Automatic Transmissions and Transaxles

Second Edition

**By Chek-Chart Publications,
a Division of
H.M. Gousha**

Michael T. Calkins, CMAT, *Editor*
Richard K. DuPuy, *Managing Editor*
Janette E. Kok, *Contributing Editor*
Robert T. Whitmore, *Contributing Editor*

HarperCollins*Publishers*

Acknowledgments

The authors have made every effort to ensure that the material in this book is as accurate and up-to-date as possible. However, neither Chek-Chart nor Harper & Row nor any related companies can be held responsible for mistakes or omissions, or for changes in procedures or specifications made by the carmakers or suppliers.

Chek-Chart is grateful to the following reviewer for his invaluable comments and suggestions:
Chane Bush, Program Supervisor, Automotive-Diesel Department, Southern Alberta Institute of Technology, Calgary, AB Canada.

At Chek-Chart, Ray Lyons managed the production of this book. Original art and photographs were produced by Gordon Agur, John Badenhop, Jim Geddes, C.J. Hepworth, Janet Jamieson, Kalton C. Lahue, and F.J. Zienty. The project is under the direction of Roger L. Fennema.

In producing this series of textbooks for automobile technicians, Chek-Chart has drawn extensively on the technical and editorial knowledge of the nation's carmakers and suppliers. Automotive design is a technical, fast-changing field, and we gratefully acknowledge the help of the following companies in allowing us to present the most up-to-date information and illustrations possible:
Borg-Warner Corporation
Chrysler Motors Corporation
Ford Motor Company
Fram Corporation, A Bendix Company
General Motors Corporation
    Hydra-matic Division
    Buick Motor Division
    Chevrolet Motor Division
    Oldsmobile Division
    Pontiac Division
Harrah's Automobile Collection
Torrington Company

AUTOMATIC TRANSMISSIONS AND TRANSAXLES, Second Edition, Classroom Manual and Shop Manual Copyright © 1989 by Chek-Chart, Simon & Schuster Inc.

Library of Congress Cataloging and Publication Data:

Chek-Chart, 1989
    Automatic Transmissions and Transaxles
    (HarperCollins /Chek-Chart Automotive Series)
v.1. Classroom Manual. v.2. Shop Manual.

ISBN: 0-06-454012-X
Library of Congress Catalog Card No.: 88-35811

Contents

On the Cover:
Front — The A4LD transmission, courtesy of Ford Motor Company.
Rear — The ATX transaxle, courtesy of Ford Motor Company; and the Torqueflite A-413 transaxle, courtesy of Chrysler Motors.

Contents

Introduction to Automatic Transmissions and Transaxles

Automatic Transmissions and Transaxles is part of the Harper & Row/Chek-Chart Automotive Series. The package for each course has two volumes, a *Classroom Manual* and a *Shop Manual*.

Other titles in this series include:
- Automotive Brake Systems
- Automotive Electrical and Electronic Systems
- Automotive Engine Repair and Rebuilding
- Fuel Systems and Emission Controls
- Engine Performance Diagnosis and Tune-Up
- Heating and Air Conditioning (due late 1989).

Each book is written to help the instructor teach students to become excellent professional automotive technicians. The two-manual texts are the core of a complete learning system that leads a student from basic theories to actual hands-on experience.

The entire series is job-oriented, especially designed for students who intend to work in the car service profession. A student will be able to use the knowledge gained from these books and from the instructor to get and keep a job. Learning the material and techniques in these volumes is a giant leap toward a satisfying, rewarding career.

The books are divided into *Classroom Manuals* and *Shop Manuals* for an improved presentation of the descriptive information and study lessons, along with the practical testing, repair, and overhaul procedures. The manuals are to be used together: the descriptive chapters in the *Classroom Manual* correspond to the application chapters in the *Shop Manual*.

Each book is divided into two parts, and each of these parts complements the other. Instructors will find the chapters to be complete, readable, and well thought-out. Students will benefit from the many learning aids included, as well as from the thoroughness of the presentation.

The series was researched and written by the editorial staff of Chek-Chart, and was produced by Harper & Row Publishers. For over 60 years, Chek-Chart has provided car and equipment manufacturers' service specifications to the automotive service field. Chek-Chart's complete, up-to-date automotive data bank was used extensively to prepare this textbook series.

Because of the comprehensive material, the hundreds of high-quality illustrations, and the inclusion of the latest automotive technology, instructors and students alike will find that these books will keep their value over the years. In fact, they will form the core of the master technician's professional library.

How
To Use
This Book

Why Are There Two Manuals?

This two-volume text — **Automatic Transmissions and Transaxles** — is not like any other textbook you've ever used before. It is actually two books, the *Classroom Manual* and the *Shop Manual*. They should be used together.

The *Classroom Manual* will teach you what you need to know about how automatic transmissions and transaxles work. The *Shop Manual* will show you how to fix and adjust the major automatic transmissions and transaxles.

The *Classroom Manual* will be valuable in class and at home, for study and for reference. It has text and pictures that you can use for years to refresh your memory about the basics of automatic transmissions and transaxles.

In the *Shop Manual,* you will learn about test procedures, troubleshooting, and overhauling the systems and parts you are studying in the *Classroom Manual.* Use the two manuals together to fully understand how transmissions and transaxles work, and how to fix them when they don't work.

What's in These Manuals?

There are several aids in the *Classroom Manual* that will help you learn more:

1. The text is broken into short bits for easier understanding and review.

2. Each chapter is fully illustrated with drawings and photographs.

3. Key words in the text are printed in **boldface type** and are defined on the same page and in a glossary at the end of the manual.

4. Review questions are included for each chapter. Use these to test your knowledge.

5. A brief summary of every chapter will help you to review for exams.

6. Every few pages you will find short blocks of "nice to know" information, in addition to the main text.

7. At the back of the *Classroom Manual* there is a sample test, similar to those given for National Institute for Automotive Service Excellence (NIASE) certification. Use it to help you study and to prepare yourself when you are ready to be certified as an expert in one of several areas of automotive technology.

The *Shop Manual* has detailed instruction on overhaul, test, and service procedures. These are easy to understand, and many have step-by-step, photo-illustrated explanations that guide you through the procedures. This is what you'll find in the *Shop Manual*:

1. Helpful information tells you how to use and maintain shop tools and test equipment.

2. Safety precautions are detailed.
3. System diagrams help you locate trouble-spots while you learn to read the diagrams.
4. Tips the professionals use are presented clearly and accurately.
5. A full index will help you quickly find what you need.
6. Test procedures and troubleshooting hints will help you work better and faster.

Where Should I Begin?

If you already know something about automatic transmissions and transaxles, how they work, and how to fix them, you may find that parts of this book are a helpful review. If you are just starting in car repair, then the subjects covered in these manuals may be all new to you.

Your instructor will design a course to take advantage of what you already know, and what facilities and equipment are available to work with. You may be asked to take certain chapters of these manuals out of order. That's fine. The important thing is to really understand each subject before you move on to the next.

Study the vocabulary words in boldface type. Use the review questions to help you understand the material. While reading in the *Classroom Manual*, refer to your *Shop Manual* to relate the descriptive text to the service procedures. And when you are working on actual car systems, look back to the *Classroom Manual* to keep the basic information fresh in your mind. Working on such a complicated piece of equipment as a modern car isn't always easy. Use the information in the *Classroom Manual*, the procedures of the *Shop Manual*, and the knowledge of your instructor to help you.

The *Shop Manual* is a good book for work, not just a good workbook. Keep it on hand while you're working on equipment. It folds flat on the workbench and under the car, and can withstand quite a bit of rough handling.

When you do test procedures and overhaul equipment, you will also need a source of accurate manufacturers' specifications. Most auto shops have either carmakers' annual shop service manuals, which list these specifications, or an independent guide, such as the **Chek-Chart Car Care Guide**. This unique book, which is updated every year, gives you the complete service instructions, electronic ignition troubleshooting tips, and tune-up information that you need to work on specific cars.

Using the Two Parts of Each Manual

Each manual is divided into two parts. In the *Classroom Manual*, Part One is titled "Transmission and Transaxle Fundamentals", and Part Two is "Specific Transmissions and Transaxles". In the *Shop Manual*, Part One is titled "General Service Operations", and Part Two is "Transmission and Transaxle Overhaul". The first part of each manual contains general information that applies to all transmissions and transaxles. Part Two of each book deals with specific transmissions and transaxles. The *Classroom Manual* tells you how they are put together and how they work. The *Shop Manual* tells you how to overhaul them.

You can begin studying or overhauling any transmission or transaxle in Part Two of either manual at any point that you wish or that your instructor assigns. You do not have to start with the first one. You will find a lot of similarities in the descriptions and overhaul procedures for different transmissions and transaxles. That's because modern automatic transmissions and transaxles are more similar to each other than they are different.

Remember, however, before you study or service the transmissions and transaxles covered in Part Two of each manual, you should read and understand the material in Part One of each book.

PART ONE

Transmission and Transaxle Fundamentals

Gears

This is a book about automatic transmissions and transaxles for modern automobiles. Automatic transmissions replace manual transmissions in rear-wheel-drive (RWD) powertrains, and relieve the driver of the need to manually shift gears. Automatic transaxles perform the same function in front-wheel-drive (FWD) vehicles (and mid- or rear-engine RWD cars), but are unique in that the differential and final drive gears are located within the transaxle housing. In this textbook, the term "transmission" will be used generically to refer to all automatic gearboxes. The term "transaxle" will be used only for specific transaxle references.

Automatic transmissions are not a recent development. Some cars built in the first decade of this century had transmissions with gears similar to those in a modern automatic. In the late 1930s, Chrysler, Ford, General Motors, Hudson, and others had automatic or semi-automatic transmissions in experimental stages or limited production. The first fully automatic transmission, the Hydramatic, was introduced in 1940 Oldsmobiles and Cadillacs, and was based on a semiautomatic unit used in 1938 Buicks. As you study modern automatic transmissions, you will find they have quite a history of development behind them.

In order to professionally service automatic transmissions, you must understand their mechanical, hydraulic, and on the latest designs, electrical systems. Part One of this book deals with basic mechanical, hydraulic, and electrical principles used by all transmissions. If you already understand these principles, the following chapters will be a good review. If these ideas are new to you, the chapters in Part One will provide the foundation for a complete understanding of automatic transmissions.

In this first chapter, we begin with a look at simple gears and how they work to transmit power. Gears are the heart of any automotive transmission, whether automatic or manual. While many other devices help make up a complete automatic transmission, steel gears are essential parts.

GEAR TYPES

Gear types used in transmissions are identified by the shape of the gear teeth and by the location of the teeth on the gear.

Spur Gears and Helical Gears

The simplest type of gear is the **spur gear**, figure 1-1. A spur gear consists of a wheel with teeth around its circumference. The teeth of a

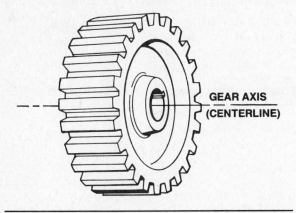

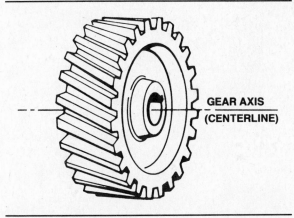

Figure 1-1. The straight-cut teeth of a spur gear are parallel to the gear's axis.

Figure 1-2. The teeth of a helical gear are at an angle to the gear's axis.

spur gear are straight cut. That is, they are parallel to the centerline, or **axis**, of the gear. As with all gears, the teeth are shaped so another set of teeth, having the same size and shape, will fit into the openings between them.

Because of the spur gear's design, only one pair of teeth are meshed at any given time, and they make contact with one another other over their full width at the same instant. These factors limit the load-carrying capacity of spur gears, and make them rather noisy in operation. Spur gears are seldom used in automatic transmissions except for reverse gear.

Although similar to a spur gear, a **helical gear**, figure 1-2, has its teeth cut at an angle to the axis of the gear. This enables one and one-half gear teeth to be in contact at all times, and allows the teeth to mesh gradually rather than all at once. As a result, helical gears are stronger and generally run more quietly than spur gears. Most of the gears used in automatic transmissions are of helical design.

Spur Gear: A gear on which the teeth are parallel with the gear's axis of rotation.

Axis: The centerline around which a gear, wheel, or shaft rotates.

Helical Gear: A gear on which the teeth are at an angle to the gear's axis of rotation.

■ Transmissions and Transaxles

While transmissions and transaxles are two distinct types of components, they perform basically the same job in vehicle powertrains; they transmit torque from the engine to the wheels and tires to move the car down the road. To do this efficiently, they use a series of gears to provide several different gear ratios. Transmissions are used primarily in rear-wheel-drive powertrains, and their output is generally routed to the driven wheels through a driveshaft, and a differential and final drive assembly at the rear axle. Transaxles are used primarily in front-wheel-drive powertrains, and because they contain the differential and final drive gears, their output is routed directly to the driven wheels.

REAR-WHEEL DRIVE

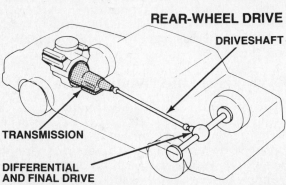

FRONT-WHEEL DRIVE

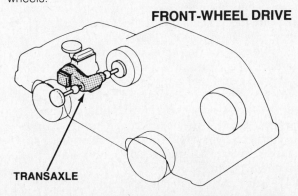

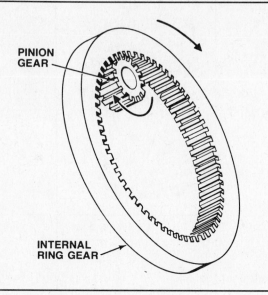

Figure 1-3. When an internal ring gear and a pinion gear are meshed, they both rotate in the same direction.

Figure 1-4. When two external gears are meshed, they rotate in opposite directions.

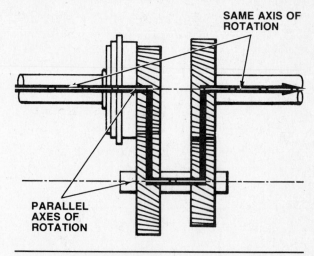

Figure 1-5. External gear axes of rotation in a manual transmission.

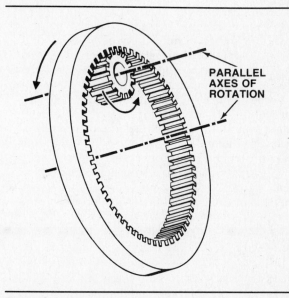

Figure 1-6. An internal ring gear and pinion gear with parallel axes of rotation.

External and Internal Gears

The teeth of the spur and helical gears shown in figures 1-1 and 1-2 are cut on the gears' outside circumferences. Gears with their teeth on the outside are called external gears. External gears are the most common type.

When a gear has teeth on its inside circumference, figure 1-3, it is called an **internal ring gear**. Like external gears, the teeth of internal ring gears may be straight-cut spur teeth (parallel to the gear axis), or helical teeth (at an angle to the axis). A smaller external gear, designed to turn on a pin while traveling around the inside of an internal ring gear, is one kind of **pinion gear**, figure 1-3.

When an external gear is meshed with an internal gear, they rotate in the same direction, figure 1-3. When two external gears are meshed, figure 1-4, they rotate in opposite directions. When internal or external gears are meshed and turned, motion is transmitted along the same axis, figure 1-5, or along parallel axes, figure 1-6.

Bevel, Worm, and Hypoid Gears

Unlike the gears discussed thus far, bevel, worm, and hypoid gearsets change the axis of rotation. In most cases, the gears' axes of rotation are 90 degrees apart, but other angles are possible depending on the design of the gears.

Simple **bevel gears** have straight-cut teeth, similar to those on a spur gear, set at an angle

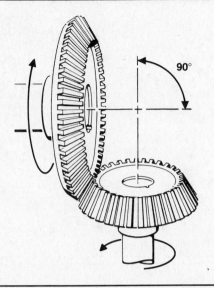

Figure 1-7. Bevel gears with their axes of rotation at a 90-degree angle.

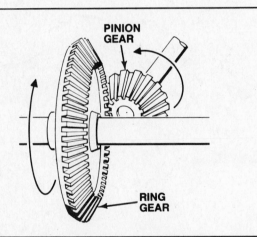

Figure 1-8. Ring and pinion bevel gears.

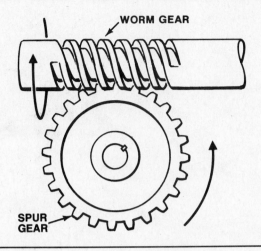

Figure 1-9. A worm gear and spur gear.

on their outside surfaces, figure 1-7. Differential spider gears are a common form of simple bevel gears. Spiral bevel gears have curved teeth, similar to those on a helical gear, for increased load-carrying ability and quieter operation. Spiral bevel gears are rare in automotive applications today.

When one gear in a pair of bevel gears is larger than the other, it is called a ring gear, figure 1-8. The smaller gear in this combination is called a pinion gear. This combination of bevel gears, or spiral bevel gears, was used for the final drive in early automotive differentials.

A **worm gear** changes the axis of rotation using a screw operating against a spur gear, figure 1-9. A unique feature of worm gears is that, while the worm can drive the spur gear, the spur gear cannot drive the worm. The most

common automotive uses of worm gears are in recirculating-ball steering boxes and speedometer cable drive mechanisms.

A **hypoid gearset**, figure 1-10, combines gear teeth that are curved and angled similarly to spiral bevel gears, with a pinion gear that is offset below the ring gear centerline similarly to a worm gear. This design provides maximum gear tooth contact for great strength, and very gradual engagement for quiet operation. Hypoid gear teeth have a great deal of sliding contact, however, and are the most difficult type to lubricate.

Hypoid gears are commonly used for the final drive in rear axle differentials where load-

Internal Ring Gear: A gear with a hole through its center and teeth cut on the inner circumference. Also called an "annulus gear."

Pinion Gear: A smaller gear that meshes with a larger gearwheel or toothed rack.

Bevel Gear: A gear with teeth cut at an angle on its outside surface. Bevel gears are commonly used to transmit power between two shafts at an angle to one another.

Worm Gear: A shaft with gear teeth cut as a continuous spiral around its outer surface. A worm gear meshes with an external gear and is used to change the axis of rotation.

Hypoid Gearset: A combination of a ring gear and pinion gear in which the pinion meshes with the ring gear below the centerline of the ring gear. Commonly used in automotive final drives.

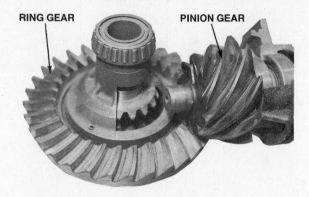

Figure 1-10. Hypoid gears as used in a RWD differential final drive.

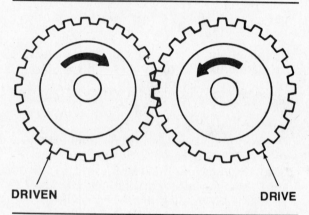

Figure 1-11. A 1 to 1 (1:1) gear ratio.

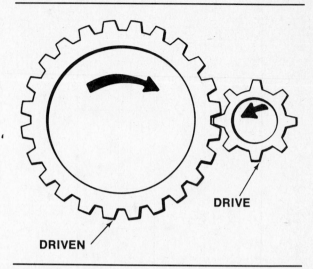

Figure 1-12. A 3 to 1 (3:1) gear reduction ratio.

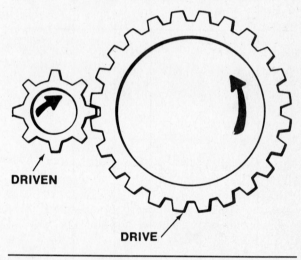

Figure 1-13. A .33:1 (1:3) overdrive gear ratio.

carrying ability and low noise are important. In addition, the offset pinion allows the driveshaft to be positioned lower in the car, reducing the size of the hump in the vehicle interior.

GEAR RATIOS

When one gear turns another, the speed at which the two turn in relation to each other is the **gear ratio**. This is expressed as the number of turns the drive gear must make in order to turn the driven gear through one revolution, and is obtained by dividing the number of teeth on the driven gear by the number of teeth on the drive gear. Gear ratios are always expressed relative to the number one.

For example, if two meshed gears are the same size and have the same number of teeth, figure 1-11, they will turn at the same speed. Since the drive gear turns once for each revolution of the driven gear, the gear ratio is one, or 1 to 1, written as 1:1. A gear ratio of one is called **direct drive**. When a transmission is in direct drive, the engine, transmission, and driveshaft (on RWD vehicles) all turn at the same speed.

Gear Reduction

If one gear drives a second gear that is three times larger and has three times the number of teeth, figure 1-12, the smaller drive gear will rotate three times in order to turn the larger driven gear through one rotation. The gear ratio in this case, obtained by dividing the teeth on the driven gear by the teeth on the drive gear, is three, 3 to 1, or 3:1. This type of gearset provides **gear reduction**.

Reduction refers to the decreased output speed of the driven gear. First gear in a transmission is called ''low'' gear because vehicle speed is low. However, low gears have numerically high ratios. That is, a 3:1 ratio is a lower gear than 2:1 or 1:1.

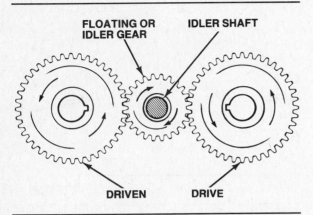

Figure 1-14. A floating or idler gear.

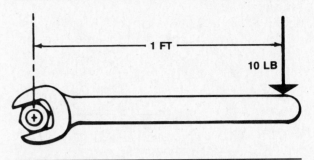

Figure 1-15. Torque is a twisting force measured by multiplying the force times its distance from a pivot point.

Overdrive

The opposite of gear reduction is **overdrive**. This occurs when the driven gear turns faster and makes more revolutions than the drive gear. In the example in figure 1-13, the driven gear turns three times for every one turn of the drive gear. This is a one third, 1 to 3, or .33:1 overdrive ratio. In actual transmissions, overdrive ratios are usually much closer to 1:1. Overdrive ratios of .89:1 and .70:1 are typical of those used in automotive applications.

Idler Gears

In a **geartrain**, gears that operate between the drive and driven gears are called floating or **idler gears**, figure 1-14. Idler gears do not affect the speed relationship of the drive or driven gears in any way. It does not matter how many idlers are in the geartrain, or how many teeth they have, the gear ratios are still computed the same way, by dividing the number of driven-gear teeth by the number of drive-gear teeth.

Idler gears do, however, affect the direction of rotation. When a drive and driven gear are meshed directly, as in figure 1-13, they rotate in opposite directions. When an idler gear is installed between the drive and driven gears, figure 1-14, they then rotate in the same direction.

TORQUE AND POWER

Gears have several different functions. Primarily, they are used to increase or decrease *torque*, and raise or lower *speeds*. In the process of doing these jobs, gears transmit *power*.

Torque is a twisting force, commonly expressed in foot-pounds, inch-pounds, or Newton-meters. Torque is calculated by multiplying the force times its distance from a center

or pivot point. For example, a 10-pound force pressing down on a wrench at a point 1 foot from the center of a bolt, figure 1-15, exerts $10 \times 1 = 10$ foot-pounds of torque at the bolt.

A gear applies torque much as a wrench does. A gear with a 2-foot radius and a force of

Gear Ratio: The number of turns, or revolutions, made by a drive gear compared to the number of turns made by a driven gear. For example, if the drive gear turns three times for one revolution of the driven gear, the gear ratio is three, 3 to 1, or 3:1. Also, the ratio between the number of teeth on two gears.

Direct Drive: A 1:1 gear ratio in which the engine, transmission, and driveshaft (on RWD vehicles) all turn at the same speed.

Gear Reduction: A condition in which the drive gear rotates faster than the driven gear. Output speed of the driven gear is reduced, while output torque is increased. A gear ratio of 3:1 is a gear reduction ratio.

Overdrive: A condition in which the drive gear rotates slower than the driven gear. Output speed of the driven gear is increased, while output torque is reduced. A gear ratio of .70:1 is an overdrive gear ratio.

Geartrain: A series of two or more gears meshed together so that power is transmitted between them.

Idler Gears: Gears that transmit movement between the drive and driven gears, but do not affect the speed relationship of those gears.

Torque: Twisting or turning force measured in terms of the distance times the amount of force applied. Commonly expressed in foot-pounds, inch-pounds, or Newton-meters.

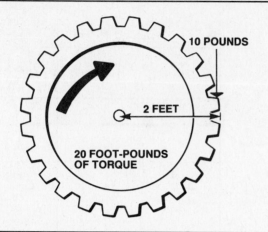

Figure 1-16. Gears apply torque in the same way a wrench does.

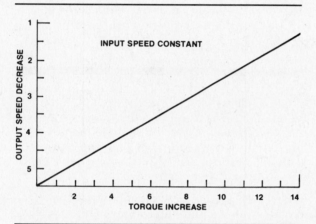

Figure 1-17. With input speed constant, torque will increase as output speed decreases.

10 pounds applied to one of its teeth, figure 1-16, exerts 20 foot-pounds of torque at the shaft to which it is attached.

Torque is *not* the same as power. **Power**, commonly expressed in horsepower or kilowatts, is calculated by multiplying torque times the speed of rotation. Therefore, power is the rate, or speed, at which torque performs **work**.

Torque and Speed Relationship

Torque and speed have an inverse relationship; that is, as one goes up, the other goes down. Assuming a constant input speed, as output speed increases, torque decreases. The opposite is also true: assuming a constant input speed, as output speed decreases, torque increases. This is shown in figure 1-17.

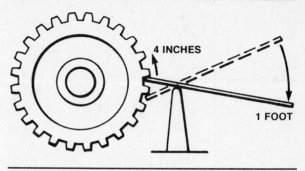

Figure 1-18. A lever can be used to multiply torque.

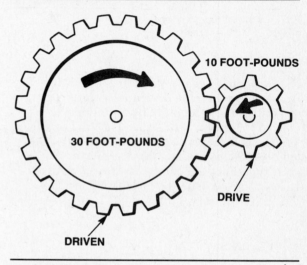

Figure 1-19. Torque multiplication of three, or 3:1.

Torque Multiplication

Simple levers can be used to increase, or multiply, torque. For example, a wheel may be too heavy for a person to turn by muscle power alone, but it *can* be turned when a lever and fulcrum are used to multiply the force of the person's arms, figure 1-18. This is possible because the person moves his or her end of the lever further than the working end moves. Distance, or speed, is traded for torque.

Gears can be used in the same way as levers to multiply torque. As we saw in figure 1-11, when two gears of the same diameter are meshed with one another, and one of them is driven, the second gear will turn at the same speed. Since there is no difference in speed, there is no difference in torque between the two gears.

If the drive gear is one-third the diameter of the driven gear with one-third as many teeth, figure 1-19, it must rotate three times for each rotation of the larger gear. As we saw earlier, this means that the larger gear will turn three

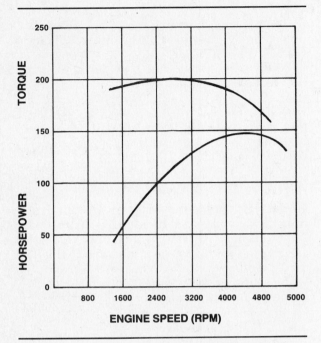

Figure 1-20. Typical engine torque and horsepower curves.

times more slowly than the smaller gear. At the same time, however, the larger gear will exert three times the twisting force of the smaller gear. In the example in figure 1-19, 10 foot-pounds of torque from the drive gear produces 30 foot-pounds of torque at the driven gear. The torque multiplying power of this system is 3:1, which is the same as the gear ratio.

It is clear that torque multiplication and gear ratios are directly related. When the gear system is in reduction, there is *more* torque available at the driven gear, but less speed. When the gear system is in overdrive, figure 1-13, there is *less* torque available at the driven gear, but greater speed. In designing mechanical devices such as transmissions, engineers choose gear ratios that will provide the torque output required to do the work.

Engine Torque Characteristics

The **torque curve** of an engine, figure 1-20, shows how much torque is available at different points in a speed range. Notice that over the rpm range shown, the torque curve is relatively flat, while the horsepower curve increases. This is because power equals torque times speed, and as long as speed increases more than torque decreases, power will increase. At the upper end of the rpm range, torque drops off so far that higher speeds can no longer compensate, and power output begins to drop as well.

The gasoline-powered internal-combustion engine has a limited torque range; high torque is available over only part of the engine's rpm range. At lower speeds, the engine does not produce enough torque at the crankshaft to move the vehicle, and soon after engine rpm exceeds the point where the torque curve peaks, the amount of power available to perform work drops off.

Because of the characteristics of gasoline engines, torque multiplication must be provided between the crankshaft and the drive axles to accelerate an automobile from low speeds or a standstill. And, once engine rpm rises beyond the torque peak, a change in gear ratios is required to bring the engine rpm back within the most efficient torque-producing range.

POWERTRAIN GEAR RATIOS

In an automobile powertrain, the job of the transmission is to transmit the torque of the engine, through various speed ranges, into the work of efficiently moving the vehicle down the road. The transmission is aided in this task by the final drive. The sections below describe how these components work together to provide a selection of gear ratios that takes maximum advantage of the engine power available.

Final Drive

The **final drive** is the last set of reduction gears the powerflow passes through on its way to the drive axles. On most FWD cars, the final drive is located in the transaxle. On most RWD cars, the final drive is located in the rear axle housing. Although the RWD axle assembly that con-

Power: The rate at which work is done or force is applied. In mechanics, power is measured as torque times speed and expressed in units such as horsepower or kilowatts.

Work: The transfer of energy from one system to another, particularly through the application of force.

Torque Curve: A graphic depiction of the amount of torque available at different engine speeds.

Final Drive: The last set of reduction gears the powerflow passes through on its way to the drive axles.

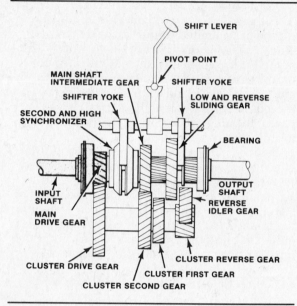

Figure 1-21. Sliding-gear manual transmission in neutral.

tains the final drive is generically called a **differential**, that term specifically refers to the mechanism that allows the wheels to turn at different speeds as the car rounds a corner. Transaxles also contain differentials. In this textbook, we will use the term "final drive" when referring to the ring and pinion gears that provide gear reduction to the drive axles.

As mentioned earlier, the final drives in RWD differentials generally use hypoid gears because the powerflow must be turned 90 degrees from the driveshaft to the drive axles. Hypoid final drive gears are also used on some FWD cars where the engine is mounted longitudinally (north/south). On FWD cars with transversely mounted (east/west) engines, the powerflow is already on a parallel axis with the drive axles, and a simple set of helical gears within the transaxle serves as the final drive.

To compensate for certain operating characteristics covered in later chapters, vehicles with automatic transmissions usually have final drive ratios closer to 1:1 than those used with manual transmissions. Common ratios for automatic transmission final drives range from 2.29 to 4:1.

Low Gear

Moving the average vehicle from a standstill requires an overall torque multiplication from the engine to the drive axles of about 12:1. The overall ratio of the engine speed (or torque) to the drive axles' speed (or torque) is the transmission gear ratio *times* the final drive gear ratio. If the final drive multiplies torque four

times (4:1), and the transmission low gear multiplies torque three times (3:1), the overall ratio is 4:1 × 3:1 = 12:1.

At the same time torque is *multiplied* by twelve, speed is *divided* by twelve. When the crankshaft is turning at 1,200 rpm, the drive wheels turn at 1,200 ÷ 12 = 100 rpm. A 12:1 gear ratio works well at low speeds, but by the time the vehicle reaches 40 mph, the engine will be turning about 4,000 rpm.

Second (Intermediate) Gear

To add flexibility to the system, a second set of transmission gears can be added, with a ratio of 2:1. Multiplying this by the final drive ratio of 4:1 gives a new overall gear ratio of 8:1. The driver can use this ratio to increase speed, while continuing to operate in favorable bands of the engine torque and horsepower curves.

Third (Direct) Gear

For cruising speeds, a third gear with a direct-drive 1:1 ratio can be built into the transmission. With a final drive ratio of 4:1, this would give an overall gear ratio of 4:1. In a 3-speed transmission, 4:1 would also be the **overall top gear ratio**.

Fourth (Overdrive) Gear

For greater fuel economy, a fourth gear with an overdrive ratio, such as .80:1, can be added to the transmission to further lower engine rpm at cruising speeds. If this ratio is multiplied times a final drive ratio of 4:1, the overall top gear ratio drops to 3.2:1.

PLANETARY GEARSETS

In a manual transmission, figure 1-21, gear ratios are obtained by the use of sliding gears in various combinations. Before a change is made from one ratio to another, the engine flywheel is disconnected from the transmission by a clutch. Then the gears are manually moved inside the transmission.

Automatic transmissions use a gear system called a **planetary gearset**, figure 1-22, that does not require gears to be shifted in order to change gear ratios. A planetary gearset consists of three primary members: the **sun gear**, the internal ring gear, and the **planet carrier assembly**. The sun gear gets its name from its position at the center of the gearset; the other gears revolve around it like planets around the sun. The outermost part of the gearset is the internal ring gear with teeth cut on its inner

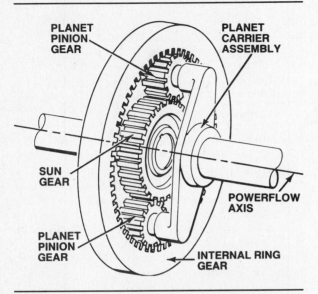

Figure 1-22. The three members of a simple planetary gearset are the sun gear, internal ring gear, and planet carrier assembly.

side. The planet carrier assembly fits between the sun gear and the internal ring gear.

The planet carrier assembly contains several **planet pinions**, held within a cage, that rotate on pins that are part of the carrier. The simple planetary gearset in figure 1-22 has only two

Differential: The assembly of a carrier and spider gears that allows the drive axles to rotate at different rates of speed as the car turns a corner.

Overall Top Gear Ratio: The ratio obtained when the gear ratio of the transmission's highest forward speed is multiplied times the final drive gear ratio.

Planetary Gearset: A system of gears consisting of a sun gear, an internal ring gear, and a planet carrier with planet pinion gears.

Sun Gear: The central gear of a planetary gearset around which the other gears rotate.

Planet Carrier Assembly: One of the members of a planetary gearset. The carrier, or bracket, on which the planet pinions are mounted.

Planet Pinions: The pinion gears mounted on the planet carrier assembly in a planetary gearset. The planet pinions rotate on the carrier and revolve around the sun gear.

planet pinions, but actual units in transmissions generally have three or four. The planet pinions are fully meshed with *both* the sun gear and the internal ring gear at all times. The gears are never disengaged to change ratios, as in a manual transmission.

Not only are the sun gear, internal ring gear, and planet pinions constant mesh, but the sun gear, internal ring gear, and planet carrier rotate on the same axis, figure 1-22. This means that all the powerflow through a planetary gearset, both input and output, occurs along a single axis.

■ **Horseless Carriage Transmissions**

The 1902 curved-dash Oldsmobile looked more like a horse-drawn buggy than a car; hence the name, "horseless carriage." However, this Merry Oldsmobile was the first car to be mass produced in the United States, so in a way, it was the ancestor of all our cars.

The curved-dash Olds also carried the ancestor of the modern automatic transmission — a planetary transmission. Of course, the transmission in the 1902 Olds didn't shift automatically, the driver had to do that with his feet, but it did have the same basic planetary gearset of sun gear, ring gear, and pinions found in modern automatics.

Other early cars that used planetary transmissions were the 1903 Buick and the 1903 Cadillac. The Buick and Cadillac transmissions were similar in concept to the Oldsmobile design, but all three differed in details. In 1903, these companies were still independent carmakers. It would be five years before they were merged into General Motors.

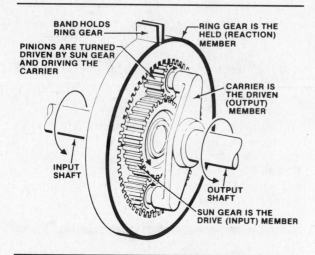

BAND HOLDS RING GEAR

PINIONS ARE TURNED DRIVEN BY SUN GEAR AND DRIVING THE CARRIER

RING GEAR IS THE HELD (REACTION) MEMBER

CARRIER IS THE DRIVEN (OUTPUT) MEMBER

INPUT SHAFT

OUTPUT SHAFT

SUN GEAR IS THE DRIVE (INPUT) MEMBER

Figure 1-23. To transmit power through a planetary gearset, one (input) member drives, another (reaction) member is held, and the third (output) member is driven.

Some of the parts in a planetary gear system are known by several different names. The internal ring gear is sometimes called the ring gear or the **annulus gear**. The planet pinion gears are often called planet gears or, simply, pinions. The planet carrier assembly is commonly referred to as simply the carrier. In this textbook, we will use the terms "sun gear," "ring gear," "pinions," and "carrier" as we describe the powerflow and operations of a planetary gearset.

PLANETARY GEARSET POWERFLOW

In a planetary gearset, the teeth of each gear are always meshed with the teeth of another. Therefore, whenever one gear is driven, all of the other gears are affected as well. To transmit power through a planetary gearset, figure 1-23, you drive one member while holding a second which causes the third to be driven. Each member of a planetary gearset can play any of the three parts in transmitting power. Various combinations of drive, held, and driven members result in a number of gear ratios, as well as changes in the direction of rotation.

When discussing powerflow through a planetary gearset, several steps are often required to get from the drive action of one member to the driven action of the last member. The terms "drive" and "driven" are convenient when describing how two gears work together. However, when three or more gears are involved, the second gear is a *driven* gear in relation to the

first, but a *drive* gear in relation to the third. For this reason, the drive member of a planetary gearset is sometimes called the **input member**, the held member is often called the **reaction member**, and the driven member may be identified as the **output member**.

Any member of the gearset that is not the drive (input), held (reaction), or driven (output) member can be described as simply being turned. The pinions in figure 1-23 are an example of this. They are driven by the sun gear, and they drive the carrier. To do this, the pinions turn on their axes and react against the stationary ring gear.

In describing planetary gearset operations in the following sections, we will use the terms "input member," "held member," and "output member." We also will describe other members of the gearset as being "turned". The devices that hold the reaction members and keep them from rotating are described at the end of this chapter and in Chapter 4.

PLANETARY GEARSET OPERATIONS

Gear reduction, overdrive, reverse, and direct drive can all be obtained by driving and holding different members of a planetary gearset. There are two ways to do each of these operations except direct drive, for which there are three possible methods. Neutral occurs when no member of the gearset is held.

Gear Reduction

If the carrier is the output member and either of the other members is the input member, with the remaining member held, the result will be gear reduction. The output rotation will be in the same direction as the input rotation. Note that in both methods of obtaining gear reduction the carrier is the output member.

Method 1
- Ring gear — input
- Sun gear — held
- Carrier — output

The internal ring gear is the input member, turning clockwise, figure 1-24. The sun gear is held. Because the ring gear rotates clockwise, it turns the pinions clockwise. And, because the pinions are meshed with the stationary sun gear, they "walk" clockwise around it, moving the carrier clockwise with them at a reduced speed. The carrier is the output member.

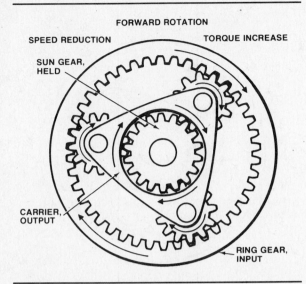

Figure 1-24. Gear reduction, method 1.

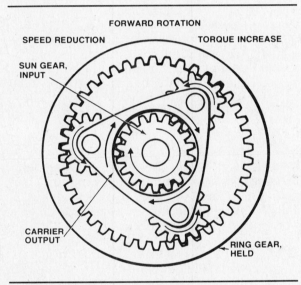

Figure 1-25. Gear reduction, method 2.

Method 2
- Sun gear — input
- Ring gear — held
- Carrier — output

If the sun gear is the input member and turns clockwise, the pinions will rotate counterclockwise, figure 1-25. Because the ring gear is held, the pinions meshed with the ring gear walk clockwise around the inside of the ring gear and take the carrier with them. The carrier is again the output member. It rotates in the same direction as the sun gear, but at a reduced speed.

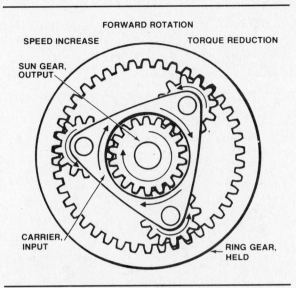

Figure 1-26. Overdrive, method 1.

Overdrive

If the planet carrier is the input member and one of the remaining members is held, the output member will rotate faster than the input member. The direction of rotation of the input and output shafts will be the same. Note that in both methods of obtaining overdrive the planet carrier is the input member.

Method 1
- Carrier — input
- Ring gear — held
- Sun gear — output

The carrier is the input member and turns clockwise, figure 1-26. The ring gear is held. The pinions turn counterclockwise inside the stationary ring gear. The pinions turn the sun gear clockwise, and it becomes the output member. The output sun gear turns in the same direction as the input carrier but at a higher speed.

Annulus Gear: Another name for the internal ring gear in a planetary gearset.

Input Member: The drive member of a planetary gearset.

Reaction Member: The member of a planetary gearset that is held in order to produce an output motion. Other members react against the stationary, held member.

Output Member: The driven member of a planetary gearset.

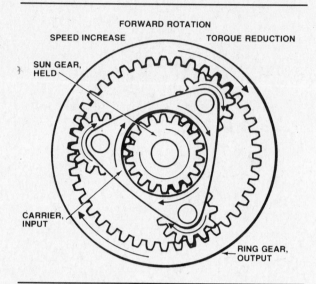

Figure 1-27. Overdrive, method 2.

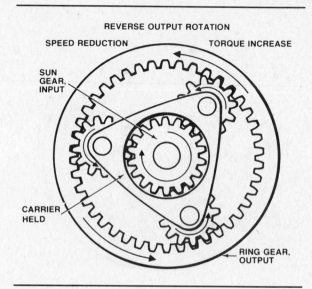

Figure 1-29. Reverse, method 2.

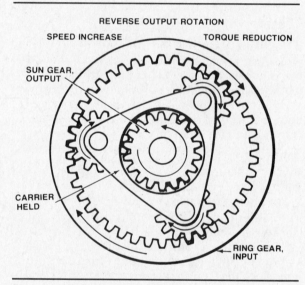

Figure 1-28. Reverse, method 1.

Method 2
- Carrier — input
- Sun gear — held
- Ring gear — output

The carrier is again the input member, turning clockwise, but the sun gear is held, figure 1-27. The pinions rotate clockwise and turn the ring gear clockwise. The ring gear becomes the output member. It rotates in the same direction as the input carrier but at a higher speed.

Reverse

If the carrier is held and either of the other two members is the input member, the remaining

member will be the output member and will turn in reverse. The rotation of the pinions on the stationary carrier reverses the direction of rotation from the input member to the output member. Note that in both methods of obtaining reverse the carrier is held.

Method 1
- Ring gear — input
- Carrier — held
- Sun gear — output

The ring gear is the input member and turns clockwise, figure 1-28. The carrier is held, but the pinions rotate clockwise on their pins. The pinions turn the sun gear counterclockwise. The output sun gear rotates faster than the input ring gear but in reverse. This is an overdrive-reverse condition.

Method 2
- Sun gear — input
- Carrier — held
- Ring gear — output

The sun gear is the input member and turns clockwise, figure 1-29. The carrier is held, but the pinions rotate counterclockwise on their pins. The pinions turn the ring gear counterclockwise. The output ring gear rotates slower than the input sun gear but in reverse. This is a gear-reduction-reverse condition.

Direct Drive

If any two members of the planetary gearset are locked together, they become a single input member; they turn in the same direction and at the same speed. In this case, the third member of the gearset is locked up by the first two and

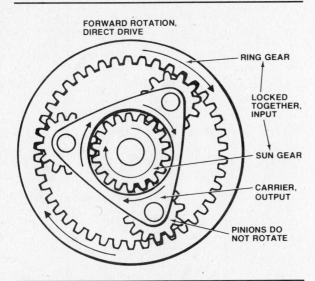

Figure 1-30. Direct drive, method 1.

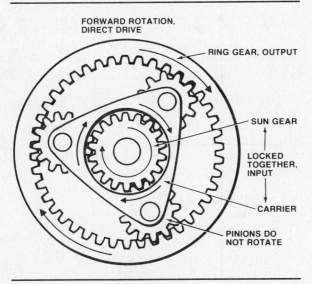

Figure 1-31. Direct drive, method 2.

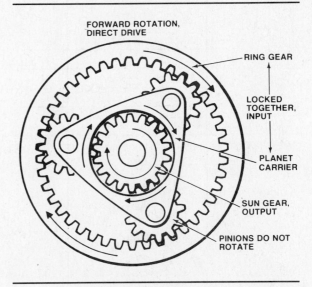

Figure 1-32. Direct drive, method 3.

Method 1
- Sun gear and ring gear — input
- Carrier — output

The sun gear and the ring gear are locked together and turned clockwise, figure 1-30. Because the sun gear and the ring gear are locked together, they hold the pinions and keep them from rotating on their pins. Because the pinions do not rotate or move in relation to the sun gear or the ring gear, the carrier is also locked. As a result, the entire planetary assembly is locked together and rotates as a unit.

Method 2
- Sun gear and carrier — input
- Ring gear — output

It is also possible to lock the sun gear and the carrier together and turn them clockwise, figure 1-31. Because the sun gear and the carrier are locked together, the pinions cannot revolve around the sun or rotate on their pins. Therefore, the pinions hold the ring gear, and again, the entire assembly rotates as a unit.

Method 3
- Ring gear and carrier — input
- Sun gear — output

If the ring gear and the carrier are locked and turned clockwise together, figure 1-32, the result is the same as when the sun gear and the carrier are locked. The pinions are again kept from rotating, and they, in turn, lock the sun gear to the rest of the assembly. The complete planetary assembly rotates in the same direction at the same speed.

is also turned in the same direction and at the same speed. The result is direct drive, a 1:1 gear ratio.

There are three ways to obtain direct drive with a simple planetary gearset. Note that in all three the third member of the gearset is locked to the first two because the pinions are kept from rotating. Because it results in the most practical design, only the first method of obtaining direct drive is commonly used in modern automatic transmissions.

	OPERATION								
	Gear Reduction		Overdrive		Reverse		Direct Drive		
	1	2	1	2	1	2	1	2	3
Sun Gear	H	I	O	H	O	I	I	I	O
Ring Gear	I	H	H	O	I	O	I	O	I
Carrier	O	O	I	I	H	H	O	I	I

I = Input O = Output H = Held

Figure 1-33. Planetary gearset operation chart.

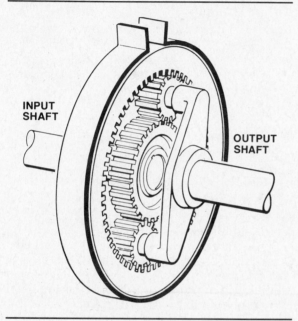

Figure 1-34. Single planetary gearset with input and output shafts.

Neutral

If any single member of a planetary gearset is turned, but none is held, there will be no output. The result is neutral. For example, if the sun gear is the input member with neither the carrier nor the ring gear being held, the pinions will rotate on their pins and carry either the carrier or the ring gear around the sun with them, depending on which has the least resistance. Regardless of which member moves with the pinions, there will be no output. The system will idle.

A similar result will occur if either the ring gear or the carrier is the input member, and no other member is held. The pinions will rotate on their pins and turn the remaining member that has the least resistance. The system will idle.

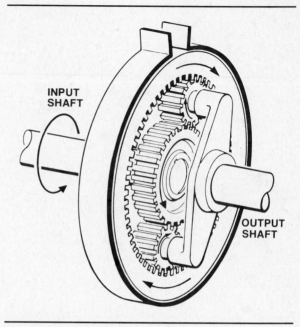

Figure 1-35. Single planetary gearset in neutral.

SIMPLE PLANETARY GEAR SYSTEMS

Simple planetary gear systems consist of one or more separate planetary gearsets. Figure 1-33 summarizes the different methods of obtaining gear reduction, overdrive, reverse, and direct drive with a single planetary gearset. The sun gear, ring gear, and carrier are identified as Input (I), Output (O), or Held (H).

Consider the possibilities for changing gear ratios without actually shifting gears in a transmission that uses a simple planetary gear system consisting of a single planetary gearset. Assume that an input shaft from the engine is attached to the sun gear, and an output shaft is attached to the carrier assembly, figure 1-34. The output shaft is connected, through the final drive, to the drive axles.

With no member held, there is no power transmission, and the gearset is in neutral, figure 1-35. When the input shaft turns, the sun gear turns, turning the planet pinions in the carrier. The carrier, which is attached to the drivetrain, has greater resistance to turning than the ring gear; therefore, the pinions turn the ring gear harmlessly opposite the direction of rotation of the sun gear.

Now assume that a contracting brake band is installed around the ring gear, to hold it from rotating, figure 1-36. Because the sun gear is the input member and the ring gear is held, the

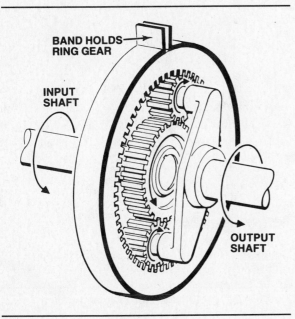

Figure 1-36. Single planetary gearset in low gear (gear reduction).

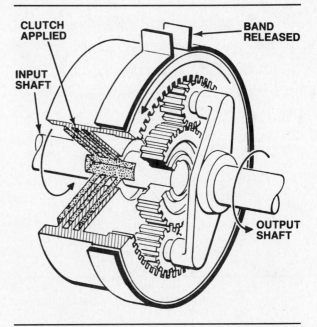

Figure 1-37. Single planetary gearset in high gear (direct drive).

carrier must be the output member. This condition results in gear reduction in the same direction of rotation as the input member. The planet carrier and output shaft rotate in the same direction as the input shaft, at reduced speed, and with increased torque.

So far, this simple transmission is capable of neutral (when none of the members of the planetary system is held), and low (when the ring gear is held).

Now, assume that a clutch is installed that can connect the input shaft to the ring gear, figure 1-37. If the band holding the ring gear is released, and the clutch is applied, the ring gear is connected to the input shaft. The ring gear then turns in the same direction and at the same speed as the sun gear, which is also connected to the input shaft. Since turning any two members together results in direct drive, the carrier, connected to the output shaft, rotates at the same speed and in the same direction as the input shaft. This is high gear.

Using a single planetary gearset, a brake band, and a clutch, we have designed an automatic transmission capable of neutral, low-, and high-gear changes, without actually shifting gears. Simple planetary gear systems like this one were used in some early automatic transmissions. The first Hydramatics used several independent planetary gearsets to provide

four speeds forward and one speed in reverse. A few late-model automatic transmissions also make limited use of simple planetary gear systems. Toyota uses a separate planetary gearset to provide overdrive fourth gear in its A40D transmission. However, most modern domestic transmissions use compound planetary gear systems.

COMPOUND PLANETARY GEAR SYSTEMS

A compound planetary gear system is an assembly of several (usually two) planetary gearsets built together to provide various overall combinations of gear reduction, direct drive, overdrive, and reverse. The most popular compound planetary gear system is the **Simpson gearset**, figure 1-38. It consists of two planetary gearsets that share a common sun gear. The methods we have just studied for obtaining various gear ratios with a simple planetary gearset also apply to the Simpson gearset.

Simpson Gearset: A compound planetary gear system consisting of two ring gears and two planet carrier assemblies that share a common sun gear.

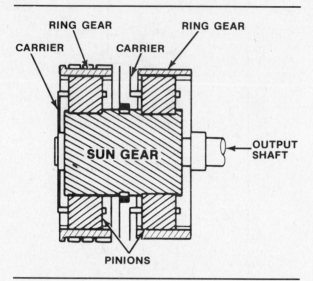

Figure 1-38. The Simpson gearset is made up of two ring gears and two planet carrier assemblies that share a common sun gear.

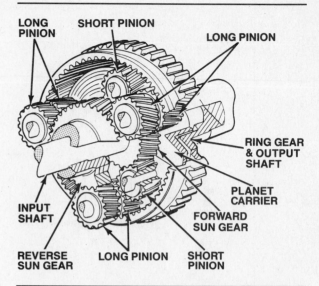

Figure 1-39. The Ravigneaux gearset is made up of two sun gears and two sets of pinions that share a common ring gear.

Another popular compound planetary gear system is called the **Ravigneaux gearset**, figure 1-39. It consists of two sun gears, two sets of planet pinions (one set longer than the other), and one ring gear. This type of gearset is easily controlled and has been used in General Motors, Ford, and some import automatic transmissions. Ford's AOD automatic overdrive transmission uses this type of geartrain, and GM used it in their Powerglide and Turbo Hydra-matic 300 gearboxes.

We will examine the operation of Simpson and Ravigneaux gearsets, and other compound planetary gear system designs, in detail in Part Two of this *Classroom Manual*, when we study specific late-model domestic automatic transmissions.

Ravigneaux Gearset: A compound planetary gear system consisting of two sun gears and two sets of planet pinions that share a common ring gear.

■ Howard Simpson and the Planetary Gearset

It's a rare occasion when major carmakers put aside their private interests and rivalries, and all adopt a single design. However, this is what happened with the Simpson geartrain that is used in most 3-speed and some 4-speed automatic transmissions.

Howard Simpson was a transmission engineer for Ford Motor Company from 1917–1938. During these years, Simpson often worked closely with Henry Ford, who had a fondness for planetary transmissions. Shortly after World War II, Simpson's doctors told him he had incurable cancer, and he retired to the Arizona desert. To pass what he thought would be the last few months of his life, he spent his days working out designs for various planetary gear arrangements.

Simpson outlived his doctors predictions; he finally died in 1963, at the age of 71. Over the course of his life, he obtained an array of patents that pretty much covered the planetary gear field. In 1953, Ford bought a license from Simpson for one of his compound gear systems that consisted of two planetary gearsets sharing a common sun gear. Because of other commitments, Ford did nothing with this license until 1964. Chrysler, however, bought a license for the same Simpson gearset to use in a new transmission for the 1956 Imperial. The result was the Chrysler Torqueflite, variations of which are still in use today.

Other modern automatic transmissions that use the Simpson geartrain are the Ford C3, C4, C5, C6, and Jatco transmissions, and the GM Turbo Hydra-matic 200, 200-R4, 250, 325, 350, and 440-T4 models. Quite a tribute to the genius of one man.

SUMMARY

Automatic transmissions replace manual gearboxes in front-engine RWD powertrains. Automatic transaxles do the same in FWD cars, and mid- or rear-engine RWD cars. Unlike transmissions, transaxles have the differential and final drive located within the transaxle housing.

Gears used in transmissions are named by the shape of their teeth and the location or relationship of the gears to other gears. Helical internal and external gears are the most common types in automatic transmissions.

Gear ratio expresses the speed relationship between two meshed gears, and is based on the number of turns the drive gear must make to turn the driven gear through one revolution. When the driven gear turns slower than the drive gear, the gearset is in gear reduction. When the driven gear turns at the same speed as the drive gear, the gearset is in direct drive. When the driven gear turns faster than the drive gear, the gearset is in overdrive.

Transmission gears are used to compensate for engine power output characteristics by regulating torque and speed. When gears are used to increase torque, output speed decreases. When gears are used to decrease torque, output speed increases.

Final drive gears provide torque multiplication in addition to that of the transmission. On front-engine RWD cars, the final drive is located in the rear axle assembly. On FWD vehicles, and mid- or rear-engine RWD cars, the final drive is located in the transaxle housing.

Planetary gearsets are used in automatic transmissions because they do not require that gears be shifted in order to change gear ratios. A simple planetary gearset consists of a sun gear, an internal ring gear, and a planet carrier assembly. To transmit power through a planetary gearset, you drive one member while holding a second which causes the third to be driven.

Simple planetary gear systems use one or more independent planetary gearsets. Most modern automatic transmissions use compound planetary gear systems that have two planetary gearsets built together. Simpson and Ravigneaux gearsets are the most common compound planetary gear systems.

Review Questions

Select the single most correct answer
Compare your answers to the correct answers on page 603

1. The teeth of a spur gear are
   _____ to the gear's axis.
   a. At a spiral
   b. At a right angle
   c. At an angle
   d. Parallel

2. Mechanic A says helical gears are
   often used to change the direction
   of the axis of rotation.
   Mechanic B says bevel gears are
   often used to change the direction
   of the axis of rotation.
   Who is right?
   a. Mechanic A
   b. Mechanic B
   c. Both mechanic A and B
   d. Neither mechanic A nor B

3. Mechanic A says the pinion centerline is below the centerline of
   the ring gear in a hypoid gear set.
   Mechanic B says the manufacturers use hypoid type final drive
   gears to lower the hump in the
   passenger compartment.
   Who is right?
   a. Mechanic A
   b. Mechanic B
   c. Both mechanic A and B
   d. Neither mechanic A nor B

4. An internal ring gear is also called:
   a. A pinion gear
   b. A center gear
   c. A sun gear
   d. An annulus gear

5. Mechanic A says overdrive is accomplished by holding the sun
   gear and driving the planet carrier.
   Mechanic B says the driven gear
   turns faster than the drive gear.
   Who is right?
   a. Mechanic A
   b. Mechanic B
   c. Both mechanic A and B
   d. Neither mechanic A nor B

6. A gear ratio of 3:1 is an example of:
   a. Overdrive
   b. Direct drive
   c. Gear reduction
   d. None of the above

7. Torque is:
   a. Calculated by multiplying force
      times distance
   b. Twisting force
   c. Sometimes expressed in
      Newton-meters
   d. All of the above

8. Power is:
   a. Torque divided by speed
   b. Torque multiplied by speed
   c. The same as torque
   d. Torque plus speed

9. As output speed increases, torque
   will:
   a. Stay the same
   b. Fluctuate
   c. Decrease
   d. Increase

10. Mechanic A says, to obtain the
    gear ratio, divide the number of
    teeth on the drive gear into the
    number of teeth on the driven
    gear.
    Mechanic B says, to obtain the
    gear ratio, multiply the number of
    teeth on the drive gear by the
    number of teeth on the driven
    gear.
    Who is right?
    a. Mechanic A
    b. Mechanic B
    c. Both mechanic A and B
    d. Neither mechanic A nor B

11. If a driven gear has three times as
    many teeth as its drive gear,
    torque will be:
    a. Divided 3 times
    b. Divided 9 times
    c. Multiplied 3 times
    d. Multiplied 9 times

12. The member of a planetary gear-
    set that is held in order to transmit
    power is often called the:
    a. Input member
    b. Output member
    c. Reaction member
    d. Turned member

13. To produce gear reduction with a
    single plantary gearset, the carrier
    is always the:
    a. Input member
    b. Reaction member
    c. Reversed member
    d. Output member

14. To produce overdrive with a single
    planetary gear set, the carrier is
    always the:
    a. Input member
    b. Reaction member
    c. Reversed member
    d. Output member

15. To produce reverse with a simple
    planetary gear set, the carrier is
    always the:
    a. Input member
    b. Reaction member
    c. Reversed member
    d. Output member

16. Mechanic A says when any two
    members of a planetary gearset
    are driven at the same speed the
    result is direct drive.
    Mechanic B says when the plane-
    tary is locked up, the result is di-
    rect drive.
    Who is right?
    a. Mechanic A
    b. Mechanic B
    c. Both mechanic A and B
    d. Neither mechanic A nor B

17. Mechanic A says Ford uses the
    Ravigneaux gearset in their
    4-speed automatic overdrive
    transmission.
    Mechanic B says GM used the
    Ravigneaux gearset in their Turbo
    Hydra-matic 300 transmission.
    Who is right?
    a. Mechanic A
    b. Mechanic B
    c. Both mechanic A and B
    d. Neither mechanic A nor B

2

Hydraulic Fundamentals

Hydraulics is the study of liquids, and their use to transmit force and motion. The term "hydraulics" comes from the Greek word "hydro," which means water. Home water systems are a common use of hydraulics, and the principles that govern the behavior of water apply to all liquids. Modern automatic transmissions use a hydraulic system to control their operation.

While all hydraulic systems use liquids, the specific liquid used in automatic transmissions is generally called transmission fluid. In this textbook, the terms "liquid" and "fluid" will be used interchangeably when referring to hydraulic systems.

HYDRAULIC PRINCIPLES

Hydraulic systems transmit force and motion through the use of pressure. **Force** is a push or pull acting on an object. Force is usually measured in pounds or Newtons. **Pressure** is force applied to a specific area. Pressure is usually measured in force per unit of area, usually pounds per square inch (psi) or kilo-pascals (kPa). One psi is equal to 6.895 kPa.

The Pascal is a unit of measure named after Blaise Pascal (1623–1662), a French mathematician and scientist recognized for his work with hydraulics. Pascal discovered some important facts about the behavior of liquids in closed systems. He found that pressure on a confined liquid is transmitted equally in all directions, and acts with equal force on equal areas. This is called Pascal's Law, and as we shall see, it is a fundamental principle that rules the operation of all hydraulic systems.

Hydraulic systems are able to use liquids to transmit force and motion because, for all practical purposes, a liquid cannot be compressed. No matter how much pressure is placed on a liquid, its volume remains the same. This allows a liquid to transfer motion much like a mechanical lever. The advantage of a liquid over a mechanical lever is that a liquid has volume, but does not have a fixed shape. Since it assumes the shape of its container, a liquid can be bent to clear any obstacle.

Hydraulic Force

A simple hydraulic system can be built using a single cylinder with an input piston and an output piston having the same surface area, figure 2-1. The space between the pistons is filled with fluid. If we apply 10 pounds of force to the input piston, 10 pounds of force will be available to do work at the output piston as well.

Now let's put the same two pistons in separate cylinders and connect the cylinders with

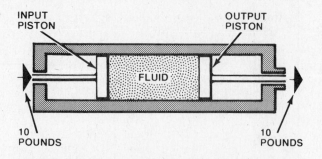

Figure 2-1. Force is transmitted from the input piston to the output piston by fluid in a closed cylinder.

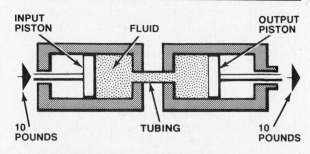

Figure 2-2. Force can be transmitted, without change, from one cylinder to another through tubing.

metal tubing, figure 2-2. Once again, the cylinders and the tubing between the pistons are filled with fluid. As before, applying a force of 10 pounds to the input piston provides a force of 10 pounds at the output piston.

Hydraulic Pressure

Thus far, we have been talking only about input and output force. The study of hydraulics gets more interesting when we examine pressures within the system. Remember that pressure is measured in force per unit of area, usually pounds per square inch (psi) or kilo-pascals (kPa). First, let's see how input force affects hydraulic system pressure.

Hydraulic system pressure is calculated by dividing the input force by the surface area of the input piston (Force ÷ Area = Pressure). In figure 2-3, ten pounds of input force divided by 1 inch of input piston area gives us a system pressure of 10 psi (10 ÷ 1 = 10). In keeping with Pascal's Law, the 10 psi pressure exists throughout the hydraulic system.

If the input force remains constant, but is applied to an input piston with more or less area, the system pressure will differ as well. For example, if we increase the input piston area to 2 inches while keeping the input force at 10 pounds, figure 2-4A, system pressure drops to

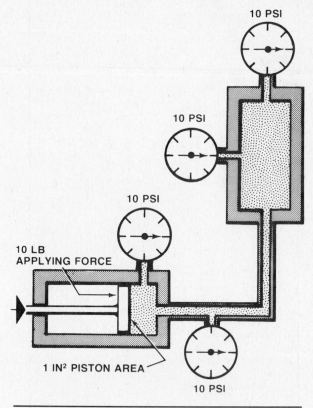

Figure 2-3. Pascal's Law dictates that hydraulic pressure is constant throughout a closed system.

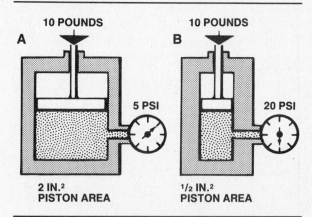

Figure 2-4. Divide the input force by the input piston area to determine system pressure (F ÷ A = P).

only 5 psi (10 ÷ 2 = 5). This also works the other way around. If we reduce the input piston area to ½ inch while keeping the input force at 10 pounds, figure 2-4B, the system pressure rises to 20 psi (10 ÷ ½ = 20).

Now let's look at how system pressure affects output force. The output force is determined by multiplying the system pressure times the output piston area (Pressure × Area = Force). In

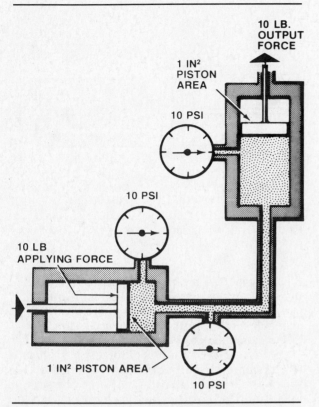

Figure 2-5. When the input and output piston areas are equal, input and output force are also equal.

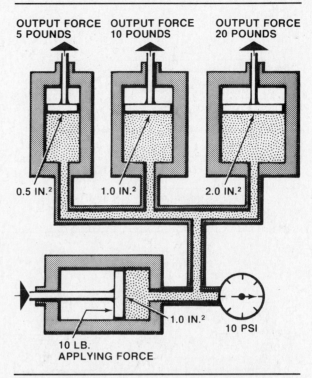

Figure 2-6. Multiply the system pressure times the area of the output piston to determine output force (P × A = F).

figure 2-5, a system pressure of 10 psi applied to a 1-square-inch piston gives an output force of 10 pounds (10 × 1 = 10). Whenever the areas of the input and output pistons are equal, input and output force are equal as well.

Figure 2-6 shows how different amounts of output force can be obtained with a constant system pressure. Apply the 10-psi of pressure to a larger 2-square-inch output piston, and the output force increases to 20 pounds (10 × 2 = 20). But if the 10 psi is applied to a smaller ½-square-inch output piston, output force drops to only 5 pounds (10 × ½ = 5).

Hydraulics: The study, or science, of liquids or fluids, and their use to transmit force and motion.

Force: A push or pull acting on an object. Force is usually measured in pounds or Newtons. Force = Pressure × Area.

Pressure: The force exerted on a given unit of area. Pressure is usually measured in pounds per square inch or kilopascals. Pressure = Force ÷ Area.

■ Torture Test

All automatic transmissions are subjected to exhaustive testing to ensure they will give car buyers dependable service. One of the hardest jobs for an automatic is when it is used to rock a car out of snow or sand. This involves shifting the transmission back and forth from Drive to Reverse several times with varying amounts of throttle.

To make sure its transmissions will stand up to this kind of use, General Motors has devised a brutal torture test. Production transmissions are selected at random and installed in test cars. The cars are then driven to an area of the GM proving grounds called "Black Lake." This is an asphalt pad, several miles square, that is an ideal place to abuse a car.

The test car is driven onto Black Lake. The driver then floors the throttle and holds it wide open while he shifts the transmission rapidly back and forth from Drive to Reverse *200 times in a row!* Although this often tears the car apart, the transmission usually lives through it. We don't recommend this torture test for your own or a customer's car, but you can bet that a transmission strong enough to survive this abuse will be dependable in normal service.

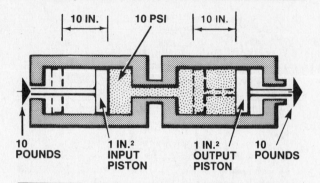

Figure 2-7. When the input and output piston areas are equal, input and output motion are also equal.

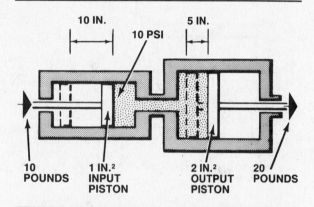

Figure 2-8. If the output piston area is doubled, output force is doubled, but output motion is halved.

Here are the two basic equations used to determine hydraulic system pressure and output force:

Pressure = Force ÷ Area (P = F ÷ A)
Force = Pressure × Area (F = P × A).

Hydraulic Motion

The previous sections discuss how hydraulic systems transmit and change *force* using hydraulic *pressure*, but what about *motion*? As we have just seen, different sized output pistons increase or decrease output force. These changes also affect output motion. The reason is described by the first law of **thermodynamics**, which states that energy cannot be destroyed, it can only be converted from one form into another. This means that whenever one form of energy is increased, another form must be decreased. Or, in even simpler terms, you don't get something for nothing.

In hydraulic systems, the tradeoff is between force and motion. Hydraulic pressure contains a limited amount of energy that must be divided between output force and output motion.

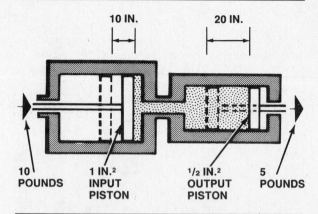

Figure 2-9. If the output piston area is halved, output force is halved, but output motion is doubled.

If energy is used for *increased* output force, output motion is *decreased*. However, if output force is *decreased*, additional energy becomes available and output motion is *increased*. Let's take a closer look at how this works.

In figure 2-7, 10 pounds of force are applied to a 1-square-inch input piston to develop a system pressure of 10 psi. The 10-psi pressure is then applied to a 1-square-inch output piston to develop 10 pounds of output force. This one-to-one relationship holds true for piston movement as well. If the input piston moves 10 inches, the output piston will also move 10 inches. As long as the input and output pistons have the same area, input and output motion will be the same.

Now consider what occurs if we hold system pressure constant at 10 psi, but double the area of the output piston to 2 square inches, figure 2-8. In this case, we double the output *force* to 20 pounds, but we cut the output *motion* in half to only 5 inches.

Just the opposite occurs if we hold the system pressure constant at 10 psi, but halve the area of the output piston to ½ square inch, figure 2-9. In this case, we halve the output *force* to 5 pounds, but we double the output *motion* to 20 inches.

Hydraulics and Work

By now, you should be starting to see that the work performed by a hydraulic system is not simply an expression of force. Although we talk about "10 pounds of force," we must also consider the distance over which the 10 pounds travels. For example, if we apply 100 pounds of force to a ½-square-inch input piston, we have created a hydraulic system pressure of 200 psi (100 ÷ ½ = 200). If the 200-psi pressure is then applied to an 8-square-inch output piston, our

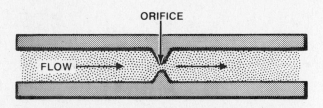

Figure 2-10. An orifice is a small opening used to regulate pressure.

input force of 100 pounds becomes an output force of 1,600 pounds (200 × 8 = 1,600).

Now assume that the ½-inch input piston is attached to the handle of a hydraulic jack, and we pump the handle so that the input piston travels 8 inches. Because we *increased* output force 16 times, we *decreased* output motion 16 times, to ⅟₁₆ of the input motion. Therefore, the hydraulic jack converts our 8-inch input piston motion into ½ inch (that is, ⅟₁₆ of 8 inches) of output piston motion that enables us to raise the 1,600-pound front end of a car.

PRINCIPAL HYDRAULIC SYSTEM PARTS

All automatic transmission hydraulic systems use a reservoir, an input source, control valving, and an output device. A **reservoir** is a tank, pan, or other container that stores fluid for use. An **input source** is a piston or pump that supplies force. **Control valving** is any device that restricts, directs, or otherwise regulates the flow of fluid. An **output device** is a piston or servo motor that transmits force created by hydraulic pressure.

Hydrodynamics

Hydrodynamics is the study of the mechanical movement and action of fluids or liquids in motion. The first thing to learn about a moving fluid is that pressure occurs only when there is resistance against the flow of the fluid.

If a pump can move fluid at 200 gallons per minute (gpm), and fluid is pushed through a hose that will carry at least 200 gpm, the fluid will flow but there will be no pressure. Pressure will not build up in the hose until some resistance is placed against the fluid.

The ability to vary the amount of resistance in a hydraulic system to achieve different degrees of pressure is the basis for the operation of all automatic transmissions. The rest of this chapter will deal with the various types of control valves used in automatic transmissions.

Transmission control valves can be divided into two general categories:
- Pressure-regulating valves — control the *amount* of pressure developed in the transmission, which is used to shift gears.
- Switching valves — control the *direction* of fluid flow, which then shifts the gears.

Some transmission valves perform both pressure regulating and switching at the same time. However, each job can be clearly seen in transmission operation.

PRESSURE-REGULATING VALVES

Orifices, pressure-relief valves, and spool valves are used to regulate hydraulic pressure in a transmission.

Orifice

An **orifice** is a small opening that functions as the simplest type of pressure-regulating valve. In an automatic transmission, an orifice might be a restriction in a line, figure 2-10, or a small hole between two fluid chambers. An orifice may also be used to restrict the flow of fluid through a passage in the transmission case.

As fluid reaches an orifice, it meets resistance and begins to build pressure. Pressure is therefore higher on the side of the orifice that has the greater amount of fluid. There will be a

Thermodynamics: The study, or science, of the relationship between mechanical and heat energy, and the laws governing the conversion of energy.

Reservoir: A tank, pan, or other container that stores fluid for use in a hydraulic system.

Input Source: The piston or pump that supplies the input force in a hydraulic system.

Control Valving: Any devices that restrict, direct, or regulate the flow of fluid in a hydraulic system.

Output Device: A piston or motor that transmits output force created by the pressure in a hydraulic system.

Hydrodynamics: The study, or science, of the mechanical movement and action of liquids or fluids in motion.

Orifice: A small opening or restriction in a line or passage that is used to regulate pressure and flow.

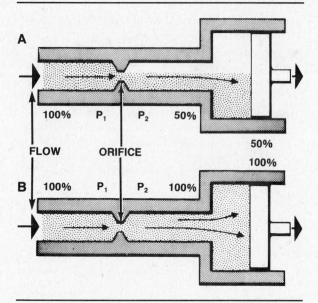

Figure 2-11. With fluid flowing, the pressure at P1 is greater than the pressure at P2 (A). When the chambers on both sides of the orifice are full and fluid no longer flows, the pressures at P1 and P2 are equal (B).

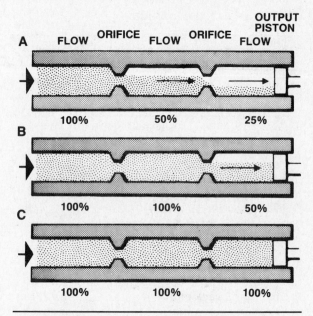

Figure 2-12. When fluid starts to flow, there is a pressure drop at each orifice (A). As the second chamber fills, the pressure equalizes in the first two chambers but still drops at the second orifice (B). When all three chambers are filled, the pressure is equal in all chambers (C).

difference in pressure between the two sides of the orifice as long as the flow of fluid through it remains constant. When the fluid flow stops, pressure becomes equal on both sides of the orifice.

In figure 2-11A, an orifice is located between two chambers of a hydraulic system. The orifice is sized to reduce the fluid pressure on one side of the orifice (P1) by one-half as it passes through to the other side (P2). This makes it possible to apply the piston on the P2 side of the orifice with half the pressure available in the system. Then, once both chambers are full and pressure has equalized, figure 2-11B, the piston is applied with full system pressure.

In the application just described, the orifice allows smooth transfer of fluid from one chamber to another, while maintaining full pressure in the rest of the hydraulic system. Only half of the available pressure in the system is used as long as fluid is flowing. When flow stops, full pressure is available on both sides of the orifice.

For even smoother transfer of fluid, two line restrictions can be used in series, with each orifice reducing the hydraulic pressure by one-half, figure 2-12A. At the start of flow, there is 100-percent pressure in the first chamber, 50-percent pressure in the second chamber, and 25-percent pressure in the third chamber that leads to an output piston.

As flow continues, the second chamber fills, and pressure in it rises to 100 percent. Pressure in the third chamber then rises to 50 percent,

and is delivered to the output piston, figure 2-12B.

When the third chamber fills, fluid flow meets complete resistance and stops. The pressure in the third chamber rises to 100 percent, and uniform pressure then exists in all three chambers. At this point full pressure acts against the output piston in the third chamber, figure 2-12C.

Pressure-Relief Valve

A **pressure-relief valve**, or pressure-limiting valve, allows fluid to escape through an outlet port once a preset level of pressure has built up in a hydraulic system. A common type of pressure-relief valve, figure 2-13, uses a piston, a spring, and an outlet port to control pressure. Fluid enters the inlet port and pushes against the piston. At the same time, the piston is pushed back against the force of the fluid by a spring. The outlet port is located on the side of the piston that has the spring.

When designing a pressure-relief valve, an engineer must specify both the size of the piston surface area and the strength of the spring. Piston area determines the amount of force that hydraulic pressure will exert against the piston. The strength of the spring controls the amount of pressure that must build up before the piston can compress the spring far enough to allow

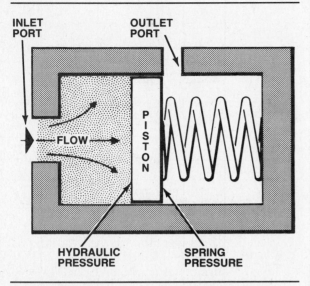

Figure 2-13. When system pressure is not high enough to overcome spring pressure, a pressure-relief valve is closed.

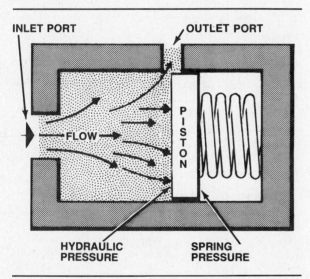

Figure 2-14. When system pressure exceeds spring pressure, a pressure-relief valve opens.

fluid to escape through the outlet port, figure 2-14.

The outlet port must be large enough to carry away all of the fluid that will be pushed through it when the pressure-relief valve opens. If the port is not large enough, it will act as an orifice, and pressure in the system will continue to increase despite the fact that the relief valve has opened.

In automatic transmissions, pressure-relief valves are used to perform a couple of jobs. First, they limit hydraulic system pressures to protect transmission components from damage that can be caused by too much pressure. Sec-

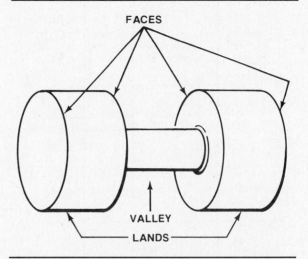

Figure 2-15. A simple spool valve resembles a spool used to wrap thread.

ond, a pressure-relief valve can be used to prevent fluid flow in a hydraulic system until a specific pressure is reached. In this case, the outlet port acts as an inlet port, and no fluid passes through until system pressure against the piston is sufficient to compress the spring and open the valve.

Spool Valves

Most of the valves used in an automatic transmission are **spool valves**, and many of these work as pressure-regulating valves. Spool valves get their name from the fact that they resemble the spools on which sewing thread is wrapped.

Simple spool valve
A simple spool valve, figure 2-15, has two **lands**, four faces, and a **valley** between the two

---

Pressure-Relief Valve: A valve that limits the maximum pressure in a hydraulic system or circuit. Also called a pressure-limiting valve.

Spool Valve: A type of sliding hydraulic control valve, made up of lands and valleys, that resembles a sewing thread spool.

Land: The large, outer circumference of a valve spool that slides against the valve bore. Most spool valves have several lands that are used to block fluid passages.

Valley: The depressions formed between the lands of a spool valve. Valleys allow the passage of fluid into and/or through the valve depending on the position of the spool.

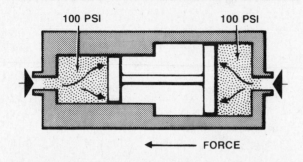

Figure 2-16. When equal pressure is applied to two different-sized faces of a spool valve, the larger face exerts more force.

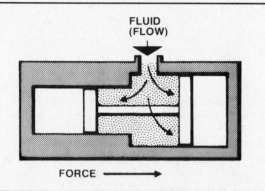

Figure 2-17. If pressure is applied to the valley of a spool valve with two different-sized faces, the larger face again exerts more force.

inner faces. Spool valves are basically a type of piston, and they slide back and forth in machined bores as hydraulic pressure is applied to their various faces.

The lands of a spool valve, and thus the areas of the faces, are not always the same size. This is done to control the movement of the valve when pressure is applied. Remember, a basic principle of hydraulics is that changes in piston (face) area allow hydraulic systems to increase or decrease force. In figure 2-16, a simple spool valve has one face that is twice as large as its other face, and the pressure acting on both faces is the same. In this example, *pressure* acting against the larger face will produce twice as much *output force*, and the valve will move toward the lesser force.

A spool valve that has equal pressure acting on unequal face areas will always move in the direction that the greater force is being applied. As the spool valve moves, it will apply a force of its own equal to the greater force minus the lesser force. For example, if 100 psi acting on the large face of the spool valve in figure 2-16 creates 200 pounds of force, and the same 100 psi acting on the small face creates 100 pounds

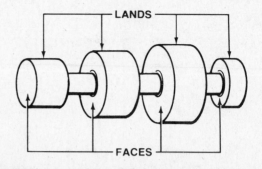

Figure 2-18. A multi-land spool valve has several lands and faces that may be of different sizes.

of force, the spool will move to the left with $(200 - 100 = 100)$ pounds of force.

The same principles apply if we feed fluid under pressure into the valley of a simple spool valve that has one face with twice the area the other face. The identical pressure acting on the larger face creates greater output force, and the spool valve moves toward the larger face (right), figure 2-17.

Multi-land spool valve
A multi-land spool valve has several lands and faces of different sizes that allow more than one force act upon it, figure 2-18. The multi-land spool valve is used in automatic transmissions to regulate hydraulic pressures for a variety of purposes.

Balance Valves

Balance valves are pressure-regulating valves that use a spool valve and spring to control hydraulic system pressures. Some balance valves also use auxiliary lever force or fluid force in doing their jobs. Each of these variations is described below.

Balance valve with spring pressure
A simple balance valve with spring pressure, figure 2-19, allows fluid to enter through an inlet port into the valley of a spool valve. Fluid does not act on the spool valve at this point because pressure is equal against both inner faces of the spool, and both faces are the same size.

A portion of the fluid exits the valley through a transfer port that routes the fluid to a chamber at the left end of the spool valve. When pressure in the left chamber develops enough force against the spool to overcome the spring pressure at the right end of the valve, the spool moves to the right. This opens an outlet port in the left chamber to feed fluid to the **hydraulic circuit** that is being regulated by the valve.

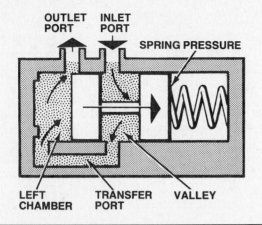

Figure 2-19. When hydraulic pressure acting on the spool valve faces overcomes spring pressure, the spool moves to open the balance valve outlet port.

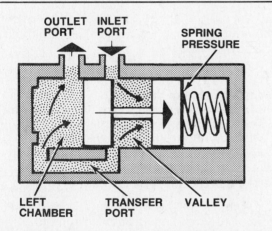

Figure 2-20. If pressure in a balance valve becomes too high, the spool moves far enough to partially block the inlet port.

As pressure in the balance valve increases and force acting on the land faces becomes greater, the spool moves so far to the right that the left land partially closes the inlet port, figure 2-20. This restricts the amount of fluid entering the system, the pressure drops, and the balance valve shifts back to the left, allowing pressure to rise again. The valve does not continually move back and forth as it might appear. Instead, it finds a fixed position that *balances* input pressure against output pressure.

Balance valve with auxiliary lever force
This type of balance valve, figure 2-21, has an auxiliary lever that enables mechanical force to be used as an input to regulate hydraulic pressure. Many of these valves also have an outlet port at the left end of the valve. In this design, the outlet port at the top of the valve acts as a

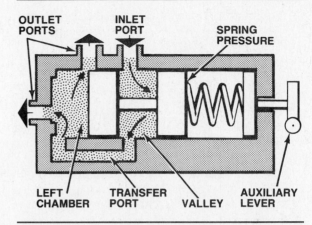

Figure 2-21. This balance valve has an auxiliary lever, and uses an output port as a variable orifice.

variable orifice that resists the flow of the fluid. Let's look at how this valve operates.

As in a basic balance valve, fluid passes through the inlet port, into the valley, then through the transfer port into the left chamber. At this point, some fluid flows through the outlet port at the left end of the valve, and some fluid pushes against the face of the spool.

As pressure builds in the left chamber, pressure on the left face of the spool valve increases, and the spool moves to the right, figure 2-21. This exposes the outlet port (orifice) at the top of the valve, which limits pressure in the outlet port at the left end of the valve.

If pressure continues to build, the spool moves farther to the right, and partially closes off the inlet port. Pressure to the outlet port at the left end of the valve is now based on:

● Pressure through the inlet port
● The restriction of the inlet port caused by the balancing effect of the valve.
● The restriction of the top outlet port (orifice)
● The area of the left-end outlet port
● The area of the left face of the spool valve
● Spring pressure at the right end of the valve.

We can increase the output pressure of the balance valve by increasing the pressure of the

Balance Valve: A spool valve and spring combination used to regulate hydraulic system pressure. Some balance valves also use auxiliary lever force or fluid force in doing their jobs.

Hydraulic Circuit: An arrangement of an input source, fluid passages, control valves, and an output device that is used to transmit motion and force to do work.

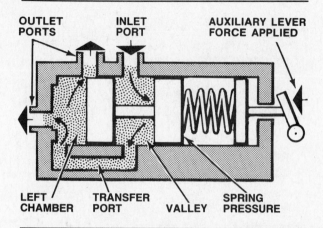

Figure 2-22. When the auxiliary lever is moved to increase spring pressure, the regulated fluid pressure increases as well.

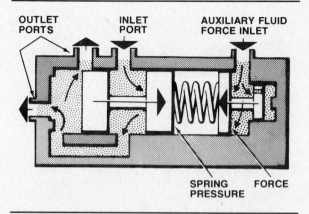

Figure 2-23. This balance valve uses auxiliary fluid force to increase spring pressure, and thus regulated hydraulic pressure.

spring. This is done by pressing on the auxiliary lever attached to the spring, figure 2-22. This increases spring pressure against the right side of the valve, which requires the development of more hydraulic pressure in the left chamber. More force must be applied to the left face of the spool valve in order to overcome the increased spring force.

The mechanically operated throttle valves in transmissions such as Chrysler's Torqueflite and GM's Turbo Hydra-matic 200 are good examples of balance valves that use auxiliary lever force. We will study the operation of these valves within a full hydraulic system in more detail in Part Two of this manual.

Balance valve with auxiliary fluid force
A balance valve can also use auxiliary fluid force in place of auxiliary lever force. In this design, figure 2-23, an auxiliary fluid force inlet is located in a valley between the inner faces of a second spool valve. The second spool is positioned in a chamber to the right of the balance valve.

In this example, the face on the left land of the auxiliary spool valve has twice the area of the face on the right land. As we have seen, this means that when equal pressure is applied to both lands, the large land will generate twice as much force as the small land. As a result, any hydraulic pressure applied to the auxiliary spool valve will cause it to move to the left. As the valve moves to the left, it increases the spring pressure against the balance-valve spool.

The factors that affect outlet pressure in a balance valve that uses auxiliary lever force also affect a balance valve using auxiliary fluid force. These things are:
* Pressure through the inlet port
* The restriction of the inlet port caused by the balancing effect of the valve.
* The restriction of the top outlet port (orifice)
* The area of the left-end outlet port
* The area of the left face of the spool valve
* Spring pressure at the right end of the valve.

SWITCHING VALVES

A switching, or directional, valve moves fluid from one hydraulic passage into another, or from one hydraulic circuit to another. A switching valve may also allow one set of passages to be used for more than one hydraulic circuit. Common types of switching valves are described in the following sections.

One-Way Check Valve

As its name implies, a **one-way check valve** allows fluid to flow through it in one direction only. A poppet valve, figure 2-24, is one type of one-way check valve. Fluid passes through the valve (in the direction flow is allowed) only after hydraulic pressure exceeds the spring pressure holding the poppet against its seat.

If a spring-loaded ball is substituted for a spring-loaded poppet in this kind of one-way check valve, the operation remains the same. Spring pressure holds the ball against its seat at the inlet port until fluid pressure becomes high enough to overcome spring pressure and lift the ball off its seat.

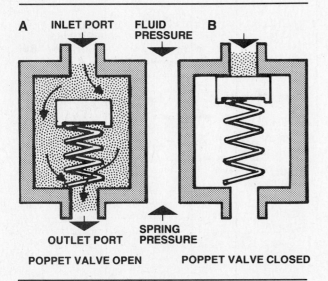

Figure 2-24. A 1-way poppet check valve allows flow when fluid pressure overcomes spring pressure (A). When fluid pressure drops below spring pressure, the valve closes to prevent flow (B).

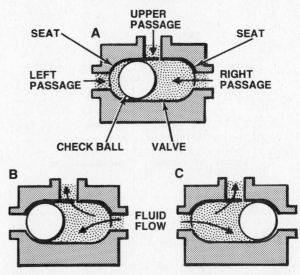

Figure 2-25. A 2-way ball-check valve has a self-sealing action that allows two hydraulic circuits to share a single passage.

In both types of spring-loaded one-way check valves, fluid cannot flow backward through the valve. When inlet pressure drops, such as during a reverse-flow or a no-flow condition, spring pressure forces the poppet or ball against its seat to close the valve. Any fluid pressure that might be applied in a reverse direction simply holds the poppet or ball against its seat that much tighter.

Two-way Check Valve

A **two-way check valve** controls fluid flow in two separate hydraulic circuits. A ball-check valve, figure 2-25A, is the simplest type of two-way check valve.

As shown in figure 2-25B, the valve operates automatically when hydraulic pressure enters the circuit. If fluid enters from the right passage, the ball moves toward the left passage. Fluid pressure then holds the ball against the seat for the left passage, and blocks fluid flow in that direction. The fluid escapes through the upper valve passage.

If fluid enters the ball-check valve from the left passage, figure 2-25C, the ball moves toward the right passage. Fluid pressure then holds the ball against the seat for the right passage. This stops fluid from entering the right passage, and flow is again directed out through the upper passage.

The automatic sealing action of a two-way ball-check valve allows two hydraulic circuits to share one passage. In automatic transmissions, fluid pressure from the two circuits is fed into

the left and right passages of the ball-check valve. The upper valve passage then becomes a shared outlet. This arrangement allows the two circuits to control the same transmission output device under different operating conditions.

Manually Operated Switching Valve

A **manually operated switching valve**, figure 2-26, is a spool valve that is moved by a mechanical lever. Depending on the position of the lever, and thus the spool in its bore, fluid pressure may be blocked from entering the valve, contained within the valve, or allowed to pass through the valve.

In figure 2-26A, the spool is positioned so that its left land blocks the inlet port. This prevents fluid from entering the switching valve.

One-Way Check Valve: A type of switching valve that allows fluid to pass in one direction only, and then only when the pressure is sufficient to unseat the valve.

Two-Way Check Valve: A type of switching valve that controls fluid flow in two separate hydraulic circuits, and allows them to share a single fluid passage.

Manually Operated Switching Valve: A spool valve whose position is controlled by a mechanical lever. Automatic transmission gear selector levers usually connect to a manually operated switching valve.

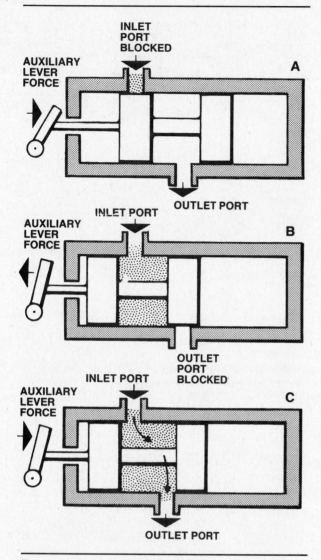

Figure 2-26. A manually operated switching valve uses a lever to change the position of a spool valve that opens and closes hydraulic circuits.

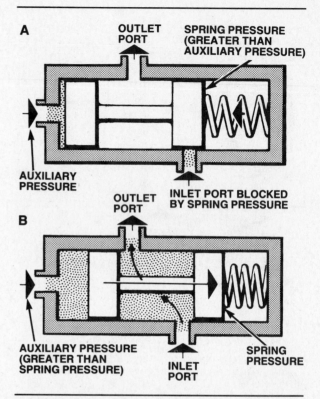

Figure 2-27. A hydraulically operated switching valve closes a circuit when spring pressure is greater than auxiliary hydraulic pressure (A), and opens the circuit when auxiliary pressure exceeds spring pressure (B).

In figure 2-26B, the lever is moved to reposition the spool so that the inlet port is open. This allows fluid to enter the valley of the spool. However, because the right land of the spool is blocking the outlet port, fluid pressure is contained within the switching valve.

In figure 2-26C, the lever is moved even further, and the spool position now allows both the input and output ports to be open. Fluid passes through the valley of the spool, and out the outlet port. Flow from the outlet port serves a specific circuit in the hydraulic system.

The shift lever of most automatic transmissions is connected to a manually operated switching valve that is used to select the desired gear ranges. Actual transmission manual switching valve assemblies have many lands

and ports, but although they are more complex, they use the same fluid control options described above.

Hydraulically Operated Switching Valve

A **hydraulically operated switching valve**, figure 2-27, is a spool valve that is moved by hydraulic pressure working against spring pressure. As on its manually operated counterpart, the lands of a hydraulically operated switching valve open and close various hydraulic circuits to control transmission operation.

In figure 2-27A, spring pressure positions the switching valve spool so that fluid cannot flow from the inlet port to the outlet port. This

Hydraulically Operated Switching Valve: A spool valve whose position is determined by hydraulic pressures acting on its faces. The shift valves that control most gearchanges in an automatic transmission are hydraulically operated switching valves.

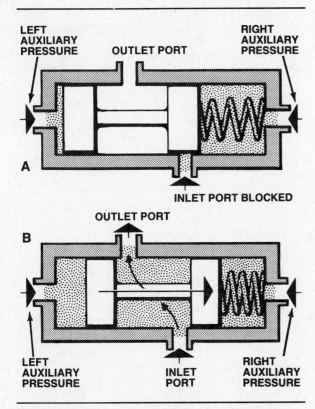

Figure 2-28. A transmission shift valve closes a circuit when spring and auxiliary hydraulic pressure on the right is greater than auxiliary hydraulic pressure on the left (A). The valve opens the circuit when left auxiliary pressure exceeds the combination of spring and auxiliary hydraulic pressure on the right (B).

blocks a circuit in the hydraulic system. However, if auxiliary hydraulic pressure at the left of the valve increases to the point where it overcomes the spring pressure, figure 2-27B, the valve will move to the right, opening the inlet port and allowing fluid pressure to feed the hydraulic circuit.

Transmission shift valve

A transmission shift valve, figure 2-28, works like a simple hydraulically operated switching valve except that *two* auxiliary hydraulic pressures are used to move the valve. One auxiliary pressure pushes against the left side of the valve, while the other pressure pushes against the right side.

As long as the combination of spring pressure and auxiliary hydraulic pressure against the right side of the valve is greater than the auxiliary hydraulic pressure against the left side of the valve, figure 2-28A, the right land of the spool valve will block the inlet port. This keeps fluid from moving through the shift valve into the hydraulic circuit.

If the auxiliary pressure at the left of the valve becomes greater than the combination of spring pressure and auxiliary hydraulic pressure at the right of the valve, figure 2-28B, the valve will move to the right. This opens the inlet port and allows fluid to pass through the valley of the spool to the outlet port. With this design, the timing of the valve movement can be varied by changing the auxiliary pressures.

Transmission shift valves are sometimes referred to as "snap valves" because they shift almost instantly when the auxiliary pressure on their ends change. We will examine shift valve operation in more detail in Chapter 4, and when we study specific transmissions in Part Two of this manual.

SUMMARY

Hydraulics is the study of liquids, or fluids, and their use of pressure to transmit force and motion. Hydraulic systems use liquids for this purpose because liquids cannot be compressed.

Hydraulic systems are governed by physical laws and principles that affect the ways they transmit and modify force. Remember these points about hydraulic systems:
1. For any given input force, hydraulic pressure is constant throughout a closed system.
2. For any given input force, a small input piston will develop more hydraulic system pressure than a large input piston.
3. For any given hydraulic system pressure, a large output piston will develop more force than a small output piston.
4. An output piston that increases output force over input force will have less travel than the input piston.
5. An output piston that reduces output force below input force will have greater travel than the input piston.

Every hydraulic system requires a fluid reservoir, an input source, control valving, and an output device. Hydraulic pressure does not exist in a system until there is resistance to fluid flow. Orifices, pressure relief valves, spool valves, and balance valves are pressure-regulating valves used to create resistance and control the resulting hydraulic pressure.

Poppet check valves, ball-check valves, manually operated switching valves, and hydraulically operated switching valves are all types of switching valves used to move hydraulic fluid from one passage to another, or from one hydraulic circuit to another. Some valves perform both pressure regulation and switching.

Review Questions

Select the single most correct answer
Compare your answers to the correct answers on page 603

1. Mechanic A says force is push or pull acting on an object.
   Mechanic B says pressure is force exerted on a given surface area.
   Who is right?
   a. Mechanic A
   b. Mechanic B
   c. Both mechanic A and B
   d. Neither mechanic A nor B

2. Mechanic A says force is calculated by multiplying pressure times surface area.
   Mechanic B says force is usually measured in foot-pounds or Newton-meters.
   Who is right?
   a. Mechanic A
   b. Mechanic B
   c. Both mechanic A and B
   d. Neither mechanic A nor B

3. Mechanic A says pressure is calculated by multiplying the force times the surface area.
   Mechanic B says pressure is usually measured in pounds per square inch or kilopascals.
   Who is right?
   a. Mechanic A
   b. Mechanic B
   c. Both mechanic A and B
   d. Neither mechanic A nor B

4. If a force of 200 pounds is applied to a 4-square-inch input piston, the resulting hydraulic pressure is:
   a. 80 psi
   b. 50 psi
   c. 20 psi
   d. 10 psi

5. If hydraulic pressure of 100 psi is applied to a 10-square-inch output piston, the output force is:
   a. 100 pounds
   b. 10 pounds
   c. 1,000 pounds
   d. 10,000 pounds

6. Mechanic A says the output force of a hydraulic system is dependent on area and pressure.
   Mechanic B says the output force of a hydraulic system is dependent on pressure only.
   Who is right?
   a. Mechanic A
   b. Mechanic B
   c. Both mechanic A and B
   d. Neither mechanic A nor B

7. If the output piston of a hydraulic system is larger than the input piston, output motion will be:
   a. Greater than input motion
   b. Less than input motion
   c. The same as input motion
   d. Any of the above depending on system pressure

8. A complete hydraulic system must have:
   a. An input source and control valving
   b. A reservoir
   c. An output device
   d. All of the above

9. Mechanic A says that pressure regulating valves control the shifts in a transmission hydraulic system.
   Mechanic B says the switching valves control the amount of pressure developed in a hydraulic system.
   Who is right?
   a. Mechanic A
   b. Mechanic B
   c. Both mechanic A and B
   d. Neither mechanic A nor B

10. Which of the following is *not* primarily a pressure-regulating valve:
    a. An orifice
    b. A shift valve
    c. A relief valve
    d. A balanced spool valve

11. Mechanic A says a mechanically operated throttle valve is a good example of a balance valve with auxiliary lever force.
    Mechanic B says a mechanically operated throttle valve uses an outlet port as a fixed orifice.
    Who is right?
    a. Mechanic A
    b. Mechanic B
    c. Both mechanic A and B
    d. Neither mechanic A nor B

12. With uniform pressure in a hydraulic system, a large output piston will develop _____ output force than a small output piston.
    a. Less
    b. More
    c. Equal
    d. None of the above

3

Transmission Hydraulic Systems

In our study of hydraulic principles in Chapter 2, we examined several kinds of pressure-regulating valves and switching valves. In this chapter, we will learn how these valves operate within automatic transmissions. We will also look at the pumps that provide automatic transmissions with three kinds of hydraulic pressure. Then, we will trace these pressures through the transmission passages and valves, and learn how they interact with one another to control upshifts and downshifts.

TRANSMISSION PRESSURES

Before we begin a detailed examination of transmission pumps and valves, we need to take a brief look at the three principle automatic transmission pressures:

- Mainline pressure
- Throttle pressure
- Governor pressure.

Remember, in order to develop hydraulic pressure, there must be a source of fluid flow and a resistance to that flow. In an automatic transmission, the pump is the source of fluid flow, and various valves and passages provide the resistance that is used to develop pressure.

Mainline Pressure

Mainline pressure is the hydraulic pressure developed by the transmission pump and controlled by the pressure-regulator valve at the pump output. Mainline pressure is used to apply the servos and clutches (explained in Chapter 4) that hold and drive various members of the planetary gearsets to provide transmission gear shifts. Mainline pressure is also the source of all other pressures in the transmission.

Some carmakers refer to mainline pressure by other names, such as "control pressure" or "drive oil." We will learn more about these different names in Part Two of this manual. In this textbook, we will use the term "mainline pressure" to refer to the regulated hydraulic pressure from the transmission pump.

Throttle Pressure

Throttle pressure is a hydraulic pressure that increases with engine load or throttle opening. Throttle pressure is developed from mainline pressure at a throttle valve that is controlled by mechanical linkage from the car's throttle, or by a vacuum diaphragm that senses engine intake manifold vacuum. Throttle pressure is used to control transmission shift points by interacting with governor pressure (see below) on shift valves.

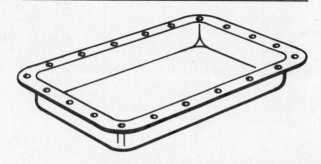

Figure 3-1. The sump, or oil pan, is the fluid reservoir of an automatic transmission hydraulic system.

Some carmakers refer to throttle pressure by other names such as "modulator pressure". We will learn more about these different names in Part Two of this manual. In this textbook, we will use the term "throttle pressure" to refer to the regulated hydraulic pressure based on engine load or throttle opening.

Governor Pressure

Governor pressure is a hydraulic pressure that increases with vehicle road speed. It is developed from mainline pressure at a centrifugally operated valve, called the governor, that is driven off the transmission output shaft. Governor pressure is used to control transmission shift points by interacting with throttle pressure on shift valves.

TRANSMISSION HYDRAULIC PUMPS

All automatic transmission operating pressures are developed from the output flow of the oil pump. The pump draws transmission fluid, through a pickup and filter, from the **sump**, figure 3-1. The sump, or oil pan, is simply a fluid reservoir that holds a supply of fluid ready for the pump at all times. Three types of pumps are currently used on domestic automatic transmissions:

• Gear
• Rotor
• Vane.

Gear and rotor pumps are called constant-displacement or **positive-displacement pumps** because they displace, or deliver, exactly the same volume of fluid with each revolution, regardless of the speed of rotation. As pump speed increases, the volume of fluid delivered *per minute* increases because the pump makes

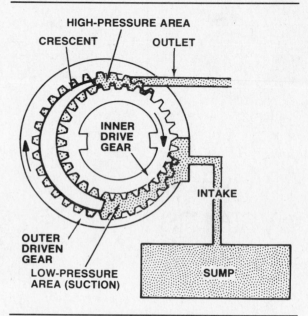

Figure 3-2. A gear pump consists of an inner drive gear and an outer driven gear separated by a stationary crescent.

more revolutions. Although the overall delivery rate increases with pump speed, the volume delivered for each revolution remains constant. The disadvantage of positive-displacement pumps is that they require increasing amounts of energy to drive as pump speed increases, even though transmission operation may not require the extra fluid flow being produced.

Simple vane pumps are positive-displacement pumps that follow the same rules of operation as the gear and rotor pumps described above. However, the vane pumps used in modern automatic transmissions are called **variable-displacement pumps** because the volume of fluid they deliver with each revolution can be changed. The pump output is automatically regulated based on the needs of the transmission, and there is no direct relationship between pump speed and fluid flow. The advantage of variable-displacement pumps is that they do not waste power producing fluid flow that is not required to operate the transmission.

Transmission pumps mount in the front of the transmission case and are driven by the torque converter. Transaxle pumps may also be driven by the torque converter, but in some applications they are driven by a separate pump driveshaft. Many older transmissions had a second pump, mounted at the rear of the case and driven by the output shaft. Rear pumps have not been used in domestic transmissions since the late 1960s.

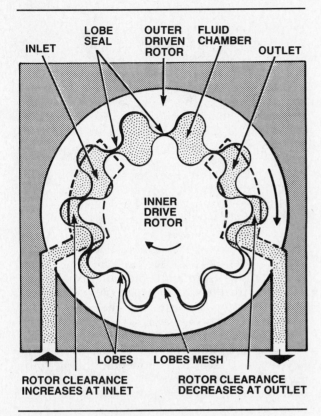

INLET LOBE SEAL OUTER DRIVEN ROTOR FLUID CHAMBER OUTLET

INNER DRIVE ROTOR

LOBES LOBES MESH

ROTOR CLEARANCE INCREASES AT INLET

ROTOR CLEARANCE DECREASES AT OUTLET

Figure 3-3. A rotor pump consists of an inner drive rotor and an outer driven rotor.

Gear Pump

A **gear pump**, figure 3-2, consists of an inner drive gear and an outer driven gear installed in a pump body. Gear pumps are built with either spur gears or helical gears, but both types work the same. The inner gear is turned at engine speed by the hub of the torque converter, and drives the outer gear. Both gears mesh at one side but are separated at the opposite side by a crescent-shaped section of the pump body. Because of this, gear pumps are sometimes called gear-and-crescent pumps.

As the gears rotate, they create a low-pressure area at the point where they separate, figure 3-2. This is called the suction side of the pump. The pressure developed at this point is lower than the atmospheric pressure pressing on the fluid in the sump. Therefore, fluid flows from the sump into the pump inlet.

As the pump gears continue to rotate, they carry the fluid in the spaces between the gear teeth past the crescent to the pump outlet. Near the outlet, the clearance between the gears decreases, forcing the fluid out of the pump. Whenever the engine is running, a gear pump is operating and delivering fluid.

Rotor Pump

A **rotor pump**, figure 3-3, operates on the same principles as a gear pump, but has inner and outer rotors rather than inner and outer gears. The inner rotor is turned at engine speed by the torque converter hub, and drives the outer rotor. Unlike gears which have teeth, rotors have lobes that mesh with one another. In a

Mainline Pressure: The pressure developed from the fluid output of the pump and controlled by the pressure regulator valve. Mainline pressure operates the apply devices in the transmission and is the source of all other pressures in the hydraulic system.

Throttle Pressure: The transmission hydraulic pressure that is directly related to engine load. Throttle pressure increases with throttle opening and engine torque output, and is one of the principle pressures used to control shift points.

Governor Pressure: The transmission hydraulic pressure that is directly related to vehicle speed. Governor pressure increases with vehicle speed and is one of the principle pressures used to control shift points.

Sump: The oil pan, or reservoir, that contains a supply of fluid for the transmission hydraulic system.

Positive-Displacement Pump: A pump that delivers the same volume of fluid with each revolution. The gear and rotor pumps used in automatic transmissions are positive-displacement pumps.

Variable-Displacement Pump: A pump whose output volume per revolution can be varied to increase or decrease fluid delivery. The vane pumps used in automatic transmissions are variable-displacement pumps.

Gear Pump: A positive-displacement pump that uses an inner drive gear and an outer driven gear, separated on one side by a crescent, to produce oil flow. Gear pumps may use either helical or spur gears, and are sometimes called gear-and-crescent pumps.

Rotor Pump: A positive-displacement pump that uses an inner drive rotor and an outer driven rotor to produce oil flow. Lobes on the rotors create fluid chambers of varying volumes, and eliminate the need for a crescent as used in a gear pump.

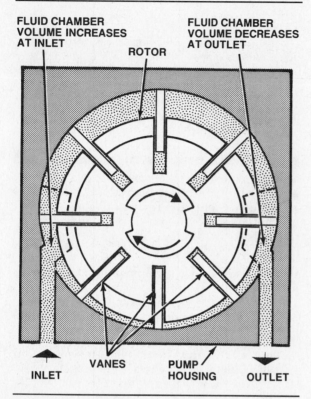

FLUID CHAMBER
VOLUME INCREASES
AT INLET

FLUID CHAMBER
VOLUME DECREASES
AT OUTLET

ROTOR

VANES   PUMP
HOUSING

INLET   OUTLET

Figure 3-4. A simple, positive displacement, vane pump.

rotor pump, the meshing action of the lobes eliminates the need for a separating crescent between the rotors.

The rotors mesh with each other at one side but separate from each other at the opposite side. Where the rotors mesh, the fluid volume between them is very small. As the rotors separate, the rounded lobes maintain a seal, creating small chambers whose volume increases as the rotors move farther apart. As the volume of the chambers increases, low-pressure areas are formed that allow fluid to flow from the sump into the pump inlet, just as in a gear pump.

As the pump rotors continue to rotate, they carry fluid in the chambers formed between the lobes. Near the pump outlet, the clearance between the rotors decreases. This reduces the volumes of the fluid chambers, and forces the fluid out of the pump. Whenever the engine is running, a rotor pump is operating and delivering fluid.

Vane Pump

In order to better understand the operation of the variable displacement vane pumps used in automatic transmissions, we will first examine the operation of a simple **vane pump**. Although

simple vane pumps are not used in automatic transmissions, they are found in other vehicle systems such as power steering.

A simple vane pump, figure 3-4, consists of a rotor, several vanes, and a pump body. The rotor is driven by the engine, and has several radial slots cut around its circumference. The vanes fit into the rotor slots, and are free to slide in and out. When the pump is turning, **centrifugal force** moves the vanes outward so their outer edges seal against the inside of the pump housing to form several fluid chambers.

The centerpoint of the rotor is offset from the centerpoint of the opening in the pump body. As a result, the rotor is closer to the pump body on one side than on the other. Where the rotor and pump body are close, the volume of the fluid chambers formed by the vanes is very small. As the distance between the rotor and pump body increases, the volume of the fluid chambers becomes greater.

As the rotor turns, the volumes of the chambers over the pump inlet increase. This forms low-pressure areas that allow fluid to flow from the sump into the pump inlet, just as in a gear or rotor pump. As the rotor continues to turn, it carries with it the fluid trapped in the chambers between the vanes.

Near the pump outlet, the clearance between the rotor and pump body decreases. This reduces the volumes of the fluid chambers, and forces the fluid out of the pump. Whenever the engine is running, a simple vane pump is operating and delivering fluid.

Variable displacement vane pump

The variable displacement vane pumps in automatic transmissions, figure 3-5, consist of a rotor, several vanes, a vane ring, a slide with a pivot pin and seal, a priming spring, and a pump body. Rather than sealing against the pump body directly, the vanes in this type of pump seal against the inside of a slide that mounts inside the pump body.

As in a simple vane pump, the centerpoint of the opening in the slide is offset from the rotor centerpoint. The difference is that the slide is able to pivot in relation to the rotor, which moves the opening centerpoint closer to or farther from that of the rotor. The pivoting slide is what gives the pump its variable displacement capability.

The rotor and vanes are basically the same as those in a simple vane pump, but a vane ring is added that fits inside the rotor and contacts the inner edges of the vanes. The vane ring limits the inward movement of the vanes, ensuring that they are always close to the inside of the slide. This prevents the vanes from sticking in

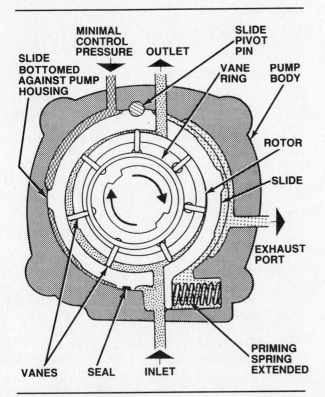

Figure 3-5. A variable displacement vane pump with its slide in the maximum output position.

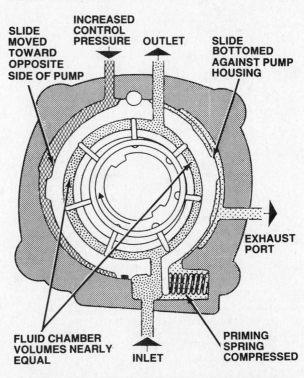

Figure 3-6. A variable displacement vane pump with its slide in the minimum output position.

the rotor slots, and minimizes the amount of fluid bypassing the vanes when pump speed is low and the amount of centrifugal force available to seal the vanes against the slide is limited.

Whenever the engine is off, figure 3-5, the priming spring holds the slide at full offset against the left side of the pump body. This position provides maximum pump output because there is the greatest variation in the sizes of the fluid chambers. It also creates a large pressure differential between atmospheric pressure acting on the fluid in the sump, and the low pressure area formed at the pump inlet. This primes the pump quickly during cranking, and produces a large volume of fluid flow to pressurize the entire transmission hydraulic system within the first few moments of engine operation.

Once the engine starts, the basic operation of a variable displacement vane pump is like that of a simple vane pump. The torque converter hub or pump driveshaft turns the rotor and vanes inside the slide. As the clearance between the rotor and slide increases on the inlet side and decreases on the outlet side, fluid is drawn into and forced out of the pump.

Under most operating conditions, however, maximum output is not required, and the pump displacement is reduced by applying a control pressure to the backside of the slide, figure 3-6. This moves the slide to the right against priming spring pressure, and brings the centerpoints of the slide opening and rotor closer together. As this happens, the variation in the volumes of the fluid chambers becomes smaller, and the output flow of the pump drops. Minimum pump output occurs when the slide is bottomed against the right side of the pump housing.

During transmission operation, the pressure regulator valve (see below) continually varies the control pressure to adjust pump output to meet immediate operating needs. To ensure accurate regulation, pressure is never allowed to build behind the right side of the slide. Any fluid that does leak past the slide pivot pin or seal into this area escapes through an exhaust port.

Vane Pump: A pump that uses a slotted rotor and sliding vanes to produce oil flow. The vane pumps used in automatic transmissions are variable-displacement pumps.

Centrifugal Force: The natural tendency of objects, when rotated, to move away from the center of rotation.

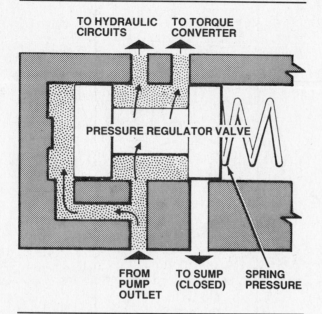

TO HYDRAULIC   TO TORQUE
CIRCUITS   CONVERTER

PRESSURE REGULATOR VALVE

FROM   TO SUMP   SPRING
PUMP   (CLOSED)   PRESSURE
OUTLET

Figure 3-7. A pressure regulator valve develops and controls mainline pressure in an automatic transmission hydraulic system by forming a variable restriction.

Remember, maximum pump output occurs when the priming spring extends fully and holds the slide against the left side of the pump body, figure 3-5. As the slide moves toward the center of the rotor, output drops until it reaches a minimum when the slide is held against the right side of the pump body, figure 3-6. We will examine the operation of several variable displacement pumps in Part Two of this manual.

MAINLINE PRESSURE

As we have noted, there must be a restriction on the pump output to develop pressure. If there is no restriction, pump output flow rate can be measured, but the pressure is zero. In reality, there are many restrictions in a transmission hydraulic system. The pump outlet port forms a partial restriction. So do the lines and passages to and from the pump control valves. The first important restriction in the system — and the one used to control mainline pressure — is the pressure regulator valve. Other valves acted on by mainline pressure include the booster valve, manual control valve, and shift valves.

Of course, mainline pressure is not the only pressure in a transmission hydraulic system. There are a number of others, but they are all based on mainline pressure, and are created by other valves with pressure-regulating functions, such as the throttle valve and the gover-

nor valve. We will see how these pressures are developed in other circuits later in this chapter.

Pressure Regulator Valve

Because pump delivery volume increases with speed, pressure would also increase with pump speed if the fluid operated against a simple fixed restriction such as an orifice. In such a system, pressure would quickly reach levels high enough to damage various transmission components. Therefore, pressure must be regulated. This is done with the **pressure regulator valve**, which forms a variable restriction. A variable restriction is an opening in a hydraulic circuit whose size can be changed in order to alter the pressure developed by the restriction.

The pressure regulator valve controls the high and low limits of mainline fluid pressure to meet various transmission operating conditions. Mainline pressure actually starts in the pressure regulator valve, but fluid diagrams often show it coming directly from the pump. This is due to the balancing effect of the valve, which makes the pressure between the pump and valve nearly equal to the pressure leaving the valve to enter the mainline circuit. A typical pressure regulator valve works as follows.

When the engine is started, fluid from the pump enters the valve, figure 3-7. At first, the fluid passes directly through the valve to fill the torque converter and mainline circuits. Once these areas are purged of air, pressure in the hydraulic system begins to build. That pressure acts on the end of the regulator valve and moves the valve against the opposing spring pressure. As long as pump pressure does not exceed spring pressure, full pump output is delivered to the transmission mainline circuit.

As engine speed increases, the volume of fluid sent to the valve by the pump also increases. This causes increased pressure on the valve, which moves it far enough against spring pressure to expose an outlet port, figure 3-8. In a transmission with a gear or rotor pump, excess pressure is then relieved by the flow of fluid from the outlet port back into the suction side of the pump. In a transmission with a vane pump, a portion of the excess pressure is routed to the backside of the pump slide to reduce the pump displacement, and thus the output flow and pressure. In either case, the transmission is protected from damage that might be caused by excessive hydraulic pressure.

In many transmissions, pressure from the pump is routed directly to the manual control valve and other valves in the hydraulic system. This is possible because all the fluid *does not* have to pass through the pressure regulator

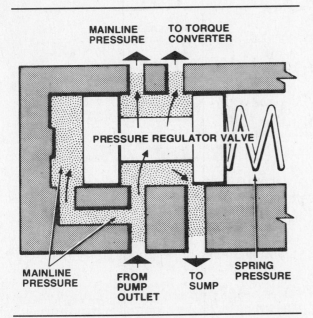

Figure 3-8. When pump pressure overcomes spring pressure, the pressure regulator valve opens to return fluid to the sump and limit mainline pressure.

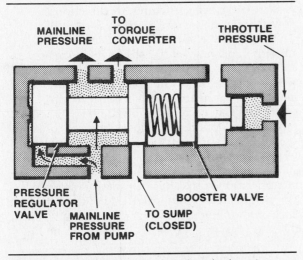

Figure 3-9. Throttle pressure against the booster valve combines with spring pressure in the pressure regulator valve to increase mainline pressure.

valve *before* going to other valves. As we noted in Chapter 2, pressure within a closed hydraulic system is the same everywhere in that system. You can think of the mainline pressure circuit as a single, closed system. No matter where the pressure regulator valve is located in that circuit, it will regulate the pressure at all points within the circuit.

Booster Valve

Under certain driving conditions, such as low-speed operation under heavy engine load,

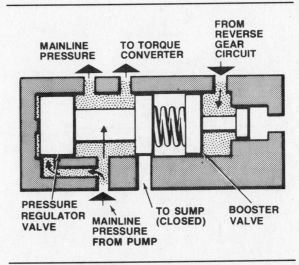

Figure 3-10. Pressure from the reverse-gear circuit can also act on the booster valve to increase mainline pressure.

transmission mainline pressure must be raised above the level allowed by the normal action of the pressure regulator valve. In these situations, higher mainline pressures increase the holding force of the apply devices. This prevents the clutches and bands from slipping under the increased torque output of the engine. To increase mainline pressure in these circumstances, most transmissions have some type of **booster valve** that acts on the pressure regulator valve.

A typical booster valve, figure 3-9, acts against the spring end of the pressure regulator valve. At full throttle, or full load, throttle pressure is routed from the throttle valve to the end of the booster valve that is farthest from the spring in the pressure regulator valve. Throttle pressure and spring pressure then combine to work against pump pressure, increasing the mainline pressure from the pressure regulator valve.

To help hold the apply devices for reverse gear, auxiliary pressure from the reverse gear circuit is fed into the valley of the booster valve, figure 3-10. Because the land of the booster valve closest to the pressure regulator valve is

Pressure Regulator Valve: The valve that regulates mainline pressure by creating a variable restriction.

Booster Valve: A valve that increases mainline pressure when driveline loads are high to prevent apply devices from slipping.

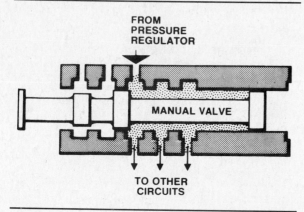

Figure 3-11. The lands and valleys of a manual valve direct mainline pressure to some circuits while blocking flow to others.

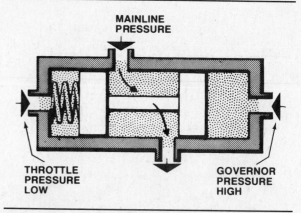

Figure 3-13. When governor pressure exceeds throttle pressure and spring pressure, the shift valve "snaps" and allows mainline pressure to pass.

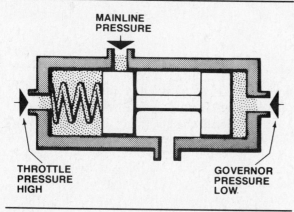

Figure 3-12. A shift valve blocks mainline pressure when throttle pressure is higher than governor pressure.

larger than the other land, greater force is developed against it. Therefore, pressure from the reverse gear circuit pushes the booster valve against the pressure regulator valve spring to increase mainline pressure.

Manual Control Valve

The manual control valve, or **manual valve**, gets its name from the fact that the driver manually moves the valve using the gear selector linkage. The manual control valve is used to shift the transmission from one driving range to another.

Mainline pressure is applied to the manual valve, figure 3-11, from which it is routed to various hydraulic circuits under different driving conditions. The lands and valleys of the manual valve direct the flow of fluid to some circuits, while blocking the flow of fluid to others. The manual valve allows fluid under pressure to be sent to the various apply devices and

valves in the transmission. All of the valves affected by the manual control valve are discussed later in this chapter.

Shift Valves

A **shift valve** is a switching, or directional, valve that uses governor pressure and throttle pressure to time shifts in an automatic transmission. For this reason, shift valves are also called shift timing valves. By sensing engine torque (by means of throttle pressure) and vehicle speed (by means of governor pressure), upshifts and downshifts can be properly timed for any driving conditions. Most transmissions have several shift valves that control the various gearchanges required.

A shift valve makes an upshift when governor pressure overcomes throttle pressure and spring pressure. In figure 3-12, governor pressure enters the right side of the shift valve and throttle pressure enters the left side of the valve. A coil spring is also located at the left side of the valve. Mainline pressure leads to the shift valve, but it is blocked by a land of the valve.

As the speed of the vehicle increases, governor pressure rises. At some point governor pressure will overcome throttle pressure and spring pressure, and move the valve to the left, figure 3-13. This allows mainline pressure to pass through the valley between the lands of the shift valve, and exit into an apply circuit. This causes an upshift. Remember, when a shift valve shifts, it must "snap," or move almost instantly and not allowed to hunt back and forth.

We will return to shift valve operation later in this chapter after we see how throttle pressure and governor pressure are developed and what their purposes are.

THROTTLE PRESSURE

As just described, throttle pressure is one of the pressures used to control the upshift and downshift points of an automatic transmission. Throttle pressure does this in relation to engine load or torque output, which directly relate to the driving conditions at any given moment.

At cruising speed or during acceleration under light load and light throttle, the engine's torque output is low. As a result, the transmission can be shifted to a higher gear sooner than it could be under heavy load. Under these conditions, throttle pressure is low, and upshifts occur at lower speeds.

During hard acceleration, hill climbing, or operation in Reverse, engine torque output is high. As a result, transmission shifts to higher gears must be delayed so they occur later than with light loads. Under these conditions, throttle pressure is high, and upshifts occur at higher speeds.

In most transmissions, throttle pressure is also used to help regulate mainline pressure. When the engine torque output is high, increased mainline pressure is required to hold the apply devices more firmly and prevent slippage. To obtain increased mainline pressure, throttle pressure (or a modified form of it) is routed through passages to the booster valve at one end of the pressure regulator valve, as described previously.

Throttle Valve

Throttle pressure is developed from mainline pressure, and is controlled by the **throttle valve**. The throttle valve senses engine torque (load) in one of two ways. It may be controlled mechanically through a linkage connected to the throttle butterfly. Or, it may be controlled by a vacuum diaphragm connected to the engine intake manifold. In either case, the throttle valve responds directly to engine load. Low load equals low throttle pressure; high load equals high throttle pressure.

When engine load (torque) increases, the throttle butterfly is open wider, manifold vacuum drops, and throttle pressure is higher. When engine load (torque) decreases, the throttle butterfly is not open as far, manifold vacuum rises, and throttle pressure is lower. The relationships between throttle pressure and engine load, manifold vacuum, and throttle position

Manual Valve: The valve that is moved manually, through the shift linkage, to select the transmission drive range. The manual valve directs and blocks fluid flow to various hydraulic circuits.

Shift Valve: A spool valve acted on by throttle pressure and governor pressure to time transmission shifts. Also called a "snap" valve or a timing valve.

Throttle Valve: The valve that regulates throttle pressure based on throttle butterfly opening or intake manifold vacuum.

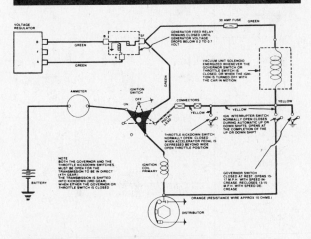

■ Hydraulics Aren't Necessary, But They Sure Are Nice

The hydraulic systems of modern automatic transmissions provide automatic upshifts and downshifts to match vehicle speeds and engine torque for smooth and efficient operation. However, a hydraulic system is not the only way to control the shifting of a transmission. Several carmakers used electric and vacuum systems to control the shifts in early semiautomatic transmissions.

Above is the electrical diagram for Chrysler's Simplimatic transmission of the 1940's. The gearbox was a 4-speed sliding-gear unit mated to a fluid coupling. The transmission shift levers were operated by vacuum servos whose vacuum supply was controlled by this electrical system.

Chrysler used variations of this transmission — under several different trade names — until 1953, when the company introduced its first fully automatic transmission.

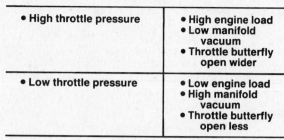

| • High throttle pressure | • High engine load
• Low manifold vacuum
• Throttle butterfly open wider |
| • Low throttle pressure | • Low engine load
• High manifold vacuum
• Throttle butterfly open less |

Figure 3-14. This table shows the relationships between throttle pressure and engine operation.

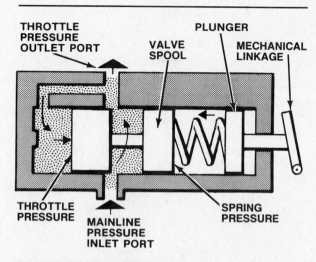

Figure 3-16. Mainline pressure enters the throttle valve where it becomes throttle pressure as fluid bleeds off through the variable restriction of the throttle valve.

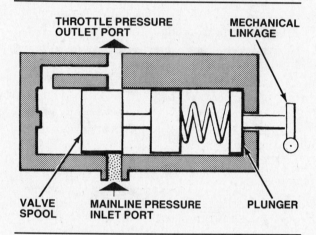

Figure 3-15. A mechanically operated throttle valve in the closed position.

are shown in figure 3-14. These relationships apply for both mechanically and vacuum-operated throttle valves.

Mechanically operated throttle valve

To learn how a throttle valve works, we will look at a simple mechanical throttle valve, figure 3-15. This throttle valve assembly consists of a valve spool, a spring, and a plunger, all contained in a bore. A mechanical linkage passes out of the valve, and is connected to the throttle butterfly on the engine. When the engine is off and the throttle is closed, figure 3-15, the left land of the throttle valve blocks the mainline pressure inlet port to the valve.

As the throttle is opened with the engine running, figure 3-16, the mechanical linkage pushes the throttle valve spool to the left. This uncovers the mainline pressure inlet port and allows fluid to enter the valve. Fluid then passes through the valley of the valve spool, and exits the throttle valve at the outlet port. Mainline pressure becomes throttle pressure at this point.

The throttle pressure at the outlet port is routed to many points in the transmission.

Some of the pressure is directed to a chamber at the left end of the throttle valve, figure 3-16. This pressure acts on the end of the valve spool, and opposes the pressure of the plunger and spring working against the other end. These opposing forces balance the position of the valve in its bore, and provide a uniform throttle pressure for any given throttle opening.

The level of throttle pressure depends on how far the throttle valve inlet and outlet ports are opened. As the engine's throttle is opened farther, the mechanical linkage pushes on the plunger to increase spring pressure on the throttle valve spool. This opens the mainline inlet port wider, and increases throttle pressure. It also increases the pressure working against the end of the throttle valve to balance the increased pressure of the spring and plunger. The net result is that throttle pressure increases as the throttle is opened wider, and stabilizes when the throttle opening is held constant.

At wide-open throttle, figure 3-17, the throttle valve is pushed as far into the valve bore as it can go. At this point, the inlet port and outlet port are both fully open. Therefore, fluid passes straight through the valve, and throttle pressure equals mainline pressure.

Some common automatic transmissions that use mechanically operated throttle valves are the GM Turbo Hydra-matic 125, 200, 200-4R, 440-T4, and 700-R4, the Chrysler Torqueflite, and the Ford AXOD. You will learn more about the throttle valve operation of these specific transmissions in Part Two of this manual.

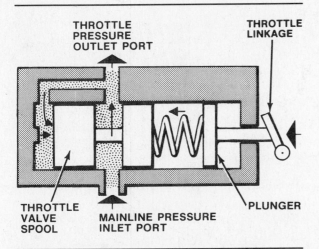

Figure 3-17. At wide-open throttle, the mechanical linkage fully opens the throttle valve, and throttle pressure equals mainline pressure.

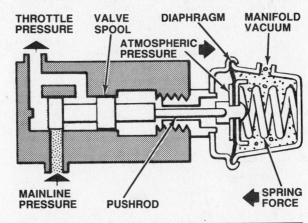

Figure 3-19. The inlet port of a vacuum-operated throttle valve is closed when engine manifold vacuum is high.

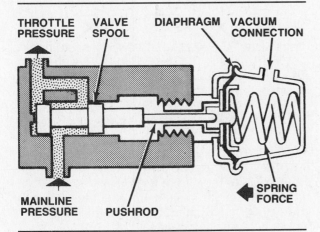

Figure 3-18. A vacuum-operated throttle valve is controlled by a vacuum modulator, but basically operates the same as a mechanical throttle valve.

Vacuum-operated throttle valve

A vacuum-operated throttle valve, figure 3-18, does the same job as a mechanically operated throttle valve, and its hydraulic valving operates in basically the same manner. However, the method used to control the valving is different.

In a vacuum-operated throttle valve, a pushrod connects the valve spool to a rubber diaphragm that controls throttle pressure in response to changes in engine intake manifold vacuum. Ford calls the component containing the vacuum diaphragm the "vacuum diaphragm control unit". General Motors refers to it as the "**vacuum modulator**". In this textbook we will use the term "vacuum modulator" to refer to the component that contains the diaphragm that regulates throttle pressure.

A vacuum modulator, figure 3-18, contains two chambers separated by a flexible rubber

diaphragm. The front chamber, on the side of the diaphragm nearest the valve spool, is open to atmospheric pressure. The rear chamber, on the side of the diaphragm farthest from the throttle valve spool, receives engine intake manifold vacuum through a line connected to the intake manifold. The rear chamber also contains a spring that acts on the back side of the diaphragm to force the valve spool into its bore.

With the engine off, manifold vacuum is zero, and there is equal (atmospheric) pressure in both chambers of the vacuum modulator. As a result, spring pressure acting on the back side of the diaphragm bottoms the throttle valve spool in its bore, figure 3-18. This fully opens both the mainline pressure inlet port and the throttle pressure outlet port.

When the engine is started, manifold vacuum is applied to the rear chamber of the modulator. If manifold vacuum is high enough, such as when the engine is idling or under light load, atmospheric pressure against the modulator diaphragm will oppose spring pressure with enough force to move the valve spool all the way to the right, closing the mainline inlet port, figure 3-19.

Vacuum Modulator: A metal housing divided into two chambers by a flexible rubber diaphragm. One chamber is open to atmospheric pressure, the other is connected to intake manifold vacuum and contains a spring. Changes in manifold vacuum cause movement of the diaphragm against spring tension. In automatic transmissions, this movement is often used to control a throttle valve.

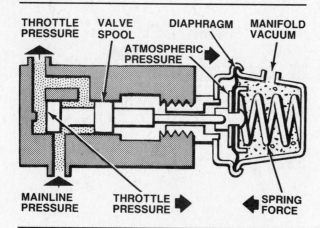

Figure 3-20. A vacuum-operated throttle valve regulating throttle pressure during engine operation.

Under most vehicle operating conditions, manifold vacuum will be insufficient to allow atmospheric pressure to cut off throttle pressure altogether. Instead, the valve spool position will only partially obstruct the mainline pressure inlet port, thus regulating throttle pressure at the outlet port, figure 3-20. During such operation, throttle pressure acting against the end of the valve spool combines with atmospheric pressure acting on the diaphragm to oppose spring pressure.

When the engine comes under heavy load, manifold vacuum drops almost to zero. At these times, pressure is nearly equal in both chambers of the vacuum modulator, and spring pressure again moves the valve spool all the way into its bore, figure 3-18. Therefore, fluid passes straight through the valve, and throttle pressure equals mainline pressure.

Vacuum-operated throttle valves are used in most early Ford automatic transmissions, and in others such as the GM Turbo Hydra-matic 250, 350, 400, and 440-T4. You will learn more about these valves in Part Two when we study specific transmissions.

Downshift Valve

Before we leave this discussion of throttle pressure, you should be aware of other types of valves that work with the throttle valve to increase pressure for forced downshifts when the car's throttle is opened wide. These may be called **downshift valves**, kickdown valves, or detent valves, but they all operate basically the same. In this textbook, we will use the term "downshift valve" to refer to valves that work with the throttle valve to force downshifts.

A simple downshift valve works like the plunger in the throttle valve bore shown in

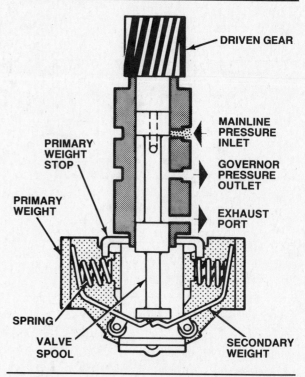

Figure 3-21. This gear-driven governor uses weights acted on by centrifugal force to convert mainline pressure into governor pressure.

figures 3-15, 3-16, and 3-17. When the throttle is opened wide, the downshift valve forces the valve spool to the end of its bore and raises throttle pressure to its maximum. Other downshift valves used with mechanical throttle valves work in the same way.

Downshift valves used with vacuum-operated throttle valves are usually installed in a separate bore. These may be operated by mechanical linkage (rods or cables) or by electric switches and solenoids. In either case, when the throttle is opened, the downshift valve boosts throttle pressure or creates an auxiliary downshift pressure. This pressure is applied to the appropriate shift valve to overcome governor pressure and force a downshift.

GOVERNOR PRESSURE

Governor pressure opposes throttle pressure at a shift valve to control an automatic transmission's upshift and downshift points in relation to vehicle road speed. Governor pressure is developed from mainline pressure, and controlled by the **governor valve**. The governor valve is driven by the output shaft of the transmission, whose rotational speed increases with vehicle speed. As the output shaft speed increases, governor pressure also increases. When output

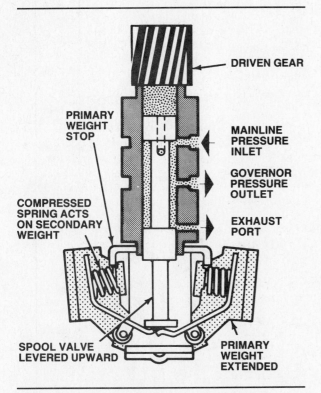

Figure 3-22. With increased governor speed, the weights lever the valve spool into its bore, closing the exhaust port and opening the mainline pressure inlet farther.

shaft speed remains constant, governor pressure also stabilizes at a fixed value for that speed.

Three types of governor valves, or governors as they are more commonly called, are used in current domestic automatic transmissions:

- Gear-driven with spool-valve
- Gear-driven with check-balls
- Shaft-mounted with spool-valve.

Gear-Driven Governor with Spool Valve

The gear-driven governor shown in figure 3-21 is mounted in the transmission extension housing. The output shaft turns the driven gear on the end of the governor, which causes the valve to rotate in the housing. The remainder of the governor consists of two sets of weights, two springs, and a spool valve.

When the vehicle is stopped, figure 3-21, the mainline pressure inlet is closed or slightly open, and both the governor pressure outlet and the exhaust port are fully open. No governor pressure is developed at this time because any pressure that leaks past the valve spool or enters through the mainline inlet escapes through the exhaust port.

As the vehicle begins to move, the transmission output shaft turns the governor. Centrifugal force throws the weights outward, which levers the valve spool farther into its bore, figure 3-22. This begins to close the exhaust port, and opens the mainline pressure inlet port farther, causing governor pressure to build.

Two sets of weights are used to increase the accuracy of pressure regulation at low vehicle speeds. The heavier primary weights move first, and act through springs against the lighter secondary weights that actually move the valve spool. In some transmissions, the spring tensions for the two primary weights differ to provide even smoother regulation of governor pressure. Once governor speed becomes great enough, the primary weights bottom against their stops, and the secondary weights alone act to move the valve spool.

Downshift Valve: An auxiliary shift valve that increases throttle pressure to force a downshift under high driveline loads. Also called a kickdown valve or detent valve.

Governor Valve: The valve that regulates governor pressure in relation to vehicle road speed. Commonly called a governor.

■ Shift Timing "Tuning"

Both mechanically and vacuum-operated throttle valves can be adjusted to fine tune shift timing, causing the transmission shifts to occur at lower or higher vehicle speeds. This is done by increasing or decreasing the force that the mechanical linkage or vacuum modulator applies to the throttle valve.

As force on the throttle valve is increased, throttle pressure increases. Therefore, more governor pressure is required to overcome throttle pressure at the shift valve, and shifts occur at a higher vehicle speed. As force on the throttle valve is decreased, throttle pressure decreases. Therefore, less governor pressure is required to overcome throttle pressure at the shift valve, and shifts occur at lower road speeds.

To make this adjustment on a mechanically operated throttle valve, the linkage is shortened or lengthened. On a vacuum-operated throttle valve, the vacuum modulator is threaded further into or out of the transmission case, or a screw inside the vacuum pipe is turned to change the spring tension inside the modulator. Automakers provide adjustment specifications for the throttle valve (TV) linkages and vacuum modulators on their vehicles to provide the best all-around shift performance.

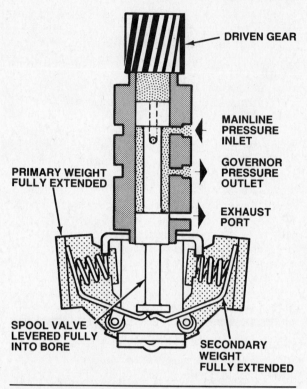

Figure 3-23. With the valve spool bottomed in its bore, governor pressure equals mainline pressure.

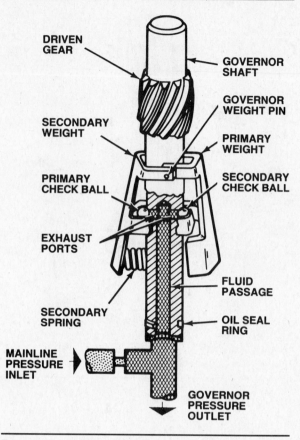

Figure 3-24. A Check-ball governor regulates governor pressure by exhausting fluid flow.

During operation, governor pressure is routed through a passage in the center of the spool valve to the end of the valve near the drive gear. This pressure acts against the end of the spool, and helps oppose the lever force applied at the opposite end of the valve. As a result of these opposing forces, governor pressure increases with vehicle speed, and stabilizes whenever vehicle speed is constant.

When vehicle speed exceeds a certain point, both sets of weights move out as far as possible, figure 3-23. At this time, the secondary weights are bottomed on the primary weights, the exhaust port is fully closed, and both the mainline pressure inlet and the governor pressure outlet are fully open. Therefore, governor pressure equals mainline pressure.

Gear-Driven Governor with Check-Balls

Some late-model automatic transmissions have a gear-driven governor that uses check balls to regulate governor pressure, figure 3-24. Like the design just described, this governor rotates in the transmission extension housing and is driven off a gear on the output shaft. The

check-ball governor consists of two check balls, a governor shaft, a primary weight, a secondary weight, a governor weight pin, a secondary spring, and an oil seal ring.

In transmissions using a gear-driven check-ball governor, mainline pressure enters the governor circuit through an orifice, and extends into the governor through a fluid passage in the center of the shaft. Two exhaust ports are drilled into the fluid passage, and governor pressure is determined by the amount of fluid allowed to flow out of these ports.

The exhaust port openings are controlled by check balls that seat in pockets formed at the ends of the ports. As the governor valve rotates, centrifugal force causes the primary weight to act on one check ball, and the secondary weight to act on the other. Each weight acts on the check ball located on the opposite side of the governor shaft. The secondary weight is assisted in its job by the secondary spring, which helps regulate pressure more smoothly as vehicle speed increases.

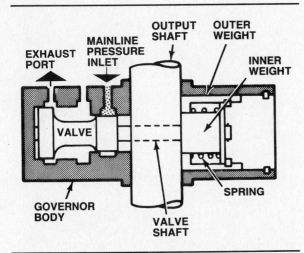

Figure 3-25. This type of governor mounts directly on the output shaft and rotates with it.

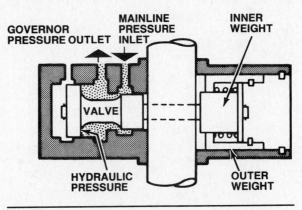

Figure 3-26. Hydraulic pressure acts against centrifugal force to regulate governor pressure in a shaft-mounted governor valve.

When the vehicle is stopped and the governor is not turning, there is no centrifugal force and all of the oil is exhausted from the governor valve. Governor pressure is zero at this time.

As the governor begins to turn, centrifugal force moves the weights outward. This force is relayed to the check balls, partially seating them in their pockets to restrict fluid flow from the exhaust ports. As governor speed increases, the forces on the check balls become greater and less fluid is allowed to escape; this causes governor pressure to increase. When the governor speed remains constant, the forces acting on the check balls are also constant, and governor pressure stabilizes.

At higher vehicle speeds, centrifugal force becomes great enough to hold both check balls fully seated so that no fluid escapes from the governor. At this point, governor pressure equals mainline pressure.

Shaft-Mounted Governor Valve

The third type of governor, figure 3-25, mounts directly on the transmission output shaft and rotates with it. The shaft-mounted governor operates similar to a gear-driven governor in that it uses centrifugal force acting on weights to control the governor pressure. This type of governor assembly consists of a valve spool, an inner weight, an outer weight, a spring, and a valve shaft that ties all these pieces together.

When the vehicle is stopped, figure 3-25, mainline pressure is blocked by the small land of the governor valve spool. Any pressure that leaks past the valve spool at this time escapes through the exhaust port that is opened by the large land of the spool. As a result, governor pressure is zero.

As the vehicle begins to move and the output shaft starts to turn, figure 3-26, the governor weights are thrown outward by centrifugal force. As the weights move, they pull on the valve shaft, which moves the valve spool inward toward the output shaft. This closes the exhaust port and opens the mainline pressure inlet. Mainline pressure then enters the governor valve where it becomes governor pressure and exits through the governor pressure outlet.

As in a gear-driven governor valve, two weights are used to regulate pressure more accurately. At low vehicle speeds, only the heavier outer weight moves. It works against spring pressure to transmit force to the valve shaft and move the spool. When the outer weight moves far enough to fully compress the spring, the inner weight begins to move as well.

Governor pressure continues to increase as centrifugal force increases, but it is modulated by pressure acting against the larger land inside the governor valve. This creates a force that tends to hold the valve back against the movement of the weights. As a result of these opposing forces, governor pressure increases with vehicle speed, and stabilizes whenever vehicle speed is constant.

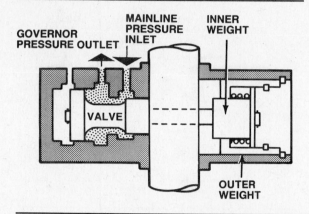

Figure 3-27. When a shaft-mounted governor valve is wide open, governor pressure equals mainline pressure.

At very high vehicle speeds, the governor valve is open as far as possible, figure 3-27. At this point, governor pressure equals mainline pressure. Governor pressure can never be higher than mainline pressure, however.

PRESSURE INTERACTION AT SHIFT VALVES

Transmission shifts must be timed so that the gear ratio selected matches the torque produced by the engine. This is necessary to provide proper performance under various driving conditions. The shifts in an automatic transmission are controlled by throttle pressure and governor pressure interacting at the shift valves. Let's look at a couple of examples of how this works.

If a vehicle is accelerated slowly from a stop, the transmission should upshift as soon as practical in order to achieve better fuel economy. Torque demand is low at this time, and the driver does not press down very far on the accelerator. This causes the throttle linkage to open the throttle butterfly only slightly, which means that manifold vacuum remains high. These actions result in low throttle pressure, which means relatively low governor pressure will be required to overcome throttle pressure at the shift valve. As a result, an upshift occurs while the car is traveling at low speed.

If the same vehicle is accelerated rapidly from a stop, or is climbing a hill, the transmission should stay in a lower gear for better performance. Torque demand is high at this time, so the driver presses the accelerator down farther. This causes the throttle linkage to open the throttle butterfly wider, which also causes intake manifold vacuum to drop. These actions

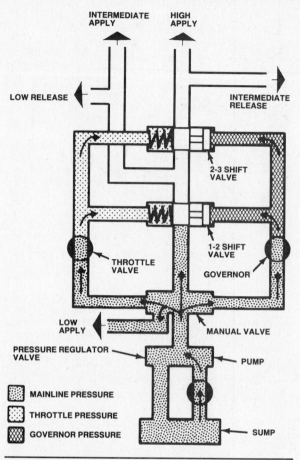

Figure 3-28. A simplified hydraulic diagram showing a 3-speed automatic transmission in first (low) gear.

result in increased throttle pressure, which means that higher governor pressure will be required to overcome throttle pressure at the shift valve. As a result, an upshift will not occur until the car reaches a higher speed.

Sample 3-Speed Automatic Transmission

Figure 3-28 shows a greatly simplified diagram of the hydraulic system for a sample 3-speed automatic. This transmission has two shift valves. The 1-2 shift valve times shifts from first to second (low to intermediate). The 2-3 shift valve times shifts from second to third (intermediate to high).

Fluid moves from the sump, through the pump, to the pressure regulator valve where it becomes mainline pressure. Mainline pressure is routed to the manual control valve which delivers it directly to the low-gear apply circuit. The manual control valve also delivers mainline pressure to the throttle valve, governor, and the 1-2 shift valve.

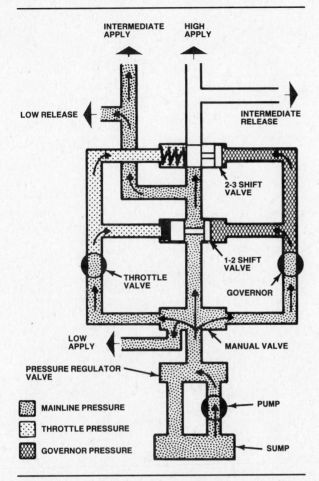

Figure 3-29. A simplified hydraulic diagram showing a 3-speed automatic transmission in second (intermediate) gear.

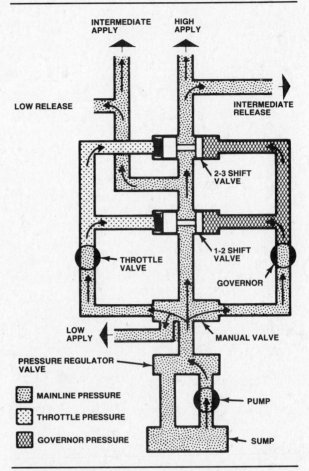

Figure 3-30. A simplified hydraulic diagram showing a 3-speed automatic transmission in third (high) gear.

When the speed of the vehicle becomes great enough, governor pressure overcomes throttle pressure at the 1-2 shift valve, figure 3-29. This moves the shift valve to the left, and routes mainline pressure to the apply devices for second gear. At the same time, pressure in the second gear apply circuit is used to release first gear. This prevents the transmission from being in two gears at once. Movement of the 1-2 shift valve also delivers mainline pressure to the 2-3 shift valve.

The auxiliary spring in the 2-3 shift valve is stronger than the spring in the 1-2 shift valve. Therefore, higher governor pressure is required to overcome the combined forces of throttle pressure and the spring. This causes the shift from second to third to occur at a higher speed than the shift from low to second.

When the speed of the vehicle becomes great enough, governor pressure overcomes throttle pressure at the 2-3 shift valve, figure 3-30. This moves the shift valve to the left, and routes

mainline pressure to the apply devices for high gear. At the same time, pressure in the third gear apply circuit is used to release second gear. This prevents the transmission from being in two gears at once.

VALVE BODIES

Most of the valves we have discussed in this chapter are housed in a casting inside the transmission called a **valve body**, figure 3-31. The valve bodies in most RWD automatic transmissions are located at the bottom of the case, just above the oil pan. In automatic transaxles, the valve body may be located in the bottom of the case, on the back side of the torque converter

Valve Body: The casting that contains most of the valves in a transmission hydraulic system. The valve body also has passages for the flow of hydraulic fluid.

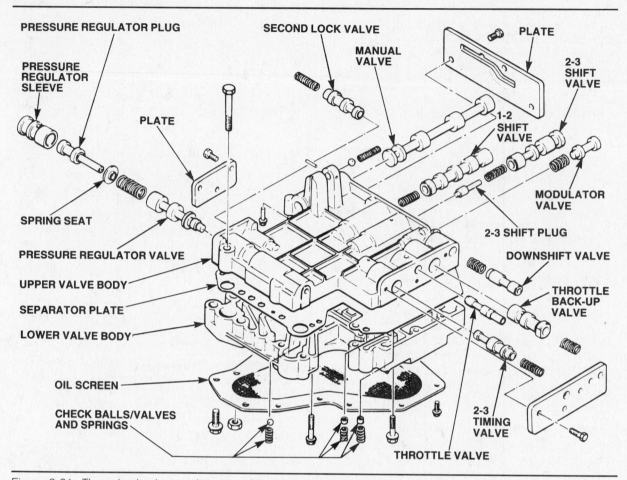

Figure 3-31. The valve body contains most of the hydraulic valves that control automatic transmission operation.

housing, or on top of the transaxle, depending on the specific design.

Valve bodies have many fluid passages for the transmission hydraulic system cast into them. Often, there are pockets that contain check balls. Sometimes, a valve body will consist of two or more sections that bolt together with a flat metal separator plate between them. The separator plate contains openings that help control fluid flow. Figure 3-31 shows a valve body with several common hydraulic valves. You will learn more about valve bodies and specialized hydraulic valves and circuits when we study specific transmissions in Part Two of this manual.

OTHER HYDRAULIC CIRCUITS

This chapter has focused on three main transmission pressures, how they are created, and how they are used to control shifting. However, the transmission pump is more than just a source of fluid for shift valve movements. It also provides fluid for the torque converter, the

transmission cooler, and various apply devices and lubrication circuits inside the transmission.

A portion of the fluid flow from the pump is used to fill the torque converter that transfers power from the engine to the transmission. If the torque converter is not completely filled, it will be unable to properly do its job. Torque converters are described in detail in Chapter 7.

Automatic transmissions create a great deal of heat in operation, so their fluid must be cooled to prevent transmission wear and damage. A portion of the fluid from the pump is sent through tubes to a transmission cooler in the vehicle radiator. The transmission cooler is a heat exchanger. Heat is transferred from the transmission fluid to the coolant in the car's radiator and then to the outside air. Cooled fluid is returned from the cooler to the transmission sump. Some cars use a transmission cooler that is separate from the radiator, and transfers heat directly to the outside air. Transmission coolers are described in detail in Chapter 5.

Finally, a portion of the fluid flow from the pump is routed through tubes and passages within the transmission case and shafts. This fluid does the very important job of lubricating the gears and other moving parts of the transmission. This fluid is also used to operate the various transmission valves and apply devices. We will look at these devices more closely in the next chapter.

SUMMARY

An automatic transmission uses a gear pump, rotor pump, or vane pump to move fluid from a sump into the transmission hydraulic circuits. Gear and rotor pumps are positive-displacement pumps whose output volume increases with pump speed. The vane pumps used in automatic transmissions are variable-displacement pumps whose output volume is controlled independent of pump speed.

The three basic pressures used in an automatic transmission are mainline pressure, throttle pressure, and governor pressure. Mainline pressure is controlled by the pressure regulator valve which forms a variable restriction in the hydraulic circuit that receives the pump output.

Mainline pressure is used to operate the apply devices in the transmission, and is the source for all other transmission hydraulic pressures. Because of this, other pressures can never be higher than mainline pressure. Although under certain driving conditions, other pressures may equal mainline pressure.

Mainline pressure is generally held at a fixed level. However, when driveline loads are high, mainline pressure may be temporarily increased using a booster valve. This prevents the transmission apply devices (clutches and bands) from slipping.

Mainline pressure is routed to the manual valve, which is moved by the shift linkage to select the transmission drive range. Mainline pressure is also routed to the shift valves, which control flow to the transmission apply devices based on throttle pressure and governor pressure.

Throttle pressure is controlled by the throttle valve which is operated by a mechanical linkage attached to the throttle butterfly, or by a vacuum modulator attached to intake manifold vacuum. The throttle valve input acts on a spool valve to restrict fluid flow. Throttle pressure increases with engine torque output, and acts against governor pressure at a shift valve to help time transmission shifts. Throttle pressure may also be used to help regulate mainline pressure.

When driveline loads are high, throttle pressure may be temporarily increased using a downshift valve. This forces the transmission to downshift, allowing the engine to operate at an rpm where greater torque is available.

Governor pressure is controlled by the governor valve which is driven off the transmission output shaft. The governor uses centrifugal force acting on weights to control a spool valve or check balls that restrict fluid flow. Governor pressure increases with vehicle road speed, and acts against throttle pressure at a shift valve to help time transmission shifts.

Throttle pressure and governor pressure interact at shift valves to cause transmission shifts. A shift occurs whenever governor pressure acting on one side of the valve exceeds throttle pressure acting on the other. The shift valve, and most other valves that control the transmission hydraulic system, are located in a casting called the valve body.

In addition to operating the shift valves, fluid from the pump travels to other areas inside and outside the transmission. Some fluid fills the torque converter. Another portion is routed through a transmission cooler. And other fluid lubricates the gears and internal components, and operates the apply devices.

Review Questions

Select the single most correct answer
Compare your answers to the correct answers on page 603

1. The pressure from which all others pressures are developed in an automatic transmission hydraulic system is:
   a. Control pressure
   b. Throttle pressure
   c. Governor pressure
   d. Mainline pressure

2. The inner drive gear of a positive-displacement oil pump in a RWD transmission is driven by the:
   a. Input shaft
   b. Governor
   c. Torque converter hub
   d. Output shaft

3. Mechanic A says a vane pump may be driven by a pump drive-shaft.
   Mechanic B says the displacement of a vane pump can be varied during transmission operation.
   Who is right?
   a. Mechanic A
   b. Mechanic B
   c. Both mechanic A and B
   d. Neither mechanic A nor B

4. Mainline pressure is developed and controlled by the:
   a. Throttle valve
   b. Modulator valve
   c. Pressure regulator valve
   d. Detent valve

5. Higher mainline pressure is necessary during:
   a. Low-speed operation under heavy load
   b. Hill climbing
   c. Reverse gear operation
   d. All of the above

6. Mechanic A says throttle pressure decreases as engine load and torque increase.
   Mechanic B says throttle pressure increases as the accelerator pedal is depressed.
   Who is right?
   a. Mechanic A
   b. Mechanic B
   c. Both mechanic A and B
   d. Neither mechanic A nor B

7. Mechanic A says governor pressure stays constant as vehicle speed increases.
   Mechanic B says as governor pressure decreases as vehicle speed increases.
   Who is right?
   a. Mechanic A
   b. Mechanic B
   c. Both mechanic A and B
   d. Neither mechanic A nor B

8. Mechanic A says throttle pressure controls upshifts and downshifts.
   Mechanic B says governor pressure controls upshifts and downshifts.
   Who is right?
   a. Mechanic A
   b. Mechanic B
   c. Both mechanic A and B
   d. Neither mechanic A nor B

9. Shift valves are sometimes referred to as:
   a. Snap valves
   b. Detent valves
   c. Modulator valves
   d. Regulator valves

10. High throttle pressure may be produced by:
    a. A wide throttle opening
    b. High engine load
    c. Low manifold vacuum
    d. All of the above

11. Mechanic A says intake manifold vacuum assists spring pressure in a vacuum operated throttle valve.
    Mechanic B says atmospheric pressure opposes spring pressure in a vacuum operated throttle valve.
    Who is right?
    a. Mechanic A
    b. Mechanic B
    c. Both mechanic A and B
    d. Neither mechanic A nor B

12. Throttle pressure is used in some transmissions to:
    a. Reduce governor pressure
    b. Increase governor pressure
    c. Reduce mainline pressure
    d. Increase mainline pressure

13. Maximum throttle pressure is never greater than:
    a. Twice mainline pressure
    b. Half mainline pressure
    c. Mainline pressure
    d. None of the above

14. Mechanic A says forced downshifts are controlled by a kickdown valve.
    Mechanic B says forced downshifts are controlled by a detent valve.
    Who is right?
    a. Mechanic A
    b. Mechanic B
    c. Both mechanic A and B
    d. Neither mechanic A nor B

15. Downshift valves may be controlled by:
    a. Rods
    b. Cables
    c. Switches and solenoids
    d. All of the above

16. Mechanic A says the weights in a governor are moved by centrifugal force.
    Mechanic B says the weights in a governor are moved by throttle pressure.
    Who is right?
    a. Mechanic A
    b. Mechanic B
    c. Both mechanic A and B
    d. Neither mechanic A nor B

17. Mechanic A says governor pressure varies in cycles at a constant vehicle speed.
    Mechanic B says governor pressure decreases as the vehicle road speed decreases.
    Who is right?
    a. Mechanic A
    b. Mechanic B
    c. Both mechanic A and B
    d. Neither mechanic A nor B

18. Mechanic A says most of the valves in a transmission hydraulic system are housed in the valve body.
    Mechanic B says the valve body can be located in various positions depending on the type of transmission.
    Who is right?
    a. Mechanic A
    b. Mechanic B
    c. Both mechanic A and B
    d. Neither mechanic A nor B

Apply Devices

In Chapter 1, we learned that in order to get output motion from a planetary gearset, you must hold one member while driving another. The mechanical devices that provide holding and driving forces to planetary gearsets are called **apply devices**. The apply devices used in automatic transmissions include bands, multiple-disc clutches, and 1-way overrunning clutches.

Automatic transmission bands are applied hydraulically by servos. A servo operates in a manner similar to the simple hydraulic cylinder discussed in Chapter 2. A band and its servo are always a holding device. The band holds the member of the planetary gearset that is the reaction member in any particular gear range.

The multiple-disc clutches used in automatic transmissions are also applied hydraulically by a clutch piston. The operation of a hydraulic clutch is again similar to that of the simple hydraulic cylinder discussed in Chapter 2. Multiple-disc clutches can be *either* holding or driving devices. Depending on its design, a hydraulic clutch may hold the reaction member, or apply input motion to the gearset.

Most automatic transmissions also use another type of clutch called a 1-way overrunning clutch, or simply, a 1-way clutch. A 1-way clutch is a mechanical device that locks up to prevent a gearset member from rotating in one direction, but unlocks and overruns to allow free rotation in the opposite direction. A 1-way clutch is considered an apply device, even though it is not hydraulically operated.

This chapter explains the general design and operation of transmission bands, servos, multiple-disc clutches, and 1-way clutches. It also describes some of the hydraulic system parts associated with the operation of these components.

TRANSMISSION BANDS

A **transmission band** stops and holds one member of a planetary gearset so that another member of the gearset can develop output motion. The band does this by tightening around the outside of a drum to keep the drum from turning, figure 4-1. The drum is engaged with the member of the gearset to be held. Remember, a transmission band is a holding device only. It is never used to drive any member of a planetary gearset.

In its simplest form, a transmission band is anchored at one end, and a force is applied against the other end. As force is applied, the band contracts around the rotating drum and

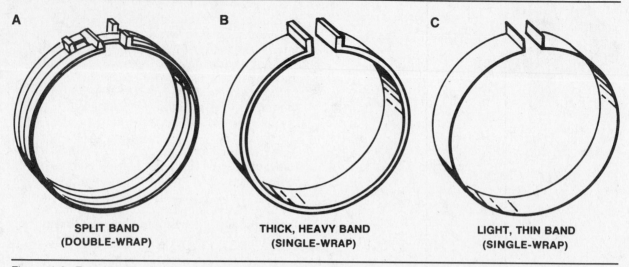

A   **B**   **C**

SPLIT BAND   THICK, HEAVY BAND   LIGHT, THIN BAND
(DOUBLE-WRAP)   (SINGLE-WRAP)   (SINGLE-WRAP)

Figure 4-2. Transmission bands come in several designs and thicknesses depending on the loads they have to carry.

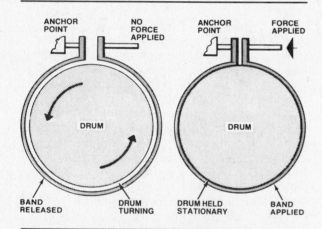

Figure 4-1. A contracting band tightens around the outside of a drum to keep the drum from turning.

applies pressure to bring it to a stop, figure 4-1. The amount of pressure generated against the drum is determined by the length and width of the band, and the amount of force applied against the band's unanchored end.

Band Designs

Transmission bands, figure 4-2, differ in size and construction depending on the amount of work they are required to do. A band that is split with overlapping ends is called a double-wrap band. A simple 1-piece band that is not split is called a single-wrap band. All transmission bands are made of flexible steel, and their inside surfaces are lined with a friction material.

A double-wrap band, figure 4-2A, conforms to the circular shape of a drum better than a single-wrap band does. As a result, it can provide greater holding power for a given application force. A double-wrap band also engages with the drum more smoothly than a single-wrap band does. This enables it to provide smoother shifts.

Single-wrap transmission bands are cheaper to make than double-wrap bands, and in many applications a single-wrap band is quite satisfactory. A single-wrap band can be large and heavy when a great deal of force and pressure must be applied to hold a particular member of a planetary gearset from rotating, figure 4-2B. However, most late-model transmissions are designed to use slim, light, flexible bands like that shown in figure 4-2C.

Bands are designed to slip slightly as they are applied so the member of the planetary gearset being acted on does not stop too quickly. A very quick stop would cause a harsh shift, and possibly damage the transmission. Too much slippage, on the other hand, causes a band to become glazed and burned.

Transmission band slippage increases as the lining wears and the clearance between the band and drum becomes greater. This means that bands require adjustment from time to time. The bands in most early automatic transmissions must be manually adjusted periodically. Due to improved band design, most newer transmissions do not require periodic adjustment.

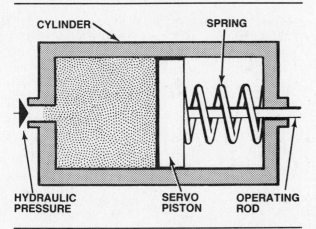

Figure 4-3. A servo assembly consists of a piston in a hydraulic cylinder. The piston applies force through an operating rod to tighten the band around a drum.

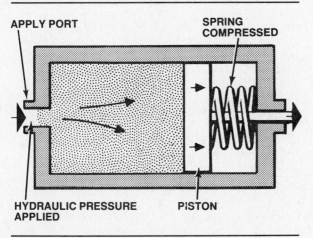

Figure 4-4. In this simple servo, hydraulic pressure moves the piston to apply the band.

HYDRAULIC SERVOS

A transmission band is applied hydraulically by a piston moving inside a cylinder. This piston and cylinder assembly is called a **servo**, figure 4-3. When the servo is not applied, the piston is held to one side by a spring. The servo operates when fluid under pressure enters the cylinder and acts against the piston, moving it against spring pressure. This transmits force through an operating rod to apply the band around the drum.

Servos operate in several ways to control the application and release of transmission bands. In the simplest design, the servo and band are applied when hydraulic pressure pushes against a piston and moves the operating rod toward the band, figure 4-4. The band and

Apply Device: Hydraulically operated bands and multi-disc clutches, and mechanically operated 1-way clutches, that drive or hold the members of a planetary gearset.

Transmission Band: A flexible steel band lined with friction material that is clamped around a circular drum to hold it from turning.

Servo: A hydraulic piston and cylinder assembly that controls the application and release of a transmission band.

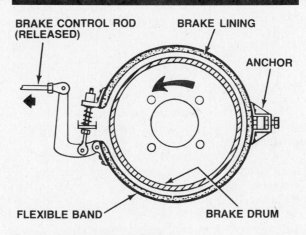

■ **Contracting-Band Brakes**

The automatic transmission band was developed from the external contracting-band wheel brake used on many early cars and trucks. In that design, a flexible metal band lined with friction material was wrapped around the outside of a brake drum attached to the wheel. The center of the band was anchored to the vehicle chassis, and the free ends were joined with a cantilever linkage that was connected, through a series of rods and cables, to the brake pedal. When the pedal was applied, the linkage clamped the brake band around the drum to slow the vehicle.

External contracting-band brakes were used for some time, but they suffered rapid wear because the lining was exposed to the elements. They could also drag and lock up if hard use caused the drum to overheat and expand. As cars got heavier and speeds increased, band-type brakes were replaced by internal expanding-shoe brakes, and later, disc brakes. However, even in recent times, contracting brake bands have been used as parking brakes on the driveshafts of Chrysler Corporation cars and many light trucks.

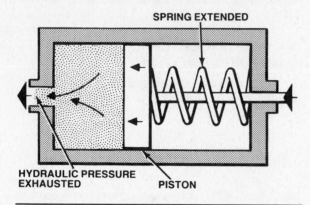

Figure 4-5. Spring force returns the servo piston to release the band when the apply pressure is cut off and exhausted.

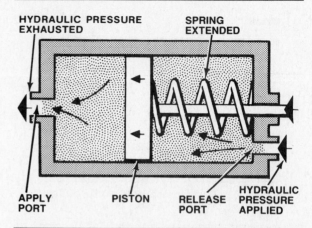

Figure 4-6. In this servo, hydraulic pressure is applied to the spring side of the piston to help release the servo and band.

servo are released when hydraulic pressure to the servo apply port is cut off and exhausted, figure 4-5. The spring on the opposite side of the piston then returns the piston to its original position.

In a more common design, the servo and band are released by a combination of spring pressure and hydraulic pressure, figure 4-6. At the same time pressure on the apply side of the servo is exhausted, pressure is applied to the side of the servo that contains the spring. The band is released and the piston moves back to its original position as soon as hydraulic pressure is equal on both sides of the piston. Once the piston has fully returned, the continued pressure on the release side of the servo ensures that the band remains unapplied.

Servo Force

The piston area of a servo is relatively large, which allows it to convert the hydraulic pressure it receives into a great deal of application force. This force is required in order to clamp the band tight enough around the spinning drum to bring the drum to a stop. In actual transmissions, servos are designed to accept varying pressures and apply different forces as required.

For example, consider a servo piston with a surface area of 3 square inches operating in a transmission circuit where the hydraulic pressure can range from 50 to 100 psi. If the pressure is 50 psi, the force applied by the piston to the operating rod equals the pressure of the hydraulic fluid times the area of the piston face (50 psi × 3 square inches), or 150 pounds. This amount of force will stop and hold a drum if there is not much input torque, or if there is not too much resistance against the output unit. If

the hydraulic pressure is increased to 100 psi, as it is under heavy engine loads, the force applied to the band doubles. This creates approximately 300 pounds of force to tighten the band around the drum.

SERVO LINKAGES

A servo is connected to a band by one of three types of operating linkages:
- Straight
- Lever
- Cantilever.

The first type of servo operating linkage uses a straight rod or strut to transfer force from the servo piston directly to the free end of the band, figure 4-7. A straight linkage is used when the transmission design allows the servo to be placed in a position where it can act directly on a band. However, a straight linkage can only be used if the servo is large enough to hold the band from slipping when maximum torque is applied to the drum.

The second type of servo operating linkage uses a lever to move the rod or strut that applies the band, figure 4-8. A lever linkage is used when the servo must be placed in an area of the transmission where it cannot act directly on the band. This type of linkage bends the force of the servo that acts against the band. In addition, a lever linkage usually increases application force because the lever is longer on the servo side of the **fulcrum** than it is on the rod or strut side. In figure 4-8, the fulcrum is the pivot pin of the apply lever.

The third type of servo operating linkage, figure 4-9, uses a lever and a **cantilever** to act on both ends of a band that is not anchored. As the servo piston applies force to the operating

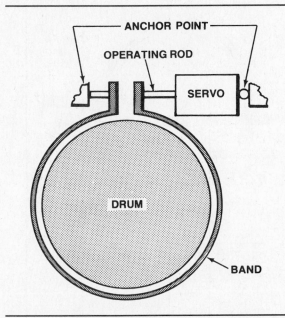

Figure 4-7. In some transmissions, a straight operating linkage delivers output force directly from a servo to the band.

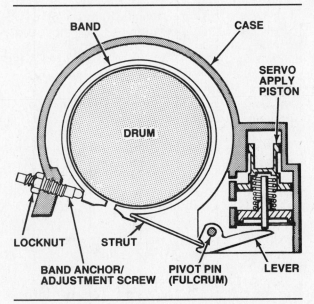

Figure 4-8. A lever linkage multiplies the application force provided by the servo.

rod, the rod moves the lever and applies force to one end of the band through a rod or strut. At the same time, the cantilever attached to the other end of the band is pulled toward the pivot pin. These actions clamp the ends of the band together, and tighten the band around the drum. A cantilever linkage increases band application force like a lever linkage. It also reduces band wear and smooths application be-

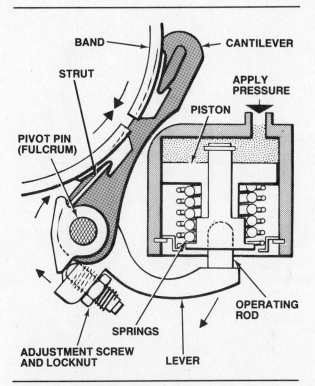

Figure 4-9. A cantilever linkage acts on both ends of a free-floating band to tighten it around a drum.

cause the band can self-center and contract more evenly around the drum.

Servo Linkage (Band) Adjustment

On most straight and lever operating linkages that require manual adjustment of the bands, the clearance between the band and drum is set using an adjustment screw that doubles as the fixed anchor for the band, figure 4-8. The anchor/adjustment screw passes through the transmission case, and is accessible from outside where it is secured with a locknut. This design makes band adjustment a relatively simple job.

Where manual band adjustment is required with a cantilever operating linkage, the band adjustment screw and locknut are located on the linkage inside the transmission case, figure 4-9. In these applications, the transmission oil

Fulcrum: The support and pivot point of a lever.

Cantilever: A lever that is anchored and supported at one end by its fulcrum, and provides an opposing force at its opposite end.

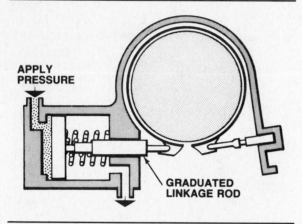

Figure 4-10. A graduated linkage rod of specific length is used to adjust band clearance in some transmissions.

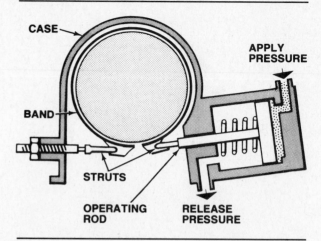

Figure 4-11. Rods and struts are vital parts of a servo operating linkage.

pan must be removed to perform the band adjustment.

One type of straight-rod linkage that does not require periodic adjustment uses graduated operating rods, figure 4-10. Adjustment is made at the time of assembly by selecting one of several different length rods. These rods are identified by either ring grooves or raised rings on the band end of the rod. When the proper length rod is installed, the servo piston moves a specific distance and applies a specific amount of force against the band.

Complete band adjustment procedures for transmissions that require them are explained in your *Shop Manual*.

Servo Rods and Struts

Servo rods and struts are operating linkage parts that apply force against the band. Rods are round metal bars; struts are flat metal plates; both are sometimes referred to as operating links. A rod or strut may be placed between the servo or lever and the band, or it may be placed between the anchor and the band. Some transmissions use rods and/or struts on both sides of the band, figure 4-11.

HYDRAULIC ACCUMULATORS

A hydraulic **accumulator** is a type of shock absorber. Accumulators are used in automatic transmissions to cushion the application of servos and hydraulic clutches. Accumulators are necessary because when hydraulic fluid accelerates or changes direction rapidly, such as when pressure is applied through a shift valve, it is subject to surging. A rapid surge of hydraulic pressure can cause an apply device to vibrate or

engage harshly. This causes rough shifts and could damage the transmission.

An accumulator cushions, or damps, hydraulic pressure surges by temporarily diverting part of the fluid in a circuit into a parallel circuit or chamber. This diversion allows pressure to increase in the main circuit more gradually, and provides the desired smooth engagement of a band or clutch.

Accumulators are classified as either piston-type or valve-type. Piston-type accumulators look like servos. In fact, some accumulator pistons share the same bore as a servo piston; we will call this design an integral accumulator. Other piston-type accumulators are installed in separate bores in the transmission case; we will call this design an independent accumulator. Both types operate in basically the same way.

Valve-type accumulators are similar to other spool valves in a hydraulic system. They do essentially the same job as a piston-type accumulator, temporarily diverting a portion of the apply pressure to a servo or clutch.

Independent Piston-Type Accumulator

A simple hydraulic circuit for an apply device that uses an independent piston-type accumulator in a separate bore is shown in figure 4-12. When the shift valve opens to apply pressure to the intermediate (second-gear) servo, part of the fluid is diverted to the 1-2 accumulator. Fluid pressure acts at the same time on both the servo piston and the accumulator piston. This causes pressure in the apply circuit to increase more gradually than it would if it were acting only on one piston.

The movement of the accumulator piston cushions application of the servo piston, and absorbs any surge created by the sudden

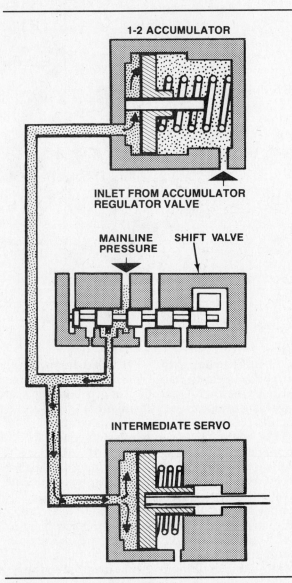

Figure 4-12. This independent piston-type accumulator cushions application of the intermediate servo.

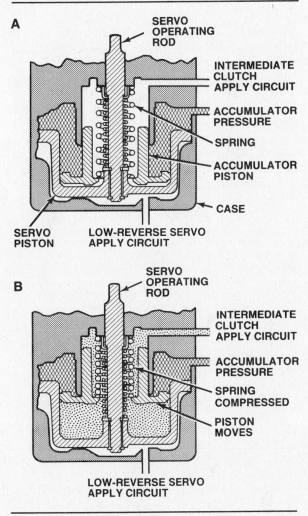

Figure 4-13. This integral piston-type accumulator is combined with a servo in a single bore.

change in pressure. However, when the accumulator piston bottoms in its bore, it can move no farther and its cushioning action is finished. Pressure in the apply circuit then increases to its maximum and firmly applies the servo and band.

The cushioning action of the accumulator can be modified under certain driving conditions by applying an auxiliary pressure to the spring side of the accumulator piston. In figure 4-12, this is shown as an inlet from an accumulator regulator valve. The auxiliary pressure alters the spring force on the piston, and allows the accumulator to provide more or less cushioning action under different operating conditions.

This type of accumulator is found in Chrysler Torqueflite and some General Motors Turbo Hydra-matic transmissions, among others. You will study these in detail in Part Two of this manual.

Integral Piston-Type Accumulator

An integral piston-type accumulator is installed in the same bore as a servo piston. In these applications, the accumulator piston is usually part of the apply circuit for a servo other than the one with which it is combined. For example, in figure 4-13, an accumulator piston for an

---

Accumulator: A device that absorbs the shock of sudden pressure surges within a hydraulic system. Accumulators are used in transmission hydraulic systems to control shift quality.

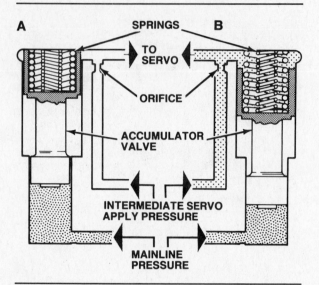

Figure 4-14. Like its piston-type counterpart, a valve-type accumulator provides hydraulic cushioning action for an apply device.

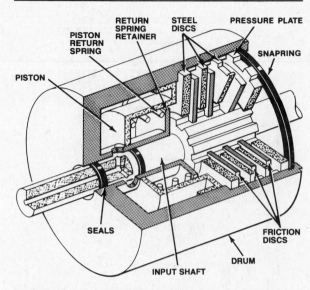

Figure 4-15. A multiple-disc hydraulic clutch assembly.

intermediate (second-gear) clutch is installed in the same bore as the servo piston for a low-reverse band.

When the low-reverse band and the intermediate clutch are both released, figure 4-13A, accumulator pressure is applied to the top side of the accumulator piston. This bottoms the accumulator piston in the servo piston and holds the servo in the released position.

When the intermediate clutch is applied, figure 4-13B, part of the fluid from that circuit is routed into the accumulator through internal passages in the accumulator piston. The fluid enters the chamber between the accumulator and servo pistons, and forces the accumulator piston to the top of its bore against spring pressure and accumulator pressure. This cushions the pressure increase for intermediate clutch application, just as with an independent accumulator.

While the accumulator is in operation, intermediate clutch pressure is applied to the top of the low-reverse servo piston, figure 4-13B. This pressure continues to hold the servo in the released position.

Integral piston-type accumulators are found inside servos in several Turbo Hydra-matic transmissions, among others. You will study these in detail in Part Two of this manual.

Valve-Type Accumulator

A valve-type accumulator provides basically the same cushioning effect as a piston-type accumulator. The intermediate (second-gear) accumulator valve in figure 4-14 works together

with a restricting orifice. Before the servo apply circuit is opened, figure 4-14A, there is no pressure in the circuit, and the valve is bottomed against spring pressure by mainline pressure acting on the small end of the valve spool.

As intermediate servo pressure enters the apply circuit, it passes through the orifice which delays the pressure buildup. Just past the orifice, the circuit splits in two with one line leading to the servo and the other to the large end of the accumulator valve spool. The combination of fluid pressure and spring force on the large end of the valve overcomes mainline pressure and moves the valve in its bore, figure 4-14B. The resulting pressure drop, combined with the delay in pressure buildup caused by the restricting orifice, cushions application of the intermediate servo.

Valve-type accumulators are used in Ford automatic transmissions, among others. You will study these in detail in Part Two of this manual.

MULTIPLE-DISC HYDRAULIC CLUTCHES

Like a transmission band, a multiple-disc hydraulic clutch is a type of apply device. The primary components in a **multiple-disc clutch**, figure 4-15, are friction discs alternated between steel discs. Friction discs have a rough friction material on their faces. Steel discs have smooth faces without friction material. Collectively, the friction discs and steel discs are called the **clutch plates**. Steel discs are sometimes also called separator plates, apply plates, or simply "steels."

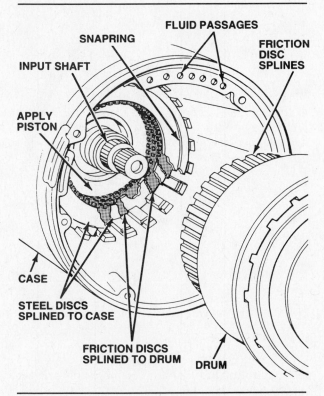

FLUID PASSAGES
SNAPRING
INPUT SHAFT
FRICTION DISC SPLINES
APPLY PISTON
CASE
STEEL DISCS SPLINED TO CASE
FRICTION DISCS SPLINED TO DRUM
DRUM

Figure 4-16. In this holding clutch, one set of discs is splined to the transmission case, the other set is splined to the drum. When the clutch is applied, the drum is locked to the case and prevented from turning.

One set of plates in a multiple-disc clutch is splined on its inner edges; the other set is splined on its outer edges. The splines on each set engage with matching splines on one of the following: a shaft, a drum, a member of the planetary gearset, or the transmission case. When the clutch is applied, the two sets of plates lock together and mechanically connect the components engaged with their splines.

A multiple-disc clutch is applied by a hydraulic piston that forces the plates together. The apply piston is fitted with at least one return spring, and some clutch pistons have several. A multiple-disc clutch also contains one or more pressure plates, seals, return spring retainers, and snaprings. A pressure plate, figure 4-15, is a heavy metal backing plate installed at one, or both, ends of the **clutch pack**. The snaprings are used to hold the parts in the clutch pack together.

Although a multiple-disc clutch is an apply device similar to a transmission band, it differs from a band because it can be used to drive members of planetary gearsets as well as hold

them. Before we examine the hydraulic operation of a multiple-disc clutch, we will look at the two variations:

● Holding clutches
● Driving clutches.

Holding Clutches

An example of a holding clutch is shown in figure 4-16. In this design, the friction discs are splined on their inner edges and engage matching splines on the outside of a drum. The friction discs are alternated between steel discs that are splined on their outer edges and engage matching splines machined into the transmission case.

As long as there is clearance between the friction discs and the steel discs, the drum is free to rotate in either direction. However, when the space between the friction discs and the steel discs is eliminated (the clutch is applied), the two sets of discs grip one another and stop the rotation of the drum. The drum is then held until the clutch is released.

The piston that applies a holding clutch operates in a bore that is located in the transmission case. Passages in the case route fluid to the apply piston.

Driving Clutches

There are two types of multiple-disc driving clutches. The first, shown in figure 4-17, has a set of clutch plates (drive discs) that are splined to the transmission input shaft and turned by it. The alternating set of steel plates (driven discs) is splined to the inside of the clutch drum.

When the clutch piston is released, the drive discs rotate with the input shaft, but do not engage the driven discs that are splined to the drum.

Multiple-Disc Clutch: A clutch that consists of alternating friction discs and steel discs that are forced together hydraulically to lock one transmission part to another.

Clutch Plates: A generic term for the friction discs and steel discs used in a multiple-disc clutch.

Clutch Pack: The assembly of clutch plates and pressure plates that provides the friction surfaces in a multiple-disc clutch.

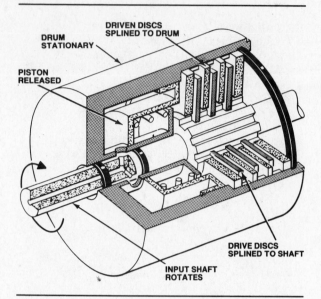

Figure 4-17. When the piston in this driving clutch is released, the input shaft can turn, but it cannot drive the clutch drum.

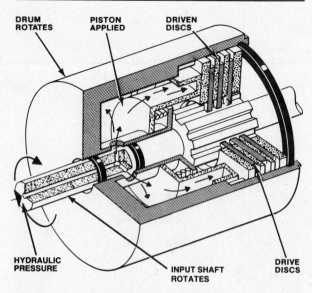

Figure 4-18. When the piston in this driving clutch is applied, the drive and driven discs are locked together to turn the drum.

When the clutch is applied, figure 4-18, the piston clamps the alternating discs together. The drive discs then engage the driven discs, which turn the drum. The drum is connected to, and drives, a member of the planetary gearset. A passage inside the input shaft carries fluid to the clutch apply piston.

The other type of multiple-disc driving clutch, figure 4-19, has the drum splined directly to the input shaft so that the drum always rotates with the shaft. The drive discs are splined to the inside of the drum. The driven discs are splined to the outside of a clutch hub that is splined to an output shaft, or attached to a member of the planetary gearset. When the clutch is applied, friction between the discs mechanically locks the drum and hub together. The input shaft then drives the output shaft, or the planetary gearset member, and both turn at the same speed.

You will learn more about specific clutch designs when we study individual transmissions in Part Two of this manual.

CLUTCH OPERATION

To explain how a hydraulic clutch operates, we will use the typical multiple-disc driving clutch shown in figure 4-19. In this design, the apply piston in the rear of the drum (piston cylinder) is held in place by return springs and a spring retainer secured by a snapring.

To apply the clutch, figure 4-19A, hydraulic fluid under pressure enters the drum through a passage in the input shaft. The fluid acts on the piston and moves it against return spring pressure to clamp the clutch plates together against the pressure plate. The friction between the discs then locks the clutch drum and hub together, causing them to turn as one unit. Because the drum is attached to a member of a planetary gearset, that member turns at shaft speed.

To release the clutch, figure 4-19B, hydraulic fluid to the piston is cut off and exhausted. The piston return spring, which was compressed as the clutch was applied, then expands to move the piston back, allowing the discs to disengage.

Clutch Vent Port and Check Ball

When a hydraulic clutch is released, fluid is thrown to the outside of the drum by centrifugal force. As the fluid reaches the outer edge of the drum, it can spread out and apply force against the piston. This can cause partial engagement of the clutch pack, which could ruin the quality of the shift, and burn, glaze, or warp the clutch discs.

To relieve any residual fluid pressure, a vent port with a check ball is built into the clutch. When full hydraulic pressure is applied to the clutch, the ball is forced against its seat and pressure is contained within the drum, figure 4-19A. When the clutch is released, and only residual hydraulic pressure pushes against the ball, centrifugal force pulls the ball from its seat and moves it to the outer edge of its retainer.

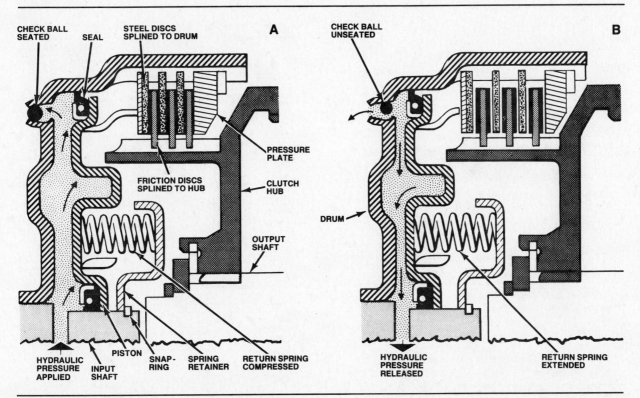

Figure 4-19. When the clutch is applied (A), hydraulic pressure seats the check ball and moves the piston against spring pressure to compress the discs. When the clutch is released (B), return springs bottom the piston in its bore, and the check ball is unseated to quickly exhaust any residual hydraulic pressure.

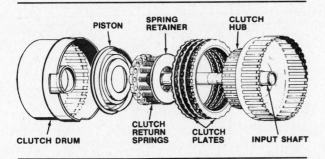

Figure 4-20. Some multiple-disc clutches use several small return springs rather than one large one.

This allows leftover fluid to escape from the drum through the open vent port, figure 4-19B.

The vent port and check ball do not necessarily have to be in the drum. In some clutches, they are located in the piston. Other transmissions have a metered orifice in either the clutch drum or the clutch piston. These do the same job as a vent port and check ball.

Clutch Spring Types

Clutch pistons can be released with one large coil spring, figure 4-15 or with several smaller springs, figure 4-20. Transmission engineers calculate the strength and number of return springs needed to release the piston quickly enough to keep the clutch from dragging, while at the same time offering the least possible resistance to the piston being applied.

Some transmission clutches may have fewer springs than there are spring pockets available. This occurs because manufacturers use the same basic clutches and transmissions in many different vehicles. Clutch springs are added or subtracted to meet the needs of specific applications.

Diaphragm spring
Another type of multiple-disc clutch return spring is the diaphragm spring, figure 4-21. This design is also called an over-center or **Belleville spring**. A diaphragm spring acts as both a clutch apply device and a piston return spring.

Belleville Spring: A diaphragm-type spring sometimes used to help apply and release a multiple-disc clutch.

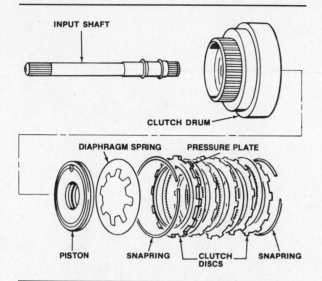

Figure 4-21. Some clutches use a diaphragm, or Belleville, return spring that also multiplies application force.

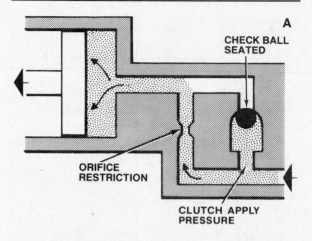

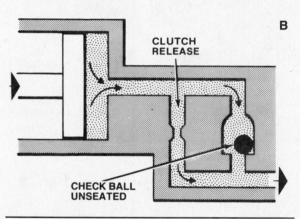

Figure 4-22. An orifice restriction and an exhaust port check ball can be combined to smooth clutch application, but still provide quick clutch release.

The outer circumference of the diaphragm spring is held in place by a snapring locked into a groove in the drum. When the clutch piston is applied, it contacts the inner ends of the diaphragm spring fingers and bends them into contact with the pressure plate to apply the clutch. Because the spring fingers contact the pressure plate near their outer edges, they also act as levers and increase clutch application force.

When hydraulic pressure to the apply piston is cut off, the diaphragm spring returns to its original shape. This pushes the piston back to its original position and releases the clutch.

CONTROLLING CLUTCH APPLICATION

Shift quality with a multiple-disc clutch is determined by the amount of pressure exerted on the piston, and the rate at which the fluid is allowed to enter the clutch assembly. Some clutch circuits include an accumulator, like those described earlier, to achieve a smooth application and yet maintain good holding power once the clutch is fully applied. However, accumulators are not the only way in which this can be done.

Some clutch pistons receive fluid through a line that has an orifice restriction, figure 4-22. The orifice reduces the initial apply pressure, but allows full pressure to act on the piston once pressure equalizes on both sides of the orifice. The orifice in our example is coupled with a check ball and an unrestricted exhaust port. During application, the check ball seats and the orifice restricts pressure to provide smooth clutch engagement, figure 4-22A. When the clutch is released, the check ball unseats and opens the unrestricted exhaust port to quickly release the clutch, figure 4-22B.

Variable Clutch Holding Force

Some multiple-disc hydraulic clutches require less holding force in high gear than they do in reverse. In these cases, application force is modified by applying the clutch piston with two separate hydraulic circuits, figure 4-23. In this design, the backside of the piston is acted on by fluid pressure in two chambers that are separated by an inner lip seal. As in all clutches, an outer lip seal contains fluid behind the piston.

In high gear, figure 4-23A, pressure from the high clutch apply circuit is routed into the high clutch chamber. This chamber allows hydraulic

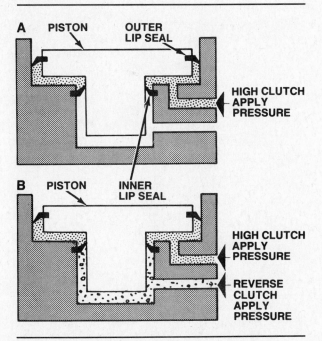

Figure 4-23. Clutch application force can be modified by applying hydraulic pressure to different areas of the piston for different driving conditions.

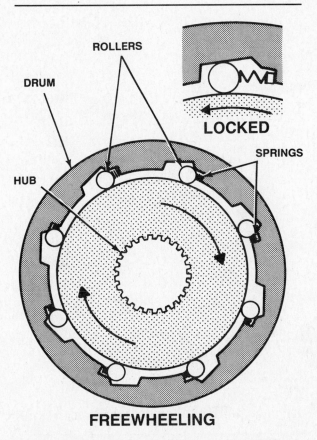

Figure 4-24. This one-way roller clutch locks when the hub rotates counterclockwise, and unlocks when it rotates clockwise. The clutch locks when the rollers wedge between the drum and hub.

pressure to act on only a portion of the total piston surface area, thus limiting application force. The 1-way sealing action of the inner lip seal (explained in the next chapter) prevents hydraulic pressure from escaping the reverse clutch chamber.

In reverse gear, figure 4-23B, additional application force is needed, and hydraulic pressure from the reverse clutch apply circuit is routed to the reverse clutch chamber. Because both circuits to the piston are now open, hydraulic pressure can act on the entire surface area of the piston to increase application force. The inner lip seal is generally ineffective at this time because hydraulic pressure is equal on both its sides. However, if pressure in the reverse clutch chamber should exceed that in the high clutch chamber, the inner lip seal will allow pressure to bypass into the high clutch chamber to ensure maximum clutch application force.

ONE-WAY CLUTCHES

Along with transmission bands and multiple-disc clutches, the **one-way clutch** is a type of apply device. Like a band, a 1-way clutch is always a holding device. One-way clutches are either roller clutches or sprag clutches. The roller clutch is more common.

One-Way Roller Clutch

A 1-way roller clutch, figure 4-24, consists of a hub, rollers, and springs surrounded by a **cam-cut drum** that contains pockets for the rollers and springs. When the clutch hub in our example rotates clockwise, the rollers move toward the large ends of the pockets, compressing the springs as they do so. This unlocks the clutch and allows the hub to freewheel. The 1-way clutch assembly allows clockwise hub rotation at all times.

One-Way Clutch: A mechanical holding device that prevents rotation in one direction, but overruns to allow it in the other. One-way clutches are either roller clutches or sprag clutches.

Cam-Cut Drum: A 1-way roller clutch drum whose inner surface is machined with a series of angled grooves into which rollers are wedged.

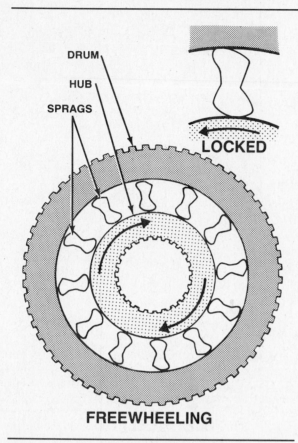

FREEWHEELING

Figure 4-25. This one-way sprag clutch locks when the hub rotates counterclockwise, and unlocks when it rotates clockwise. The clutch locks when the sprags tilt and wedge between the hub and drum.

If the clutch drum is held, the moment the hub rotates counterclockwise the rollers are wedged into the small ends of the pockets cut in the drum, locking the drum and hub together. The springs help in this wedging action. The 1-way clutch assembly locks up and prevents counterclockwise hub rotation as soon as the rollers wedge into their pockets.

One-Way Sprag Clutch

A 1-way sprag clutch consists of a hub and a drum separated by figure-eight-shaped metal pieces called **sprags**, figure 4-25. The sprags lock and unlock the clutch in a manner similar to the rollers in a 1-way roller clutch. When the clutch hub in our example rotates clockwise, the sprags tilt and open a space between themselves, the hub, and the drum. This unlocks the clutch and allows the hub to freewheel. The 1-way clutch assembly allows clockwise hub rotation at all times.

If the clutch drum is held, the moment the hub rotates counterclockwise the sprags tilt the

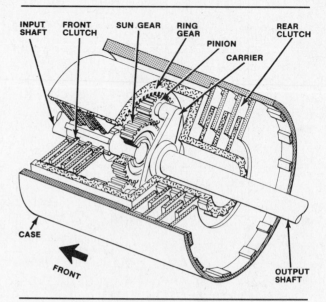

Figure 4-26. A single planetary gearset with two multiple-disc clutches. With the clutches released, the gearset is in neutral; there is no output motion.

other way and are wedged between the hub and drum, locking those two parts together. The 1-way clutch assembly locks up and prevents counterclockwise hub rotation as soon as the sprags wedge between the hub and drum.

One-Way Clutch Applications

One-way clutches have some advantages over other types of apply devices. First, they do not require hydraulic force to operate. Second, they apply and release almost instantly, unlike bands and multiple-disc clutches. And finally, a 1-way clutch that holds the reaction member of a planetary gearset can release that member automatically as soon as the reaction member turns faster than (overruns) its respective drive member, or tries to rotate in the opposite direction.

INTERACTION OF BANDS, MULTIPLE-DISC CLUTCHES, AND ONE-WAY CLUTCHES

Automatic transmissions use planetary gearsets controlled by various combinations of bands, multiple-disc clutches, and 1-way clutches to provide forward and reverse gear ranges. Precisely how this is done in each specific transmission will be covered in Part Two of this manual. The following sections explain some basic ways in which apply devices interact within a transmission to control planetary gearset operation.

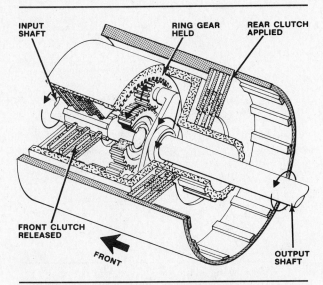

Figure 4-27. With the front clutch released and the rear clutch applied, the ring gear is held and the gearset produces output motion in gear reduction.

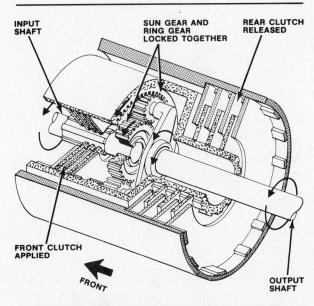

Figure 4-28. With the front clutch applied and the rear clutch released, the sun gear and ring gear are locked together, and the gearset produces output motion in a direct, 1:1 high-gear ratio.

Simple Two-Speed Transmission — Type 1

If you combine two multiple-disc clutches with a single planetary gearset, figure 4-26, it is possible to build a simple 2-speed automatic transmission. The two forward gear ranges are obtained by holding the ring gear (locking it to the transmission case), and by turning the ring gear with the input shaft. To do this, a front clutch is installed between the input shaft and the inside of the ring gear. Also, a rear clutch is installed between the outside of the ring gear and the inside of the transmission case.

In neutral, figure 4-26, both clutches are released. The input shaft and the sun gear splined to it rotate clockwise as viewed from the front. Resistance from the rear wheels holds the output shaft and the carrier splined to it. Therefore the planet pinions (driven by the sun gear) rotate counterclockwise and drive the ring gear counterclockwise. However, because the ring gear is not connected to the output shaft, there is no output from the transmission.

In low gear, figure 4-27, the front clutch is released and the rear clutch is applied. The rear clutch holds the ring gear by locking it to the transmission case. The planet pinions are still driven counterclockwise by the sun gear. However, because the ring gear is held, the pinions walk around the inside of the ring gear and drive the carrier clockwise. The carrier then drives the output shaft in gear reduction, or low gear.

In high gear, figure 4-28, the front clutch is applied and the rear clutch is released. The

front clutch locks the sun gear and the ring gear together. With two members of the planetary gearset locked together, the entire gearset turns clockwise as a unit. The pinions then turn the carrier, which drives the output shaft in direct, or high gear.

The simple transmission just described provides only two speeds forward, and no reverse. It does, however, provide a good basic example of how multiple-disc clutches are used as driving and holding devices for a planetary gearset.

Simple Two-Speed Transmission — Type 2

We can take the same single planetary gearset — still providing only two forward speeds — and give it additional capabilities by adding two more apply devices and rearranging the multiple-disc clutches. The transmission shown in figure 4-29 contains: a front clutch between the input shaft and the inside of the ring gear; a rear clutch between the input shaft and the inside of a clutch drum splined to the sun gear; a band around the outside of the ring gear; and a 1-way roller clutch.

Sprag: A figure-eight-shaped locking element of a 1-way sprag clutch.

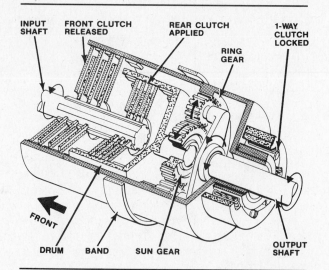

Figure 4-29. With the sun gear driven by the applied rear clutch, and the ring gear held from counterclockwise rotation by the one-way clutch, the gearset produces output motion in gear reduction.

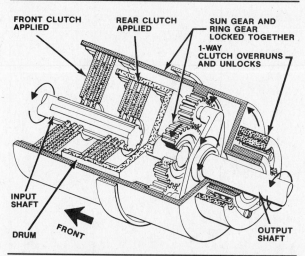

Figure 4-30. With both the front and rear clutches applied, and the one-way clutch allowing the ring gear to turn clockwise, the gearset produces output motion in a direct, 1:1 high-gear ratio.

The inner hub of the 1-way clutch is attached to the ring gear. The outer drum of the 1-way clutch is attached to the inside of the transmission case. The clutch allows the ring gear to rotate clockwise, as viewed from the front. However, the clutch locks and holds the ring gear if the ring gear attempts to rotate counterclockwise.

In this transmission, as with any planetary gearset, it is necessary to turn one member of the gearset while holding another member in order to get output motion. In neutral, none of the members is turned because both clutches are released. The input shaft rotates, but it is not connected to any member of the planetary gearset. As a result, there is no output motion.

In low gear, figure 4-29, the rear clutch is applied. This connects the input shaft to the sun gear through the rear clutch drum. The sun gear then turns clockwise. Resistance from the rear wheels holds the output shaft and the carrier splined to it, so the pinion gears turn counterclockwise and attempt to drive the ring gear in the same direction. The ring gear, however, is held from turning counterclockwise by the 1-way clutch. Therefore, the pinions walk around the inside of the ring gear and drive the carrier and output shaft in gear reduction, or low gear.

In high gear, figure 4-30, both the front clutch and the rear clutch are applied. This connects the sun gear and the ring gear to the input shaft, locking them together. With two members of the planetary gearset locked together, the entire gearset turns clockwise as a

unit. The pinions then turn the carrier, which drives the output shaft in direct, or high gear.

The advantage of the 1-way clutch in this design is that it eliminates timing problems when the transmission is shifted into high gear. Unlike the first type of simple two-speed transmission described in the previous section, there is

■ Crunch Gears

The Ford Motor Company might never have used sliding-gear manual transmissions in its cars if Old Henry had had his way. Henry Ford Sr. never did learn to shift a manual 3-speed transmission properly, and he was adamantly opposed to using one in the car that would replace his beloved Model T. The elder Ford wanted a planetary transmission that would shift automatically through the use of a hydraulic system.

After much argument, Edsel Ford Sr. and others convinced Henry to adopt a 3-speed manual gearbox for the new Model A in 1928. When he finally consented to a manual transmission — "crunch gears," he called it — he reportedly said to one of his engineers, "If the public wants a sliding-gear transmission, let them have it. Let them find out what a contraption it is."

Henry never did understand why the Model A's transmission was so popular. He continued experiments with automatic planetary transmissions until 1941. But by that time, General Motors had launched its Hydra-matic and beaten Henry to the marketplace with the first fully automatic transmission. It wasn't until 1951 — four years after Henry's death — that the first Ford was built with an automatic transmission.

no need to exhaust hydraulic fluid from the apply device holding the ring gear. The moment the front clutch is applied, the 1-way clutch releases automatically and allows the ring gear to rotate clockwise.

Engine braking

In normal low gear operation, the 1-way clutch holds the ring gear only when the engine provides torque to the sun gear, and the ring gear attempts to turn counterclockwise. During closed- or trailing-throttle coasting, the rear wheels provide the torque to the transmission. This makes the carrier the input member of the planetary gearset, and causes the pinions to walk around the sun gear and drive the ring gear clockwise. The 1-way clutch then overruns, and there is no longer a direct connection between the engine and the rear wheels. As a result, engine compression cannot be used for braking.

The band around the ring gear in our second type of simple two-speed transmission is used to shift into manual low gear. In manual low gear, the band is applied to hold the ring gear from rotating in *either* direction. This maintains a direct connection between the engine and rear wheels, and allows the driver to use engine compression for braking.

Modern Transmission Designs

Most modern 3- and 4-speed automatics use compound planetary gear systems that are controlled by one or two bands, two to four clutches, and one or two 1-way clutches. Although the powerflow through these transmissions is more complex than that of the two simple transmissions we have just described, the interactions of their apply devices are basically the same.

In Part Two of this manual we will examine the more popular domestic automatic transmissions in detail. As you study each transmission, memorize the apply devices used in each gear. This is information you need to know in order to accurately diagnose transmission problems.

SUMMARY

The parts used to hold and drive planetary gearset members are called apply devices. The apply devices used in automatic transmissions include bands, multiple-disc clutches, and 1-way overrunning clutches. Bands and 1-way clutches are holding devices only; multiple-disc clutches can be either holding or driving devices.

Bands may be either double-wrap or single wrap, and come in several thicknesses depending on the loads they have to carry. All bands are applied by hydraulic servos acting through a linkage. A straight linkage transmits servo force directly to the band; lever and cantilever linkages transmit servo force indirectly and can increase the amount of force.

Some servo hydraulic circuits contain an accumulator that smooths pressure buildup and damps pressure surges. This is done by temporarily diverting part of the circuit's fluid into a parallel circuit or chamber. Accumulators may be either piston-type or valve type. Piston-type accumulators are further broken down into independent designs that fit in their own bore, or integral designs that share a bore with a servo piston.

Multiple-disc hydraulic clutches consist of friction discs alternated between steel discs. One set of discs is splined on its inner edge, the other set is splined on its outer edge. When the discs are forced together by a hydraulic piston, the clutch locks up and connects the two components that are engaged with the discs' splines.

Multiple-disc clutch application may be controlled with accumulators, restricting orifices, and/or dual hydraulic circuits for the clutch apply piston. Clutch release is aided by fluid vent ports, check balls, and one or more return springs.

One-way clutches are either roller type or sprag type. Both designs freewheel in one direction, and lock up in the other. One-way clutches are purely mechanical devices and require no hydraulic pressure to operate.

Review Questions

Select the single most correct answer
Compare your answers to the correct answers on page 603

1. Apply devices consist of:
   a. Hydraulic clutches
   b. Hydraulic servos
   c. 1-way clutches
   d. All of the above

2. Mechanic A says a band is used as a driving device only.
   Mechanic B says a band is used as a holding device only.
   Who is right?
   a. Mechanic A
   b. Mechanic B
   c. Both mechanic A and B
   d. Neither mechanic A nor B

3. Mechanic A says a double-wrap band provides greater holding force than a single-wrap band.
   Mechanic B says a double-wrap band costs more to make but matches the circumference of a drum more closely than a single-wrap band.
   Who is right?
   a. Mechanic A
   b. Mechanic B
   c. Both mechanic A and B
   d. Neither mechanic A nor B

4. The application force of a servo equals the system pressure times the area of the servo piston. Therefore, 125 psi on a 4-square-inch piston exerts _____ pounds of force:
   a. 300
   b. 350
   c. 500
   d. 550

5. Mechanic A says cantilever servo linkage is used with a band that is anchored at one end.
   Mechanic B says cantilever servo linkage is used on the Torqueflite transmission.
   Who is right?
   a. Mechanic A
   b. Mechanic B
   c. Both mechanic A and B
   d. Neither mechanic A nor B

6. Periodic transmission band adjustment is *not* required when the servo linkage contains a:
   a. Cantilever
   b. Lever
   c. Strut
   d. Graduated rod

7. An accumulator in a transmission hydraulic system:
   a. Delays pressure buildup on an apply device
   b. Cushions the shock of sudden pressure changes
   c. Provides a smooth shift
   d. All of the above

8. Mechanic A says an accumulator may be either a spool valve in the valve body or piston in a servo bore.
   Mechanic B says accumulators may be installed in separate bores in the transmission case.
   Who is right?
   a. Mechanic A
   b. Mechanic B
   c. Both mechanic A and B
   d. Neither mechanic A nor B

9. Mechanic A says when an accumulator reaches the end of its stroke, the pressure on the apply device increases to its full value.
   Mechanic B says when an accumulator reaches the end of its stroke, the pressure on the apply device increases to twice the level at the shift valve.
   Who is right?
   a. Mechanic A
   b. Mechanic B
   c. Both mechanic A and B
   d. Neither mechanic A nor B

10. Mechanic A says a multiple-disc holding clutch is used to lock the input and output shafts together.
    Mechanic B says a multiple-disc driving clutch is used to lock the planetary ring gear to the transmission case.
    Who is right?
    a. Mechanic A
    b. Mechanic B
    c. Both mechanic A and B
    d. Neither mechanic A nor B

11. The diaphragm return spring used in some multiple-disc clutches is also called a:
    a. Coil spring
    b. Belleville spring
    c. Ackerman spring
    d. Roller spring

12. Mechanic A says a check ball is used in some clutch drums to provide an escape for any oil remaining in the drum when the clutch is released.
    Mechanic B says a check ball is used in some clutch drums to prevent the clutch from releasing when fluid is exhausted.
    Who is right?
    a. Mechanic A
    b. Mechanic B
    c. Both mechanic A and B
    d. Neither mechanic A nor B

13. The most common type of 1-way clutch is the _____ type:
    a. Sprag
    b. Roller
    c. Belleville
    d. Bendix

14. Mechanic A says a 1-way clutch locks up whenever it overruns.
    Mechanic B says an advantage of a 1-way clutch is that it requires less hydraulic pressure to apply.
    Who is right?
    a. Mechanic A
    b. Mechanic B
    c. Both mechanic A and B
    d. Neither mechanic A nor B

5

Transmission Fluids, Filters, and Coolers

Automatic transmission hydraulic systems require special fluids in order for the transmission to operate properly and have a long service life. Filters are used to keep these fluids clean, and prevent wear or damage to the transmission internal components. Cars with automatic transmissions also have coolers that lower transmission operating temperatures by removing excess heat from the fluid.

AUTOMATIC TRANSMISSION FLUID

Automatic transmission fluid (ATF) is the vital operating fluid of every automatic transmission. In this role, it is called upon to perform a number of jobs including:

- Power transmission
- Cooling
- Lubrication
- Cleaning
- Shift control
- Apply device operation.

Power Transmission

With a manual transmission, the engine crankshaft is mechanically connected to the transmission input shaft through a clutch operated by the driver. When the transmission is shifted, or the vehicle is stopped in gear with the engine running, the clutch must be disengaged to isolate the engine from the transmission. This unloads the gears when shifting, and prevents the engine from stalling as the car comes to a stop.

As discussed in Chapter 1, the planetary gearsets in an automatic transmission do not require a driver-operated clutch for shifting. A clutch is also unnecessary when stopping the car in gear because an automatic transmission is *not* mechanically connected to the crankshaft. Instead, ATF circulating through a **torque converter**, figure 5-1, transfers engine power to the transmission hydraulically. Unless the vehicle is equipped with a lockup torque converter, ATF provides the *only* connection between the engine crankshaft and the transmission input shaft. Both conventional and lockup torque converters are covered in detail in Chapter 7.

Cooling, Lubrication, and Cleaning

The torque converter and the transmission gears, clutches, and bands create a great deal of heat during operation. The ATF circulating through and over these parts cools them by absorbing much of the heat created. Heated fluid

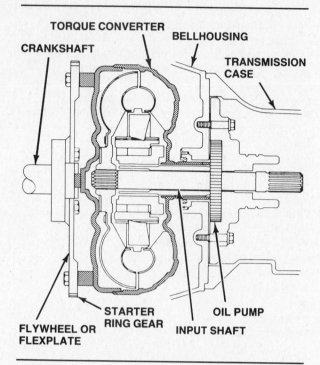

TORQUE CONVERTER
CRANKSHAFT
BELLHOUSING
TRANSMISSION CASE
FLYWHEEL OR FLEXPLATE
STARTER RING GEAR
OIL PUMP
INPUT SHAFT

Figure 5-1. ATF inside the torque converter transmits engine power to the transmission.

is then circulated through a transmission cooler and returned to the transmission. These coolers are described later in the chapter.

The ATF also lubricates the transmission as it passes through and over the various internal components. In some applications, cooled fluid returned to the transmission goes directly into lubricating the bushings, bearings, and gears. In other transmissions, the oil returns to the sump where it is picked up by the pump, directed to the pressure regulator valve, and then sent throughout the transmission.

As it cools and lubricates the transmission, ATF works to keep internal parts clean. Special additives in the fluid help prevent varnish buildup, and work to keep various contaminants in suspension so they can be trapped by the transmission filter.

Shift Control and Apply Device Operation

The ATF circulates under pressure through the transmission valve body to move the valves that control transmission shifting. As described in Chapter 3, hydraulic pressure is developed by the transmission pump and controlled by regulating valves that sense engine torque and road speed. Fluid flow and pressure from these valves operates the switching valves that time the transmission gear ratio changes.

Finally, ATF operates the apply devices in the transmission. The clutches and bands are applied, and often released, by fluid pressure. This happens when the switching valves direct fluid through the hydraulic circuits to these apply devices.

GENERAL ATF PROPERTIES

Automatic transmission fluid can be a petroleum-base oil, a 100-percent synthetic lubricant, or a combination of these two types. As of the 1988 model year, all factory-fill ATFs used by the big three domestic automakers are petroleum-based fluids. However, some import automakers are using semi-synthetic fluids, and both partial- and full-synthetic ATFs are available from aftermarket suppliers.

Under normal conditions, petroleum-based ATF has a clear red color that helps technicians determine whether an under-car leak is from the transmission or some other powertrain component. ATF that is contaminated with water or engine coolant, most likely as the result of a transmission cooler leak, has a milky pink color. Overheated petroleum-base ATF smells burnt, and contaminated petroleum-base fluids turn reddish-brown, brown, or brown-black. Either or both of these conditions indicate that the transmission needs to be serviced.

Some partially or fully synthetic ATFs have a darker red initial color, and these fluids often smell burnt or turn reddish-brown, brown, or even brown-black soon after being put into service. Smell and color (except for the milky pink color that indicates coolant contamination) are *not* accurate indicators of fluid condition when using part- or full-synthetic ATF.

In general, ATF is similar to engine oil, but it does have some important differences that allow for the unique requirements of automatic transmissions. Before it can be certified for use, an ATF must pass a series of stringent tests established by the transmission and/or vehicle manufacturer. For example, figure 5-2 lists the 16 tests General Motors currently requires an ATF to pass before it can be labeled as DEXRON®-II. The paragraphs below describe typical tests used to determine several important fluid properties.

Miscibility Tests

Although ATFs from several manufacturers may meet the same set of specifications, this does not mean they are exactly the same in their chemical makeup. To ensure compatibility between different brands of fluid, every fluid must pass a **miscibility** test. In a typical test of

Test	Criteria Measured
A	Miscibility
B	Viscosity at 210°F (100°C)
C	Flash Point
D	Fire Point
E	Low Temperature Fluidity at −10°F (−23°C) at −40°F (−40°C)
F	Copper Staining
G	Foaming at 200°F (93°C) at 275°F (135°C)
H	Corrosion or Rusting
I	Effects on Seals
J	Odor
K	Oxidation Resistance and Thermal Stability
L	Low-Energy Transmission Cycling, Friction-Retention, and Oxidation
M	High-Energy Transmission Cycling, Friction-Retention, and Oxidation
N	Nylon Compatibility
O	Fluid Friction Characteristics
P	Performance in Transmission Under Service Conditions

Figure 5-2. These are the tests required by GM for DEXRON®-II transmission fluids.

this type, fluid samples are mixed with several reference fluids, and the mixtures are subjected to high and sub-freezing temperatures. In order to pass, there must be no fluid separation or color change at the end of the test.

Viscosity Tests

The resistance to flow, or **viscosity**, of ATF is roughly the same as that of a 0W-20 engine oil. However, just as with engine oils, an ATF's viscosity is greatly affected by temperature.

When ATF is cold, its viscosity increases and the fluid does not flow as well. If an ATF's viscosity is too high at low temperatures, there will be a lag between the time the engine is started and when fluid under pressure actually reaches the control valves, apply devices, and heavily loaded transmission parts. Shift action will be delayed, and there will be increased slippage of the bands and clutches as they are applied. Additives called pour point depressants are used in ATF to give it better low temperature flowability.

When ATF is hot, it thins out and flows quite easily. If an ATF's viscosity is too low at high temperatures, the lubricating film between critical parts may break down and allow them to make contact, causing wear and possibly damage. Additives called Viscosity Index (VI) improvers are used in ATF to enable it to maintain its viscosity at elevated temperatures.

To make sure an ATF is capable of providing adequate lubrication throughout the full range

of transmission operating conditions, the fluid is tested for viscosity at both high and low temperatures. Typically, an ATF will be tested at the temperature extremes of −40°F (−40°C) and 210°F (100°C), and sometimes at other designated temperatures in between.

Flash and Flame Point Tests

Safety is a concern in any vehicle system, but especially so when dealing with hot volatile fluids such as ATF. Typically, an ATF must meet minimum specifications for both flash point and fire point. The flash point is the temperature at which vapors from heated ATF will ignite when exposed to an open flame. The fire point is the slightly higher temperature at which heated ATF will begin to burn when exposed to an open flame. Generally, both temperatures are above 300°F (149°C) for petroleum-base ATF, and in excess of 400°F (204°C) for synthetic fluids.

Foaming Test

The inside of an automatic transmission in operation is filled with rapidly spinning components that can agitate the ATF and churn it into foam. When this happens, air bubbles in the fluid reduce lubrication, and the transmission hydraulic system is unable to function properly because the air in the fluid can be compressed. In cases of minor foaming, shift action is delayed or erratic; in more extreme situations, slippage of the clutches and bands can cause extreme heat, wear, and fluid breakdown that will lead to the need for a transmission overhaul.

To avoid these problems, ATF contains antifoaming agents that help prevent air bubbles and limit the lifespan of those that do form. The effectiveness of these additives is checked in laboratory tests that agitate the fluid in a controlled manner for specified times at different

Torque Converter: A type of fluid coupling used to connect the engine crankshaft to the automatic transmission input shaft. Torque converters multiply the available engine torque under certain operating conditions.

Miscibility: The property that allows one fluid to blend with other fluids.

Viscosity: The tendency of a liquid, such as an oil or hydraulic fluid, to resist flowing.

Fluid Type	Friction Modified	Non-Modified	Obsolete
Type A	X		X
Type A, Suffix A	X		
DEXRON®-B	X		
DEXRON®-IIC	X		X
DEXRON®-IID	X		
Type F			
1P-XXXXXX		X	X
2P-XXXXXX		X	
Type G		X	
Type CJ	X		X
Type H	X		X
MERCON®	X		

Figure 5-3. Many types of ATF have been recommended by vehicle manufacturers over the years.

temperatures. Depending on the exact test being performed, there must be either no sign of foam on the surface of the fluid, or the amount of foam and the time it takes to subside must fall within certain limits.

Oxidation Resistance Test

ATF generally operates at higher temperatures than engine oil; under heavy loads it can run at *much* higher temperatures. Because of this, ATF is subject to greater **oxidation** of its oil molecules. Oxidation causes the formation of varnish, sludge, and acids that can result in transmission damage. To combat such heat-related fluid breakdown, additives called oxidation inhibitors are used in modern ATF.

A fluid's effectiveness in resisting oxidation is measured by installing the fluid in a test transmission, then operating the transmission through a strictly controlled test cycle. Afterwards, the fluid is chemically analyzed for breakdown, and the transmission is disassembled and its parts inspected for physical damage and deposits.

Rust, Corrosion, and Compatibility Tests

Modern ATFs also contain additives called corrosion inhibitors that prevent the rusting or etching of metal components. These properties are tested by immersing strips and pins of various metals in heated ATF for a given period of time. At the end of this time, the metal samples are removed from the fluid, weighed to check for metal loss, and visually inspected for signs of staining or physical damage.

ATF must also be compatible with rubber and nylon seals, and nylon speedometer drive gears. As with the rust and corrosion tests, these properties are checked by immersing

samples of the seal and gear materials in heated fluid. After a specified time, rubber samples are measured to check for excessive swelling or softening of the material. Nylon parts must show no signs of physical deterioration.

Friction and Wear Tests

Friction and wear tests determine how well an ATF protects the transmission over an extended period of operation, and how well the fluid maintains its frictional properties. Zinc, sulphur, and phosphorous anti-wear additives in the fluid combat friction. Detergent additives help keep the transmission clean. Dispersant additives hold contaminants suspended in the fluid so they can be trapped by the filter. The frictional properties of an ATF are determined by additives called **friction modifiers**. These are discussed in greater detail in the next section.

During the friction and wear tests, the fluid is installed in a newly rebuilt test transmission that is then subjected to several hundred hours of controlled cycle operation. For the duration of the test, shift times must remain within specified limits. Afterwards, the fluid is tested to ensure its frictional properties have not deteriorated, and the transmission is torn down and inspected for cleanliness and physical condition.

ATF FRICTIONAL PROPERTIES

All ATFs fall into two groups, figure 5-3, those that contain friction modifiers and those that do not. Fluid Types A, CJ, H, DEXRON®, DEXRON®-II, and MERCON® are friction modified. Types F and G fluid are non-modified. The two types of ATF act in opposite ways within a transmission, but to explain how they differ, we must first define two terms, **static friction** and **dynamic friction**.

Static friction is the **coefficient of friction** between two surfaces that are in fixed, or nearly fixed, contact with one another. Dynamic, or kinetic, friction is the coefficient of friction between two surfaces that have relative motion between them — such as when they are sliding against one another. Under normal conditions, the coefficient of static friction is always higher than that of dynamic friction. This is why it is harder to *start* an object moving than it is to *keep* it moving.

A non-modified ATF like Type F or G provides a certain coefficient of dynamic friction between multiple-disc clutch plates, or between a transmission band and drum, when there is relative motion between the parts. As the clutch or band locks up and the relative motion

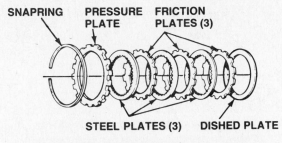

SNAPRING   PRESSURE   FRICTION
                    PLATE   PLATES (3)

STEEL PLATES (3)   DISHED PLATE

JATCO LOW-REVERSE CLUTCH

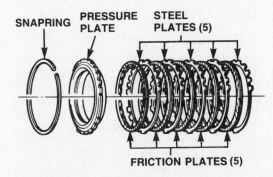

SNAPRING   PRESSURE   STEEL
                    PLATE   PLATES (5)

FRICTION PLATES (5)

THM 250 LOW-REVERSE CLUTCH

Figure 5-4. These low-reverse clutches are from a Ford Jatco transmission that uses non-modified Type F ATF, and a THM 250 that uses friction modified DEXRON®-II fluid. Note the differences in the size and number of clutch plates.

becomes zero, a non-modified ATF provides an *increasing* coefficient of static friction that is ultimately *greater* than the dynamic friction present before lockup.

The friction modifiers in fluids such as DEXRON®-II or MERCON® cause these ATFs to work in the opposite fashion. When there is relative motion between the elements of a clutch or band assembly, the fluid provides a certain coefficient of dynamic friction. However, as the clutch or band locks up, a friction-modified fluid reacts to actually *decrease* the coefficient of static friction. In other words, with a friction-modified fluid, the coefficient of static friction is *less* than the coefficient of dynamic friction.

Fluid Effects on Transmission Design

When engineers design an automatic transmission and calibrate its shift action, they take the friction properties of the ATF into account. For example, non-modified fluids with their increasing coefficient of static friction allow the transmission engineer to use smaller bands, or

fewer plates in a clutch pack, figure 5-4, yet still achieve good lockup.

The problem with non-modified fluids is that their lower coefficient of dynamic friction requires that the bands and clutches be applied quickly and with high hydraulic pressure in order to avoid excessive slippage that causes wear and heat buildup. Ford transmissions that use non-modified Type F fluid are known for their firm shifting with distinct rpm changes between gears.

Friction-modified ATFs require a very different type of transmission construction and shift programming. The benefit of these fluids is their relatively high coefficient of dynamic friction which, when combined with lower clutch and band application pressures, provides a very smooth shift feel. General Motors transmissions that use friction-modified DEXRON®-II fluid are noted for their smooth seamless flow of power with almost undetectable shifts in some models.

The problem with friction-modified fluids is that their decreasing coefficient of static friction can lead to excessive slippage and heat buildup as the clutch or band locks up. As a result, transmissions designed to operate with friction-modified fluids generally have larger bands and more plates in their clutch packs, figure 5-4, than do transmissions that use non-modified fluids.

Oxidation: The process in which one element of a compound is combined with oxygen in a chemical process that produces another compound. Oxidation generates heat as a byproduct.

Friction Modifiers: Additives that enable lubricants to maintain their viscosity over a wide range of temperatures.

Static Friction: The coefficient of friction between two surfaces that are in fixed, or nearly fixed, contact with one another.

Dynamic Friction: The coefficient of friction between two surfaces that have relative motion between them. Also called kinetic friction.

Coefficient Of Friction: A numerical value expressing the amount of friction between two surfaces. The coefficient of friction is obtained by dividing the force required to slide the surfaces across one another by the pressure holding the surfaces together.

Improper Fluid Use

The consequences of using the wrong ATF in a transmission vary depending on the type of fluid originally specified by the vehicle manufacturer. When friction-modified ATF is used in an automatic transmission designed for non-modified fluid, excessive slippage results. The slippage is especially prevalent when high torque is generated during trailer towing or hill climbing. This slippage causes accelerated wear of the friction materials on the clutches and bands, and shortens the transmission service life.

If non-modified ATF is used in an automatic transmission designed for friction-modified fluid, harsh or abrupt shifts will result. While this increases the loads on certain transmission components, and may be objectionable to the vehicle owner, it is less likely to create long-term problems that will affect the service life of the transmission.

"Universal" Transmission Fluids

Certain oil companies have developed full-synthetic ATFs that have good performance characteristics in a wide variety of automatic transmissions. At least one formulation has been tested to meet or exceed certain performance requirements for Dexron®-II, Type F, and Type CJ ATF. These synthetic fluids are sometimes called "universal" ATFs, and their manufacturers claim they may be used in any automatic transmission.

Ford Motor Company, the main manufacturer of transmissions that require non-modified ATFs, disagrees. In a letter to the American Petroleum Institute, reprinted in the July, 1987 issue of Motorcraft *Shop Tips*, Ford made its position clear. "With the information available as of the date of this letter, Ford Motor Company can state that no single product formulation exists which can meet *all* of the current Ford material specifications for Automatic Transmission Fluid (ATF)."

Ford's statement is based on the fact that no single ATF can be both friction-modified and non-modified. As explained earlier, these two types of fluids operate in fundamentally opposite ways. A synthetic fluid formulation may meet many of the requirements for both types of fluid, and may even work acceptably well in certain driving conditions; however, it cannot defy the laws of chemistry — or meet *all* of Ford's specifications at this time.

SPECIFIC ATF TYPES

The first ATF was a straight mineral engine oil dyed red to help detect leaks. Limited additives were used to improve low-temperature performance, counteract sludge and varnish formation, and resolve compatibility problems with metals, rubbers, and the friction materials on clutch discs and transmission bands. Over time, more advanced ATFs have been developed to meet the needs of modern automatic transmissions. The most popular of these are listed in figure 5-3, and described in detail in the following sections.

Type A Fluids

The Type A fluid specification was developed by General Motors and was recommended for all American automatic transmissions manufactured prior to 1956. Type A fluid was friction modified and had special additives that allowed it to maintain proper viscosity, resist oxidation and foaming, and be compatible with clutch and band friction materials.

Type A fluid is now obsolete and unavailable. All approved Type A fluids had a qualification number, AQ-ATF-XXX, on the container. The Xs represent the three-digit number assigned to the fluid manufactured by each specific oil company.

Type A, Suffix A fluid

In 1957 the Type A specification was upgraded, and the resulting fluids are known as Type A, Suffix A. The primary difference from Type A fluids is higher oxidation limits.

Type A, Suffix A fluids were used in passenger cars through the late 1960s. This fluid is now obsolete for on-road vehicles, but it is still recommended for limited use in the transmissions and hydraulic systems of industrial equipment.

All approved Type A, Suffix A fluids have a qualification number, AQ-ATF-XXXX, on the container. The Xs represent the four-digit number assigned to the fluid manufactured by each specific oil company.

DEXRON® Fluids

In 1967 General Motors introduced DEXRON® fluid to compensate for the increased operating temperatures of newer automatic transmissions being used with more powerful engines. DEXRON® ATF is friction modified like Type A fluids, but it has a lower static coefficient of friction. It also has superior performance at low temperatures, does a better job of resisting oxidation, and retains these properties much longer than do Type A fluids.

An approved DEXRON® fluid has a qualification number, B-XXXXX, on its container; the Xs represent the five-digit number assigned to the fluid manufactured by each specific oil company. Because a "B" replaces the "A" in the qualification number, this fluid formulation is sometimes called DEXRON®-B.

DEXRON®-B ATF may be used in older GM transmissions that originally required Type A fluids. However, DEXRON®-B fluid does *not* necessarily supersede Type A fluids in transmissions built by other manufacturers. Although it has been superseded by DEXRON®-II in later GM applications, DEXRON®-B remains the primary recommendation of a number of import automakers, and is widely used in off-road and industrial equipment transmissions.

DEXRON®-IIC fluid

In 1973 General Motors introduced DEXRON®-II fluid. This fluid had an even lower static coefficient of friction than DEXRON®-B ATF. It also worked better at low temperatures, had increased oxidation resistance, and exhibited higher frictional stability. The first DEXRON®-II fluids carried a qualification number, C-XXXXX, on their containers; the Xs represent the five-digit number assigned to the fluid manufactured by each specific oil company. Because a "C" replaced the "B" in the qualification number, this fluid formulation was sometimes called DEXRON®-IIC.

A couple years after its introduction, it was discovered that DEXRON®-IIC was allowing corrosion of the transmission cooler line fittings on some cars. To correct this problem, DEXRON®-IIC was superseded by DEXRON®-IID fluid. Today, DEXRON®-IIC ATF is obsolete and unavailable.

DEXRON®-IID fluid

In 1976 the DEXRON®-II fluid formulation was changed. This ATF passes the same tests as the C-XXXXX fluids, but has special additives to prevent the corrosion of transmission cooler line fittings. These fluids are marked with a D-XXXXX qualification number on their containers; the Xs represent the five-digit number assigned to the fluid manufactured by each specific oil company. Because a "D" replaces the "C" in the qualification number, this fluid formulation is sometimes called DEXRON®-IID.

DEXRON®-IID ATF can be used in older GM transmissions that originally required DEXRON®-IIC, DEXRON®-B, or Type A fluids.

However, DEXRON®-IID does *not* necessarily supersede these fluids in transmissions built by other manufacturers. Some import automakers allow the use of DEXRON®-IID for top up or fluid changes, but require DEXRON®-B or a proprietary fluid when a new or rebuilt automatic transmission is installed.

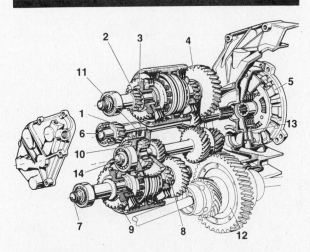

LEGEND:
1. REVERSE PINION
2. 1ST SPEED PINION
3. 4TH SPEED PINION
4. 2ND SPEED PINION
5. CENTRIFUGAL CLUTCH
6. INPUT SHAFT
7. LOWER COUNTERSHAFT
8. 3RD SPEED WITH CLUTCH
9. 5TH SPEED WITH CLUTCH
10. 1ST SPEED WITH FREEWHEEL
11. SLIDING DOG ON OUTPUT SHAFT
12. DIFFERENTIAL GEAR
13. OUTPUT PINION
14. OUTPUT PINION

■ An Automated Transmission for Fuel Economy

Engineers have always known that the most efficient automatic transmission design would be one with infinitely variable gear ratios. However, because most transmissions use mechanical gears with fixed ratios, this has not been practical.

In a step toward the ideal of a stepless transmission, Fiat of Italy developed a transmission that used constant-mesh helical gears — not planetary gearsets — that were engaged and disengaged by hydraulic multiple-disc clutches. Power from the engine was transferred to the gearbox by a centrifugal clutch.

The tricky part of this transmission was the electrical system that regulated engine speed as well as shifting. The accelerator linkage was connected to a variable resistor that controlled a servo motor to open and close the throttle at the carburetor. The electronic control box matched the engine speed to the car speed, and controlled upshifts and downshifts using solenoids and an electro-hydraulic brake on the input shaft.

Type F Fluid

The Ford Motor Company developed non-modified Type F fluid in 1959. Type F fluid had improved viscosity and greater resistance to oxidation than the Type A, Suffix A fluid Ford had been recommending previously. Like DEXRON®-B, Type F fluid was designed to cope with the increased heat produced by newer transmissions teamed with more powerful engines. However, unlike DEXRON®-B, Type F fluid did this by providing faster lockup of the bands and clutches.

Early approved Type F fluids (specification ESW-M2C33-D) came in containers marked with the qualification number 1P-XXXXXX; the Xs represented the six-digit number assigned to the fluid manufactured by each specific oil company. This early fluid formulation was suitable for use in selected automatic transmissions manufactured prior to 1967. This fluid is now obsolete and unavailable.

An improved Type F fluid (specification ESW-M2C33-F) was introduced in 1967. It carries the qualification number, 2P-XXXXXX, and has special frictional characteristics and better resistance to oxidation. This fluid is now recommended for all transmissions requiring Type F fluid.

Type G Fluid

Ford of Europe introduced an upgraded version of Type F fluid in 1972 under the name Type G fluid (specification ESP-M2C33-G); no qualification numbers are used. This non-modified ATF has greater high-temperature stability than Type F fluid to better cope with European high-speed driving conditions.

Some import vehicle manufacturers have used Type G fluid as the factory fill in their automatic transmissions. When these cars are sold in the United States where Type G fluid is usually unavailable, Type F fluid is commonly specified for top up or refill.

Type CJ Fluid

In 1977 Ford introduced Type CJ fluid (specification M2C138-CJ); no qualification numbers are used. Type CJ fluid was used in the redesigned C6 automatic transmission, and was also recommended for Jatco transmissions in passenger cars from 1977-79. Type CJ fluid was a friction-modified ATF similar to DEXRON®-II, and it was required for compatibility with the friction materials on the bands and clutches of the transmissions mentioned above.

Type CJ fluid was discontinued from service availability in 1980 and, at that time, Ford recommended DEXRON®-II be used in its place. Ford's latest recommendation is that transmissions originally requiring Type CJ fluid now be filled with MERCON®.

Type H Fluid

In 1982, Ford introduced Type H fluid (specification ESP-M2C166-H); no qualification numbers are used. Type H fluid was used in the C5 automatic transmission with lockup torque converter. The specification for Type H fluid has an extra friction-retention requirement not included for Type CJ or DEXRON®-II fluids. Unless Type H fluid is used in the C5, a "shudder" will occur as the torque converter locks up.

Type H fluid was discontinued from service availability in 1988. Ford's latest recommendation is that transmissions originally requiring this fluid now be filled with MERCON®.

MERCON® Fluid

In 1988, Ford introduced a new multi-purpose friction-modified ATF called MERCON® (specification WSP-M2C185-A). Approved MERCON® fluids come in containers marked with the qualification number M-XXXXXX; the Xs represented the six-digit number assigned to the fluid manufactured by each specific oil company.

MERCON® was developed to reduce the number of ATFs needed to service Ford automatic transmissions, and this fluid is also used in certain Jeep/Eagle automatic transmissions. MERCON® replaces DEXRON®-II, Type CJ, and Type H fluids in Ford's recommendations. It specifically *does not* replace Type F or G fluids.

Proprietary ATF Types

The previous sections describe the most common types of ATF, but these are not the only fluids available. In some cases, a vehicle manufacturer will require a special fluid that meets additional requirements unique to a particular application. Ford's Type H fluid for the C5 transmission was one such fluid, and Chrysler currently specifies an "ATF+" for their automatic transmissions. Certain import automakers, such as Mercedes-Benz and Peugeot, recommend only their own brand of ATF for use in their automatic transmissions.

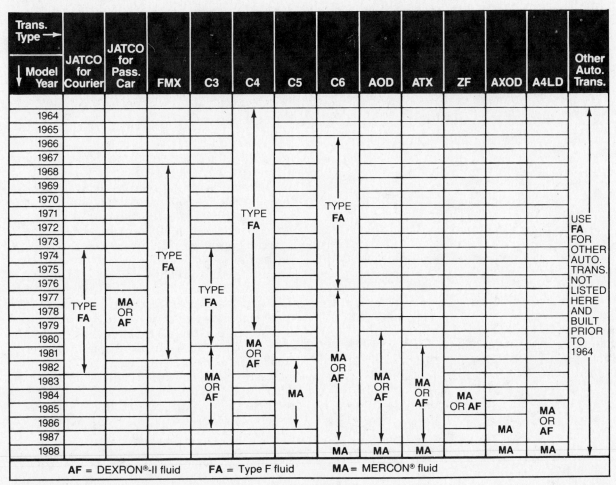

Figure 5-5. Ford automatic transmission fluid recommendations.

MANUFACTURERS' ATF RECOMMENDATIONS

American Motors Corporation required DEXRON®-IID fluid for all automatic transmissions in vehicles built through 1986. Following the purchase of AMC by Chrysler, the fluid recommendations for Jeep/Eagle models are as follows. The 1987-88 Eagle (AMC) 4WD wagon, Eagle Premier, and Eagle Medallion require MERCON®. The 1987-88 Jeep models continue to use DEXRON®-IID.

Chrysler Motors recommended DEXRON®-IID fluid for all its 1986 and earlier automatic transmissions. For 1987-88 automatics, Chrysler specifies a special ATF+ Type 6167 (Part No. 4318077). This fluid is also recommended for earlier Chrysler automatics with locking torque converters to help eliminate lockup shudder.

Ford Motor Company ATF recommendations have changed several times over the years. As mentioned in the fluid descriptions, certain Ford automatics use two different fluid types depending on the vehicle in which they are installed, or the year in which the transmission was manufactured. The Ford ATF recommendations in effect as of the 1988 model year are shown on the chart in figure 5-5.

There are some generalizations that can be made about Ford ATF recommendations. First, *all* Ford automatics in 1976 and earlier cars use Type F fluid. Second, MERCON® is the preferred fluid for *all* 1983-87 Ford automatics, although DEXRON®-II remains a secondary recommendation for *some* transmissions built in these years. Finally, the use of MERCON® is a warranty *requirement* for all 1988 and later Ford automatic transmissions manufactured in North America, and Ford Motor Company vehicles sold in North America.

General Motors Corporation has had the most consistent ATF recommendations of the big three automakers. As of the 1988 model year, all GM automatic transmissions require DEXRON®-IID fluid.

TRANSMISSION FILTERS

Transmission filters remove dirt, metal, and friction material particles from the ATF so that

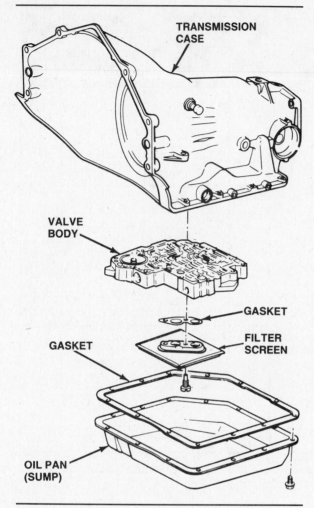

Figure 5-6. A typical automatic transmission filter installation.

that replace the standard parts to provide greater oil capacity for increased cooling. Transmissions equipped with such pans require special filters that are longer and extend down to the bottom of the pan to pick up the oil.

Screen Filters

The oldest type of transmission filter is the metal screen filter, which is nothing more than a fine wire mesh designed to trap contaminants. A screen filter should be cleaned every time the transmission is serviced, and should be replaced if it is torn, broken, or clogged. A screen with varnish build up on its mesh surface can be soaked in carburetor cleaner, flushed in solvent, and reused. If the varnish cannot be removed in this manner, the filter must be replaced.

Newer synthetic screen filters are similar to the metal screen design, but the mesh is made of nylon or polyester. These types of mesh are finer than a metal screen, and able to filter out smaller particles. A synthetic filter screen can be washed with solvent to remove trapped particles; however, it cannot be soaked in carburetor cleaner to remove varnish. If a synthetic screen filter is broken, torn, clogged, or shows any signs of varnish buildup, it must be replaced.

Metal and synthetic screen filters are **surface filters**. A surface filter traps contaminants on its surface, and as the particles collect, fluid flow is reduced. Because of this, the mesh openings in screen filters must be large enough to allow smaller particles to pass through or the filter will quickly become clogged. Most metal screen filters have mesh openings in the area of 130 microns (.005 in.); typical synthetic screen filter openings are roughly 100 microns (.004 in.) in size. Even larger particles with irregular shapes, such as slivers, can pass through these relatively large mesh openings.

Paper Filters

Paper transmission filters are made of an oil-resistant fabric similar to that in an engine oil filter. The fabric may be made of a natural material such as cellulose, but many newer designs use a synthetic fiber such as dacron. Paper filters cannot be cleaned, and should be replaced whenever the transmission is serviced. Generally this will be at least every 25,000 miles (40,000 km) for vehicles operated in severe service.

these contaminants do not circulate through the transmission where they can increase wear or cause valves to stick. If the transmission filler tube is left open, dirt may enter from outside the transmission. Metal and friction-material particles, however, are generally the result of normal wear or specific problems within the gearbox.

The transmission filter is located inside the transmission case between the pickup for the oil pump and the bottom of the sump, figure 5-6. The three types of filters used in automatic transmissions are:

- Screen filters
- Paper filters
- Felt filters.

Common examples of these filters are shown in figure 5-7. Note that some automatic transmissions are available with optional deep pans

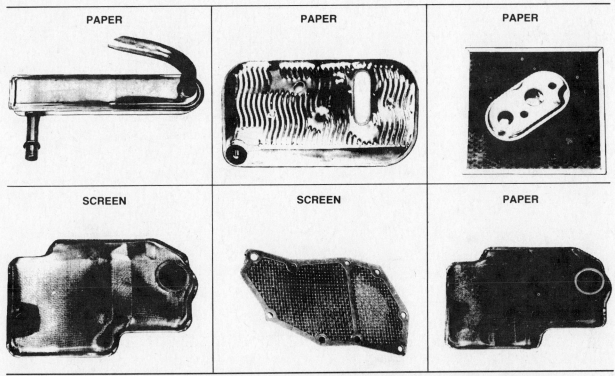

| PAPER | PAPER | PAPER |
| SCREEN | SCREEN | PAPER |

Figure 5-7. Screen, paper, and felt filters in automatic transmissions must be cleaned or replaced periodically.

Paper filters are more efficient than screen filters, and trap a greater number and smaller size of particles for a given fluid pressure drop. However, paper filters are still surface filters, and they ultimately suffer from the same limitations as screen filters. To prevent a loss of pressure should the filter become clogged, a dirty oil bypass is used in some transmissions with paper filters. However, such a bypass is unacceptable in the latest transmission designs which are built to close tolerances and contain sensitive solenoid valves as part of their electronic controls.

Felt Filters

Felt filters are made from specially treated polyester fibers that are randomly spaced and not woven. Like paper filters, felt filters cannot be cleaned, and should be replaced whenever the transmission is serviced. Generally, this is usually at least every 25,000 miles (40,000 km) for vehicles operated in severe service.

Unlike screen and paper filters, felt filters are **depth filters**. Depth filters trap contaminant particles within the matrix of the filter material rather than on its surface. As a result, they are able to trap finer particles, and hold larger quantities of contaminants, with little or no restriction in fluid flow. A quality felt filter can trap particles as small as 59 microns (.0025 in.), and all of the latest domestic transmissions use filters of this type.

Secondary Filters

Most automatic transmissions have one or more small screen filters located in a passage, hole, or slot somewhere in the transmission. These are called secondary filters, and they are used to help keep foreign materials out of critical components such as pumps, valves, governors, and solenoids.

For example, in THM 325-4L transaxles, a secondary filter screen located in the pump oil

Surface Filter: A filter that traps contaminants on its surface. Screen and paper filters are types of surface filters.

Depth Filter: A filter that traps contaminants within the matrix of the filter material. A felt filter is a type of depth filter.

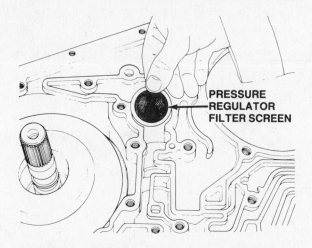

Figure 5-8. A secondary filter in a THM 325-4L transmission.

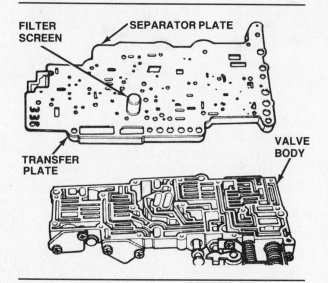

Figure 5-9. A secondary filter in a Torqueflite transmission.

ATF Operating Temperature		Projected ATF Service Life	
°F	°C	Miles	KM
175	80	100,000	160,000
195	90	50,000	80,000
215	100	25,000	40,000
235	115	2,000	20,000
255	125	6,250	10,000
275	135	3,000	5,000
295	145	1,500	2,500
315	155	750	1,200
335	170	325	500
355	180	160	250
375	190	80	125
390	200	40	65
410	210	20	32

Figure 5-10. Temperature versus service life in both miles and kilometers for petroleum-based automatic transmission fluid.

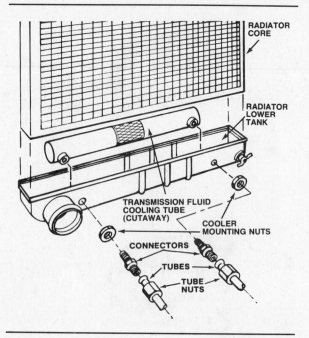

Figure 5-11. A typical oil-to-water transmission cooler installation in a radiator.

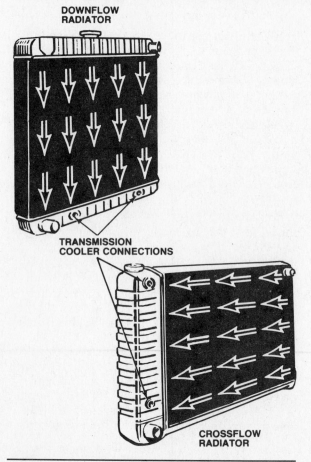

Figure 5-12. Transmission oil coolers are located in the lower tanks of downflow radiators, and the outlet tanks of crossflow radiators.

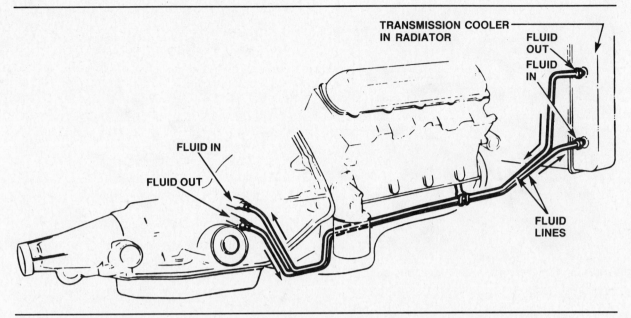

Figure 5-13. The transmission hydraulic system is connected to the cooler by two steel lines.

circuit in the case protects the pressure regulator valve, figure 5-8. Torqueflite transmissions have secondary screen filters in the valve body at the mainline pressure outlet and in the governor casting to protect the governor from contamination, figure 5-9. Secondary filters are generally serviced only during a transmission or valve body overhaul.

TRANSMISSION COOLERS

During transmission operation, ATF undergoes a great deal of stress and strain that can shorten its life. The greatest enemy of ATF is heat that breaks down the fluid and allows varnish to form. Clutch and band slippage creates some of the heat, friction between moving parts adds more, but most of the heat is generated by **fluid shear** in the torque converter. This is especially true during severe service operation such as trailer towing. Continued use of overheated fluid will result in poor lubrication, and varnish formation that can lead to sticking valves, poor shift quality, burned apply device friction materials, and the need for a transmission overhaul.

The performance specifications for petroleum-based ATF are established using an operating temperature of about 175°F (80°C). At this temperature, the service life of a typical ATF is roughly 100,000 miles (160,000 km). As shown in figure 5-10, raising the temperature only 20°F (10°C) to 195°F (90°C) cuts the fluid life in half. And should the temperature reach 410°F (210°C) the fluid will be completely destroyed in only 20 miles (32 km) of driving!

To maintain ATF operating temperatures within safe limits, two types of transmission coolers are available:
- Oil-to-water
- Oil-to-air.

A single cooler of either type may be used to cool an automatic transmission. In heavy-duty applications, one of each type cooler may be plumbed in series for added cooling capacity.

Oil-to-Water Coolers

Most original equipment (OE) transmission coolers are oil-to-water designs built into the engine radiator, figure 5-11. The cooler is always located in bottom tank of a downflow radiator, or the outlet tank of a crossflow radiator, figure 5-12. Hot fluid from the transmission is pumped to the cooler where heat from the ATF is transferred into the engine coolant. Heat from the coolant is then transferred into the radiator fins where it is carried away by the passing airflow. Finally, the cooled fluid is returned to the transmission.

Two steel lines connect the transmission hydraulic system to the cooler, figure 5-13. The

Fluid Shear: The internal friction that occurs within a fluid when its rate or direction of flow is suddenly changed. Fluid shear in the torque converter is the major source of transmission fluid heat buildup.

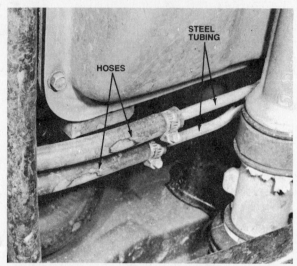

Figure 5-14. Sections of rubber hose may be used in the cooler lines to allow for the small amount of transmission movement.

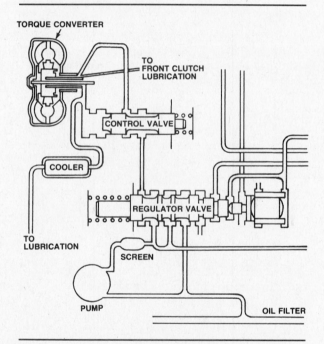

Figure 5-15. A typical transmission cooler oil circuit.

lines end in fittings that thread into matching fittings on the transmission case and the cooler assembly. In some applications short lengths of high-pressure reinforced rubber hose are used to make the connections between the steel lines and the cooler, figure 5-14. These hoses allow for the small amount of relative movement between the engine and transmission on their mounts, and the radiator which is solidly attached to the body or frame.

Fluid flows to the transmission cooler from the torque converter circuit, figure 5-15, be-

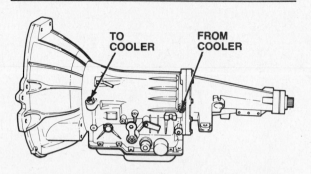

Figure 5-16. Typical cooler line connection locations on a RWD automatic transmission.

cause that is where most of the heat is generated. Therefore, on a RWD transmission, the fitting for the line running *to* the cooler is generally at the front of the transmission, and the fitting for the return line is at the rear, figure 5-16, although many GM cars are exceptions to this rule. On transaxle installations, the cooler line fitting locations vary with the particular application.

The advantages of oil-to water transmission coolers are their compact size and relatively low cost. Also, because engine coolant warms to operating temperature faster than ATF when the car is first started, coolant heat is transferred *into* the ATF at this time to speed warmup of the transmission. One disadvantage of the oil-to-water design is that a leaking cooler contaminates both the engine cooling system and the transmission hydraulic system. In addition, this design is generally less efficient at reducing fluid temperatures than an oil-to-air cooler.

Oil-to-Air Coolers

Vehicles equipped with heavy-duty cooling systems or trailer-towing packages often have an auxiliary oil-to-air ATF cooler, figure 5-17. These coolers are also available as aftermarket add-ons. Oil-to-air coolers mount in front of the radiator or air conditioning condenser, or elsewhere in an unobstructed airstream. An air-to-oil transmission cooler is essentially a small radiator; heat from the ATF is transferred directly into the cooler fins where it is carried away by the passing airflow.

Although oil-to-air coolers are more expensive than oil-to-water designs, they are also able to provide greater reductions in ATF temperature. This is possible because the **temperature differential** between the hot ATF and the relatively cool outside air is much greater than the differential between the hot ATF and the similarly heated engine coolant.

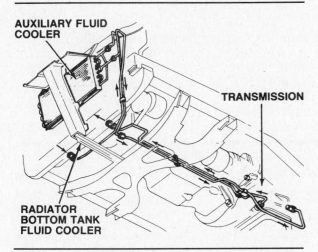

Figure 5-17. An auxiliary oil-to-air transmission cooler installation.

Oil-to-air transmission coolers used as auxiliary units are plumbed in series with the standard in-radiator cooler. Fluid flows first from the transmission to the in-radiator cooler, then to the auxiliary cooler, and finally back to the transmission.

In rare instances, an oil-to-air unit may be used as the sole transmission cooler. This type of installation eliminates the possibility of fluid leaks between the engine cooling system and the transmission hydraulic system. However, it prevents coolant heat from aiding in transmission warmup. Oil-to-air coolers are also inefficient in heavy traffic where there is little airflow over the cooler.

ATF CHANGE INTERVALS

Even with transmission coolers, ATF operating temperatures often exceed the design limit of 175°F (80°C). This can occur at points inside the transmission where friction creates momentary hot spots, but it is most likely to result from severe service such as trailer towing or rocking a car back and forth to free it from a snow bank. These operations make fluid temperatures skyrocket, and result in fluid breakdown that eventually leads to burned bands and clutches, and slippage during shifts.

Temperature Differential: The difference in temperature between two objects or areas. The greater the temperature differential, the greater the rate at which heat will be transferred between the objects or areas.

Because of fluid breakdown, all vehicle manufacturers once recommended that the transmission fluid be changed at periodic intervals. As fluid and transmission technologies improved, however, these intervals were gradually made longer and longer. Today, some automakers claim that in normal use the ATF is good for the life of the car and need never be changed. However, these same manufacturers recommend ATF change intervals as short as 15,000 miles (24,000 km) when the vehicle is used in severe service!

Field experience shows that the normal fluid operating temperature in most cars is actually about 190° to 195°F (88° to 90°C). This means that whenever petroleum-based ATF is used, as it is in all domestic automatic transmissions, fluid breakdown is occurring continuously whenever the car is driven. To prevent damage from this breakdown, most transmission specialists suggest that the fluid be changed in accordance with the vehicle manufacturers' severe service recommendations.

SUMMARY

Automatic transmission fluid performs a number of important jobs including: power transmission, cooling, lubrication, cleaning, shift control, and apply device operation. On vehicles without a lockup torque converter, the ATF transmits all of the power from the engine to the transmission.

ATF can be a petroleum-base oil, a 100-percent synthetic lubricant, or a combination of these two types. Fluids now in common use include Type F, DEXRON®-II, MERCON®, and several proprietary fluids specified by individual vehicle manufacturers.

All ATFs are either friction modified or nonmodified. As a clutch or band locks up, nonmodified ATF provides an increasing coefficient of static friction that is ultimately *greater* than the coefficient dynamic friction that existed before lockup. A friction-modified fluid provides a decreasing coefficient of static friction that is ultimately *less than* the coefficient of dynamic friction that existed before lockup.

Transmission filters remove dirt, metal, and friction material particles from the ATF. Screen filters use a wire or nylon mesh to trap particles. Paper filters use a fabric made of cellulose or synthetic fibers such as Dacron. Most automatic transmissions have one or more smaller secondary screen filters that keep foreign materials out of critical components.

Transmission coolers help keep ATF operating temperatures within design limits to prevent fluid breakdown. Most OE coolers are oil-to-water designs built into the engine radiator. Oil-to-air coolers, used for additional cooling, mount in front of the radiator and are plumbed in series with the main cooler in the radiator. Steel lines and reinforced rubber hoses connect the transmission hydraulic system to the coolers.

The normal transmission fluid operating temperature in most cars is above the design temperature for petroleum-base ATF. As a result, fluid breakdown occurs whenever the car is driven. To prevent damage from deteriorated fluid, the ATF should be changed at regular intervals.

Review Questions

Select the single most correct answer
Compare your answers to the correct answers on page 603

1. Mechanic A says there is no mechanical coupling between the engine and the transmission when the car is idling in gear at a stoplight.
   Mechanic B says a lockup torque converter can provide a mechanical coupling between the engine and the transmission under certain operating conditions.
   Who is right?
   a. Mechanic A
   b. Mechanic B
   c. Both mechanic A and B
   d. Neither mechanic A nor B

2. Mechanic A says ATF is used to clean, cool, and lubricate the transmission.
   Mechanic B says ATF operates the switching valves that time the transmission gear ratio changes.
   Who is right?
   a. Mechanic A
   b. Mechanic B
   c. Both mechanic A and B
   d. Neither mechanic A nor B

3. Most ATF is dyed _____ in color.
   a. Red
   b. Brown
   c. Black
   d. Brown-black

4. Mechanic A says operating temperature does not affect the service life of ATF.
   Mechanic B says age is the main factor that causes ATF to oxidize.
   Who is right?
   a. Mechanic A
   b. Mechanic B
   c. Both mechanic A and B
   d. Neither mechanic A nor B

5. Mechanic A says ATF that is brown-black is contaminated with engine coolant.
   Mechanic B says that smell alone cannot be used to determine the condition of synthetic ATF.
   Who is right?
   a. Mechanic A
   b. Mechanic B
   c. Both mechanic A and B
   d. Neither mechanic A nor B

6. Mechanic A says DEXRON® ATF is friction modified.
   Mechanic B says Type H fluid replaces an ATF that is obsolete.
   Who is right?
   a. Mechanic A
   b. Mechanic B
   c. Both mechanic A and B
   d. Neither mechanic A nor B

7. Mechanic A says transmissions that use friction modified ATF's are known for their firm shifts.
   Mechanic B says a non-modified ATF provides a greater amount of static friction than dynamic friction.
   Who is right?
   a. Mechanic A
   b. Mechanic B
   c. Both mechanic A and B
   d. Neither mechanic A nor B

8. MERCON® ATF was developed by:
   a. Ford
   b. Chrysler
   c. General Motors
   d. American Motors

9. Mechanic A says Type G fluid was a further development of Type F.
   Mechanic B says Ford recommends a universal ATF for Type G applications in the United States.
   Who is right?
   a. Mechanic A
   b. Mechanic B
   c. Both mechanic A and B
   d. Neither mechanic A nor B

10. Transmission filters are made of:
    a. Synthetic fibers
    b. Paper
    c. Metal screens
    d. All of the above

11. Mechanic A says secondary filters should be changed whenever the transmission is serviced.
    Mechanic B says synthetic filter screens should be soaked in carburetor cleaner to remove varnish deposits.
    Who is right?
    a. Mechanic A
    b. Mechanic B
    c. Both mechanic A and B
    d. Neither mechanic A nor B

12. The greatest amount of heat in a transmission is generated by fluid shear in the:
    a. Valve body
    b. Gear train
    c. Torque converter
    d. Pump

13. The most common type of original equipment transmission cooler is the _____ design.
    a. Crossflow
    b. Oil-to-air
    c. Downflow
    d. Oil-to-water

14. Mechanic A says ATF temperature is one factor that would affect when to change the fluid.
    Mechanic B says poor shift quality could indicate the need to change the transmission fluid.
    Who is right?
    a. Mechanic A
    b. Mechanic B
    c. Both mechanic A and B
    d. Neither mechanic A nor B

6

Gaskets, Seals, Bushings, Bearings, Thrust Washers, and Snaprings

In addition to the components discussed in earlier chapters, automatic transmissions contain a number of other parts that play important roles in their operation. Gaskets and seals contain fluid within the transmission case, and prevent pressurized fluid from leaking out of the various hydraulic circuits. Bushings, bearings, and thrust washers reduce friction between moving parts and control play within the transmission. Snaprings hold various assemblies together and in position inside the transmission.

GASKETS

A gasket seals the space between two parts with irregular surfaces. In an automatic transmission, some gaskets seal parts together in order to contain ATF; other gaskets help channel fluid from one part of the transmission to another. The gaskets used in automatic transmissions are made of paper, cork, rubber, plastic, or synthetic materials. A gasket that combines two or more of these materials is called a composition gasket.

Gasket Compressibility

All gaskets used in automatic transmissions are resistant to ATF. However, depending on the specific application, transmission gaskets require varying amounts of **compressibility**. A gasket's compressibility is its ability to conform to irregularities in the sealing surfaces. Gaskets that can be compressed less than 20 percent in thickness are called "hard" gaskets, figure 6-1A; a plain paper gasket is a common type of hard gasket. Gaskets with greater than 20 percent compressibility are called "soft" gaskets, figure 6-1B, a cork/rubber composition gasket is a common type of soft gasket.

Hard gaskets do a good job of sealing where the surfaces to be mated are smooth and rigid. A good example is a cast valve body that is mounted to a cast transmission case. Hard gaskets also allow good torque retention; they do not "relax" over time and allow the fasteners holding the assembly together to loosen.

Soft gaskets do an excellent job of sealing where one or both surfaces to be mated are irregular or subject to distortion when tightened. A good example is a stamped steel oil pan that is mounted to a cast transmission case. The tradeoff for high compressibility is that soft gaskets are often weak in the area of torque retention; they tend to relax over time, reducing the amount of joining force provided by the fasteners that hold the assembly together.

A

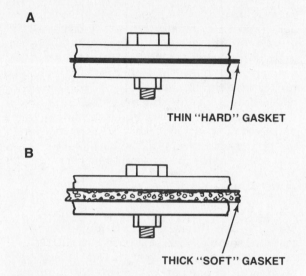

THIN "HARD" GASKET

B

THICK "SOFT" GASKET

Figure 6-1. "Hard" and "soft" gaskets have differing degrees of compressibility.

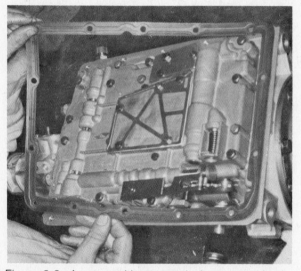

Figure 6-2. A composition transmission pan gasket is a common example of a "soft" gasket.

Two basic types of gaskets are used in modern automatic transmissions:
- Pan gaskets
- Mating assembly gaskets.

Pan Gaskets

The pan gasket on most automatic transmissions, figure 6-2, is a "soft" gasket that is 1/16 to 1/8 inch (1.5 to 3 millimeters) thick. Early pan gaskets were made entirely of cork. However, while cork is a good sealing material, used alone it is unstable and will shrink and crack as it ages. Cork can also act as a wick, pulling

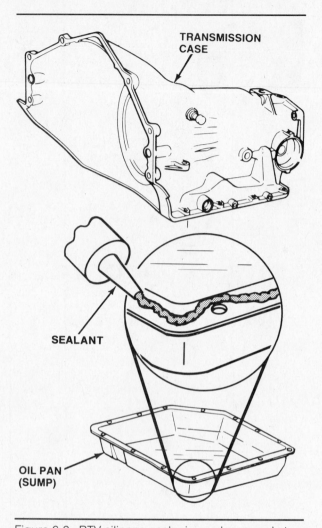

TRANSMISSION CASE

SEALANT

OIL PAN (SUMP)

Figure 6-3. RTV silicone sealer is used as a gasket on some transmission pans.

transmission fluid through the gasket and giving the appearance of a leak. To overcome these problems, the typical modern pan gasket is a composition design made of cork mixed with rubber for added stability.

Some of the latest aftermarket pan gaskets use a "sandwich" construction with a stiff reinforcing layer in the center. Others have metal inserts surrounding the pan bolt openings to prevent overtightening. Yet others are made of synthetic rubber compounds that offer added sealing benefits. A few transmissions do not use a pan gasket; instead, a bead of room temperature vulcanizing (RTV) silicone sealant is placed between the pan and transmission case to create an oil-tight seal, figure 6-3. Aftermarket gaskets are available for many applications that originally used RTV.

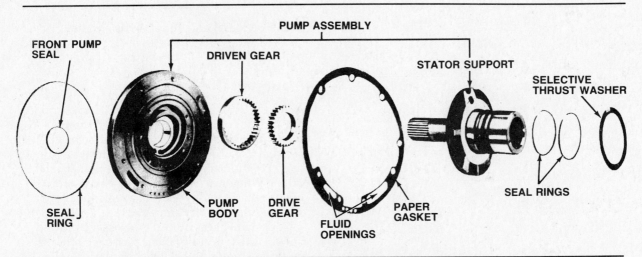

Figure 6-4. A paper gasket is used between the oil pump body and the transmission case.

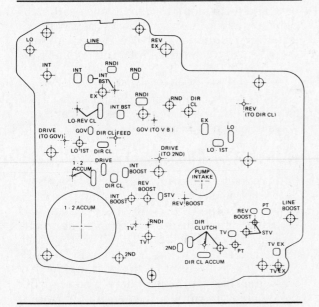

Figure 6-5. The holes in a valve body gasket allow ATF to pass through the separator plate between the two halves of the valve body, or between the valve body and the case.

Because most leaks from automatic transmissions originate from the pan gasket, it is important to use the proper type of gasket and/or sealant when servicing a transmission. When there is a particular leakage problem, vehicle manufacturers sometimes release updated gasket designs, and even new oil pans, to improve transmission sealing.

Mating Assembly Gaskets

The gaskets used to seal smooth and rigid mating surfaces, such as those where the extension housing, valve body, and pump attach to the transmission case, figure 6-4, are usually "hard" gaskets made of thin cellulose paper. Some servo and governor covers also use paper gaskets. Thin paper gaskets are desirable in these applications because they are unlikely to compress and allow the retaining nuts or bolts to loosen over time.

Paper gaskets between the separator plate and the valve body halves, figure 6-5, help direct the flow of fluid from one half of the valve body, through the separator plate, to different circuits in the other half of the valve body. Generally, one gasket is used on each side of the separator plate, although the Torqueflite valve body has no gaskets, and the C4, C5, and C6 use only a single valve body gasket.

Manufacturers often use the same transmission case and valve body castings in several different engine/vehicle applications. Because of this, a valve body gasket that *blocks* certain circuits between the valve body halves in one application may *open* those same circuits in another application. Compare the valve body gaskets shown in figures 6-5 and 6-6.

Some valve body gaskets have markings that help technicians identify their proper location. For example, General Motors gaskets usually have either C or VB printed on or cut into them; the C means the gasket goes between separator plate and case; the VB means the gasket goes between valve body and separator plate.

Compressibility: The ability of a gasket to conform to surface irregularities. "Soft" gaskets are more compressible than "hard" gaskets.

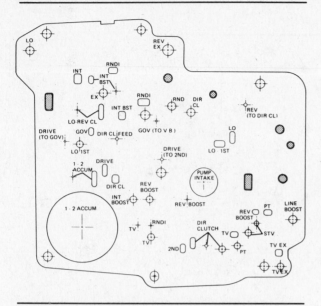

Figure 6-6. This gasket is similar to the one in figure 6-5, but it has extra fluid holes for a different valve body.

Gasket Sealants

In servicing, repairing, or overhauling automatic transmissions, gasket sealants are generally not recommended except for specific applications. In most situations, a quality gasket is all that is needed to effect a seal, providing the surfaces to be joined are flat and clean.

Gasket sealants should not be used on pump and valve body gaskets because excess sealant may get into the hydraulic system and block passages or cause valves and check balls to stick. Gasket sealants are not recommended for highly compressible pan gaskets, such as cork ones, because the sealant may act as a lubricant and cause the gasket to squeeze out of the joint.

Using RTV sealant in place of a pan gasket is acceptable in some applications, as is a light application of nonhardening sealant on the thin paper gaskets used for cast transmission pans and extension housings. Nonhardening sealant may also be used on the outer surface of a metal-clad lip seal. If a metal-clad seal has a rubber coating on its outer circumference, however, no sealant is necessary.

TRANSMISSION SEALS

Three types of rubber seals and three types of metal and Teflon seals are used in automatic transmissions. The rubber seals are the O-ring seal, the square-cut (lathe-cut) seal, and the lip seal. The metal and Teflon seals are the open-end seal, the butt-end seal, and the locking-end seal.

Static and Dynamic Seals

All of the seals listed above fall into one or both of two categories, static seals and dynamic seals. A **static seal** blocks the passage of fluid between two or more parts that are always in fixed positions relative to one another. There can be little or no movement of either part when a static seal is used or a leak will occur and the seal may be damaged.

A **dynamic seal** blocks the passage of fluid between two or more parts that have relative motion between them. Dynamic seals are required where there is axial or rotational motion between the parts. **Axial motion** is movement back and forth along the length of a shaft, or parallel to the centerline (axis) of the shaft. **Rotational motion** is present when one of the parts being sealed turns (rotates) in relation to the other.

Several types of static and dynamic seals are used in modern automatic transmissions. The following sections describe these seals and the ways in which they are used.

RUBBER SEALS

Early automatic transmissions required as few as five rubber seals, but the typical modern design has more than 20. Rubber seal materials must have three main characteristics: abrasion resistance, compatibility with ATF, and sealing ability across a wide range of operating temperatures. If a seal cannot resist abrasion, it will wear quickly — particulary during occasional periods of "dry running" when limited lubrication is available. When a seal is incompatible with ATF, it softens and swells. And when a seal is forced to operate at temperatures below or above its design range, it becomes brittle and may crack. All of these circumstances result in seal deterioration that can lead to fluid leaks, poor shift quality, or complete transmission failure.

Rubber Compounds

Because natural rubber is not nearly tough enough, all automatic transmission seals are made of synthetic rubber. Over the years, seal designs have been simplified and seal materials have been improved to the point where the service life of the typical seal now exceeds that of most other transmission parts! These advances in seal life are primarily the result of the

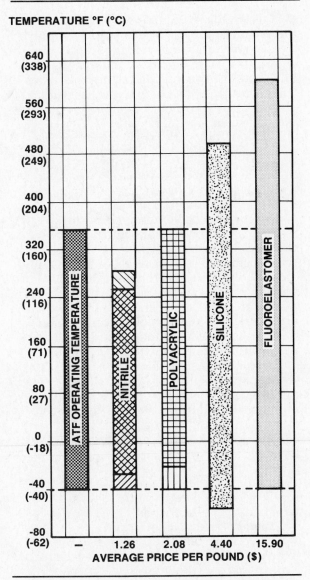

Figure 6-7. This chart shows the operating temperature ranges and relative costs for synthetic rubber seal materials.

ongoing research into synthetic rubber compounds.

Early seals were made of a synthetic rubber **polymer** called neoprene. Compared to natural rubber, neoprene was less brittle and more resistant to oil, heat, and oxidation. But compared to the more advanced synthetic rubber **copolymers** now available, neoprene is only a "fair" seal material and is seldom used today. The synthetic rubbers used for modern transmission seals break down into four groups:

- Nitriles
- Polyacrylics
- Silicones
- Fluoroelastomers.

Nitrile rubbers
Nitrile rubbers such as Buna N have excellent abrasion resistance and very good compatibility with ATF. Nitrile rubbers can also be formulated to work well in temperatures from −40°F (−40°C) to 300°F (149°C), figure 6-7. However, when a nitrile-rubber seal is designed to be durable at high temperatures, its low-temperature performance suffers. Similarly, when a nitrile-rubber seal is designed to remain flexible at low temperatures, its high-temperature durability is reduced. Nitrile rubbers are usually gray-black or black in color.

Although nitrile rubbers are among the least expensive seal materials, their limited operating temperature range makes them only "good" overall. Nitrile-rubber transmission seals were used almost universally a few years ago. However, in late-model cars, nitrile seals are being replaced by newer synthetics that offer better high temperature performance.

Polyacrylic rubbers
Polyacrylic rubbers such as polyacrylate or Vamac® have only fair abrasion resistance, but very good compatibility with ATF. The advantage of polyacrylics over nitrile rubbers is that they work better at higher temperatures, operating across a range from −40°F (−40°C) to 350°F (177°C). Like nitrile rubbers, polyacrylic seal materials are usually gray-black or black in color.

Static Seal: A seal that prevents fluid passage between two parts that are in fixed positions relative to one another.

Dynamic Seal: A seal that prevents fluid passage between two parts that are in motion relative to one another.

Axial Motion: Movement along, or parallel to, the centerline (axis) of a shaft. A dynamic seal is required to contain fluids where axial motion is present.

Rotational Motion: Movement that occurs when a shaft turns (rotates) on its axis. A dynamic seal is required to contain fluids where rotational motion is present.

Polymer: A substance made of giant molecules formed from smaller molecules of the same substance.

Copolymer: A substance made of giant molecules formed from the smaller molecules of two or more unlike substances.

Figure 6-8. A burn test may be the only way to assess the quality of unidentified seal rubbers.

The disadvantages of polyacrylic rubbers are their higher cost, poor dry-running ability, and limited low-temperature performance with certain formulations. Despite these drawbacks, polyacrylic rubbers are rated "very good" for use in transmission seals, and are frequently used for pump seals, piston seals, and temperature-critical applications.

Silicone rubbers

Silicone rubbers are soft **elastomers** that have relatively poor resistance to abrasion, and offer only fair compatibility with ATF. The saving grace of silicone seals is that they work extremely well over a very wide range of temperatures from −65°F (−54°C) to 500°F (260°C). Silicone rubbers are usually red or orange, but may also be gray or blue.

The disadvantages of silicone-rubber include its high cost (greater than nitrile or polyacrylic rubbers), low tolerance of certain oxidized oils, and relative softness. This last factor contributes to the poor abrasion resistance of silicone-rubber seals, and makes them prone to handling and installation damage. Because of these limitations, silicone rubber seals are primarily used only for pump seals in automatic transmissions at this time.

Fluoroelastomer rubbers

Fluoroelastomer rubbers such as Viton® offer good compatibility with ATF and excellent abrasion resistance. They also have the best high-temperature performance, remaining flexible from −40°F (−40°C) all the way up to 600°F

(316°C). Fluoroelastomer rubbers are generally brown or black, but may also be blue or green.

The main disadvantage of fluoroelastomer rubbers is their extreme cost in comparison to all other seal materials. Because of this expense, current use of fluoroelastomer seals is limited to rare instances in which the demands of the application cannot be met by any other seal material. Typical automatic transmission applications include check balls and speedometer drive seals.

Identifying Seal Materials

Unless the material is identified by the manufacturer, it can be difficult to determine the type of rubber used in a seal. Color is not a reliable guide to seal materials because manufacturers establish seal colors with dyes. For example, some orange-colored seals now on the market are made of nitrile rubber rather than the more costly and durable silicone materials. Certain other orange and blue seals are blends of silicone and nitrile rubber.

Part numbers are also not an assurance of seal quality, material, or performance. Anyone can use part numbers.

The manufacturing processes used to make rubber O-ring and lathe-cut seals make it virtually impossible to mold-in identification. With these seals, a burn test can be used to determine the type of rubber. This test is destructive and cannot be used on a part that will be put into service; however, it can be used to check whether a parts supplier is delivering quality seal materials suitable for the job.

To perform a burn test, figure 6-8, use a match to light the seal on fire. Allow the rubber to burn for two or three seconds, then extinguish it. If the smoke smells like burning tires, the seal is made of nitrile rubber. A sweet smell identifies a polyacrylic rubber seal. Silicone rubber will have a white ash on its surface after it is extinguished.

RUBBER SEAL TYPES

As mentioned earlier, three types of synthetic rubber seals are used in automatic transmissions:

● O-ring
● Square-cut
● Lip.

O-Ring Seals

The O-ring seals used in automatic transmissions are round and have a circular cross

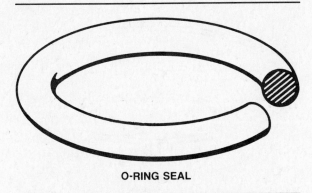

O-RING SEAL

Figure 6-9. An O-ring seal is not used where rotational force or a great deal of axial force might be placed on it.

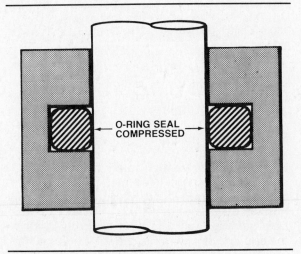

O-RING SEAL COMPRESSED

Figure 6-10. An O-ring creates a seal when it is compressed between two parts.

section, figure 6-9. An O-ring seal works by having its shape distorted. In figure 6-10, one metal part is slipped over another, and an O-ring is installed in a groove cut into the outer part. Because the groove is not as deep as the thickness of the O-ring, the ring is compressed between the groove and the inner metal part. This pressure distorts the ring and forms a tight seal between the two parts.

O-ring seals are commonly used as static seals, and also see limited service as dynamic seals. An O-ring seal like that in figure 6-10 is an example of a static seal. Once the O-ring is compressed, it remains in position and does not move.

An O-ring seal in a groove on a clutch piston that moves inside a cylinder or drum is an example of a dynamic seal. An O-ring will maintain a seal under these conditions, providing

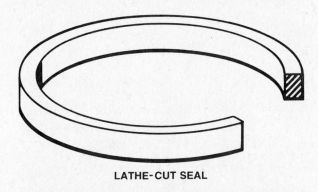

LATHE-CUT SEAL

Figure 6-11. A square-cut seal is often used between parts affected by axial movement.

the amount of axial motion is not too great. If the piston moves too far, the O-ring will roll in its groove, become damaged, and not seal properly.

O-ring seals are never used where rotational motion occurs. As in the case of excessive axial motion, rotation will cause an O-ring seal to bunch up and be damaged, making it unable to maintain a seal.

Square-Cut Seals

The square-cut, also called lathe-cut, seals used in automatic transmissions are circular with a square or rectangular cross section, figure 6-11. Like O-ring seals, square-cut seals are used in both static and dynamic seal applications. However, a square-cut seal can maintain a seal over a larger range of axial motion than can an O-ring seal. This is possible because the shape of the square-cut seal prevents it from rolling over in its groove and becoming damaged.

The action of a square-cut seal in a dynamic application is shown in figure 6-12. When hydraulic pressure moves the piston through the cylinder, the outer edge of the seal slides along the cylinder wall, but does not move as far as the inner edge that is mounted in the piston. When hydraulic pressure against the piston is released, the bent edge of the square-cut seal moves back to its original position and helps draw the piston back into the cylinder.

Elastomer: A synthetic polymer or copolymer that has elastic properties similar to those of rubber.

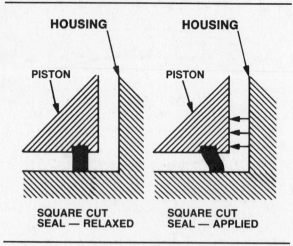

Figure 6-12. A square-cut seal compressed between a piston and a cylinder flexes in the direction of the moving piston.

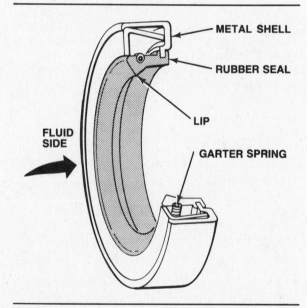

Figure 6-13. A shaft seal is used where the sealing area is subject to high rotational movement. The lip is installed facing the fluid to be contained.

Like O-ring seals, square-cut seals are not used where rotational motion can occur. Rotational motion causes a square-cut seal to bunch up in its groove and become damaged in much the same manner as an O-ring seal.

Lip Seals

Lip seals are circular and made of rubber like O-ring and square-cut seals. However, the actual sealing is done by a thin, flexible lip that

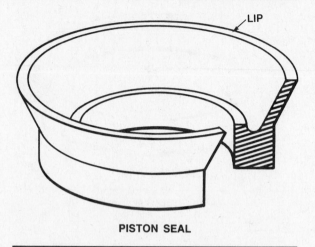

Figure 6-14. Piston seals are a type of lip seal that can contain high hydraulic pressures.

is molded as part of the seal. Two types of lip seals are used in automatic transmissions:
- Shaft seals
- Piston seals.

Shaft seals

Shaft seals, figure 6-13, are a type of dynamic lip seal used where high rotational motion is present. A shaft seal has a metal outer shell that press fits into a bore machined in the transmission case. The rubber seal is bonded to the inside of this shell, and a garter spring holds the seal lip firmly against the rotating part. Because the seal lip is firmly attached to the rigid metal shell, and the area that receives the rotational motion is very small, a shaft seal will not distort the way an O-ring or lathe-cut seal would.

Shaft seals are used for the front and rear transmission seals that ride on the torque converter hub and the transmission output shaft. The job of these seals is to keep ATF within the transmission. To do this, a shaft seal is always installed with its lip facing the fluid to be contained. Shaft seals are *not* designed to contain highly pressurized fluids, and can rarely contain more than about 15 psi (103 kPa) of pressure.

Piston seals

Piston seals, figure 6-14, are a type of dynamic lip seal used where large axial movements are present. A piston seal is made entirely of rubber, and is similar to a square-cut seal with a lip molded on. These types of seals are commonly used as internal and external seals on transmission servo and clutch pistons.

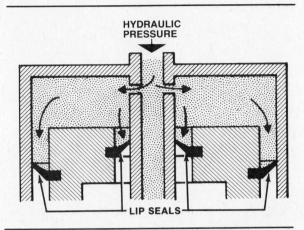

Figure 6-15. Hydraulic pressure on piston seal lips presses them tightly against the cylinder walls.

Unlike shaft seals, piston seals are designed to contain high hydraulic pressures. To do this, piston seals are always installed with the lip facing the source of hydraulic pressure. When pressure is applied, figure 6-15, the seal lip flares and is pressed harder against the cylinder wall for a tighter seal. When pressure against the lip is relieved, the seal slides easily against the cylinder in either direction.

It is important to realize that lip seals as used on pistons are effective in one direction only. If fluid pressure is applied to the backside of the sealing lip, the lip will collapse and allow fluid to bypass the seal.

METAL SEALS

The metal seals used in automatic transmissions are open-end seals, butt-end seals, or locking-end seals, figure 6-16. Open-end seals have a small space between the ends of the seal when it is installed. The square-cut ends of butt-end seals touch, or butt against each other. Locking-end seals have small lips on the seal ends that are interlocked for better sealing after the seal is installed. Both types of metal seals are commonly called steel rings, even though they are made of cast iron which is softer for better sealing.

Because the designs of these seals allow for a small amount of fluid seepage past the ends of the steel rings, they are used only where an absolutely fluid-tight seal is not required. However, the strength and rigidity of such seals allows their use where both axial and rotational motion are present.

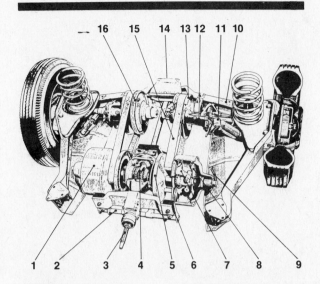

LEGEND:
1. VACUUM CYLINDER (PRIMARY GROUP VARIOMATIC)
2. VARIOMATIC ADJUSTMENT FRAME
3. TRANSMISSION SHAFT, POWER DIVIDER
4. CONTROL MECHANISM FOR FORWARD AND REVERSE POSITIONS
5. FIXED FRONT PULLEY HALF
6. SILENTBLOCK
7. CENTRIFUGAL WEIGHTS
8. PARTITION
9. AIR VACUUM CONNECTIONS
10. TRIANGULAR LINK
11. SHOCK ABSORBER
12. REDUCTION CASE
13. FIXED REAR PULLEY HALF
14. ADJUSTING BOLD
15. V-BELT
16. REAR PULLEY

■ A Belt-Driven Transmission

Over the years, several automatic transmissions have been built that do not use gears to transmit power to the wheels. However, most of these infinitely variable, or stepless, transmissions have been inefficient because of slippage or frictional power losses. One of the more successful attempts was the Variomatic transmission produced by DAF of Holland and used in the Daffodil car imported to the U.S. in the early 1960s.

A driveshaft connected the Daffodil's front-mounted, air-cooled, 2-cylinder, horizontally opposed engine to a pair of centrifugal clutches. The clutches operated variable pulleys that were connected by V-belts to fixed pulleys on the independent rear axles. The belts slipped on the variable pulleys at idle, but as engine speed increased, the clutches tightened the pulleys against the belts to drive the rear axles.

As vehicle and engine speeds increased, the clutches further tightened the pulleys, effectively increasing their diameters. This changed the ratios between the variable and fixed pulleys, and was equivalent to shifting into higher gears. The Variomatic was capable of an infinite range of ratios from 16.4:1 to 3.9:1. While the car was underway, the centrifugal clutches constantly varied the ratio to provide the best balance of speed and power for driver demands and the prevailing road conditions.

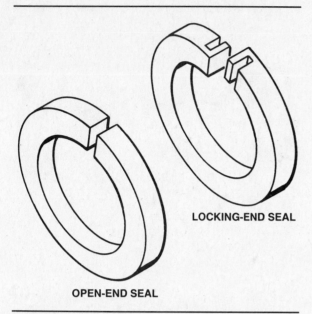

LOCKING-END SEAL

OPEN-END SEAL

Figure 6-16. Metal seal rings are used where an absolutely fluid-tight seal is not necessary. Shown are the locking- and butt-end types.

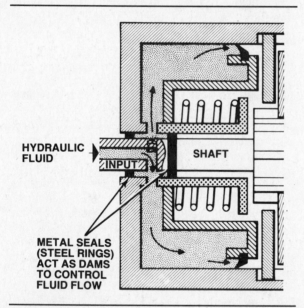

HYDRAULIC FLUID

INPUT

SHAFT

METAL SEALS (STEEL RINGS) ACT AS DAMS TO CONTROL FLUID FLOW

Figure 6-17. The seal rings in this clutch installation direct fluid flow from a passage in the shaft into the clutch drum.

Metal seals are often used as shaft seals, where they act as dams to direct fluid pressure from oil passages in the shaft or pump to clutch drums, figure 6-17. Steel rings also are used as seals on some accumulator and servo pistons.

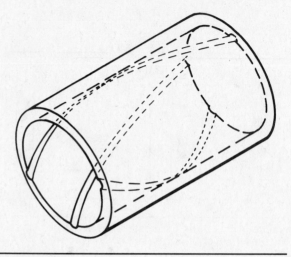

Figure 6-18. Many metal bushings used in transmissions have oil grooves to help carry fluid for lubrication.

TEFLON SEALS

Teflon seals have basically the same construction as metal seals, and do the same jobs in the same locations. The angled ends of Teflon locking-end seals vary slightly from the hooked ends of metal ones. The Teflon type of locking-end seal is generally called a "scarf-cut ring". Teflon seals are popular with transmission manufacturers because they are less expensive than metal seals, and when new, a Teflon seal provides excellent sealing.

The problem with Teflon seals is that they are relatively soft, similar to plastic, and therefore susceptible to scratches from contaminants in the ATF. These scratches decrease sealing ability. In addition, metal particles can become impregnated in the face of a Teflon seal, causing increased wear of the surface on which the seal rides. For these reasons, many manufacturers recommend that Teflon seals be replaced with steel seals when the transmission is overhauled.

BUSHINGS, BEARINGS, AND THRUST WASHERS

The bushings, bearings, and thrust washers in an automatic transmission absorb the radial and axial play from the gears and drums mounted on the transmission shafts. Such play must be kept to a minimum to ensure proper operation and limit component wear.

Radial play is the side-to-side movement of a gear, gearset, hub, or drum on a shaft. Because automatic transmissions use planetary gears that are continuously meshed, there is much less radial play, or side thrust, than in a

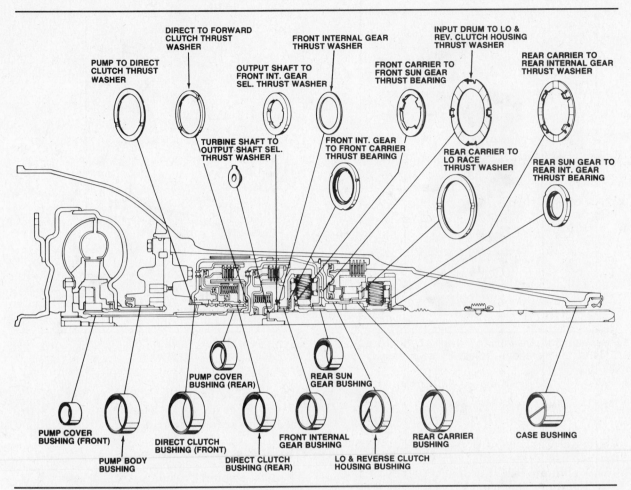

Figure 6-19. The bushings, thrust bearings, and thrust washers in this Turbo Hydra-matic 200 are typical of those used in modern transmissions.

manual transmission which uses sliding gears. As a result, very few ball or roller bearings are used in automatic transmissions; bushings are used instead.

Axial play is the back and forth movement of a gear, gearset, hub, or drum along the axis of a shaft. A small amount of axial play is necessary in an automatic transmission for several reasons. First, it allows metal parts to expand without binding as the transmission reaches operating temperature. Second, it allows clutch drums and other parts to expand under hydraulic pressure. And finally, it allows fluid to escape between the parts when a clutch is released.

Bushings

A bushing, figure 6-18, is a metal sleeve or cylinder placed on a shaft or installed in the transmission case to act as a bearing between gears, drums, shafts and other parts. Bushings support the rotating parts, and may also act as internal seals to restrict ATF flow from one area of the transmission to another.

Because the bushings in an automatic transmission do not have to absorb very much radial play, they wear very slowly. Where a bushing is not used as a seal, it may have helical or diagonal oil grooves cut into its inner surface for better lubrication. Figure 6-19 shows the locations of several bushings used in a typical automatic transmission.

Radial Play: Movement at a right angle to the axis of rotation of a shaft, or along the radius of a circle or shaft. Also called side thrust or sideplay.

Axial Play: Movement along, or parallel to, the centerline (axis) of a shaft. Also called end thrust or endplay.

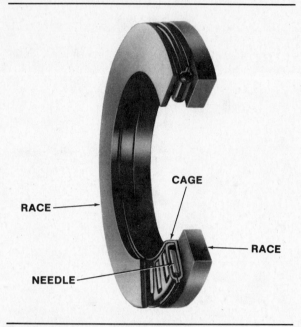

Figure 6-20. A roller thrust (Torrington) bearing consists of rollers, or needles, in a cage between two races.

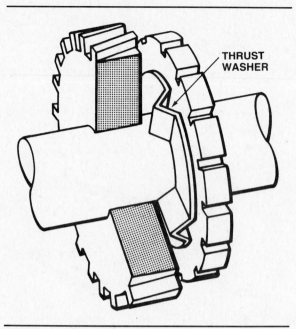

Figure 6-21. Thrust washers control axial play, and keep parts from rubbing together and causing wear.

Roller Thrust Bearings

Roller thrust bearings, sometimes called Torrington bearings, are rollers inside a cage that are used to separate parts of the transmission, figure 6-20. In some applications, the cage with the rollers and the two races are three separate pieces; in other designs, the bearings and races are assembled as a unit.

Roller thrust bearings are used in automatic transmissions to reduce friction and prevent wear between drums and gears on shafts, and between drums and planetary gearset carriers. Together with the thrust washers described in the next section, roller thrust bearings also help control axial play. Figure 6-19 shows the locations of several roller thrust bearings used in a typical automatic transmission.

Thrust Washers

Thrust washers, figure 6-21, absorb axial play and prevent automatic transmission parts from rubbing together and wearing out. Instead, the thrust washers suffer the wear themselves. To help them in this job, thrust washers are generally made of soft steel with copper facings, and are well lubricated.

Like roller thrust bearings, thrust washers are placed between drums and gears on transmission shafts, and between drums and planetary gearset carriers. Because thrust washers wear in normal use, they are usually replaced

during an overhaul. Figure 6-19 shows the locations of several thrust bearings used in a typical automatic transmission.

Transmission Endplay

The total amount of axial play in an assembled transmission is called the **endplay**. Manufacturers provide endplay specifications for each of their transmissions. For example, GM specifies an endplay of .010 to .044 in. (.25 to 1.12 mm) for their THM 350 transmission. As long as the endplay is less than the maximum value, the transmission will not lose hydraulic pressure when it is cold. As long as the endplay is greater than the minimum value, the transmission will not seize when it warms up.

Endplay is measured using a special gauge called a dial indicator, figure 6-22. The measurement is taken before a transmission is disassembled for overhaul so that the technician will be able to adjust the endplay during reassembly if needed. Endplay is also measured after the transmission is assembled to ensure that it is within the specified limits. These procedures are covered in detail in the Shop Manual.

Selective-fit thrust washers

Transmission endplay is usually adjusted with selective-fit thrust washers that are installed at certain locations on the shafts, although some transmissions also use steel shims at the pump body or extension housing for this purpose. The exact locations of the washers vary with

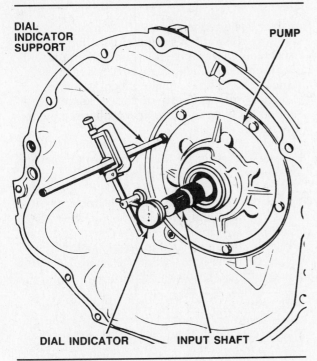

Figure 6-22. Transmission endplay is typically measured at the input shaft with a dial indicator.

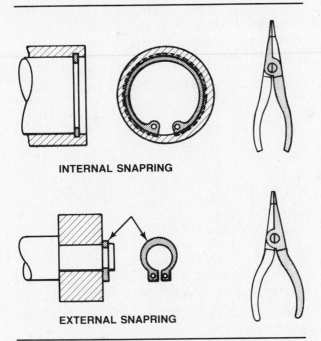

Figure 6-23. Typical internal and external snaprings, and the special pliers used for their installation.

each model of transmission. Transmission manufacturers provide selective-fit washers in many thicknesses so that the proper sizes can be selected to establish the endplay within the

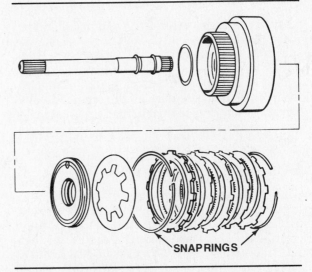

Figure 6-24. Large snaprings are often used to hold a clutch assembly together.

specified limits. However, when a transmission is overhauled, adjust the endplay near the minimum measurement because endplay increases with wear.

SNAPRINGS

Internal and external snaprings, figure 6-23, are retaining devices that are used in many places in automatic transmissions. External snaprings are commonly used to hold clutch and gear assemblies in place on transmission shafts. Internal snaprings are often used to hold servo piston retainers in their bores, and to hold clutch assemblies together, figure 6-24.

The snaprings for some clutch assemblies come in several different thicknesses. Like selective thrust washers, this allows the technician to adjust the total amount of clearance between the clutch plates. A thick snapring in a clutch provides less clearance for the discs than a thin snapring does. Some clutches also use a wave-type snapring that smooths clutch application.

Snaprings with holes in their ends, like those in figure 6-23, are called Tru-arc® snaprings and are removed and installed with special pliers. However, many snaprings in automatic transmissions have simple square or tapered ends;

Endplay: The total amount of axial play in an automatic transmission. Endplay is typically measured at the input shaft.

these snaprings are removed and installed by hand, sometimes with the aid of a small screwdriver or special pliers.

All snaprings are stamped from steel sheet stock, and as a result, the edges of the ring are sharper on one side than on the other. Always install a snapring so the sharper-edged side faces away from any force that may be exerted on the ring.

SUMMARY

Transmission gaskets made of paper, cork, rubber, plastic, and synthetic materials are used to contain fluid and direct fluid to specific hydraulic circuits. Hard gaskets seal surfaces that are smooth and rigid. Soft gaskets seal joints that are irregular or subject to distortion. Pan gaskets are typically soft gaskets. Mating assembly gaskets are typically hard gaskets.

All transmission seals are either static seals that block the passage of fluid between parts in fixed positions relative to one another, or dynamic seals that block the passage of fluid between parts that have relative motion.

Rubber, metal, and Teflon seals are used in transmissions. O-ring, square-cut, and lip seals are made of synthetic rubber from one of four groups: nitriles, polyacrylics, silicones, and fluoroelastomers.

O-ring seals are round, have a circular cross section, and work by having their shape distorted. O-ring seals are commonly used as static seals, and see limited service as dynamic seals. O-ring seals are never used where rotational force occurs.

Square-cut seals are circular, have a square or rectangular cross section, and are used in both static and dynamic sealing applications. A square-cut seal can operate over a larger range of axial movement than an O-ring seal, but is also never used where rotational force occurs.

Shaft and piston lip seals are circular like O-ring and square-cut seals, but the sealing is done by a thin, flexible lip molded into the seal. These seals are always installed with the lip facing the fluid to be contained.

Shaft seals are dynamic seals used where high rotational forces are present. Shaft seals are not designed to contain fluid under high pressure. Piston seals are dynamic lip seals used where large axial movements are present, and high hydraulic pressure must be contained. Piston lip seals are effective in one direction only; pressure on the backside of the sealing lip will cause the lip to collapse and allow fluid to bypass.

Metal and Teflon open-end, butt-end, and locking-end seals are used where rotational and axial movement may be present. Open-end seals allow a small space between their ends. The ends of butt-end seals meet. Locking-end seals interlock at their ends. Because they allow fluid seepage, these types of seals are used only where totally fluid-tight joints are not required. Teflon seals are usually replaced with steel seals when a transmission is overhauled.

Bushings, bearings, and thrust washers absorb radial and axial play. Radial play is side-to-side movement of a part on a shaft. Axial play is forward-and-back movement of a part along the axis of a shaft.

Bushings are metal sleeves or cylinders placed on a shaft or installed in the transmission case to act as a bearing and/or seal between gears, drums, shafts and other parts.

Roller thrust bearings and thrust washers control axial play, and reduce friction and wear between drums and gears on shafts, and between drums and planetary gearset carriers. Selective-fit thrust washers are used to adjust the transmission endplay within specified limits so the gearbox will not lose hydraulic pressure when cold, or seize when warmed up.

Internal and external snaprings are retaining devices used to hold clutch and gear assemblies in place on transmission shafts, to hold servo piston retainers in their bores, and to hold clutch assemblies together. Tru-arc® snaprings with holes in their ends are removed and installed with special pliers. Snaprings with square or tapered ends are removed and installed by hand with the aid of a small screwdriver. All snaprings are installed with their sharp-edged side facing away from any force that may be exerted on the ring.

Review Questions

Select the single most correct answer
Compare your answers to the correct answers on page 603

1. Mechanic A says a hard gasket is used to seal a joint that is subject to distortion.
   Mechanic B says a cork and rubber gasket is called a composition gasket.
   Who is right?
   a. Mechanic A
   b. Mechanic B
   c. Both mechanic A and B
   d. Neither mechanic A nor B

2. A gasket with good compressibility:
   a. Sometimes has poor torque retention
   b. Is preferred for sealing smooth, rigid surfaces
   c. Can be replaced by paper gasket if desired
   d. All of the above

3. Mechanic A says Chrysler Torqueflite transmissions do not use valve body gaskets.
   Mechanic B says some valve body gaskets are marked to indicate their positions in the transmission.
   Who is right?
   a. Mechanic A
   b. Mechanic B
   c. Both mechanic A and B
   d. Neither mechanic A nor B

4. Mechanic A says a nonhardening sealant should be used on the outside of metal-clad seals that are coated with rubber.
   Mechanic B says RTV sealant is recommended for valve body gaskets.
   Who is right?
   a. Mechanic A
   b. Mechanic B
   c. Both mechanic A and B
   d. Neither mechanic A nor B

5. A(n) _____ seal is used to prevent the passage of fluid between parts that are moving against one another.
   a. Axial
   b. Static
   c. Rotational
   d. Dynamic

6. Which of the following is *not* a necessary characteristic of rubber seal materials?
   a. Compatible with ATF
   b. Resistant to abrasion
   c. Unlimited dry-running ability
   d. Wide temperature operating range

7. Mechanic A says a polyacrylic rubber seal may be used where fluid temperatures reach 350°F (177°C).
   Mechanic B says most modern transmission seals are made of nitrile rubber.
   Who is right?
   a. Mechanic A
   b. Mechanic B
   c. Both mechanic A and B
   d. Neither mechanic A nor B

8. Mechanic A says a silicone rubber seal will be coated with white ash following a burn test.
   Mechanic B says the RMA assigns identifying initials to seal manufacturers.
   Who is right?
   a. Mechanic A
   b. Mechanic B
   c. Both mechanic A and B
   d. Neither mechanic A nor B

9. Mechanic A says O-ring seals are commonly used as static seals.
   Mechanic B says O-ring seals may also be used in limited dynamic sealing applications.
   Who is right?
   a. Mechanic A
   b. Mechanic B
   c. Both mechanic A and B
   d. Neither mechanic A nor B

10. Mechanic A says square-cut seals are better than O-ring seals when rotational motion is present.
    Mechanic B says square-cut seals flex to help apply the piston.
    Who is right?
    a. Mechanic A
    b. Mechanic B
    c. Both mechanic A and B
    d. Neither mechanic A nor B

11. Which of the following is *not* true of shaft seals?
    a. They are used where rotational motion is present
    b. They are designed to contain high fluid pressures
    c. They are dynamic seals
    d. They have a seal lip attached to a metal shell

12. A lip-type piston seal:
    a. Is always installed with the lip facing the fluid to be contained
    b. Is able to seal in only one direction
    c. Forms a tighter seal as hydraulic pressure against it increases
    d. All of the above

13. Mechanic A says metal seals and Teflon seals are basically the same except for the materials from which they are made.
    Mechanic B says open-end seals provide better sealing than locking-end seals.
    Who is right?
    a. Mechanic A
    b. Mechanic B
    c. Both mechanic A and B
    d. Neither mechanic A nor B

14. Mechanic A says radial play is the back and forth movement of a part along the axis of a shaft.
    Mechanic B says transmission endplay is adjusted with selective-fit thrust bearings.
    Who is right?
    a. Mechanic A
    b. Mechanic B
    c. Both mechanic A and B
    d. Neither mechanic A nor B

15. Mechanic A says bushings can serve as a type of internal transmission seal.
    Mechanic B says Torrington bearings are used to prevent wear and reduce friction.
    Who is right?
    a. Mechanic A
    b. Mechanic B
    c. Both mechanic A and B
    d. Neither mechanic A nor B

16. Transmission endplay is:
    a. Checked with a micrometer
    b. Not the same for all transmissions
    c. Usually adjusted with steel shims
    d. Measured only after the transmission is overhauled

17. Mechanic A says Tru-arc® snaprings are stamped out of sheet steel and installed with the aid of a screwdriver.
    Mechanic B says selective-fit snaprings may be used to adjust the clearance of a clutch pack.
    Who is right?
    a. Mechanic A
    b. Mechanic B
    c. Both mechanic A and B
    d. Neither mechanic A nor B

7

Fluid Couplings and Torque Converters

In earlier chapters, we learned how automatic transmission hydraulic systems are used to hold and drive various members of planetary gearsets to change gear ratios. These changes are made automatically without any manual gear shifting. A transmission is not fully automatic, however, unless it includes a mechanism to automatically couple and uncouple the engine and the gearbox. Two devices that do this are fluid couplings and torque converters; both transfer engine torque to the transmission, but a torque converter can increase torque while a fluid coupling cannot.

This chapter explains the operation of both fluid couplings and torque converters. Fluid couplings are simpler devices than torque converters, and have not been used on domestic automatic transmissions for over 20 years. However, because fluid couplings operate on much the same principles as torque converters, an understanding of their operation will make the study of torque converters easier.

FLUID COUPLINGS

A fluid coupling, figure 7-1, consists of an impeller and a turbine with internal vanes that face each other. The impeller, sometimes called the pump, is attached to the engine flywheel or flexplate, and the turbine is attached to the transmission input shaft. The impeller is the driving member, and the turbine is the driven member.

Both the impeller and turbine are enclosed in the fluid coupling housing. Fluid is pumped into the coupling housing by the transmission pump. As the impeller is turned by the engine, its vanes pick up fluid and pump it toward the turbine.

The moving fluid in the coupling travels in two paths, **rotary flow** and **vortex flow**. Rotary flow is the movement of the fluid in the same clockwise circular path as the rotation of the impeller. In other words, the fluid flows in a circle around the axis of the engine crankshaft and the transmission input shaft, figure 7-2.

As the fluid moves in the rotary flow path, centrifugal force throws it to the outer edge of the impeller. Because the impeller is curved, the fluid turns as it strikes the outer edge of the impeller and flows toward the turbine. The fluid then moves in a second circular flow path at a right angle to the rotary flow path. This is called vortex flow, figure 7-2.

The fluid in a fluid coupling moves through both the rotary flow and the vortex flow paths at the same time. The rotary flow, caused by the impeller, carries with it the rotating torque of the engine. However, the torque cannot be

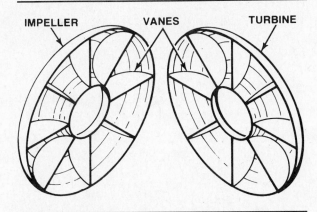

IMPELLER   VANES   TURBINE

Figure 7-1. A fluid coupling consists of an impeller and a turbine.

Rotary Flow: The oil flow path, in a fluid coupling or torque converter, that is in the same circular direction as the rotation of the impeller.

Vortex Flow: The oil flow path, in a fluid coupling or torque converter, that is at a right angle to the rotation of the impeller and to rotary flow.

Resultant Force: The combined force and oil flow direction produced by rotary flow and vortex flow in a fluid coupling or torque converter.

Speed Ratio: The number of revolutions made by the turbine for each revolution of the impeller. Turbine (output) speed is divided by impeller (input) speed and expressed as a percentage.

transferred to the transmission, without the vortex flow that moves the fluid *from* the impeller *to* the turbine. The rotating force of the impeller vanes is applied to the turbine vanes by the combination of rotary and vortex fluid flow.

Fluid that leaves a moving impeller to enter a turbine does not exit as only rotary flow or only vortex flow, but as a combination of both, figure 7-3. The combined flow paths create a **resultant force** that causes the fluid from the impeller to strike the vanes of the turbine at an angle. When the force of the fluid hitting the turbine vanes becomes great enough, the turbine rotates and turns the transmission input shaft, figure 7-4.

If the speed of one member of the fluid coupling is much faster than the speed of the other member, oil flow turbulence results, figure 7-5. This causes oil to swirl in all directions inside the fluid coupling, and drastically reduces the force provided by rotary and vortex fluid flow. To limit turbulence, a split guide ring can be added to the fluid coupling, figure 7-6. This ring guides and directs the moving oil to limit the force of any flow counter to that of the rotary and vortex flow paths.

Speed Ratio

Oil flow in a fluid coupling is directly related to the **speed ratio** of the coupling. The speed ratio is a measure of coupling efficiency expressed as a percentage. Speed ratio is determined by comparing how far the turbine rotates for each revolution of the impeller. For example, if the impeller rotates at 1,000 rpm and the turbine rotates at 900 rpm, the speed ratio of the fluid coupling is 90 percent. When the impeller rotates but the turbine does not, such as when the car is stopped and idling in gear, the speed ratio is zero.

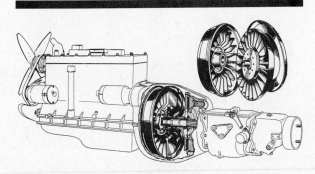

■ **A Metallic Grapefruit**

In 1938, Chrysler introduced the first fluid coupling for a transmission. The company described its Fluid Drive as being "similar to two halves of a metallic grapefruit from which the fruit has been removed without damage to the sectional membrane." The impeller half of the coupling was attached to the engine flywheel, while the turbine half was attached to a conventional single-disc clutch. The clutch, in turn, drove the input shaft of the standard, sliding-gear manual transmission.

Although Fluid Drive didn't replace the manual clutch and shift lever with an automatic shifting system, it did allow the engine to idle in gear with the clutch engaged. Fluid Drive also reduced the number of times the driver had to shift, and shifting could be eliminated almost altogether in the forward speeds if the driver simply left the transmission in high gear. Of course, acceleration from a stop was leisurely when operating the system in this manner, so Chrysler's 1940 shop manual advised that, "The transmission and clutch may be used in a conventional manner for shifting gears when flashing acceleration is desired...!"

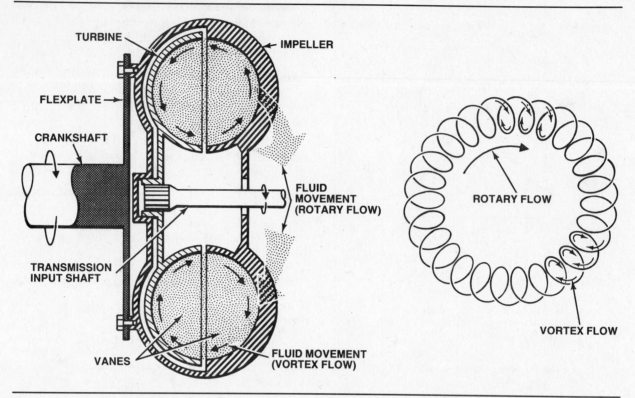

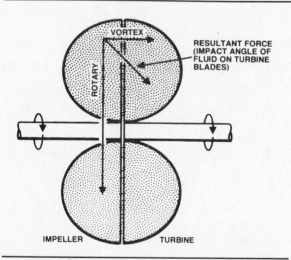

Figure 7-2. Fluid pumped into the turbine by the impeller creates rotary and vortex flow paths.

Figure 7-3. Fluid leaves the rotating impeller at an angle that is a combination of the rotary and vortex flow paths.

As we saw in figure 7-3, rotary flow and vortex flow in a fluid coupling combine to create a resultant force. At a zero speed ratio, a great deal of vortex flow moves through the turbine vanes, creating cross circulation between the impeller and turbine. Both the fixed turbine

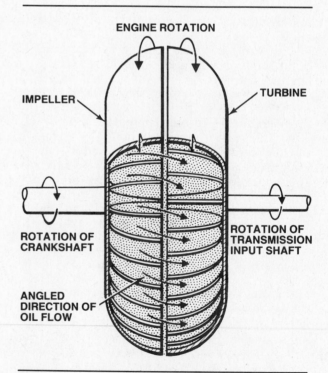

Figure 7-4. The turbine of a fluid coupling begins to rotate when fluid from the impeller exerts enough force on the turbine vanes.

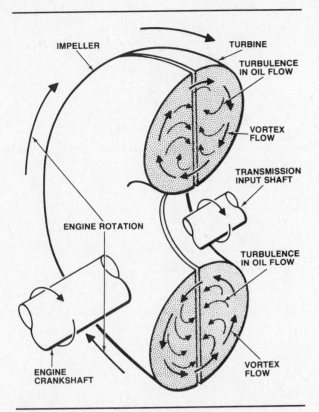

Figure 7-5. Oil flow turbulence is created in a fluid coupling when there is a large speed difference between the impeller and the turbine.

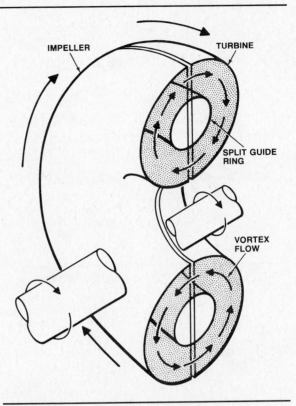

Figure 7-6. A split guide ring reduces oil flow turbulence in a fluid coupling. One half of the ring is attached to the impeller vanes, the other half is attached to the turbine vanes.

and the high level of vortex flow work to oppose rotary flow. As a result, the direction of the resultant force is close to the direction of vortex flow, figure 7-7.

Once the vehicle is moving, the turbine speed approaches that of the impeller and the speed ratio is higher. At this time, rotary flow increases and vortex flow decreases. The direction of the resultant force created by oil flow then moves closer to the direction of rotary flow, figure 7-8. At the point where the resultant force becomes almost completely rotary, the turbine is said to be coupled to the impeller.

Fluid flow can never be completely rotary, because without some vortex flow between the impeller and turbine there would be no transfer of torque through the coupling. In addition, the oil circulation provided by vortex flow is necessary to keep the impeller and turbine vanes filled with oil. If the turbine ever turned at exactly the same speed as the impeller, coupling action would be lost. Because the turbine always turns slower than the impeller, there is always some slippage in a fluid coupling.

Fluid Coupling Operation

When the engine is turning at idle speed, the impeller of a fluid coupling does not create enough oil flow to drive the turbine and transmit torque to the transmission. As a result, the gearbox is automatically uncoupled from the engine.

As engine speed rises, so does the oil flow produced by the impeller. The resultant force of combined rotary and vortex flow then drives the turbine, coupling the engine to the transmission, and transmitting the torque required to move the vehicle.

Torque transfer between the impeller and turbine is made possible by the internal friction that exists between the fluid in each half of the coupling. At highway speeds, constant torque transfer is maintained because of the oil's natural resistance to the shearing action between the two halves of the coupling.

A fluid coupling also keeps the drive wheels of the vehicle coupled to the engine to create a braking action. When the car decelerates, the turbine acts as an impeller to send oil back

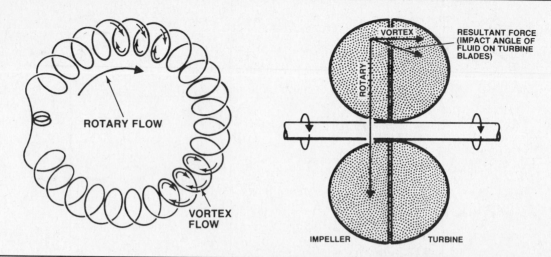

Figure 7-7. At a zero speed ratio, there is a great deal of vortex flow between the impeller and turbine. Therefore, the resultant force angle of the combined rotary and vortex flows is close to the direction of vortex flow.

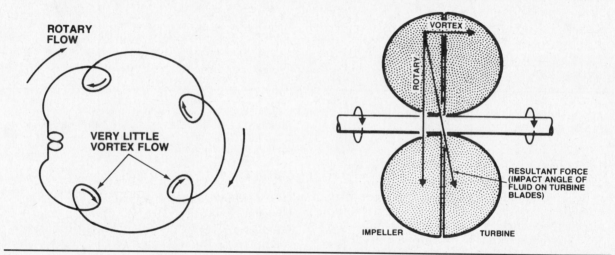

Figure 7-8. At high speed ratios, there is very little vortex flow between the impeller and turbine. Therefore, the resultant force angle of the combined rotary and vortex flows is close to the direction of rotary flow.

against the vanes of the actual impeller which then functions as a turbine. Torque transmitted back through the coupling in this manner works against crankshaft rotation and engine compression to slow the vehicle.

Fluid couplings have one big drawback; an efficient transfer of torque from the impeller to the turbine is only possible when the turbine moves almost as fast as the flow of fluid through it, in other words, when the coupling's speed ratio is high. When the speed ratio is low, a fluid coupling does not create enough fluid flow to transfer torque efficiently. Torque converters replaced fluid couplings because they are able to overcome this problem.

TORQUE CONVERTERS

The torque converter was developed in the early part of the 20th century by marine engineers who sought a method of achieving both gear reduction and speed reduction by using multiple turbines. Torque converters are more efficient than fluid couplings because they are able to increase fluid flow, and thereby multiply engine torque, at low speed ratios. Although fluid couplings were still used with some imported automatic transmissions into the 1960s, torque converters have been in use since 1948, and since 1964 all American automatics have used torque converters.

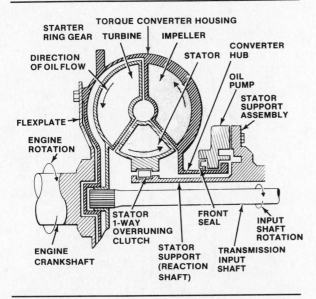

Figure 7-9. This cutaway of half of a typical torque converter installation shows the three elements of the converter (impeller, turbine, stator) and other related parts.

Torque Converter Construction

A modern torque converter, figure 7-9, consists of three elements that are contained in a housing that is filled with fluid by the transmission pump:

● An impeller (driving member)
● A turbine (driven member)
● A stator (reaction member).

A torque converter also contains a split guide ring to reduce oil flow turbulence.

Just as in a fluid coupling, the impeller (sometimes called the pump) is the driving member of a torque converter. The vanes of the impeller pick up fluid in the converter housing and direct it toward the turbine. The turbine is the driven member whose vanes receive the oil flow from the impeller. When there is adequate flow, the turbine rotates to turn the transmission input shaft. The stator (sometimes called the reactor) is the reaction member whose vanes multiply force by redirecting fluid flow from the turbine back to the impeller.

The vanes used in 3-element torque converters are curved, figure 7-10, rather than straight like the vanes in a fluid coupling. Impeller vanes curve backward to accelerate the oil flow as it leaves the impeller, figure 7-11 — the higher the fluid velocity, the greater the torque transfer to the turbine.

The inlet sides of the turbine vanes are curved back toward the impeller to absorb as much energy as possible from the fluid flow. Turbine vanes are also curved to absorb shock

and limit the loss of power that can occur when there is a sudden change in oil flow. The curve of turbine vanes takes advantage of a basic hydraulic principle: *the more the direction of a moving fluid is diverted, the greater the force exerted by the fluid on the diverting surface.*

■ Multi-Element Torque Converters

Although each element in a torque converter is an independent hydrodynamic member (driving, driven, reaction), each member may consist of more than one element. For example, the first torque converter on a domestic car, used with the 1948 Buick Dynaflow transmission, had a torque converter containing five elements: the driving member consisted of two impellers; the reaction member consisted of two stators; and a single turbine served as the driven member.

The gearbox of the Dynaflow transmission had only a single forward speed instead of the two, three, or four found in more modern designs. In place of using several different fixed gear ratios, the multiple elements in the Dynaflow's torque converter provided highly variable torque multiplication to allow for a wide range of driving conditions.

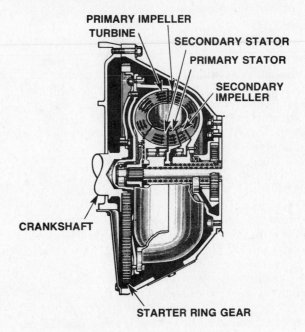

The Dynaflow transmission with its multiple-element torque converter was made possible, if not entirely practical, by the large-displacement, high-torque engines of its day. Although the Dynaflow worked acceptably well it was very inefficient — both performance and fuel economy suffered in comparison to similar vehicles equipped with more conventional transmissions.

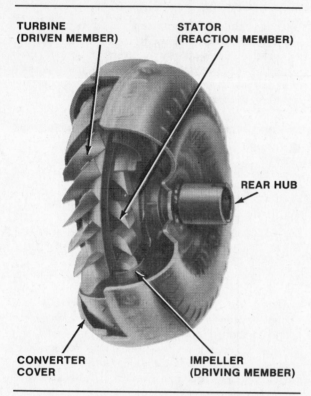

Figure 7-10. The vanes in this modern 3-element torque converter are curved. Note the welded construction of the converter cover.

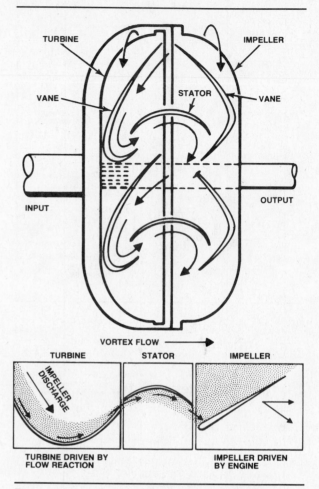

Figure 7-11. The curvature of the various vanes affects the fluid flow pattern in a torque converter.

Stator vanes curve opposite the direction of the impeller and turbine vanes. Because the stator is located between the impeller and turbine, its curved vanes redirect the force returning to the impeller from the turbine. We will discuss this in more detail later in the chapter.

The earliest torque converters were made of several pieces that bolted together, figure 7-12. Multipiece converters can be disassembled for service and repair of their component parts. All late-model torque converters are welded assemblies, figure 7-10, that cannot be disassembled or repaired. These converters must be replaced as a unit if they fail.

Older torque converters have drain plugs, figure 7-13, that allow the fluid trapped in the converter to be changed when the transmission is serviced. Most newer converters, however, do not have drain plugs. The fluid in these converters cannot be drained during normal service, which makes it impossible to completely change the fluid in the transmission hydraulic system.

General Motors eliminated torque converter drain plugs from its transmissions in the mid-1960s. Plugs were eliminated from Chrysler

and AMC converters in 1978. Ford was the last to abandon drain plugs, with a few models still having them into the early 1980s. Aftermarket kits are available to install drain plugs in some newer converters not originally equipped with plugs.

Torque Converter Attachment to Engine And Transmission

The torque converter generally attaches to the engine through a **flexplate** that is mounted on the engine crankshaft, figure 7-9. The flexplate replaces the heavy flywheel that smoothes engine power pulses when a manual transmission is fitted. A conventional flywheel is usually not required with an automatic transmission because the torque converter full of fluid provides sufficient mass to damp any vibrations.

An external ring gear is sometimes attached to the outer rim of the flexplate. The starter motor pinion gear engages this ring gear to crank

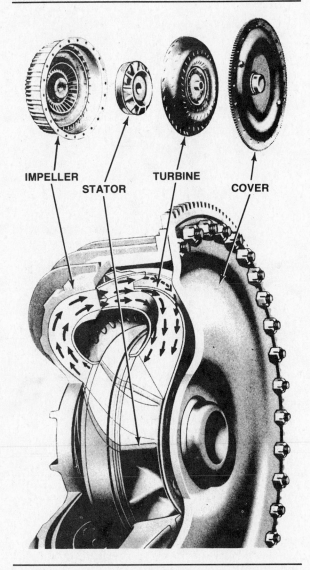

IMPELLER   TURBINE

STATOR   COVER

Figure 7-12. Early multipiece torque converters could be taken apart and repaired.

the engine during starting. In other applications, the starter ring gear is welded to the outside of the torque converter housing.

The torque converter attaches to the transmission in several ways, figure 7-9. First, the turbine is splined to the input shaft. Second, the stator is mounted on a 1-way overrunning clutch that is splined to a stationary extension of the oil pump assembly called the stator support (sometimes called the reaction shaft). And finally, the hub at the rear of the torque converter housing passes over the stator support and through the front seal into the oil pump.

Tangs on the converter hub engage the oil pump drive gear, rotor, or vane ring, and turn it to operate the pump and provide trans-

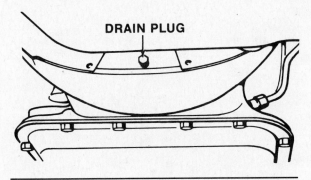

DRAIN PLUG

Figure 7-13. A typical torque converter drain plug location.

mission hydraulic pressure. Because the converter hub is part of the converter housing (it is attached to the back of the impeller) which is bolted to the engine, the oil pump is turning and pumping fluid whenever the engine is running.

Torque Converter Operation

Fluid sent to the torque converter from the transmission oil pump is picked up by the spinning impeller vanes and transferred into the turbine vanes through the same basic rotary flow and vortex flow paths as in a fluid coupling. The major difference in the oil flow of a torque converter compared to that of a fluid coupling is that torque multiplication takes place in a converter when the speed ratio is low.

Torque multiplication phase

Torque multiplication occurs when fluid leaving the turbine strikes the concave (front) sides of the stator vanes. These vanes redirect the fluid so that it joins the fluid being pumped from the impeller to the next turbine vane, figure 7-14. The force of the oil flow from the stator accelerates the flow of oil from the impeller, increasing the amount of torque transferred from the impeller to the turbine. This is called the converter **torque multiplication phase**.

Flexplate: The thin metal plate, used in place of a flywheel, that joins the engine crankshaft to the fluid coupling or torque converter.

Torque Multiplication Phase: The period of torque converter operation when vortex oil flow is redirected through the stator to accelerate impeller flow to the turbine and increase engine torque.

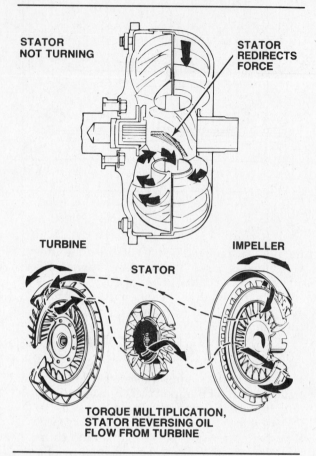

Figure 7-14. Torque multiplication occurs when fluid leaving the turbine strikes the fronts of the stator vanes and is redirected to the impeller.

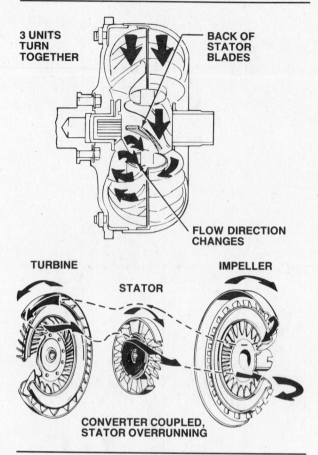

Figure 7-15. Converter coupling occurs when fluid leaving the turbine strikes the backs of the stator vanes, causing the stator to unlock and turn with the impeller and turbine.

The stator is able to redirect the fluid flow as just described because it remains stationary on the stator support. Remember, the stator is mounted on a 1-way roller clutch like those we studied in Chapter 4. When the force of the fluid from the turbine strikes the concave (front) sides of the stator vanes and tries to drive the stator counterclockwise, the 1-way clutch locks and holds the stator stationary. The stator can then redirect the fluid back to the impeller.

Torque multiplication occurs whenever *vortex* oil flow makes a full cycle from the impeller to the turbine and back through the stator to the impeller again. This means the torque converter multiplies engine torque in relation to the speed ratio between the impeller and turbine. At low speed ratios when the impeller is turning very fast, but the turbine is turning slowly, there is high vortex flow and high torque multiplication. As the turbine spins faster and approaches the speed of the impeller, rotary flow increases, and both vortex flow

and torque multiplication are reduced. As the speed ratio approaches 90 percent, torque multiplication is minimal.

The stator is considered the reaction member of the torque converter because it is held stationary so that the fluid force generated by the impeller and turbine can react against it. In the next section we will see how the stator unlocks and rotates under certain operating conditions.

Coupling phase

As the speed ratio of the impeller and turbine approaches 90 percent, the fluid flow in the torque converter becomes mostly rotary, and the angle of flow from the turbine to the stator becomes steeper. Eventually, the fluid strikes the convex (back) sides of the stator vanes rather than the concave front sides, figure 7-15. When the speed of the fluid striking the back-sides of the vanes becomes great enough to drive the stator clockwise, the 1-way clutch overruns, and the impeller, turbine, and stator all rotate in the same direction at approximately the same speed. This is called the converter **coupling phase**.

If the stator clutch did not unlock when the fluid flow angle changed enough to drive the stator clockwise, fluid would bounce off the vanes in a direction that would *oppose* the flow from the impeller to turbine. This would cause the converter to work against itself, and greatly reduce efficiency.

All torque multiplication ceases with the coupling phase because the fluid flow is no longer being redirected by the stator to increase the force of the fluid flowing from the impeller to the turbine. When the converter is coupled, rotary flow is much greater than vortex flow, the turbine turns at a slightly slower speed than the impeller, and the torque converter operates basically like a fluid coupling.

STALL SPEED

To better comprehend the principles of torque multiplication and coupling in a torque converter, it is helpful to understand the principle of stall. Stall is the condition that exists when the impeller moves but the turbine does not. The greatest amount of stall occurs when the impeller is driven at the maximum speed possible without moving the turbine. The speed at which this occurs is called the torque converter **stall speed**.

To determine the stall speed of a converter in a vehicle, set the parking brake, apply the service brake pedal firmly, place the transmission in gear, and apply full throttle for *no more than five seconds*. Engine rpm will rise until vortex fluid flow from the turbine strikes the impeller vanes with a force equal to that being applied by the engine. This causes the engine speed to stabilize at the converter stall speed.

Because the difference between impeller speed and turbine speed is greatest at the converter stall speed, this is also the point of the lowest possible speed ratio, and thus the greatest possible torque multiplication. The opposite extreme is when the converter enters the coupling phase where the speed ratio is at its maximum and there is no torque multiplication.

The stall speed of a torque converter is related to its outside diameter, and to the angle of its stator blades. If a small-diameter converter and a large-diameter converter with the same stator blade angle are turned at the same speed, the smaller converter will create less centrifugal force to move any fluid inside than will the larger converter. As a result, small-diameter converters: have higher stall speeds, multiply torque at higher engine rpm, and do not couple

until the vehicle reaches higher speeds. Large-diameter converters: have lower stall speeds, multiply torque at lower engine rpm, and couple at somewhat lower vehicle speeds.

If two converters of equal diameter are turned at the same speed, but one converter has a sharp stator blade angle and the other has a low (nearly straight) stator blade angle, the stator blades with the sharp angle will move less fluid. As a result, converters with sharply-angled stator blades: have higher stall speeds, multiply torque at higher engine rpm, and do not couple until the vehicle reaches higher speeds. Converters with low-angled stator blades: have lower stall speeds, multiply torque at lower engine rpm, and couple at somewhat lower vehicle speeds.

Vehicle manufacturers select the torque converter size to match the demands of each particular application. Cars with big engines that produce lots of torque at low rpm typically have torque converters that couple at relatively low speeds for greater fuel economy. Cars with smaller engines that produce less torque at low rpm typically have converters that allow the engine to operate higher in its rpm band where there is more power available to move the vehicle. High-performance vehicles with automatic transmissions use small-diameter converters for this same reason.

CONVERTER CAPACITY

Another factor that engineers consider when selecting the torque converter for an application is capacity. **Torque converter capacity** is the ability of a converter to absorb and transmit engine torque in relation to the amount of slippage it allows while doing so.

A high-capacity converter has a low stall speed and transmits torque to the transmission

---

Coupling Phase: The period of torque converter operation when there is no torque multiplication and rotary flow causes the stator to unlock and rotate with the impeller and turbine at approximately the same speed.

Stall Speed: The maximum possible engine and torque converter impeller speed, measured in rpm, with the turbine held stationary and the engine throttle wide open.

Torque Converter Capacity: The ability of a torque converter to absorb and transmit engine torque in relation to the amount of slippage in the converter.

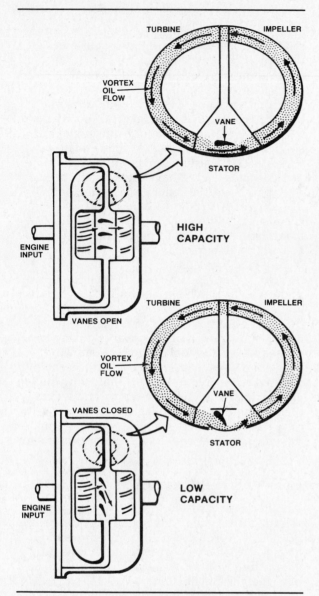

HIGH CAPACITY

LOW CAPACITY

Figure 7-16. The vanes of a variable-pitch stator can be opened or closed to alter the speed of the fluid being redirected from the turbine to the impeller.

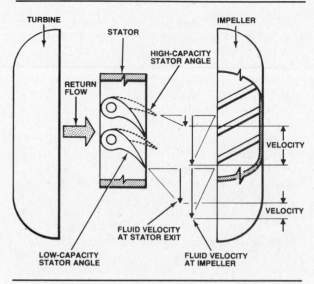

Figure 7-17. A variable-pitch stator creates low- and high-capacity conditions in a torque converter.

Torque converter capacity limits engine speed as a result of the loads placed on the impeller as it transmits torque to the turbine. These loads are high when the vehicle first starts off, but as the car gains momentum and turbine speed approaches that of the impeller, the torque required to drive the impeller falls off and engine speed naturally rises.

In order to get the best combination of acceleration and fuel economy, it is important that the engine and the torque converter be properly matched. A converter whose capacity is too low for a given application will allow the engine rpm to exceed the point of maximum torque output. A converter whose capacity is too high for an application will prevent the engine from reaching the rpm at which maximum torque is produced. The normal practice is to use a converter whose capacity allows the stall speed to occur at the engine rpm where maximum torque is produced.

Variable-Pitch Stators

In order to gain the benefits of both high- and low-capacity designs, some manufacturers have built torque converters with variable-pitch stators. A variable-pitch stator has vanes that can be opened (low angle) or closed (high angle) to change the speed of the fluid flowing from the turbine to the impeller, figure 7-16. The variable-pitch stator was first used with the 1956 Buick Dynaflow transmission. A more recent application was the variable-pitch stator used with the Turbo Hydra-matic 400 transmission from 1964 through 1967.

with a minimum of slippage. High-capacity torque converters are efficient and give good fuel economy at highway speeds, but they generally operate at higher speed ratios and provide less torque multiplication.

A low-capacity converter has a high stall speed and allows a greater amount of slippage as torque is transmitted. Low-capacity torque converters are less efficient, but they generally operate at lower speed ratios and provide greater torque multiplication for better acceleration during stop-and-go driving.

When the vanes of a variable-pitch stator are closed, the speed of oil flow from the stator to the impeller is increased. This creates a low-capacity converter stage, figure 7-17. The low-capacity stage allows engine rpm to increase for more rapid acceleration, and extends the torque multiplication range to higher speeds.

When the vanes of a variable-pitch stator are open, the speed of oil flow from the stator to the impeller is decreased. This creates a high-capacity converter stage, figure 7-17. A high-capacity converter stage allows torque transfer and coupling at lower engine speeds for better fuel economy.

LOCKUP TORQUE CONVERTERS

Even when a torque converter is in the coupling phase, there is still a small amount of slippage — about three to six percent. If this slippage can be eliminated, fuel economy will improve up to five percent during highway cruising. Transmission operating temperatures will also be lowered because fluid shear within the converter is eliminated. Lockup torque converters prevent any slippage by using a torque converter clutch (TCC) inside the converter cover to lock the impeller to the turbine. This creates a direct 1:1 mechanical connection between the engine and the transmission.

Some early automatic transmissions provided for mechanical lockup of the torque converter. Both the 1949 Packard Ultramatic and the 1950 Studebaker automatic had a torque converter clutch that locked the turbine to the impeller in high gear. However, because these early lockup converters added to the cost of the vehicle and were unnecessary considering the low fuel prices of the day, they were dropped after only a few years.

The concept of a lockup torque converter was revived by Chrysler in 1978 in response to the need for better fuel economy. Today, virtually all new cars with automatic transmissions have lockup torque converters. Two types are now in use. The first uses hydraulic pressure to lock the converter; the second relies on centrifugal force.

HYDRAULICALLY LOCKED CONVERTERS

Hydraulically locked torque converters use the oil pressure in the transmission hydraulic system to apply the TCC and lock the impeller and

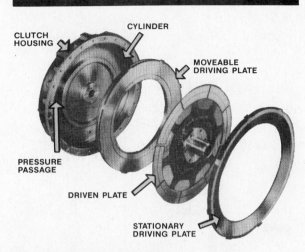

■ Lockup is Nothing New

When Chrysler introduced its lockup torque converter in 1978, it was not the first time this idea had been used. Back in 1949, the Packard Ultramatic transmission had a torque converter with a mechanical clutch that locked up for direct drive, and Studebaker automatics in the early 1950s had a similar lockup clutch.

Packard's Ultramatic transmission had a 4-element torque converter with two turbines. The lockup clutch was located at the front of the converter housing, and its driven plate was bolted to the first turbine. When hydraulic fluid was routed to the clutch housing, a moveable driving plate locked the driven plate to the converter housing. This is exactly the same principle that was reintroduced by Chrysler almost 30 years later.

The Ultramatic clutch was controlled by applying governor pressure and throttle pressure to opposite ends of a shift valve. Below 15 mph, the converter was always unlocked. Above 15 mph, governor pressure moved the shift valve to lock the clutch when throttle pressure dropped as the car reached cruising speed. If the driver pressed hard on the accelerator for passing power, increased throttle pressure moved the shift valve against governor pressure to exhaust fluid from the clutch housing and unlock the converter.

Packard and other manufacturers dropped early lockup torque converters because they were expensive, gasoline was cheap, and car owners didn't appreciate their value. Today, gasoline is more expensive, consumers are more sophisticated, and automakers must meet Federal minimum fuel economy standards. For these reasons, Chrysler and others have revived an older, better idea.

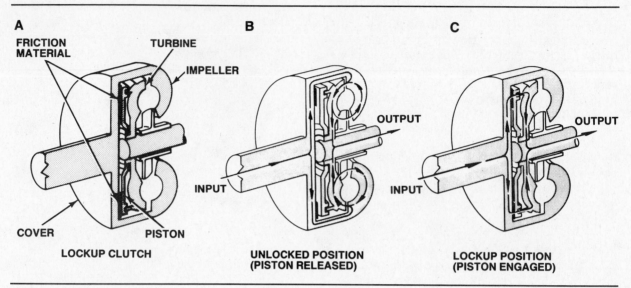

Figure 7-18. A hydraulically locked converter uses a piston as a clutch plate to lock the impeller to the turbine and eliminate slippage.

turbine together. The clutch consists of a piston, which serves as a clutch plate, placed between the turbine and the converter cover, figure 7-18A. The piston is splined to the turbine, and the cover is attached to the impeller.

When the converter is unlocked, figure 7-18B, it functions in exactly the same manner as a non-locking torque converter. Fluid flow from the impeller, to the turbine, and back through the stator provides the only connection between the engine and the transmission.

To lock the converter, figure 7-18C, hydraulic pressure is routed between the turbine and piston. This forces the piston forward and holds it tightly against the converter cover, locking the impeller to the turbine. Power is then transmitted directly (mechanically) from the cover, through the piston, to the turbine; fluid flow plays no part in transmitting the power.

Hydraulic Converter Clutch Controls

The first hydraulically locked torque converters were controlled entirely by hydraulic pressures and spool valves within the transmission. In addition to spool valves, some later designs employed simple electrical switches and a TCC solenoid to control the oil pressure that applied the clutch. All current hydraulic torque converter clutches use an electronic control system to switch the TCC solenoid.

Many early torque converter clutches were engaged only in high gear. Lockup was not used in lower gears because it would eliminate the torque multiplication necessary for maximum acceleration. However, the sophisticated

electronic controls of most newer hydraulically locked converters allow them to lock in lower gears under certain operating conditions.

Although all hydraulically locked converters operate in the same basic manner, there are minor variations in their mechanical construction. The sections below describe the lockup converters used by the major domestic vehicle manufacturers. The electrical and electronic controls used to regulate hydraulic converter clutches are covered in the next chapter. The hydraulic system modifications and controls are described in later chapters that deal with specific transmissions.

Chrysler Lockup Converter

In the Chrysler lockup torque converter, figure 7-19, the inner surface of the converter cover is lined with friction material. The piston is attached to the front outer edge of the turbine through ten torsional damping springs that cushion initial engagement of the converter clutch, and damp engine power pulses once the converter is locked.

Early Chrysler lockup converters are controlled entirely by hydraulic pressures acting on spool valves in the transmission valve body. In lower gears, converter pressure is routed to the chamber in front of the piston to keep the converter unlocked. When the transmission shifts into high gear and a minimum vehicle speed is reached, converter pressure is vented from the front chamber, and mainline pressure is directed between the turbine and the lockup piston. This moves the piston into contact with the

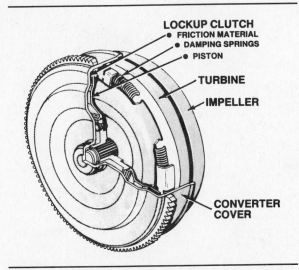

Figure 7-19. In Chrysler's lockup converter, the clutch piston attaches to the turbine through 10 torsional damping springs.

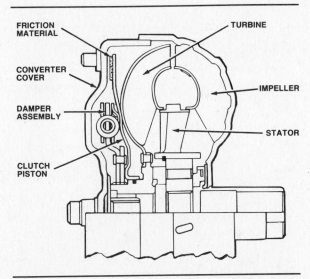

Figure 7-20. In the Ford lockup torque converter, the clutch piston is splined directly to the turbine.

converter cover, and locks the turbine to the impeller. The lockup clutch disengages automatically during part- or full-throttle downshifts, or if vehicle speed falls below the preset limit.

The converter clutches on later Chrysler models operate in the same basic manner as just described. However, beginning with selected 1986 truck applications, the pressures that apply and release the clutch piston are controlled by an electric solenoid valve that is switched on and off by the electronic engine control system. We will study Chrysler's lockup torque converter in greater detail in Chapter 9.

Ford Lockup Converter

The Ford hydraulic lockup torque converter design, figure 7-20, operates in the same basic manner as the Chrysler design, with the following construction differences. The piston has a band of friction material around its outer edge. In addition, the clutch piston is splined directly to the turbine. To absorb the shock of clutch application, and damp engine power pulses once the converter is locked, a spring-type torsional damper is built into the piston hub.

Some Ford converter clutches are operated by hydraulic pressures and spool valves whose actions can be overridden by an electric solenoid valve that is switched on and off by the electronic engine control system. In other applications, the lockup clutch is entirely controlled by the engine electronics. We will study these lockup torque converters in more detail in later chapters covering specific Ford transmissions.

General Motors Lockup Converter

The General Motors lockup torque converter, figure 7-21, was introduced on selected models beginning in the 1980 model year. The GM converter clutch design operates in the same basic manner as the Chrysler and Ford designs, with the following construction differences. The piston has a one-inch band of friction material around its outer edge, and additional center blocks of friction material that support the piston evenly as it is held against the converter cover under pressure. There is no friction material inside the converter cover.

The clutch piston is splined directly to the turbine. To absorb the shock of clutch application, and damp engine power pulses once the converter is locked, a spring-type torsional damper is built onto the back of the piston, figure 7-22. The damper allows the piston to rotate independently of the turbine up to 45 degrees in either direction.

On models with diesel engines (except the 6.2-liter diesel), which have very high compression, the damper assembly includes a pair of poppet-type pressure release valves. If engine braking becomes strong enough to rotate the damper assembly past a certain point, ramps on the outer edge of the damper lift the poppet valves off their seats. This opens a pair of orifices that allow pressure to equalize on both sides of the piston, releasing the clutch to lessen the effect of engine braking.

All GM lockup torque converters are controlled by a converter clutch solenoid. On early models, the clutch operates only in high gear, and the solenoid is controlled by a series of

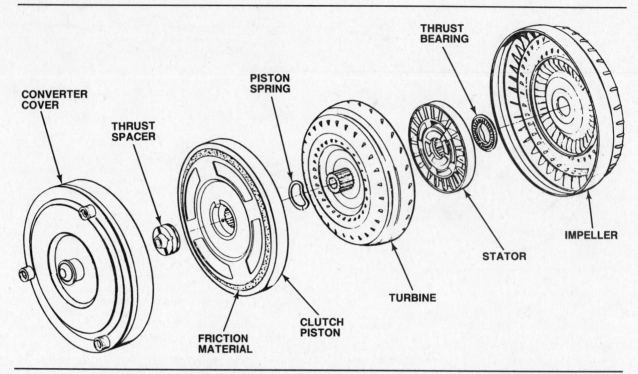

Figure 7-21. An exploded view of the GM lockup torque converter.

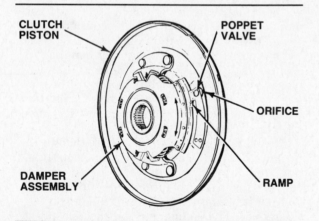

Figure 7-22. The shock damper is built onto the back of the GM converter clutch piston (diesel version shown).

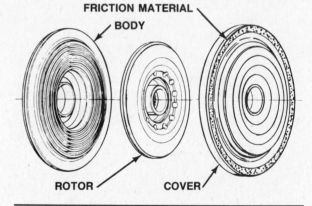

Figure 7-23. A viscous converter clutch is used in place of a spring-type torsion damper on some GM models.

simple switches. On later models, the solenoid is controlled by the electronic engine control system, and the clutch is applied in selected lower gears on some models. We will study these lockup torque converters in more detail in later chapters covering specific GM transmissions.

Viscous converter clutch
Introduced on 1985 Cadillac models with the Turbo Hydra-matic 440-T4 transaxle, a viscous

converter clutch is used in some GM applications where maximum smoothness of power transmission is desired. The viscous clutch piston assembly, figure 7-23, primarily consists of an outer cover and body, and an inner rotor. The clutch friction material is attached to the piston cover, which is joined to the piston body. The rotor is splined to the converter turbine. The body and rotor have a series of interlocking grooves, and the inside of the assembly is entirely filled with a special silicone fluid.

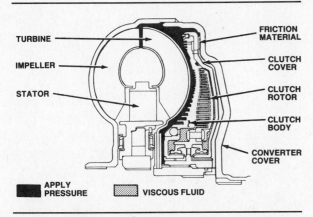

Figure 7-24. The viscous converter clutch piston assembly contains a special silicone fluid that allows a small amount of shock-damping slippage.

When the converter clutch is engaged, figure 7-24, power is passed from the piston cover to the body, through the silicone fluid to the rotor, and then to the turbine and transmission input shaft. The silicone fluid allows a small amount of slippage between the body and rotor to absorb the shock of engagement and damp engine power pulses once the converter is locked.

Technically, the viscous clutch does not completely lock the converter. However, the amount of slippage that occurs is kept to a minimum by the silicone fluid whose viscosity

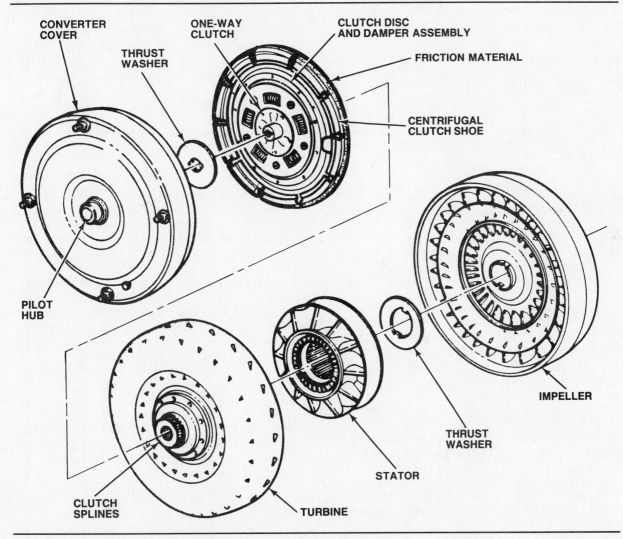

Figure 7-25. Ford's centrifugally locking torque converter is effective in all forward gears.

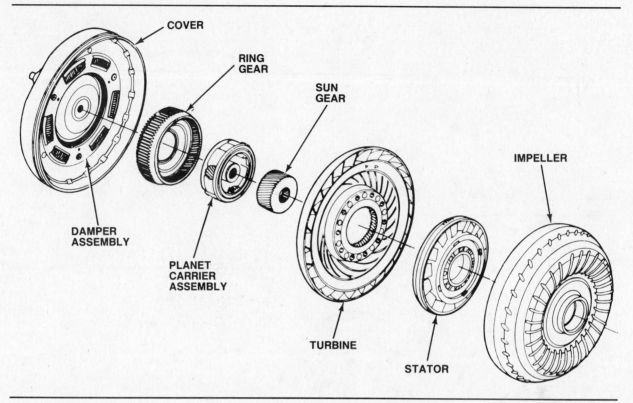

Figure 7-26. The Ford splitter-gear converter allows engine torque to be transmitted both hydraulically and mechanically.

automatically increases with the **speed differential** between the body and rotor. In actual use, viscous clutch slippage amounts to only about 40 rpm at 60 mph, which still allows good fuel economy with maximum smoothness.

CENTRIFUGALLY LOCKED CONVERTERS

Ford is the only domestic manufacturer to use a centrifugally locked torque converter. This type of unit has been used for several years with the C5 transmission, and for at least one year with the ATX transaxle.

A centrifugally locked converter, figure 7-25, contains a clutch disc that is splined to the turbine through a 1-way overrunning clutch. A spring-type torsional damper is built into the disc to damp engine power pulses once the clutch is engaged. The outside of the clutch disc supports a series of centrifugal clutch shoes held in place by return springs. The outer faces of the shoes are lined with friction material.

As the car accelerates and turbine speed builds, centrifugal force acting on the clutch shoes pulls them outward toward the converter cover. When turbine speed becomes great

enough, the shoes' friction materials contact the inside of the converter cover and lock the turbine to the impeller. When vehicle speed, and thus turbine speed, drops below a preset level, the return springs retract the clutch shoes and unlock the converter.

When torque multiplication for acceleration or increased pulling power is required *after* the converter has locked, the clutch shoes are designed to slip on the inside of the converter cover. To limit engine braking for maximum fuel economy, the 1-way overrunning clutch allows the turbine to turn faster than the impeller during trailing throttle conditions.

A centrifugally locked converter is simpler than any of the hydraulically locked designs, and it works in any forward gear when the turbine speed rises above the designed-in lockup speed. However, a centrifugally locked converter cannot be controlled as precisely as a hydraulically locked unit, and both lockup speed and locking force will vary as the friction

Speed Differential: The difference in speed between two objects.

material on the clutch shoes wears. Finally, the greater amount of slippage that occurs with this type of lockup mechanism releases more friction material particles to contaminate the transmission fluid.

SPLITTER-GEAR TORQUE CONVERTERS

The splitter-gear torque converter is used with certain Ford ATX automatic transaxles to increase fuel economy. A splitter-gear converter, figure 7-26, is similar to other torque converters except that it contains a planetary gearset that allows engine torque to be transmitted by a combination of hydraulic and mechanical means depending on the gear selected.

In Reverse and first gears, torque is transmitted 100 percent hydraulically as in any other torque converter. In second gear, torque is transmitted 38 percent hydraulically and 62 percent mechanically. In third gear, torque is transmitted 7 percent hydraulically and 93 percent mechanically. Because a portion of the engine power is transmitted by mechanical means in second and third gears, a spring-type torsion damper is welded into the converter cover to absorb engine power pulsations in these gears.

The methods used to obtain these various torque splits involve planetary gearflow and the applications of various hydraulic clutches within the transaxle. These processes will be described in detail in Chapter 11 which deals exclusively with the ATX transaxle.

SUMMARY

Fluid couplings and torque converters couple and uncouple the engine and the transmission. A torque converter can multiply engine torque, a fluid coupling cannot.

Couplings and converters use vaned impellers and turbines to develop the fluid flow that transfers engine torque. Fluid moving through a coupling or converter travels in rotary and vortex flow paths. Rotary flow carries engine torque; vortex flow transfers torque from the impeller to the turbine. Combined rotary and vortex flows create a resultant force that drives the turbine.

The speed ratio is a comparison of impeller and turbine speed, expressed as a percentage, used to measure coupling or converter efficiency. At low speed ratios: vortex flow is high, rotary flow is low, and the coupling or converter is less efficient. At high speed ratios: vortex flow is low, rotary flow is high, and the coupling or converter is more efficient.

Torque converters are more efficient than fluid couplings at low speed ratios because they multiply engine torque using a stator between the turbine and impeller. Torque multiplication occurs at lower speed ratios when vortex flow is high and fluid leaving the turbine strikes the front sides of the stator vanes which redirect and accelerate the fluid so it joins the fluid being pumped from the impeller to the next turbine vane.

Both couplings and converters are most efficient when coupled. Torque converter coupling occurs at speed ratios approaching 90 percent when rotary fluid flow is high and fluid leaving the turbine strikes the back sides of the stator vanes causing the stator 1-way clutch to overrun and allowing the stator to rotate with the impeller and turbine.

The stall speed of a torque converter is the maximum speed at which the impeller can be driven without turning the turbine. The stall speed is also the speed at which maximum torque multiplication occurs. Small-diameter converters have higher stall speeds than large-diameter converters.

Torque converter capacity is the ability of a converter to transmit torque compared to the amount of slippage it allows while doing so. High-capacity converters have low stall speeds and transmit torque with minimal slippage, but they operate at higher speed ratios and provide less torque multiplication. Low-capacity converters have high stall speeds and allow greater slippage, but they operate at lower speed ratios and provide greater torque multiplication. A variable-pitch stator allows a converter to provide some benefits of both high- and low-capacity designs.

Lockup torque converters mechanically connect the turbine to the impeller to eliminate slippage for improved fuel economy. Hydraulically locked converters use oil pressure to apply a clutch piston that is splined to the turbine against the inside of the converter cover. Centrifugally locked converters use centrifugal force to expand a series of clutch shoes that are attached to a disc splined to the turbine outward against the inside of the converter cover. A spring-type damper or viscous coupling smoothes clutch engagement and absorbs engine power pulses when the converter is locked.

Hydraulically locked converters may be applied in high gear only, or in several gears depending on the system. Clutch application can be controlled solely by hydraulic means, or by a combination of hydraulic, electrical, and electronic components. Centrifugally locked converters require no external controls, and apply automatically in any forward gear whenever turbine speed reaches the designed-in lockup speed.

Splitter-gear torque converters contain a planetary gearset and damper assembly that allow engine torque to be transmitted by a combination of hydraulic and mechanical means, depending on the gear selected.

Review Questions

Select the single most correct answer
Compare your answers to the correct answers on page 603

1. Mechanic A says the first fluid coupling on a domestic passenger car was used by Chrysler in 1938. Mechanic B says the first torque converter used on a production car was used by Buick in 1948. Who is right?
   a. Mechanic A
   b. Mechanic B
   c. Both mechanic A and B
   d. Neither mechanic A nor B

2. Mechanic A says the biggest disadvantage of a fluid coupling compared to a torque converter is that a coupling must be larger than a converter. Mechanic B says the biggest disadvantage of a fluid coupling compared to a torque converter is that a coupling cannot provide engine compression braking. Who is right?
   a. Mechanic A
   b. Mechanic B
   c. Both mechanic A and B
   d. Neither mechanic A nor B

3. Rotary oil flow in a fluid coupling or torque converter is:
   a. At a right angle to impeller rotation
   b. In the same direction as impeller rotation
   c. At a variable angle in relation to impeller rotation
   d. Perpendicular to impeller rotation

4. The combination of rotary flow and vortex flow produces a:
   a. Toroid force
   b. Multiplication force
   c. Coupling force
   d. Resultant force

5. Mechanic A says vortex flow in a fluid coupling or torque converter is at a right angle to the direction of impeller rotation. Mechanic B says a guide ring in a fluid coupling or torque converter is used to prevent the buildup of resultant force. Who is right?
   a. Mechanic A
   b. Mechanic B
   c. Both mechanic A and B
   d. Neither mechanic A nor B

6. Mechanic A says the speed ratio of a fluid coupling or torque converter is turbine speed divided by impeller speed. Mechanic B says the speed ratio of a coupling or converter is high when there is more vortex flow than rotary flow. Who is right?
   a. Mechanic A
   b. Mechanic B
   c. Both mechanic A and B
   d. Neither mechanic A nor B

7. Mechanic A says only vortex oil flow is needed to keep the impeller and turbine vanes filled with fluid. Mechanic B says fluid is supplied to the torque converter by the transmission oil pump. Who is right?
   a. Mechanic A
   b. Mechanic B
   c. Both mechanic A and B
   d. Neither mechanic A nor B

8. Mechanic A says that the impeller of a fluid coupling or torque converter acts as a turbine during deceleration to provide engine braking. Mechanic B says torque converters have replaced fluid couplings because they are more efficient at high speed ratios. Who is right?
   a. Mechanic A
   b. Mechanic B
   c. Both mechanic A and B
   d. Neither mechanic A nor B

9. In a torque convertor, the stator is the _____ member.
   a. Drive
   b. Reaction
   c. Driven
   d. Pump

10. Mechanic A says the vanes in 3-element torque converters are curved to reduce oil flow turbulence. Mechanic B says the impeller vanes curve opposite the direction of the turbine and stator vanes to increase torque. Who is right?
    a. Mechanic A
    b. Mechanic B
    c. Both mechanic A and B
    d. Neither mechanic A nor B

11. Mechanic A says most new domestic torque converters have no drain plugs and must be replaced as a unit if they fail. Mechanic B says a flywheel is not usually required with an automatic transmission because the starter ring gear is attached to the outside of the torque converter. Who is right?
    a. Mechanic A
    b. Mechanic B
    c. Both mechanic A and B
    d. Neither mechanic A nor B

12. Which of the following is *not* a common connection between the torque converter and the transmission?
    a. The turbine is splined to the input shaft
    b. The converter hub drives the oil pump
    c. The impeller is splined to the reaction shaft
    d. The stator overrunning clutch is mounted on an extension of the oil pump

13. Mechanic A says the torque multiplication phase is made possible, in part, by lockup of the stator clutch.
    Mechanic B says torque multiplication occurs when rotary oil flow makes a complete circuit from the impeller to the turbine, and back through the stator to the impeller.
    Who is right?
    a. Mechanic A
    b. Mechanic B
    c. Both mechanic A and B
    d. Neither mechanic A nor B

14. Mechanic A says maximum torque multiplication occurs at the converter stall speed.
    Mechanic B says converter coupling occurs when fluid flow strikes the concave faces of the stator vanes.
    Who is right?
    a. Mechanic A
    b. Mechanic B
    c. Both mechanic A and B
    d. Neither mechanic A nor B

15. When the stator clutch in a torque converter overruns:
    a. The stator turns counter-clockwise
    b. Vortex flow is greater than rotary flow
    c. Fluid is redirected by the stator to increase torque
    d. The impeller turns faster than the turbine

16. Mechanic A says that larger-diameter torque converters have higher stall speeds.
    Mechanic B says there is no torque multiplication during the converter coupling phase.
    Who is right?
    a. Mechanic A
    b. Mechanic B
    c. Both mechanic A and B
    d. Neither mechanic A nor B

17. Mechanic A says a low-capacity torque converter also has a low stall speed.
    Mechanic B says vehicle manufacturers specify a torque converter capacity that allows maximum engine torque and the converter stall speed to occur at the same rpm.
    Who is right?
    a. Mechanic A
    b. Mechanic B
    c. Both mechanic A and B
    d. Neither mechanic A nor B

18. A variable-pitch stator:
    a. Provides some benefits of both low- and high-capacity torque converters
    b. Has low capacity when the vanes are open
    c. Has high capacity when the vanes are closed
    d. All of the above

19. Mechanic A says a lockup torque converter helps lower transmission operating temperatures.
    Mechanic B says the turbine is locked to the converter cover in a lockup torque converter.
    Who is right?
    a. Mechanic A
    b. Mechanic B
    c. Both mechanic A and B
    d. Neither mechanic A nor B

20. Mechanic A says a piston is used to apply a hydraulically locked torque converter.
    Mechanic B says hydraulically locked torque converters without electronic controls are only applied in high gear.
    Who is right?
    a. Mechanic A
    b. Mechanic B
    c. Both mechanic A and B
    d. Neither mechanic A nor B

21. Mechanic A says silicone fluid is used to apply the GM viscous converter clutch.
    Mechanic B says centrifugal force causes the silicone fluid in the GM viscous converter clutch to change viscosity.
    Who is right?
    a. Mechanic A
    b. Mechanic B
    c. Both mechanic A and B
    d. Neither mechanic A nor B

22. Which of the following is *not* true of the Ford centrifugally locked torque converter?
    a. It is applied by shoes lined with friction material that contact the inside of the converter cover
    b. Lockup is controlled by the turbine speed
    c. Torque multiplication is not possible after the converter locks
    d. An overrunning clutch allows the turbine to turn faster than the impeller during deceleration

23. Mechanic A says the Ford splitter-gear torque converter has a torsion damper welded into the converter cover.
    Mechanic B says that in Low gear 100 percent of the engine's torque is transmitted mechanically through a splitter-gear torque converter.
    Who is right?
    a. Mechanic A
    b. Mechanic B
    c. Both mechanic A and B
    d. Neither mechanic A nor B

8

Electrical and Electronic Transmission Controls

Beginning with the earliest designs, most automatic transmissions have been operated primarily by mechanical and hydraulic means. The driver selects the gear range by moving the shift lever, which is part of a mechanical linkage that positions the manual valve inside the transmission. Once the gear range is chosen, shift action is regulated by hydraulic throttle and governor pressures that interact on the shift valves. When a forced downshift is desired, a mechanical linkage, or a simple switch and electrical solenoid, are used to boost hydraulic pressure and move the appropriate shift valve.

The first lockup torque converters were also controlled by hydraulics and basic electrical circuits. However, as the use of torque converter clutches increased, it soon became apparent that these simple lockup controls were inadequate. Factors outside the transmission, such as engine temperature and brake application, had to be considered when determining whether to engage the converter clutch. In 1981, General Motors introduced electronic torque converter clutch controls. Other manufacturers soon followed, and today, all lockup torque converters are controlled electronically.

Torque converter clutches are not the only application of electronics for automatic transmissions, however. In the late 1970s, imported vehicles from Toyota and Renault began using computer controlled transmissions that employed electronics to regulate shifting. In the mid-1980s, American manufacturers began to introduce automatic gearboxes with partial electronic shift control. And by 1988, BMW, Mazda, Mitsubishi, and others were also using electronic shift controls.

In this chapter, we will examine the operation of basic electrical and electronic control systems. Then we will look at how these systems are used to regulate torque converter lockup and transmission shifting.

BASIC TORQUE CONVERTER CLUTCH CONTROL

As discussed in the last chapter, a TCC provides a direct 1:1 mechanical coupling between the engine and transmission. A few applications use centrifugally applied converter clutches, but in most cases, lockup is achieved using a piston placed between the turbine and the converter housing (impeller). When hydraulic oil pressure is routed between the turbine and the piston, the piston moves forward and contacts the inside of the housing to lock the turbine to the impeller. This eliminates the

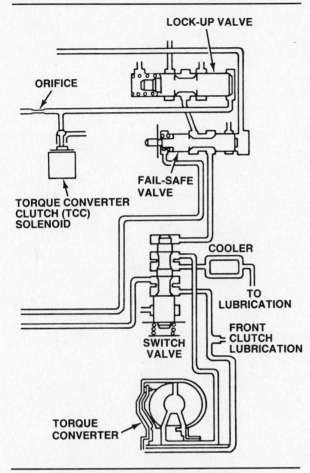

Figure 8-1. This circuit diagram of Chrysler's electronically controlled torque converter clutch shows the spool valves and solenoid used to lock and unlock the converter clutch.

slippage that occurs in a conventional torque converter, and results in better fuel economy and reduced transmission temperatures.

Electrical and electronic torque converter clutch controls are used only with hydraulically applied converter clutches. In these systems, the application of the clutch piston is regulated by an electric solenoid and one or more spool valves, figure 8-1. The solenoid controls a pressure that signals the spool valves when to lock and unlock the converter clutch assembly. The spool valves control the various pressures that actually apply and release the clutch piston. This chapter focuses on the operation of the electrical solenoid. The hydraulic valves involved vary with the application, and will be detailed in later chapters that cover specific transmissions.

The solenoid that regulates converter clutch operation is the part of the system that is electrically or electronically controlled. This torque

converter clutch solenoid is called different names by different manufacturers; in this text we will call it the TCC solenoid.

Torque Converter Clutch Application Restrictions

Before we look at the various electrical and electronic means used to control the TCC solenoid, it is important to understand when the converter clutch can and cannot be applied, and

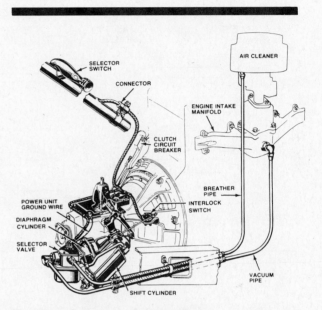

■ Black Boxes Are Nothing New

Every year, new cars seem to have more electronic "black boxes" that control various areas of vehicle operation — including the transmission. Actually, electrical transmission controls are not all that new; Hudson had them in the 1930's. The Hudson Electric Hand transmission didn't have a solid-state computer, but it did use an array of switches, circuit breakers, solenoids, and vacuum servos to shift a conventional 3-speed sliding gear transmission.

In Hudson's system, the driver moved a small fingertip lever on the steering column through a standard H-pattern to select three speeds forward and reverse. The lever operated a switch that controlled several solenoids, which in turn, directed engine manifold vacuum through a selector valve to the vacuum servos in a shift cylinder. The servos moved the shift levers on the transmission to change the gear ratios.

As a safety measure, a circuit breaker attached to the clutch linkage prevented the transmission from shifting unless the clutch was depressed. An interlock switch was also used to keep the vacuum servos from trying to engage two gears at once.

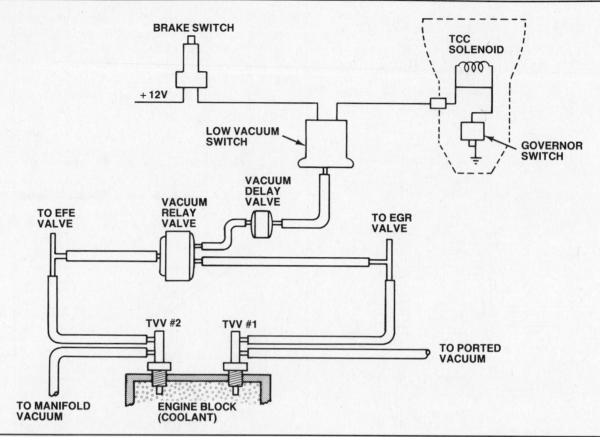

Figure 8-2. Early GM converter clutches were operated by simple electrical switches and several vacuum control devices.

why these restrictions are necessary. Many driveability complaints today are caused by converter clutches that lock or unlock at inappropriate times.

The converter clutch should apply only after several conditions are met. First, the engine should be warm enough to handle the extra load without stumbling. Second, the vehicle speed should be high enough to allow smooth power transmission without detonation or noticeable engine power pulses. And finally, the converter clutch should only engage when engine and vehicle speed (impeller and turbine speed) are such that lockup will not create an objectionable surge or shudder.

The converter clutch should not lock, or should release if it is already locked, when certain other conditions exist. The clutch should unlock whenever torque multiplication is needed in the converter for faster acceleration. The clutch should release during closed-throttle deceleration because leaving it engaged would increase exhaust emissions and lower fuel economy. The clutch should unlock during braking to prevent killing the engine if the

brakes lock. If the clutch can apply in more than one gear, it should unlock during shifts between those gears to smooth the flow of power. And finally, the clutch must not be applied when the vehicle comes to a stop or the engine will stall.

ELECTRICAL TORQUE CONVERTER CLUTCH CONTROLS

Before electronic torque converter clutch control systems became commonplace, simple electrical circuits and switches were used to energize the TCC solenoid on many General Motors vehicles. The first such system was introduced on cars with both gasoline and diesel engines in the 1980 model year. Although most gasoline-engine applications were discontinued after the 1981 model year, variations of these systems continued to be used on later diesel-powered vehicles.

Electrical converter clutch control systems share some parts with later electronically regulated systems, and they perform the same job based on the same requirements discussed in

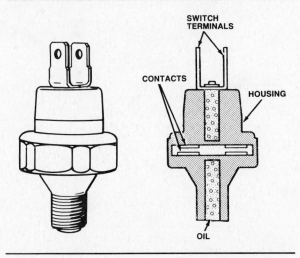

Figure 8-3. The governor switch grounds the TCC solenoid when governor pressure rises high enough to close the switch contacts.

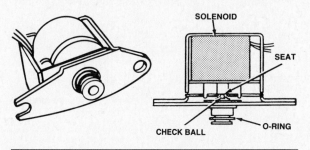

Figure 8-4. A typical TCC solenoid.

the previous section. Understanding the components used in electrical converter clutch control systems will make it easier to grasp the operation of computer-controlled converter clutches.

GM GASOLINE-ENGINE ELECTRICAL TORQUE CONVERTER CLUTCH CONTROLS

The 1980 GM torque converter clutch system used with gasoline engines, figure 8-2, has three electrical switches that control the operation of the TCC solenoid:

- Brake switch
- Low vacuum switch
- Governor switch.

The brake and low vacuum switches are installed in the power supply circuit to the TCC solenoid. When these switches are closed, the TCC solenoid receives 12 volts. The solenoid ground is controlled by the governor switch. When governor pressure becomes high enough, this switch closes and grounds the TCC solenoid. Whenever all three switches in the converter clutch circuit are closed, the solenoid is energized to lock the converter clutch.

The system shown in figure 8-2 is one of many electrical converter clutch control systems used on gasoline-powered GM vehicles. There are too many variations to discuss each one in this text; however, the following sections describe the different system components in more detail. A particular system may use many of the parts described, but no one system contains them all.

Governor Switch

The governor switch, figure 8-3, is a normally open (NO) switch located on the transmission valve body in the governor pressure circuit. As vehicle speed increases, so does the governor pressure. When governor pressure reaches a predetermined level (between 35 and 50 mph or 55 and 80 kph road speed, depending on the application), the governor switch contacts close to ground the TCC solenoid.

TCC Solenoid

The TCC solenoid, figure 8-4, contains a ball and seat check valve. This type of assembly is sometimes called a **solenoid ball valve**. The TCC solenoid is located in the valve body or front pump depending on the transmission.

When the solenoid is not energized, figure 8-5A, oil pressure in the converter clutch apply circuit passes through the check valve and is vented, or exhausted, to the transmission sump. When the solenoid is energized, figure 8-5B, the check ball seats to close off the exhaust passage, and oil pressure is directed to the spool valves that apply the torque converter clutch.

A restricting orifice is located in the apply pressure line to the TCC solenoid. The orifice limits the volume of fluid flow in the apply circuit to ensure that all the fluid can be vented through the check valve. This makes it impossible for the converter clutch to apply when the solenoid is not energized.

Solenoid Ball Valve: An assembly consisting of a ball and seat check valve controlled by an electrical solenoid. When the solenoid is not energized, fluid flows through the check valve. When the solenoid is energized, fluid flow through the check valve is blocked.

A

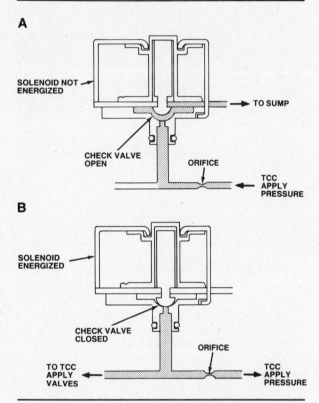

Figure 8-5. The TCC solenoid either vents or seals the converter clutch apply circuit to control clutch lockup.

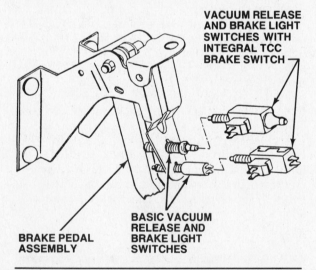

Figure 8-6. The converter clutch brake switch on GM vehicles is usually combined with the brake light switch or cruise control vacuum release switch.

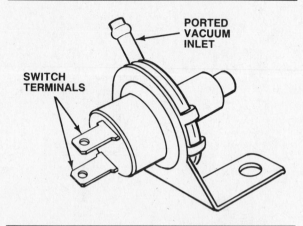

Figure 8-7. When regulated (diesel engines) or ported vacuum drops below a certain level, the low vacuum switch cuts power to the TCC solenoid.

Brake Switch

The brake switch, figure 8-6, is a normally closed (NC) switch located under the instrument panel on the pedal assembly. This switch is often combined into either the brake light switch or the cruise control vacuum release switch. When the brake pedal is depressed, the switch contacts open and cut off power to the TCC solenoid. This disengages the converter clutch to prevent killing the engine if the brakes lock.

Low Vacuum Switch

The low vacuum switch, figure 8-7, is a normally open switch often located on an inner fender housing. The switch receives a **ported vacuum** signal by way of several control devices that are explained in the next section. When the throttle is partially open (ported vacuum is high), the low vacuum switch closes to help complete the power supply circuit to the TCC solenoid.

When the throttle is closed or open fairly wide (ported vacuum is low), the low vacuum switch opens and cuts power to the TCC solenoid. During closed-throttle deceleration, the low vacuum switch unlocks the converter clutch to reduce exhaust emissions and improve fuel economy. During large throttle openings, the low vacuum switch unlocks the converter clutch to allow torque multiplication in the converter for faster acceleration.

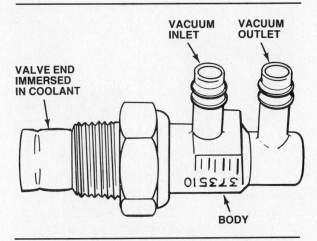

Figure 8-8. Thermal vacuum valves are used to prevent converter clutch lockup until the engine coolant reaches a certain temperature.

In early GM converter clutch systems, figure 8-2, the ported vacuum signal to the low vacuum switch is controlled by a pair of thermal vacuum valves and a vacuum relay valve. These valves work together to cut off the flow of vacuum to the low vacuum switch, and thus prevent converter clutch lockup, when the engine is cold. The same controls are also used to regulate the flow of vacuum to the **early fuel evaporation** (EFE) and **exhaust gas recirculation** (EGR) systems as explained below.

Thermal vacuum valves
A thermal vacuum valve (TVV), figure 8-8, blocks or allows the flow of vacuum based on temperature. A TVV threads into the engine cooling system so that the end of the valve is immersed in the coolant. Vacuum inlet and outlet hoses attach to fittings on the body of the valve. The two TVV's used in early GM TCC control systems operate in different ways. For purposes of explanation, we will call them TVV #1 and TVV #2.

TVV #1 is installed between the ported vacuum source on the throttle body and two components, the EGR valve and the 2-port side of the vacuum relay valve. When coolant temperature is below the TVV switch point, vacuum flow is blocked. When the temperature rises above the TVV switch point, ported vacuum flows to the EGR valve and the 2-port side of the vacuum relay valve.

TVV #2 is installed between an **intake manifold vacuum** source and two components, the EFE valve and the 1-port side of the vacuum relay valve. The operation of TVV #2 is opposite that of TVV #1. When the coolant temperature is below the TVV switch point, manifold vacuum flows to the EFE valve and the

1-port side of the vacuum relay valve. When the temperature rises above the switch point, manifold vacuum flow to these parts is blocked.

Vacuum relay valve
The vacuum relay valve allows vacuum to flow through the 2-port side of the valve to the low vacuum switch only when no vacuum is applied to the 1-port side of the valve. This happens only after the coolant temperature has risen above the switch points of *both* TVV's. A typical sequence of events is as follows.

When a cold engine is started, TVV #1 *blocks* ported vacuum flow to the EGR valve and the 2-port side of the vacuum relay valve. TVV #2 *allows* manifold vacuum flow to the EFE valve and the 1-port side of the vacuum relay valve. As the engine warms up, the TVV's reach their switch points. Depending on the valve calibrations, either valve may operate first.

If TVV #1 switches first, ported vacuum is routed to the EGR valve and the 2-port side of the vacuum relay valve. The EGR system begins to operate immediately, but the vacuum relay valve blocks vacuum flow to the low vacuum switch because manifold vacuum is still being applied to the 1-port side of the valve.

If TVV #2 switches first, manifold vacuum to the EFE valve and the 1-port side of the vacuum relay valve is blocked. This opens the vacuum relay valve, but because TVV #1 has not yet opened, no ported vacuum is available for routing to the low vacuum switch.

Once both TVV's have switched, ported vacuum is routed through the open vacuum relay valve to the low vacuum switch. The ported vacuum is then used to open and close the switch contacts in the TCC solenoid power supply circuit as already described.

Ported Vacuum: The low pressure area (vacuum) located just above the throttle butterfly valve in a carburetor or fuel injection throttle body.

Early Fuel Evaporation (EFE): A system that preheats the incoming air-fuel mixture to improve driveability when the engine is cold.

Exhaust Gas Recirculation (EGR): A pollution control system that injects a small quantity of exhaust gasses into the air-fuel mixture to lower combustion temperatures and limit the formation of oxides of nitrogen (NO_x).

Intake Manifold Vacuum: The low pressure area (vacuum) located below the throttle butterfly valve in the engine intake manifold.

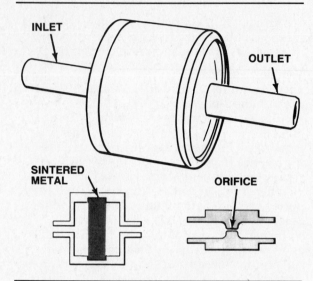

Figure 8-9. Vacuum delay valves are used to damp sudden changes in ported vacuum.

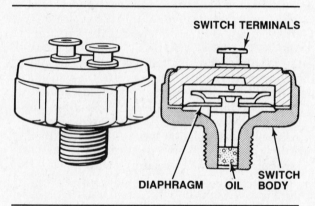

Figure 8-10. Third and fourth gear switches are used in a variety of ways depending on the application.

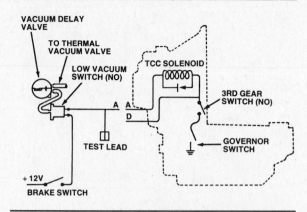

Figure 8-11. The third gear switch in this transaxle ensures that the converter clutch engages only in high gear.

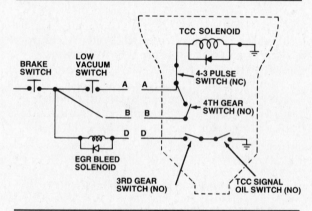

Figure 8-12. This transmission uses several switches on the valve body to help control the torque converter clutch and related systems.

Vacuum delay valve

A vacuum delay valve, figure 8-9, is installed between the vacuum relay valve and the low vacuum switch in some torque converter clutch systems. The vacuum delay valve contains a **sintered** metal disc or very small orifice that slows the flow of vacuum from one side of the valve to the other. This enables the valve to damp fluctuations in ported vacuum to the low vacuum switch, thus preventing repeated locking and unlocking of the converter clutch when the throttle is momentarily opened or closed.

Third Gear Switch

The third gear switch, figure 8-10, is a normally open switch installed in the third gear circuit in the transmission valve body. When the transmission shifts into third gear, oil pressure acting on a diaphragm closes the switch contacts to complete the power supply circuit to the TCC solenoid.

In some 3-speed applications, figure 8-11, the third gear switch is wired in series with the governor switch to ensure that the converter clutch engages only in high gear. In 4-speed transmissions with electrical converter clutch controls, figure 8-12, the third gear switch works with a torque converter clutch switch to control the EGR bleed solenoid.

Fourth Gear Switch

The fourth gear switch is a normally open switch installed in the fourth gear circuit in the transmission valve body. The fourth gear switch looks and operates like a third gear

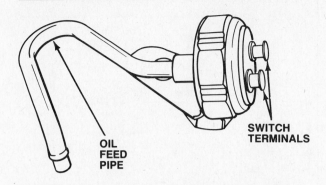

Figure 8-13. The 4-3 pulse switch is used to unlock the converter clutch during a downshift.

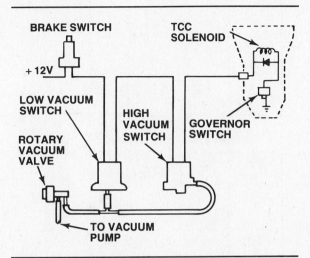

Figure 8-14. Because of their operating characteristics, diesel engines require different torque converter clutch controls than gasoline engines.

switch except that its contacts close when the transmission is in fourth gear. When a fourth gear switch is used in an application with basic electrical converter clutch controls, figure 8-12, the governor switch is eliminated and the TCC solenoid is grounded at all times. When the transmission shifts into high gear, the fourth gear switch bypasses the low vacuum switch and completes the power supply circuit to lock the converter clutch regardless of intake manifold vacuum.

4-3 Pulse Switch

The 4-3 pulse switch, figure 8-13, is a normally closed (NC) switch located on the transmission valve body that momentarily opens the power supply circuit to the converter clutch on a 4-3 downshift. This reduces downshift harshness by allowing normal torque converter slippage to occur during the shift.

EGR Bleed Solenoid

The EGR bleed solenoid, figure 8-12, operates a normally closed air bleed in the EGR control vacuum passage. The bleed solenoid is usually located on the right inner fender housing. Because less EGR is needed at higher engine speeds, the EGR bleed solenoid is energized at the same time as the TCC solenoid. This opens the air bleed in the EGR control vacuum passage, which reduces or eliminates EGR to help prevent engine surging while the converter clutch is engaged.

EARLY GM DIESEL-ENGINE ELECTRICAL TORQUE CONVERTER CLUTCH CONTROLS

The electrically controlled TCC system used with 1980 GM diesel engines, figure 8-14, is similar to the first gasoline-engine systems.

In fact, the TCC solenoid, brake switch, and governor switch operate in exactly the same manner. However, there are also several differences because diesel engines have little or no intake manifold vacuum. As a result, a **vacuum pump** and rotary vacuum valve are used to produce the vacuum signals that help control converter clutch operation. The sections below describe the parts whose design or operation differs from those on gasoline-powered GM vehicles.

Rotary Vacuum Valve

The rotary vacuum valve is mounted on the end of the throttle shaft on the fuel injection pump. The valve receives a constant high vacuum signal from the engine vacuum pump, and modifies it into a variable vacuum signal that is high at idle but decreases with engine speed. The output from the rotary vacuum valve is connected to both the low vacuum and high vacuum switches.

Sintered: Metal Particles fused together under high heat and pressure. Sintering is used to make the porous metal discs in some vacuum delay valves.

Vacuum Pump: A mechanically or electrically driven pump that provides a source of vacuum. Vacuum pumps are commonly used with diesel engines which have little or no intake manifold vacuum.

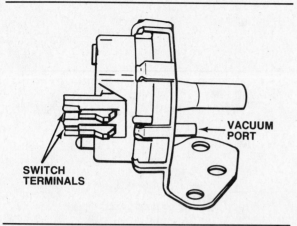

Figure 8-15. A high vacuum switch is used to help control the converter clutch on early GM vehicles with diesel engines.

Low Vacuum Switch

The low vacuum switch is a normally open switch that receives its vacuum signal through a hose from the rotary vacuum valve. When the engine is running and the throttle is closed or partially open, the high vacuum signal from the rotary vacuum valve closes the electrical contacts in the low vacuum switch. This allows engagement of the converter clutch, providing all other switches in the converter clutch power supply circuit are also closed.

When the throttle is opened beyond a certain point, usually just before a part-throttle downshift, the vacuum signal from the rotary vacuum valve drops far enough that the low vacuum switch contacts open. This cuts off power to the TCC solenoid, unlocks the converter clutch, and allows torque multiplication in the converter for a smooth downshift and faster acceleration.

High Vacuum Switch

The high vacuum switch, figure 8-15, is a normally closed (NC) switch that receives its vacuum signal through a hose from the rotary vacuum valve. When the engine is running and the throttle is closed or only slightly open, such as during closed-throttle deceleration, the high vacuum signal from the rotary vacuum valve opens the electrical contacts in the low vacuum switch. This cuts off power to the TCC solenoid, and unlocks the converter clutch to reduce exhaust emissions and limit the feeling of engine power pulses during coastdown.

The high vacuum switch was eliminated on 1981 and later diesel models. In its place, poppet valves were installed on the clutch piston

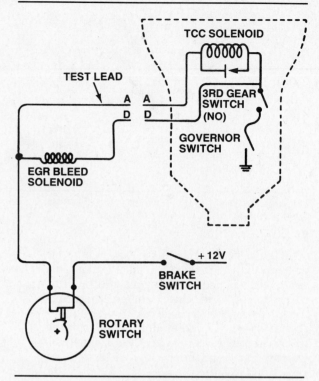

Figure 8-16. Newer GM diesel converter clutch systems use mechanical switches in place of vacuum controls.

to release the converter clutch under hard deceleration. The operation of these valves is described in Chapter 7.

LATE GM DIESEL-ENGINE ELECTRICAL TORQUE CONVERTER CLUTCH CONTROLS

Like gasoline-engine electrical converter clutch control systems, the diesel versions have gone through changes over the years. Beginning in 1981, vacuum was eliminated as a means of converter clutch control. In place of vacuum switches, a number of mechanical switches were adopted, figure 8-16. Once again, there are too many different systems to cover in this text; however, the components used in several representative diesel converter clutch control systems are described below. Once again, keep in mind that no one system contains all of the parts described.

Rotary Switch

The rotary switch replaced the rotary vacuum valve on 1981 diesel engine applications with a torque converter clutch. The rotary switch is mounted on the end of the throttle shaft on the fuel injection pump. Two types of rotary

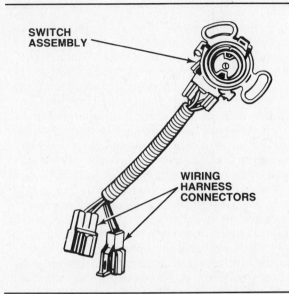

Figure 8-17. This rotary switch contains two sets of contacts that are used to control converter clutch lockup.

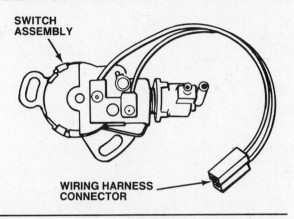

Figure 8-18. The vacuum regulator valve replaced the rotary switches and rotary vacuum valves used in earlier diesel converter clutch control systems.

■ **The V-SAC Converter Clutch Control**

switches are used. The first contains a set of electrical contacts that are normally closed, figure 8-16. When the accelerator is at or near full throttle, a cam opens the switch contacts. This cuts off power to the TCC solenoid, unlocks the converter clutch, and allows torque multiplication in the converter for a smooth downshift and faster acceleration.

Dual-contact rotary switch
The second type of rotary switch, figure 8-17, uses two sets of contacts wired in series. The first contact set is normally open when the engine is at or near idle speed. This cuts the power supply to the TCC solenoid during closed-throttle deceleration, which reduces emissions and the feeling of engine power pulses.

When the throttle is opened partway, the first set of contacts closes to complete the power supply circuit to the TCC solenoid. Because the second set of contacts is normally closed, the converter clutch will then lock up, providing all other switches in the circuit are also closed.

The second set of contacts opens when the accelerator is at or near full throttle. This cuts off power to the TCC solenoid, unlocks the converter clutch, and allows torque multiplication in the converter for a smooth downshift and faster acceleration.

Vacuum Regulator Valve

The vacuum regulator valve (VRV), figure 8-18, replaced the rotary switch on 1982 diesel en-

Although any converter clutch helps improve fuel economy, not all torque converter clutch control systems are equally refined. Because of this, drivers may complain that a converter clutch locks or unlocks harshly, or that it behaves annoyingly by repeatedly engaging and disengaging in response to small throttle movements. In some cars, erratic converter clutch operation can even give an impression that the engine has a miss.

The simple way to solve these problems might be to cut the wires to the TCC solenoid. This will definitely prevent lockup, although with a slight penalty in fuel economy. Unfortunately, the truth is not so simple. On transmissions like the THM 200-4R and 700-R4, disconnecting the converter clutch will burn out fourth gear and ruin the front pump. In addition, torque converters that contain clutches generally have lower capacities than those that do not. As a result, disconnecting the converter clutch will allow excessive slippage in higher gears, and cause dangerously high transmission temperatures.

To resolve complaints of bothersome converter clutch operation without sacrificing transmission durability, the Component Research Corporation of McLeansboro, Illinois developed the V-SAC system. This unit overrides the stock torque converter clutch controls, and regulates the converter clutch based solely on vehicle speed. The V-SAC package consists of a speed sensor that installs in the speedometer cable, a computer unit that fits under the dash, and all necessary wiring. A knob on the computer unit allows the driver to set the speed at which lockup occurs; 42 mph or 68 kph is suggested for average use.

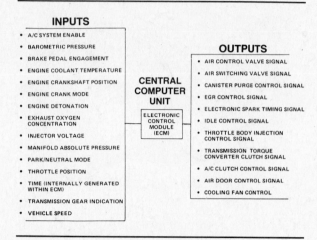

INPUTS

- A/C SYSTEM ENABLE
- BAROMETRIC PRESSURE
- BRAKE PEDAL ENGAGEMENT
- ENGINE COOLANT TEMPERATURE
- ENGINE CRANKSHAFT POSITION
- ENGINE CRANK MODE
- ENGINE DETONATION
- EXHAUST OXYGEN CONCENTRATION
- INJECTOR VOLTAGE
- MANIFOLD ABSOLUTE PRESSURE
- PARK/NEUTRAL MODE
- THROTTLE POSITION
- TIME (INTERNALLY GENERATED WITHIN ECM)
- TRANSMISSION GEAR INDICATION
- VEHICLE SPEED

CENTRAL COMPUTER UNIT

ELECTRONIC CONTROL MODULE (ECM)

OUTPUTS

- AIR CONTROL VALVE SIGNAL
- AIR SWITCHING VALVE SIGNAL
- CANISTER PURGE CONTROL SIGNAL
- EGR CONTROL SIGNAL
- ELECTRONIC SPARK TIMING SIGNAL
- IDLE CONTROL SIGNAL
- THROTTLE BODY INJECTION CONTROL SIGNAL
- TRANSMISSION TORQUE CONVERTER CLUTCH SIGNAL
- A/C CLUTCH CONTROL SIGNAL
- AIR DOOR CONTROL SIGNAL
- COOLING FAN CONTROL

Figure 8-19. All electronic control systems are made up of the same three groups of components.

gines. Like its predecessors, the VRV is positioned on the end of the throttle shaft on the fuel injection pump. Like the dual-contact rotary switch described above, the VRV contains two sets of contacts that are wired in series. Operation of the VRV in controlling the TCC solenoid is identical to that of the dual-contact rotary switch.

BASIC ELECTRONIC CONTROL SYSTEMS

Most current electronic torque converter clutch and transmission shift controls are integrated into larger and more comprehensive electronic engine control systems. The first electronic engine control systems were introduced in the late 1970s to eliminate mechanical and vacuum advance mechanisms from the distributor. These components were replaced by a computer that continuously varied the spark timing for optimum combustion.

Once a computer was installed in the car, engineers quickly developed other uses for it. Today, the typical electronic engine control system regulates spark timing, fuel delivery, engine idle speed, emission control devices, and many other systems and components including the torque converter clutch, and in some applications, transmission shifting.

Although each manufacturer has a different name for its electronic engine control systems, and the abilities of the systems vary from one application to another, they all function in essentially the same manner. Every system is made up of three distinct groups of components, figure 8-19:

- Input devices
- A central computer
- Output devices.

Input devices tell the central computer what is happening in several areas of vehicle operation at any given moment. Input devices may be either switches, which provide an on/off signal, or sensors, which provide a variable signal. Typical input devices supply information on ignition switching, engine rpm, manifold and barometric pressure, throttle position, coolant and intake air temperature, vehicle speed, intake air volume or mass, transmission gear position, and exhaust oxygen content.

The *central computer* analyzes information from the various input sensors and switches according to instructions contained in its program. It then provides signals to the output devices that control the operation of various engine and vehicle systems. Many central computer units contain a replaceable chip called a Programmable Read Only Memory (PROM) or calibration module. This chip contains the specialized instructions for that particular application. Revised PROMs are often made available to resolve driveability problems.

Output devices receive electrical signals from the computer once it has calculated which engine and vehicle systems must be adjusted to meet the demands of the moment. Among other things, the output devices may control the fuel pump relay, idle speed, spark timing, mixture control solenoid or fuel injectors, TCC solenoid, transmission shifting, and the system self-test output.

The following sections describe the electronic torque converter clutch control systems used by the major domestic manufacturers. In all cases, the central computer is the engine control system computer, and the output device being controlled is the TCC solenoid. Only the input devices will vary, although most systems use basically the same inputs to control converter clutch lockup. The descriptions do not list *all* of the sensors in every system, only those used to control the converter clutch. As you study these systems, note how the different input devices are used to perform the same jobs done by electrical switches and vacuum controls in earlier systems.

CHRYSLER ELECTRONIC TCC CONTROLS

Chrysler introduced an electronically controlled torque converter clutch with its 3-speed Torque-flite transmission in selected 1986 trucks and vans, and includes it on some 1987 and later passenger cars. In this system, figure 8-20, the TCC solenoid is controlled by the engine control computer through a converter clutch relay.

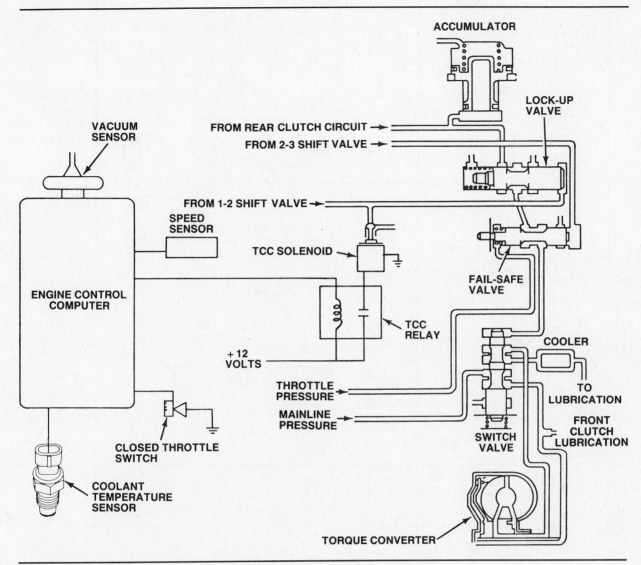

Figure 8-20. The Chrysler electronic torque converter clutch control system.

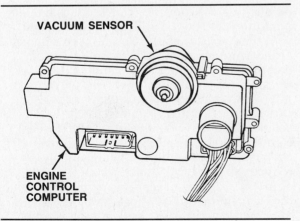

Figure 8-21. The Chrysler intake manifold vacuum sensor is located on the control computer body.

The computer determines when to lock or unlock the converter clutch based on information from four input devices:

- Coolant sensor
- Vacuum sensor
- Speed sensor
- Closed throttle switch.

The coolant sensor is located in the engine block and provides the computer with a variable signal based on engine temperature. The vacuum sensor, figure 8-21, is located on the computer body and is connected to an intake manifold vacuum source through a length of hose. The vacuum sensor provides the computer with a variable signal based on intake manifold vacuum (engine load). The speed

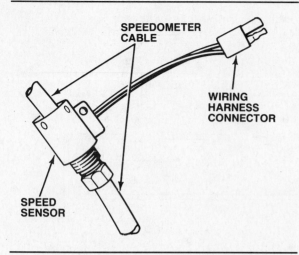

Figure 8-22. The Chrysler vehicle speed sensor is mounted in the speedometer cable.

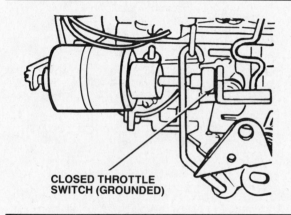

Figure 8-23. The Chrysler closed throttle switch is located on the throttle linkage.

sensor, figure 8-22, is located in the speedometer cable and provides the computer with a variable signal based on vehicle speed. The closed throttle switch, figure 8-23, is located on the throttle linkage and sends a ground signal to the computer when the throttle butterfly is closed.

The TCC solenoid, figure 8-24, mounts on the valve body. A threaded electrical connector installed through the transmission case provides the wiring connection to the solenoid. When the solenoid is not energized, it vents the pressure used to apply the converter clutch, thus preventing the clutch from locking. When the solenoid is energized, it closes the vent and directs mainline pressure to the converter clutch lockup valve which causes the converter clutch to lockup.

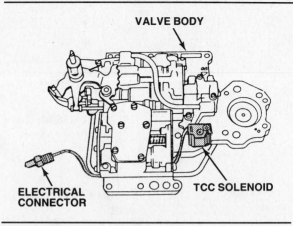

Figure 8-24. The Chrysler TCC solenoid is installed on the valve body.

The next section describes how the electronic controls work together to switch the TCC solenoid, and thus lock and unlock the converter clutch. The hydraulic valves and circuits that actually apply and release the clutch piston are described in Chapter 9 which covers Torqueflite transmissions and transaxles.

System Operation

Lockup of the Chrysler electronically controlled TCC will occur only when:
- Engine coolant temperature is above 150°F (66°C)
- The throttle is open
- Vehicle speed is above 40 mph (64 kph)
- Intake manifold vacuum is between approximately 8 and 20 in-Hg.

As long as the coolant temperature sensor indicates that the engine is below 150°F (66°C), the computer will not energize the TCC solenoid to lock the converter clutch. Once the temperature rises above 150°F (66°C), the computer uses the signals from the closed throttle switch, speed sensor, and vacuum sensor to decide whether the converter clutch should be locked or unlocked.

Whenever the closed throttle switch is closed, the computer de-energizes the TCC solenoid and unlocks the clutch. The closed throttle switch is used to unlock the converter clutch during closed throttle deceleration to reduce exhaust emissions and increase fuel economy. To prevent repeated engagement and disengagement during short-duration throttle openings and closings, the computer is programmed to delay converter clutch application for a certain period of time after the closed throttle switch opens.

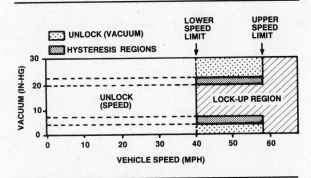

Figure 8-25. Once coolant temperature reaches 150°F (66°C), the engine control computer locks and unlocks the Chrysler converter clutch based primarily on manifold vacuum and vehicle speed.

Vehicle speed and intake manifold vacuum are the primary factors in determining converter clutch operation, figure 8-25. If the vehicle speed is below the lower limit, the computer will not energize the TCC solenoid. If the road speed is above the lower limit but below the upper mph limit, the computer will use the signal from the vacuum sensor to determine whether the converter clutch should be locked or unlocked. Once vehicle speed exceeds the upper limit, the computer will energize the TCC solenoid and lock the clutch regardless of the vacuum sensor signal.

When vehicle speed is between the upper and lower limits, the computer's decision to lock or unlock the converter clutch is determined by a set of vacuum versus speed curves that are programmed into memory. For the computer to lock the TCC, intake manifold vacuum must be within a relatively narrow range of values — about 8 to 20 in-Hg as shown in figure 8-25. Limiting lockup to within this range smoothes converter clutch engagement by ensuring that the vehicle is at a steady cruise, not accelerating hard or under heavy load, or decelerating under light throttle. To help prevent momentary converter clutch engagement during transient throttle openings, lockup is delayed for a programmed period of time after the vacuum level enters the lockup range.

Once the converter clutch is locked, manifold vacuum must fall outside of a different, somewhat wider, range of values before the computer will unlock the converter clutch — about 4 to 23 in-Hg as shown in figure 8-25. Allowing this wider range of values improves driveability by preventing the TCC from repeatedly unlocking and relocking as the manifold vacuum level moves outside of the narrow lockup limits during small changes in engine

load. The areas between the locking and unlocking curves are called **hysteresis bands**.

FORD ELECTRONIC TORQUE CONVERTER CLUTCH CONTROLS

Ford uses an electronically controlled torque converter clutch on both its AXOD transaxle and the A4LD transmission. Because the systems do not work in the same manner, they are explained separately below.

Ford AXOD Lockup Converter

In the 4-speed AXOD transaxle, the converter clutch may be locked in either third or fourth gear. Operation of the TCC solenoid, called the "converter clutch bypass solenoid" by Ford, is regulated by a fourth-generation Electronic Engine Control (EEC-IV) system based on signals from a number of input devices:

- Engine coolant sensor
- Throttle position sensor
- Vehicle speed sensor
- 3-2 pressure switch
- 4-3 pressure switch
- Barometric pressure sensor
- Brake switch.

The coolant sensor is located in the engine block, and provides the computer with a variable signal based on engine temperature. The throttle position sensor is located on the throttle body and provides the computer with a variable signal based on the percentage of throttle opening. The speed sensor provides the computer with a variable signal based on vehicle road speed. The 3-2 and 4-3 pressure switches are located on the transmission valve body, figure 8-26, and provide the computer with on/off signals that indicate specific shift conditions. The barometric pressure sensor is located in the engine compartment and provides the computer with a variable signal based on atmospheric pressure. The brake switch is located on the pedal assembly and provides the computer with an on/off signal based on brake application.

Hysteresis Bands: The area between two curves on a graph that indicate when a TCC locks and when it unlocks. Hysteresis bands also exist between graph curves that indicate when a transmission upshifts and when it downshifts between two gears.

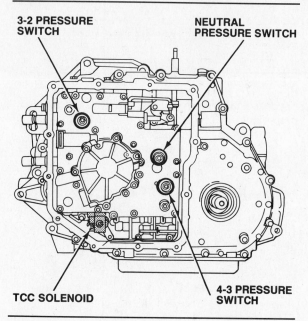

Figure 8-26. Pressure switches are among the inputs that help regulate the TCC solenoid in Ford's AXOD transaxle.

The TCC solenoid, figure 8-26, mounts on the valve body. An electrical connector installed through the transmission case provides the wiring connection to the solenoid. When the solenoid is not energized, it vents the pressure used to apply the converter clutch, thus preventing the clutch from locking. When the solenoid is energized, it closes the vent and directs mainline pressure to the spool valves that apply the converter clutch.

The next section describes how the electronic controls work together to switch the TCC solenoid, and thus lock and unlock the converter clutch. The hydraulic valves and circuits that actually apply and release the clutch piston are described in Chapter 13, which covers the Ford AXOD transaxle.

System operation

Based on sensor inputs, the AXOD converter clutch *will not* engage if any of the following conditions exist:

- Coolant temperature is below 75°F (24°C)
- The throttle is fully closed
- The throttle is suddenly opened or closed
- During 4-3 downshifts
- During 2-3 upshifts
- The brakes are applied
- Barometric pressure below 20 in-Hg.

The TCC also *may not* engage if the EEC-IV system is operating in Failure Mode Management or Limited Operation Strategy (LOS) due to the malfunction of an input device (switch or sensor) critical to the function of the EEC-IV system.

As long as the coolant temperature sensor indicates that the engine is below 75°F or 24°C the computer will not energize the TCC solenoid to lock the converter clutch. Once the temperature rises above 75°F or 24°C, the throttle position sensor and vehicle speed sensor are the two main inputs used to determine when the converter clutch locks and unlocks. The EEC-IV computer makes its decision based on two sets of throttle-opening versus vehicle-speed curves, figure 8-27.

In third gear, figure 8-27A, the converter clutch can apply at speeds as low as 27 mph or 43 kph if the throttle opening is small. At larger throttle openings, lockup may not occur until vehicle speed is as high as 55 mph or 89 kph. While in third gear, the TCC can be applied only when the throttle opening (from the hot idle position) is greater than 8 percent and less than 59 percent.

In fourth gear, figure 8-27B, the converter clutch can apply at speeds as low as 35 mph or 56 kph if the throttle opening is small. At larger throttle openings, lockup may not occur until vehicle speed is as high as 55 mph or 89 kph. While in fourth gear the converter clutch can be applied only when the throttle opening (from the hot idle position) is between 5 and 45 percent.

In a manner similar to the manifold-vacuum versus vehicle-speed curves used in Chrysler's converter clutch system, the Ford throttle-opening versus vehicle-speed curves create hysteresis bands between when the converter clutch locks and when it unlocks. Smaller throttle openings and higher vehicle speeds are required before the converter clutch will lock. However, once the clutch has locked, larger throttle openings and lower vehicle speeds are possible before the clutch will unlock. Allowing these wider ranges of values improves driveability by preventing repeated unlocking and relocking of the converter clutch as throttle position and vehicle speed move outside the lockup limits during minor changes in vehicle operation.

Whenever the throttle position sensor indicates the throttle is fully closed, the computer de-energizes the TCC solenoid. This unlocks the converter clutch during closed-throttle deceleration to reduce exhaust emissions and increase fuel economy. To prevent repeated converter clutch engagement and disengagement during short-duration throttle openings and closings, the computer is programmed to delay

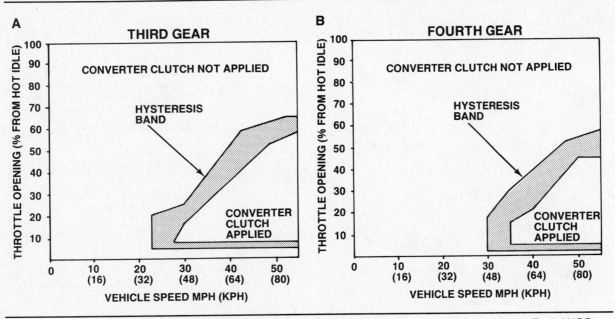

Figure 8-27. Once the coolant temperature reaches 75°F (24°C), the EEC-IV computer controls the Ford AXOD converter clutch based primarily on throttle opening and vehicle speed.

any change in the converter clutch status for a certain period of time after a significant change in throttle position.

During both upshifts and downshifts, 3-2 and 4-3 pressure switches signal the computer to de-energize the TCC solenoid and unlock the converter clutch. This smoothes the shifts by allowing normal slippage within the converter. The computer will also de-energize the TCC solenoid whenever the signal from the brake switch indicates that the brakes are applied. This prevents killing the engine if the brakes should lock the wheels.

Finally, converter clutch lockup is prevented whenever the barometric pressure sensor signal indicates that the atmospheric pressure is less than 20 in-Hg. This prevents converter clutch lockup at higher altitudes where engine power is reduced and torque multiplication in the converter is desirable to help maintain performance.

Ford A4LD Lockup Converter

Operation of the A4LD lockup converter is also regulated by the EEC-IV system. However, the TCC solenoid in this design, called an "override solenoid" by Ford, operates differently than any of the others discussed in this chapter. In the A4LD, the converter clutch is engaged solely by hydraulic pressures — much like the early Chrysler system. The TCC solenoid is used to inhibit, or override, the hydraulic programming when it is desirable to unlock the converter clutch.

Some of the EEC-IV input devices that provide signals to regulate the TCC solenoid in the A4LD are the same as those used in the AXOD lockup system, but there are differences. Switches and sensors used in the A4LD application include the:

- Engine coolant sensor
- Throttle position sensor
- Ignition pickup
- Manifold absolute pressure sensor
- Brake switch.

The coolant sensor is located in the engine block, and provides the computer with a variable signal based on engine temperature. The throttle position sensor is located on the throttle body and provides the computer with a variable signal based on the percentage of throttle opening. The ignition pickup is located in the distributor, and provides the computer with a variable signal based on engine rpm. The manifold absolute pressure (MAP) sensor, located in the engine compartment and connected to intake manifold vacuum, provides the computer with a variable signal based on intake manifold vacuum (engine load). The brake switch is located on the pedal assembly and provides the computer with an on/off signal based on brake application.

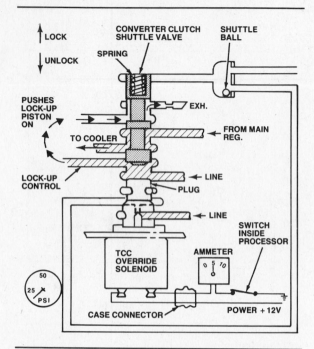

Figure 8-28. When the TCC solenoid in the Ford A4LD transmission is energized, mainline pressure is blocked from acting on the plug in the converter clutch shuttle valve.

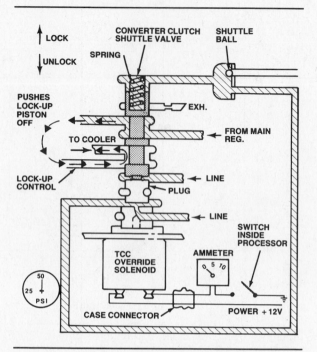

Figure 8-29. When the TCC solenoid in the Ford A4LD transmission is deenergized, mainline pressure moves the plug and hydraulic pressure shifts the shuttle valve to unlock the converter clutch.

The TCC solenoid in the A4LD works in an opposite manner from those we have studied thus far. The typical TCC solenoid is normally *de-energized* to vent fluid flow and prevent clutch lockup; when lockup is desired, the solenoid is energized to direct fluid flow to the appropriate valves. In the A4LD system, the TCC solenoid is normally *energized* to block fluid flow that is used to unlock the converter clutch, figure 8-28. This allows the hydraulic valving to lock the converter whenever operating conditions are right. When the A4LD TCC solenoid is de-energized, figure 8-29, hydraulic pressure flows to the end of the converter clutch shuttle valve to release the converter clutch.

The hydraulic valves and circuits that apply the converter clutch are described in Chapter 10, which covers the A4LD transmission in detail. The next section describes how the electronic controls work together to switch the TCC solenoid and unlock the converter clutch.

System operation

In the A4LD system, The EEC-IV computer de-energizes the TCC solenoid to prevent lockup, or unlock the converter clutch, whenever:

- Coolant temperature is below 128°F (53°C) or above 240°F (116°C)
- The throttle is fully closed
- The accelerator is at heavy or wide-open throttle

- The throttle is suddenly opened or closed
- Engine speed is below a programmed value at lower intake manifold vacuum levels
- The brakes are applied.

If the coolant temperature sensor indicates that the engine is below 128°F or 53°C, the computer will not energize the TCC solenoid. This prevents converter clutch lockup before the engine is warm enough to smoothly handle the added load. When coolant temperature is between 128°F and 240°F (53°C and 116°C), the computer energizes the TCC solenoid, providing none of the other conditions described below are present. Should the coolant temperature rise above 240°F or 116°C, the computer will again de-energize the TCC solenoid to reduce driveline loads and help cool the engine.

Once the coolant temperature is within allowable limits, the computer will monitor the throttle position sensor. Whenever the throttle position sensor indicates the throttle is fully closed, the computer de-energizes the TCC solenoid. This unlocks the converter clutch during closed-throttle deceleration to reduce exhaust emissions and increase fuel economy.

Whenever the throttle position sensor indicates the throttle is heavily applied or wide open, the computer de-energizes the TCC solenoid to allow torque multiplication in the converter for faster acceleration. To prevent re-

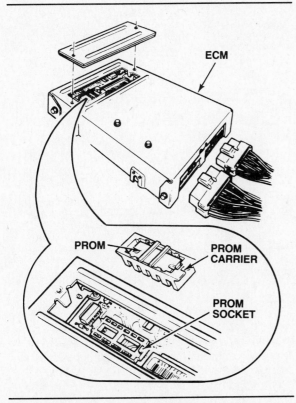

Figure 8-30. The PROM in the ECM contains instructions on when to energize and de-energize the TCC solenoid.

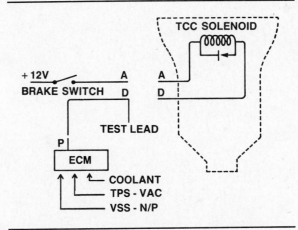

Figure 8-31. This is the most basic version of GM's electronic torque converter clutch control system.

peated converter clutch engagement and disengagement during short-duration throttle openings and closings, the computer is programmed to delay any change in the converter clutch status for a certain period of time after a significant change in throttle position.

Using the respective sensors, the computer also monitors the balance between engine rpm and intake manifold vacuum. When rpm drops below a certain value at lower vacuum levels (based on a curve programmed into the computer), the computer de-energizes the TCC solenoid. This ensures that 4-3 downshifts occur with the converter unlocked for maximum smoothness.

Finally, the computer will also de-energize the TCC solenoid whenever the signal from the brake switch indicates that the brakes are applied. This prevents killing the engine if the brakes should lock the wheels.

GM ELECTRONIC TORQUE CONVERTER CLUTCH CONTROLS

General Motors torque converter clutch systems have gone through two stages of development. The first generation controls were introduced in the 1980 model year, and used electrical and vacuum switches to regulate the TCC solenoid. We looked at these systems at the beginning of this chapter. The second generation controls were introduced in the 1981 model year, and regulate the TCC solenoid using the Electronic Control Module (ECM), figure 8-30, of the Computer Command Control (CCC) system. The ECM controls the converter clutch on most 1982 and later gasoline-powered GM cars and trucks. All diesel-powered vehicles continue to use variations of the earlier electrical and vacuum controls.

GM Torque Converter Clutch Control Sensors

When the converter clutch is regulated by the CCC system, figure 8-31, the TCC solenoid ground is provided through the ECM. In essence, the ECM ground replaces the governor switch of earlier systems. In all GM electronic converter clutch control systems, the ECM decides when to ground the TCC solenoid based on the input signals from four sensors:

- Coolant temperature sensor
- Throttle position sensor
- Vacuum or manifold absolute pressure sensor
- Vehicle speed sensor.

Coolant temperature sensor

The coolant temperature sensor (CTS), figure 8-32, is a variable resistor whose resistance changes with temperature. The sensor threads into the engine cooling system so that the end of the unit is immersed in coolant. Although the CTS may be located on the engine in a number of positions, it is usually installed near the thermostat housing.

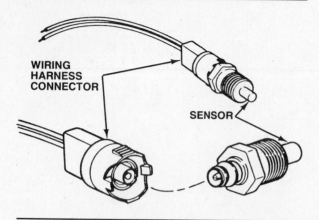

Figure 8-32. The GM coolant temperature sensor and its wiring harness connector.

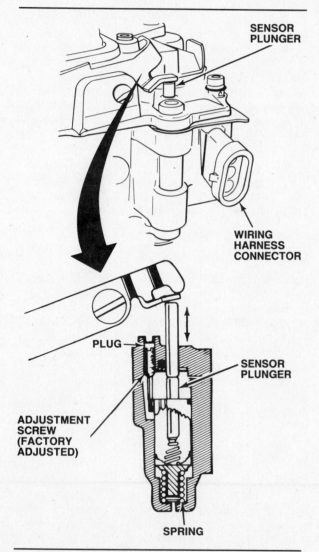

Figure 8-33. The TPS is located inside the carburetor on some GM vehicles.

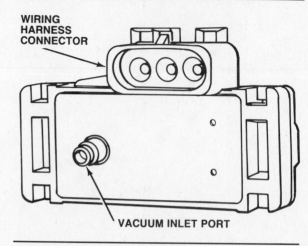

Figure 8-34. Vacuum and MAP sensors measure changes in intake manifold pressure (vacuum).

During engine operation, the ECM delivers a five-volt **reference voltage** to the CTS. When the coolant is cold, the sensor has a high resistance; when the coolant is at normal operating temperature, sensor resistance is low. The resistance of the sensor reduces the reference voltage, and the lower voltage is then returned to the ECM. The difference between the reference voltage and the return voltage represents the coolant temperature to the ECM.

Throttle position sensor

The throttle position sensor (TPS) is a variable resistor whose resistance changes with the degree of throttle opening. The TPS is located in the carburetor on carbureted engines, figure 8-33, or on the throttle body on fuel-injected engines.

During engine operation, the ECM delivers a five-volt reference voltage to the TPS. When the throttle is closed, the sensor has a high resistance; when the throttle is wide open, sensor resistance is low. The resistance of the sensor reduces the reference voltage, and the lower voltage is then returned to the ECM. The difference between the reference voltage and the return voltage represents the throttle position to the ECM.

Vacuum or manifold absolute pressure sensor

The vacuum sensor (VS) or manifold absolute pressure (MAP) sensor, figure 8-34, is a variable resistor whose resistance changes in response to changes in pressure. The sensor location varies with the application, but these units are often mounted on one of the front inner fender housings. A hose connects the VS or MAP sensor to an intake manifold vacuum source.

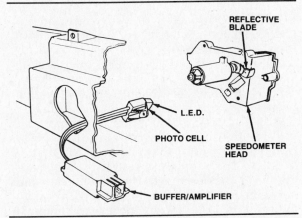

Figure 8-35. The VSS on most GM cars attaches to the back of the speedometer head.

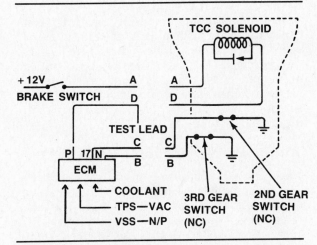

Figure 8-36. This GM electronic torque converter clutch control system uses second and third gear switches as inputs to the ECM.

Vacuum sensors, used in most early systems, measure the manifold vacuum level relative to ambient atmospheric pressure. MAP sensors, used on most newer systems, measure the manifold vacuum level relative to a fixed, known pressure. Although the two parts are not interchangeable, they work in the same manner.

During engine operation, the ECM delivers a five-volt reference voltage to the VS or MAP sensor. When manifold vacuum is high (pressure is low), the sensor has a high resistance; when manifold vacuum is low (pressure is high), sensor resistance is low. The resistance of the sensor reduces the reference voltage, and the lower voltage is then returned to the ECM. The difference between the reference voltage and the return voltage represents the manifold vacuum to the ECM.

Vehicle speed sensor
The vehicle speed sensor (VSS), figure 8-35, generates a signal whose frequency increases with vehicle speed. The VSS is located on the speedometer head behind the instrument panel, and consists of a light emitting diode (LED), a photo cell, and a buffer/amplifier. The VSS also uses a reflective blade that is built into the speedometer head assembly.

When the vehicle is in motion, the speedometer cable and the reflective blade attached to it rotate. As the blade enters the light beam provided by the LED, light is reflected back at the photo cell. The photo cell "sees" the light and sends a low power electrical impulse to the buffer/amplifier. The buffer/amplifier conditions and strengthens this signal, then sends it to the ECM. The number of electrical impulses generated in a fixed period of time (the signal frequency) indicates road speed to the ECM.

GM Torque Converter Clutch Control Switches

In addition to the four sensors just described, one or more switches may be used to help control the TCC solenoid or related systems, figure 8-36. Keep in mind that not all these switches are used in every system, and a particular switch may be used for different purposes in different systems:

- Brake switch
- Second gear switch
- Third gear switch
- Fourth gear switch
- 4-3 pulse switch.

A brake switch is used in all GM electronically controlled converter clutch systems, and it operates in exactly the same manner as the brake switch in the first generation GM converter clutch controls. The switch is located under the dash on the pedal assembly, and opens the power supply circuit to the TCC solenoid when the brakes are applied.

The gear switches also operate in basically the same manner as those in the early GM converter clutch systems; they install in hydraulic circuits, and open or close when pressure is applied or released in those circuits. However, the

Reference Voltage: The basic operating voltage in an electronic engine control system. The reference voltage is often modified by sensors to provide the computer with information on engine and vehicle operation.

switches in GM electronic converter clutch control systems are installed and used in several different ways. Not all the switches are located on the valve body; some thread into the outside of the transmission case. In addition, while many switches are wired directly into the TCC solenoid ground circuit, others are used as input devices to the ECM.

Torque Converter Clutch Relay

A converter clutch relay is used to control the TCC solenoid in certain GM applications that use an EGR bleed solenoid. The relay isolates the ECM from potentially damaging power surges. It also allows a single ECM circuit to control both solenoids without overloading the switching transistor. The converter clutch relay is usually located in the engine compartment near the EGR bleed solenoid.

GM ELECTRONIC TORQUE CONVERTER CLUTCH SYSTEM OPERATION

As with early GM electrical TCC solenoid controls, there are too many variations of the electronic systems to cover them all in this text. The section below describes how the ECM uses the four sensor inputs to control converter clutch operation. The section following describes several different system configurations. Consult the appropriate shop manual for a list of the components used in a specific application.

GM Torque Converter Clutch Control Strategy

The ECM will provide a ground for the TCC solenoid whenever four conditions are met:
- Coolant temperature is above a predetermined value
- The throttle is open
- The engine is not under heavy load
- Vehicle speed is above a predetermined mph or kph.

As long as the coolant sensor indicates that the engine is below a predetermined temperature, usually around 150°F or 66°C, the ECM will not energize the TCC solenoid to lock the converter clutch. This prevents TCC engagement before the engine is warm enough to handle the extra load. Once the coolant temperature rises above the predetermined value, the ECM uses signals from the other sensors to decide when to lock and unlock the converter clutch. To prevent driveability problems, the

ECM is programmed to prevent converter clutch lockup for a fixed period of time following a hot engine restart.

Chrysler and Ford electronic converter clutch systems use control curves based primarily on *two* sensors to regulate lockup once the engine has warmed to a minimum temperature. The GM system programming is more complex, and uses multiple curves that integrate the signals from *three* sensors (TPS, vacuum or MAP, and VSS) into an overall control strategy. The paragraphs below describe the general relationships between the sensors, and how the ECM uses their inputs. Once the engine is warm enough, three sets of conditions must exist before the ECM will energize the TCC solenoid.

First, the TPS must indicate that the throttle is open more than 2 percent, but less than 80 percent. Within this range, converter clutch lockup occurs earlier at smaller throttle openings, and later at larger throttle openings. When the throttle opening is 2 percent or less, the ECM unlocks the converter clutch to reduce exhaust emissions and improve fuel economy during coastdown. When the throttle opening is 80 percent or more, the ECM unlocks the converter clutch to allow torque multiplication in the converter for faster acceleration.

Second, the vacuum or MAP sensor must indicate a minimum level of intake manifold vacuum, usually between 9 and 14 in-Hg depending on throttle position and vehicle speed. Within this range, converter clutch lockup occurs earlier at higher vacuum levels, and later at lower vacuum levels. This smoothes converter clutch engagement by ensuring that the vehicle is not accelerating, decelerating, or under heavy load when lockup occurs. The ECM also uses the signal from the vacuum or MAP sensor, along with that from the TPS, to determine when to unlock the converter clutch during deceleration (high vacuum) or heavy load (low vacuum) conditions.

Third, the VSS must indicate that the vehicle speed is above a certain level. The exact speed varies with throttle position and intake manifold vacuum, but usually falls between 27 and 51 mph (43 and 82 kph). Once again, this smoothes converter clutch engagement by ensuring that impeller and turbine speeds in the torque converter are approximately matched.

The various curves used by the ECM in deciding when to lock and unlock the converter clutch create hysteresis bands much like those in the Chrysler and Ford systems. Basically, the ranges of throttle opening, intake manifold vacuum, and vehicle speed that are required before the converter clutch will lock are more

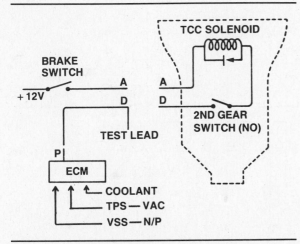

Figure 8-37. Gear switches may be used to limit converter clutch operation to higher gears only.

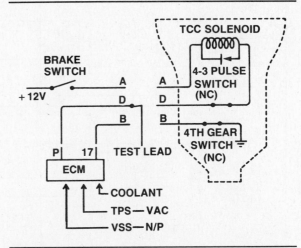

Figure 8-38. In this system, a pulse switch smoothes 4-3 downshifts, and a fourth gear switch alters the control curve used by the ECM to regulate converter clutch operation.

limited than those that are necessary for the converter clutch to unlock. Allowing wider ranges of these values before the converter clutch will unlock prevents repeated converter clutch unlocking and relocking as the signals from the TPS, vacuum or MAP sensor, and VSS move outside the lockup limits during minor changes in vehicle operation.

GM Torque Converter Clutch Switch Operation

As already stated, switches are used in a variety of ways in GM electronic torque converter clutch control systems. In the 3-speed transmission shown in figure 8-36, the second and third gear switches are used as inputs for the ECM. The signals from these switches are used to alter the converter lockup points to compensate for the different powertrain loads in each gear.

Figure 8-37 shows a 3-speed transmission in which a second gear switch is used to help complete the ground circuit to the ECM. The switch prevents converter clutch lockup unless the transmission is in second or third gear. In other applications, a third gear switch is used in a similar manner to allow lockup in high gear only.

In figure 8-38, a 4-3 pulse switch is used to open the converter clutch ground circuit and momentarily unlock the converter clutch during downshifts. This smoothes the shift by allowing normal slippage within the converter. In some 3-speed transmissions, a 3-2 switch is used for the same purpose. The fourth (high) gear switch in the system shown in figure 8-38

signals the ECM to allow a wider range of TPS movement before unlocking the converter clutch.

The system shown in figure 8-39 uses two gear switches that operate in exactly the same manner as those in figure 8-38. The difference is that the system in figure 8-39 uses a torque converter clutch relay to operate the TCC solenoid and the EGR bleed solenoid.

ELECTRONIC TRANSMISSION SHIFT CONTROLS

In addition to torque converter clutch control, electronics are now being used to regulate automatic transmission shifting, figure 8-40. One reason is the increasingly complex valve bodies required to control modern 4-speed transmissions with lockup torque converters. The valve body castings are expensive to manufacture, and the many valves, springs, check balls, and orifices they contain are prone to wear, clogging, and sticking. Computerized shift controls replace many of these parts with solenoid ball valves.

Another reason for the introduction of electronic shift controls is the greater programming flexibility they provide. As we saw with converter clutch controls, traditional mechanical and hydraulic mechanisms are limited in the number of factors they can consider when making control decisions. Electronic regulation allows many additional operating conditions to be taken into account.

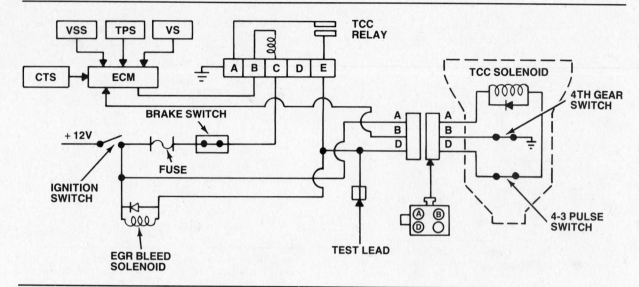

Figure 8-39. When an EGR bleed solenoid is used along with a TCC solenoid, the ECM grounds a converter clutch relay to provide the ground for both solenoids.

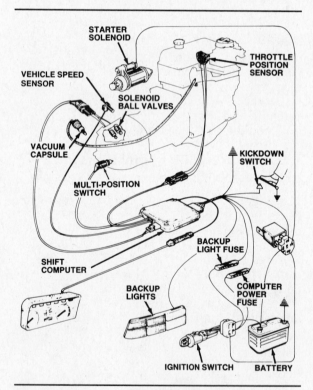

Figure 8-40. The components in this Renault Alliance electronic shift control system are typical of those in other such systems.

The operation of an electronically shifted transmission is similar to that of an electronically controlled converter clutch. First, several input devices send signals to the shift computer, which may be either a dedicated unit or the engine control computer. The computer then analyzes the input signals by comparing them to shift instructions programmed into its memory. Finally, the computer switches the appropriate solenoid ball valves on or off as needed to provide the proper gear.

Electronic Shift Control Inputs

Many of the same inputs used to control the torque converter clutch are also used in electronic shift control systems. The most common signals sent to the computer report on:

- Engine temperature
- Engine rpm
- Vehicle speed
- Throttle position
- Gear selector position.

These are only the basic control inputs. Many systems use other sensors and switches to provide additional control functions.

The engine temperature signal comes from a coolant temperature sensor located in the cooling system. When the engine is cold, the computer generally upshifts the transmission later, and downshifts earlier, to provide better driveability and faster warmup. If the transmission is equipped with a torque converter clutch, the computer will also prevent clutch lockup when the engine is cold.

The engine rpm signal is usually taken from the ignition pickup located in the distributor, or on the flywheel or damper pulley. The computer uses the engine rpm signal to help time the upshifts and downshifts.

The vehicle speed signal comes from a speed sensor located in the speedometer head or on the transmission. The speed sensor signal is the electronic equivalent of governor pressure in a hydraulically controlled transmission. The computer uses the vehicle speed signal to help time the upshifts and downshifts.

The throttle position signal comes from a throttle position sensor mounted on the carburetor, or on the throttle body of a fuel injected engine. The throttle position sensor signal is the electronic equivalent of throttle pressure in a hydraulically controlled transmission. The computer uses the throttle position signal as an indicator of engine load to help time the upshifts and downshifts.

The gear selector position signal is provided by a multi-position switch located on the shift mechanism or transmission. The gear selector switch signal is based on the position of the manual valve in the transmission, and informs the computer which gear range the driver has chosen. The gear selector switch may also be used to operate the backup lights when the vehicle is in reverse, and in some applications it serves as the neutral safety switch to prevent the engine from starting when the transmission is in gear.

Solenoid Ball Valve Usage

Once the computer has analyzed all of the input signals, it sends output signals to the solenoid ball valves to engage the proper gear for the existing driving conditions. The solenoid ball valves used for shifting a transmission operate in much the same manner as a TCC solenoid. When voltage is applied to the solenoid windings, the ball seats and blocks fluid flow. When no voltage is applied, the ball unseats, and fluid is free to flow through the valve.

Most electronically shifted transmissions are designed and built under the same basic patents, and have two solenoid ball valves that control shift action. However, there are a number of different ways the valves can be used to change the gear ratios. For example, in some Renault 3-speed transmissions, one solenoid is energized for first gear, both are energized for second, and neither is energized for third. In other Renault transmissions both solenoids are energized for first gear, one is energized for second, and once again, neither is energized for third.

Electronically shifted 4-speed transmissions also vary in how the solenoids are used. Certain Toyota transmissions energize one solenoid for first, the other for third, both solenoids for second, and neither for high. Mitsubishi

transmissions energize both solenoids for first, neither for third, one solenoid for second, and the other for high.

Simple Electronic Shift Control System

Figure 8-41 shows the hydraulic components of a greatly simplified electronic shift control system from an imaginary 4-speed automatic transmission; actual electronic shift controls are far more complex. We will use this simple design to examine how solenoid ball valves can be used to shift the forward gears.

In our imaginary transmission, the pressure used to apply the clutches is routed from the manual valve through several passages to the shift control valve. The left end of the shift control valve is acted on by a spring, the right end of the valve is acted on by mainline pressure in a circuit that includes the two solenoid ball valves.

The mainline pressure that acts on the shift control valve must pass through a restricting orifice. The orifice ensures that all of the pressure acting on the valve can be exhausted when both solenoid ball valves are open. When just one of the solenoid ball valves is open, only part of the pressure is vented — how much depends on which valve is open. In this example, solenoid ball valve #1 has a smaller bleed opening, and thus releases less pressure, than solenoid ball valve #2.

In first gear, figure 8-41A, the computer opens solenoid #1 to exhaust a small portion of the restricted mainline pressure acting on the right end of the shift control valve. This allows the spring acting on the left end of the valve to move the valve a specific distance to the right. A valley in the valve then opens the passage that allows hydraulic pressure to pass through and apply the forward clutch for first gear.

In second gear, figure 8-41B, the computer opens solenoid #2 to exhaust a greater portion of the restricted mainline pressure acting on the right end of the shift control valve. This allows the spring acting on the left end of the valve to move the valve a further specific distance to the right. The shape of the valve allows the forward clutch to remain applied, and another valley in the valve opens the passage that allows hydraulic pressure to pass through and apply the intermediate clutch. Together, the two clutches provide second gear.

In third gear, figure 8-41C, the computer opens both solenoids to exhaust all of the restricted mainline pressure acting on the right end of the shift control valve. This allows the spring acting on the left end of the valve to

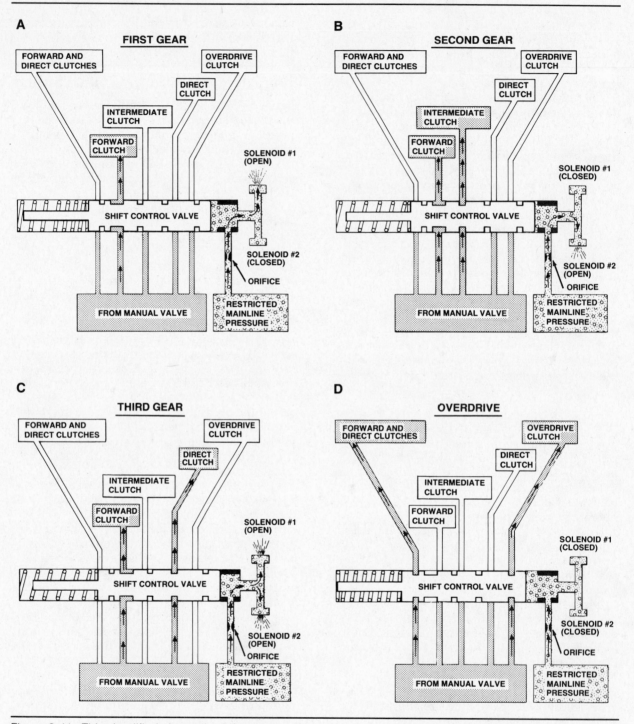

Figure 8-41. This simplified electronic shift control system shows how a spool valve and two solenoid ball valves can be used to control a 4-speed automatic transmission.

move the valve fully to the right. The shape of the valve again allows the forward clutch to remain applied, and another valley in the valve opens the passage that allows hydraulic pressure to pass through and apply the direct clutch. Together, the two clutches provide third gear.

In fourth gear, figure 8-41D, the computer closes both solenoids, which allows full mainline pressure to act on the right end of the shift control valve. Mainline pressure overcomes the spring acting on the left end of the valve, and moves the valve fully to the left. Two valleys

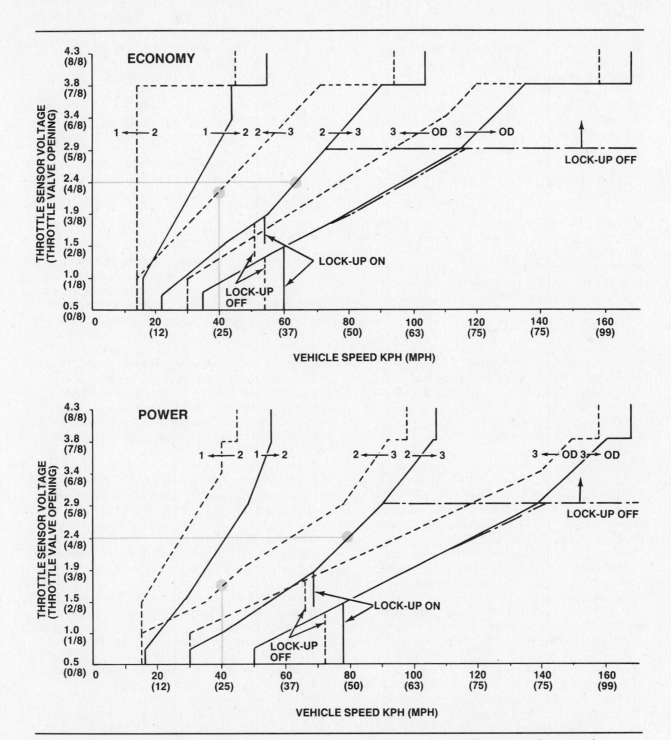

Figure 8-42. These charts show the throttle-opening versus vehicle-speed curves that are used to control transmission shifting in two different shift programs.

in the valve then open passages that allows hydraulic pressure to pass through and apply the forward, direct, and overdrive clutches. Together, the three clutches provide fourth gear.

SHIFT CONTROL OPTIONS

An added advantage of using a computer to regulate shifting is that it allows several shift control options not possible or practical with hydraulic shift controls. These options generally give the driver more control over transmission operation, or offer improvements in vehicle driveability.

One electronic shift control system offers a "hold" feature that allows the driver to lock the transmission in the selected gear by pressing a button on the shift lever. This gives the driver full manual shift control over the transmission, and it permits the use of a higher gear for added traction when starting off on slippery surfaces.

To reduce creep, some electronic shift control systems place the transmission in second gear when the vehicle is stopped in gear with the engine idling. As soon as the throttle position sensor indicates the driver has depressed the accelerator, the transmission is immediately shifted back into low gear.

Selectable Shift Programs

One of the more common options on vehicles with electronic shift controls is a choice between two or more **shift programs** that the driver selects using a switch on the instrument panel or console. The different shift programs alter the shifting behavior of the transmission to achieve specific results.

Figure 8-42 shows the shift curves for a vehicle that offers a choice between "economy" and "power" shift programs. In the "economy" program designed to conserve fuel, the transmission upshifts at lower vehicle speeds, and downshifts only under larger throttle openings (heavier engine loads). In the "power" program designed for greater performance, the transmission upshifts at higher vehicle speeds, and downshifts in response to smaller throttle openings (lighter engine loads).

Some comparisons of the charts will show how the different programs alter transmission shifting. Suppose the vehicle is accelerated from a stop at half (4/8) throttle. In the "economy" program, the 2-3 upshift occurs at approximately 40 mph (64 kph), but in the "power" program the transmission remains in second gear until over 50 mph or 80 kph for faster acceleration.

Shift Program: The set of instructions that the computer in an electronic shift control system uses to decide when to upshift and downshift the transmission. Many systems offer a choice of two or more shift programs that provide different performance characteristics.

Now suppose that the vehicle is driven at a steady 25 mph or 40 kph, and the throttle is depressed to accelerate around a slow moving vehicle. In the "economy" program, the throttle must be opened almost half (4/8) way before a 3-2 downshift takes place, and a 2-1 downshift does not occur until the throttle is open almost all (7/8) the way. In the "power" program, however, the 3-2 downshift takes place at just over one quarter (2/8) throttle, and a 2-1 downshift takes place at three quarters (6/8) throttle.

SUMMARY

Lockup torque converters were the first transmission components to be electrically and electronically controlled. These controls were adopted because vehicle systems outside the transmission needed to be considered when deciding to lock or unlock the converter clutch.

In general, a torque converter clutch should not lock until: the engine is warm enough to handle the extra load, the vehicle speed is high enough to allow smooth power transmission, and the engine and vehicle speeds are matched so that lockup will not create a surge or shudder. A torque converter clutch should unlock: during braking, during downshifts, whenever torque multiplication is needed in the converter, when leaving it engaged would increase exhaust emissions or lower fuel economy, and when the vehicle comes to a stop.

All electrically and electronically controlled torque converter clutches are locked and unlocked using a TCC solenoid. General Motors introduced the first electrical converter clutch controls which used various mechanical and vacuum switches to operate the TCC solenoid. All late model converter clutch systems have electronic controls.

Most electronic torque converter clutch and transmission shift controls are integrated into overall electronic engine control systems that regulate many vehicle systems and components. All electronic control systems operate in basically the same way, and they are made up of three groups of components: input devices, a central computer, and output devices.

Input devices are switches or sensors that tell the computer what is happening in several areas of vehicle operation. The central computer analyzes information from the input switches and sensors, according to instructions contained in its program, and provides signals to the output devices. The output devices convert the electrical signals from the computer into adjustments of engine, transmission, and other vehicle systems to meet the demands of the moment.

Electronic shift controls increase shift programming flexibility and reduce valve body complexity. They also allow such shift control options as selectable shift programs. In an electronically shifted transmission, spool valves are replaced with solenoid ball valves that are used in various combinations to control the hydraulic pressures that effect gear changes.

The operation of an electronically shifted transmission is similar to that of an electronically controlled torque converter clutch. Several input devices send signals to a shift computer. The computer analyzes the input signals by comparing them to shift curves in its memory. Then, the computer switches the solenoid ball valves on or off as needed to provide the proper gear.

Review Questions

Select the single most correct answer
Compare your answers to the correct answers on page 603

1. Mechanic A says early Chrysler lockup torque converters were controlled entirely by hydraulics. Mechanic B says electrical controls were never used in transmissions until the introduction of the torque converter clutch. Who is right?
   a. Mechanic A
   b. Mechanic B
   c. Both mechanic A and B
   d. Neither mechanic A nor B

2. Mechanic A says spool valves and a solenoid may be used to control a centrifugally locking converter clutch. Mechanic B says the converter clutch piston is electrically switched to control converter lockup. Who is right?
   a. Mechanic A
   b. Mechanic B
   c. Both mechanic A and B
   d. Neither mechanic A nor B

3. Mechanic A says a converter clutch should not lock until the engine is warm enough to handle the added load. Mechanic B says a converter clutch should unlock under certain conditions to improve acceleration and reduce exhaust emissions. Who is right?
   a. Mechanic A
   b. Mechanic B
   c. Both mechanic A and B
   d. Neither mechanic A nor B

4. Which of the following switches is *not* used to help control the TCC solenoid in early GM electrical TCC systems?
   a. Low vacuum switch
   b. Brake switch
   c. Pulse switch
   d. Governor switch

5. Mechanic A says the TCC solenoid is a type of solenoid ball valve. Mechanic B says a GM high vacuum switch operates on ported vacuum. Who is right?
   a. Mechanic A
   b. Mechanic B
   c. Both mechanic A and B
   d. Neither mechanic A nor B

6. Mechanic A says the vacuum relay valve damps pulsations in ported vacuum flow. Mechanic B says that, depending on the application, a thermal vacuum valve may be either open or closed when the engine is cold. Who is right?
   a. Mechanic A
   b. Mechanic B
   c. Both mechanic A and B
   d. Neither mechanic A nor B

7. Mechanic A says a fourth gear switch may be used to bypass the low vacuum switch in high gear. Mechanic B says when an EGR bleed solenoid is used, it is often energized at the same time as the TCC solenoid. Who is right?
   a. Mechanic A
   b. Mechanic B
   c. Both mechanic A and B
   d. Neither mechanic A nor B

8. Mechanic A says pulse switches unlock the torque converter clutch to smooth shifts between gears. Mechanic B says the EGR bleed solenoid is used to control an air bleed in a vacuum passage. Who is right?
   a. Mechanic A
   b. Mechanic B
   c. Both mechanic A and B
   d. Neither mechanic A nor B

9. Mechanic A says the low vacuum switch in early GM electrical converter clutch systems was replaced by poppet valves in the torque converter. Mechanic B says electrical switches on the shaft of the vacuum pump may be used to control the converter clutch in some GM diesel applications. Who is right?
   a. Mechanic A
   b. Mechanic B
   c. Both mechanic A and B
   d. Neither mechanic A nor B

10. In an electronic control system, the instructions on when to lock and unlock the torque converter clutch may be contained in
   a. An input device
   b. A calibration module
   c. An output device
   d. None of the above

11. Mechanic A says the Chrysler lockup torque converter was introduced on trucks and vans in 1986. Mechanic B says the Chrysler lockup torque converter is released on deceleration by a closed throttle switch. Who is right?
   a. Mechanic A
   b. Mechanic B
   c. Both mechanic A and B
   d. Neither mechanic A nor B

12. Mechanic A says the Chrysler lockup converter is controlled primarily by throttle position and intake manifold vacuum. Mechanic B says hysteresis bands separate the conditions under which the Chrysler converter clutch locks and unlocks. Who is right?
   a. Mechanic A
   b. Mechanic B
   c. Both mechanic A and B
   d. Neither mechanic A nor B

13. Mechanic A says Ford has two different electronically controlled torque converter clutch systems. Mechanic B says the converter clutch used with the Ford AXOD transaxle will not lock if the car is above a certain altitude.
    Who is right?
    a. Mechanic A
    b. Mechanic B
    c. Both mechanic A and B
    d. Neither mechanic A nor B

14. The EEC-IV computer will not energize the TCC solenoid used with the AXOD transaxle if:
    a. The throttle is closed in third or fourth gear
    b. Coolant temperature is above 75°F (24°C)
    c. The brakes are released
    d. Vehicle speed is above 55 mph (89 kph)

15. Mechanic A says the torque converter clutch used with the Ford A4LD transmission is applied entirely by hydraulic means. Mechanic B says the TCC solenoid used with the A4LD lockup converter is normally deenergized to exhaust fluid flow.
    Who is right?
    a. Mechanic A
    b. Mechanic B
    c. Both mechanic A and B
    d. Neither mechanic A nor B

16. Mechanic A says GM introduced the electronically controlled converter clutch in 1981 model year. Mechanic B says the TCC solenoid on all 1982 and later GM vehicles is controlled by the ECM.
    Who is right?
    a. Mechanic A
    b. Mechanic B
    c. Both mechanic A and B
    d. Neither mechanic A nor B

17. Mechanic A says the GM ECM uses at least five sensors to control the ground of the TCC solenoid. Mechanic B says the resistance of the GM vehicle speed sensor changes with vehicle road speed.
    Who is right?
    a. Mechanic A
    b. Mechanic B
    c. Both mechanic A and B
    d. Neither mechanic A nor B

18. Mechanic A says the reference voltage at the GM coolant temperature sensor (CTS) increases as the coolant temperature increases.
    Mechanic B says when the coolant is cold, the CTS has high resistance.
    Who is right?
    a. Mechanic A
    b. Mechanic B
    c. Both mechanic A and B
    d. Neither mechanic A nor B

19. Mechanic A says the GM throttle position sensor (TPS) is sometimes mounted inside the carburetor.
    Mechanic B says the TPS is a variable resistor that converts the degree of throttle plate opening to an electrical signal.
    Who is right?
    a. Mechanic A
    b. Mechanic B
    c. Both mechanic A and B
    d. Neither mechanic A nor B

20. Mechanic A says the manifold absolute pressure sensor supplies a voltage signal to the ECM relative to the strength of atmospheric pressure.
    Mechanic B says a vacuum sensor is connected through a hose to a ported vacuum source.
    Who is right?
    a. Mechanic A
    b. Mechanic B
    c. Both mechanic A and B
    d. Neither mechanic A nor B

21. Mechanic A says a brake switch is used in all GM electronic converter clutch control systems.
    Mechanic B says the gear switches in GM electronically controlled converter clutch systems are sometimes used as ECM inputs.
    Who is right?
    a. Mechanic A
    b. Mechanic B
    c. Both mechanic A and B
    d. Neither mechanic A nor B

22. Lockup of an electronically controlled GM torque converter clutch is based on:
    a. Vehicle speed and throttle opening
    b. Throttle opening and manifold vacuum
    c. Manifold vacuum and vehicle speed
    d. All of the above

23. Mechanic A says electronic shift controls operate in the same basic manner as electronic converter clutch controls.
    Mechanic B says electronic shift controls complicate valve body construction by adding solenoid ball valves.
    Who is right?
    a. Mechanic A
    b. Mechanic B
    c. Both mechanic A and B
    d. Neither mechanic A nor B

24. Mechanic A says the signal from the vehicle speed sensor in an electronically controlled transmission is the equivalent of governor pressure in a hydraulically controlled gearbox.
    Mechanic B says the rpm signal in an electronically controlled transmission provides the computer with an indication of engine load.
    Who is right?
    a. Mechanic A
    b. Mechanic B
    c. Both mechanic A and B
    d. Neither mechanic A nor B

25. Mechanic A says a "power" shift program will upshift the transmission at a lower vehicle speed than an "economy" shift program.
    Mechanic B says electronic shift controls use solenoid ball valves to contain and bleed off hydraulic pressure.
    Who is right?
    a. Mechanic A
    b. Mechanic B
    c. Both mechanic A and B
    d. Neither mechanic A nor B

PART TWO

Specific Transmissions and Transaxles

Chapter

9

Torqueflite Transmissions and Transaxles

HISTORY AND MODEL VARIATIONS

Chrysler Corporation introduced the Torqueflite transmission in the 1956 Imperial. The Torqueflite was the first really modern 3-speed automatic transmission with a torque converter. It was also the first to use the Simpson 2-planetary compound geartrain. The original Torqueflite was called the A-466, and had a cast-iron case with separate aluminum castings for the bellhousing and rear extension housing. Late-model Torqueflites have an aluminum 1-piece case that incorporates the bellhousing; the rear extension housing remains a separate part to ease assembly.

There are two basic versions of the aluminum Torqueflite RWD transmission, the A-904 and the A-727. Introduced in 1959, the A-904 is the light-duty version, sometimes called the Torqueflite 6. Introduced in 1961, the A-727 is the heavy-duty version, sometimes called the Torqueflite 8. The Torqueflite 6 and 8 names are misleading, however, because some 6-cylinder engines use the A-727 and some V-8 engines use the A-904.

The basic A-904, figure 9-1, is used with most 6-cylinder engines in domestic passenger cars, with 4-cylinder engines in Simcas built by Chrysler of Europe, and with 4-cylinder engines in Mitsubishi-built Colt, Arrow, and Challenger/Sapporo models. Other versions of the A-904 are used with some small-block 273- and 318-cid V-8 engines. These are known as the A-904-1 and the A-904-A or -LA. Introduced in the 1974 model year, the A-998 and A-999 are two other versions of the A-904 that have been used with 318- and 360-cid V-8s. In 1981, the A-904T transmission was introduced for 6-cylinder truck and heavy-duty passenger car applications.

The A-727, figure 9-2, is used with most large-block V-8 engines, including the 361-, 383-, 400-, 413-, 426-, and 440-cid powerplants. This transmission is also used with all 340-cid, and some 318- and 360-cid, small-block V-8s. Finally, the A-727 is used in all Chrysler Corporation light- and medium-duty trucks with automatic transmissions, and in some 6-cylinder passenger cars built for police and taxi service.

Chrysler Torqueflite transmissions were also used in some 1972-88 American Motors (AMC) and 1972-80 International Harvester (IHC) vehicles. AMC used both Torqueflite versions, but dropped the "A" prefixes and identified the transmissions as the Torque-Command 904, 998, 999, and 727 models. The 904 was used with AMC 4- and 6-cylinder engines. The 998

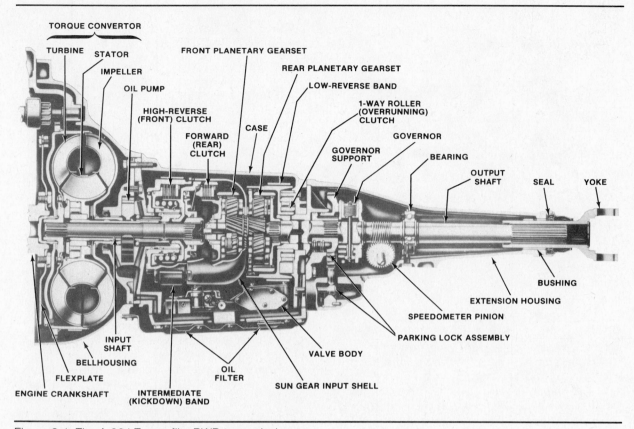

Figure 9-1. The A-904 Torqueflite RWD transmission.

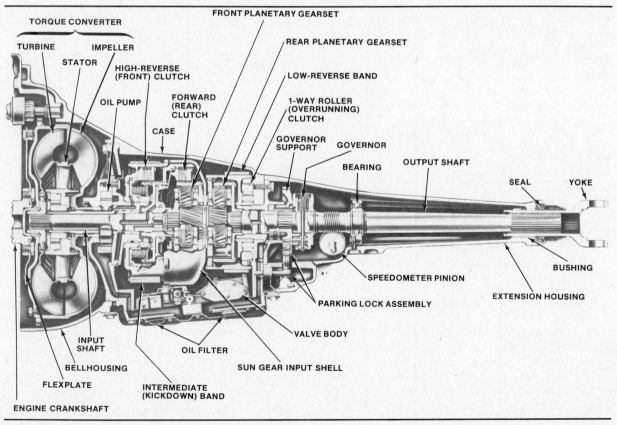

Figure 9-2. The A-727 Torqueflite RWD transmission.

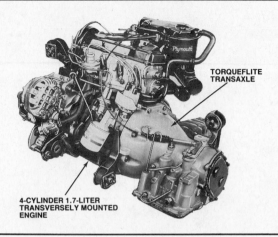

Figure 9-3. The first Torqueflite FWD transaxles were installed in Omni/Horizon hatchbacks with Volkswagen engines.

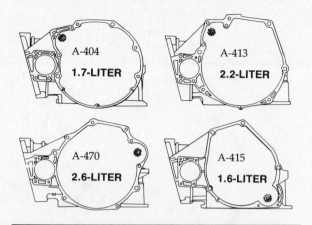

Figure 9-4. The bellhousing shapes of Torqueflite FWD transaxles vary to accommodate different engines and starter locations.

and 999 were used with some of AMC's large 6-cylinder and small V-8 engines. The 727 was used with most large AMC V-8s, and with 6-cylinder engines in heavy-duty applications. IHC bought only the A-727 Torqueflite, renamed it the T-407, and used it with 4-, 6-, and 8-cylinder engines in light- and medium-duty vehicles.

The AMC and IHC versions of Torqueflite transmissions were essentially identical to the Chrysler models. Minor internal changes are made to match the transmissions to the weights, axle ratios, and engine torque characteristics of the AMC and IHC vehicles. Externally, the bellhousings are redesigned to mate with AMC and IHC engines, and the extension housings and output yokes are altered to fit AMC and IHC chassis and drivelines.

Torqueflite Transaxles

In 1978, the Torqueflite transmission was extensively re-engineered for use as a FWD transaxle, figure 9-3. Torqueflite transaxles contain essentially the same parts as an A-904 transmission, with the addition of a transfer shaft, final drive gears, and a differential assembly. Torqueflite transaxles have been built in five versions. All share the same basic internal construction, but the case design varies for each individual engine and starter mounting location, figure 9-4.

The first Torqueflite transaxle was the A-404, figure 9-5, used with 1.7-liter Volkswagen-built engines in 1978 and later Omni/Horizon models. In 1982, the A-413 and the A-470 were added to the lineup — the A-413 for use with domestic 2.2- and 2.5-liter engines, and the A-470 for use with Mitsubishi-built 2.6-liter

engines. In 1983, the A-415 was added for use with the Peugeot-built 1.6-liter engine in the Omni/Horizon. Finally, in 1987, the A-670 was added for use with the Mitsubishi-built 3.0-liter V-6 engine. The A-404 was discontinued in 1983, and the A-415 was dropped in 1986. The A-413, A-470, and A-670 continue in Chrysler FWD applications through the 1988 model year.

Fluid Recommendations

Torqueflite transmissions in AMC and IHC vehicles use DEXRON®-IID fluid for models built through 1986. Following the purchase of AMC by Chrysler, the 1987-88 Eagle 4WD wagon transmission requires MERCON® ATF. The 1987-88 Jeep models continue to use DEXRON®-IID.

Chrysler recommended DEXRON®-IID fluid for all of its 1986 and earlier Torqueflite transmissions. For 1987-88 automatics, Chrysler specifies a special ATF+ Type 6167 (Part No. 4318077). This fluid is also recommended for earlier Chrysler automatics with locking torque converters to help eliminate lockup shudder.

Transmission Identification

You can identify a RWD Torqueflite transmission by the model number cast into the case on the lower left-hand side of the bellhousing, figure 9-6. As described earlier, transaxle model numbers are determined by the engine application.

You need more than just a model number when you order parts to repair a specific transmission, however. The most important identification number for a technician is the transmission part number. On Torqueflite

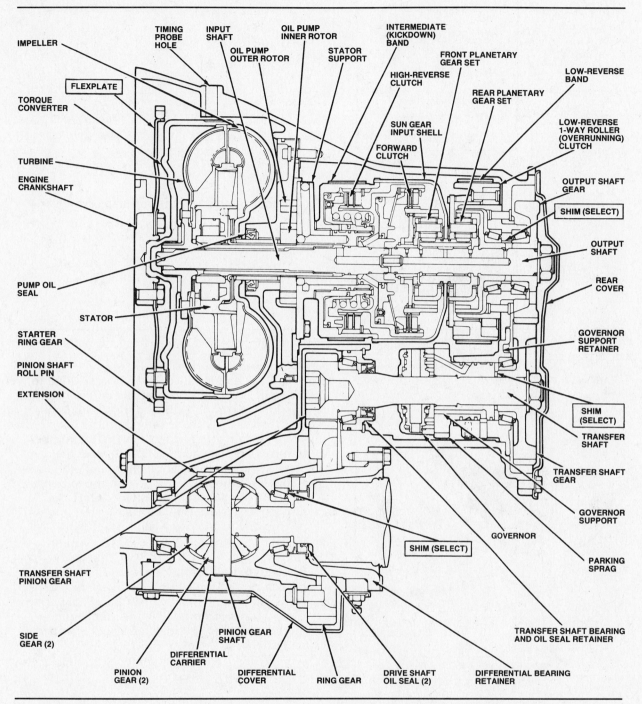

Figure 9-5. This A-404 cutaway shows the internal construction of a typical Torqueflite FWD transaxle.

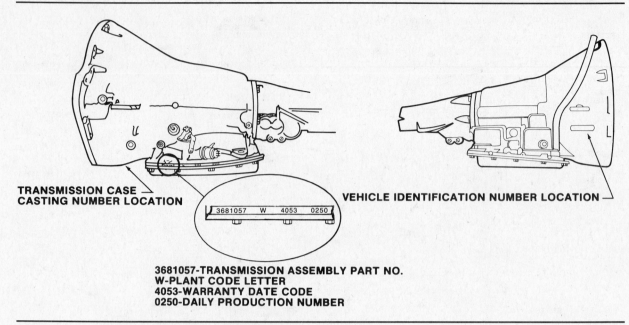

TRANSMISSION CASE CASTING NUMBER LOCATION

3681057 W 4053 0250

VEHICLE IDENTIFICATION NUMBER LOCATION

3681057-TRANSMISSION ASSEMBLY PART NO.
W-PLANT CODE LETTER
4053-WARRANTY DATE CODE
0250-DAILY PRODUCTION NUMBER

Figure 9-6. Torqueflite RWD transmission identification numbers.

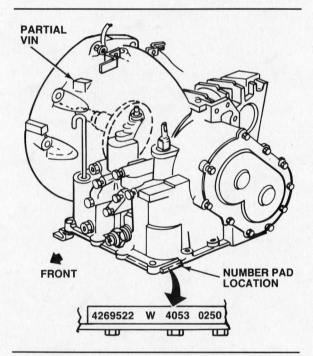

PARTIAL VIN

FRONT

NUMBER PAD LOCATION

4269522 W 4053 0250

Figure 9-7. Torqueflite FWD transaxle identification numbers.

transmissions and transaxles, this 7-digit number is stamped into the oil pan flange on the left side of the gearbox case, figures 9-6 and 9-7. Other numbers that appear in this area are codes for the manufacturing plant, the warranty date, and the daily production number.

In 1969, an identification pad was added to the right side of the bellhousing on RWD transmissions to comply with Federal regulations that require the vehicle serial number to be on all major components, figure 9-6. The number stamped on this pad is a short form of the complete Vehicle Identification Number (VIN) that is found on the plate mounted to the top of the dashboard under the windshield. On Torqueflite transaxles, the VIN pad is cast into the upper forward side of the bellhousing, figure 9-7.

An example of the VIN on a Torqueflite transmission is 6 A 100001. The first character, the digit 6, indicates model year 1976. The second character, the letter A, is a code that indicates the assembly plant. The last six digits are the vehicle serial number. Chrysler recommends that rebuilt transmissions be stamped with a letter "R" before the VIN.

MAJOR CHANGES

Early Torqueflite transmissions used push-button shift controls and a driveshaft-mounted parking brake. Along with the change from a cast-iron to an aluminum case, one of the earliest changes was to replace the external parking brake with an internal parking gear on the output shaft. In 1965, the push-button shift controls were replaced by conventional column- or console-mounted shift levers.

Before 1966, Torqueflite transmissions had a rear oil pump, driven by the output shaft. This pump supplied oil to the governor and provided working pressure for the transmission during push starts. Additionally, the rear pump cut into the main hydraulic system and took over the duties of the front pump at a designated speed. The rear pump was eliminated from Torqueflite transmissions at the beginning of the 1966 model year.

Listed below are many other changes made to Torqueflite transmissions over the years. Note that only the *major* changes are described; it is impossible to list every one of the dozens of running changes that take place every year. Always consult shop manuals, service bulletins, and aftermarket parts suppliers for the latest information on the transmission being serviced.

1966 • Rear pump eliminated.
    • Output shaft seal and yoke redesigned.
1967 • Low-reverse servo redesigned.
    • A-727 modified for high performance use with 426- and 440-cid engines. Changes included the use of 4-pinion carrier assemblies and high-strength shafts.
1968 • A-904 modified for use with 318-cid V-8 engine.
    • Lubrication changes made to extend clutch life.
1969 • Combination neutral-start switch and backup light switch added.
    • Vehicle Identification Number (VIN) added.
1971 • A-904 modified for use with 1.6- and 1.8-liter Simca engines.
    • Transmission cases and hydraulic components redesigned to improve shift quality. Some parts not interchangeable with earlier models.
1972 • Chrysler began supplying Torqueflites to AMC and IHC.
    • High-reverse clutch seals and converter hub seals modified.
1973 • Oil filter area increased 50 percent for all models.
    • A-727 intermediate band changed to a flexible type.
1974 • Lubrication circuits redesigned for better cooling.
    • Hydraulic components and clutch parts altered to improve shifts into Drive or Reverse from Neutral.
    • Insert-type oil filters added to governor feed circuit and pressure regulator circuit.
    • A-904 modified for use in Mitsubishi-built import vehicles.
    • 1-way roller clutch in A-904 enlarged.
    • A-999 version of A-904 transmission developed for use with 360-cid V-8.

    • 1-piece low-reverse band lever developed for the A-904. This change altered the band adjustment specifications.
    • A-998 version of the A-904 transmission developed for use with 318-cid V-8.
1975 • Part-throttle kickdown feature modified to reduce sensitivity in relation to carburetor opening.
1976 • A-727 output shaft and front carrier strengthened.
    • A-904 thrust washers redesigned.
    • Teflon seal rings replaced cast-iron rings as original equipment in A-904 transmissions. Cast-iron rings continued for service replacement.
1978 • Lockup torque converter introduced on RWD transmissions. This involved many new and revised components including the input shaft, oil pump, and valve body.
    • Drain plugs eliminated from all torque converters.
    • Chip collector magnet incorporated into all oil pans except on transmissions in import models.
    • A-404 FWD transaxle introduced on Omni/Horizon.
1979 • Intermediate servo of A-904 modified to improve 2-3 and 3-2 shift quality.
    • Lockup torque converter added to some AMC Torque-Command transmissions.
    • Select-fit #3 thrust washer introduced to adjust A-404 endplay.
    • A-998 and A-999 valve bodies modified to eliminate reverse squawk.
1980 • A-904 and A-998 wide-ratio gear set introduced to improve performance.
    • A-404 valve body and other components modified at mid-year for better shift quality, less internal leakage, and lower line pressures.
    • A-904 family gets new design intermediate flex-band.
1981 • Counter-bore added to low-reverse servo feed passage of A-904 for improved Neutral-Reverse shift quality.
    • A-413 and A-470 transaxles introduced.
    • Wide-ratio gearset and lower final-drive ratio introduced on all transaxles.
    • Forward clutch changed from 2-disc to 3-disc design on A-413 and A-470.
1982 • AMC versions of A-904 get wide-ratio gearing and modified valve body to match shift quality to the gearset.
    • Teflon seals adopted for A-413 intermediate and accumulator pistons, and input shaft.

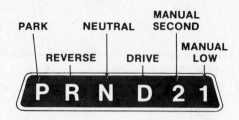

Figure 9-8. Torqueflite gear selector positions.
Pre-1965 models used pushbuttons.

1983 • Revised rotor-type pump and matching torque converter adopted on 4-cylinder A-904 applications.
• A-904 oil pan gasket and pan revised to reduce leaks at rear of pan.
• A-415 and A-470 transaxles introduced.
• Common transmission and differential oil sump introduced on all transaxles to reduce possibility of transfer shaft bearing failure due to low oil level. New, longer dipstick required with common oil sump.
• Strengthened differential assembly, transfer shaft, final drive gears, and planetary gearset introduced for transaxles used in heavy-duty service.
• New 4-disc forward clutch for heavy-duty transaxle applications.

1984 • Transaxle differential bearings modified for improved durability.
• Transaxle 3-disc forward clutch redesigned to provide for common pressure and reaction plates with the 4-disc clutch introduced for heavy-duty applications in 1983.
• Use of heavy-duty transaxles expanded to 2.2-liter Turbo engines, mini-vans, and fleet applications.

1985 • Cooler line fittings on transaxles redesigned with longer hose nipple to permit using a wider hose clamp.

1986 • Part-throttle torque converter unlocking added to A-999 and A-904T.
• Transaxle cases revised with one threaded and two drilled holes for easier starter removal and installation.
• Transaxle inner oil pump gear and converter hub revised to incorporate flats instead of two lugs.
• Input shaft and stator structure changed on all transaxles.
• Lockup torque converter introduced on limited transaxle applications.

1987 • A-670 transaxle introduced for V-6 applications.
• Lockup torque converter use expanded to include all transaxles except those used with turbocharged engines.

GEAR RANGES AND RATIOS

Regardless of whether it controls a transmission or a transaxle, every Torqueflite gear selector has six positions, figure 9-8. They are:
P — Park
R — Reverse
N — Neutral
D — Drive
2 — Manual second
1 — Manual low.
Torqueflite RWD transmission gear ratios are:
• Low (first) — 2.45:1 w/standard gearset
• Low (first) — 2.67:1 w/wide-ratio gearset
• Second (intermediate) — 1.45:1 w/standard gearset
• Second (intermediate) — 1.52:1 w/wide-ratio gearset
• Third (direct drive) — 1.00:1
• Reverse — 2.20:1 or 2.21:1
Torqueflite FWD transaxle gear ratios are:
• Low (first) — 2.48:1 w/standard gearset
• Low (first) — 2.69:1 w/wide-ratio gearset
• Second (intermediate) — 1.48:1 w/standard gearset
• Second (intermediate) — 1.55:1 w/wide-ratio gearset
• Third (direct drive) — 1.00:1
• Reverse — 2.10:1
The overall top gear ratio in a Torqueflite FWD transaxle is determined by the transfer shaft and final drive gearing. The range of overall top gear ratios varies widely depending on model year, engine displacement, transaxle type, emission requirements, and the locale where the vehicle is designed to be sold. Ratios that have been used include 2.78, 3.02, 3.22, 3.48, 3.67, and 3.74:1.

Neutral and Park

The engine can be started only when the gear selector is in N or P (Neutral or Park). No clutches or bands are applied in these positions, and there is no powerflow through the transmission. In Neutral, the output shaft is free to turn, and the car can be pushed. In Park, a lever is mechanically engaged with a large gear on the output shaft to lock the transmission and prevent the vehicle from moving.

Drive

When the vehicle is stationary and the gear selector is placed in D (Drive), the transmission is automatically in low gear. As the vehicle accelerates from a standstill, the transmission upshifts automatically to second and then to high gear. The speeds at which these upshifts occur

CHRYSLER TERM	TEXTBOOK TERM
Overrunning clutch	1-way roller clutch
Rear clutch	Forward clutch
Front clutch	High-reverse clutch
Kickdown, or front, band	Intermediate band
Front servo	Intermediate servo
Rear band	Low-reverse band
Rear servo	Low-reverse servo
Annulus gear	Ring gear
Clutch retainer	Clutch drum
Sun gear driving shell	Input shell
Reaction shaft support	Stator support
Breakaway	Low, or first, gear
Kickdown	Second, or intermediate, gear
Line pressure	Mainline pressure

Figure 9-9. Chrysler transmission nomenclature table.

vary and are controlled by vehicle speed and engine load.

The driver can force downshifts in Drive by partially or completely opening the throttle. Up to about 35 or 40 mph (60 to 65 kph), the transmission makes a 3-1 downshift when the throttle is opened wide. Up to approximately 45 to 50 mph (70 to 80 kph), the transmission makes a 3-2 downshift when the throttle is partially opened from a cruising position. From about 50 to 75 mph (80 to 120 kph), the transmission makes a 3-2 downshift only when the throttle is opened completely. Above 75 mph, the transmission does not downshift, even when the throttle is fully open.

When the car decelerates in Drive to a full stop, the transmission automatically downshifts only once. This is called a coasting downshift, and it is a shift directly from high to low, bypassing second. A coasting downshift occurs just before the vehicle comes to a complete stop.

Manual Second

When the gear selector is placed in 2 (manual second) and the vehicle is accelerated from a standstill, the transmission automatically upshifts from low to second but not from second to high. The 1-2 upshift will occur at the same speed as it would if the gear selector were in Drive. Once in second, the driver can force a downshift back to low by opening the throttle completely. During deceleration, the transmission automatically downshifts to low as the vehicle speed drops below 10 to 15 mph (16 to 24 kph).

When the vehicle is cruising in high gear and greater engine braking is desired, the driver can manually downshift from third to second by moving the selector level from D to 2. In this case, the transmission shifts into manual second gear regardless of throttle position or vehicle speed.

Manual Low

When the gear selector is placed in 1 (manual low) and the vehicle is accelerated from a standstill, the transmission stays in low gear. It does not upshift to second. Because different apply devices are used in manual low than are used in Drive low, this gear range can be used to provide engine compression braking to slow the vehicle.

If the driver moves the gear selector from Drive to the 1 position at high speed, the transmission does not shift directly from third to low gear. To avoid engine or driveline damage, the transmission first downshifts to second and then automatically downshifts to low when vehicle speed drops below about 20 mph or 30 kph.

Reverse

When the gear selector is moved to R (Reverse), the transmission shifts into Reverse. This should be done only when the vehicle is standing still. On older Torqueflites with rear oil pumps and pushbutton shift controls, the design of the valve body prevents the transmission from being accidentally or intentionally shifted into Reverse while the car is moving forward. On newer models, the lockout feature of the shift gate helps to prevent accidental shifting into Reverse. However, if the driver intentionally moves the gear selector to R while the vehicle is moving forward, the transmission *will* shift into Reverse and damage may result.

CHRYSLER TRANSMISSION NOMENCLATURE

For the sake of clarity, this text uses consistent terms for the different transmission parts and functions. However, the various vehicle manufacturers may use different terms. Therefore, before we examine the buildup of the Torqueflite and the operation of its geartrain and hydraulic system, you should be aware of the unique names that Chrysler uses for some transmission parts and operations.

Knowing the manufacturer's names will make it easier in the future if you look in a Chrysler manual and see, for example, a reference to the "kickdown band". You will then realize this is the intermediate band, and that it is similar to the same part in any other transmission that uses a comparable geartrain and apply devices. Figure 9-9 lists Chrysler nomen-

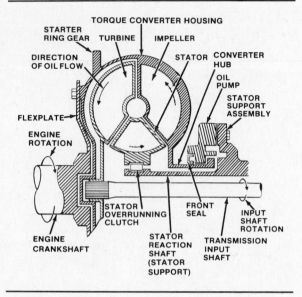

Figure 9-10. Torqueflite torque converter and oil pump installation.

support, figure 9-10. The splines in the stator 1-way clutch engage the splines on the stator support, and the splines in the turbine engage the transmission input shaft. The torque converter hub, at the rear of the converter housing, engages the drive lugs on the oil pump to drive the pump. The converter housing includes the impeller and is bolted to the engine flexplate or flywheel so that the oil pump turns and pumps fluid whenever the engine is running.

Torqueflite RWD transmissions use a rotor-type oil pump mounted in the front of the transmission case, figure 9-11, while FWD transaxles use a gear-type pump, figure 9-12. The pump's inner rotor or gear is driven by the torque converter hub. The pump draws fluid from the sump, through the filter, and delivers it to the pressure regulator valve and the manual valve.

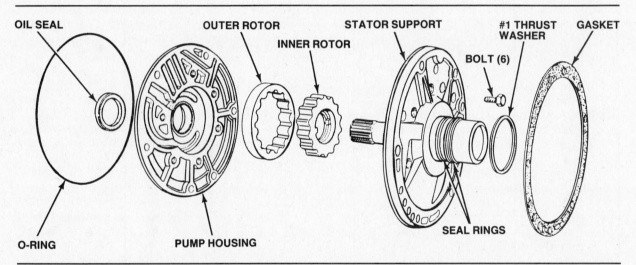

Figure 9-11. Torqueflite RWD transmission rotor-type oil pump assembly.

clature and the textbook terms for some transmission parts and operations.

TORQUEFLITE BUILDUP

The Torqueflite uses a torque converter that is a welded unit and cannot be disassembled. The torque converter is bolted to the flexplate or flywheel that is attached to the engine crankshaft, figure 9-10. On Chrysler RWD transmissions, the starter motor ring gear is welded to the converter housing. On Chrysler FWD transaxles and AMC products, the starter ring gear is on the flexplate or flywheel.

The torque converter is installed in the transmission by sliding it onto the stationary stator

The transmission input shaft is turned clockwise by the torque converter turbine. The shaft passes through the stator support into the transmission case. The high-reverse clutch hub and forward clutch drum are splined to the end of the input shaft, figure 9-13, and rotate clockwise with it whenever the torque converter turbine is being driven by the impeller. The hub and drum are splined together, or built as a single assembly, depending on the application.

Refer to figure 9-14 while studying the relationship of the clutches and bands to the gear-train. The inner circumference of each high-reverse clutch friction disc is splined to the outer circumference of the high-reverse clutch

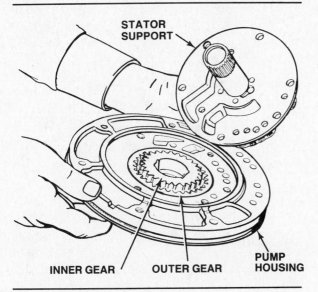

Figure 9-12. Torqueflite FWD transaxle gear-type oil pump assembly.

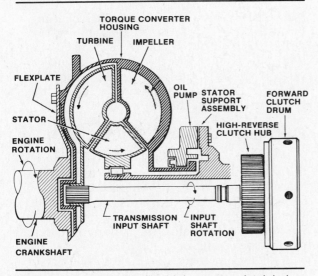

Figure 9-13. The Torqueflite high-reverse clutch hub and forward clutch drum are splined to the end of the input shaft.

hub and turned by it. The outer circumference of each high-reverse clutch steel separator disc is splined to the inner circumference of the high-reverse clutch drum. When the high-reverse clutch is applied, the drum is rotated clockwise by the input shaft.

The high-reverse clutch drum engages the sun gear input shell. The sun gear rides on bushings on the output shaft but is not splined to the shaft. Both the front and rear pinions mesh with the sun gear. When the high-reverse clutch is applied (in high or Reverse), the drum, the input shell, and the sun gear are rotated clockwise by the input shaft.

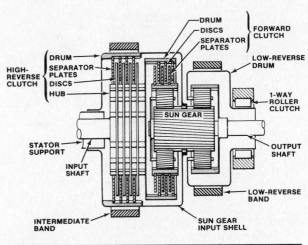

Figure 9-14. Torqueflite geartrain, clutches, and bands.

The intermediate band is wrapped around the outside of the high-reverse clutch drum. When the intermediate band is applied, the drum, the input shell, and the sun gear are all held stationary. This occurs only in second gear.

The outer circumference of each forward clutch steel disc is splined to the inner circumference of the forward clutch drum, which is attached to the input shaft. The inner circumference of each forward clutch friction disc is splined to the outer circumference of the front ring gear in the planetary gearset. Therefore, whenever the forward clutch is applied, the front ring gear is rotated clockwise.

None of the members of the planetary gearset is attached directly to the input shaft. However, two members of the gearset, the front carrier and the rear ring gear, are splined to the output shaft, figure 9-15. This means that one or the other of these units is always the final driving member of the gearset.

The rear carrier is supported by the rear pinions and engages the inside of the low-reverse drum. Therefore, the low-reverse drum and rear carrier are locked together. The inside of the low-reverse drum hub rotates on the governor oil delivery and output shaft support that protrudes inside the rear of the transmission case. The outside of the low-reverse drum hub is splined to the inner race of the 1-way roller clutch at the rear of the transmission case.

The outer race of the 1-way roller clutch is a splined press fit into the case on 727 models and is riveted into the case on 904 models. Replacement units are bolted to the case. When the inner race of the clutch, along with the low-reverse drum and rear carrier, tries to rotate

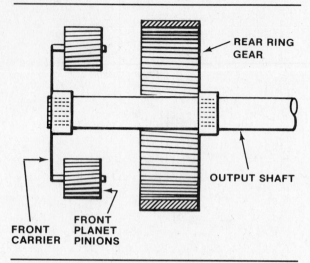

Figure 9-15. The front carrier and rear ring gear are splined to the Torqueflite output shaft.

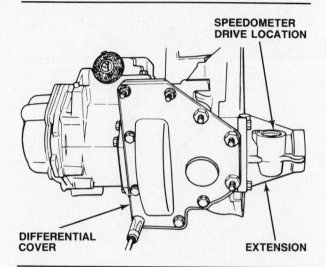

Figure 9-16. The speedometer drive is located in an extension housing on the right side of Torqueflite transaxles.

counterclockwise, the clutch locks up. This action occurs in Drive low to hold the rear carrier. When the inner race of the clutch tries to rotate clockwise, the clutch unlocks and allows the low-reverse drum and carrier to freewheel.

The low-reverse band wraps around the low-reverse drum at the rear of the transmission case. When the low-reverse band is applied, the drum and rear carrier are held. This occurs in Reverse and manual low. Unlike the 1-way clutch, the low-reverse band holds the drum and carrier from rotating in either direction.

Up to this point, the buildup of both Torqueflite RWD transmissions and FWD transaxles is essentially the same. From here on, however, there are substantial differences. The next two sections describe the remainder of the buildup for the respective units.

RWD Transmission Final Buildup

In RWD Torqueflite transmissions, the output shaft passes through the rear of the transmission case. Just inside the extension housing, the governor support is splined to the output shaft, figures 9-1 and 9-2. The governor is bolted to the governor support and turned by the output shaft whenever the vehicle is moving. The outer circumference of the governor support forms the parking gear. To hold the car from moving, this gear is locked to the transmission case by a pawl when the transmission is in Park.

The speedometer drive pinion is just behind the governor in the extension housing. The drive pinion is driven by a worm gear cut into the output shaft. To the rear of the speedometer pinion, the output shaft is supported by a

ball bearing retained in the extension housing by a snapring. The output yoke is splined to the output shaft at the rear of the transmission. The outer circumference of the output yoke rides in a bushing pressed into the rear of the extension housing. This, in turn, supports the rear of the output shaft.

FWD Transaxle Final Buildup

In FWD transaxles, figure 9-5, power from the output shaft is routed through two helical gears to a transfer shaft positioned behind, below, and parallel to the transaxle mainshaft. The governor support plate, governor, and parking gear (also called the parking sprag), are mounted on the transfer shaft.

A helical pinion gear on the opposite end of the transfer shaft drives the differential ring gear and carrier. The differential assembly, through its pinion and side gears, then drives the axles that power the wheels. The speedometer drive assembly is installed in a small extension housing that bolts to the transaxle case where the right driveshaft exits, figure 9-16.

As mentioned earlier, the overall top gear ratio in a Torqueflite transaxle is determined by the combined gear ratios of the transfer shaft drive gears and the final drive ring and pinion gears.

TORQUEFLITE POWER FLOW

Now that we have taken a look at how a Torqueflite is assembled, we can trace the flow of power through the apply devices and the planetary gearset in each of the gear ranges.

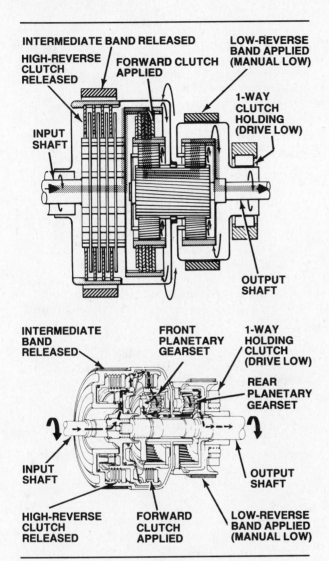

Figure 9-17. Torqueflite powerflow in Drive low or manual low.

Park and Neutral

When the gear selector is in Park or Neutral, no clutches or bands are applied. Because no member of the planetary gearset is held, there can be no output motion. The torque converter impeller and the oil pump rotate clockwise at engine speed. The torque converter turbine and the transmission input shaft also rotate clockwise at engine speed.

In Neutral, the planetary gearset idles. The output shaft is free to rotate in either direction so the car can be pushed or pulled backwards or forwards.

In Park, a mechanical pawl is engaged with the parking gear or sprag that is splined to the output shaft (RWD) or transfer shaft (FWD). This locks the output shaft (RWD) or differential assembly (FWD) in position, prevents the driveshaft or halfshafts from turning, and stops the car from moving in either direction.

Drive Low

When the gear selector lever is moved to Drive, the car starts to move with the transmission automatically in low gear. The forward clutch is applied, and it is the only hydraulic apply device that operates in Drive low, figure 9-17. The forward clutch locks the front ring gear to the input shaft so that the ring gear rotates clockwise at turbine speed.

Before the car begins to move, the transmission output shaft is held stationary by the driveline and wheels. This also holds the front carrier and rear ring gear stationary. This means the clockwise rotation of the front ring gear turns the front pinions clockwise on the stationary carrier. The front pinion rotation then causes the sun gear to turn counterclockwise.

Counterclockwise rotation of the sun gear turns the rear pinions clockwise. The rear pinions then try to walk clockwise around the inside of the rear ring gear, which is splined to the stationary output shaft. This action causes the rear carrier to try to rotate counterclockwise. However, the rear carrier is attached to the low-reverse drum, which is splined to the inner race of the 1-way roller clutch at the rear of the case.

As the rear carrier tries to rotate counterclockwise, the 1-way clutch locks up and prevents rotation. Therefore, the torque from the rear pinions is transferred to the rear ring gear and output shaft instead of the rear carrier. Because the 1-way clutch holds the carrier more securely than the drive wheels can hold the output shaft and ring gear, the output shaft and ring gear begin to rotate clockwise in gear reduction.

As long as the car is accelerating in Drive low, torque is transferred from the engine, through the planetary gearset, to the output shaft. If the driver lets up on the throttle, the engine speed drops while the drive wheels try to keep turning at the same speed. This removes the clockwise-rotating torque load from the 1-way clutch, and it unlocks. This allows the rear planetary gearset to freewheel, so no engine compression braking is available. This is a major difference between Drive low and manual low, as we will see later.

Depending on whether the gearbox is fitted with a standard or wide-ratio gearset, the low gear ratio at the output shaft is either 2.45 or 2.67:1 with a RWD transmission, or 2.48 or 2.69:1 with a FWD transaxle. This reduction is

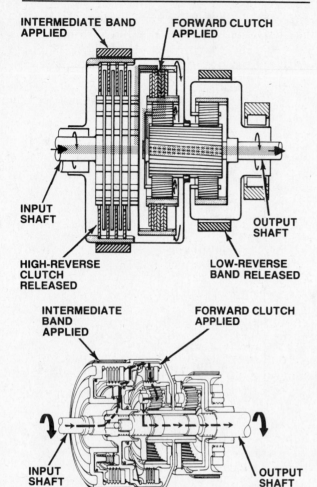

INTERMEDIATE BAND
APPLIED

FORWARD CLUTCH
APPLIED

INPUT
SHAFT

OUTPUT
SHAFT

HIGH-REVERSE
CLUTCH
RELEASED

LOW-REVERSE
BAND RELEASED

INTERMEDIATE
BAND
APPLIED

FORWARD CLUTCH
APPLIED

INPUT
SHAFT

OUTPUT
SHAFT

Figure 9-18. Torqueflite powerflow in Drive second or manual second.

produced by a combination of the front and rear planetary gearsets.

Remember, there are only two apply devices working in Drive low, the forward clutch and the 1-way roller clutch. The forward clutch is the only hydraulic unit that is working, and it is the input device. The 1-way roller clutch is the holding device.

Drive Second

As the car continues to accelerate in Drive, the transmission hydraulic system produces an automatic upshift to second gear by applying the intermediate band, figure 9-18. The forward clutch remains applied, and power from the engine is still delivered to the front ring gear. The front ring gear rotates clockwise and turns the front pinions clockwise.

At the same time, the intermediate band locks the high-reverse clutch drum and sun gear input shell to the transmission case. This holds the sun gear stationary. As the front ring gear turns the front pinions clockwise, they walk around the stationary sun gear. This causes the front carrier to rotate clockwise. The front carrier is splined to the output shaft and drives it in gear reduction.

Depending on whether the gearbox is fitted with a standard or wide-ratio gearset, the second gear ratio at the output shaft is either 1.45 or 1.52:1 with a RWD transmission, or 1.48 or 1.55:1 with a FWD transaxle. All of the gear reduction for second gear is produced by the front planetary gearset. The rear gearset simply idles. The 1-way clutch that held the rear carrier in Drive low overruns in Drive second because the pinions in the rear carrier are meshed with the common sun gear that is stationary. Therefore, the rear pinions are forced to walk around the sun gear, causing the rear carrier to rotate clockwise and the 1-way clutch inner race to rotate clockwise. The rear ring gear, which is splined to the output shaft, continues to rotate with the shaft, but freewheels because no holding device is applied to the rear gearset.

In second gear, there are again only two apply devices working, the forward clutch and the intermediate band. The forward clutch is the input device. The intermediate band is the holding device.

Drive Third

As the car continues to accelerate in Drive, the transmission automatically upshifts from second to third gear. To make this happen, the forward clutch remains applied, the intermediate band is released, and the high-reverse clutch is applied, figure 9-19.

When the high-reverse clutch is applied, the high-reverse clutch drum and the sun gear input shell are locked to the transmission input shaft. Because the sun gear input shell is splined to the sun gear, this causes the sun gear to revolve clockwise with the input shaft at turbine speed unless the converter clutch is applied.

front ring gear to the input shaft. This means that two members of the planetary gearset, the sun gear and the ring gear, are locked to the input shaft, and thus to each other. With two members of the gearset locked together, the entire gearset turns as a unit.

With the front planetary gearset locked, there is no relative motion between the sun gear, ring gear, carrier, or the pinions. As a result, input torque to the sun and ring gears is

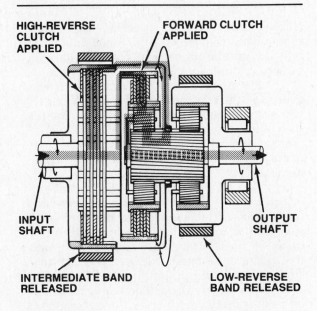

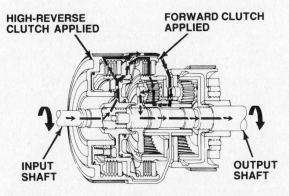

Figure 9-19. Torqueflite powerflow in direct drive (high gear).

transmitted through the planet carrier, which is splined to the output shaft, to drive the output shaft.

In third gear in both RWD transmissions and FWD transaxles, there is a direct 1:1 gear ratio between the input shaft speed and the output shaft speed. There is no reduction or increase in either speed or torque through the transmission.

In third gear there are again two apply devices working, the forward clutch and the high-reverse clutch. Both clutches are input or driving devices.

Manual Low

In manual low (1 on the gear indicator), the flow of power through the planetary gearset is the same as in Drive low. The only difference is that the low-reverse band is applied in manual

low to hold the rear carrier, figure 9-17. In Drive low, the 1-way clutch holds the rear carrier only during acceleration; the clutch freewheels on deceleration. In manual low, the low-reverse band holds the rear carrier during both acceleration and deceleration. This provides engine braking on deceleration. Even though the 1-way clutch acts to help hold the rear carrier during acceleration in manual low, its action is secondary so long as the low-reverse band is applied.

As long as the gear selector remains in the 1 position, the hydraulic system keeps the low-reverse band applied and prevents the intermediate band from applying. If the driver moves the gear selector from D to 1 while the car is moving at low to medium speeds, the transmission downshifts directly from high to low gear by releasing the high-reverse clutch and applying the low-reverse band. If the driver moves the gear selector from D to 1 while the car is moving at higher speeds, the transmission first shifts into second by releasing the high-reverse clutch and applying the intermediate band. The transmission then downshifts to low at about 20 mph or 30 kph by releasing the intermediate band and applying the low-reverse band.

In manual low, two apply devices are again operating. The forward clutch is the input device, and the low-reverse band is the holding device.

Manual Second

The only difference between manual second (2 on the gear selector) and Drive is that the hydraulic system prevents the transmission from upshifting into high gear. The flow of power through the planetary gearsets in first and second gears is identical, and the same apply devices are used.

When the gear selector is at 2 and the car accelerates from a standstill, the transmission starts in Drive low, figure 9-17. The forward clutch drives the front ring gear, and the 1-way clutch holds the rear carrier. As the vehicle continues to accelerate, the transmission automatically shifts to second by applying the intermediate band to hold the sun gear, figure 9-18.

If the driver moves the gear selector to 2 while the car is moving in Drive third, the hydraulic system will downshift the transmission to second gear by releasing the high-reverse clutch and applying the intermediate band. This capability provides the driver with a controlled downshift for engine braking or improved acceleration at any speed or throttle position.

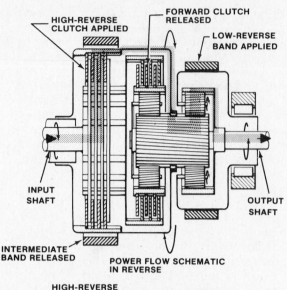

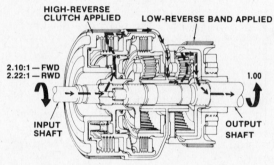

Figure 9-20. Torqueflite powerflow in Reverse.

3-2 kickdown

The driver can also cause a 3-2 downshift by opening the throttle more than a certain amount based on vehicle speed. We will see how this is controlled when we examine the Torqueflite hydraulic system later in this chapter. This type of 3-2 downshift is called a "kickdown". The effect on the transmission gearset and apply devices is the same as a manual downshift from Drive to 2. The high-reverse clutch is released, and the intermediate band is applied.

Reverse

When the gear selector is moved to Reverse, the transmission must change clockwise input shaft rotation to counterclockwise output shaft rotation. To do this, the high-reverse clutch and low-reverse band are applied, figure 9-20.

When the high-reverse clutch is applied, the high-reverse clutch drum and the sun gear input shell are locked to the transmission input shaft. Because the sun gear input shell is

splined to the sun gear, this causes the sun gear to revolve clockwise with the input shaft at turbine speed. The sun gear then drives the rear pinions counterclockwise.

The low-reverse band holds the low-reverse drum and the rear carrier stationary. With the rear carrier held, the rear pinions drive the rear ring gear counterclockwise. The rear ring gear is splined to the output shaft, and turns it counterclockwise in gear reduction to provide Reverse.

The Reverse gear ratio in RWD transmissions is 2.20:1 or 2.21:1, depending on the year of the transmission. The Reverse gear ratio in FWD transaxles is 2.10:1. All gear reduction is produced by the rear planetary gearset. Although there is input to the front gearset through the sun gear and the front pinions, no other member of the gearset is held. Therefore, there can be no power output from the front gearset — it simply freewheels.

In Reverse, two apply devices are used. The high-reverse clutch is the input device. The low-reverse band is the holding device.

Torqueflite Clutch and Band Applications

Before we look at the Torqueflite hydraulic system, let's review the apply devices used for each gear range in this transmission. The clutch and band application chart in figure 9-21 summarizes the apply devices used in each gear. Also, remember these facts about Torqueflite apply devices:

- The multiple-disc clutches are input devices.
- The bands and the 1-way clutch are holding devices.
- The forward clutch is applied in all forward gear ranges.
- The intermediate band is used only in second gear.
- The high-reverse clutch and the low-reverse band are each used in Reverse and in one forward gear.

THE TORQUEFLITE HYDRAULIC SYSTEM

The Torqueflite hydraulic system controls shifting under varying vehicle loads and speeds, as well as in response to manual gear selection by the driver. Figure 9-22 is a diagram of the complete hydraulic system for a late-model Torqueflite RWD transmission. Figure 9-23 is a diagram of the complete hydraulic system for a

GEAR RANGES	CLUTCHES			BANDS	
	1-Way Roller (Overrunning)	Forward (Rear)	High-Reverse (Front)	Intermediate (Kickdown, or Front)	Low-Reverse (Rear)
Drive Low	●	●			
Drive Second		●		●	
Drive High		●	●		
Manual Low	*	●			●
Manual Second		●		●	
Reverse			●		●

★ The 1-way roller clutch will hold in manual low if the low-reverse band fails but will not provide engine braking.

Figure 9-21. This chart indicates the clutches and bands that are applied, or working, in each gear range of a Torqueflite transmission or transaxle.

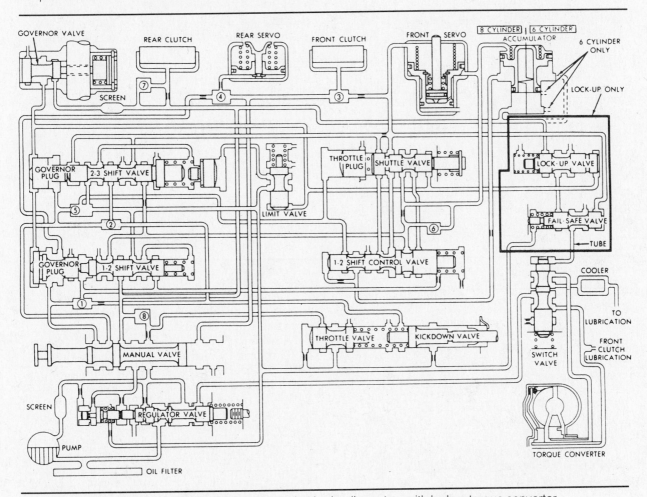

Figure 9-22. Late-model Torqueflite RWD transmission hydraulic system with lockup torque converter.

Torqueflite FWD transaxle. Note that the transaxle hydraulic system does not include controls for a lockup torque converter. You will find it helpful to refer to these diagrams as you read the following paragraphs.

The hydraulic systems in RWD and FWD Torqueflites are basically the same, although there are some notable differences. All late-model RWD transmissions have a 1-2 shift control valve, a limit valve, and a throttle plug at

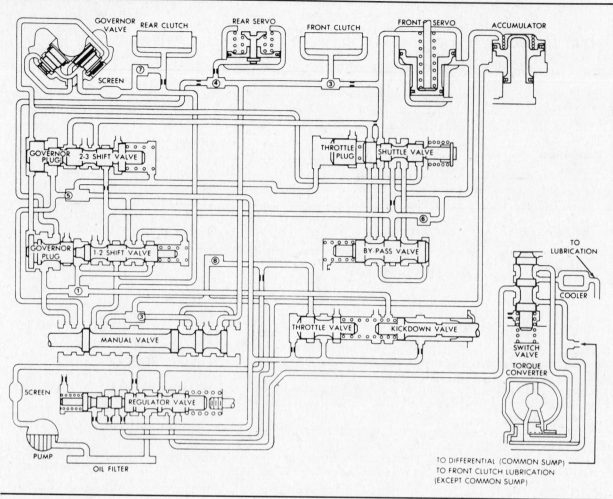

Figure 9-23. Torqueflite FWD transaxle hydraulic system without dockup torque converter.

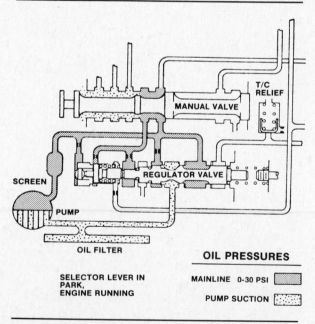

OIL PRESSURES

SELECTOR LEVER IN
PARK,
ENGINE RUNNING

MAINLINE 0-30 PSI

PUMP SUCTION

Figure 9-24. The oil pump delivers hydraulic fluid to the pressure regulator valve and the manual valve.

the end of the 2-3 shift valve. These parts are not used in FWD transaxles, or in early 6-cylinder RWD transmissions. In turn, the FWD transaxle hydraulic system has a bypass valve not found in RWD applications.

All RWD transmissions and FWD transaxles with lockup torque converters have a lockup valve and a fail-safe valve that other Torqueflites do not have. In addition, those with electronic torque converter clutch controls have a TCC solenoid. The lockup torque converter and its hydraulic circuits are discussed at the end of this section.

Hydraulic Pump

We will begin our study of the Torqueflite hydraulic system with the oil pump, figure 9-24. Fluid is stored in the pan (sump) at the bottom of the transmission. The oil pump draws fluid through the filter attached to the bottom of the valve body. Fluid flows through a passage in

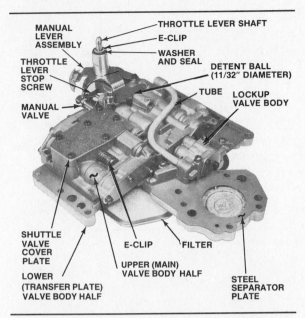

Figure 9-25. This Torqueflite valve body is from a 1978 RWD transmission equipped with a lockup torque converter.

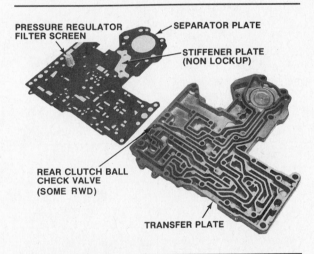

Figure 9-26. The transfer plate and separator plate from a RWD Torqueflite transmission.

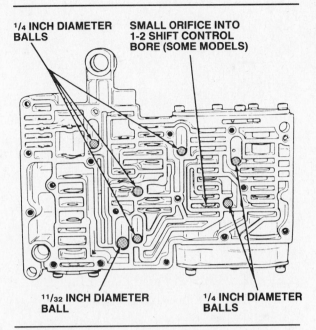

Figure 9-27. Typical Torqueflite RWD transmission ball-check valve locations.

Valve Body

All of the valves in the Torqueflite hydraulic system, except the governor valve, are housed in the valve body, figure 9-25. There are two main parts to the Torqueflite valve body, the upper half and the lower half; these are separated by a steel separator plate. The upper valve body half (the main valve body) contains most of the spool valves, valve plugs, and ball-check valves of the hydraulic system.

The lower valve body half (the transfer plate), figure 9-26, contains passages that carry fluid from one valve to another. In RWD transmissions, an extension of the transfer plate serves as the bottom of the 1-2 accumulator cylinder in the transmission case. The transfer plate also houses a forward clutch ball-check valve in some RWD applications.

Torqueflites with lockup torque converters have a small auxiliary valve body, figure 9-25, that attaches to the transfer plate in a location occupied by a stiffener plate, figure 9-26, in non-lockup applications. The auxiliary valve body houses the lockup and fail-safe valves that control the torque converter clutch. An external tube routes fluid from the main valve body to the auxiliary valve body.

Ball-check valves

Most RWD Torqueflites have seven ball-check valves in the upper valve body, figure 9-27, and one in the transfer plate, figure 9-26. The FWD transaxles have eight ball-check valves in the

the valve body and then through another passage in the front of the transmission case to the pump.

Torqueflite RWD transmissions use a rotor-type oil pump, figure 9-11, while FWD transaxles use a gear-type pump, figure 9-12. The pump's inner rotor or gear is driven by the torque converter hub. Whenever the engine is running, the pump delivers hydraulic fluid to the pressure regulator valve and the manual valve. From these valves, fluid flows to the rest of the valves in the hydraulic system.

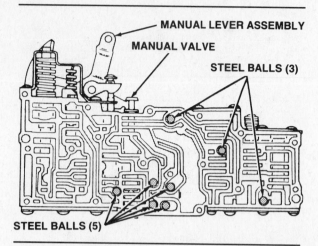

Figure 9-28. Typical Torqueflite FWD transaxle ball-check valve locations.

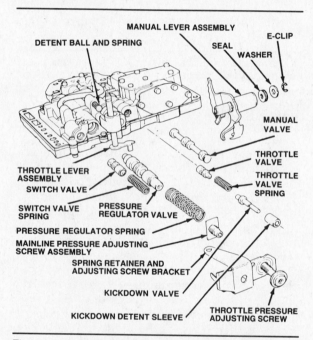

Figure 9-29. The valve body of a RWD Torqueflite transmission showing the pressure regulator valve, the manual valve, the throttle valve, and the kickdown valve assembly.

upper valve body, figure 9-28, and none in the transfer plate. In all Torqueflites, each clutch also has a ball-check valve to vent residual fluid after the clutch is released.

The valve body check balls differ in number, diameter, and location depending on the application. One of the check balls in some valve bodies is a high-pressure relief valve that protects the transmission from too much pressure from the pump. A spring behind this ball seats it against a hole in the separator plate.

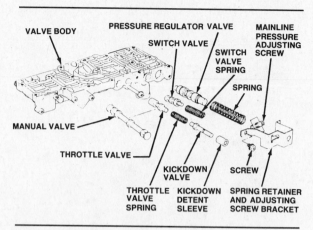

Figure 9-30. The valve body of a FWD Torqueflite transaxle showing the pressure regulator valve, the manual valve, the throttle valve, and the kickdown valve assembly.

Torqueflite Hydraulic Valves

The hydraulic system of a Torqueflite transmission or transaxle will contain some, but not all, of these valves:

1. Pressure regulator valve
2. Torque converter control valve (early RWD only)
3. Switch valve (late RWD, all FWD)
4. Manual valve
5. Throttle valve
6. Kickdown valve
7. Governor valve
8. 1-2 shift valve
9. 1-2 shift control valve (some RWD only)
10. 2-3 shift valve
11. Limit valve (some RWD only)
12. Shuttle valve
13. Kickdown valve
14. Bypass valve (FWD only)
15. Lockup valve (with lockup torque converter only)
16. Fail-safe valve (with lockup torque converter only).

As indicated, the use of certain valves is limited to early- or late-model Torqueflites. Other valves are used only in RWD or FWD applications. And some valves are found only in gearboxes equipped with a lockup torque converter. The following paragraphs describe the operation of the Torqueflite hydraulic system valves.

Pressure Regulator Valve

The pressure regulator valve receives fluid from the pump, then delivers fluid at the desired pressure to other valves and parts in the hydraulic system. The pressure regulator valve is

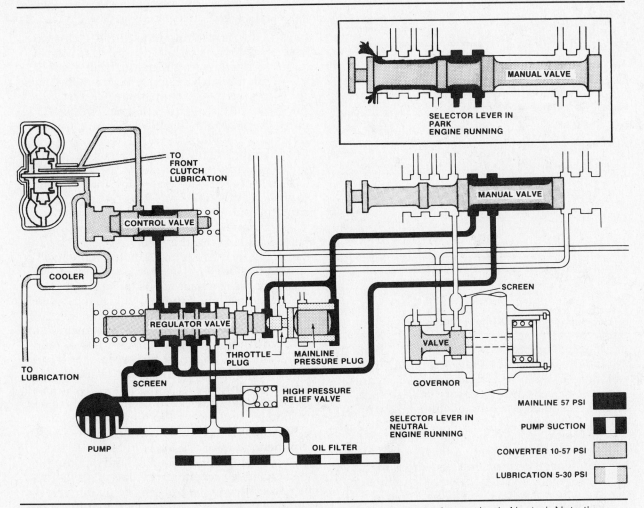

Figure 9-31. Operation of the early Torqueflite RWD transmission pressure regulator valve in Neutral. Note the different position of the manual valve in Park (inset).

near the larger end of the valve body as shown in figures 9-29 and 9-30.

Three different pressure regulator valve designs are used in Torqueflite gearboxes. Early RWD transmissions through 1977 use one design, late RWD transmissions from 1978 on use a second design, and all FWD transaxles use a third design. Although all three pressure regulator valves do the same job and operate on the same principles, there are significant differences between them. These are called out in the following paragraphs which describe the operation of the pressure regulator valve within the hydraulic system.

Pressure regulation — early RWD transmission
The pressure regulator valve assembly in a 1977 or earlier RWD transmission, figure 9-31, consists of: the pressure regulator valve, a throttle pressure plug, a mainline pressure plug, and a large spring.

Fluid from the pump is delivered directly to the center of the regulator valve through two ports. Fluid is also routed to the regulator valve assembly through two ports from the manual valve. One line from the manual valve enters the regulator valve assembly at the head of the mainline pressure plug. The other line from the manual valve enters between the throttle pressure plug and the end of the regulator valve.

When the engine is stopped and the pump is not producing any fluid flow, the spring holds the pressure regulator valve to the right. When the engine is started, pressure from the manual valve builds against the mainline pressure plug and the regulator valve, causing the regulator valve to move left against spring tension. When the regulator valve moves far enough, it uncovers the port to the torque converter control valve. Regulated mainline pressure then flows into the torque converter circuit.

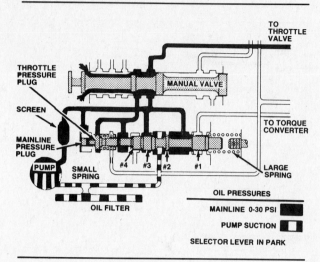

Figure 9-32. Operation of the late Torqueflite RWD transmission pressure regulator valve in Park.

As pressure continues to increase and the regulator valve moves farther to the left, an exhaust port is opened to return a portion of the fluid to the suction side of the pump. As the exhaust port opens and fluid is vented, hydraulic pressure and spring pressure balance each other to maintain a steady mainline pressure.

Auxiliary hydraulic pressures are used to further control the pressure regulator valve in some operating modes, but not in Park and Neutral. However, mainline pressure is higher in Neutral than in Park because of the different manual valve positions. In Neutral, fluid is trapped between two large lands on the manual valve, figure 9-31. This allows some pressure to build up before fluid is routed to the regulator valve. In Park, fluid is routed between one large land and one small land on the manual valve, figure 9-31 (inset). The clearance between the small land and the valve bore allows some fluid to escape from the valve body into the sump. This reduces the amount of pressure build up before fluid is routed to the pressure regulator valve.

Pressure regulation — late RWD transmission

The pressure regulator valve assembly in a 1978 or later RWD transmission, figure 9-32, consists of: the pressure regulator valve, a throttle pressure plug, a mainline pressure plug, a large spring, and a small spring. These are basically the same parts as the early design, but the routing of hydraulic fluid is different, the throttle pressure plug and mainline pressure plug sizes are altered, and a small spring is added around the throttle pressure plug.

Fluid from the pump is delivered directly to the head of the mainline pressure plug, and

through two ports to the center of the regulator valve. Fluid is also routed to the regulator valve assembly through two ports from the manual valve. One line from the manual valve enters the regulator valve assembly between the mainline pressure plug and the throttle pressure plug. The other line from the manual valve enters between the throttle pressure plug and the end of the regulator valve.

When the engine is stopped and the pump is not producing any fluid flow, the springs hold the pressure regulator valve and the two plugs to the left. When the engine is started, pressure from the pump and manual valve builds against the mainline pressure plug, the throttle pressure plug, and the regulator valve, causing the regulator valve to move to the right against spring tension.

When the transmission is in Park, figure 9-32, the position of the manual valve allows some of the fluid to escape to the transmission sump. Because this reduces the volume of fluid flow, the pressure regulator valve moves only a small amount, and most fluid recirculates back through the manual valve and pressure regulator valve into the transmission sump. A small amount of fluid passes through the groove next to land #1 of the pressure regulator valve, and slowly fills the torque converter circuit.

When the transmission is in Neutral or any gear position, figure 9-33, the lands on the manual valve prevent the escape of fluid. This increases flow to the pressure regulator valve, and moves the valve further to the right. Mainline pressure then passes through the groove next to land #1, and flows into the torque converter circuit leading to the switch valve.

As pressure continues to increase and the regulator valve moves farther to the right, land #2 opens an exhaust port that allows a portion of the fluid to return through the pressure regulator valve to the suction side on the pump. As the exhaust port opens and fluid is vented, hydraulic pressure and spring pressure balance each other to maintain a steady mainline pressure.

Note that as the position of the pressure regulator valve changes and fluid is exhausted from between lands #1 and #2, the pressure of the fluid bypassing land #1 into the torque converter circuit is also affected. In addition to mainline pressure, the pressure regulator valve in this design also regulates converter circuit pressure. This allows the torque converter control valve of early RWD transmissions to be replaced by a switch valve. Both of these valves are described in detail later in the chapter.

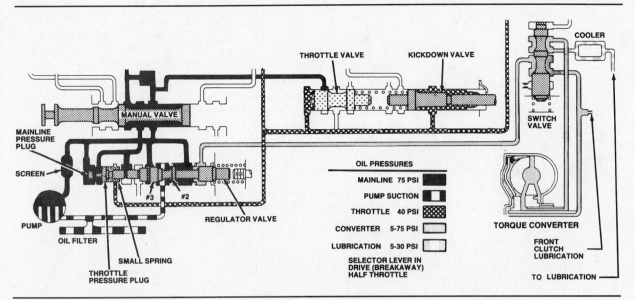

Figure 9-33. Operation of the late Torqueflite RWD transmission pressure regulator valve in Drive.

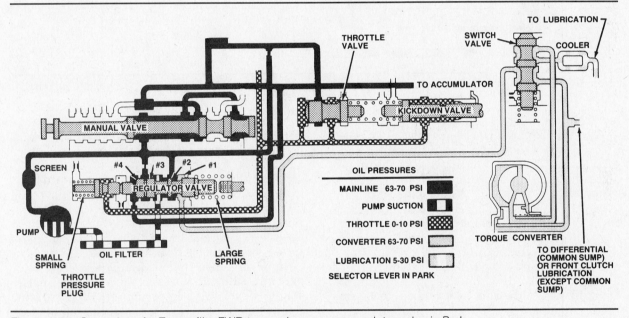

Figure 9-34. Operation of a Torqueflite FWD transaxle pressure regulator valve in Park.

Pressure regulation — FWD transaxles

The pressure regulator valve assembly in a FWD transaxle, figure 9-34, differs from those in RWD transmissions and consists of: the pressure regulator valve, a throttle pressure plug, a large spring, and a small spring. The routing of hydraulic fluid is again different, there is no mainline pressure plug, and the small spring acts differently on the throttle pressure plug.

Fluid from the pump is delivered directly to the regulator valve through two ports. Fluid is also routed from the manual valve to the left of land #4 on the regulator valve.

When the engine is stopped and the pump is not producing any fluid flow, the pressure regulator valve is held in a balanced position between the two springs acting on its ends. When the engine is started, pump pressure from the manual valve builds against the left side of land #4, and together with the small throttle pressure plug spring, moves the regulator valve to the right against the tension of the larger spring. When the regulator valve moves far enough, it uncovers the port to the switch valve. Fluid then flows into the torque converter circuit and to the switch valve.

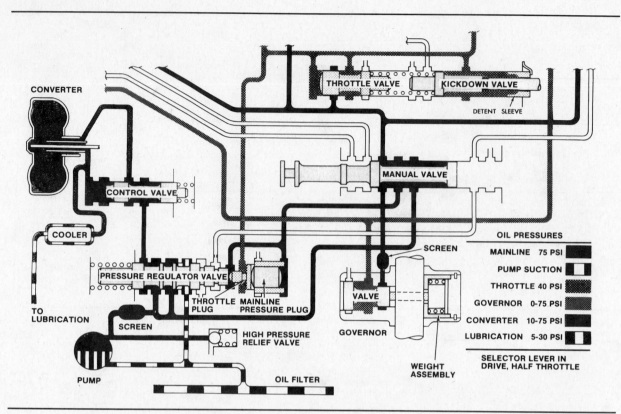

Figure 9-35. Throttle pressure, acting on the throttle pressure plug, raises mainline pressure by working against the mainline pressure plug in the pressure regulator valve assembly.

As pressure continues to increase and the regulator valve moves farther to the right, land #2 opens an exhaust port that allows a portion of the fluid to return through the pressure regulator valve to the sump. As the exhaust port opens and fluid is vented, hydraulic pressure and spring pressure balance each other to maintain a steady mainline pressure.

Note that as the position of the pressure regulator valve changes and fluid is exhausted from between lands #1 and #2, the pressure of the fluid bypassing land #1 into the torque converter circuit is also affected. As in late-model RWD transmissions, the pressure regulator valve in this design also regulates converter circuit pressure. As a result, all FWD transaxles use a switch valve to regulate fluid flow to the converter. This valve is described in detail later in the chapter.

Throttle valve effect on pressure regulation
Torqueflite pressure regulation is not controlled only by the regulator valve and the manual valve. The throttle valve affects mainline pressure, and the torque converter control valve on early RWD transmissions is also a pressure-regulating valve. We will examine the torque converter control valve and throttle valve in

detail later. At this point, we will look only at the throttle valve's effect on the pressure regulator valve.

When the gear selector is in any forward gear position, mainline pressure is routed to the throttle valve where it becomes throttle pressure. One of several passages routes the throttle pressure to the pressure regulator valve where it acts on the throttle pressure plug.

In RWD transmissions, figure 9-35, throttle pressure enters the pressure regulator valve assembly between lands of different diameters on the throttle plug. Throttle pressure against the larger land of the throttle plug works against mainline pressure on the smaller mainline pressure plug. As throttle pressure increases, the throttle plug moves the mainline pressure plug back into its bore. The spring then moves the regulator valve to restrict the exhaust port to the sump, thus increasing mainline pressure.

In FWD transaxles, figure 9-34, throttle pressure against the throttle pressure plug compresses the small spring acting on the left end of the pressure regulator valve. This reduces the force of the small spring, which allows the larger spring on the right end of the pressure regulator valve to move the valve further to the left. This restricts the exhaust port to the sump, and increases mainline pressure.

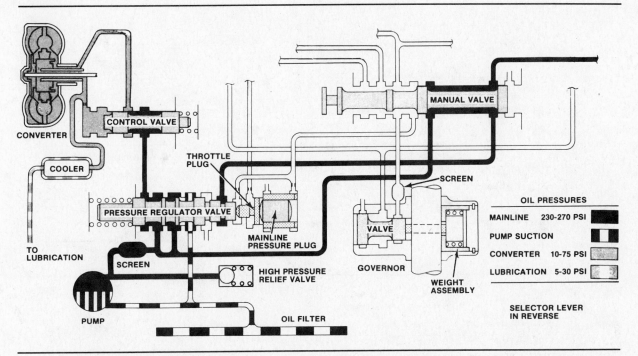

Figure 9-36. Reverse gear pressure regulation in early RWD Torqueflite transmissions.

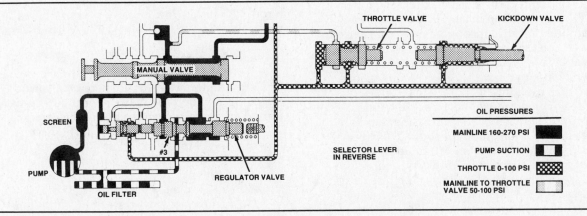

Figure 9-37. Reverse gear pressure regulation in late RWD Torqueflite transmissions.

Reverse gear effect on pressure regulation

In Reverse, mainline pressure must be high to hold clutches and bands securely under high-torque conditions. Once again, this is done in three different ways depending on the model of gearbox.

When an early (pre-1978) Torqueflite RWD transmission is put in Reverse, figure 9-36, the manual valve blocks the fluid passage to the mainline pressure plug and the right end of the regulator valve. Instead, fluid flows from the manual valve to the second chamber of the regulator valve, which offers less surface area for the fluid to act on. As a result, the spring moves the pressure regulator valve to the right

to restrict the exhaust port to the sump, and mainline pressure increases. In early RWD transmissions, reverse mainline pressure is about 260 psi or 1,790 kPa at all throttle positions.

When a 1978 or later RWD transmission is put in Reverse, figure 9-37, mainline pressure is still applied to the mainline pressure plug and to the reaction area on the left side of land #3, but the manual valve blocks fluid flow to the areas between the mainline pressure plug and the throttle pressure plug, and between the throttle pressure plug and the pressure regulator valve. As a result, the large spring moves

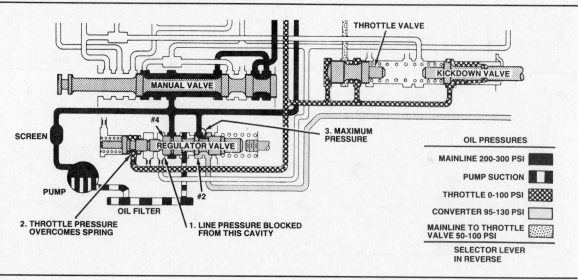

Figure 9-38. Reverse gear pressure regulation in FWD Torqueflite transaxles.

the pressure regulator valve to the left to restrict the exhaust port to the sump. This increases mainline pressure to approximately 160 psi or 1,100 kPa at idle.

Unlike the earlier design, the late-model RWD transmission hydraulic system routes mainline pressure through a metered orifice to the throttle valve when in Reverse. As the throttle is opened, throttle pressure acting on the throttle pressure plug forces the mainline pressure plug into its bore. This enables the large spring to move the pressure regulator valve further to the left, increasing mainline pressure to a maximum of about 270 psi or 1,860 kPa at full throttle. Modulating Reverse gear mainline pressure in this manner reduces the loads on the high-reverse clutch seals and increases their lifespan.

When a FWD transaxle is put in Reverse, figure 9-38, the manual valve blocks the flow of mainline pressure to the reaction area left of land #4 on the pressure regulator valve. As a result, the large spring moves the pressure regulator valve to the left to restrict the exhaust port to the sump. This increases mainline pressure to approximately 200 psi or 1,380 at idle.

As in late-model RWD transmissions, mainline pressure is supplied through a metered orifice to the throttle valve. As the throttle is opened, throttle pressure acting on the throttle pressure plug compresses the small spring acting on the left side of the pressure regulator valve. This enables the large spring to move the pressure regulator valve further to the left, increasing mainline pressure to a maximum of about 300 psi or 2,070 kPa at full throttle.

Torque Converter Control Valve

The torque converter control valve, figure 9-35, is a small pressure-regulating valve used in 1977 and earlier Torqueflite transmissions. The torque converter control valve is located next to the pressure regulator valve in the valve body, and controls oil pressure and flow into and out of the torque converter.

Before the engine is started, the control valve is pushed all the way into its bore by a spring on the right end of the valve. Once the engine starts and the pump produces sufficient pressure to open the converter feed port in the pressure regulator valve, fluid flows from the pressure regulator valve to the control valve. At first, fluid flows unrestricted through the control valve to fill the torque converter.

After the torque converter is full, fluid flows through a restriction to the left end of the control valve. As pressure starts to build, it moves the control valve to the right against spring tension to restrict the fluid passage to the converter. This limits the fluid volume and pressure to the desired level. If converter pressure drops, the control valve spring moves the valve to the left, opening the passage farther to increase the fluid supply to the converter.

From the converter, fluid flows to the transmission cooler in the car's radiator. The cooler provides a restriction that helps keep converter pressure within the desired operating range. Fluid leaves the cooler under a low pressure of 5 to 30 psi (35 to 207 kPa) and returns to the transmission case to lubricate gears, bushings, and other moving parts before returning to the sump.

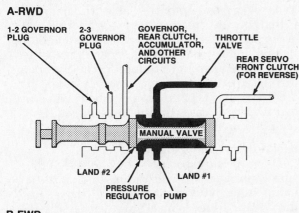

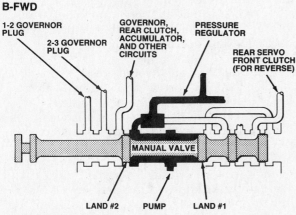

Figure 9-39. The manual valves used in Torqueflite RWD transmissions and FWD transaxles.

Switch Valve

In 1978 and later Torqueflite RWD transmissions, and in all FWD transaxles, torque converter circuit pressure is controlled by the pressure regulator valve as described earlier. In these applications, the torque converter control valve is replaced by a switch valve as shown in figures 9-33 and 9-34.

On models without a lockup torque converter, the switch valve simply directs fluid flow to the converter. Fluid returning from the converter is then routed back through the switch valve and from there to the transmission cooler.

On models with lockup torque converters, pressure applied to the end of the switch valve opposite the spring moves the valve and changes the direction of fluid flow to lock the converter clutch. Movement of the switch valve also affects the flow path of fluid to the transmission cooler. Switch valve operation is covered in detail in the section that explains the lockup and fail-safe valves.

Manual Valve

The Torqueflite manual valve is a directional valve. Its interaction with the pressure regulator valve has already been discussed, and in addition, it routes fluid to the clutches, servos, shift valves, and other valves of the hydraulic system figures 9-22 and 9-23.

The manual valve in RWD transmissions, figure 9-39A, has two large lands of equal diameter and one small land. The valley between the two large lands connects the pump and regulator circuits with the circuits for the various driving ranges. The valley between the large center land and the small land vents the forward drive circuits that are not in use. The manual valve in FWD transaxles, figure 9-39B, has a slightly different design, but operates in essentially the same manner.

The manual valve is operated by the manual valve lever at one side of the valve body, figures 9-28 and 9-29. The manual valve lever is connected by linkage or cable to the gear selector lever inside the vehicle. A spring-loaded detent ball in the valve body engages the "rooster comb" arm of the manual valve lever to hold the valve in the selected position. Another arm of the lever engages a groove in the end of the manual valve that sticks out of the valve body. A third arm of the manual valve lever has two contacts that operate the starting safety switch and backup light switch.

Throttle Valve and Kickdown Valve

Throttle pressure provides engine load information that is used to help determine when the transmission upshifts and downshifts. Without throttle pressure, the transmission would upshift and downshift at exactly the same speed regardless of throttle opening. The Torqueflite throttle valve produces throttle pressure by sensing engine torque through a mechanical linkage connected to the engine's accelerator linkage.

The throttle valve and kickdown valve are located next to the manual valve in the valve body, figures 9-29 and 9-30. The valve assembly consists of the throttle valve, the kickdown valve and detent sleeve, and a spring that separates the two valves. The throttle lever bears against the end of the kickdown valve that extends out of the valve body.

When the engine throttle is closed, the spring between the kickdown and throttle valves tends to push each valve to its end of the

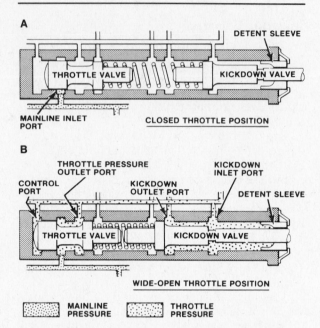

Figure 9-40. Torqueflite throttle valve and kickdown valve operation.

bore, figure 9-40A. The throttle valve inlet port for mainline pressure from the manual valve is closed. As the driver depresses the accelerator, the kickdown valve is pushed towards the throttle valve, and the spring between the valves begins to compress and transmit motion to the throttle valve. The throttle valve then starts to uncover the mainline pressure inlet port.

Fluid enters the mainline port and passes through the valley of the throttle valve to the outlet port. The mainline pressure becomes throttle pressure as it leaves the valve and enters the throttle pressure circuit. The throttle pressure circuit has several branches that lead to various other transmission valves. One of the branches leads back to a control port at the left end of the throttle valve itself, figure 9-40B.

As accelerator pedal force moves the kickdown valve farther into the valve bore, the throttle valve opens the mainline inlet port wider to increase throttle pressure. The increased throttle pressure is applied through the control port at the left end of the throttle valve, where it opposes the kickdown valve and spring pressure working on the right end of the throttle valve. These opposing forces balance the position of the throttle valve to provide a fixed throttle pressure for any given engine throttle opening.

As the engine throttle is opened and closed, the balance between the opposing pressures acting on the throttle valve constantly changes

to alter the position of the valve. The net result is that throttle pressure increases as the throttle is opened wider, stabilizes as the engine throttle opening is held constant, and decreases as the throttle opening is reduced. At wide-open throttle, the kickdown valve and throttle valve are pressed as far into the valve body as they can go. In this case, throttle pressure equals mainline pressure.

Kickdown valve operation

We have already discussed how throttle pressure is used as an auxiliary pressure on the regulator valve to boost mainline pressure as the throttle is opened wider. Throttle pressure is also used as an auxiliary pressure to delay full-throttle upshifts, and force full-throttle downshifts. The kickdown valve provides this control.

One branch of the throttle pressure circuit leads to the inlet port of the kickdown valve, figure 9-40B. When the engine throttle is opened wide and the accelerator linkage forces the kickdown valve all the way into its bore, the outlet port of the kickdown valve is opened. Throttle pressure, which is equal to mainline pressure at this time, then passes through the kickdown valve and into the auxiliary throttle pressure circuit. Depending on the application and operating conditions, fluid in the auxiliary throttle pressure circuit acts on the 1-2 shift valve, the 1-2 shift control valve, and/or the 2-3 shift valve. The operation of these valves is explained later in this chapter.

Governor Valve

The governor valve, or governor, senses vehicle speed from the rotation of the output shaft (RWD) or transfer shaft (FWD), and creates governor pressure that is used to help time transmission upshifts and downshifts. Governor pressure acts on one end of each shift valve, while throttle pressure and spring force act on the other. When governor pressure overcomes throttle pressure and spring force, the transmission upshifts. The governor design varies between RWD and FWD gearboxes, so they are explained separately below.

RWD governor operation

The RWD transmission governor assembly, figure 9-41, consists of the governor valve, an inner weight, an outer weight, and a spring. All of these are contained in a governor body that is mounted on the governor support. The governor support is splined to the output shaft, and the shaft passes through the center of the governor body. Fluid flows from the manual valve through the rear of the transmission case

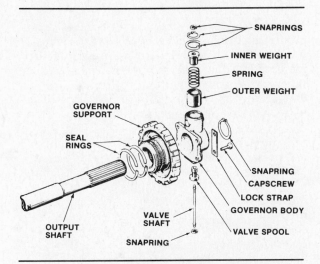

Figure 9-41. The governor assembly mounts on the output shaft in Torqueflite RWD transmissions.

and a passage in the output shaft to the governor, where it becomes governor pressure.

In Park, Reverse, and Neutral, the manual valve blocks mainline pressure from reaching the governor. In all forward driving ranges, mainline pressure is directed to the mainline (inlet) port of the governor, figures 9-42.

When the vehicle is stopped, figure 9-42A, mainline pressure is blocked by the small land of the governor valve spool. Any pressure that leaks past the valve spool at this time escapes through the exhaust port that is opened by the large land of the spool. As a result, governor pressure is zero.

As the vehicle begins to move and the output shaft starts to turn, figure 9-42B, the governor weights are thrown outward by centrifugal force. As the weights move, they pull on the valve shaft, which moves the valve spool inward toward the output shaft. This closes the exhaust port and opens the mainline pressure inlet. Mainline pressure then enters the governor valve where it becomes governor pressure and exits through the governor pressure outlet.

Two weights are used to regulate pressure accurately. At low vehicle speeds, only the heavier outer weight moves. It works against spring pressure to transmit force to the valve shaft and move the spool. When the outer weight moves far enough to fully compress the spring, the inner weight begins to move as well.

Governor pressure continues to increase as centrifugal force increases, but it is modulated by pressure acting against the larger land inside the governor valve. This creates a force that tends to hold the valve back against the move-

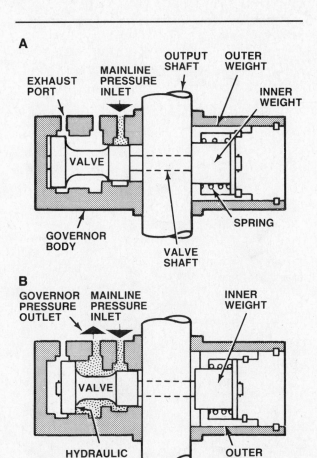

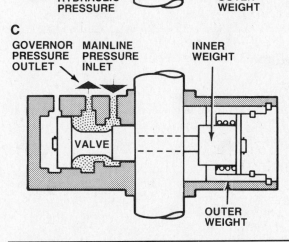

Figure 9-42. Torqueflite RWD transmission governor operation.

ment of the weights. As a result of these opposing forces, governor pressure increases with vehicle speed, and stabilizes whenever vehicle speed is constant.

At very high vehicle speeds, the governor valve is open as far as possible, figure 9-42C. At this point, governor pressure equals mainline pressure. Governor pressure can never be higher than mainline pressure, however.

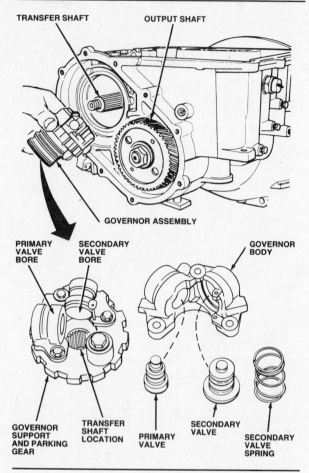

TRANSFER SHAFT   OUTPUT SHAFT

GOVERNOR ASSEMBLY

PRIMARY   SECONDARY   GOVERNOR
VALVE   VALVE   BODY
BORE   BORE

GOVERNOR   TRANSFER   SECONDARY
SUPPORT   SHAFT   PRIMARY   VALVE   SECONDARY
AND PARKING   LOCATION   VALVE   VALVE
GEAR   SPRING

Figure 9-43. The Torqueflite governor assembly mounts on the transfer shaft in FWD transaxles.

FWD governor operation

The FWD transaxle governor assembly, figure 9-43, mounts on the transfer shaft, and consists of a primary valve, a secondary valve, and a secondary valve spring. All of these are contained in a governor body that is mounted on the governor support. The governor support is splined to the transfer shaft, which passes through its center.

In Park, Reverse, and Neutral, the manual valve blocks mainline pressure from reaching the governor. In all forward driving ranges, mainline pressure is directed through an inlet port to the centers of both the primary and secondary governor valves, figure 9-44. When the vehicle is stopped in a forward gear, mainline pressure moves both valves fully inward toward the transfer shaft. In addition, the secondary valve spring helps hold the secondary valve inward.

Once the car begins to move, transfer shaft rotation spins the governor. Actual pressure regulation occurs in two stages. At lower

speeds (the primary stage), centrifugal force moves the primary valve away from the transfer shaft, which allows mainline pressure to be metered into the governor pressure chamber where it becomes governor pressure, figure 9-44A.

As transfer shaft speed continues to increase, mainline pressure and centrifugal force also increase. To regulate the increase in governor pressure, the primary valve has a reaction area in the governor pressure chamber. Governor pressure acting against this area opposes outward movement of the primary valve. As a result of the opposing forces, governor pressure increases with vehicle speed, and stabilizes whenever vehicle speed is constant.

When transfer shaft rotation exceeds a certain higher speed (the secondary stage), centrifugal force overcomes the hydraulic and spring pressures acting on the secondary valve, figure 9-44B. The secondary valve then moves outward, and mainline pressure is metered through the valve where it becomes secondary pressure. The secondary pressure is then routed to the reaction area on the outside of the innermost land of the primary valve. This further limits the rate of governor pressure increase.

At very high car speeds, the primary valve is open as far as possible, and governor pressure equals mainline pressure. Governor pressure can never exceed mainline pressure, however.

Accumulator

The accumulator in the Torqueflite RWD transmission, figure 9-45, cushions engagement of the forward clutch when the gear selector is moved to any forward driving range. The accumulator also cushions application of the intermediate band on a 1-2 upshift. In FWD transaxles, the accumulator cushions only the application of the intermediate band on the 1-2 upshift. The accumulator in both RWD and FWD gearboxes is located in the case between two servos. In RWD transmissions, the lower end of the accumulator is covered by the valve body transfer plate. In FWD transaxles, the accumulator has a separate cover plate.

The accumulator uses a piston with two different diameters that looks something like a spool valve. The piston fits within a 2-diameter cylinder in the case. A metal ring seals each diameter of the piston to its bore. The larger end of the piston is at the lower end of the cylinder. A spring is placed between the larger end of the piston and the valve body transfer plate (RWD) or the accumulator cover plate (FWD).

A — PRIMARY STAGE

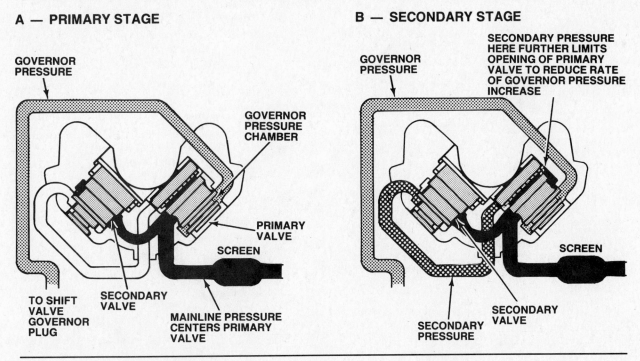

GOVERNOR PRESSURE

GOVERNOR PRESSURE CHAMBER

PRIMARY VALVE

SCREEN

TO SHIFT VALVE GOVERNOR PLUG

SECONDARY VALVE

MAINLINE PRESSURE CENTERS PRIMARY VALVE

B — SECONDARY STAGE

GOVERNOR PRESSURE

SECONDARY PRESSURE HERE FURTHER LIMITS OPENING OF PRIMARY VALVE TO REDUCE RATE OF GOVERNOR PRESSURE INCREASE

SCREEN

SECONDARY PRESSURE

SECONDARY VALVE

Figure 9-44. Torqueflite FWD transaxle governor operation.

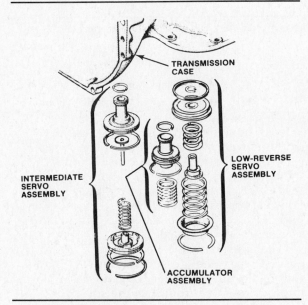

TRANSMISSION CASE

LOW-REVERSE SERVO ASSEMBLY

INTERMEDIATE SERVO ASSEMBLY

ACCUMULATOR ASSEMBLY

Figure 9-45. Accumulator and servo installations in the Torqueflite RWD transmission.

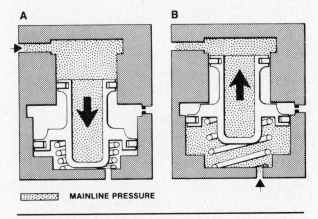

MAINLINE PRESSURE

Figure 9-46. Torqueflite accumulator operation.

Accumulator operation is shown in figure 9-46. When the manual valve is moved to any forward driving position in a RWD transmission, mainline pressure flows to the 1-2 shift valve, the forward clutch, and the accumulator. Fluid enters the accumulator at the small, upper end of the cylinder on 8-cylinder transmissions, or at the center of the cylinder between the two

piston diameters on 6-cylinder transmissions. In either case, the piston is pushed downward in its bore and compresses the spring. Because mainline pressure in this same circuit flows to the forward clutch, movement of the accumulator piston causes clutch apply pressure to build up gradually, which smoothes clutch engagement. The forward clutch is not fully applied until the accumulator piston is fully seated against the transfer plate.

The accumulators in both RWD and FWD Torqueflites cushion application of the intermediate band. When the governor plug of the

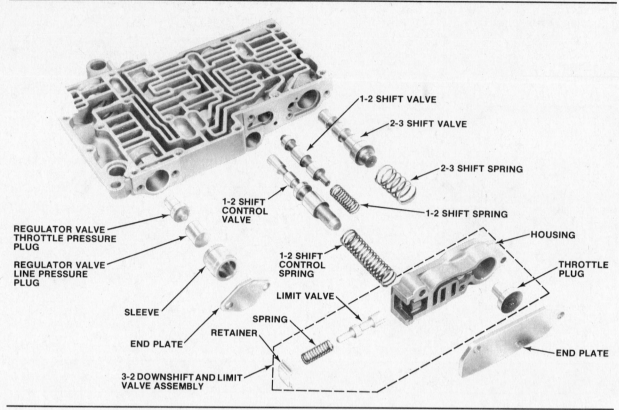

Figure 9-47. Torqueflite RWD transmission shift valves and pressure regulator valve plugs for an 8-cylinder application.

1-2 shift valve moves the valve for the 1-2 up-shift, the shift valve directs mainline pressure to the intermediate servo and the spring end of the accumulator. Mainline pressure and spring force then move the accumulator piston upward. Because fluid is diverted to the accumulator, pressure builds gradually in the intermediate servo to smooth band application. The intermediate band is not fully applied until the accumulator reaches the top of its bore.

In an 8-cylinder Torqueflite RWD transmission with a 1-2 shift control valve, shift control pressure is applied to the center chamber of the accumulator at the same time mainline pressure is applied to the small end. Shift control pressure is throttle pressure that has been metered by the 1-2 shift control valve. As the transmission makes the 1-2 upshift, mainline pressure enters the large end of the accumulator to move the piston upward. The combination of shift control pressure in the center chamber of the accumulator, and mainline pressure on the small end, makes application of the intermediate band smoother than it would be with mainline pressure acting on the small end alone.

1-2 Shift Valve

The 1-2 shift valve controls the upshift from low to second, and the downshift from second to low. The complete 1-2 shift valve assembly is located toward the narrow end of the upper valve body, figure 9-47, and consists of the 1-2 shift valve, a spring at one end, and a governor plug at the other. The governor plug is inserted from the opposite side of the valve body, figure 9-48. Figure 9-49 shows the same components as used in FWD transaxles.

When the manual valve is in any forward position, mainline pressure is routed to the 1-2 shift valve. As the accelerator is depressed, throttle pressure is routed to the spring end of the shift valve, figure 9-50A. Throttle pressure and spring force hold the valve so that mainline pressure is blocked by one of its lands.

As the car accelerates, governor pressure builds up and is applied to the governor plug at the other end of the shift valve. When governor pressure overcomes throttle pressure and spring force, figure 9-50B, the governor plug moves the shift valve towards the spring. This opens the mainline port so that mainline pressure can pass through the valve to cause the 1-2 upshift.

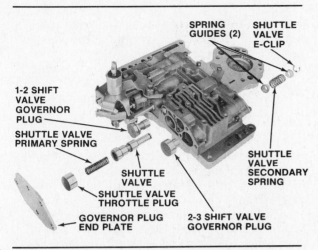

Figure 9-48. Torqueflite RWD transmission shuttle valve and shift valve governor plugs for an 8-cylinder application.

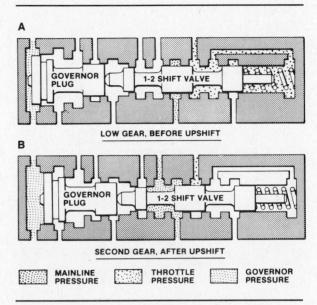

Figure 9-50. Torqueflite 1-2 shift valve operation.

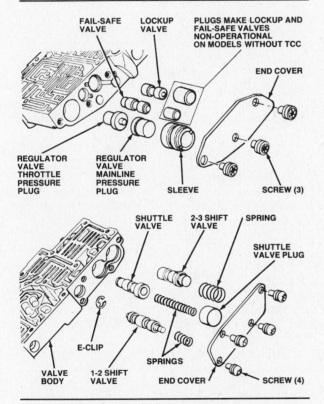

Figure 9-49. Torqueflite FWD transaxle shift valves, shuttle valve, and shift valve governor plugs.

As the shift valve moves to open the mainline port, it also closes the throttle pressure port. Any leftover throttle pressure inside the valve is vented to the sump. Once the shift valve shifts, governor pressure has only spring pressure to oppose it. Therefore, governor pressure can vary within a given range as vehicle speed changes slightly, without the shift valve snapping back and forth.

1-2 shift control valve

Torqueflite RWD transmissions for 8-cylinder cars have a 1-2 shift control valve not used in other applications. The 1-2 shift control valve is located next to the 1-2 shift valve in the valve body, figure 9-47, and has two functions. First, it helps control the quality of 1-2 upshifts. Second, it helps control the quality and timing of part-throttle and full-throttle 3-2 downshifts.

When the transmission is in Drive low or Drive second, the 1-2 shift control valve sends shift control pressure (modulated throttle pressure) to the center chamber of the accumulator as described earlier. This pressure works with mainline pressure in the accumulator to cushion application of the intermediate band and smooth the 1-2 upshift. The shift control valve regulates the cushioning effect of the accumulator.

For a full-throttle downshift, kickdown throttle pressure is routed between the kickdown valve and the 1-2 shift control valve to create an additional pressure at the spring end of the 1-2 shift valve. This boosts throttle pressure at full throttle for a somewhat firmer and later 2-3 upshift.

As the 2-3 upshift is completed, mainline pressure is directed from the shuttle valve to the center chamber of the shift control valve. The shift control valve then moves against its spring to vent shift control pressure and close the port from the kickdown valve.

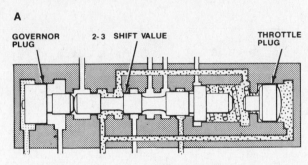

SECOND GEAR, BEFORE UPSHIFT

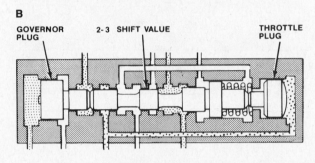

HIGH GEAR, AFTER UPSHIFT

Figure 9-51. Torqueflite 2-3 shift valve operation.

2-3 Shift Valve

The 2-3 shift valve controls the upshift from second to third, and the downshift from third to second. The complete 2-3 shift valve assembly is located toward the narrow end of the upper valve body, figure 9-47. The assembly consists of the 2-3 shift valve, a spring, a throttle plug (8-cylinder applications only), and a governor plug that is inserted from the opposite side of the valve body, figure 9-48. Figure 9-49 shows the same components as used in FWD transaxles.

Once the transmission is in Drive second, mainline pressure from the manual valve passes through the 1-2 shift valve. One branch of this circuit routes mainline fluid to the 2-3 shift valve where it is blocked by a land on the valve, figure 9-51A.

In second gear, throttle pressure is routed to the 2-3 shift valve where it passes through the valve between two lands at one end, and then through a passage to a chamber at the spring end. In 8-cylinder applications, throttle pressure is also routed around the valve to a chamber behind the throttle plug. Throttle pressure and spring pressure combine to hold the 2-3

shift valve in the position that blocks the passage of mainline pressure for the 2-3 upshift.

As vehicle speed increases, governor pressure increases against the head of the governor plug at one end of the 2-3 shift valve. When governor pressure is high enough, the governor plug moves the shift valve against spring and throttle pressure, figure 9-51B. As the shift valve moves, it closes the throttle pressure inlet port and opens the mainline port. Mainline pressure then passes through the valve to apply the high-reverse clutch and release the intermediate servo.

Mainline pressure that passes through the 2-3 shift valve is also routed to the governor end of the shift valve where it acts equally on the ends of the valve and the governor plug. As vehicle speed and governor pressure drop, this mainline pressure helps keep the shift valve open until governor pressure is low enough to permit the 1-2 shift valve to close.

In 8-cylinder RWD transmissions, throttle pressure is still applied to the throttle plug at the spring end of the 2-3 shift valve after the upshift to high gear. Throttle pressure acting on the throttle plug moves the 2-3 shift valve to provide a faster part-throttle downshift at low speeds when the throttle is opened fairly wide.

Six-cylinder RWD transmissions and all FWD transaxles do not have a throttle plug at the end of the 2-3 shift valve. In these gearboxes, throttle pressure is not completely cut off from the end of the shift valve after the shift into high gear — it is simply reduced. This allows governor pressure to move the valve for an upshift and provides enough leftover throttle pressure for a part-throttle downshift.

For a full-throttle downshift, the kickdown valve opens another throttle pressure passage to both the 1-2 and 2-3 shift valves. This auxiliary throttle pressure can overcome governor pressure at either or both of the shift valves for the required 3-2, 2-1, or 3-1 downshift.

When the transmission is in manual low or manual second, mainline pressure is routed to the center chamber of the 2-3 governor plug. This holds the plug to the left against governor pressure, and locks the 2-3 shift valve in the second-gear position to prevent an upshift.

Limit Valve

The limit valve, figure 9-52, is used in 8-cylinder Torqueflite RWD transmissions to prevent a part-throttle 2-3 downshift at medium to high speeds. The limit valve consists of a valve spool and a spring. Throttle pressure passes through the center chamber of the limit on the way to the throttle plug of the

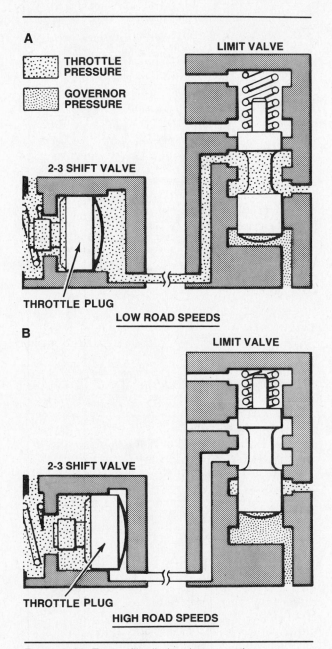

Figure 9-52. Torqueflite limit valve operation.

2-3 shift valve. Governor pressure acts on one end of the limit valve, and spring pressure acts on the other.

At low-to-medium speeds, the limit valve is pushed against the end of its bore by the spring, figure 9-52A. Governor pressure is present at the other end of the limit valve, but it is not strong enough to overcome spring pressure. Throttle pressure flows through the limit valve to the 2-3 shift valve throttle plug. If the accelerator is depressed part way, throttle pressure is free to move the throttle plug for a 3-2 downshift.

At higher speeds, governor pressure increases against the end of the limit valve and overcomes spring pressure. This shifts the limit valve to block throttle pressure to the 2-3 throttle plug, figure 9-52B. Then, if the accelerator is depressed part way, throttle pressure cannot move the throttle plug so there is no 3-2 downshift.

Shuttle Valve

The shuttle valve, figures 9-48 and 9-49, is located in the valve body above the shift valves, and works with the accumulator to cushion application of the intermediate servo and band. The valve assembly consists of the shuttle valve spool, a throttle plug, and a spring. The shuttle valve in 8-cylinder RWD transmissions has a second spring. One end of the valve spool sticks through the valve body and is retained on the outside by an E-clip that limits its inward travel.

During a part- or full-throttle 1-2 upshift, throttle pressure on the throttle plug holds the shuttle valve toward the retainer end of its bore. This opens a bypass circuit for mainline pressure on the way to the intermediate servo. Pressure in the bypass circuit helps the accumulator cushion application of the intermediate servo.

Once the servo is fully applied and the accumulator is at the top of its cylinder, mainline pressure builds in the shuttle valve and moves the valve against throttle pressure to close the bypass circuit. Mainline pressure closes this circuit quickly in case of a lift-foot upshift. This provides a quick release of the band to prevent any engine braking during the lift-foot upshift.

Bypass Valve

The bypass valve, used in FWD transaxles, provides quick venting of the high-reverse clutch and intermediate servo release circuit during a 3-2 downshift; this helps prevent clutch slippage. Use of the bypass valve also allows using a smaller orifice in the apply circuit to the high-reverse clutch, which improves the quality of the 2-3 upshift.

With the transaxle in second gear, figure 9-53, the bypass valve spring seats the valve to the right. When the 2-3 upshift occurs, mainline pressure from the shift valve first fills the passage to the orifice and shuttle valve, and then the circuit through the shuttle valve to the bypass valve. A transfer passage in the bypass valve supplies pressure to the reaction area on the right end of the valve, which moves the bypass valve to the left against spring pressure,

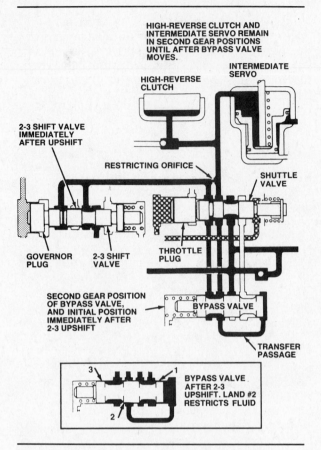

Figure 9-53. Bypass valve operation during a 2-3 upshift.

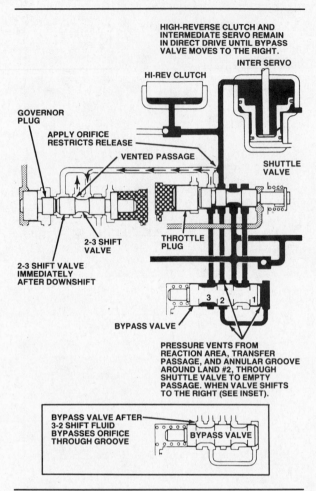

Figure 9-54. Bypass valve operation during a 3-2 downshift.

figure 9-53 (inset). The second land of the bypass valve then blocks the free flow of fluid through the bypass valve to the high-reverse clutch and intermediate servo. As a result, all of the apply pressure for these components must pass through the restricting orifice, which slows the buildup of pressure and improves 2-3 upshift quality.

When throttle pressure forces a 3-2 downshift, figure 9-54, the 2-3 shift valve immediately vents fluid from the passage to the restricting orifice and shuttle valve. However, the orifice initially restricts venting of fluid from the high-reverse clutch and intermediate servo to maintain direct drive. Fluid also flows from the reaction area and transfer passage of the bypass valve, back to the shuttle valve, where it continues on to the 2-3 shift valve and is vented. Spring pressure moves the bypass valve back to the right, opening the bypass groove to the high-reverse clutch and intermediate servo release circuit, figure 9-54 (inset). The bypass groove rapidly vents fluid from the clutch and servo, allowing a quick 3-2 downshift.

Lockup and Fail-Safe Valves

In 1978, Chrysler introduced a lockup torque converter for the Torqueflite RWD transmission. In 1986, electronic control of the torque converter clutch was added in some RWD applications. Selected FWD transaxles received the electronically controlled converter clutch in 1987. The construction of the lockup torque converter is described in Chapter 7. This section deals only with the hydraulic valves that control converter clutch operation.

In Torqueflites with a lockup torque converter, an auxiliary valve body contains the lockup and fail-safe valves that control the converter clutch, figure 9-55. The lockup valve controls the pressure used to lock and unlock the clutch. The fail-safe valve prevents lockup unless the gearbox is in high gear.

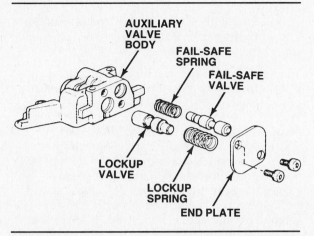

Figure 9-55. The lockup and fail-safe valves for lockup torque converters are located in an auxiliary valve body.

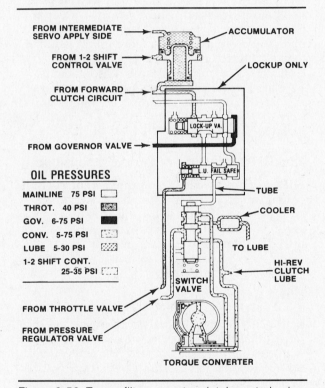

Figure 9-56. Torqueflite converter clutch control valve positions in Drive second, before lockup.

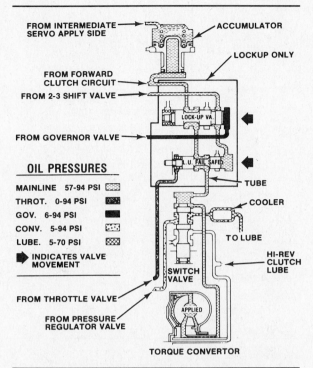

Figure 9-57. Torqueflite converter clutch control valve positions in Drive high, after lockup.

When the transmission is placed in Drive, figure 9-56, mainline pressure from the forward clutch circuit is routed to the lockup valve. However, the pressure cannot enter because the valve is held to the right by spring pressure, and a land on the valve blocks the inlet port. At the same time, converter pressure from the regulator valve flows through the switch valve to the torque converter where it forces the clutch piston away from the converter cover to keep the converter clutch unlocked. Pressure from the converter is routed back through the switch valve, and from there to the transmission cooler.

As the car begins to move, governor pressure is applied to the right side of the upshift valve. When the vehicle speed reaches approximately 27 mph (43 kph) for 8-cylinder cars, or 31 mph (50 kph) for 6-cylinder vehicles, governor pressure overcomes spring pressure and moves the lockup valve to the left. This allows mainline pressure to pass through the lockup valve to the fail-safe valve. The converter clutch still does not apply, however, unless the transmission is in high gear.

Whenever the accelerator is depressed, throttle pressure is routed to the spring end of the fail-safe valve to help hold the valve to the right. When the transmission shifts into high gear, figure 9-57, mainline pressure from the 2-3 shift valve is routed to the right side of the fail-safe valve. This moves the fail-safe valve to the left against spring pressure and throttle pressure, which opens a passage from the upshift valve to the switch valve.

Once both the upshift valve and the failsafe valve have been shifted, mainline pressure moves the switch valve downward. The switch valve then blocks the flow of converter pressure

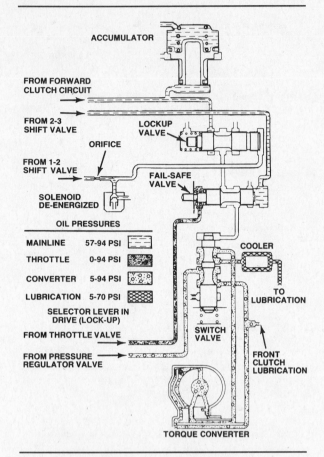

ACCUMULATOR

FROM FORWARD
CLUTCH CIRCUIT

FROM 2-3
SHIFT VALVE   LOCKUP
                                VALVE
        ORIFICE

FROM 1-2
SHIFT VALVE
                    FAIL-SAFE
                    VALVE
SOLENOID
DE-ENERGIZED

OIL PRESSURES

MAINLINE	57-94 PSI	
THROTTLE	0-94 PSI	
CONVERTER	5-94 PSI	
LUBRICATION	5-70 PSI	

SELECTOR LEVER IN
DRIVE (LOCK-UP)

FROM THROTTLE VALVE

FROM PRESSURE
REGULATOR VALVE

COOLER

TO
LUBRICATION

SWITCH
VALVE

FRONT
CLUTCH
LUBRICATION

TORQUE CONVERTER

Figure 9-58. Torqueflite electronic converter clutch control valve positions in Drive high, before lockup. Note that the TCC solenoid is venting pressure that will be used to shift the lockup valve.

that was holding the converter clutch in the un-applied position, and routes mainline pressure between the clutch piston and turbine to apply the clutch. At the same time, the switch valve reroutes the flow of converter pressure from the regulator valve directly to the transmission cooler.

During wide-open throttle conditions, a combination of high throttle pressure and spring pressure moves the fail-safe valve back to the right. This cuts off mainline pressure to the switch valve, and unlocks the converter clutch to allow torque multiplication in the converter for faster acceleration. If there is a forced 3-2 downshift, mainline pressure to the right side of the fail-safe valve is lost, and the valve again moves to the right to unlock the converter clutch. Similarly, if the vehicle speed in high gear drops below the minimum lockup speed, spring pressure will overcome governor pressure on the lockup valve, moving the valve to the right and unlocking the converter clutch. As you can see, lockup cannot occur or be

maintained without the proper combination of governor pressure, throttle pressure, and mainline pressure from the 2-3 shift valve.

Electronic torque converter clutch control

Torqueflites with electronic torque converter clutch control continue to use a lockup valve and fail-safe valve, figure 9-58, and the basic hydraulic operation is similar to the earlier design with a few exceptions. First, mainline pressure from the 2-3 shift valve to the fail-safe valve is routed through a separate passage, rather than through the lockup valve. Second, mainline pressure, not governor pressure, is used to upshift the lockup valve. Finally, the flow of mainline pressure to the lockup valve is controlled by a TCC solenoid mounted on the valve body.

When the solenoid is not energized, it vents the mainline pressure and prevents converter clutch lockup. When the solenoid is energized, mainline pressure is no longer vented, and the lockup valve shifts to apply the converter clutch. The operation of the TCC solenoid is regulated by the electronic engine control system based on several vehicle operating conditions. These electronic controls are explained in detail in Chapter 8.

Multiple-Disc Clutches And Servos

The preceding sections examined all of the individual valves in the Torqueflite hydraulic system. Before we go on to tracing the complete system operation in each driving range, including lockup torque converter operation, we will take a quick look at the two clutches and the servos of the Torqueflite.

Forward clutch

The forward (rear) clutch, figure 9-59, is applied to transmit driving torque in all forward gears. The number of discs and separator plates varies with different vehicle and engine applications. The forward clutch uses a Belleville spring to transmit force from the clutch piston to the clutch pack. The Belleville spring multiplies the applying force of the piston and acts as a piston return spring.

Hydraulic pressure is supplied to the forward clutch piston through the input shaft. When the clutch is released, a vent and a ball-check valve in the drum discharges fluid quickly to prevent clutch drag.

High-reverse clutch

The high-reverse (front) clutch, figure 9-59, is applied in third and Reverse gears to drive the sun gear. Depending on the application, the high-reverse clutch uses one large coil spring,

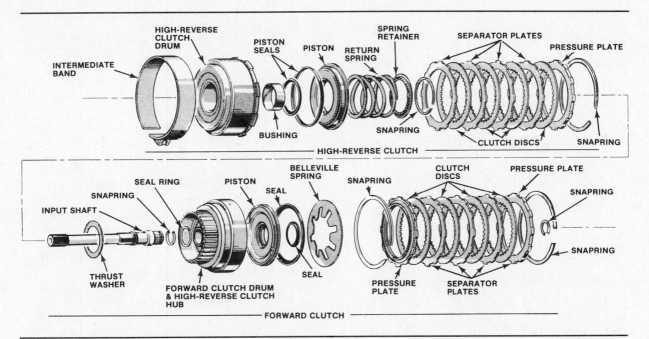

Figure 9-59. These Torqueflite clutch assemblies are from an A-904 RWD transmission which uses a single large return spring for the high-reverse clutch. Torqueflite FWD transaxles are similar, but the A-727 RWD transmission has several small springs.

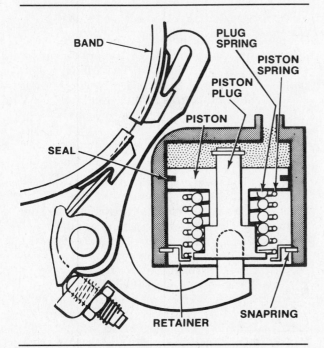

Figure 9-60. Torqueflite low-reverse servo operation.

or several smaller coil springs, to return the piston when hydraulic pressure is released.

Fluid is supplied to the clutch piston through passages in the gearbox side of the stator support, which is bolted to the pump housing. The high-reverse clutch also uses a vent and a ball-check valve to discharge fluid quickly when the clutch is released. A fiber thrust washer in front of the high-reverse clutch hub absorbs any thrust loads between the forward and the high-reverse clutches.

Low-reverse servo

The low-reverse (rear) servo, figure 9-60, applies the low-reverse band in Reverse and manual low. The servo is a single-diameter piston located in a bore in the right rear of the transmission case. Fluid is routed to the top of the servo to apply the band, and the servo is released by spring pressure.

When the servo and band are released, the piston is held at the top of the bore by a piston spring. The piston spring is held at the lower end of the bore by a retainer and snap ring. A plug in the center of the piston is held in an extended position by a plug spring. A retaining clip at the top of the plug keeps it from coming out of the piston.

As the servo is first applied, the piston and plug move down together to take up the clearance in the cantilever linkage that applies the band. Once the slack is absorbed, plug movement is restricted and further piston movement compresses the plug spring until the piston seats against a shoulder on the plug. Until the piston seats, the band is tightened gradually by increasing spring tension — this cushions the shift. Once bottomed, the piston and plug move together to fully apply the band.

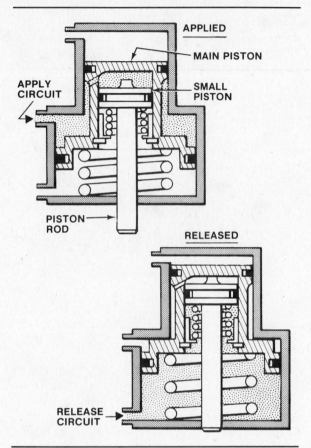

Figure 9-61. Torqueflite controlled-load intermediate servo operation.

Intermediate servo

The intermediate (kickdown or front) servo, figure 9-61, applies the intermediate band for second gear. The servo is a 2-diameter piston in a bore at the right front of the transmission case. The piston is sealed in the bore with metal seal rings. Servo apply pressure is routed to the center chamber between the two piston lands. Servo release pressure is applied to the lower, larger chamber. The upper end of the servo bore is vented to the case. A combination cover plate and piston rod guide closes the bottom of the servo bore and is held in place with a snapring.

Most passenger car Torqueflites have a controlled-load servo like that shown in figure 9-61. Inside the main servo piston is a smaller inner piston attached to the piston rod. The inner piston is held in position by a spring and retainer. A passage in the main piston allows fluid flow to the top of the inner piston.

To prevent slippage on a 2-3 upshift, the intermediate band must hold until the high-reverse clutch begins to apply. However, the band must also release almost instantly before the clutch is fully applied. Servo release timing

is critical. The inner piston in a controlled-load intermediate servo provides added cushioning for band application, and also allows faster band release.

When the servo is applied on a 1-2 upshift, fluid fills the center chamber of the piston, and flows through the passage to the top of the inner piston. The inner piston moves first to take up clearance in the band and provide smooth application. Once the inner piston seats on a ledge in its bore, the main piston moves to fully apply the band.

On a 2-3 upshift or a 2-1 downshift, fluid is supplied to the lower side of the main piston to quickly release the servo. On a 3-2 downshift, fluid trapped between the main piston and the inner piston cushions the application of the intermediate band.

Some heavy-duty and high-performance Torqueflites do not have the controlled-load intermediate servo. These transmissions use an intermediate servo with no inner piston. This type of servo does not have the cushioning effect and provides a firmer band application.

HYDRAULIC SYSTEM SUMMARY

We have examined the individual valves, clutches, and servos in the hydraulic system and discussed their function. The following paragraphs summarize their combined operation in the various gear ranges. Refer to figures 9-22 and 9-23 to trace the hydraulic system operation.

Park and Neutral

When the gear selector is moved to Park or Neutral, oil flows from the pump to the pressure regulator valve and the manual valve. The manual valve blocks oil flow in Park and Neutral. The pressure regulator valve sends pressure to: the torque converter control valve or switch valve, the throttle valve, the kickdown valve, and the accumulator (FWD).

The torque converter control valve or switch valve sends oil to fill the converter, oil that returns through the transmission cooler and lubrication circuits. In a lockup torque converter, the switch valve sends oil through the passage that unlocks the converter.

The throttle valve regulates mainline pressure according to engine load and routes throttle pressure to the spring ends of the 1-2 and 2-3 shift valves to oppose upshifts. These throttle pressure applications are made in all gear ranges.

In Park and Neutral, all clutches and bands are released. Valve positions are determined by spring force and throttle pressure alone. In

Park, a mechanical linkage engages a pawl with the lugs of the parking gear (sprag) on the governor support to lock the transmission output shaft.

Drive Low

When the gear selector is moved to Drive, mainline pressure is routed through the manual valve to: the governor valve, the apply area of the forward clutch, the 1-2 shift valve, the accumulator (RWD), and the lockup valve (in transmissions so equipped).

The governor valve regulates mainline pressure in relation to vehicle speed and then applies governor pressure to the 1-2 and 2-3 governor plugs, and the spring end of the shuttle valve. In RWD vehicles, governor pressure also flows to ends of the limit and lockup valves opposite spring force. These governor pressure applications are made in all forward gear ranges.

Mainline pressure applies the forward clutch and is blocked at the downshifted 1-2 shift valve. In RWD vehicles, the forward clutch application is cushioned by the accumulator. The lockup valve does not operate in low.

Drive Second

As governor pressure increases, it upshifts the 1-2 shift valve. Mainline pressure flows through the upshifted valve to: the apply side of the intermediate servo, the 2-3 shift valve, and the shuttle valve.

The intermediate servo applies the intermediate band. The accumulator and shuttle valve cushion the band application. Mainline pressure is blocked at the downshifted 2-3 shift valve.

Drive Third

As governor pressure increases further, it upshifts the 2-3 shift valve. Mainline pressure then flows through the 2-3 shift valve to: the shuttle valve, the release side of the intermediate servo, the fail-safe valve, and the apply area of the high-reverse clutch.

With the help of a spring, the intermediate servo releases. Mainline pressure applies the high-reverse clutch, and opens the fail-safe valve. The lockup torque converter may or may not be applied, depending on vehicle speed or TCC solenoid operation.

Drive Range Forced Downshifts

A 3-2 forced downshift occurs at speeds below 65 to 70 mph (105 to 113) when the throttle is opened wide and increased throttle pressure moves the throttle valve to its extreme inner position. This allows throttle pressure into the kickdown circuits where it combines with spring force to overcome mainline and governor pressure and downshift the 2-3 shift valve. A full-throttle 3-1 downshift occurs in the same way at speeds below 25 to 35 mph (40 to 48 kph).

A 2-1 part-throttle downshift can be forced at speeds from 16 to 30 mph (26 to 48 kph). Throttle pressure increases enough to overcome governor pressure at the 1-2 shift valve and produce the downshift.

Manual Second

When the gear selector is placed in manual second, mainline pressure from the manual valve is routed to the 2-3 shift valve governor plug, and the spring end of the 2-3 shift valve. This moves the valve to provide a 3-2 downshift, and prevents an upshift at any road speed. When the gear selector is placed in manual second and the car is accelerated from a standstill, the transmission starts in Drive low and then upshifts to manual second.

Manual Low

When the gear selector is placed in manual low, mainline pressure from the manual valve is directed between the two lands of the 1-2 shift valve governor plug, to the spring end of the 1-2 shift valve, and to the apply side of the low-reverse servo. The servo applies the low-reverse band. Mainline pressure on the governor plug and the end of the 1-2 shift valve moves the valve to provide a 2-1 downshift and prevent an upshift at any road speed.

If the driver shifts from Drive third to manual low at high speed, the transmission does not downshift directly from high to low because governor pressure is still too strong. The downshift pressure first overcomes governor pressure at the 2-3 shift valve, causing a 3-2 downshift. When road speed drops to about 20 mph or 30 kph, the downshift to low takes place.

Reverse

When the gear selector is placed in Reverse, mainline pressure from the manual is routed to: the apply area of the high-reverse clutch, the apply side of the low-reverse servo, and the pressure regulator valve. The high-reverse clutch and low-reverse band apply, and the pressure regulator valve raises mainline pressure to provide increased clutch and band holding power.

Review Questions

Select the single most correct answer
Compare your answers to the correct answers on page 603

1. Mechanic A says the A-904 Torqueflite was introduced in the 1956 Imperial.
   Mechanic B says the A-727 Torqueflite is sometimes called the Torqueflite 6.
   Who is right?
   a. Mechanic A
   b. Mechanic B
   c. Both mechanic A and B
   d. Neither mechanic A nor B

2. Mechanic A says International Harvester used a version of the Torqueflite called the T-904.
   Mechanic B says American Motors used versions of the A-727 Torqueflite.
   Who is right?
   a. Mechanic A
   b. Mechanic B
   c. Both mechanic A and B
   d. Neither mechanic A nor B

3. Mechanic A says the Torqueflite FWD transaxle was introduced in 1978.
   Mechanic B says the first Torqueflite FWD transaxle was the A-999 model.
   Who is right?
   a. Mechanic A
   b. Mechanic B
   c. Both mechanic A and B
   d. Neither mechanic A nor B

4. Mechanic A says early Torqueflite FWD transaxles had a separate sump for the final drive.
   Mechanic B says Chrysler recommends a special ATF+ for 1987 and later Torqueflites, and earlier RWD transmissions with lockup torque converters.
   Who is right?
   a. Mechanic A
   b. Mechanic B
   c. Both mechanic A and B
   d. Neither mechanic A nor B

5. Mechanic A says Torqueflite FWD transaxles have a rotor-type oil pump.
   Mechanic B says that Torqueflite RWD transmissions have a gear-type oil pump.
   Who is right?
   a. Mechanic A
   b. Mechanic B
   c. Both mechanic A and B
   d. Neither mechanic A nor B

6. Mechanic A says Torqueflite gear selectors have 6 positions.
   Mechanic B says the Torqueflite rear pump was eliminated in 1966.
   Who is right?
   a. Mechanic A
   b. Mechanic B
   c. Both mechanic A and B
   d. Neither mechanic A nor B

7. Mechanic A says the Torqueflite RWD transmission part number is stamped onto a tag riveted to the rear servo cover.
   Mechanic B says the Torqueflite FWD transaxle models are identified by the shape of the bellhousing.
   Who is right?
   a. Mechanic A
   b. Mechanic B
   c. Both mechanic A and B
   d. Neither mechanic A nor B

8. The Torqueflite lockup torque converter was introduced in:
   a. 1976
   b. 1977
   c. 1978
   d. 1979

9. Mechanic A says Chrysler refers to a planetary ring gear as an annulus gear.
   Mechanic B says Chrysler refers to a clutch drum as a clutch housing.
   Who is right?
   a. Mechanic A
   b. Mechanic B
   c. Both mechanic A and B
   d. Neither mechanic A nor B

10. Mechanic A says Chrysler refers to the intermediate band as the manual second band.
    Mechanic B says Chrysler refers to the front band as the kickdown band.
    Who is right?
    a. Mechanic A
    b. Mechanic B
    c. Both mechanic A and B
    d. Neither mechanic A nor B

11. Mechanic A says Chrysler refers to the input shell as the sun gear driving shell.
    Mechanic B says *breakaway* is Chrysler's name for manual second.
    Who is right?
    a. Mechanic A
    b. Mechanic B
    c. Both mechanic A and B
    d. Neither mechanic A nor B

12. Mechanic A says the Torqueflite kickdown band is wrapped around the forward clutch drum.
    Mechanic B says the Torqueflite kickdown band is wrapped around the high-reverse clutch drum.
    Who is right?
    a. Mechanic A
    b. Mechanic B
    c. Both mechanic A and B
    d. Neither mechanic A nor B

13. Mechanic A says that when the Torqueflite gear selector is in Park or Neutral, no clutches or bands are applied.
    Mechanic B says that when the Torqueflite gear selector is in Park, the output shaft is locked to the case.
    Who is right?
    a. Mechanic A
    b. Mechanic B
    c. Both mechanic A and B
    d. Neither mechanic A nor B

14. Mechanic A says that when the gear selector is placed in manual second, a Torqueflite starts in second gear.
    Mechanic B says that when the Torqueflite gear selector is moved from Drive to manual low at higher speeds, the transmission will immediately shift into second gear.
    Who is right?
    a. Mechanic A
    b. Mechanic B
    c. Both mechanic A and B
    d. Neither mechanic A nor B

15. The member of the Simpson planetary gearset that is splined directly to the input shaft is the:
    a. Front carrier
    b. Front ring gear
    c. Rear ring gear
    d. None of the above

16. Mechanic A says the front carrier in the Torqueflite is splined to the output shaft.
    Mechanic B says the front clutch in a Torqueflite is the forward clutch.
    Who is right?
    a. Mechanic A
    b. Mechanic B
    c. Both mechanic A and B
    d. Neither mechanic A nor B

17. Mechanic A says the rear band in a Torqueflite is applied in manual low.
    Mechanic B says the front clutch in a Torqueflite is applied in Reverse.
    Who is right?
    a. Mechanic A
    b. Mechanic B
    c. Both mechanic A and B
    d. Neither mechanic A nor B

18. Torqueflite governors are mounted:
    a. On the transfer shaft
    b. In the extension housing
    c. On the output shaft
    d. All of the above

19. Mechanic A says throttle pressure in a Torqueflite FWD transaxle is controlled by a vacuum modulator.
    Mechanic B says throttle pressure in a Torqueflite RWD transmission is controlled by mechanical linkage.
    Who is right?
    a. Mechanic A
    b. Mechanic B
    c. Both mechanic A and B
    d. Neither mechanic A nor B

20. The two members of the Simpson geartrain that are splined to the output shaft are the:
    a. Sun gear and rear carrier
    b. Sun gear and rear ring gear
    c. Front ring gear and rear carrier
    d. Front carrier and rear ring gear

21. Mechanic A says the Simpson geartrain uses the sun gear as an input member in Reverse.
    Mechanic B says the Simpson geartrain uses the front ring gear as an input member in Drive.
    Who is right?
    a. Mechanic A
    b. Mechanic B
    c. Both mechanic A and B
    d. Neither mechanic A nor B

22. The Torqueflite forward clutch is the input device in:
    a. Drive low and Reverse
    b. Drive low only
    c. Manual low only
    d. All forward gears

23. Mechanic A says the manual valves for Torqueflite RWD transmissions and FWD transaxles are the same.
    Mechanic B says Torqueflite FWD transaxles have a bypass valve that the RWD transmissions do not.
    Who is right?
    a. Mechanic A
    b. Mechanic B
    c. Both mechanic A and B
    d. Neither mechanic A nor B

24. Mechanic A says the upper half of the Torqueflite valve body contains most of the valves that control the hydraulic system.
    Mechanic B says RWD Torqueflite transmissions with lockup torque converters have an auxiliary valve body that contains the lockup valve and the shuttle valve.
    Who is right?
    a. Mechanic A
    b. Mechanic B
    c. Both mechanic A and B
    d. Neither mechanic A nor B

25. Mechanic A says 1978 and earlier Torqueflites without lockup converters use a torque converter control valve to regulate oil flow and pressure into and out of the converter.
    Mechanic B says a switch valve does these same jobs on later Torqueflites.
    Who is right?
    a. Mechanic A
    b. Mechanic B
    c. Both mechanic A and B
    d. Neither mechanic A nor B

10

Ford C3, C4, C5, C6, A4LD, and Jatco Transmissions

HISTORY AND MODEL VARIATIONS

The Ford Motor Company has six transmissions that use the Simpson gear train and are quite similar to each other. These are the C3, C4, C5, C6, Jatco, and A4LD transmissions. The most obvious difference in these transmissions is that the A4LD is a 4-speed while all of the others are 3-speeds. Another difference is that the C3, C4, C5, and A4LD have a low-reverse band, while the C6 and Jatco have a multiple-disc clutch to do the same job. All the transmissions use the same basic apply devices and have similar hydraulic systems.

The oldest of these transmissions is the C4, figure 10-1, which was introduced in 1964 for use with small- and medium-sized 6-cylinder and V-8 engines. The C4 continued through 1982, but was superseded in later vehicles by the C5, which was basically an updated C4 with a lockup torque converter. The C5, figure 10-2, debuted in 1981 and continued in selected models through 1986.

The C6 transmission, figure 10-3, was introduced in 1966 as a heavy-duty gearbox for use with larger V-8 engines. The C6 was not used in cars after 1980, but its use continues in light trucks through the 1988 model year.

In the early 1970s, the Ford Motor Company entered into a cooperative effort with the Japan Automatic Transmission Company (Jatco) to develop a gearbox for light- and medium-duty use. The result was the Jatco transmission, figure 10-4, which was used in Mazda vehicles beginning in 1972, and in Ford Courier trucks from 1974 through 1982. Ford also used the Jatco transmission in certain 1977-79 domestic passenger cars with 6-cylinder engines. The Jatco was also used extensively by Nissan (Datsun).

The C3 transmission, figure 10-5, was a light-duty unit used from 1974 through 1986 with certain 4-cylinder, V-6, and small V-8 engines. The 4-speed A4LD, figure 10-6, is a stretched version of the C3 with an additional gearset to provide an overdrive fourth gear. The A4LD was introduced in 1985 in the Ranger, Bronco II, Aerostar van, and selected light trucks. The A4LD continues in use through the 1988 model year and is the first Ford automatic transmission to have electronic controls.

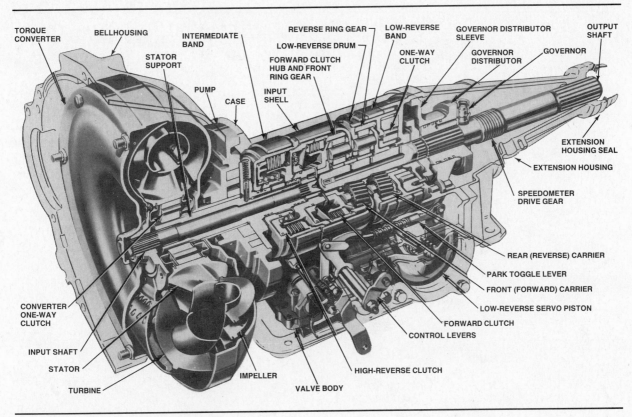

Figure 10-1. The Ford C4 transmission.

Fluid Recommendations

Ford Motor Company ATF recommendations for the transmissions discussed in this chapter have changed several times over the years. In the past, some of these transmissions used fluids that are now obsolete and unavailable. In addition, certain of these automatics use two different fluid types depending on the vehicle in which they are installed, or the year that the transmission was manufactured. The Ford ATF recommendations in effect as of the 1988 model year are shown on the chart in figure 5-5, which is located in Chapter 5 of this text. Specific recommendations are also called out in the *Major Changes* section below.

There are some generalizations that can be made about Ford ATF recommendations. First, *all* Ford automatic transmissions in 1976 and earlier vehicles use Type F fluid. Second, MERCON® is the preferred fluid for *all* 1983-87 Ford automatics, although DEXRON®-II remains a secondary recommendation for *some* transmissions built in these years. Finally, the use of MERCON® is a warranty *requirement* for all 1988 and later Ford automatic transmissions manufactured in North America, and Ford Motor Company vehicles sold in North America.

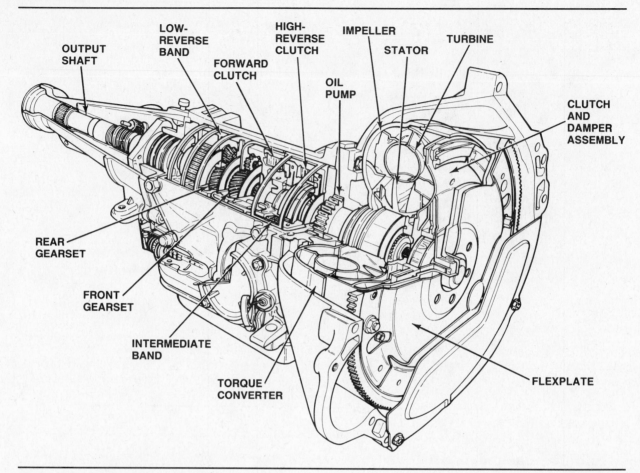

Figure 10-2. The Ford C5 transmission.

Transmission Identification

Identifying a Ford automatic transmission is a 2-step process. First, you determine the model of transmission. All Ford vehicles have an identification plate on the rear of the left front door, figure 10-7. In addition to the vehicle identification number and other data, this plate has a code letter that identifies the automatic transmission model; manual transmissions are identified by numbers rather than letters.

Figure 10-8 lists the Ford automatic transmission code letters and the models they represent. Note that the letter "T" is used for several models, Ford uses the "T" to represent all of their 4-speed automatics. In these cases, the shop manual for the vehicle gives further information, such as engine type, that makes it possible to precisely identify the transmission model.

In addition to the transmission model, you must know the exact transmission assembly part number and/or other identifying characters to obtain repair parts. This information is stamped on a tag attached to the transmission. The location of the tag and the type of information it contains varies with the transmission. Figure 10-9 gives the tag location for each transmission covered in this chapter, and explains the meaning of the data stamped on the tag. As required by Federal regulations, Ford transmissions also carry the serial number of the vehicle in which they were originally installed.

Because of ongoing improvements and updates, all of the information on the transmission tag must be available when ordering repair parts. For example, the C6 tag shown in figure 10-9 contains a model code, an assembly part number prefix and suffix, a serial number,

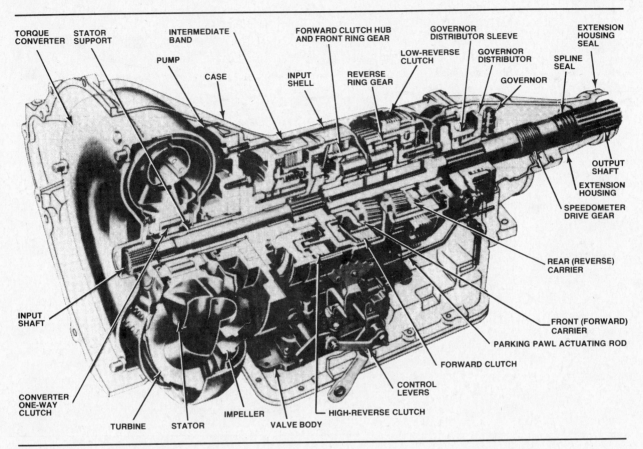

Figure 10-3. The Ford C6 transmission.

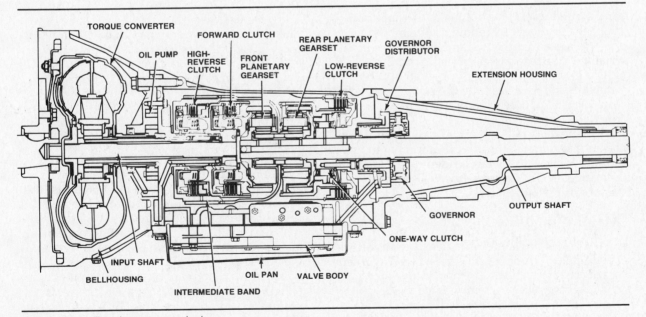

Figure 10-4. The Jatco transmission.

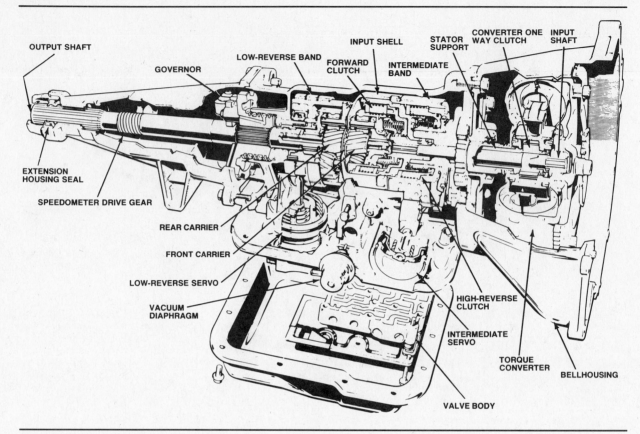

Figure 10-5. The Ford C3 transmission.

and a build-date code. In this particular application, the number "2" following the model number indicates that the transmission has internal changes compared to earlier units marked PJD-AN, or PJD-AN 1. Although all of these transmissions are basically the same, some service parts would be slightly different for each one.

Whenever you overhaul a Ford transmission, be sure to reinstall the identification tag in its correct location. Without it, identification of the transmission for future service will be difficult, if not impossible.

Major Changes

Listed below are some of the changes that have been made to the Ford transmissions discussed in this chapter. The nature and extent of the changes vary with each individual model. Note that only the *major* changes are described; it is impossible to list every one of the dozens of running changes that take place every year. Always consult appropriate shop manuals, service bulletins, and aftermarket parts suppliers for the latest information on the transmission being serviced.

C3 transmissions

The C3 did not undergo any major changes or redesigns during its use from 1974-86. However, the governor was revised in 1978. All 1974-80 C3 transmissions require Type F fluid. MERCON® is the preferred fluid for 1981-86 C3 transmissions, although DEXRON®-IID is acceptable.

C4 transmissions

The C4 transmission was first used in some 1964-model cars. When the gear selector is placed in manual second on 1964-66 units, the transmission starts in second gear and upshifts to third as road speed increases. When the gear selector is placed in manual second on 1967 and later units, the transmission starts and remains in second gear regardless of vehicle speed.

The C4S was a semi-automatic version of the C4 used only in the 1970 Maverick. It was a manually operated transmission that does not have a vacuum modulator, a governor, a throttle rod, or a downshift lever assembly.

In 1970, the C4 valve body was re-designed to add a throttle limit valve in the lower section. This small valve and its spring are held in the lower valve body by a tab on the fluid filter

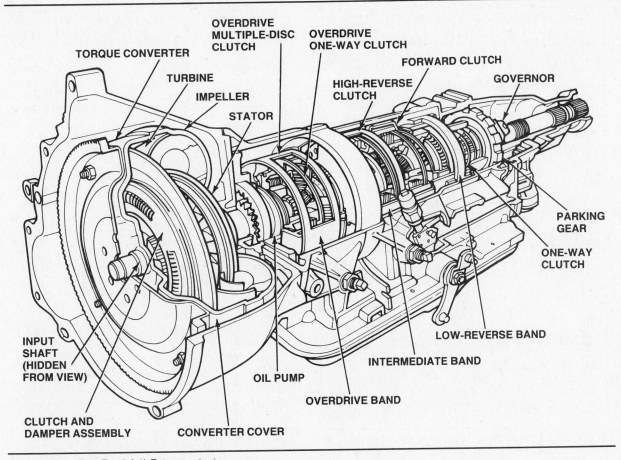

Figure 10-6. The Ford A4LD transmission.

screen, figure 10-10. Take care when removing the fluid screen on these models because the valve pops out and is easily lost.

You should also be aware that some C4 transmissions have a deep oil pan that holds additional fluid for better cooling. The filters used with the standard and deep pans have the same bolt pattern, but they are not interchangeable.

All 1964-79 C4 transmissions require Type F fluid. MERCON® is the preferred fluid for 1980-82 C4 transmissions, although DEXRON®-IID is acceptable.

C5 transmissions

The C5 transmission superseded the C4 and has a somewhat different hydraulic system. Several new valves and a new timing valve body are added. Also, the converter relief valve is moved from the stator support (in the pump assembly) to the timing valve body. As a result, the C4 oil pump assembly cannot be used to service the C5.

The C5 also differs from the C4 in that it has a 12-inch torque converter with a centrifugally locking torque converter clutch. Because of this

converter clutch, Ford initially required a special Type H transmission fluid for the C5. Type H fluid has since become obsolete, and Ford now recommends MERCON® fluid for these transmissions.

C6 transmissions

Like early C4 transmissions, the first C6 units in 1966 provided an automatic upshift from second to third gear when the gear selector lever was in manual second. This was changed in late 1966 and early 1967 so that the transmission starts and remains in second gear when the selector is in manual second. Except for a few late 1976 transmissions described below, all 1966-76 C6 gearboxes require Type F fluid.

In 1976, the C6 transmission was extensively re-engineered to provide smoother shifts and reduced backlash. Some redesigned transmissions were used on late 1976 vehicles, and revised models became the standard C6 transmission for 1977 and later vehicles. Most internal parts were changed during the C6 redesign, and many are not interchangeable with parts from earlier C6 units.

```
MFD. BY FORD MOTOR CO. IN U.S.A.

DATE: 8/77          GVWR 6538
GAWR: FRONT 3222, REAR 3375

THIS VEHICLE CONFORMS TO ALL
APPLICABLE FEDERAL MOTOR
VEHICLE SAFETY STANDARDS IN
EFFECT ON THE DATE OF MANU-
FACTURE SHOWN ABOVE.

                      F0012/R0165
8S63H100001           PASSENGER
VEH. IDENT. NO.          TYPE

BODY | COLOR | TRIM | TRANS. | AXLE | DSO
53H  | 1C-YA | DD   |   X    |  6   | 48
```

```
MFD. BY FORD MOTOR CO. IN U.S.A.

DATE: 09-85                    GVWR: 5347 LB - 2425 KG
FRONT GAWR: 2714 LB            REAR GAWR: 2683 LB
         1231 KG                          1216 KG

THIS VEHICLE CONFORMS TO ALL APPLICABLE FEDERAL MOTOR VEHICLE SAFETY AND
BUMPER STANDARDS IN EFFECT ON THE DATE OF MANUFACTURE SHOWN ABOVE.

VEH. IDENT. NO. 1FABP43MZGX100001      F0276
TYPE PASSENGER                         R0141
2A                                     482450
EXTERIOR PAINT COLORS                  DSO

BODY | VR | MLDG. | INT. TRIM | A/C | R | S | AX | TR
54K  | YP | 59P   | GG        | A   | 2 | B | 8  | TBBB
```

Figure 10-7. These are examples of the identification plates located on the rear of the left front door on Ford vehicles. The transmission code identifies the transmission model.

TRANSMISSION CODE	TRANSMISSION MODEL	YEAR
C	C5	1982-86
S	Jatco	1977-79
T	A4LD	1985-88
T	AOD	1980-88
T	AXOD	1986-88
U	C6	1966-88
V	C4S (Semiautomatic, Maverick only)	1970
V	C3	1974-86
W	C4	1964-82
X	FMX	1968-81
X	FX	1967
Y	MX	1967-68
Y	CW	1973-74
Y	FMX	1975-76
Z	C6 (Police)	1966-88

Figure 10-8. Ford automatic transmission codes, 1964-88.

One of the most important changes made in the revised C6 transmission was a new clutch friction material that required a friction-modified Type CJ transmission fluid to provide smoother shifts. Early transmissions that required Type CJ fluid had the ESP-M2C138-CJ specification number inscribed on the dipstick, figure 10-11.

Type CJ fluid was made obsolete in 1980, and at that time Ford recommended DEXRON®-IID fluid as a replacement in the C6 transmission. Ford now recommends MERCON® as the preferred fluid for late-1976 through 1987 C6 transmissions, although DEXRON®-IID is acceptable. The use of MERCON® is a warranty requirement for 1988 and later C6 transmissions.

Jatco transmissions

The earliest Jatco transmissions were used in Mazda vehicles and the Ford Courier truck. The intermediate servo adjustment screw of these units is inside the transmission case. In 1977, the case and servo were redesigned to place the adjustment screw on the outside of the case, at the right front. All Jatco transmissions used in 1977 and later domestic Ford vehicles have the externally adjustable servo.

Jatco transmissions used in Mazda vehicles, and *all* Ford Couriers, require Type F fluid. Jatco transmissions in 1977-79 domestic Ford passenger cars used the same friction material as in the re-engineered C6. As a result, these transmissions first required Type CJ fluid, then DEXRON®-IID fluid when Type CJ became obsolete. Ford now recommends MERCON® as the preferred fluid for passenger car Jatco transmissions, although DEXRON®-II is acceptable.

A4LD transmissions

The A4LD automatic is similar to the C3 transmission from the high-reverse clutch drum rearward. An additional gearset is added ahead of this point to provide the overdrive fourth gear. There have been no major changes in the A4LD since its introduction.

MERCON® is the preferred fluid for 1985-87 A4LD transmissions, although DEXRON®-IID is acceptable. The use of MERCON® is a warranty requirement for 1988 and later A4LD transmissions.

GEAR RANGES

Ford 3-speed automatic transmission gear selectors have six positions, figure 10-12A:

P — Park
R — Reverse
N — Neutral
D — Drive
2 — Manual second
1 — Manual low.

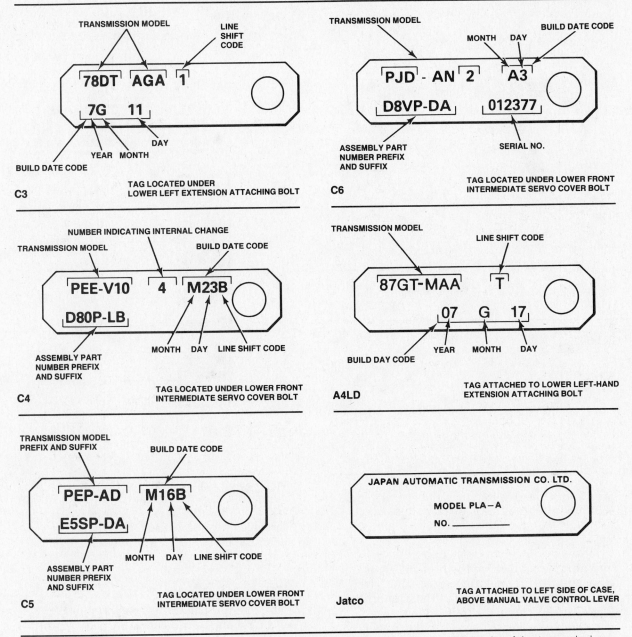

Figure 10-9. These tags, attached to Ford automatics, give exact information on the identity of the transmissions.

On 1966 and earlier cars with transmissions that provide a 2-3 upshift when the gear selector lever is in manual second, the selector is marked with D1 for Drive, D2 for manual second, and L for manual low. The numbers in the D1 and D2 markings indicate the gear that the transmission starts in for each Drive range.

The floor-mounted gear selector of the 4-speed A4LD transmission has seven positions, figure 10-12B:

P — Park
R — Reverse
N — Neutral
Ⓓ — Overdrive
D — Drive
2 — Manual second
1 — Manual low.

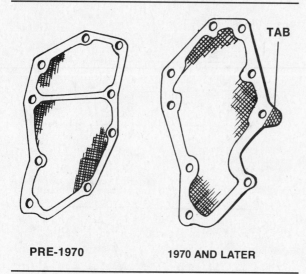

PRE-1970   **1970 AND LATER**

Figure 10-10. The tab on 1970 and later C4 filters holds the throttle pressure limit valve in the valve body.

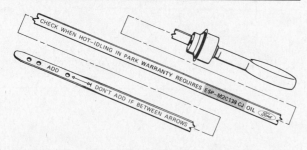

Figure 10-11. The Type CJ specification number is inscribed on the dipstick of some re-engineered C6 transmissions beginning in late 1976.

Gear Ratios

The gear ratios for all of the transmissions discussed in this chapter are very similar. The ratios for the C5 are typical of those in other Ford 3-speed automatic transmissions:

- First (low) — 2.46:1
- Second (intermediate) — 1.46:1
- Third (direct drive) — 1.00:1
- Reverse — 2.19:1

The ratios in Reverse and the lower three gears of the 4-speed A4LD transmission are similar to those above. The fourth gear (overdrive) gear ratio is 0.75:1.

Neutral and Park

The engine can be started only when the transmission is in Neutral or Park. No clutches or bands are applied in these selector positions, and there is no flow of power through the

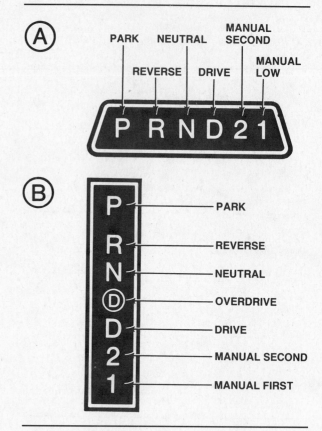

Figure 10-12. Ford automatic transmission gear selector positions.

transmission. In Neutral, the output shaft is free to turn, and the car can be pushed. In Park, a mechanical lever, rod, or pawl engages a large gear on the output shaft to lock the transmission and prevent the vehicle from being moved.

Overdrive (A4LD)

This is the normal driving position for the A4LD. When the vehicle is stopped and the gear selector is placed in Overdrive (Ⓓ), the transmission is automatically in low gear. The vehicle starts to move in low gear and automatically upshifts to second, and then to third gear. Once in third gear, the converter clutch may engage, depending on driving conditions. As the driver eases up on the accelerator to allow the vehicle to cruise, the transmission upshifts into fourth gear, or overdrive, which reduces engine rpm by 25 percent. The converter clutch will also engage in fourth gear if powertrain loads are light.

Depressing the accelerator for added power will first unlock the converter clutch, and then force a 4-3 downshift if the power demand is great enough. The cruise control system can

FORD TERM	TEXTBOOK TERM
Clutch cylinder	Clutch drum
Internal gear	Ring gear
Turbine shaft	Input shaft
Reactor	Stator
Line or control pressure	Mainline pressure
Main pressure regulator valve	Pressure regulator valve
T.V. pressure	Throttle pressure
Throttle control pressure relief valve	Throttle limit valve
Overrunning clutch	1-way roller clutch
Rear clutch	Forward clutch
Front clutch	High-reverse clutch
Kickdown, or front, band	Intermediate band
Kickdown, or front, servo	Intermediate servo
Rear band	Low-reverse band
Rear servo	Low-reverse servo

Figure 10-13. Ford transmission nomenclature table.

also force a part-throttle 4-3 downshift to maintain cruising speed on upgrades or into a headwind. Downshifts to lower gears are also possible, and occur as described in the Drive section below.

Drive

This is the normal driving position for all of the transmissions discussed in this chapter except the A4LD. When the vehicle is stopped and the gear selector is placed in Drive, the transmission is automatically in low gear. The vehicle starts to move in low gear and automatically upshifts to second, and then to third gear. The speeds at which upshifts occur vary and are controlled by vehicle speed and engine load.

When the A4LD gear selector is in Drive, the transmission will not upshift to fourth, although the converter clutch can still lock in third gear, depending on driving conditions. In the C5 transmission, the converter clutch will lock whenever turbine speed exceeds a certain rpm.

The driver can force a 3-2 downshift in Drive by depressing the accelerator partially or completely. The speeds at which the downshift occurs vary for different transmission, engine, axle ratio, tire size, and vehicle weight combinations. A forced 3-1 downshift in Drive is possible with the C3, A4LD, and Jatco transmissions at speeds below approximately 27 mph (43 kph). With the C4, C5, and C6 gearboxes, a forced 3-1 downshift in Drive is possible at speeds below approximately 48 mph (77 kph).

When the car decelerates in Drive to a full stop, the transmission downshifts automatically only once. This is called coasting downshift and is a shift directly from third to low, bypassing second. It occurs at a speed below 10 mph or (16 kph), just before the vehicle comes to a full stop.

Manual Second

The manual second range on these Ford transmissions is different from the same range on Chrysler and GM 3-speed transmissions. On 1964-66 C4 transmissions, and early 1966 C6 units, the gearbox starts in second and upshifts to third when the gear selector is placed in manual second. On later models of these transmissions, and all other Ford transmissions discussed in this chapter, the vehicle starts in second gear and remains there when the gear selector is placed in manual second. The transmission neither upshifts nor downshifts regardless of throttle opening and road speed.

Manual Low

When the vehicle is stopped and the gear selector is placed in 1 or L (manual low), the transmission starts and stays in low gear. It does not upshift to second. In addition, because different apply devices are used in manual low than are used in Drive low, this gear range can be used to provide engine compression braking to slow the vehicle.

If the driver shifts to the 1 or L position at high speed, the transmission does not shift directly from high to low gear. To avoid engine or driveline damage, the transmission first downshifts to second, and then automatically downshifts to low when the speed drops below approximately 25 mph (40 kph). The transmission then remains in low gear until the selector is moved to another position.

Reverse

When the gear selector is moved to R, the transmission shifts to Reverse. The transmission should be shifted to Reverse only when the vehicle is standing still. The lockout feature of the shift gate should prevent accidental shifting into Reverse while the vehicle is moving forward. However, if the driver intentionally moves the selector to R while moving forward, the transmission *will* shift into Reverse and damage may result.

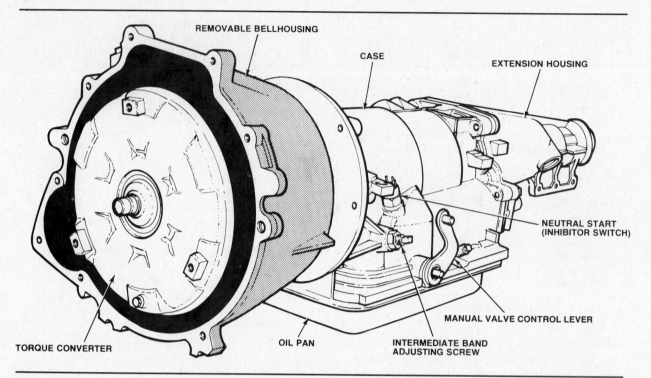

Figure 10-14. The bellhousings on C3, C4, C5, A4LD, and Jatco transmissions are removable from the main case.

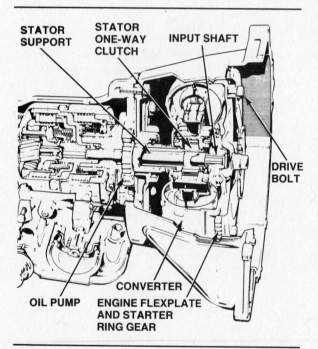

Figure 10-15. Torque converter and oil pump installation.

FORD TRANSMISSION NOMENCLATURE

For the sake of clarity, this text uses consistent terms for transmission parts and functions. However, the various manufacturers may use different terms. Therefore, before we examine the buildup of the C3, C4, C5, C6, A4LD, and Jatco, along with the operation of their geartrains and hydraulic systems, you should be aware of the unique names that Ford uses for some transmission parts and operations. This will make it easier in the future if you should look in a Ford manual and see a reference to the "reactor". You will realize that this is simply the stator, and is similar to the same part in any other transmission. Figure 10-13 lists Ford nomenclature and the textbook terms for some transmission parts and operations.

C3, C4, C5, C6, A4LD, AND JATCO BUILDUP

On C3, C4, C5, A4LD, and Jatco transmissions, the torque converter bellhousing is a separate casting bolted to the front of the main transmission case, figure 10-14. On C6 transmissions, the bellhousing and main case are a 1-piece casting. All of these transmissions have separate, removable rear extension housings.

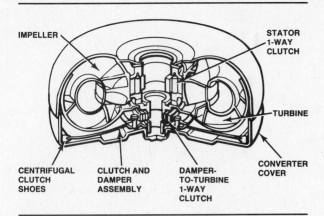

Figure 10-16. The C5 torque converter with its centrifugally locking clutch assembly.

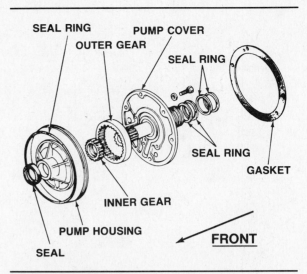

Figure 10-17. This Jatco oil pump is typical of those used in Ford transmissions.

The torque converter bolts to the engine flexplate, which has the starter ring gear around its outer diameter. The torque converter is a welded assembly and cannot be disassembled. It is installed in the transmission by sliding it onto the stationary stator support. The splines in the stator 1-way clutch engage the splines on the stator support, and the splines in the turbine hub engage the input shaft, figure 10-15.

The C5 torque converter contains a centrifugally operated clutch that includes friction shoes, a damper assembly, and a 1-way clutch, figure 10-16. The operation of this type of converter is described in Chapter 7. The A4LD torque converter contains a hydraulically operated clutch that is partially controlled by the Electronic Engine Control (EEC-IV) system. The electronic torque converter clutch controls are explained in Chapter 8; we will study the hydraulic control components later in this chapter.

The torque converter hub is at the rear of the converter housing and engages the drive lugs on the transmission oil pump, figure 10-17. Because the converter cover is bolted to the engine flexplate, the oil pump turns and pumps fluid whenever the engine is running. All of the Ford transmissions discussed in this chapter use a gear-type pump mounted in the front of the transmission case. The pump's inner gear is driven by the torque converter hub. The pump draws fluid from the sump, through the filter, and delivers it to the pressure regulator valve, the manual valve, and other valves in the hydraulic system.

The input shaft is turned clockwise by the torque converter turbine. The shaft passes

through the stator support into the transmission case. The input shaft is supported by bushings in the stator support, and by the forward clutch drum (C3, C4, C5, C6, and Jatco) or overdrive carrier (A4LD).

A4LD Overdrive Assembly

In the A4LD transmission, an overdrive gearset and several apply devices are located between the torque converter and the Simpson geartrain, figure 10-18. We will study this unique overdrive assembly before we go on to the rest of the geartrain, which is common to all the transmissions discussed in this chapter. The A4LD overdrive assembly includes a simple planetary gearset, a multiple-disc clutch, a 1-way clutch, and a band. Ford uses the term "overdrive" in the names of all components in the overdrive assembly.

The input shaft from the torque converter turbine passes through the overdrive sun gear and is splined to the overdrive carrier. The carrier contains the overdrive pinions that turn between the sun gear and the ring gear. The front end of the overdrive center shaft is splined to the ring gear, the back end of the center shaft is splined to the forward clutch drum and high-reverse clutch hub assembly. Essentially, the center shaft acts as the input shaft to the Simpson geartrain. The 1-way clutch fits between an extension on the back end of the carrier, and the inside of the center shaft.

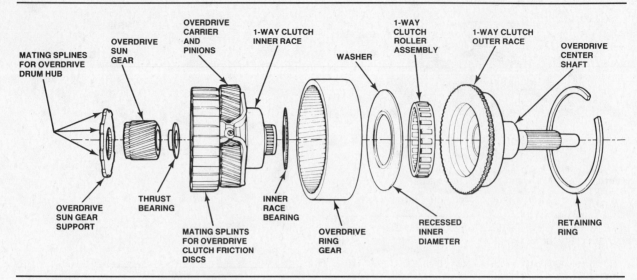

Figure 10-18. The A4LD overdrive assembly.

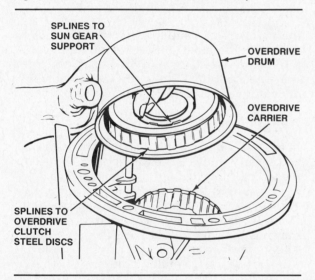

Figure 10-19. The A4LD overdrive clutch and clutch drum assembly.

The overdrive drum fits over the gearset, figure 10-19, and is surrounded by the overdrive band. The hub of the drum is splined to the overdrive sun gear support, which is splined to the sun gear. The inside of the drum is splined to the outer edges of the overdrive clutch steel discs. The inner edges of the overdrive clutch friction discs are splined to the outside of an extension on the front end of the overdrive carrier.

A center support helps locate the overdrive assembly in the case, and separates it from the rest of the geartrain. From this point on, the A4LD buildup is similar to that of all the other transmissions discussed in this chapter.

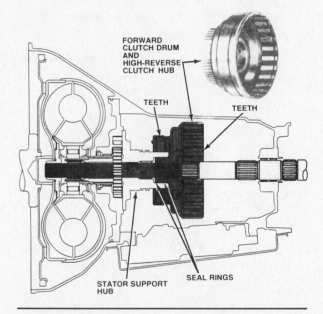

Figure 10-20. Forward clutch drum and high-reverse clutch hub installation in the C6 transmission.

C3, C4, C5, C6, A4LD, And Jatco Simpson Geartrain

The input shaft (overdrive center shaft in the A4LD) is splined to the forward clutch drum and high-reverse clutch hub assembly, figure 10-20. In all the transmissions covered in this chapter, the forward clutch drum and high-reverse clutch hub are one part. The drum-and-hub assembly is piloted inside the stator support hub, and seal rings are used to seal the oil passages to the forward clutch. The Jatco has a

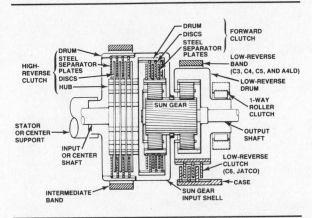

Figure 10-21. All the transmissions discussed in this chapter use the Simpson geartrain. The one major mechanical difference is the use of a low-reverse band in some transmissions (C3, C4, C5, and A4LD), compared to a low-reverse clutch in others (C6 and Jatco).

thrust bearing between the drum-and-hub assembly and the stator support.

Refer to figure 10-21 while studying the relationship of the clutches and bands to the geartrain. The inner circumference of each high-reverse clutch friction disc is splined to the outer circumference of the high-reverse clutch hub and turned by it. The outer circumference of each high-reverse clutch steel disc is splined to the inner circumference of the high-reverse clutch drum. When the high-reverse clutch is applied, the drum is rotated clockwise by the input shaft.

The high-reverse clutch drum engages the sun gear input shell. The input shell is splined to the center of the sun gear and secured with snap rings. The sun gear rides on a bushing on the output shaft but is not splined to the shaft. Both the front and rear pinions mesh with the sun gear. When the high-reverse clutch is applied (in high or Reverse), the drum, the input shell, and the sun gear are rotated clockwise by the input shaft.

The intermediate band is wrapped around the outside of the high-reverse clutch drum. When the intermediate band is applied, the drum, the input shell, and the sun gear are all held stationary. This occurs only in second gear.

The outer circumference of each forward clutch friction disc is splined to the inner circumference of the forward clutch drum, which is attached to the input shaft. The inner circumference of each forward clutch steel separator disc is splined to the outer circumference of the forward clutch hub. In the C3 and C6, the outside of the front ring gear in the planetary gearset serves as the forward clutch hub. In the C4,

C5, A4LD and Jatco, the forward clutch hub is splined to the front ring gear and held in place by a snapring. Whenever the forward clutch is applied, the front ring gear is rotated clockwise.

The front ring gear meshes with the front pinions. The front pinions are supported and turn in the front carrier, and mesh with the sun gear. The sun gear meshes with the rear pinions, which are supported and turn in the rear carrier, and mesh with the rear ring gear.

The outside of the rear carrier is splined to the inside of the low-reverse drum, figure 10-22, and retained by a snapring. A 1-way roller clutch is attached to the back end of the low-reverse drum. In C3, C6, A4LD, and Jatco transmissions, the outer race of the 1-way clutch is locked to the low-reverse drum, and the inner race is bolted to the case, figure 10-23. In the C4 and C5, the outer race is bolted to the case, and the inner race is splined to the drum. In either case, when the low-reverse drum tries to rotate counterclockwise, the clutch locks up. This action occurs in Drive low to hold the rear carrier. When the drum rotates clockwise, the 1-way clutch overruns.

The low-reverse drum can also be held by a low-reverse band or a multiple-disc clutch. In the C3, C4, C5, and A4LD, figure 10-23, a low-reverse band wraps around the low-reverse drum. When the band is applied, the drum and rear carrier are held. In the C6 and Jatco, figure 10-22, the friction discs of a low-reverse clutch are splined to the outside of the low-reverse drum. The steel discs of the clutch are splined to the transmission case. When the low-reverse clutch is applied, the drum and rear carrier are again held.

The low-reverse band or clutch is applied in Reverse and manual low. Unlike the 1-way clutch, the low-reverse band or clutch holds the drum and rear carrier from rotating in either direction. Although the designs of these apply devices differ, they serve the same function.

No member of the planetary gearset is attached directly to the input shaft, but two members of the gearset *are* splined to the output shaft, the front carrier and the rear ring gear, figure 10-24. This means that one or the other of these two units is always the final driving member of the planetary gearset. The front carrier is splined directly to the output shaft. The rear ring gear is splined to a drive hub or flange that is splined to the output shaft. The ring gear is retained on the hub or flange by a snapring. Both the front carrier and the rear ring gear can either drive the output shaft or be driven by it, depending on which bands and clutches are applied.

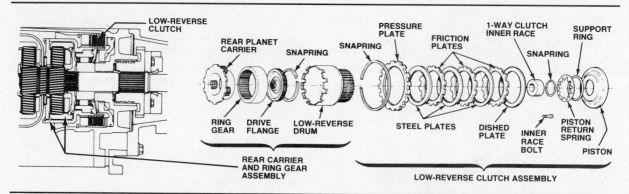

Figure 10-22. Jatco rear ring gear, low-reverse drum, and low-reverse clutch installation.

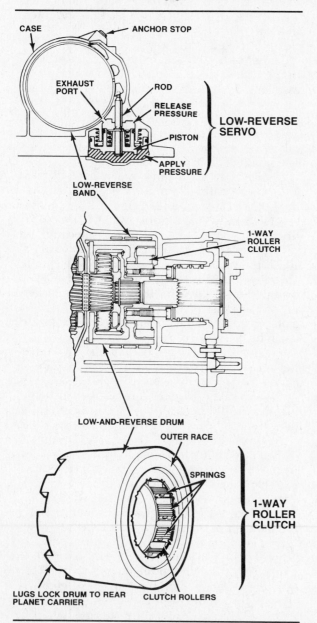

Figure 10-23. C3 low-reverse band, servo, and 1-way clutch installation.

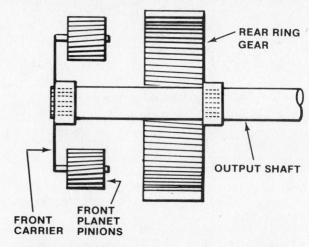

Figure 10-24. The front carrier and the rear ring gear are splined to the output shaft.

Figure 10-25 shows the alignment of the various transmission shafts. The output shaft passes through the 1-way clutch and the rear of the main transmission case where it is supported by a bushing. The output shaft extends from the rear of the extension housing. The output yoke is splined to the end of the shaft, and is supported in the extension housing by a bushing. A lip seal at the rear of the housing contains fluid within the transmission.

The parking gear and the governor distributor are splined to the output shaft, figure 10-26. The governor distributor body is located at the front of the extension housing and rotates inside a sleeve. The space between the distributor and sleeve is taken up by oil seal rings. The distributor routes fluid to the governor. The governor fits around the output shaft and is bolted to the distributor body.

The speedometer drive pinion is located just behind the governor in the extension housing and is driven by a gear on the output shaft.

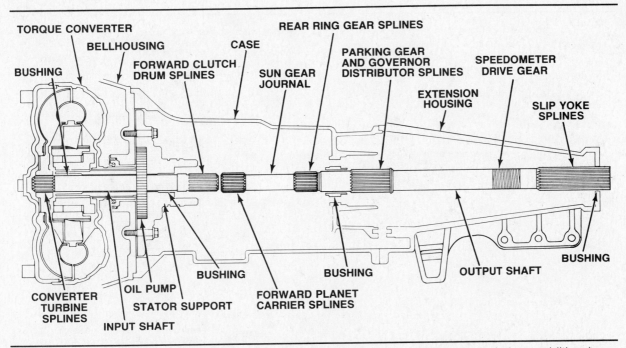

Figure 10-25. Typical Ford input and output shaft installation from a C3. An A4LD would include an additional overdrive center shaft between the input and output shafts.

C3, C4, C5, C6, A4LD, AND JATCO POWER FLOW

Now that we have taken a general look at how these transmissions are assembled, we can trace the flow of power through the apply devices and the planetary gearset in each of the gear ranges.

Park and Neutral

When the gear selector is in Park or Neutral, no clutches or bands are applied, except in the A4LD where the overdrive 1-way clutch locks and transmits input shaft motion to the center shaft. Because no member of the Simpson planetary gearset is held, there is no output motion. The torque converter impeller and the oil pump rotate clockwise at engine speed. The torque converter turbine, the transmission input shaft, and the center shaft in the A4LD also rotate, but at less than engine speed.

In Neutral, the planetary gearset idles. The output shaft is free to rotate in either direction so the car can be pushed or pulled backwards or forwards.

In Park, a mechanical pawl is engaged with the parking gear that is splined to the output shaft. This locks the output shaft in position, prevents the driveshaft from turning, and stops the car from moving in either direction.

A4LD Overdrive and Drive Ranges

Before we look at the powerflow through these transmissions in the forward gear ranges, it is necessary to first examine the operation of the A4LD overdrive assembly. The powerflow through the overdrive gearset varies depending on whether the gear selector is in the Overdrive or Drive position.

When the gear selector is in Overdrive, neither the overdrive multiple-disc clutch nor the overdrive band is applied. The overdrive carrier is driven clockwise by the input shaft, which causes the overdrive 1-way clutch to lock up and drive the center shaft at turbine speed. Power from the torque converter passes directly through the overdrive assembly to the Simpson geartrain. If the center shaft turns faster than the overdrive carrier and input shaft, such as during deceleration, the 1-way clutch overruns and the overdrive gearset freewheels.

With the selector in any position except Overdrive, Park, or Neutral, the overdrive multiple-disc clutch is applied. This locks the overdrive carrier to the overdrive sun gear, causing the overdrive gearset to turn as a unit in direct drive. The overdrive ring gear then turns the center shaft clockwise to transmit power to the Simpson geartrain. The overdrive 1-way clutch also helps transmit power at this time, but its action is secondary to that of the

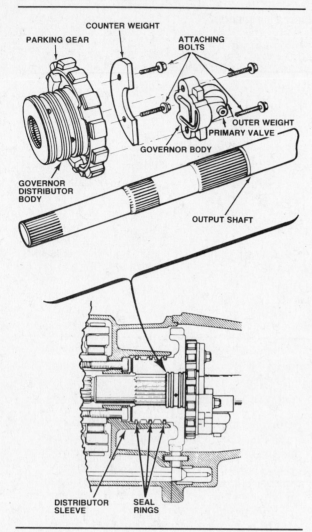

Figure 10-26. C3 governor and parking gear installation.

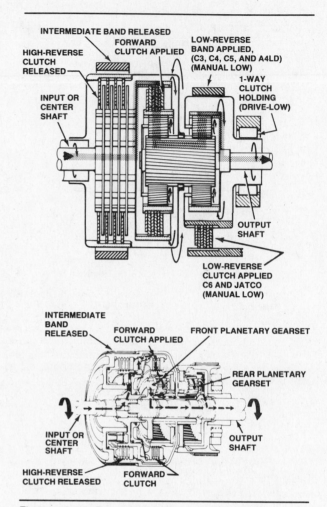

Figure 10-27. Powerflow in Overdrive (A4LD) and Drive low, and in Manual low.

overdrive multiple-disc clutch. Locking the overdrive gearset in this manner makes engine braking possible in reverse, and under certain conditions, in the lower three forward gears. How this happens is explained below.

Overdrive (A4LD) and Drive Low

When the gear selector is moved to Drive or Overdrive, the car starts to move with the transmission automatically in low gear. The forward clutch is applied, figure 10-27, and locks the front ring gear to the input shaft or A4LD center shaft. This causes the ring gear to rotate clockwise at turbine speed. In all but the A4LD transmission, the forward clutch is the only hydraulic apply device that operates in Drive low. In the A4LD, only the forward clutch operates in Overdrive low, but in Drive low the overdrive multiple-disc clutch is also applied.

Before the car begins to move, the transmission output shaft is held stationary by the driveline and wheels. This also holds the front carrier and rear ring gear stationary. This means the clockwise rotation of the front ring gear turns the front pinions clockwise in the stationary carrier. The front pinion rotation then causes the sun gear to turn counterclockwise.

Counterclockwise rotation of the sun gear turns the rear pinions clockwise. The rear pinions then try to walk clockwise around the inside of the rear ring gear, which is splined to the stationary output shaft. This action causes the rear carrier to try to rotate counterclockwise. However, the rear carrier is attached to the low-reverse drum (or clutch hub), which is splined to the 1-way clutch at the rear of the case.

As the rear carrier tries to rotate counterclockwise, the 1-way clutch locks up and prevents rotation. Therefore, the torque from the rear pinions is transferred to the rear ring gear

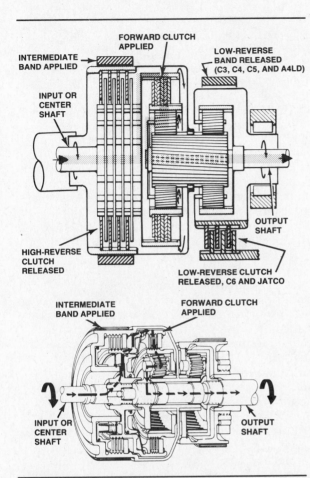

Figure 10-28. Powerflow in Overdrive (A4LD) and Drive second, and in Manual second.

and output shaft instead of the rear carrier. Because the 1-way clutch holds the carrier more securely than the drive wheels can hold the output shaft and ring gear, the output shaft and rear ring gear begin to rotate clockwise in gear reduction.

As long as the car is accelerating in Drive or Overdrive low, torque is transferred from the engine, through the planetary gearset, to the output shaft. If the driver lets up on the throttle, the engine speed drops while the drive wheels try to keep turning at the same speed. This removes the clockwise-rotating torque load from the 1-way clutch, which then unlocks and allows the rear planetary gearset to freewheel so that no engine compression braking is available. This is a major difference between Drive or Overdrive low, and manual low, as we will see later.

Note that even when the A4LD is in Drive low, with the overdrive clutch applied and the overdrive gearset locked, there is no engine braking. This is because the 1-way clutch in the

Simpson gearset allows the rear planetary gearset to freewheel.

In Drive or Overdrive low, gear reduction is produced by a combination of the front and rear planetary gearsets. The front ring gear is the input member; the rear carrier is the held member; and the rear ring gear is the output member.

In all but the A4LD transmission, there are only two apply devices working in Drive low, the forward clutch and the 1-way clutch. The forward clutch is the input device. The 1-way clutch is the holding device. In the A4LD, one or two additional apply devices are used, depending on the gear selector position. In Overdrive low, the overdrive 1-way clutch is an input device. In Drive low, the overdrive clutch is also used as an input device.

Overdrive (A4LD) and Drive Second

As the car continues to accelerate in Drive, the transmission hydraulic system produces an automatic upshift to second gear by applying the intermediate band, figure 10-28. The forward clutch remains applied, and power from the engine is still delivered to the front ring gear. The front ring gear rotates clockwise and turns the front pinions clockwise.

At the same time, the intermediate band locks the high-reverse clutch drum and sun gear input shell to the transmission case. This holds the sun gear stationary. As the front ring gear turns the front pinions clockwise, they walk around the stationary sun gear. This causes the front carrier to rotate clockwise. The front carrier is splined to the output shaft and drives it in gear reduction.

As long as the car is accelerating in Drive or Overdrive second, torque is transferred from the engine, through the planetary gearset, to the output shaft. As vehicle speed rises and the driver lets up on the accelerator pedal, the transmission upshifts to third — even when an A4LD is in Drive second, with the overdrive clutch applied and the overdrive gearset locked. Because of this upshift, there is no engine compression braking. This is a major difference between Drive or Overdrive second, and manual second, as we will see later.

All of the gear reduction for second gear is produced by the front planetary gearset. The rear gearset simply idles. The 1-way clutch that held the rear carrier in Drive or Overdrive low overruns in second because the sun gear is held, forcing the rear pinions to walk around the stationary sun gear and the carrier to turn clockwise. The rear ring gear, which is splined to the output shaft, continues to rotate with the

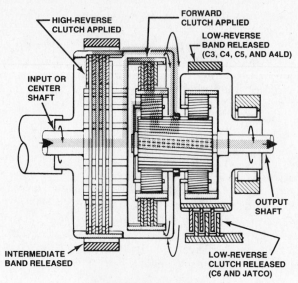

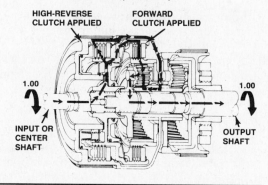

Figure 10-29. Powerflow in Overdrive (A4LD) and Drive third.

shaft, but freewheels because no holding device is applied to the rear gearset.

In all but the A4LD transmission, there are again only two apply devices working in Drive second, the forward clutch and the intermediate band. The forward clutch is the input device. The intermediate band is the holding device. In the A4LD, one or two additional apply devices are used, depending on the gear selector position. In Overdrive second, the overdrive 1-way clutch is an input device. In Drive second, the overdrive clutch is also used as an input device.

Overdrive (A4LD) and Drive Third

As the car continues to accelerate in Drive or Overdrive, the transmission automatically upshifts from second to third gear. For this to happen, the forward clutch remains applied, the intermediate band is released, and the high-reverse clutch is applied, figure 10-29.

When the high-reverse clutch is applied, the high-reverse clutch drum and the sun gear input shell are locked to the transmission input shaft. Because the sun gear input shell is splined to the sun gear, this causes the sun gear to revolve clockwise with the input shaft at engine speed.

The forward clutch continues to lock the front ring gear to the input shaft. This means that two members of the planetary gearset, the sun gear and the ring gear, are locked to the input shaft, and thus to each other. With two members of the gearset locked together, the entire gearset turns as a unit.

With the front planetary gearset locked, there is no relative motion between the sun gear, ring gear, carrier, or the pinions. As a result, input torque to the sun and ring gears is transmitted through the planet carrier, which is splined to the output shaft, to drive the output shaft. In third gear, there is a direct 1:1 gear ratio between the input or center shaft speed and the output shaft speed. There is no reduction or increase in either speed or torque through the transmission.

Engine compression braking is available in Drive third because there is a direct mechanical connection between the input shaft and the output shaft; no 1-way clutches are involved. In the A4LD, the hydraulic system prevents an upshift to fourth, and the overdrive multiple-disc clutch is applied to lock the gearset and bypass the overdrive 1-way clutch. Compression braking is not possible when an A4LD is in Overdrive third, however, because at higher speeds the transmission will upshift to fourth, or at lower speeds the overdrive 1-way clutch will overrun because the overdrive multiple-disc clutch is not applied.

In third gear, the input and driven members of the compound planetary gearset are the sun gear and the front ring gear; the front carrier is the output member. In all but A4LD transmissions, there are again only two apply devices working in Drive third, the forward clutch and the high-reverse clutch. Both clutches are input and holding devices equally. In the A4LD, one or two additional apply devices are used, depending on the gear selector position. In Overdrive third, the overdrive 1-way clutch is an input device. In Drive third, the overdrive clutch is also used as an input device.

Overdrive Fourth (A4LD)

As the vehicle continues to accelerate in Overdrive, the transmission shifts into fourth gear.

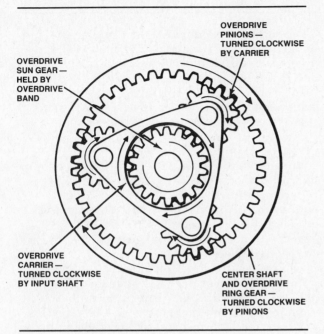

Figure 10-30. Operation of the A4LD overdrive planetary gearset in Overdrive fourth.

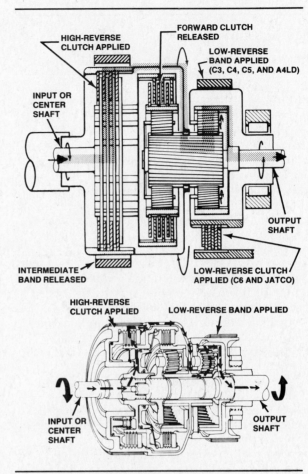

Figure 10-31. Powerflow in Reverse.

To do this, the overdrive multiple-disc clutch is released and the overdrive band is applied. This locks the overdrive drum, and the overdrive sun gear splined to it, to the transmission case. The overdrive pinions then walk around the sun gear and overdrive the ring gear and center shaft at faster-than-turbine speed, figure 10-30.

In Overdrive fourth, the overdrive 1-way clutch overruns because its outer race turns with the center shaft faster than its inner race turns with the overdrive carrier. The center shaft provides the input to the Simpson compound planetary gearset, which is still in direct drive, figure 10-29. The Simpson gearset then transmits the overdrive speed to the output shaft.

The input member of the overdrive gearset is the overdrive carrier; the held member is the overdrive sun gear; and the output member is the overdrive ring gear. There is no input device because the input shaft is splined directly to the overdrive carrier. The holding device is the overdrive band. The input and output members and devices of the Simpson gearset are the same as in Overdrive or Drive third.

Manual Second

Except on 1964-66 C4 and early 1966 C6 transmissions, when the gear selector is in manual second (2 on the gear selector), the transmission starts and remains in second gear; the hydraulic system will not allow the transmission to downshift to low or upshift to third. Other than this, there are no differences between manual second and Drive second. The flow of power through the planetary gearsets is identical, and the same apply devices are used, figure 10-28. Note that in the A4LD the overdrive multiple-disc clutch is applied in Drive second.

At any time, the driver can manually downshift from third to second by moving the gear selector from D to 2. In this case, the high-reverse clutch is released, and the intermediate band is applied. In the A4LD, the driver can also manually downshift from Ⓓ to 2. In this case, the high-reverse clutch and overdrive band are released, and the overdrive multiple-disc clutch and intermediate band are applied. These capabilities provide the driver with a controlled downshift for engine braking or improved acceleration at any speed or throttle position.

Manual Low

In manual low (1 or L on the gear indicator), the flow of power through the planetary gearset is basically the same as in Drive low. Note that in the A4LD transmission the overdrive multiple-disc clutch is applied in Drive low. The only difference between Drive low and manual low is the use of one apply device.

In manual low, a low-reverse band (C3, C4, C5, and A4LD) or multiple-disc clutch (C6 and Jatco) is applied to hold the rear carrier, figure 10-27. In Drive low, the 1-way clutch does this job, but it holds the rear carrier only during acceleration; the 1-way clutch freewheels during deceleration. In manual low, the low-reverse band or multiple-disc clutch holds the rear carrier during acceleration *and* deceleration, which provides engine braking on deceleration. The 1-way clutch still helps hold the rear carrier during acceleration in manual low, but its action is secondary as long as the low-reverse band or multiple-disc clutch is applied.

The driver can move the gear selector from Ⓓ or D to 1 or L while the car is moving. At lower speeds, the transmission will downshift directly to low gear by releasing the appropriate apply devices and applying the low-reverse band or multiple-disc clutch. At higher speeds, the transmission first shifts into second by releasing the appropriate apply devices and applying the intermediate band. The transmission will then automatically downshift to low at about 25 mph (40 kph) by releasing the intermediate band and applying the low-reverse band or multiple-disc clutch.

As long as the gear selector remains in the 1 or L position, the hydraulic system keeps the low-reverse band or multiple-disc clutch applied, and prevents the intermediate band from being applied. In manual low, two apply devices are again operating. The forward clutch is the input device, and the low-reverse band is the holding device.

Reverse

When the gear selector is moved to Reverse, the input shaft is still turned clockwise by the engine. The transmission must reverse this rotation to a counterclockwise direction.

In Reverse, the forward clutch is released and the high-reverse clutch is applied, figure 10-31. This connects the high-reverse clutch drum and the sun gear input shell to the input shaft so that the input shaft drives the sun gear clockwise. The sun gear drives the rear pinions counterclockwise.

The other apply device used in Reverse is the low-reverse band (C3, C4, C5, and A4LD) or multiple-disc clutch (C6 and Jatco). This device holds the low-reverse drum and the rear carrier. Because the rear carrier is held stationary, the rear pinions drive the rear ring gear counterclockwise. The rear ring gear is splined to the output shaft and turns it counterclockwise to provide Reverse.

All gear reduction in Reverse is produced by the rear planetary gearset. Although there is input to the front gearset through the sun gear and the front pinions, no other member of the gearset is held. Therefore, there can be no power output from the front gearset, and it simply freewheels.

In Reverse, the input member of the compound planetary gearset is the sun gear; the held member is the rear carrier; and the output member is the rear ring gear. Two apply devices are used: the high-reverse clutch is the input device, and the low-reverse band or multiple-disc clutch is the holding device. In the A4LD transmission, the overdrive multiple-disc clutch is also applied in Reverse. This provides engine compression braking by preventing the overdrive 1-way clutch from overrunning when the throttle is released.

C3, C4, C5, C6, A4LD and Jatco Clutch and Band Applications

Before we take a closer look at these transmissions' hydraulic systems, we should review the apply devices that are used in each of the gear ranges. Figure 10-32 is a clutch and band application chart for the C3, C4, and C5. Figure 10-33 is a chart for the C6 and Jatco. Figure 10-34 is a chart for the A4LD. Also, remember these facts about the apply devices in these transmissions:

• The forward and high-reverse clutches are input devices.
• The bands, 1-way roller clutch, and low-reverse clutch (C6 and Jatco) are holding devices.
• The forward clutch is applied in all forward gear ranges.
• The intermediate band is applied only in second gear.
• The high-reverse clutch is applied in third, fourth (A4LD), and Reverse.
• The low-reverse band or clutch is applied in Reverse and manual low.
• The overdrive band is applied in fourth gear only.

GEAR RANGES	CLUTCHES			BANDS	
	1-Way Roller (Overrunning)	Forward (Rear)	High-Reverse (Front)	Intermediate (Kickdown, or Front)	Low-Reverse (Rear)
Drive Low	●	●			
Drive Second		●		●	
Drive High		●	●		
Manual Low	*	●			●
Manual Second		●		●	
Reverse			●		●

* The 1-way roller clutch will hold in manual low if the low-reverse band fails, but will not provide engine braking.

Figure 10-32. Ford C3, C4, and C5 clutch and band application chart.

GEAR RANGES	CLUTCHES				BANDS
	1-Way Roller (Overrunning)	Forward (Rear)	High-Reverse (Front, or Direct)	Low-Reverse (Rear)	Intermediate (Kickdown, or Front)
Drive Low	●	●			
Drive Second		●			●
Drive High		●	●		
Manual Low	*	●		●	
Manual Second		●			●
Reverse			●	●	

* The 1-way roller clutch will hold in manual low if the low-reverse clutch fails, but will not provide engine braking.

Figure 10-33. Ford C6 and Jatco clutch and band application chart.

GEAR RANGES	CLUTCHES					BANDS		
	1-Way Roller (Over-running)	Forward (Rear)	High-Reverse (Front)	Overdrive 1-Way Roller (Over-running)	Overdrive Multiple-Disc	Inter-mediate (Kickdown, or Front)	Low-Reverse (Rear)	Over-drive
Overdrive Low	●	●		●				
Overdrive Second		●		●		●		
Overdrive Third		●	●	●				
Overdrive Fourth		●	●					●
Drive Low	●	●		*	●			
Drive Second		●		*	●	●		
Drive Third		●	●	*	●			
Manual Low	†	●		*	●		●	
Manual Second		●		*	●	●		
Reverse			●	*	●		●	

* The overdrive 1-way clutch will hold in these ranges if the overdrive multiple-disc clutch fails, but will not provide engine braking.
† The 1-way roller clutch will hold in manual low if the low-reverse band fails, but will not provide engine braking.

Figure 10-34. Ford A4LD clutch and band application chart.

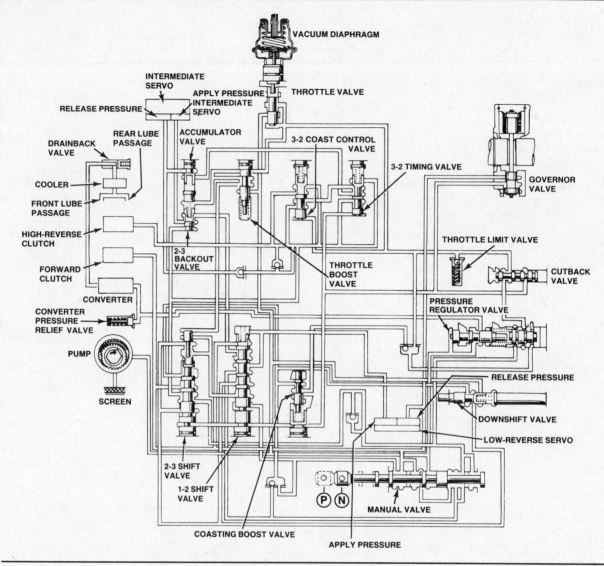

Figure 10-35. C3 transmission hydraulic system.

• The overdrive multiple-disc clutch is applied in all gear selector ranges except Park, Neutral, and Overdrive.
• The overdrive 1-way clutch holds in all gears except fourth.

THE C3, C4, C5, C6, A4LD AND JATCO HYDRAULIC SYSTEMS

The hydraulic systems of these Ford transmissions control shifting under varying vehicle loads and speeds, as well as in response to manual gear selection by the driver. The systems of all six transmissions are similar, with some variations for the different apply devices and downshift controls.

Figures 10-35, 10-36, 10-37, 10-38, 10-39, 10-40, and 10-41 diagram the complete hydraulic systems of the transmissions that are covered in this chapter. We will begin our study of these systems with the pump.

Hydraulic Pump

All of these Ford transmissions use gear-type oil pumps, figure 10-17. The rotation of the gears creates suction that draws stored fluid from the oil pan at the bottom of the transmission. The fluid is drawn through the filter attached to the bottom of the valve body, flows through a passage in the valve body, and then is routed through another passage in the front of the transmission case to the pump.

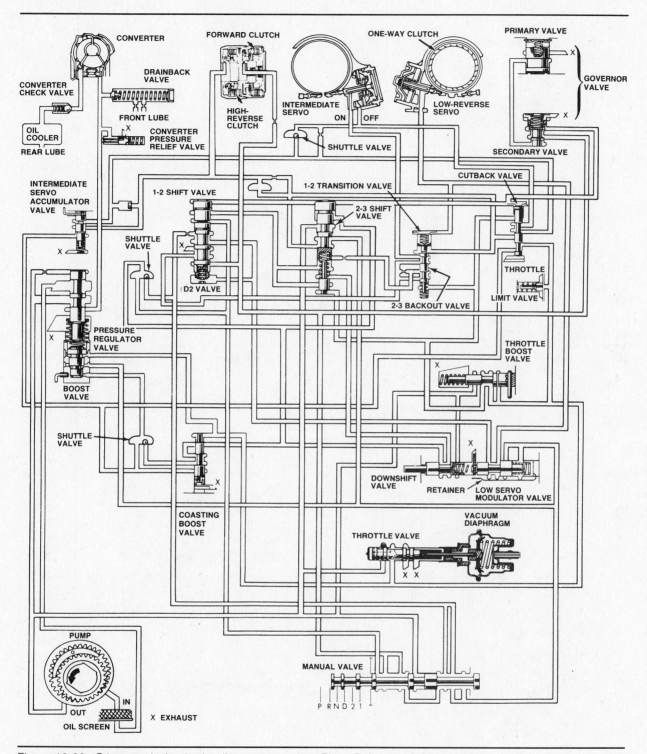

Figure 10-36. C4 transmission hydraulic system, except Pinto, Bobcat, and Mustang II.

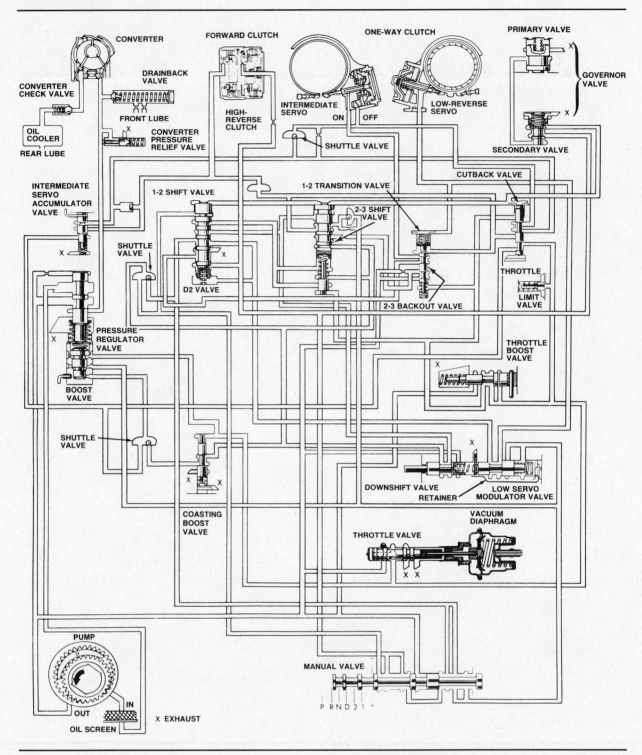

Figure 10-37. C4 transmission hydraulic system for Pinto, Bobcat, and Mustang II.

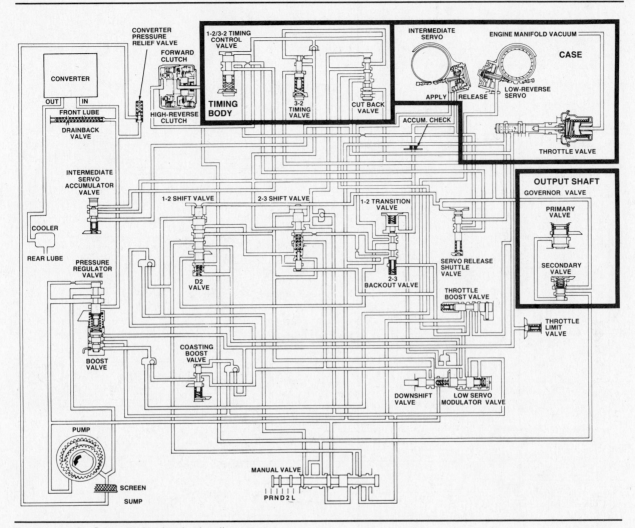

Figure 10-38. C5 transmission hydraulic system.

The pump delivers fluid to the pressure regulator valve, the manual valve, and the throttle valve. Fluid also flows directly from the pump to the:

- Throttle boost valve (C3, C4, C5, C6, A4LD)
- Downshift valve (C3, C4, C5, A4LD, Jatco)
- Intermediate servo accumulator valve (C6)
- 2-3 shift valve (C4)
- 2-3 backout valve (C5 and C4 in Pinto, Bobcat, and Mustang II)
- Coasting boost valve (C6)
- 4-3 torque demand valve (A4LD)
- Converter clutch shuttle valve (A4LD).

Hydraulic Valves

The C3, C4, C5, C6, Jatco, and A4LD transmissions contain the following valves:

1. Pressure regulator valve
2. Converter pressure relief valve
3. Converter check valve (C4, C6)
4. Drainback valve (C3, C4, C6, Jatco)
5. Manual valve
6. Throttle valve
7. Throttle limit valve
8. Throttle boost valve
9. Cutback valve
10. Governor valve
11. Downshift valve
12. 1-2 shift valve
13. Second lock valve (Jatco)
14. Manual low 2-1 scheduling valve (C6)
15. Intermediate servo modulator valve (C6)
16. Intermediate servo accumulator valve (C3, C4, C5, C6, A4LD)
17. 1-2 transition valve (C4, C5, A4LD)
18. 2-3 shift valve
19. 2-3 backout valve (C3, C4, C5, C6, A4LD)
20. 2-3 timing valve (Jatco)
21. 3-2 timing valve (C3, C5, A4LD)
22. 1-2/3-2 timing control valve (C5)
23. Servo release shuttle valve (C5)

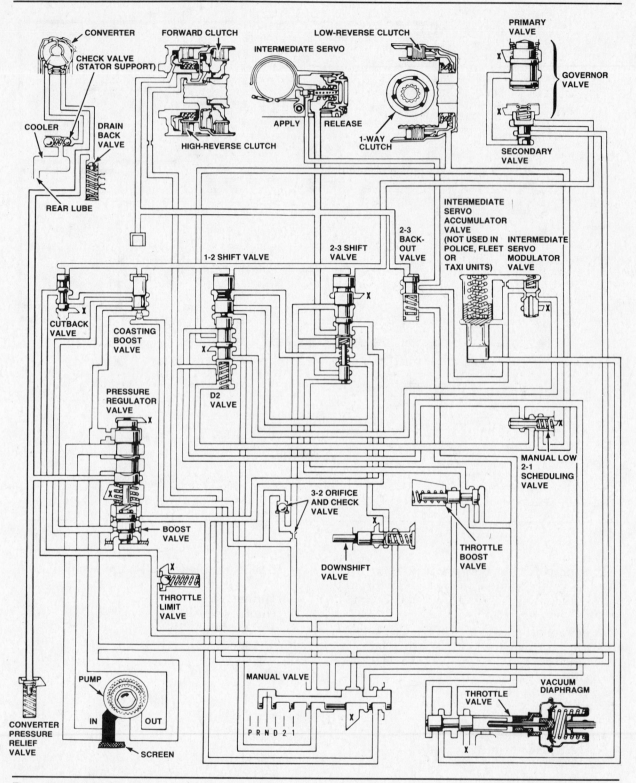

Figure 10-39. C6 transmission hydraulic system.

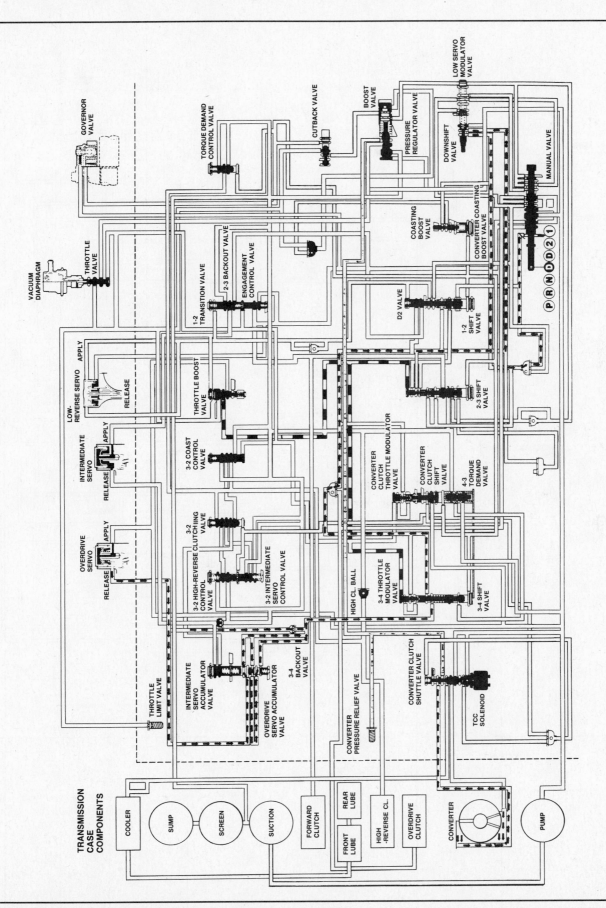

Figure 10-40. A4LD transmission hydraulic system.

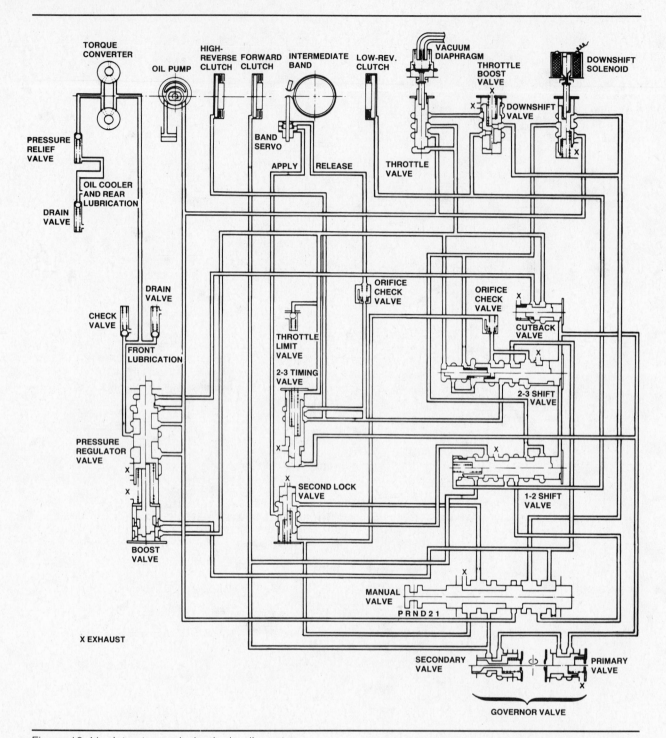

Figure 10-41. Jatco transmission hydraulic system.

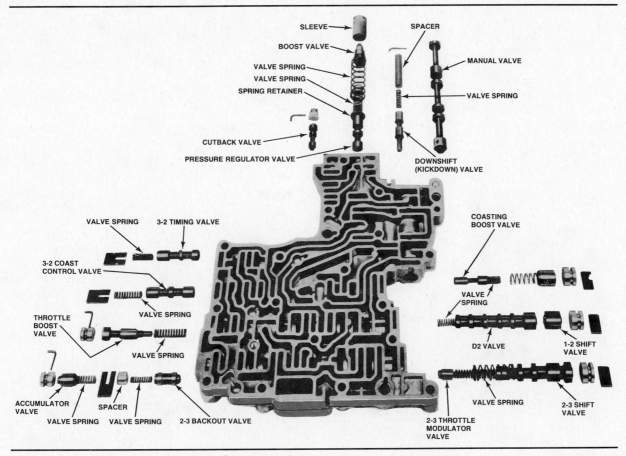

Figure 10-42. An exploded view of the C3 valve body.

24. 3-2 coast control valve (C3, A4LD)
25. Coasting boost valve (C3, C4, C5, C6, A4LD)
26. Low servo modulator valve (C4, C5, A4LD).

The following valves are unique to the A4LD hydraulic system:

1. Engagement control valve
2. 3-2 high-reverse clutch control valve
3. 3-2 intermediate servo control valve
4. Converter coasting boost valve
5. 3-4 shift valve
6. Overdrive servo accumulator valve
7. Torque demand control valve
8. 4-3 torque demand valve
9. 3-4 backout valve
10. Converter clutch throttle modulator valve
11. Converter clutch shift valve
12. Converter clutch shuttle valve.

Valve body and ball-check valves

In the C3, C4, C5, C6, and A4LD, all valves except the governor and throttle valve are housed in the valve body. In the Jatco, the throttle valve is also in the valve body. Typical valve bodies for these transmissions are shown in figures 10-42, 10-43, 10-44, 10-45, 10-46 and 10-47. Note that the C5 has an auxiliary timing body that contains three valves: the cutback valve, the 1-2/3-2 shift timing valve, and the 3-2 timing valve.

The upper half of the valve body for each transmission contains fluid passages, most of the hydraulic valves, and some of the ball-check valves. The lower half of the valve body contains passages that carry fluid from one valve to another. Some check balls are also housed in the lower valve body. The upper and lower valve bodies are separated by a steel separator plate.

The number of check balls varies among these transmissions, and even among different versions of the same transmission in different applications. Some ball-check valves are spring-loaded relief valves, while others are single-action or double-action valves that control the direction of oil flow within the valve body passages.

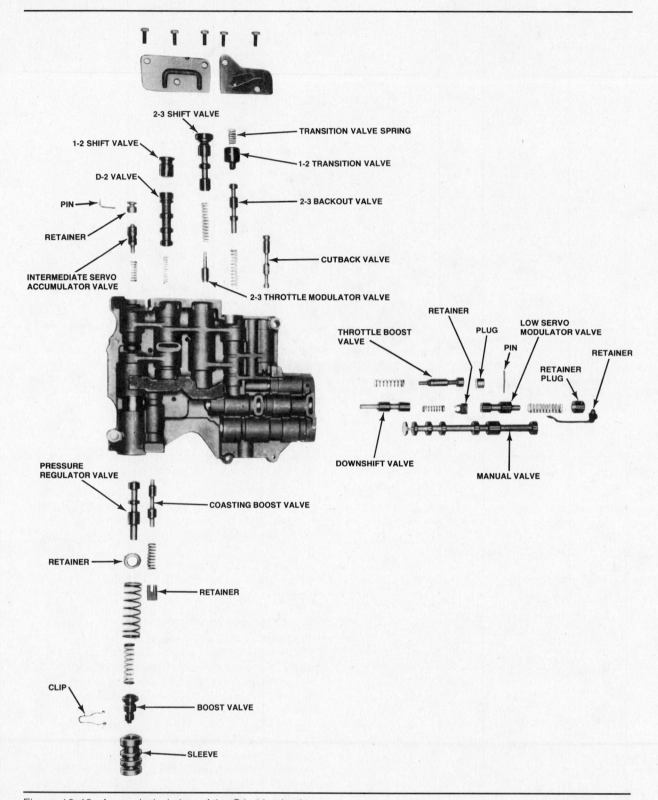

Figure 10-43. An exploded view of the C4 valve body.

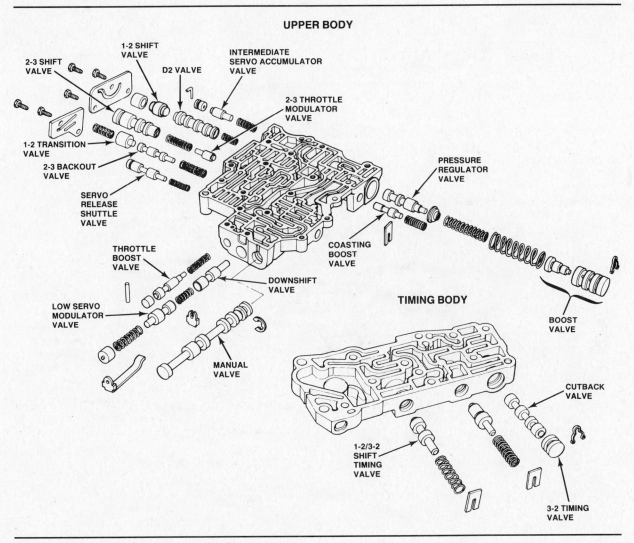

Figure 10-44. An exploded view of the C5 valve body.

Pressure regulator valve

The pump delivers fluid to two passages at the pressure regulator valve. In the C3, C4, C5, and A4LD, the first of these passages leads to the end of the valve opposite spring force. In the C6 and Jatco, the first passage directs the fluid between two lands of different diameters, where fluid pressure against the larger land moves the valve against spring force. The second passage routes fluid to the center of the regulator valve where it enters between two lands of equal diameter and so does not act on the valve in either direction. In the Jatco, there are two passages to the center of the valve, rather than one.

As the hydraulic system first fills with fluid from the pump, there is little resistance to flow in the system, so pressure does not build up immediately, figure 10-48A. In this stage, springs hold the regulator valve closed. Even though there is no pressure in the system when the pump is not turning, the lines do not drain. Therefore, the initial pressure buildup occurs almost instantly. The system needs only enough fluid from the pump to make up for slight internal leakage.

As the pressure rises and reaches about 55 to 60 psi (186 to 203 kPa), the regulator valve moves against the springs. This opens another port at the center passage that leads to the

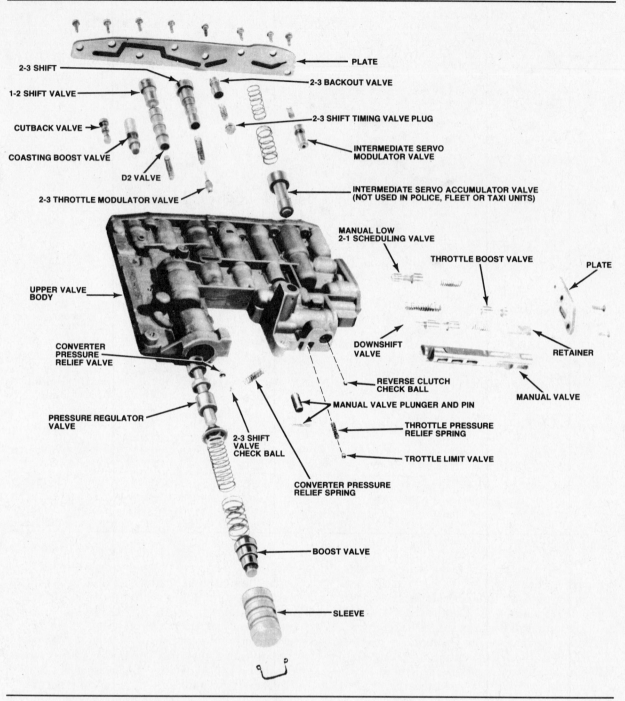

2-3 SHIFT

1-2 SHIFT VALVE

CUTBACK VALVE

COASTING BOOST VALVE

D2 VALVE

2-3 THROTTLE MODULATOR VALVE

PLATE

2-3 BACKOUT VALVE

2-3 SHIFT TIMING VALVE PLUG

INTERMEDIATE SERVO MODULATOR VALVE

INTERMEDIATE SERVO ACCUMULATOR VALVE (NOT USED IN POLICE, FLEET OR TAXI UNITS)

MANUAL LOW 2-1 SCHEDULING VALVE

THROTTLE BOOST VALVE

PLATE

UPPER VALVE BODY

CONVERTER PRESSURE RELIEF VALVE

PRESSURE REGULATOR VALVE

2-3 SHIFT VALVE CHECK BALL

CONVERTER PRESSURE RELIEF SPRING

DOWNSHIFT VALVE

RETAINER

REVERSE CLUTCH CHECK BALL

MANUAL VALVE

MANUAL VALVE PLUNGER AND PIN

THROTTLE PRESSURE RELIEF SPRING

TROTTLE LIMIT VALVE

BOOST VALVE

SLEEVE

Figure 10-45. An exploded view of the C6 valve body.

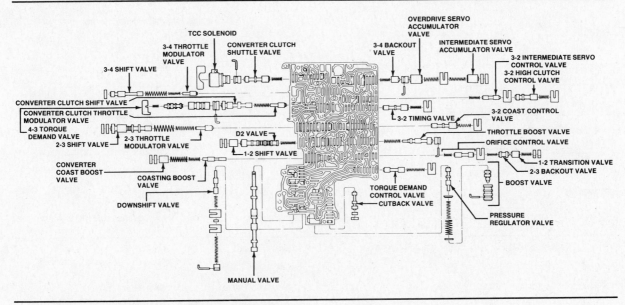

Figure 10-46. An exploded view of the A4LD valve body.

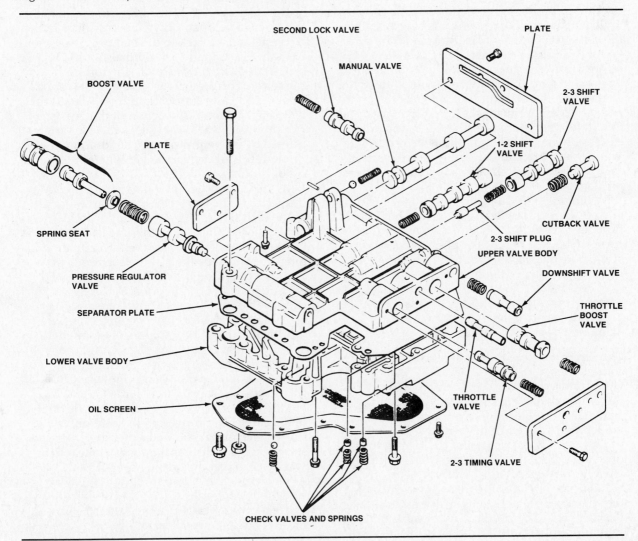

Figure 10-47. An exploded view of the Jatco valve body.

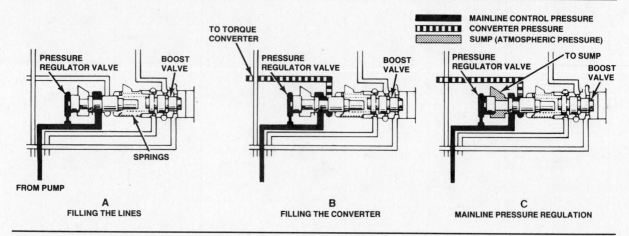

Figure 10-48. Pressure regulator valve operation occurs in three stages.

torque converter, figure 10-48B. The converter is pressurized almost immediately because it also remains filled when the transmission is not operating. In the A4LD, the direction of oil flow to the lockup torque converter is controlled by the converter clutch shuttle valve and the converter clutch shift valve as described later.

Pressure continues to build up at the regulator valve after the converter is pressurized. This moves the valve farther against the spring and opens an exhaust port to return excess fluid to the sump, figure 10-48C. As the exhaust port is opened and fluid is vented, hydraulic pressure and spring pressure balance each other to maintain a steady mainline pressure.

When engine load is high, additional mainline pressure is needed to apply and hold the clutches and bands. To increase mainline pressure, throttle pressure is applied to two lands of the boost valve. At about 17 inches of vacuum, or 10 psi (34 kPa) of throttle pressure, the boost valve is forced against the regulator valve spring. As a result, more pressure is needed to balance the regulator valve, and mainline pressure increases.

Converter pressure relief valve, converter check valve, and drainback valve

Fluid from the pressure regulator valve enters the torque converter through passages in the stator support. Depending on the transmission model, there are up to three valves that help control oil flow in this part of the circuit: the pressure relief valve, the converter check valve, and one or more drainback valves.

Oil from the regulator valve flows through the converter pressure relief valve on its way to the converter. The pressure relief valve limits converter pressure to a fixed amount, no matter

how high mainline pressure rises. In the C4, C5, and C6, converter pressure is about 90 psi or 305 kPa. In the C3, Jatco, and A4LD, it is about 115 psi or 390 kPa. At higher pressures, the relief valve opens to vent excess fluid in the converter circuit back to the sump.

In the C4, C5, and C6, fluid also flows to a drainback valve on its way to the torque converter. The drainback valve opens at about 5 psi or 15 kPa to supply fluid to the circuit that lubricates the forward portion of the geartrain. The drainback valve also keeps the converter primed by preventing fluid from draining out through the lubrication circuit when the engine is not running.

In the C4 and C6, fluid flows from the torque converter to the converter check valve located in the case side of the stator support. The converter check valve maintains at least 10 psi or 34 kPa in the converter when the engine is running. It also keeps fluid in the converter from draining out through the cooler circuit when the engine is not running. From the converter check valve, fluid flows to the transmission cooler in the radiator. After cooling, the fluid returns to the rear of the transmission where it enters the rear lubrication circuit, and then returns to the sump.

The Jatco torque converter and lubrication circuits are similar to those in the C4 and C6, but the converter pressure relief valve is on the outlet side of the converter. In addition to controlling maximum converter pressure, the pressure relief valve routes fluid to the oil cooler and rear lubrication circuit. A drain valve returns excess fluid to the sump. On the inlet side of the converter, a check valve routes fluid to the front lubrication circuit, and a drain valve returns excess fluid to the sump.

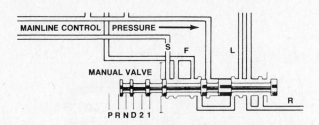

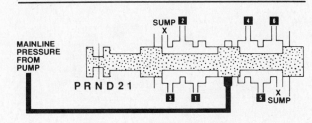

PASSAGE	FUNCTION
S	First gear lockout
F	Forward clutch and governor
L	Low-reverse apply and second gear lockout
R	Reverse pressure booster and high-reverse apply

Figure 10-49. C3, C4, and C5 manual valve passages.

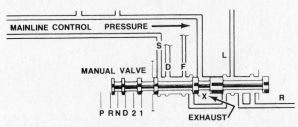

PASSAGE	FUNCTION
D	Shift control supply — Drive only
S	First gear lockout
F	Forward clutch and governor
L	Low-reverse apply and second gear lockout
R	Reverse pressure booster and high-reverse apply

Figure 10-50. C6 manual valve passages.

PASSAGE	FUNCTION
1	Forward clutch and governor; also 1-2 upshift in Drive
2	Second lock and servo apply in manual second
3	High clutch apply and servo release on 2-3 upshift; also overcome first gear lockout
4	Downshift supply in 1 & 2 ranges; throttle pressure vent in Drive and boosted throttle pressure in 2
5	Low-reverse apply and second gear lockout
6	Reverse pressure booster and high-reverse apply

Figure 10-51. Jatco manual valve passages.

The C3 converter fill circuit is also similar to that of the C4 and C6, except that the drainback valve is on the outlet side of the converter. The drainback valve keeps the converter primed and controls fluid flow to the cooler. From the cooler, fluid returns to *both* the front and rear lubrication circuits.

In the A4LD the converter feed circuit includes a number of valves that control the application and release of the converter clutch. We will study those valves later in this section.

Manual valve

The manual valve is a directional valve that receives fluid from the pump and directs it to apply devices and other valves in the hydraulic system in order to provide both automatic and manual upshifts and downshifts. In the C3, C4, and C5, there are four passages leading from the manual valve, figure 10-49. In the C6, there are five passages, figure 10-50. In the Jatco, there are six passages, figure 10-51. In the A4LD, figure 10-40, there are seven.

A mechanical linkage from the gear selector controls the manual valve position. The valve is held in each position by a spring-loaded lever and plunger. In Park and Neutral, mainline pressure is blocked by the manual valve and cannot flow to other circuits from the valve. In the other gear selector positions, various circuits are opened by the manual valve.

In all forward gear ranges (Ⓓ, D, 2, and 1) the manual valve directs fluid to apply the forward clutch, figure 10-52. Fluid is also supplied to the governor through the same circuit. In the C3, C4, C5, C6, and A4LD, this circuit also routes fluid to the D2 (DR2 or Drive 2) valve which is part of the 1-2 shift valve assembly. The Jatco does not have a D2 valve, and simply routes fluid to the 1-2 shift valve. When the 1-2 shift occurs, oil from the D2 valve or the 1-2 shift valve charges the apply side of the intermediate servo.

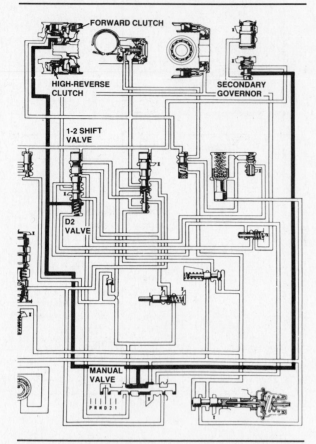

Figure 10-52. C6 forward clutch and governor circuit.

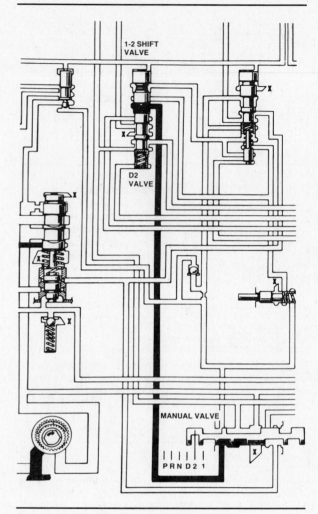

Figure 10-53. C6 first gear lockout circuit.

The first gear lockout circuit in the C3, C4, C5, C6, and A4LD prevents these transmissions from shifting to low gear when the gear selector is in the manual second position, figure 10-53. In the C4, C5, and C6, pressure is applied between the D2 valve and the 1-2 shift valve. This bottoms the 1-2 shift valve and forces the D2 valve back against spring pressure to prevent a downshift. In the C3 and A4LD, the same action is applied to the D2 valve, while the 2-3 shift valve is locked in the downshift position to prevent an upshift to third. The Jatco has a second-lock valve that routes oil to the intermediate servo while blocking it from the intermediate servo release and high-reverse clutch apply circuits, figure 10-41.

The low-reverse circuit in the C3, C4, C5, C6, and A4LD supplies mainline pressure to the low-reverse clutch or servo when the transmission is in manual low or Reverse, figure 10-54. Fluid passes through the downshifted 1-2 shift valve on the way to the servo or clutch piston and is applied to the spring end of the D2 valve to keep the 1-2 shift valve downshifted. In the C4, C5, and A4LD, pressure also is applied to the 1-2 transition valve to block

fluid from the intermediate servo and the 2-3 shift valve. In the C3 and A4LD, pressure is applied to the 2-3 shift valve to keep it downshifted. In the Jatco, there is no D2 valve, so pressure is applied to the spring end of the 1-2 shift valve to keep it downshifted. In all the transmissions, pressure is also applied to the throttle boost valve.

When the transmission is in Reverse, pressure is applied through the R passage (number 6 passage in a Jatco) to two circuits: the high-reverse clutch and the boost valve, figure 10-55. The high-reverse clutch circuit directs pressure to apply the clutch and release the intermediate servo. The pressure boost circuit sends mainline pressure to the boost valve to force it against the pressure regulator valve spring to increase mainline pressure. Higher pressure is needed in Reverse because of the greater torque loads on the transmission.

In addition to the valves already mentioned, the A4LD has a fourth gear lockout circuit that

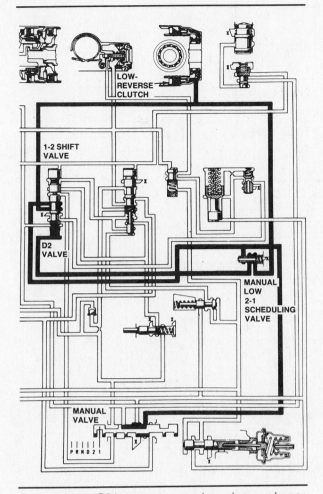

Figure 10-54. C6 low-reverse apply and second gear lockout circuit.

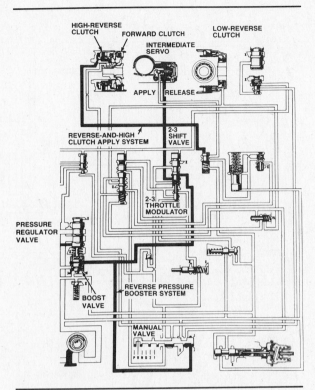

Figure 10-55. C6 reverse pressure booster and high-reverse apply circuit.

receives fluid in every gear range except Overdrive. Oil in this circuit applies the overdrive multiple-disc clutch, and prevents the 3-4 shift valve from upshifting and applying the overdrive band.

Throttle valve

All the transmissions covered in this chapter use a vacuum-modulated throttle valve to develop throttle pressure, figure 10-56. Throttle pressure indicates engine torque and helps determine the timing of the transmission upshifts and downshifts. Without throttle pressure, the transmission would upshift and downshift at exactly the same speed for different throttle openings and different load conditions.

The throttle valve senses engine torque through a vacuum diaphragm mounted on the transmission case and connected to intake manifold vacuum. In the C3, C4, C5, C6, and A4LD, the throttle valve is in the transmission

case. In the Jatco, it is in the valve body. In all models, a pushrod connects the valve to the diaphragm.

Without the diaphragm, the throttle valve would be a balanced valve, with throttle pressure balancing spring force. When mainline pressure enters the throttle valve, it is routed to the end of the valve as throttle pressure. As pressure rises, it tries to push the throttle valve against the spring. This tends to close the valve's inlet port, causing throttle pressure to drop. As throttle pressure drops, the spring pushes the valve back to open the port, allowing throttle pressure to rise again, figure 10-57.

Manifold vacuum is applied to the spring side of the diaphragm, and atmospheric pressure is applied to the opposite side. Under light engine loads, high manifold vacuum reduces the spring force and thus the throttle pressure. Under heavy engine loads, vacuum drops and more spring force is allowed to work against the atmospheric pressure and throttle pressure acting on the end of the valve. As a result, throttle pressure rises.

Some C6 transmissions have an altitude-compensating diaphragm. This unit has an evacuated, aneroid bellows that is sensitive to atmospheric pressure, which drops by about

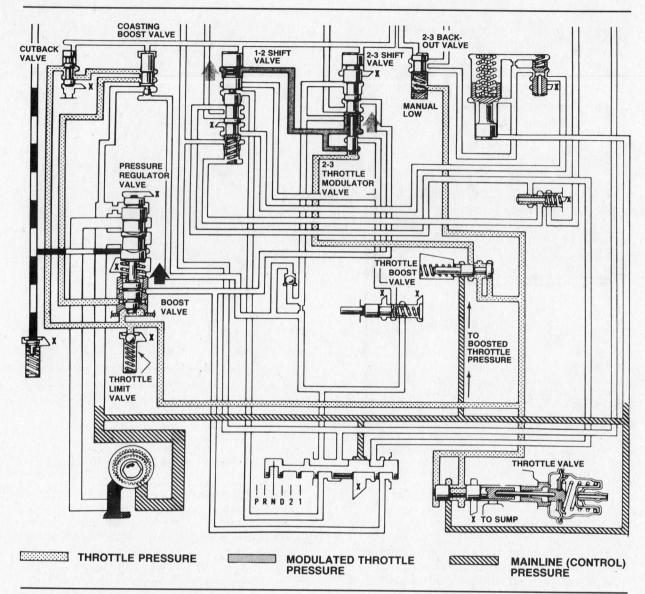

THROTTLE PRESSURE   MODULATED THROTTLE   MAINLINE (CONTROL)
                          PRESSURE   PRESSURE

Figure 10-56. C6 throttle pressure system.

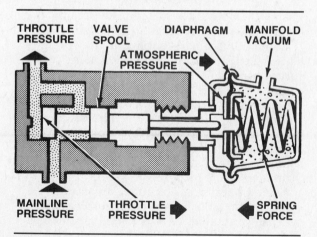

Figure 10-57. A vacuum operated throttle valve regulating throttle pressure during engine operation.

one inch of mercury for every 1,000 feet of altitude. The aneroid bellows decreases throttle pressure at higher altitudes to maintain shift points comparable to those at low altitudes.

Throttle limit valve

The throttle limit valve is part of the throttle pressure circuit, figure 10-56. It is a safety valve that keeps throttle pressure at a safe level. Should throttle pressure exceed the valve's spring force, the valve opens to vent excess pressure to the sump.

Throttle boost valve — Jatco

In C3, C4, C5, C6, and A4LD transmissions, the throttle valve dumps excess fluid directly

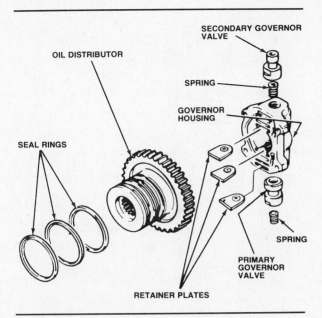

Figure 10-58. This Jatco governor is typical of C4, C5, and C6 models as well.

back to the sump. In the Jatco, figure 10-41, the throttle valve sends excess fluid through a throttle boost valve on the way to the sump. When the low-reverse clutch is applied, the throttle boost valve is forced against its spring by clutch apply pressure and vents fluid directly to the sump. When the low-reverse clutch is released, the boost valve vents through the manual valve.

In manual second, mainline pressure is applied to the throttle boost valve through the manual valve. At the same time, spring force holds the boost valve so that the direct vent to the sump is closed. Therefore, throttle pressure becomes equal to boost valve pressure. This is the only condition in which the Jatco develops boosted throttle pressure.

Throttle boost valve — C3, C4, C5, C6, and A4LD

Throttle pressure increases more or less proportionately as engine manifold vacuum decreases. However, at more than half throttle, the rate of vacuum change is slower than it is at less than half throttle. To compensate for this, the throttle boost valve in these transmissions helps increase throttle pressure at larger throttle openings.

Up to about 60 psi (205 kPa) throttle pressure in C4, C5, and C6 transmissions, or 24 psi (80 kPa) in the C3 and A4LD gearboxes, throttle pressure passes around and through the boost

valve without acting on it, figure 10-56. At 6 inches of vacuum (3 inches in the C3 and A4LD), throttle pressure on the end of the boost valve moves it against its spring. This blocks the throttle pressure passage through the valve and opens the center chamber to mainline pressure. The throttle boost valve then applies this boosted throttle pressure, which is really mainline pressure, to the 2-3 throttle modulator valve. This causes a higher modulated throttle pressure, which delays the upshifts.

Cutback valve

Throttle pressure also passes through the cutback valve on its way to the boost valve in the pressure regulator valve assembly, figure 10-56. Throttle pressure at the cutback valve opposes governor pressure. Between 10 and 30 mph (16 and 48 kph), governor pressure overcomes throttle pressure and moves the cutback valve. Under light throttle, this shift occurs at the low end of the speed range. Under heavier throttle, it occurs at the higher end of the range. When the cutback valve shifts, throttle pressure is cut off from the center passage of the boost valve. This reduces mainline pressure at cruising speeds.

Governor valve

We have been discussing the development and control of throttle pressure, which is one of the two auxiliary pressures that work on the shift valves. Before we get to the shift valve operation, we must explain the governor valve (or governor), and how it creates governor pressure.

The governor is mounted on the transmission output shaft and rotates with it. In the C4, C5, C6, and Jatco, the governor contains two spool valves in a housing splined to the shaft, figure 10-58. In early C3 transmissions, the assembly contains one spool and two spring-loaded weights. In 1978 and later C3 models, and in the A4LD, it contains one spring-loaded spool. The governor housing is bolted to the distributor sleeve. Oil is delivered to the governor through passages in the case and the distributor sleeve on the output shaft.

In the C4, C5, C6, and Jatco governors, the primary valve prevents governor pressure from developing at speeds under approximately 10 mph (15 kph). The secondary valve acts as a balance valve and regulates governor pressure in proportion to output shaft speed. When the car is stationary in a forward gear with the

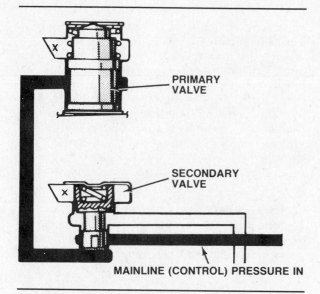

Figure 10-59. Governor operation below 10 mph (16 kph).

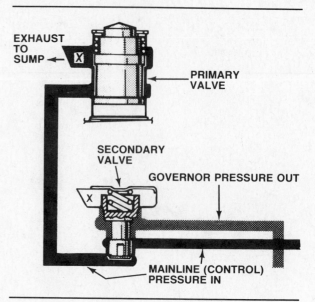

Figure 10-60. Governor operation above 10 mph (16 kph).

engine running, figure 10-59, mainline pressure acts first on the large land of the secondary valve. This moves the secondary valve inward and opens a passage to the primary valve. Mainline pressure is blocked between the two lands of the primary valve. Pressure builds up against the outer end of the secondary valve and holds it in against its spring.

As the car accelerates to about 10 mph (15 kph), centrifugal force overcomes primary valve spring force and moves the valve outward, figure 10-60. This opens a passage to relieve mainline pressure to the sump. With this pressure relieved, the secondary valve moves out in its bore and opens a passage for mainline pressure to go through the governor pressure outlet port. The secondary valve then becomes a balanced, pressure-regulating valve. Governor pressure tries to push the valve inward while spring force and centrifugal force try to move it outward. The result is that governor pressure increases as car speed increases. At very high vehicle speeds, the governor is open as far as possible, and governor pressure equals mainline pressure.

The operation of the C3 and A4LD governors is similar to governor operation in the other transmissions. However, pressure development and regulation is done with one spool valve instead of two.

Downshift valve

The downshift, or kickdown, valve overrides governor control of shift speeds when the accelerator is depressed. In the C3, C4, C5, C6, and A4LD, the valve is operated by mechanical linkage to the accelerator pedal. The end of the valve extends from the case and is connected to the linkage. In the Jatco, the downshift valve is electrically operated. The accelerator linkage closes a switch when the pedal is depressed. The switch energizes a solenoid on the transmission case, which operates the valve. In either case, the downshift valve forces downshifts at higher speeds, or delays upshifts longer than throttle pressure alone would. Hydraulic pressure through the downshift valve acts on the shift valves in the same way as throttle pressure, figure 10-61.

In the C3, C6, A4LD, and Jatco transmissions, the downshift valve receives mainline pressure directly from the manual valve in Drive. In the C4 and C5, the downshift valve receives boosted throttle pressure from the throttle boost valve. Before the downshift valve is actuated, it is held in position in its bore by a spring, and one of its lands blocks mainline or boosted throttle pressure. When the accelerator linkage moves the valve, it opens a passage to apply pressure to the spring ends of the 3-4 (A4LD only) and 2-3 shift valves, and to a differential area of the 1-2 shift valve. Depending on the levels of governor pressure and mainline or boosted throttle pressure, either or both of the shift valves is moved for a downshift.

In the C4, C5, and C6 transmissions, the driver can force a 3-2 downshift at approximately 45 to 90 mph (70 to 145 kph), and a 3-1 downshift at any speed up to approximately 45

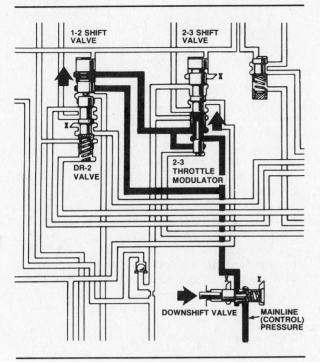

Figure 10-61. C6 downshift valve circuit.

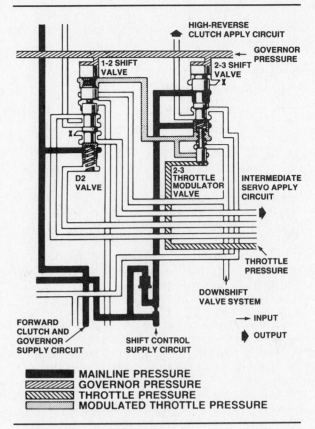

MAINLINE PRESSURE
GOVERNOR PRESSURE
THROTTLE PRESSURE
MODULATED THROTTLE PRESSURE

Figure 10-62. C6 shift valves.

mph (72 kph). The exact speeds will vary based on tire size and axle ratio. In the C3 and A4LD, the 3-2 downshift can be forced from 27 to 65 mph (43 to 105 kph), and a 3-1 shift can be forced below 27 mph (43 kph).

1-2 shift valve assembly

The 1-2 shift valve assembly controls upshifts from low to second, and downshifts from second to low. In all but the Jatco transmission, the 1-2 shift valve assembly includes the 1-2 shift valve and the D2 valve. The Jatco has only the 1-2 shift valve.

In Overdrive, Drive, and manual low, the manual valve sends mainline pressure to a differential area of the D2 valve. At low speeds, the valve assembly is downshifted by modulated throttle pressure and D2 valve spring force. In manual low, mainline pressure is applied to the spring end of the D2 valve to prevent an upshift at any speed. In the C6, which has no D2 valve, spring force acts directly on the 1-2 shift valve. While the shift valve assembly is downshifted, mainline pressure is blocked and the intermediate servo apply passage exhausts to the sump, figure 10-62.

As the car accelerates, governor pressure builds up against the end of the 1-2 shift valve opposite the D2 valve or spring. When governor pressure overcomes throttle pressure and spring force, the 1-2 valve shifts. This moves

the D2 valve to block the servo apply exhaust passage and route mainline pressure to the apply side of the intermediate servo. The servo applies the band, and the transmission shifts into second gear. In the Jatco, this action is controlled by the 1-2 valve alone.

When the gear selector is placed in manual second, mainline pressure from the manual valve keeps the 1-2 shift valve downshifted, but fluid flows through the D2 valve to apply the intermediate band. This provides a second gear start. In the Jatco, the second-lock valve applies the band for manual second.

Each transmission has several other valves in the intermediate servo apply circuit to tailor the shifts for various driving conditions. The C6 has a capacity modulator valve and an accumulator valve. The C4, C5, and A4LD have a 1-2 transition valve and an intermediate accumulator valve. The Jatco has a second-lock valve. In all of these transmissions, the 2-3 backout valve is also in the servo apply circuit, but that valve works on a 2-3 shift only.

Second-lock valve

When a Jatco transmission is placed in manual second, the second-lock valve receives mainline pressure from the manual valve and directs the pressure to the apply side of the intermediate servo. This applies the intermediate band without upshifting the 1-2 shift valve, thus providing a second gear start.

Manual low 2-1 scheduling valve

When a C6 transmission is placed in manual low, the manual valve sends mainline pressure to the end of the manual low 2-1 scheduling valve, where it opposes spring force. The pressure flows through the 2-1 scheduling valve to the spring chamber of the D2 valve. Meanwhile, the mainline pressure at the opposite end of the D2 valve flows through that valve to the spring end of the manual low 2-1 scheduling valve where it moves the valve and flows into the passage to the D2 spring chamber. These mainline applications make a 1-2 upshift impossible.

Intermediate servo modulator and accumulator valves

In the C6, the intermediate servo modulator valve works with the intermediate servo accumulator valve to cushion the 1-2 upshift by controlling the rate at which the intermediate band applies. Before the servo apply circuit is charged, mainline pressure from the manual valve is applied only to the accumulator valve. When the D2 valve moves, servo apply oil is routed through the modulator valve. This oil also is routed through a drilled passage in the modulator valve to the bottom of the valve, and to the differential areas between the lands of the accumulator valve. The oil then flows to the intermediate servo.

Another branch of the intermediate servo apply circuit leads through an orifice to the spring ends of the modulator valve and the accumulator valve. The orifice delays pressure buildup at the spring ends of the valves for an instant. This is time enough for the band to contact the drum. The pressure at the spring ends is lower than the pressure in the rest of this circuit because the fluid passes through the orifice, and pressure in the circuit is lower than mainline pressure because of the accumulator action.

When the accumulator is all the way down, no more fluid can pass through the orifice. Then, the entire apply circuit is equalized at full mainline pressure.

The C3, C4, C5, and A4LD intermediate servo accumulator valve works similarly to the C6 accumulator, but without the capacity modulator valve.

1-2 transition valve

In the C4, C5, and A4LD transmissions, the 1-2 transition valve routes mainline pressure to the apply side of the intermediate servo during a 1-2 upshift and to the release side during a 2-3 upshift. The valve also controls intermediate servo apply and release pressures during downshifts, and blocks mainline pressure in manual low and Reverse. On the A4LD, the 1-2 transition valve and the 2-3 backout valve prevent an intermediate to low-reverse band tie-up on a manual 1-2 shift.

When the 1-2 shift valve and the D2 valve move for an upshift, mainline pressure flows through the 1-2 transition valve to the 2-3 shift valve, where it is blocked. Modulated throttle pressure and spring force hold the 1-2 transition valve so that the passage to the servo and the 2-3 shift valve remain open. The higher the throttle pressure, the further the valve is held open.

When the transmission is in manual low or Reverse, mainline pressure pushes the 1-2 transition valve back, blocking the passage to the intermediate servo and the 2-3 shift valve. In the C4 and C5, the low servo modulator valve provides the pressure that closes the 1-2 transition valve, and the 1-2 transition valve mechanically closes the 2-3 backout valve. Closing the 1-2 transition valve helps lock out second gear.

2-3 shift valve

The 2-3 shift valve assembly, figure 10-62, controls upshifts from second to third, and downshifts from third to second. The assembly includes the 2-3 shift valve and the 2-3 throttle modulator valve.

The 2-3 shift valve receives mainline pressure from the manual valve in Drive and Overdrive. Governor pressure is applied to the end of the shift valve opposite the spring and the 2-3 throttle modulator valve. Throttle pressure is applied to the 2-3 throttle modulator valve to oppose governor pressure. In addition, the throttle modulator valve produces a modulated throttle pressure that is applied to the spring chambers of the 2-3 and 1-2 shift valves. In the C4, C5, and A4LD, modulated throttle pressure is also applied to the 1-2 transition valve. The 2-3 valve is kept downshifted by a combination of throttle pressure, modulated throttle pressure, and spring force.

As the car accelerates, governor pressure increases on the other end of the valve. When governor pressure becomes high enough, it moves the shift valve to open the circuit to the high-reverse clutch. This circuit also includes the release side of the intermediate servo. The surface area of the release side of the servo is larger than the area of the apply side, so the servo releases the band as the high-reverse clutch is applied.

2-3 backout valve

If the driver releases the throttle after the 2-3 upshift has started but before it is complete, the shift will be harsh. With the sudden reduction in engine torque, there might be enough pressure to apply the high-reverse clutch but not to release the intermediate servo and band. In the C3, C4, C5, C6, and A4LD, the 2-3 backout valve keeps the clutch and band from fighting each other and trying to engage two gears at once.

For a power-on shift, throttle pressure holds the backout valve closed. With zero throttle pressure, a servo-release and clutch-apply pressure of about 10 psi or 30 kPa will push the valve down against its spring. As the backout valve moves, it cuts off mainline pressure to the apply side of the servo. It also opens a passage so that apply pressure trapped between the backout valve and the servo can mix with the release pressure. This equalizes servo apply and release pressures, and the spring can release the servo regardless of the hydraulic pressure.

In manual low and Reverse in the C4 and C5, low-reverse servo apply pressure from the low servo modulator valve closes the 1-2 transition valve, and the 1-2 transition valve mechanically closes the 2-3 backout valve.

2-3 timing valve

The Jatco transmission uses a 2-3 timing valve and an orifice to control full- and light-throttle 2-3 upshifts, figure 10-41. For a full-throttle upshift, throttle pressure and spring force hold the timing valve down against governor pressure. This opens a direct passage from the 2-3 shift valve to the clutch-apply and servo-release circuit. The clutch engages instantly as the band releases.

For a light-throttle 2-3 upshift, governor pressure overcomes the reduced throttle pressure on the 2-3 timing valve. This moves the valve to close the direct passage to the clutch apply circuit. Fluid then flows through an orifice to delay the clutch engagement. Full pressure is still applied to the servo-release circuit to release the servo faster than the clutch is applied.

3-2 timing valve

The 3-2 timing valve in the C3, C5, and A4LD regulates the rate at which a 3-2 downshift takes place based on vehicle speed. The 3-2 timing valve is controlled by a spring at one end

and governor pressure at the other. While high-reverse clutch apply and intermediate servo release oil exhaust, this valve regulates the flow so that the servo can completely apply the intermediate band before the high-reverse clutch is completely released. When the clutch circuit is fully exhausted, the 3-2 timing valve stops regulating.

In the A4LD, the 3-2 timing valve routes the exhausting clutch and servo release fluid differently according to road speed. At low speeds, the fluid is sent to the torque demand control valve. At higher speeds, the fluid is sent to the 3-2 intermediate servo control valve and the 3-2 high-reverse clutch control valve.

1-2/3-2 timing control valve

The 1-2/3-2 timing control valve in the C5 transmission, figure 10-38, controls fluid exhausting from the release side of the intermediate servo. On 1-2 upshifts, it routes intermediate servo release exhaust oil to the intermediate servo accumulator valve; on 3-2 downshifts it combines high-reverse clutch exhaust oil with that from the servo release.

On a 1-2 upshift below 30 mph (48 kph), intermediate servo release exhaust oil flows around the servo release shuttle valve to the 1-2/3-2 timing control valve, which then sends the fluid to the top of the intermediate servo accumulator valve. When the pressure of this fluid overcomes throttle and spring pressure acting on the opposite end of the intermediate servo accumulator valve, it pushes the valve down. The fluid then flows to the 2-3 shift valve, and exhausts through the manual valve.

On a 3-2 downshift, the 1-2/3-2 timing control valve slows high-reverse clutch oil exhausting through the 2-3 shift valve, and joins this fluid to that from the servo release exhaust. The high- and low-speed orifices in the servo release circuit meter the pressure, so that intermediate servo pressure is not the same as high-reverse clutch pressure.

Servo release shuttle valve

The servo release shuttle valve in the C5 transmission, figure 10-38, selects intermediate servo release exhaust orifices during a 3-2 downshift based on road speed. The valve is controlled at one end by a spring, and at the other end by governor pressure. At lower speeds, spring force opens the low-speed orifice; at wide-open-throttle and high road speeds, governor pressure opens the high-speed orifice. When governor pressure overcomes spring force, the valve routes servo release exhaust oil through the high-speed orifice to join high-reverse clutch exhaust.

3-2 coast control valve

The 3-2 coast control valve in the C3 and A4LD controls the exhaust of high-reverse clutch apply and intermediate servo release oil during a coasting (zero-throttle) 3-2 downshift.

Coasting boost valve

The coasting boost valve in the C3, C4, C5, C6, and A4LD provides higher mainline pressure when the transmission is shifted from Drive or Overdrive to manual low or manual second. When a shift to manual low is made above 35 mph (56 kph), the transmission first shifts to second rather than low because downshift pressure cannot move the 1-2 shift valve against governor pressure at higher speeds. Downshift pressure is high enough to move the 2-3 shift valve, however. As speed drops below 35 mph, and governor pressure is reduced, the transmission will automatically downshift from second to low.

In either case, when the transmission downshifts from third to second at speeds above 35 mph (56 kph), it creates a problem with the intermediate band. In third, the drum that the band engages turns clockwise. This rotation tends to loosen the band by forcing the apply strut back against the servo. When the 1-2 upshift is made, this is not a problem because the drum is turning counterclockwise and tends to tighten the band. To compensate for the loosening effect of clockwise drum rotation, particularly at higher speeds, mainline pressure must be increased.

When the manual valve is moved from ⒟ or D to 1, the pressure that has been holding the coasting boost valve in place is exhausted. The valve then opens a circuit that routes mainline pressure to the pressure boost valve. The pressure boost valve acts on the pressure regulator valve to increase mainline pressure. In the C6, mainline pressure holds the coasting boost valve in place. In the C3, C4, C5, and A4LD, throttle pressure holds the valve in place.

The coasting boost valve is not necessary for a full-throttle 3-2 kickdown shift because increased throttle pressure acts on the pressure regulator valve to raise mainline pressure at that time.

Low servo modulator valve

The low servo modulator valve in the C4, C5, and A4LD applies boosted mainline pressure to the low-reverse servo for a standing start in manual low. Mainline pressure from the manual valve goes through the 1-2 shift valve to the low servo modulator valve where it is directed to the low-reverse servo. Fluid from the low servo modulator valve is also routed to the 1-2 transition valve, forcing it down.

Additional A4LD Valves

The valves described in the following sections are unique to the A4LD 4-speed automatic transmission. Refer to figure 10-40 while reading these descriptions.

Engagement control valve

To provide smooth lockup without excessive slippage, the engagement control/orifice control valve feeds oil to the forward clutch through a small orifice at low throttle pressure, and through a larger orifice at higher throttle pressure. A balance between spring force and throttle pressure controls the valve.

3-2 high-reverse clutch control valve

The 3-2 high-reverse clutch control valve restricts clutch fluid exhaust during a 3-2 downshift to delay clutch release until the intermediate band is applied.

3-2 intermediate servo control valve

The 3-2 intermediate servo control valve acts during higher speed 3-2 downshifts to control the rate at which intermediate servo release oil exhausts.

Converter coasting boost valve

The converter coasting boost valve is part of the coasting boost valve assembly. The converter coasting boost valve acts on the coasting boost valve to increase mainline pressure when the torque converter clutch is applied. This helps hold the apply devices more securely under the higher driveline loads that occur when the converter is locked.

3-4 shift valve

The 3-4 shift valve controls 3-4 upshifts and 4-3 downshifts. It is located in the valve body next to the 4-3 torque demand valve and a bushing. The assembly consists of the 3-4 shift valve, two springs, and the 3-4 throttle modulator valve. Governor pressure at one end of the 3-4 shift valve overcomes spring force and throttle pressure to upshift the valve. This allows mainline pressure through the valve to the overdrive servo which applies the overdrive band.

To delay 3-4 upshifts, the 3-4 throttle modulator valve provides modulated throttle pressure to the 3-4 shift valve when throttle pressure is high. As throttle pressure increases, it pushes the throttle modulator valve and spring against the shift valve, so that a higher governor pressure is needed to move the shift valve and effect the upshift.

Overdrive servo accumulator valve

The overdrive servo accumulator valve controls overdrive band application on 3-4 upshifts. Spring force holds the accumulator closed until overdrive servo release exhaust pushes down against spring force, giving the overdrive band a smooth apply. The servo release oil exhausts through the 3-4 shift valve, the kickdown valve, a ball-check orifice, and finally, the manual valve.

Torque demand control valve

The torque demand control valve smoothes low-speed 3-2 downshifts by maintaining a balance between spring force at one end and governor pressure at the other. By doing this, the valve controls the rate at which high-reverse clutch apply oil, and intermediate servo release oil, exhaust during low-speed 3-2 downshifts. The higher the vehicle speed, and thus the governor pressure, the greater restriction the valve places on the exhaust flow.

4-3 torque demand valve

The 4-3 torque demand valve has a spring acting on one end, and throttle pressure acting on the other. Under heavy load, throttle pressure overcomes spring force and moves the valve. High-reverse clutch oil then flows through the valve to the 3-4 shift valve where it acts on the shift valve to allow for 4-3 torque demand downshifts at higher road speeds.

3-4 backout valve

The 3-4 backout valve is a restricting orifice that controls the rate at which overdrive servo release oil exhausts during a 3-4 backout upshift. A portion of the exhaust oil flows through the 3-4 backout valve, bypassing the overdrive servo accumulator valve. The 3-4 backout valve restricts the servo release exhaust so that exhaust pressure builds sufficiently to overcome spring force in the overdrive servo accumulator valve.

A4LD Converter Clutch Control Valves

The basic operation of the lockup torque converter used with the A4LD is controlled hydraulically through three valves:

- Converter clutch throttle modulator valve
- Converter clutch shift valve
- Converter clutch shuttle valve.

Although these valves have the primary responsibility for controlling converter clutch operation, they can be overridden by the Electronic Engine Control (EEC-IV) system which controls a normally energized TCC solenoid. If the EEC-IV computer determines the converter

should not be locked at any given moment, it will de-energize the TCC solenoid to unlock the clutch. The electronic parts of this system are described in Chapter 8; the hydraulic controls are covered below.

Converter clutch throttle modulator valve

The converter clutch throttle modulator valve, figure 10-40, is controlled by spring force at one end, and throttle pressure at the other. When throttle pressure rises high enough to overcome spring force, the valve opens and routes modulated throttle pressure to the spring end of the converter clutch shift valve. This pressure helps delay converter clutch engagement at larger throttle openings.

Converter clutch shift valve and shuttle valve

The converter clutch shift valve works with the converter clutch shuttle valve and the TCC solenoid to control the converter clutch, figure 10-63. The converter clutch shift valve prevents converter clutch engagement below a minimum speed. Spring force and modulated throttle pressure act on one end of the shift valve, governor pressure acts on the opposite end.

The converter clutch shuttle valve controls the flow of fluid to apply and release the converter clutch. Spring force and mainline pressure act on one end of the shuttle valve, mainline pressure alone acts on the other end. Mainline pressure to the end of the shuttle valve without a spring can come from two passages. One enters between the shuttle valve and the lockup inhibitor plug, the other passage is controlled by the TCC solenoid and enters between the solenoid and the lockup inhibitor plug.

At lower vehicle speeds, figure 10-63, modulated throttle pressure and spring force combine to hold the shift valve in its downshifted position. In this position, the valve sends mainline pressure to the spring chamber of the shuttle valve. A two-way ball check valve seals a passage in the same circuit that leads to the space between the TCC solenoid and the lockup inhibitor plug. The TCC solenoid is also energized to prevent the flow of mainline pressure into this area.

The combination of spring force and mainline pressure acting on the shuttle valve overcomes mainline pressure alone on the other end, and holds the valve in the unlock position. Converter pressure then passes through the shuttle valve and into a passage that holds the TCC in the released position. Fluid returning from the converter flows through a valley in the shuttle valve, and is routed to the transmission cooler.

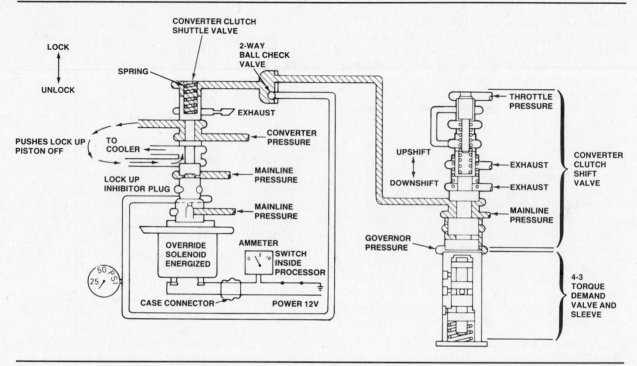

Figure 10-63. The A4LD converter clutch unlocked by the transmission hydraulic controls.

When vehicle speed becomes high enough, governor pressure overcomes spring and modulated throttle pressure, and upshifts the shift valve, figure 10-64. This cuts off the flow of mainline pressure to the spring end of the shuttle valve, and vents the circuit to the sump. With only spring pressure now acting on the upper end of the shuttle valve, mainline pressure entering between the valve and the lockup inhibitor plug moves the shuttle valve to the lockup position.

In the lockup position, mainline pressure is routed through the shuttle valve into a passage that holds the converter clutch in the locked position. Fluid returning from the converter is routed through a valley in the shuttle valve, and then through a restricting orifice where it is exhausted to the sump. While the converter clutch is locked, normal converter pressure is routed directly through the shuttle valve to the transmission cooler. Note that the converter clutch is still under hydraulic control; the TCC solenoid remains energized, and the lockup inhibitor plug has not moved.

Once the converter clutch is locked, the *hydraulic* controls can unlock it only if spring and modulated throttle pressure overcome governor pressure at the shift valve. However, the EEC-IV *electronic* controls can unlock the converter clutch under a number of other conditions simply by de-energizing the TCC solenoid, figure 10-65, which opens a mainline fluid

passage. Mainline pressure then enters between the solenoid and the lockup inhibitor plug, where it moves the plug and is routed to the spring end of the shuttle valve. The 2-way ball check valve seats to prevent fluid in this circuit from venting through the shift valve.

With equal mainline pressure on both ends of the shuttle valve, the spring moves the valve downward to unlock the converter clutch. Converter pressure then flows through the shuttle valve into a passage that holds the converter clutch in the released position. As in figure 10-63, returning fluid passes through a valley in the shuttle valve and then is routed to the transmission cooler.

When the EEC-IV computer determines that the converter can again be locked, it energizes the solenoid to cut off mainline pressure to both ends of the shuttle valve. Mainline pressure entering between the lockup inhibitor plug and the valve then returns the valve to the locked position, and oil flows through the valve is as shown in figure 10-64.

MULTIPLE-DISC CLUTCHES AND SERVOS

The preceding paragraphs examined the individual valves in these transmissions' hydraulic systems. We will now take a quick look at the multiple-disc clutches and servos in these transmissions.

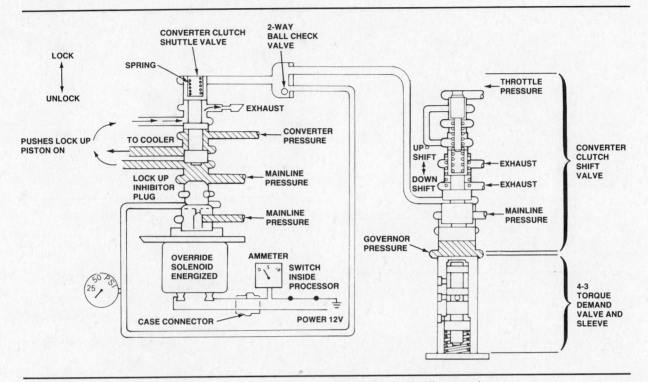

Figure 10-64. The A4LD converter clutch locked by the transmission hydraulic controls.

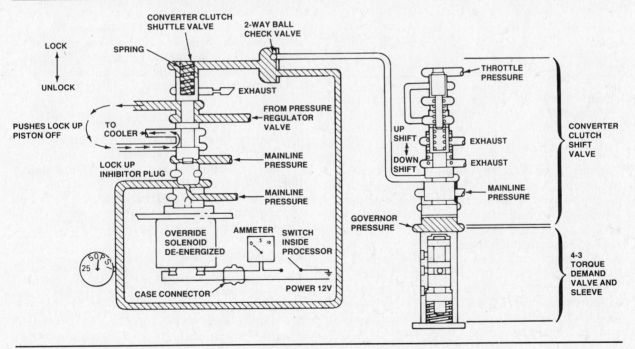

Figure 10-65. The A4LD converter clutch unlocked by the EEC-IV system electronic controls.

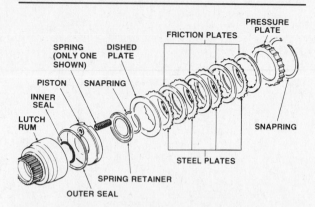

Figure 10-66. This Jatco forward clutch is typical of Ford clutches that use several small springs to return the piston.

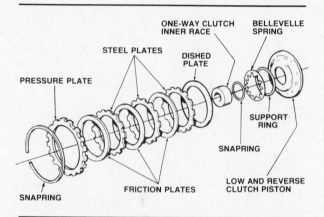

Figure 10-67. This Jatco low-reverse clutch is typical of Ford clutches that use a single Belleville spring to return the piston.

Forward Clutch

The forward clutch is applied to transmit driving torque in all forward gears. The number of discs and separator plates in this clutch varies with different transmissions. Except in the A4LD, hydraulic pressure is supplied to the clutch piston through the stator support. In the A4LD, the case center support supplies the hydraulic pressure. The C3, C6, A4LD, and Jatco forward clutches use several coil springs to return the clutch piston, figure 10-66; C4 and C5 forward clutches use a single Belleville spring.

High-Reverse Clutch

The high-reverse clutch is applied in third, fourth, and Reverse gears to hold or drive the sun gear. Except in the A4LD, fluid is supplied to the clutch through the stator support. In the A4LD, the case center support supplies the

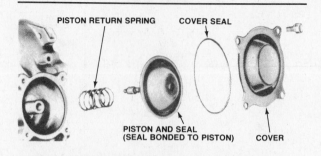

Figure 10-68. A typical C4 low-reverse servo.

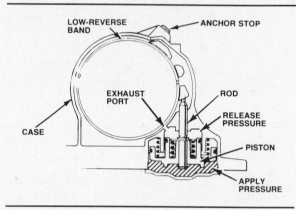

Figure 10-69. A C3 low-reverse servo installation.

hydraulic pressure. The high-reverse clutches in all of these transmissions use several small coil springs to return the clutch piston. Some C4 models use a single large coil spring.

Low-Reverse Clutch

The low-reverse clutch in C6 and Jatco transmissions holds the low-reverse drum and rear carrier in manual low and Reverse. Fluid is supplied to the clutch through passages in the transmission case. The Jatco clutch uses a Belleville spring to release the clutch, figure 10-67. The C6 clutch uses several small coil springs.

Low-Reverse Servo

The low-reverse servo in the C3, C4, C5, and A4LD applies the low-reverse band to hold the low-reverse drum and rear carrier in manual low and Reverse. The C4 and C5 servos are accessible from the outside of the case, at the left rear, figure 10-68. The C3 and A4LD servos are located inside the case, figure 10-69, and the oil pan must be removed for access. The C4 and C5 low-reverse bands are adjustable. C3 and A4LD band adjustments are controlled by selective piston rod lengths that are chosen during initial transmission assembly.

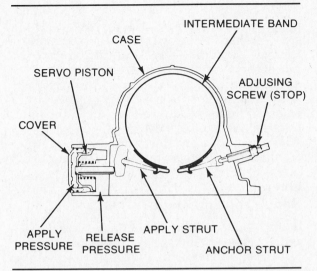

Figure 10-70. This C3 intermediate servo installation is typical of other Ford transmissions as well.

Intermediate Servo

The intermediate servo, figure 10-70, applies the intermediate band in second gear. The intermediate servo is accessible from outside the case. The servo cover is at the right front of the case. The intermediate band is adjustable.

HYDRAULIC SYSTEM SUMMARY

We have examined the individual valves, clutches, bands, and servos in the hydraulic systems and discussed their functions. The following paragraphs summarize their combined operations in the various gear ranges. Refer to figures 10-35, 10-36, 10-37, 10-38, 10-39, 10-40, and 10-41 to trace the hydraulic system operation.

Park and Neutral

The hydraulic flow in Park and Neutral is the same. Oil flows from the pump to the pressure regulator valve, the manual valve, and the throttle valve. The pressure regulator valve feeds the converter, which routes oil to the cooler and the lubrication circuits. The manual valve has no open ports in Neutral and Park. The throttle valve develops throttle pressure and sends it to the spring ends of the shift valves, the pressure boost valve, and the 2-3 and 3-4 (A4LD) throttle modulator valves.

The pressure boost valve acts with spring force on the pressure regulator valve to increase mainline pressure in relation to throttle opening. The 2-3 throttle modulator valve routes modulated throttle pressure to the spring ends of the shift valves (and the 1-2 transition valve in the C4, C5, and A4LD) to help delay upshifts and force downshifts. In the A4LD, the 3-4 throttle modulator valve also receives throttle pressure to control upshifts and downshifts.

Throttle pressure also is routed through the cutback valve to: the spring end of the intermediate servo accumulator (C4, C5, A4LD); the intermediate servo modulator valve and accumulator (C6); the 2-3 timing valve (Jatco); and the end of the coasting boost valve opposite spring force (C3, C4, C5, C6, A4LD). These throttle valve applications are made in all gear ranges.

In Park and Neutral, no clutches or bands are applied, and all valve positions are determined by spring force alone. In Park, the parking pawl engages the parking gear attached to the output shaft to prevent the car from being moved.

Overdrive (A4LD) and Drive Low

In all forward gear ranges, the manual valve directs mainline pressure to the forward clutch apply circuit, the shift valves, and the governor. The forward clutch is applied. In the A4LD, the engagement control valve regulates forward clutch apply oil in relation to throttle opening. Mainline pressure at the downshifted shift valves is blocked. The governor regulates mainline pressure in relation to road speed, producing governor pressure.

Governor pressure is routed to the shift valves and the cutback valve. Governor pressure also flows to: the 1-2/3-2 timing control and servo release shuttle valves (C5); the 3-2 timing valve (C3, C5, and A4LD); the 3-2 high-reverse clutch control valve, 3-2 intermediate servo control valve, and the torque demand control valve (A4LD); and to the coasting boost valve (C3, C4, C5, C6, and A4LD).

When the A4LD transmission is in any gear range except Overdrive, the manual valve also routes oil to the overdrive multiple-disc clutch, and to the spring end of the 3-4 shift valve. This pressure applies the overdrive clutch and prevents the 3-4 shift valve from upshifting.

The forward clutch and the 1-way roller clutch provide low gear. If the A4LD is in Drive low, the overdrive multiple-disc clutch is also applied. If the A4LD is in Overdrive low, the overdrive multiple-disc clutch is released, but the overdrive 1-way clutch is locked during acceleration.

Overdrive (A4LD) and Drive Second

As governor pressure increases, it pushes the 1-2 shift valve against throttle pressure and spring force, and mainline pressure flows through the D2 valve to the apply side of the intermediate servo, and to the 2-3 backout valve (except Jatco). In the Jatco, there is no D2 valve, and servo apply pressure comes from the 1-2 shift valve. In the C4, C5, and A4LD, servo apply oil travels through the 1-2 transition valve. In the C5 and A4LD, the intermediate servo accumulator cushions the upshift. In the C5, the 1-2/3-2 timing control valve provides servo release exhaust to the accumulator to time the upshift. In the C6, the intermediate servo modulator valve and accumulator cushion band application.

The forward clutch remains applied, and together with the intermediate band, provides second gear. If the A4LD is in Drive second, the overdrive multiple-disc clutch is also applied. If the A4LD is in Overdrive second, the overdrive multiple-disc clutch is released, but the overdrive 1-way clutch is locked during acceleration.

Overdrive (A4LD) and Drive Third

As governor pressure continues to increase, it overcomes throttle pressure and spring force to upshift the 2-3 shift valve. The upshift is delayed at large throttle openings by the 2-3 throttle modulator valve. Mainline pressure flows through the upshifted shift valve to the release side of the intermediate servo and the apply area of the high-reverse clutch. At closed throttle, the 2-3 backout valve (except Jatco) helps time the shift. In the Jatco, the 2-3 timing valve controls band release and clutch apply oil in relation to throttle opening.

Two apply devices provide third. The forward clutch remains applied, and the high-reverse clutch is applied. If the A4LD is in Drive third, the overdrive multiple-disc clutch is also applied. If the A4LD is in Overdrive third, the overdrive multiple-disc clutch is released, but the overdrive 1-way clutch is locked during acceleration.

Overdrive Fourth (A4LD)

As governor pressure continues to increase, it overcomes throttle pressure and spring force to upshift the 3-4 shift valve. At large throttle openings, the upshift is delayed by the 3-4 throttle modulator valve. Mainline pressure flows through the upshifted shift valve to the apply side of the overdrive servo. The apply pressure causes overdrive servo release pressure to exhaust. The overdrive servo accumulator valve and the 3-4 backout valve control the rate of servo release exhaust, and therefore the rate of band application. From the overdrive servo accumulator valve, servo release exhaust empties into the kickdown channel and exhausts at the manual valve.

The overdrive band, forward clutch, and high-reverse clutch are applied in fourth gear.

Overdrive and Drive Range Forced Downshifts

In the A4LD, pressing down on the accelerator pedal acts on the linkage to the downshift valve, so that the valve sends mainline pressure to downshift the 3-4 shift valve and release the converter clutch. The 4-3 torque demand valve helps cause the downshift at low vacuum.

Detent (full-throttle) 3-2 downshifts in all of these transmissions can be forced at speeds between 30 and 65 mph (48 and 105 kph). In the C3, C4, C5, C6, and A4LD, the linkage from the accelerator moves the downshift valve to its extreme inner position. The Jatco uses an electric solenoid, energized by a switch on the carburetor at full throttle, to operate the downshift valve. The downshift valve directs mainline (C3, C6, A4LD, and Jatco) or boosted throttle (C4 and C5) pressure into the part-throttle and detent passages. Mainline and boosted throttle pressure then combine with spring force to downshift the 3-2 shift valve. At speeds below 30 mph (48 kph), both the 2-3 and the 1-2 shift valves are moved for a 3-1 downshift.

Manual Low

In manual low, the manual valve sends mainline pressure to the spring end of the 1-2 shift valve to prevent it from upshifting, and also to the low-reverse servo or clutch. In the C3, C6, and Jatco, mainline pressure is directed from the manual valve to the 1-2 shift valve and D2 valve to keep the transmission in low gear and apply the low-reverse band (C3) or clutch (C6 and Jatco). In the C4, C5, and A4LD, mainline pressure is directed from the manual valve to the low servo modulator valve where it becomes low-reverse servo apply oil. This oil is then sent to close the 1-2 transition valve, which causes the 1-2 transition valve to mechanically close the 2-3 backout valve.

If the transmission is in high gear and is manually shifted to manual low at speeds above 35 mph (56 kph), the transmission first shifts into second gear. High-reverse clutch oil and intermediate band release oil exhaust at the manual valve so the intermediate band applies. In the C3, C4, C5, and C6, the coasting boost valve opens a passage to the pressure boost valve, so that mainline pressure increases and the intermediate band holds firmly. When vehicle speed drops below 35 mph, mainline pressure and spring force overcome governor pressure to downshift the 1-2 shift valve and keep it from upshifting.

The forward clutch and the low-reverse band or clutch provide low gear in manual low. In the A4LD, the overdrive multiple-disc clutch is also applied.

Manual Second

In manual second, all of these transmissions start and stay in second gear. The manual valve routes mainline pressure to the spring end of the shift valves assembly to prevent their upshifting. Pressure from the manual valve also acts on the pressure boost valve to boost mainline pressure.

In the C3, C4, C5, and A4LD, the D2 valve feeds the intermediate servo, so the band is applied even though mainline pressure holds the 1-2 shift valve closed. In the C6, mainline pressure from the D2 valve is routed through the intermediate servo modulator valve and servo accumulator to cushion the band apply, while the 1-2 and 2-3 shift valves are held closed by boosted throttle pressure. In the Jatco, mainline pressure from the manual valve is routed through a second lock valve, to apply the intermediate band by filling the apply servo. The Jatco's two shift valves are held closed by mainline pressure.

The forward clutch and the intermediate band provide second gear. In the A4LD, the overdrive clutch is also applied.

Reverse

In Reverse, the manual valve opens the passage to apply mainline pressure to the reverse circuits. The C3, C4, and C5 have L and R circuits open in Reverse. Pressure in the L circuit applies the low-reverse band through the D2 valve, and charges the downshift passages. In the C4, C5, and A4LD, this pressure also closes the 1-2 transition valve. In the C4 and C5, mainline pressure passes through the low servo modulator valve and is modulated before closing the 1-2 transition valve, which mechanically closes the 2-3 backout valve.

The R circuit includes the pressure boost valve to increase mainline pressure at the spring end of the pressure regulator. The R passage also routes oil through the 2-3 shift valve to the high-reverse clutch piston.

The C6 and Jatco have a low-reverse clutch instead of a band, but the oil flow from the manual valve is very similar to that just described. The A4LD reverse circuit is also similar, with the addition that the overdrive clutch is also applied through the manual valve.

Review Questions

Select the single most correct answer
Compare your answers to the correct answers on page 603

1. Ford Motor Company has _____ transmissions that use the 3-speed Simpson gear train.
   a. 3
   b. 4
   c. 5
   d. 6

2. Mechanic A says C3 and C4 transmissions have a low-reverse band. Mechanic B says C5 and A4LD transmissions have a low-reverse band.
   Who is right?
   a. Mechanic A
   b. Mechanic B
   c. Both mechanic A and B
   d. Neither mechanic A nor B

3. Mechanic A says the C5 transmission has a hydraulically locked torque converter. Mechanic B says the C6 and Jatco transmissions have a low-reverse clutch.
   Who is right?
   a. Mechanic A
   b. Mechanic B
   c. Both mechanic A and B
   d. Neither mechanic A nor B

4. Mechanic A says Ford transmission code letters are listed on the identification plate on the left front door. Mechanic B says the code letter "L" is used to identify all Ford 4-speed automatic transmissions.
   Who is right?
   a. Mechanic A
   b. Mechanic B
   c. Both mechanic A and B
   d. Neither mechanic A nor B

5. Mechanic A says 1964-66 C4 transmissions will not start in second gear. Mechanic B says the C5 transmission introduced in 1980 is basically an updated C4.
   Who is right?
   a. Mechanic A
   b. Mechanic B
   c. Both mechanic A and B
   d. Neither mechanic A nor B

6. Mechanic A says the C6 transmission was introduced as a heavy-duty unit for use with larger engines in 1970. Mechanic B says the C6 transmission is not used in passenger cars after 1980.
   Who is right?
   a. Mechanic A
   b. Mechanic B
   c. Both mechanic A and B
   d. Neither mechanic A nor B

7. Mechanic A says Type F fluid is used in all 1976 and earlier Ford automatic transmissions. Mechanic B says MERCON® is the preferred fluid for all 1983-87 Ford automatic transmissions. Who is right?
   a. Mechanic A
   b. Mechanic B
   c. Both mechanic A and B
   d. Neither mechanic A nor B

8. Mechanic A says the fluid screen holds a small valve into the valve body on 1970 and later C4 transmissions. Mechanic B says 3-2 downshifts are controlled by an electric solenoid on the A4LD transmission. Who is right?
   a. Mechanic A
   b. Mechanic B
   c. Both mechanic A and B
   d. Neither mechanic A nor B

9. Mechanic A says Ford refers to mainline pressure as control pressure. Mechanic B says Ford refers to the stator as the rotator. Who is right?
   a. Mechanic A
   b. Mechanic B
   c. Both mechanic A and B
   d. Neither mechanic A nor B

10. The oil pumps used in C3, C4, C5, C6, A4LD, and Jatco transmissions are all:
    a. Vane type
    b. Rotor type
    c. Gear type
    d. Piston type

11. Mechanic A says the front clutch in the C3, C4, C5, C6, and Jatco transmissions is the high-reverse clutch. Mechanic B says the rear clutch in the A4LD is the low-reverse clutch. Who is right?
    a. Mechanic A
    b. Mechanic B
    c. Both mechanic A and B
    d. Neither mechanic A nor B

12. Mechanic A says that in C3, C4, C5, C6, and Jatco transmissions, the input shaft is splined directly to the front ring gear. Mechanic B says that in the A4LD transmission, the input shaft is splined directly to the overdrive sun gear. Who is right?
    a. Mechanic A
    b. Mechanic B
    c. Both mechanic A and B
    d. Neither mechanic A nor B

13. Mechanic A says the intermediate band in the C3, C4, C5, C6, Jatco, and A4LD transmissions is applied in manual low and manual second. Mechanic B says the overdrive band in the A4LD is applied in all gear selector positions except Park, Neutral, and Overdrive Ⓓ. Who is right?
    a. Mechanic A
    b. Mechanic B
    c. Both mechanic A and B
    d. Neither mechanic A nor B

14. Mechanic A says the sun gear is the input member of the Simpson geartrain in Reverse. Mechanic B says the sun gear is the input member of the Simpson geartrain in high gear. Who is right?
    a. Mechanic A
    b. Mechanic B
    c. Both mechanic A and B
    d. Neither mechanic A nor B

15. Mechanic A says the forward clutch drives the sun gear of the Simpson geartrain. Mechanic B says the high-reverse clutch drives the ring gear of the Simpson geartrain. Who is right?
    a. Mechanic A
    b. Mechanic B
    c. Both mechanic A and B
    d. Neither mechanic A nor B

16. The forward clutch is the input device to the Simpson geartrain in:
    a. Drive low and Reverse
    b. All forward gears
    c. Manual low only
    d. Drive low only

17. In high gear, the front ring gear and the _____ are locked to the input shaft by apply devices.
    a. Rear carrier
    b. Front carrier
    c. Sun gear
    d. Rear ring gear

18. The drainback valve prevents:
    a. The governor from draining
    b. The pump from draining
    c. The torque converter from draining
    d. All of the above

19. Mechanic A says C6 and Jatco transmissions use a servo to hold the rear drum in low gear. Mechanic B says C3, C4, C5, C6, Jatco, and A4LD transmissions use a vacuum-modulated throttle valve. Who is right?
    a. Mechanic A
    b. Mechanic B
    c. Both mechanic A and B
    d. Neither mechanic A nor B

20. A D2 valve is *not* used in the _____ transmission.
    a. C3
    b. Jatco
    c. C6
    d. A4LD

Chapter

11

The Ford ATX Transaxle

design is used in Escort/Lynx ATX transaxles through the 1988 model year, and in 1984–86 Tempo/Topaz ATX transaxles.

In 1986–87 Taurus/Sable models with the ATX, the split-torque converter is replaced by converter with a centrifugally locking converter (CLC) similar to that used with the C5 RWD

HISTORY AND MODEL VARIATIONS

The Ford ATX automatic transaxle, figure 11-1, was introduced in 1980 in the Escort/Lynx. In 1984, Tempo/Topaz models were added to the list of cars using the ATX. And in 1986, The ATX was adopted for use in 4-cylinder Taurus/ Sable models. The ATX is a 3-speed automatic transaxle that uses the Ravigneaux compound planetary geartrain. Unusual features of the ATX include a valve body mounted on top of the case, an oil pump installed at the opposite end of the transaxle from the torque converter, and a parking gear located on the final drive output gear.

The original version of the ATX transaxle also uses a unique splitter gear torque converter that improves efficiency. The splitter gear converter contains a planetary gearset that divides input power delivery between mechanical and hydraulic means in second and third gears. To accomplish this, two concentric shafts transfer power from the converter to the geartrain. This design is used in Escort/Lynx ATX transaxles through the 1988 model year, and in 1984-86 Tempo/Topaz ATX transaxles.

In 1986-87 Taurus/Sable models with the ATX, the split-torque converter is replaced by a converter with a centrifugally locking converter (CLC) similar to that used with the C5 RWD transmission. The CLC torque converter contains a centrifugally operated clutch that includes friction shoes, a damper assembly, and a 1-way clutch. The operation of this converter clutch design is described in Chapter 7.

In 1987 and later Tempo/Topaz models, and in 1988 and later Taurus/Sable applications, the ATX uses a fluidically linked torque converter (FLC) in place of the previous converter designs. The FLC is a conventional 3-element torque converter without internal locking devices.

Aside from their torque converters, all ATX transaxles are essentially the same. Therefore, the bulk of this chapter will cover the original ATX transaxle equipped with the splitter gear torque converter. The minor differences for transaxles with CLC and FLC converters will be pointed out where necessary.

Fluid Recommendations

Ford initially recommended Type CJ fluid for the ATX transaxle. When that fluid became obsolete, they changed the recommendation to DEXRON®-IID fluid. Ford's latest recommendations, in effect as of the 1988 model year, are

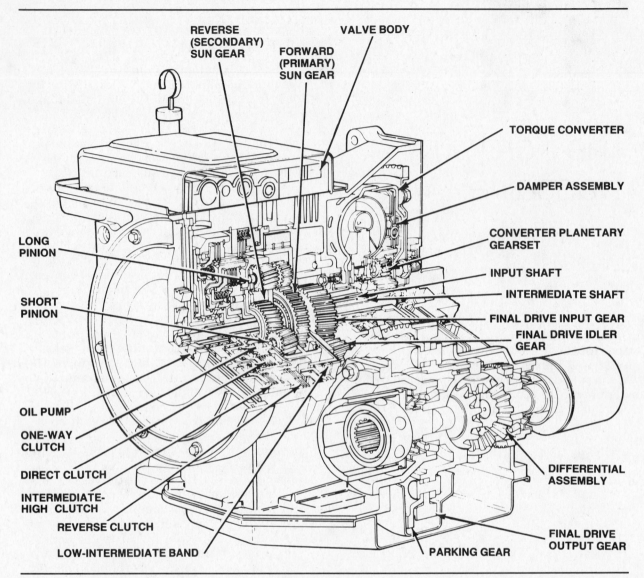

Figure 11-1. The ATX transaxle.

that MERCON® is the preferred fluid for 1980-87 ATX transaxles, although DEXRON®-IID is acceptable. The use of MERCON® is a warranty *requirement* for all 1988 and later ATX transaxles.

Transaxle Identification

Identifying a Ford automatic transaxle is a 2-step process. First, you determine the model of transaxle. All Ford vehicles have an identification plate on the rear of the left front door. In addition to the vehicle identification number and other data, this plate has a code letter that identifies the automatic transaxle model; the ATX is identified by the letter B.

In addition to the transaxle model, you must know the exact transaxle assembly part number

and/or other identifying characters to obtain repair parts. This information is stamped on a tag attached to the transaxle. On the ATX, the tag is retained under one of the valve body cover attaching bolts. Figure 11-2 shows a typical ATX identification tag and explains the meaning of the data stamped on the tag. As required by Federal regulations, Ford transaxles also carry the serial number of the vehicle in which they were originally installed.

Major Changes

Listed below are the major changes made to the ATX transaxle, and the years in which they occurred:

1981 • Steel seal rings replace plastic rings on second and third clutch hubs.

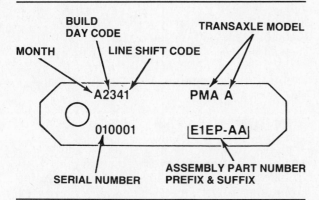

Figure 11-2. The ATX transaxle identification tag is located under one of the valve body cover attaching bolts.

1985 • Direct clutch piston seals revised.
1986 • Planetary gearset design changes reduce second gear noise.
• CLC introduced in Taurus/Sable applications.
1987 • FLC replaces splitter gear converter in Tempo/Topaz.
• Finer-pitch final drive gears introduced.
1988 • FLC replaces CLC in Taurus/Sable.
• Low-reverse brake hub snap ring eliminated in Tracer.

GEAR RANGES

The ATX gear selector, which may be either floor or column mounted depending on the application, has six positions, figure 11-3:
P — Park
R — Reverse
N — Neutral
D — Drive
2 — Manual second
1 — Manual low.
The ATX gear ratios are:
• Low (first) — 2.79:1
• Second (intermediate) — 1.61:1
• Third (direct drive) — 1.00:1
• Reverse — 1.97:1.

Neutral and Park

The engine can be started only when the transaxle is in Neutral or Park. No clutches or bands are applied in these selector positions, and there is no flow of power through the transaxle. In Neutral, the differential assembly is free to turn, and the car can be pushed. In Park, a mechanical pawl engages a large gear on the final drive output gear to lock the differential assembly to the case and prevent the vehicle from being moved.

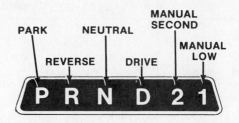

Figure 11-3. ATX gear selector positions.

Drive

When the vehicle is stopped and the gear selector is placed in Drive, the transaxle is automatically in low gear. The vehicle starts to move in low gear and automatically upshifts to second, and then to third gear. The speeds at which upshifts occur vary and are controlled by vehicle speed and engine load.

Once the transaxle is in high gear at a moderate road speed and light throttle, the driver can force a 3-2 downshift by depressing the accelerator partially or completely. A forced 3-1 downshift in Drive is also possible, provided that the vehicle road speed is low enough.

As the vehicle speed decreases, the transaxle automatically downshifts. When the car decelerates in Drive to a full stop, the transaxle makes two downshifts, first from high to second, then from second to low.

Manual Second

Unlike many Ford transmissions, when the gear selector of an ATX is placed in 2 (manual second), and the car is accelerated from a standstill, the transaxle starts in Drive low and then upshifts to second. It does not upshift to third. When vehicle speed drops below a certain point in second, the transaxle downshifts to low.

If the driver moves the gear selector from Drive high to manual second at *any* speed, the transaxle will immediately downshift to provide engine braking. In addition, if the vehicle speed is low enough while the transaxle is in manual second, the driver can force a downshift from second to low by partially or completely depressing the accelerator.

Manual Low

When the vehicle is stopped and the gear selector is placed in 1 (manual low), the transaxle starts and stays in low gear. It does not upshift to second. Because different apply devices are used in manual low than are used in Drive low,

FORD TERM	TEXTBOOK TERM
Reactor	Stator
Clutch cylinder	Clutch drum
Turbine shaft	Input shaft
Line pressure	Mainline pressure
Main oil pressure regulator valve	Pressure regulator valve
T.V. pressure	Throttle pressure
T.V. limit valve	Throttle limit valve
1-2 shift T.V. modulator valve	1-2 throttle modulator valve
2-3 shift T.V. modulator valve	2-3 throttle modulator valve
Planetaries	Pinions
Reverse sun gear	Secondary sun gear
Forward sun gear	Primary sun gear
Intermediate clutch	Intermediate-high clutch

Figure 11-4. Ford ATX transaxle nomenclature table.

this gear range can be used to provide engine compression braking to slow the vehicle.

The driver may also downshift from manual second to manual low for increased engine braking. If the driver shifts to manual low at high speed, the transaxle will not shift directly from high to low gear. To avoid engine or driveline damage, the transaxle first downshifts to second, and then automatically downshifts to low when the speed drops below approximately 20 to 25 mph (30 to 40 kph). The transaxle then remains in low gear until the gear selector is moved to another position.

Reverse

When the gear selector is moved to R, the transaxle shifts into Reverse. The transaxle should be shifted to Reverse only when the vehicle is standing still. The lockout feature of the shift gate should prevent accidental shifting into Reverse while the vehicle is moving forward. However, if the driver intentionally moves the selector to R while moving forward, the transaxle *will* shift into Reverse, and damage may result.

FORD TRANSAXLE NOMENCLATURE

For the sake of clarity, this text uses consistent terms for transaxle parts and functions. However, the various manufacturers may use different terms. Therefore, before we examine the buildup of the ATX, along with the operation of its geartrain and hydraulic system, you should be aware of the unique names that Ford uses for some transaxle parts and operations. This will make it easier in the future if you should

look in a Ford manual and see a reference to the "reactor". You will realize that this is simply the stator, and is similar to the same part in any other transaxle. Figure 11-4 lists Ford nomenclature and the textbook terms for some transaxle parts and operations.

ATX BUILDUP

The ATX transaxle case is a 1-piece aluminum casting. The torque converter bolts to the engine flexplate, which has the starter ring gear around its outer diameter. The torque converter is a welded assembly and cannot be disassembled. It is installed in the transaxle by sliding it onto the stationary stator support. The splines in the stator 1-way clutch engage the splines on the stator support.

The splitter gear ATX torque converter contains a planetary gearset that controls whether power is transmitted through the torque converter hydraulically or mechanically, figure 11-5. The outside of the converter sun gear is splined to the turbine, the inside of the sun gear is splined to the input shaft. Therefore, the sun gear turns with the turbine and drives the input shaft.

The input shaft passes through the stator support and into the transaxle case where it is splined to the inner race of the 1-way clutch assembly on the direct clutch drum. The shaft is supported by a bushing on the far side of the drum, and drives the direct clutch drum at turbine speed through the 1-way clutch assembly. The input shaft provides only hydraulic input to the geartrain.

The outside of the converter ring gear is splined to a damper assembly inside the converter cover. The inside of the ring gear meshes with the converter pinions, which mesh with the sun gear and turn on pinion shafts attached to the converter carrier. The carrier is splined to the intermediate shaft. Input torque is delivered to the carrier and intermediate shaft mechanically through the ring gear and pinions, and hydraulically through the sun gear and pinions. The percentage of torque delivered through each path varies with the relative speeds of the turbine and converter cover.

The intermediate shaft is located inside the input shaft, and passes through the stator into the transaxle case where it is splined to the intermediate-high clutch drum. Therefore, the intermediate shaft drives the intermediate-high clutch drum by way of the converter planetary gearset. The intermediate shaft provides both hydraulic and mechanical input to the geartrain.

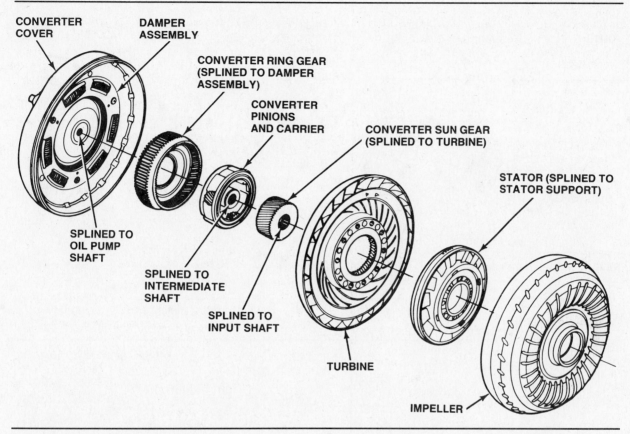

CONVERTER COVER

DAMPER ASSEMBLY

CONVERTER RING GEAR (SPLINED TO DAMPER ASSEMBLY)

CONVERTER PINIONS AND CARRIER

CONVERTER SUN GEAR (SPLINED TO TURBINE)

STATOR (SPLINED TO STATOR SUPPORT)

SPLINED TO OIL PUMP SHAFT

SPLINED TO INTERMEDIATE SHAFT

SPLINED TO INPUT SHAFT

TURBINE

IMPELLER

Figure 11-5. The ATX splitter gear torque converter contains a planetary gearset.

The CLC and FLC torque converters used in some ATX transaxle applications contain no planetary gearset. In these units, the turbine is splined to both the input shaft and the intermediate shaft. Input torque delivery to the geartrain is 100 percent hydraulic, except under those conditions when the CLC converter locks the converter cover to the turbine.

The ATX uses a gear-type oil pump located at the end of the transaxle case opposite the torque converter. The pump is driven by a shaft splined to the converter cover. The oil pump shaft is inside the intermediate shaft, which in turn is inside the input shaft. None of the shafts touches any other.

The pump draws fluid from the sump through a screen, and routes the fluid through passages in the case to the pressure regulator valve, manual valve, manual downshift modulator valve, and throttle valve of the hydraulic system. Because the converter cover is bolted to the engine flexplate, the oil pump shaft turns and the oil pump pumps fluid whenever the engine is running.

The end of the input shaft opposite the torque converter is attached to the inner race of the 1-way clutch on the direct clutch drum, figure 11-6. The outer race of the 1-way clutch is located inside the primary (forward) sun gear of the Ravigneaux geartrain. The input shaft is also attached to the direct clutch drum. The outer edges of the direct clutch steel discs are splined to the inside of the clutch drum. The inner edges of the direct clutch friction discs are splined to the outside of the 1-way clutch outer race and primary (forward) sun gear assembly.

The intermediate shaft passes through the inside of the input shaft and attaches to the intermediate-high clutch drum, figure 11-7. The outer edges of the intermediate-high clutch steel discs are splined to the inside of the clutch drum. The inner edges of the intermediate-high clutch friction discs are splined to the outside of the Ravigneaux geartrain ring gear. The outside of the ring gear is also splined to the inner edges of the reverse clutch friction discs. The outer edges of the reverse clutch steel plates are splined to the transaxle case.

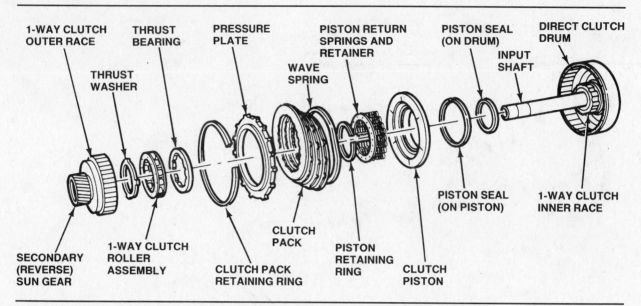

Figure 11-6. The ATX input shaft assembly consists of the 1-way clutch, the direct clutch, and the secondary (reverse) sun gear.

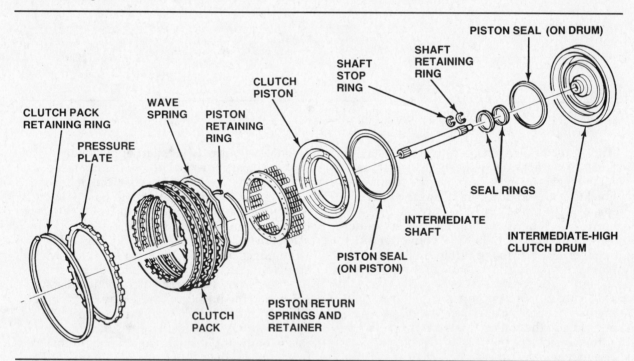

Figure 11-7. The ATX intermediate shaft drives the intermediate-high clutch assembly.

The ring gear surrounds the Ravigneaux planetary gearset, and meshes with the long pinions, figure 11-8. The long pinions mesh with the short pinions, and with the secondary (reverse) sun gear. The secondary sun gear is part of an input shell that is surrounded by the low-intermediate band. The short pinions mesh with the primary (forward) sun gear. Both the

long and short pinions turn in the carrier, which is splined to the final drive input gear. As a result, the planetary gearset carrier is always the output member of the geartrain.

The final drive assembly, figure 11-9, consists of an input gear, an idler gear, and an output gear. The input gear is driven by the Ravigneaux gearset carrier and drives the idler

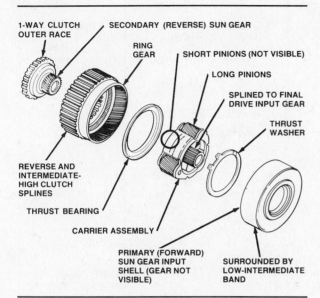

Figure 11-8. The ATX Ravigneaux geartrain.

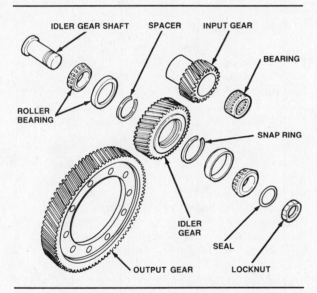

Figure 11-9. The ATX final drive gear assembly.

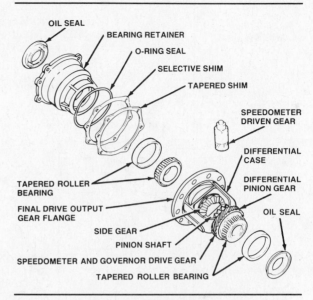

Figure 11-10. The ATX differential assembly.

gear, which turns on a shaft supported by the transaxle case and the transfer gear housing. The idler gear then drives the output gear which, together with the parking gear, is riveted to the differential case. Because of the idler gear between them, the final drive input and output gears turn in the same direction.

The differential case, figure 11-10, turns with the final drive output gear, and is supported by tapered roller bearings on each side. The driver's side bearing is supported by a retainer bolted to the transaxle case. A single gear, attached to the differential case, drives both the

governor and the speedometer cable. The differential case contains four bevel gears: two pinion gears and two side gears. The pinion gears are installed on a pinion shaft that is retained by a pin in the differential case. The pinion gears transmit power from the case to the side gears. Each side gear is splined to one of the halfshafts that transmit power to the front wheels.

ATX POWER FLOW

Now that we have taken a quick look at the general way in which an ATX transaxle is assembled, we can trace the flow of power through the apply devices and the compound planetary gearset in each of the gear ranges.

Park and Neutral

When the gear selector is in Park or Neutral, the converter cover, the oil pump drive shaft, and the converter ring gear all rotate clockwise at engine speed. The oil pump drive shaft turns the oil pump. The turbine, the converter sun gear, and the input shaft also rotate, but at less than engine speed.

The input shaft turns the 1-way roller clutch, causing it to lock up and turn the primary (forward) sun gear. The converter ring gear is turning slightly faster than the converter sun gear at this time, so the converter pinions walk slowly around the sun gear, turning the carrier and the intermediate shaft. Because no member of the planetary gearset is held in Park or Neutral, there is no output to the final drive gears and differential.

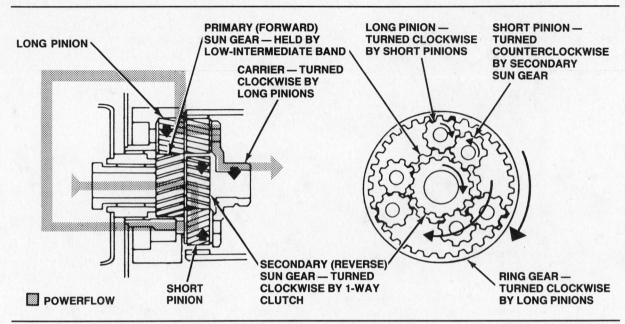

LONG PINION

PRIMARY (FORWARD) SUN GEAR — HELD BY LOW-INTERMEDIATE BAND

CARRIER — TURNED CLOCKWISE BY LONG PINIONS

LONG PINION — TURNED CLOCKWISE BY SHORT PINIONS

SHORT PINION — TURNED COUNTERCLOCKWISE BY SECONDARY SUN GEAR

SECONDARY (REVERSE) SUN GEAR — TURNED CLOCKWISE BY 1-WAY CLUTCH

SHORT PINION

RING GEAR — TURNED CLOCKWISE BY LONG PINIONS

■ POWERFLOW

Figure 11-11. ATX powerflow in Drive low. Input to the geartrain is 100 percent hydraulic.

In Neutral, the differential assembly is free to turn, and the vehicle can be pushed or pulled. In Park, the differential is locked to the case by a pawl that engages the parking gear.

Drive Low

When the gear selector is moved to Drive, the car starts with the transaxle automatically in low gear, figure 11-11. The hydraulic system applies the low-intermediate band to lock the secondary (reverse) sun gear and input shell to the transaxle case.

As the input shaft turns, the 1-way roller clutch locks up and drives the primary (forward) sun gear clockwise. The primary sun gear then turns the short pinions counterclockwise, and the short pinions drive the long pinions clockwise. The long pinions walk clockwise around the stationary secondary (reverse) sun gear, driving the carrier in gear reduction. The ring gear is also turned clockwise by the long pinions, but it simply freewheels at this time.

The carrier turns the final drive input gear clockwise, which turns final drive idler gear counterclockwise, which turns the final drive output gear and the differential case clockwise. The differential pinion gears then transmit the motion of the differential case to the differential side gears, which drive the axle halfshafts and the front wheels. The power flow through the final drive gears and differential assembly is the same for all forward gears.

If the vehicle coasts in Drive low, the primary (forward) sun gear turns faster than the input shaft, causing the 1-way roller clutch to overrun. Because of this, there is no engine braking in Drive low. This is the main difference between Drive low and manual low, as we will see.

Because only the input shaft is linked to the gearset in Drive Low, power input is 100 percent hydraulic. The input member of the gearset is the primary (forward) sun gear. The held member is the secondary (reverse) sun gear, and the output member is the carrier. The input device is the 1-way roller clutch, and the holding device is the low-intermediate band.

Drive Second

As the car continues to accelerate in Drive, the hydraulic system causes an upshift to second gear, figure 11-12, by applying the intermediate-high clutch, which transmits power from the intermediate shaft to the ring gear of the Ravigneaux geartrain. The low-intermediate band remains applied to lock the secondary (reverse) sun gear to the transaxle case.

The intermediate shaft turns the ring gear clockwise, and the ring gear then drives the long pinions. The long pinions walk clockwise around the stationary secondary (reverse) sun gear, driving the carrier in gear reduction. The clockwise-rotating carrier provides input to the final drive gears and differential assembly as in Drive low.

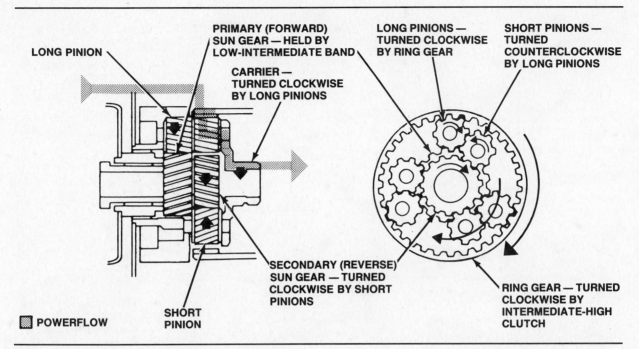

Figure 11-12. ATX powerflow in Drive second. Input to the geartrain is 62 percent mechanical and 38 percent hydraulic.

In second gear, the long pinions also drive the short pinions counterclockwise, and the short pinions drive the primary (forward) sun gear/1-way clutch outer race assembly clockwise. However, because the primary (forward) sun gear/1-way clutch outer race assembly is turned faster than the input shaft and the 1-way clutch inner race, the 1-way roller clutch overruns. Thus, there is no power input from the input shaft; the intermediate shaft is the only link to the gearset.

The planetary gearset in the splitter gear torque converter automatically divides the torque applied to the intermediate shaft between mechanical and hydraulic means. The torque split takes place as follows. The converter cover drives the converter cover (impeller) and ring gear mechanically. Mechanical input torque is transmitted from the ring gear, through the pinions and carrier, to the intermediate shaft. The impeller drives the turbine and converter sun gear hydraulically. Hydraulic input torque is transmitted from the turbine, through the sun gear, pinions, and carrier, to the intermediate shaft.

Assuming the converter is in the torque multiplication phase (as it generally is in second gear), the impeller and ring gear are turning faster than the turbine and sun gear. The majority of the power transmitted to the carrier and intermediate shaft still comes from the sun gear, but the faster turning ring gear causes the

converter pinions to walk around the sun gear and transmit a portion of the driving torque to the carrier. Power input to the Ravigneaux geartrain under these conditions is approximately 62 percent hydraulic and 38 percent mechanical. As we will see in the next section, the closer the speeds of the sun gear and ring gear become, the greater the percentage of torque that is transmitted mechanically.

In second gear, the input member of the Ravigneaux gearset is the ring gear. The held member is the secondary (reverse) sun gear, and the output member is the carrier. The input device is the intermediate-high clutch, and the holding device is the low-intermediate band.

Drive High

As the car continues to accelerate in Drive, the hydraulic system produces an upshift to third, figure 11-13, by releasing the low-intermediate band and applying the direct clutch. The direct clutch transmits engine power from the input shaft to the primary (forward) sun gear. The intermediate-high clutch remains applied and transmits power from the intermediate shaft to the ring gear.

Assuming that the torque converter is in its coupling phase, the input shaft and the intermediate shaft are turning at about the same speed. As a result, both the primary (forward) sun gear and the ring gear rotate clockwise at

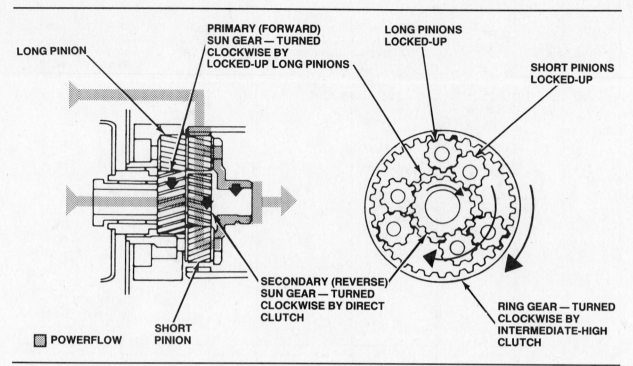

LONG PINION

PRIMARY (FORWARD) SUN GEAR — TURNED CLOCKWISE BY LOCKED-UP LONG PINIONS

LONG PINIONS LOCKED-UP

SHORT PINIONS LOCKED-UP

SECONDARY (REVERSE) SUN GEAR — TURNED CLOCKWISE BY DIRECT CLUTCH

RING GEAR — TURNED CLOCKWISE BY INTERMEDIATE-HIGH CLUTCH

▨ **POWERFLOW**   **SHORT PINION**

Figure 11-13. ATX powerflow in Drive high. Input to the geartrain is 93 percent mechanical and 7 percent hydraulic.

the same speed. This locks up the planetary gearset, which then rotates clockwise as a unit in direct drive. The clockwise-rotating carrier provides input to the final drive gears and differential assembly as in Drive low and Drive second.

When the splitter gear torque converter is in the coupling phase, the converter cover and turbine, and thus the planetary gearset ring gear and sun gear, rotate at nearly the same speed. This causes the gearset to almost completely lock. The converter pinions no longer walk around the sun gear as rapidly as they did in the converter torque multiplication phase (second gear), and therefore, they transmit a greater amount of torque from the ring gear to the carrier and intermediate shaft. Once the converter enters the coupling phase in high gear, torque transmission is approximately 93 percent mechanical and only 7 percent hydraulic.

In high gear, the two input members to the Ravigneaux gearset are the primary (forward) sun gear and the ring gear. The output member is the carrier. The two input devices are the direct clutch and the intermediate-high clutch.

Manual Second

In manual second, the hydraulic system prevents the transaxle from upshifting into high

gear. When the gear selector is placed in the 2 position and the car is accelerated from a standstill, the transaxle starts in Drive low and upshifts to second. If the vehicle slows down, the transaxle downshifts to Drive low. The power flow and apply devices used in manual second are exactly the same as in Drive second, figure 11-12.

Manual Low

In manual low, the power flow through the transaxle is the same as in Drive low, figure 11-11, but the hydraulic system prevents an upshift. The main difference between Drive low and manual low is the use of one apply device.

In Drive low, the 1-way clutch locks and drives the primary (forward) sun gear during acceleration; however, the clutch overruns when the vehicle decelerates. In manual low, the direct clutch is applied to transmit power from the input shaft to the primary (forward) sun gear. If the vehicle decelerates, causing the sun gear to turn faster than the input shaft, the direct clutch transmits power from the sun gear to the input shaft to provide engine braking. The 1-way roller clutch still acts to drive the primary (forward) sun gear during acceleration, but its action is secondary to that of the direct clutch.

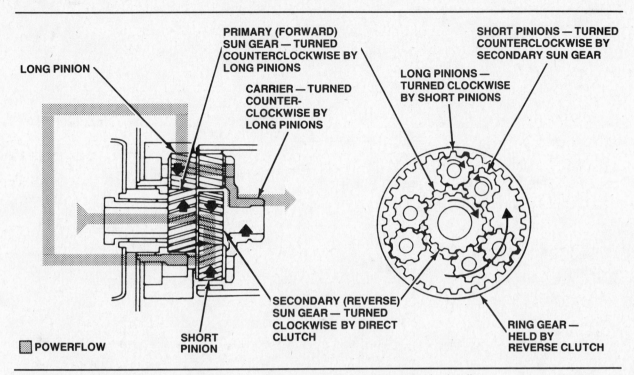

Figure 11-14. ATX powerflow in Reverse. Input to the geartrain is 100 percent hydraulic.

GEAR RANGES	CLUTCHES				BAND
	1-Way	Intermediate-High	Direct	Reverse	Low-Intermediate
Drive Low	●				●
Drive Second		●			●
Drive High		●	●		
Manual Low	*		●		●
Manual Second		●			●
Reverse	*		●	●	

* The 1-way clutch will provide input in these gear ranges if the direct clutch fails, but it will not provide engine braking.

Figure 11-15. Ford ATX transaxle clutch and band application chart.

Reverse

In Reverse, figure 11-14, the hydraulic system applies the direct clutch and the reverse clutch. The direct clutch transmits power from the input shaft to the primary (forward) sun gear. The reverse clutch locks the ring gear of the Ravigneaux geartrain to the transaxle case.

The primary (forward) sun gear rotates clockwise and drives the short pinions counterclockwise. The short pinions drive the long pinions clockwise. Because the ring gear is locked to the case, the long pinions walk around the inside of it, driving the carrier counterclockwise. The carrier then drives the final drive gears and differential assembly as in the forward gears, except that the rotation of each component is reversed. The secondary (reverse) sun gear is also turned counterclockwise by the long pinions, but it simply freewheels at this time.

In Reverse, torque input to the gear train is 100 percent hydraulic. The input member of the Ravigneaux gearset is the primary (forward) sun gear. The held member is the ring gear, and the output member is the carrier. The input device is the direct clutch. The holding device is the reverse clutch.

ATX Clutch and Band Applications

Before we move on to a closer look at the ATX hydraulic system, we should review the apply devices used in each gear range in the ATX

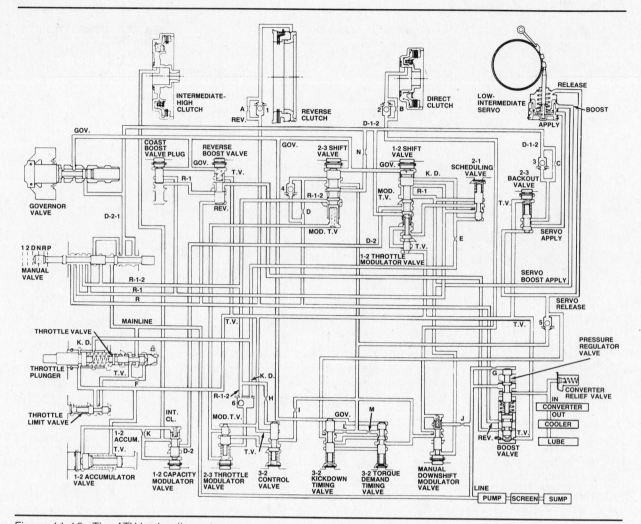

Figure 11-16. The ATX hydraulic system.

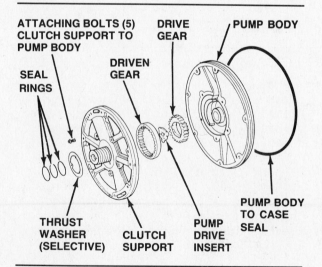

Figure 11-17. The ATX oil pump assembly.

transaxle. Figure 11-15 is a clutch and band application chart. Also, remember these facts about the apply devices in the ATX transaxle:

• The 1-way clutch, intermediate-high clutch, and direct clutch are input devices.

• The low-intermediate band and reverse clutch are holding devices.

• The low-intermediate band is applied in first and second gears.

• The intermediate-high clutch is applied in second and third gears.

• The direct clutch is applied in manual first, third, and Reverse gears.

• The reverse clutch is applied only in Reverse.

THE ATX HYDRAULIC SYSTEM

The hydraulic system of the Ford ATX transaxle controls shifting under varying vehicle loads and speeds, as well as in response to manual gear selection by the driver. Figure 11-16 is a

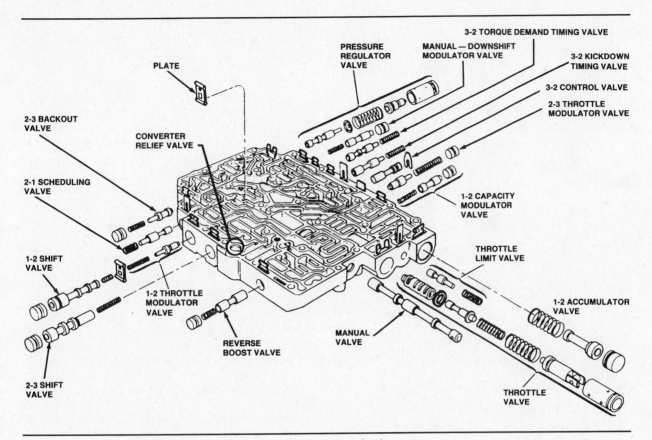

Figure 11-18. The ATX valve body and hydraulic system control valves.

diagram of the complete hydraulic systems of the ATX. We will begin our study of these hydraulic systems with the pump.

Hydraulic Pump

The ATX uses a gear-type oil pump located at the end of the transaxle opposite the torque converter. The pump gears are installed in the pump body, which is bolted to a clutch support, figure 11-17. The clutch support has passages cast in it through which direct and intermediate-high clutch apply oil flows. A drive insert in the center of the pump drive gear is turned by the oil pump drive shaft which is splined to the center of the torque converter cover.

Whenever the engine is running, the pump drive shaft turns the drive insert and drive gear. The rotation of the pump gears creates suction that draws fluid through a screen from the sump at the bottom of the transaxle case. Oil from the pump flows to the pressure regulator valve, the manual valve, the throttle valve, and the manual downshift modulator valve.

Hydraulic Valves

The ATX hydraulic system contains the following valves:

1. Pressure regulator valve
2. Converter relief valve
3. Manual valve
4. Throttle valve
5. Throttle limit valve
6. Governor valve
7. 1-2 shift valve
8. 1-2 throttle modulator valve
9. 1-2 accumulator valve
10. 1-2 capacity modulator valve
11. 2-1 scheduling valve
12. 2-3 shift valve
13. 2-3 throttle modulator valve
14. 2-3 backout valve
15. 3-2 control valve
16. 3-2 kickdown timing valve
17. 3-2 torque demand timing valve
18. Manual downshift modulator valve
19. Reverse boost valve.

Valve body and check valves

The ATX valve body, figure 11-18, contains all the valves of the hydraulic system except the

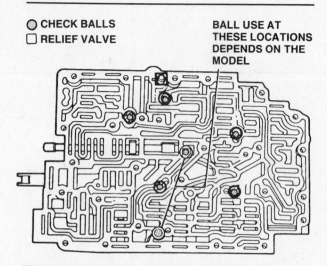

Figure 11-19. ATX check ball locations.

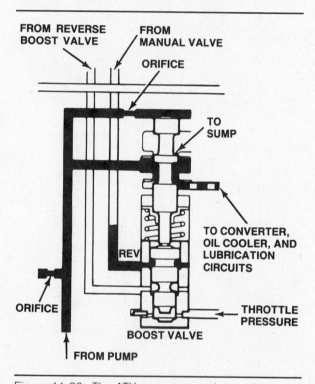

Figure 11-20. The ATX pressure regulator valve.

governor. The valve body casting also contains passages for fluid to flow through. Additional passages are cast into the top of the transaxle case where the valve body is mounted. A separator plate and gaskets are fitted between the valve body from the case. The ATX valve body contains five to seven check balls, depending on the model, figure 11-19.

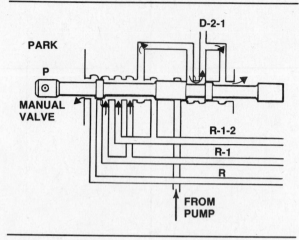

Figure 11-21. The ATX manual valve in Park.

Pressure regulator valve assembly

The pressure regulator valve, figure 11-20, regulates mainline pressure by balancing pump pressure against spring force. When the pump first starts to pump oil, the spring holds the regulator valve closed, and pump pressure flows through two passages to the valve. The passage at the end of the valve is the control passage. When pressure builds up here, it moves the valve against spring force. As the valve moves, it opens a vent to the sump and bleeds off excess pressure. Oil pressure and spring force then balance one another to provide a constant mainline pressure.

The passage at the center of the pressure regulator valve routes oil to the converter circuit. When the converter is full, oil flows to the transaxle cooler. Cooler return fluid is routed through the lubrication circuits and then returned to the sump.

The boost valve is located against the spring end of the pressure regulator valve. Throttle pressure applied to the end of the boost valve is used to increase mainline pressure at larger throttle openings. Two other passages to the boost valve, a reverse passage from the manual valve, and a passage from the reverse boost valve, provide pressure to increase mainline pressure when the transaxle is in Reverse. Pressure at any one of the three passages to the boost valve moves the boost valve so that it increases the tension of the regulator valve spring.

Converter relief valve

The converter relief valve prevents excessive pressure build-up in the torque converter, oil cooler, and transaxle lubrication circuits. The relief valve is a check ball that is held seated by a spring. If torque converter pressure gets too high, the relief valve unseats to exhaust excess

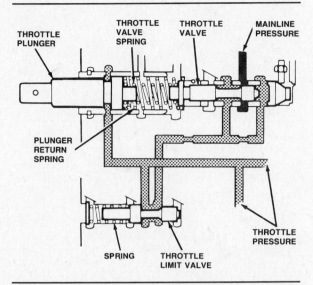

Figure 11-22. The ATX throttle valve at closed throttle.

oil to the sump. When converter pressure drops, the valve seats again.

Manual valve

The manual valve is a directional valve that is moved by a linkage from the gear selector lever. The manual valve receives fluid from the pump and directs it to apply devices and other valves in the hydraulic system in order to provide both automatic and manual upshifts and downshifts.

Four passages may be charged with mainline pressure from the manual valve. In Neutral, the lands on the manual valve block oil from any of the passages. In Park, one land on the manual valve blocks the oil from entering the valve altogether, figure 11-21. Ford refers to mainline pressure from the manual valve by the name of the passage it flows through.

The D-2-1 passage is charged in all forward driving ranges; it routes pressure to the:

- Governor valve
- 2-3 backout valve
- 1-2 shift valve
- Apply side of the low-intermediate servo
- 2-3 shift valve.

The R-1-2 passage is charged in Reverse, manual low, and manual second; it routes pressure to the:

- 2-3 shift valve
- 2-3 throttle modulator valve
- 3-2 control valve.

The R-1 passage is charged in Reverse and manual low; it routes pressure to the:

- Coast boost valve plug
- 1-2 shift valve

- Boost area of the low-intermediate servo
- 2-1 scheduling valve
- Manual downshift modulator valve.

The R passage is charged in Reverse only; it routes pressure to the:

- Reverse boost valve
- 2-3 shift valve
- Boost valve
- Manual downshift modulator valve
- Apply area of the reverse clutch
- Apply area of the direct clutch.

Throttle valve

The ATX transaxle develops throttle pressure using a throttle valve operated by a mechanical linkage, figure 11-22. Throttle pressure indicates engine torque and helps time the transaxle upshifts and downshifts. Without throttle pressure, the transaxle would upshift and downshift at exactly the same speed for different throttle openings and load conditions.

The throttle valve is controlled by opposition between hydraulic pressure and spring force. The hydraulic pressure is mainline pressure that is routed to the end of the valve opposite the spring. The amount of spring force acting on the opposite end of the valve is controlled by the throttle plunger. The plunger is linked to the accelerator so that its position is determined by throttle opening. If the link between the accelerator and the throttle plunger is broken, the plunger moves to the wide-open-throttle position to create maximum throttle pressure. Although the transaxle will not upshift and downshift at the appropriate times, this protects the clutches and bands from slipping and burning.

At closed throttle, figure 11-22, the throttle plunger exerts no force on the throttle valve spring, and the throttle valve allows only enough oil through to create about 10 psi or 70 kPa of throttle pressure. As the accelerator is depressed, the linkage moves the throttle plunger to compress the throttle valve spring. This moves the throttle valve to allow additional pressure into the throttle pressure passages. Thus, throttle pressure increases with throttle opening.

At wide-open throttle, figure 11-23, the throttle plunger is pushed into the throttle valve as far as it can be. The plunger completely compresses the spring, contacts the throttle valve spool, and bottoms the valve to allow mainline pressure into the throttle pressure passage. At the same time, the plunger opens the kickdown passage to route throttle pressure from the throttle limit valve to the

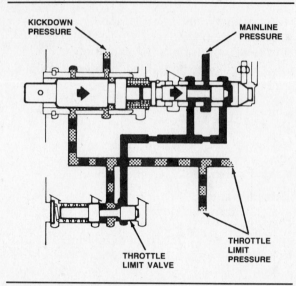

Figure 11-23. The ATX throttle valve at wide-open throttle.

2-3 throttle modulator valve, the 2-1 scheduling valve, and the 1-2 shift valve to force a downshift.

Some throttle pressure is sent through the throttle limit valve to a groove in the throttle plunger to help regulate throttle pressure. The rest is sent through the throttle limit valve (see next section) to the hydraulic system. Throttle pressure leaving the throttle limit valve flows to the:

- Reverse boost valve
- 2-3 throttle modulator valve
- 3-2 control valve
- 1-2 throttle modulator valve
- 2-3 backout valve
- 1-2 accumulator valve
- Boost valve.

Throttle limit valve
Throttle pressure flows through the throttle limit valve, figure 11-23, on its way to the hydraulic system. The throttle limit valve is controlled by spring force, and acts to limit throttle pressure to a maximum of 85 psi (586 kPa). When the throttle limit valve is regulating, the pressure in the throttle valve circuits is called throttle limit pressure. Throttle limit pressure that is routed back through the throttle valve past the throttle plunger is called kickdown pressure.

Governor valve
We have been discussing the development and control of throttle pressure, which is one of the two auxiliary pressures that work on the shift valves. Before we get to the shift valve operation, we must explain the governor valve (or

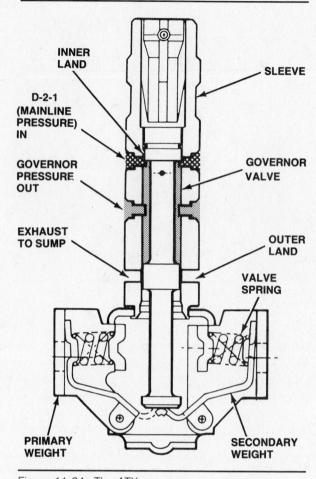

Figure 11-24. The ATX governor.

governor), and how it creates governor pressure. Governor pressure provides a road speed signal to the hydraulic system that causes automatic upshifts to occur as road speed increases, and permits automatic downshifts with decreased road speed.

The governor, figure 11-24, is located in a sleeve in the transaxle case, and is driven by a gear on the differential case. The governor assembly includes a: primary weight and spring, secondary weight and spring, and a spool valve.

The governor receives D, 2, 1 oil from the manual valve at one port. When the vehicle is stopped, the governor does not turn and the valve position allows pressure to exhaust through another port to the sump. When the vehicle moves, the governor rotates at a speed proportional to road speed, and centrifugal force pushes on the weights and springs. The weights move the valve to close the exhaust port and open a port that leads to the hydraulic system. The pressure that exits through this port is governor pressure.

The opening of the governor pressure port varies with the action of the weights so that governor pressure increases and decreases with vehicle speed, and remains constant at a fixed vehicle speed. Above a certain speed, the exhaust port is completely closed and the governor pressure port is completely open. At this point, governor pressure is equal to mainline pressure.

Governor pressure is produced in all forward gear ranges when the vehicle is moving. Governor pressure is applied to five valves:

- Reverse boost valve
- 1-2 shift valve
- 2-3 shift valve
- 3-2 kickdown timing valve
- 3-2 torque demand timing valve.

1-2 shift valve and 1-2 throttle modulator valve
The 1-2 shift valve controls upshifts from low to second, and downshifts from second to low. The assembly includes the 1-2 shift valve, a spring, and the 1-2 throttle modulator valve, figure 11-25. The 1-2 throttle modulator valve receives throttle pressure and acts on the 1-2 shift valve spring to oppose upshifts. However, it also can produce a modulated throttle pressure that is sent to the spring chamber of the shift valve assembly.

In all forward gear ranges, D-2-1 oil is routed to the 1-2 shift valve. When the valve is downshifted, it blocks this pressure. Throttle pressure helps keep the valve downshifted by acting on the 1-2 throttle modulator valve, which moves to increase the force of the shift valve spring. If the throttle pressure is high enough, the 1-2 throttle modulator valve moves so far that it opens the modulated throttle pressure port. This allows modulated throttle pressure to fill the spring chamber of the shift valve assembly and help oppose an upshift.

Governor pressure is applied to the end of the 1-2 shift valve opposite the spring. As the vehicle accelerates, governor pressure increases until it overcomes throttle pressure and spring force to upshift the valve. Once the shift valve has been upshifted, the mainline pressure that was blocked is allowed to flow to the 1-2 capacity modulator valve, which works with the 1-2 accumulator valve to apply the intermediate-high clutch and shift the transaxle into second gear.

In manual low and Reverse, R-1 oil is sent to the 1-2 shift valve, and acts on it to prevent an upshift. From there it flows to the 2-1 scheduling valve, the manual downshift modulator

valve, and through the servo boost apply passage to the low-intermediate servo where it assists servo apply pressure.

As the vehicle slows, throttle and spring pressure acting on the 1-2 shift valve overcome governor pressure to downshift the valve. At wide-open throttle, kickdown pressure is sent to the 1-2 shift valve to help force a downshift.

**1-2 accumulator valve and
1-2 capacity modulator valve**
The 1-2 accumulator valve and 1-2 capacity modulator valve, figure 11-26, work together to cushion application of the intermediate-high clutch. When the 1-2 shift valve upshifts, it routes mainline pressure to the intermediate-high clutch apply circuit. Pressure in this circuit is routed to the ends of the 1-2 accumulator valve and 1-2 capacity modulator valve that are opposite springs.

The spring chamber of the 1-2 accumulator valve receives throttle pressure that helps the spring hold the valve in the extended position. When the 1-2 upshift occurs, a portion of the mainline pressure in the clutch apply circuit is diverted through an orifice to bottom the 1-2 accumulator valve against spring force and throttle pressure. The time required to do this varies with the throttle opening.

Another portion of the mainline pressure in the clutch apply circuit is routed to the end of the 1-2 capacity modulator valve opposite the spring end. This moves the valve against spring force to limit the rate at which pressure enters the clutch apply circuit. As pressure on the 1-2 accumulator side of the orifice increases, it combines with spring force to move the 1-2 capacity modulator valve and allow greater amounts of pressure into the clutch apply circuit.

When the 1-2 accumulator valve is bottomed, and the 1-2 capacity modulator valve is fully open, the pressure in the intermediate-high clutch apply circuit reaches mainline pressure. Together, the two valves ensure that clutch apply pressure builds up gradually so the clutch applies smoothly.

2-1 scheduling valve
The 2-1 scheduling valve, figure 11-25, helps time manual downshifts to low. When the gear selector is moved to the 1 position, the manual valve sends R-1 oil to the 1-2 shift valve where it acts on the valve to cause the downshift. As the shift valve moves against governor pressure, it opens a passage that routes R-1 oil through the 2-1 scheduling valve to the spring

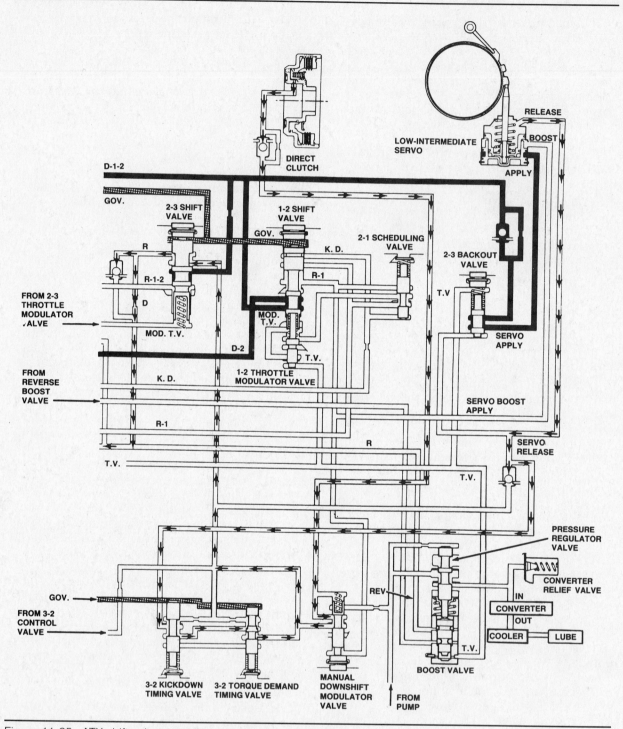

Figure 11-25. ATX shift valve operation during a 3-2 downshift.

chamber of the 1-2 shift valve. This additional pressure helps downshift the shift valve and lock out second gear.

If the throttle is wide open, the end of the 2-1 scheduling valve opposite spring force receives kickdown pressure. This moves the valve against spring force, and opens a passage that leads to the 1-2 shift valve spring chamber. If a manual downshift to low occurs at this time, R-1 oil is immediately routed through this passage to the 1-2 shift valve spring chamber to provide a faster downshift.

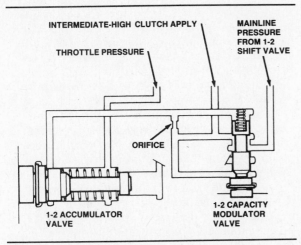

Figure 11-26. Operation of the 1-2 accumulator valve and 1-2 capacity modulator valve.

2-3 shift valve

The 2-3 shift valve, figure 11-25, controls upshifts from second to third, and downshifts from third to second. In all forward ranges, D-1-2 oil is sent to the 2-3 shift valve. When the valve is downshifted, it blocks this pressure. Modulated throttle pressure from the 2-3 throttle modulator valve is routed to the spring chamber of the 2-3 shift valve to help keep the valve downshifted. Note that the 2-3 throttle modulator valve is not in the same bore as the 2-3 shift valve.

Governor pressure is applied to the end of the 2-3 shift valve opposite the spring. As the vehicle accelerates, governor pressure increases until it overcomes modulated throttle pressure and spring force to upshift the valve. Once the shift valve has been upshifted, the mainline pressure that was blocked is allowed to flow to the 3-2 kickdown timing valve and the 3-2 torque demand timing valve. These valves act with others to apply the direct clutch and release the low-intermediate servo, and thus shift the transaxle into third gear.

In manual second, R-1-2 oil is sent from the manual valve to the 2-3 shift valve to keep it downshifted. The manual valve R passage is also connected to the 2-3 shift valve. When the transaxle downshifts from third to second, direct clutch apply oil exhausts through the shift valve and R passage. In addition, when the transaxle is in Reverse, R oil from the manual valve can flow through the downshifted 2-3 shift valve into the low-intermediate servo release passage.

2-3 throttle modulator valve

The 2-3 throttle modulator valve, figure 11-16, receives throttle pressure at the end opposite spring force. The throttle pressure moves the valve until it opens a port into the modulated throttle pressure passage. Some of the pressure in this circuit is applied to the spring end of the 2-3 throttle modulator valve to help it regulate. The rest of the modulated throttle pressure is sent to the spring end of the 2-3 shift valve to help oppose upshifts.

Another passage to the 2-3 throttle modulator valve spring chamber can be charged with either kickdown pressure or R-1-2 oil. When pressure is sent through this passage, the 2-3 throttle modulator valve stops regulating, and either kickdown pressure or mainline pressure flows into the modulated throttle pressure circuit to oppose upshifts at the 2-3 shift valve.

Finally, there is a passage to the 2-3 throttle modulator valve from the 3-2 control valve. Pressure sent through this passage moves the throttle modulator valve against spring force to help it regulate modulated throttle pressure.

2-3 backout valve

If the driver releases the throttle after the 2-3 upshift has started but before it is complete, the shift will be harsh. With the sudden reduction in engine torque, there might be enough pressure to apply the direct clutch, but not to release the low-intermediate servo and band. The 2-3 backout valve, figure 11-25, keeps the clutch and band from fighting each other and trying to engage two gears at once.

In all forward gear ranges, the 2-3 backout valve receives D-1-2 oil from the manual valve and sends it to the apply side of the low-intermediate servo where it overcomes spring force to apply the servo and band. Throttle pressure routed to the spring chamber of the 2-3 backout valve helps keep the valve bottomed so the servo apply pressure passage is fully open. When the transaxle upshifts to third, servo release pressure is routed from the 2-3 shift valve to the release side of the servo, and to the end of the backout valve opposite the spring chamber.

If throttle pressure is low, servo release pressure moves the 2-3 backout valve against throttle pressure and spring force. This restricts the apply pressure passage and slows the flow of oil out of the apply side of the servo. This slows the release of the low-intermediate band, and gives the direct clutch more time to engage. Note that servo apply oil is not completely blocked, nor is it exhausted. The servo releases because the combination of spring force and servo release pressure is enough to overcome servo apply pressure alone.

If throttle pressure is high, indicating greater driveline loads, the direct clutch will take less time to engage and servo release pressure will be unable to move the 2-3 backout valve as far to restrict the flow of apply oil out of the servo. In this case, the servo and band will release more rapidly to match the timing of the direct clutch to ensure a smoother shift.

3-2 control valve

The 3-2 control valve, figure 11-16, is part of the direct clutch apply circuit. Spring force at one end of the valve opposes throttle pressure in the center of the valve. When the throttle pressure is low and spring force controls the valve, mainline pressure from the clutch apply circuit flows through the 3-2 control valve to the 2-3 throttle modulator valve.

When throttle pressure is high, the 3-2 control valve moves against spring force and limits the flow of mainline pressure to the 2-3 throttle modulator valve. At wide open throttle, kickdown pressure is applied to the end of the 3-2 control valve opposite spring force. This moves the 3-2 control valve far enough to completely block the flow of mainline pressure to 2-3 throttle modulator valve.

3-2 kickdown timing valve and 3-2 torque demand timing valve

The 3-2 kickdown timing valve and 3-2 torque demand timing valve are part of the direct clutch apply circuit, figure 11-25. They are both controlled by governor pressure opposing spring force.

When the transaxle is in high gear, the circuits to and from these valves are charged with mainline pressure. Oil that flows through the 3-2 kickdown timing valve enters the passage to the 2-3 backout valve and the release side of the low-intermediate servo. Oil that flows through the 3-2 torque demand timing valve enters a passage to the manual downshift modulator valve, from which it travels to the direct clutch.

The most important job of these valves is to control the rate of low-intermediate servo release oil exhaust in relation to vehicle speed during a 3-2 downshift. When the servo release oil is vented, it seats a check ball and must exhaust through the 3-2 kickdown timing valve and 3-2 torque demand timing valve. At low governor pressure, the valves are controlled by spring force and their ports are unrestricted; servo release oil exhausts quickly through both valves.

If governor pressure is higher, it moves the 3-2 torque demand timing valve against spring force to block the passage to the manual downshift modulator valve. After servo release exhaust oil flows into the 3-2 kickdown timing

valve, some vents through a passage containing a restricting orifice, and some flows to the 3-2 torque demand timing valve where it vents through a second passage containing a restricting orifice. The reduced exhaust flow through the two orifices slows servo and band release.

If governor pressure is very high, it also moves the 3-2 kickdown timing valve against spring force to block the passage with the restricting orifice that leads from that valve. All of the servo release exhaust oil must then flow to the 3-2 torque demand timing valve where it exhausts through the passage with the restricting orifice that leads from that valve. The reduced exhaust flow through the single orifice causes the servo and band to release very slowly.

Manual downshift modulator valve

The manual downshift modulator valve, figure 11-25, is controlled by opposition between spring force and R-1 oil from the manual valve. The valve also receives pressure from the pump through a restricting orifice.

In all gear ranges except Reverse and manual low, spring force bottoms the valve. In this position, the valve blocks the pump pressure port, and opens the direct clutch apply passage to mainline pressure from the 2-3 shift valve and the 3-2 torque demand timing valve. In third gear, mainline pressure flows through the manual downshift modulator valve into the clutch apply passage. Some of the apply oil is routed to the spring chamber of the valve to help keep it bottomed.

In Reverse and manual low, the manual valve sends R-1 oil to the bottom of the manual downshift modulator valve, moving it against spring force to close the inlet from the 2-3 torque demand timing valve. In this position, pressure from the pump flows through the valve, which modulates the pressure and sends it to apply the direct clutch.

Reverse boost valve

The reverse boost valve, figure 11-16, receives throttle pressure and can be controlled by governor pressure and spring force, or by R oil. In all forward gears, governor pressure and spring force bottom the valve so it blocks the throttle pressure port.

In Reverse, there is no governor pressure, and the manual valve sends R oil to the end of the reverse boost valve opposite the spring. This moves the valve against spring pressure and allows throttle pressure to flow through the valve and into a passage that leads to the boost valve in the pressure regulator valve

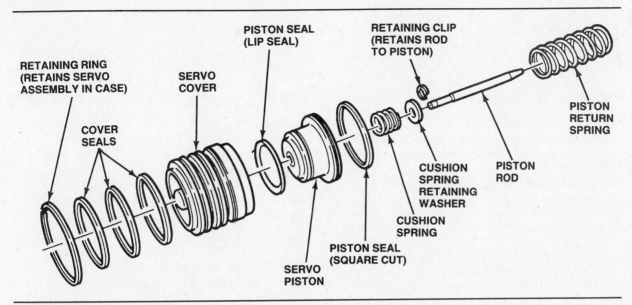

Figure 11-27. The ATX low-intermediate servo.

assembly. This increases mainline pressure in Reverse to hold the apply devices more tightly.

MULTIPLE-DISC CLUTCHES AND SERVOS

The preceding paragraphs examined the individual valves in the ATX transaxle hydraulic system. We will now take a quick look at the multiple-disc clutches and servo in this transaxle.

Direct Clutch

To apply the direct clutch, figure 11-6, hydraulic pressure is routed behind the clutch piston through a passage in the clutch support. The direct clutch is applied in Drive third, manual low, and Reverse. To release the clutch, hydraulic pressure is exhausted through the manual downshift modulator valve, and several small coil springs force the piston back into the clutch drum.

Intermediate-High Clutch

To apply the intermediate-high clutch, figure 11-7, hydraulic pressure is routed behind the clutch piston through a passage in the clutch support. The intermediate-high clutch is applied Drive second, Drive third, and manual second. When pressure is exhausted to release the clutch, several small coil springs force the piston back into the clutch drum.

Reverse Clutch

The reverse clutch is installed in a pocket in the transaxle case, instead of in a rotating drum as are the other clutches. A hydraulic passage for clutch apply oil is drilled into the case directly from the valve body mounting surface. Pressure in this passage moves the piston to apply the clutch. The reverse clutch is applied in Reverse only. When reverse clutch apply pressure exhausts through the manual valve, several small coil springs and a wave spring force the piston back into its bore in the case.

Low-Intermediate Servo

The low-intermediate servo assembly, figure 11-27, is installed into a machined bore from outside the transaxle case. The servo applies the low-intermediate band which holds the secondary (reverse) sun gear input shell in all forward gear ranges except Drive third. The low-intermediate band is not adjustable.

HYDRAULIC SYSTEM SUMMARY

We have examined the individual valves, clutches, and servos in the hydraulic system and discussed their functions. The following paragraphs summarize their combined operations in the various gear ranges. Refer to figure 11-16 to trace the hydraulic system operation.

Park and Neutral

In Park and Neutral, oil flows from the pump to the pressure regulator valve, the throttle valve, the manual valve, and the manual downshift modulator valve. Oil flows to these valves in all gear ranges. The oil that flows to the manual downshift modulator valve is blocked.

The pressure regulator valve regulates mainline pressure and charges the converter-cooler-lubrication circuits. Excess fluid is vented to the sump. When the manual valve is in Neutral, mainline pressure is trapped between two lands of the valve. When the manual valve is in Park, mainline pressure is blocked from entering the valve bore.

The throttle valve modifies mainline pressure to create a throttle pressure that increases with throttle opening. Throttle pressure is applied to the 1-2 accumulator valve, the 1-2 throttle modulator valve, the 2-3 throttle modulator valve, the 2-3 backout valve, the 3-2 control valve, the reverse boost valve, and the boost valve. These throttle pressure applications are made in all gear ranges.

The 1-2 and 2-3 throttle modulator valves produce modulated throttle pressures that are applied to the spring chambers of their respective shift valves. If throttle pressure rises too high, it is limited to 85 psi or 586 kPa by the throttle limit valve.

In Park and Neutral, no clutches or bands are applied, and all valve positions are determined by spring force alone. In Park, a pawl attached to the transaxle case engages the parking gear attached to the final drive output gear to prevent the car from being moved.

Drive Low

When the manual valve is moved to the D position, mainline pressure is routed through the D-2-1 passage to the governor, the 1-2 shift valve, the 2-3 shift valve, and the 2-3 backout valve. At the 1-2 and 2-3 shift valves, D-2-1 oil is blocked by the downshifted valves. The governor modifies D-2-1 oil to create a governor pressure that rises with road speed. Governor pressure is sent to the shift valves, but is not yet great enough to upshift them. The 2-3 backout valve sends mainline pressure to the apply side of the low-intermediate servo, and controls the rate of oil flow in relation to throttle pressure.

The low-intermediate band is the only hydraulic device applied in Drive low.

Drive Second

As vehicle speed and governor pressure increase, governor pressure upshifts the 1-2 shift valve. The upshifted valve allows mainline pressure to flow to the 1-2 capacity modulator valve and the 1-2 accumulator valve. These valves act together to cushion application of the intermediate-high clutch by producing an appropriate clutch apply pressure.

In Drive second, the low-intermediate band remains applied, and works together with the intermediate-high clutch to provide second gear.

Drive High

As vehicle speed and governor pressure continue to increase, governor pressure upshifts the 2-3 shift valve. The upshifted valve allows mainline pressure to flow to the servo release circuit, the 3-2 control valve, the 3-2 kickdown timing valve, the 3-2 torque demand timing valve, and the manual downshift modulator valve.

The 3-2 control valve regulates the mainline pressure in relation to throttle pressure, and applies it to the 2-3 throttle modulator valve. The 3-2 kickdown timing valve sends mainline pressure to the low-intermediate servo release circuit. This pressure is applied to the 3-2 backout valve, and to the release side of the servo. The 3-2 torque demand timing valve sends mainline pressure to the manual downshift modulator valve. The manual downshift modulator valve sends oil to apply the direct clutch.

In Drive high, the low-intermediate band is released, the intermediate-high clutch remains applied, and the direct clutch is applied to provide third gear.

Drive Range Forced Downshifts

A part-throttle forced downshift can take place at moderate speeds. As the driver presses down on the accelerator, the throttle plunger moves to increase throttle pressure up to a maximum of 85 psi or 586 kPa. Throttle pressure applied to the 1-2 and 2-3 throttle modulator valves creates modulated throttle pressure that is applied to the spring chambers of the shift valves. If the combination of modulated throttle pressure and spring force is great enough, it overcomes governor pressure to downshift one or both of the shift valves. During part-throttle 3-2 downshifts, the 3-2 torque demand timing valve moves to slow exhausting servo release oil, so the low-intermediate band does not apply until the direct clutch is released.

Full-throttle downshifts can be forced at higher speeds if the driver depresses the accelerator completely. Throttle pressure then increases to full throttle limit pressure (85 psi or 586 kPa), and the throttle plunger moves far enough to open the kickdown circuit to throttle limit pressure.

Kickdown pressure is applied the 2-3 throttle modulator valve and the 1-2 shift valve. Kickdown pressure causes the 2-3 throttle modulator valve to stop modulating and send throttle limit pressure to the spring chamber of the 2-3 shift valve. Kickdown pressure moves the 1-2 shift valve to cause a downshift. Depending on governor pressure, a 3-2, 2-1, or 3-1 downshift can be forced.

During a full-throttle downshift, governor pressure moves the 3-2 kickdown timing valve and the 3-2 torque demand timing valve to slow the servo release exhaust. This ensures that the low-intermediate band is not applied until the direct clutch is released.

Manual Second

In manual second, the oil flow and clutch and band applications are the same as in Drive second, except that the manual valve also opens the R-1-2 passage. Mainline pressure in this passage is applied to the 3-2 control valve, the 2-3 throttle modulator valve, and the 2-3 shift valve.

Mainline pressure pushes the 3-2 control valve down and fills the spring chamber of the 2-3 throttle modulator valve so that it stops regulating. Mainline pressure then flows through the 2-3 throttle modulator valve into the modulated throttle pressure passage to fill the spring chamber of the 2-3 shift valve and oppose an upshift. The application of R-1-2 oil to the shift valve itself also opposes an upshift. All of these pressure applications work to lock out an upshift to third gear.

Manual Low

In manual low, the oil flow is the same as in drive low except that the manual valve also opens the R-1 passage. Mainline pressure in this passage is applied to the 1-2 shift valve and the 2-1 scheduling valve. Mainline pressure on the 1-2 shift valve opposes an upshift.

The 2-1 scheduling valve sends R-1 oil to the spring chamber of the 1-2 shift valve, the servo boost apply area, and to the end of the manual downshift modulator valve opposite the spring. Mainline pressure in the 1-2 shift valve spring chamber also fills the modulated throttle pressure passage from the 1-2 throttle modulator valve. All of these applications in the 1-2 shift valve assembly work to lock out an upshift to second gear.

The R-1 oil sent to the servo boost area gives the servo extra apply force. R-1 oil at the manual downshift modulator valve pushes the valve up to open the port from the pump circuit and send a regulated mainline pressure to the apply area of the direct clutch.

In manual low, the low-intermediate band is applied, just as in Drive low. However, the direct clutch is also applied to provide engine braking that is not available in Drive low.

Reverse

In Reverse, the manual valve opens the R-1-2 passage, the R-1 passage, and the R passage. The D-2-1 passage is closed, and the lack of oil in this circuit means there is no governor pressure to upshift the shift valves.

R-1-2 oil is applied to the 2-3 shift valve and spring chamber, as described for manual second, ensuring that the 2-3 shift valve is in its downshifted position. R-1 oil is applied as described for manual low to help apply the direct clutch. R oil flows to the reverse clutch apply area and the reverse boost valve. The reverse boost valve sends throttle pressure to the boost valve in the pressure regulator system to increase mainline pressure in Reverse. R oil also flows through the downshifted 2-3 shift valve into the direct clutch apply/low-intermediate servo release circuit.

In Reverse, the direct clutch and the reverse clutch are applied.

Review Questions

Select the single most correct answer
Compare your answers to the correct answers on page 603

1. Mechanic A says the ATX was first used in the Escort/Lynx.
   Mechanic B says the Escort/Lynx ATX has a splitter gear torque converter that provides 100 percent mechanical power delivery in high gear.
   Who is right?
   a. Mechanic A
   b. Mechanic B
   c. Both mechanic A and B
   d. Neither mechanic A nor B

2. Mechanic A says the ATX CLC converter in the Taurus/Sable is similar to the converter used with the C5 transmission.
   Mechanic B says the ATX FLC converter was first used in the Tempo/Topaz in 1987.
   Who is right?
   a. Mechanic A
   b. Mechanic B
   c. Both mechanic A and B
   d. Neither mechanic A nor B

3. Mechanic A says MERCON® is the preferred fluid for all ATX transaxles.
   Mechanic B says the use of DEXRON®-IID fluid is acceptable in 1987 and earlier ATX transaxles.
   Who is right?
   a. Mechanic A
   b. Mechanic B
   c. Both mechanic A and B
   d. Neither mechanic A nor B

4. The ATX differential case contains the:
   a. Planetary gears
   b. Damper assembly
   c. Final drive gears
   d. Pinion and side gears

5. Mechanic A says the ATX speedometer cable is driven by a gear on the output shaft.
   Mechanic B says the ATX governor is driven by a gear on the differential case.
   Who is right?
   a. Mechanic A
   b. Mechanic B
   c. Both mechanic A and B
   d. Neither mechanic A nor B

6. Mechanic A says the ATX gear selector has six positions.
   Mechanic B says the ATX starts and remains in second gear when the selector is placed in 2.
   Who is right?
   a. Mechanic A
   b. Mechanic B
   c. Both mechanic A and B
   d. Neither mechanic A nor B

7. Mechanic A says the ATX oil pump is driven by the innermost shaft from the converter cover.
   Mechanic B says the inner race of the ATX 1-way clutch is driven by the intermediate shaft from the torque converter.
   Who is right?
   a. Mechanic A
   b. Mechanic B
   c. Both mechanic A and B
   d. Neither mechanic A nor B

8. Mechanic A says the ATX converter sun gear is splined to the turbine, and transmits hydraulic input torque to the input shaft.
   Mechanic B says the ATX converter ring gear is splined to the damper in the converter cover, and transmits mechanical input torque to the converter pinions and carrier.
   Who is right?
   a. Mechanic A
   b. Mechanic B
   c. Both mechanic A and B
   d. Neither mechanic A nor B

9. The valve body of the ATX transaxle is located:
   a. In the oil pan
   b. On top of the case
   c. At the back of the case
   d. None of the above

10. Mechanic A says the ATX final drive input gear is driven by the Ravigneaux geartrain carrier.
    Mechanic B says the ATX final drive idler and output gears rotate in the same direction.
    Who is right?
    a. Mechanic A
    b. Mechanic B
    c. Both mechanic A and B
    d. Neither mechanic A nor B

11. Mechanic A says the ATX direct clutch is applied in all forward gears.
    Mechanic B says the ATX reverse clutch is applied in reverse and manual low gears.
    Who is right?
    a. Mechanic A
    b. Mechanic B
    c. Both mechanic A and B
    d. Neither mechanic A nor B

12. Mechanic A says the powerflow through the ATX geartrain in Drive low is: secondary (reverse) sun gear, to short pinions, to long pinions, to carrier.
    Mechanic B says the primary (forward) sun gear is held and the ring gear freewheels in Drive low.
    Who is right?
    a. Mechanic A
    b. Mechanic B
    c. Both mechanic A and B
    d. Neither mechanic A nor B

13. Mechanic A says the ATX low-intermediate band is applied in Reverse.
    Mechanic B says the ATX direct clutch is applied to provide engine braking in manual low.
    Who is right?
    a. Mechanic A
    b. Mechanic B
    c. Both mechanic A and B
    d. Neither mechanic A nor B

14. Mechanic A says the powerflow through the ATX geartrain in Reverse is: secondary (reverse) sun gear, to short pinions, to long pinions, to carrier.
    Mechanic B says the primary (forward) sun gear freewheels and the ring gear is held in Reverse.
    Who is right?
    a. Mechanic A
    b. Mechanic B
    c. Both mechanic A and B
    d. Neither mechanic A nor B

15. Mechanic A says apply pressure is routed to the low-intermediate servo in all forward gear ranges.
    Mechanic B says low-intermediate apply pressure is not exhausted when the servo is released.
    Who is right?
    a. Mechanic A
    b. Mechanic B
    c. Both mechanic A and B
    d. Neither mechanic A nor B

12

The Ford AOD Transmission

HISTORY AND MODEL VARIATIONS

The Ford Automatic Overdrive (AOD) transmission, figure 12-1, was introduced in 1980 to replace the C6 in cars, and the FMX in both cars and trucks. The AOD was first used in 1980 Cougars and Thunderbirds, and in 1981, Ford began using it in some F-series trucks. In 1982, use of the AOD was expanded to include the Lincoln Continental, Mark VI, and Town Car. In 1984, selected Econoline vans and wagons, and Mustang/Capri models got the AOD. Since 1984, the AOD has become standard or optional equipment in most Ford Motor Company RWD cars and trucks.

The AOD is a 4-speed automatic transmission that uses the Ravigneaux planetary gearset. In addition to the usual input shaft splined to the turbine, the AOD torque converter contains a damper assembly that turns a direct drive shaft to provide input torque to the geartrain in third and fourth gears. The planetary gearset is similar to that in the earlier 3-speed FMX, with basically the same power flow in first, second, and third gears. The hydraulic system uses an oil pump and stator support similar to those in the C6. The servo pockets and the valve body upper passages cast into the case are similar to those in the C3.

Fluid Recommendations

Ford initially recommended Type CJ fluid for AOD transmissions. When that fluid became obsolete, they changed the recommendation to DEXRON®-IID fluid. Ford's latest recommendations, in effect as of the 1988 model year, are that MERCON® is the preferred fluid for 1980-87 AOD transmissions, although DEXRON®-IID is acceptable. The use of MERCON® is a warranty *requirement* for all 1988 and later AOD transmissions.

Transmission Identification

Identifying a Ford automatic transmission is a 2-step process. First, you determine the model of transmission. All Ford vehicles have an identification plate on the rear of the left front door. In addition to the vehicle identification number and other data, this plate has a code letter that identifies the transmission model. Like all other Ford 4-speed automatics, the AOD is identified by the letter T.

In addition to the transmission model, you must know the exact transmission assembly part number and/or other identifying characters to obtain repair parts. This information is

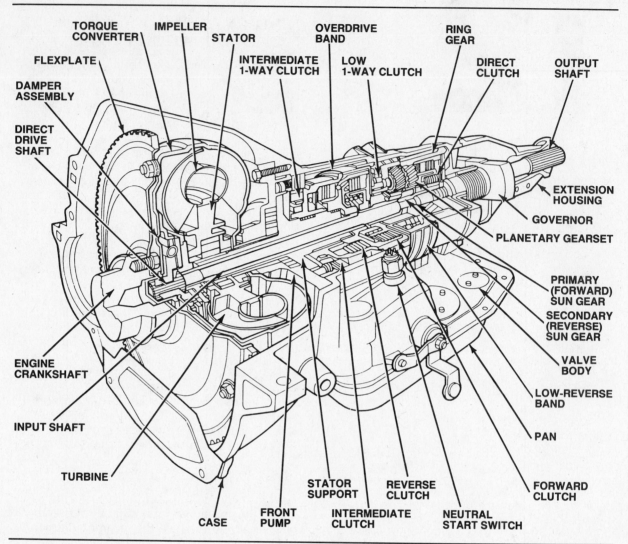

Figure 12-1. The AOD transmission.

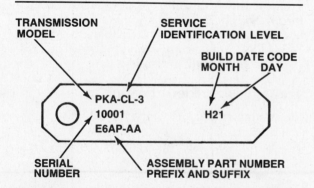

Figure 12-2. The AOD transmission identification tag is located under one of the extension housing-to-case bolts.

stamped on a tag attached to the transmission. On the AOD, the tag is retained under the upper right-hand extension housing-to-case bolt. Figure 12-2 shows a typical AOD identification tag and explains the meaning of the data stamped on the tag. As required by Federal regulations, Ford transmissions also carry the serial number of the vehicle in which they were originally installed.

Major Changes

Through the 1988 model year, there have been no major changes to the AOD transmission. Early 1980 transmissions had a case casting problem that was corrected before the end of the model year.

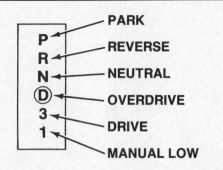

Figure 12-3. The AOD gear selector positions.

GEAR RANGES

The AOD gear selector, which may be either floor or column mounted, has six positions, figure 12-3:
P — Park
R — Reverse
N — Neutral
Ⓓ — Overdrive
3 — Drive
1 — Manual low.
The AOD gear ratios are:
- First (low) — 2.40:1
- Second (intermediate) — 1.47:1
- Third (direct drive) — 1.00:1
- Fourth (overdrive) — 0.667:1
- Reverse — 2.00:1.

Park and Neutral

The engine can be started only when the transmission is in Park or Neutral. No clutches or bands are applied in Neutral; the low-reverse band is applied in Park but does not cause any drive condition. In both Park and Neutral, there is no flow of power through the transmission.

In Neutral, the output shaft is free to turn, and the vehicle can be pushed or pulled. In Park, a pawl is mechanically engaged with a large gear on the outside of the planetary gearset ring gear. This locks the output shaft to the transmission case, and prevents the vehicle from being moved.

Overdrive

Overdrive is the normal driving position for the AOD transmission. When the vehicle is stopped and the gear selector is moved to Ⓓ, the transmission is automatically in low gear. The vehicle starts in low gear, and automatically upshifts to second, third, and fourth gears. The speeds at which the upshifts occur vary and are controlled by vehicle speed and throttle position.

In fourth gear, the driver can force a 4-3 or 4-2 part-throttle (torque-demand) downshift simply by depressing the accelerator far enough. A 4-1 part-throttle downshift is not possible with the AOD. The driver can also force a 4-3, 4-2, or 4-1 kickdown (through-detent) downshift by depressing the accelerator completely. The number of gears that the transmission will downshift depends on the vehicle speed. However, the transmission will *always* downshift at least to third; it cannot stay in fourth gear at wide-open throttle.

As vehicle speed decreases, the transmission automatically downshifts. The AOD downshifts through all of the gears — third, second, and first — as the vehicle coasts down to a stop.

Drive

When the vehicle is stopped and the gear selector is placed in 3 (Drive), the transmission starts in low as it did in Overdrive. However, in Drive, the transmission upshifts only through second and third. It will not shift into fourth. The speeds at which the upshifts occur vary and are controlled by vehicle speed and throttle position.

In third gear, the driver can force a 3-2 part-throttle downshift simply by depressing the accelerator far enough. A 3-1 part-throttle downshift is not possible with the AOD. The driver can also force a 3-2 or 3-1 kickdown downshift by depressing the accelerator completely. The number of gears that the transmission will downshift depends on the vehicle speed.

As vehicle speed decreases, the transmission automatically downshifts. In Drive, the AOD downshifts into second and then into first as the vehicle coasts down to a stop.

Manual Low

When the gear selector is placed in 1 (manual low), the transmission starts and stays in low gear; it will not upshift to second. Because the low-reverse band is applied, manual low can be used to provide engine braking.

If the gear selector lever is moved to 1 from Ⓓ or 3 at speeds above approximately 25 mph (40 kph), the transmission will downshift to second gear for engine braking. The transmission then downshifts to first as road speed drops below about 25 mph. Once in low, the transmission will not upshift until the gear selector is moved to another position.

FORD TERM	TEXTBOOK TERM
Reactor	Stator
Clutch cylinder	Clutch drum
Converter pump	Impeller
Turbine shaft	Input shaft
Line pressure	Mainline pressure
Main oil pressure regulator valve	Pressure regulator valve
T.V. pressure	Throttle pressure
T.V. limit valve	Throttle limit valve
2-3 T.V. modulator valve	2-3 throttle modulator valve
3-4 T.V. modulator valve	3-4 throttle modulator valve
Planetary 1-way clutch	Low 1-way clutch
Low-intermediate drum	Reverse sun gear input shell
Forward sun gear	Primary sun gear
Reverse sun gear	Secondary sun gear
Overdrive lockout range	Drive range

Figure 12-4. Ford transmission nomenclature table.

Reverse

When the gear selector is moved to R, the transmission shifts into Reverse. The transmission should be shifted into Reverse only when the vehicle is standing still. The lockout feature of the shift gate should prevent accidental shifting into Reverse while the vehicle is moving forward. However, if the driver intentionally moves the shift lever to R while moving forward, the transmission *will* shift into Reverse and damage may result.

FORD TRANSMISSION NOMENCLATURE

For the sake of clarity, this text uses consistent terms for transmission parts and functions. However, the various manufacturers may use different terms. Therefore, before we examine the buildup of the AOD and the operation of its geartrain and hydraulic system, you should be aware of the unique names that Ford uses for some transmission parts and operations. This will make it easier in the future if you should look in a Ford manual and see a reference to the "clutch cylinder". You will realize that this is simply a clutch drum, and is similar to the same part in any other transmission. Figure 12-4 lists Ford nomenclature and the textbook terms for some transmission parts and operations.

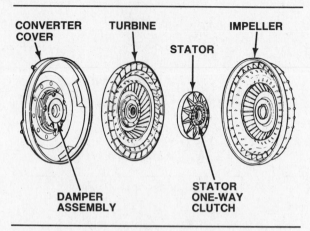

Figure 12-5. The AOD torque converter contains a damper assembly that turns the direct drive shaft.

AOD BUILDUP

The AOD transmission has a 1-piece, cast-aluminum bellhousing and case. The extension housing bolts to the rear of the case.

The AOD torque converter is a welded assembly and cannot be disassembled. The converter bolts to the engine flexplate, which has the starter ring gear around its outer edge. In addition to an impeller, turbine, and stator, the AOD torque converter contains a damper assembly in the converter cover, figure 12-5.

The hub at the rear of the torque converter housing engages the drive lugs on the transmission oil pump, and drives the pump whenever the engine is running. The AOD uses a gear-type oil pump mounted at the front of the transmission case. The pump draws fluid from the sump through a filter and delivers it to the pressure regulator valve, the manual valve, and other valves in the hydraulic system.

The AOD torque converter differs from most converters in having two power inputs to the gear train. The input shaft is splined to the turbine hub, and provides hydraulic input to the geartrain. Inside the input shaft is a direct drive shaft that is splined to the converter cover damper assembly. The direct drive shaft provides mechanical input to the geartrain.

In first, second, and Reverse gears, only the input shaft delivers torque to the geartrain; input torque delivery is 100 percent hydraulic. In third gear, both the input shaft and the direct drive shaft deliver torque to the geartrain. The hydraulic torque input through the input shaft is about 40 percent, and the mechanical input

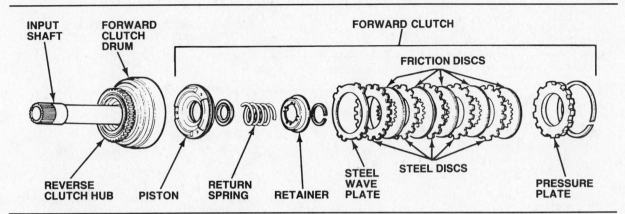

Figure 12-6. The AOD input shaft and forward clutch assembly.

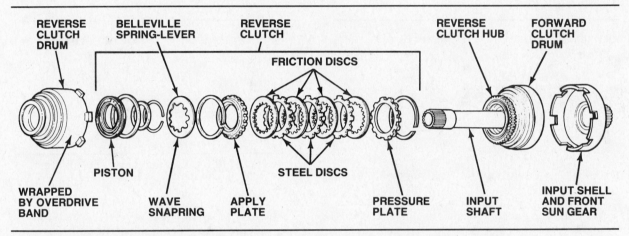

Figure 12-7. The AOD reverse clutch assembly, input shell, and reverse (secondary) sun gear.

through the direct drive shaft is about 60 percent. In fourth gear, only the direct drive shaft delivers torque to the geartrain, and input torque delivery is 100 percent mechanical. How these shafts are used will become clear as we study the buildup and power flow of the AOD transmission.

The torque converter is installed in the transmission by sliding it onto the stator support. The splines in the stator 1-way clutch engage the splines on the stator support. The splines in the turbine hub engage the input shaft. The other end of the input shaft is splined to the forward clutch drum/reverse clutch hub assembly, figure 12-6. The input shaft is supported by one bushing in the stator support, and another on the direct drive shaft.

The forward clutch drum and the reverse clutch hub are one assembly, figure 12-6. Both are turned by the input shaft at turbine speed. The outside of each forward clutch steel disc is splined to the forward clutch drum. The inside of each friction disc is splined to the forward

clutch hub. The hub is splined to the primary (forward) sun gear. When the forward clutch is applied, it transmits engine power from the input shaft to the primary (forward) sun gear.

The inside of each reverse clutch friction disc, figure 12-7, is splined to the reverse clutch hub and is turned by it. The outside of each steel disc is splined to the reverse clutch drum. The reverse clutch drum has lugs on it that interlock with notches in the input shell attached to the secondary (reverse) sun gear. When the reverse clutch is applied, it transmits engine power from the input shaft to the secondary (reverse) sun gear.

The intermediate 1-way clutch inner race is located on the front of the reverse clutch drum, figure 12-8. The intermediate 1-way clutch outer race forms the intermediate clutch hub. The inside of each intermediate clutch friction disc is splined to this race/hub. The outside of each steel disc is splined to the transmission case. The intermediate clutch apply piston is located in a bore machined into the backside of the oil pump housing.

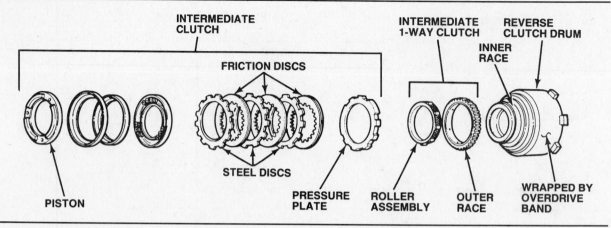

Figure 12-8. The AOD intermediate clutch and intermediate 1-way clutch assembly.

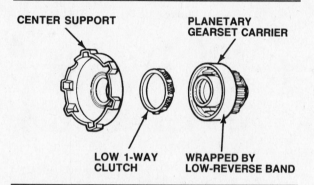

Figure 12-9. The AOD center support, low 1-way clutch, and planetary gearset carrier.

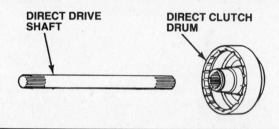

Figure 12-10. The AOD direct drive shaft turns the direct clutch drum.

When the intermediate clutch is applied, it prevents the reverse clutch and reverse sun gear from rotating counterclockwise through the intermediate 1-way clutch, because the reverse clutch drum is splined to the input shell. The intermediate 1-way clutch prevents counterclockwise rotation of the reverse clutch drum, but if the reverse clutch drum turns clockwise, the 1-way clutch overruns.

The reverse clutch drum is also surrounded by the overdrive band. When the band is applied, it prevents the reverse clutch drum, the input shell, and the secondary (reverse) sun gear from turning in *either* direction.

The AOD transmission has a center support that is splined to the case, figure 12-9. All of the components discussed thus far (forward, reverse, and intermediate clutches, intermediate 1-way clutch, and overdrive band) are located in front of the center support. All of the components in the remainder of the buildup are located behind the center support.

The inner race of the low 1-way clutch is located on the back side of the center support. The outer race of the low 1-way clutch is inside the front of the planetary gearset carrier. The low 1-way clutch holds the carrier from turning counterclockwise. If the carrier turns clockwise, the low 1-way clutch overruns.

The planetary gearset carrier is surrounded by the low-reverse band. When the band is applied, it prevents the carrier from turning in *either* direction.

As mentioned earlier, the forward end of the direct drive shaft is splined to the damper assembly in the converter cover. The back end of the shaft is splined to the direct clutch drum, figure 12-10, and turns the drum at engine speed. The direct clutch drum is supported by a bushing on the output shaft.

The outside of each direct clutch steel disc, figure 12-11, is splined to the inside of the direct clutch drum and turned by it. The inside of each friction disc is splined to the outside of the direct clutch hub. The inside of the clutch hub is splined to the rear of the planetary gearset carrier. When the direct clutch is applied, it transmits engine power from the direct drive shaft to the gearset carrier.

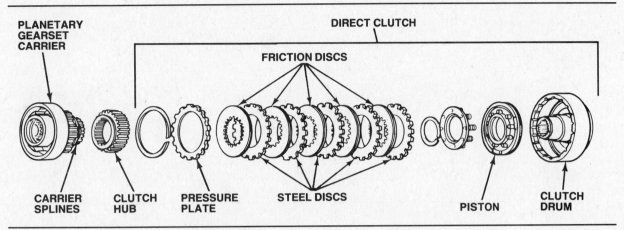

Figure 12-11. The AOD direct clutch assembly.

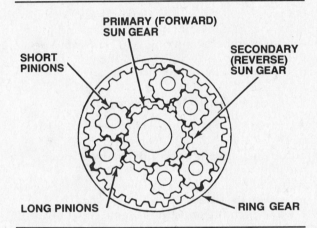

Figure 12-12. The AOD Ravigneaux planetary gearset.

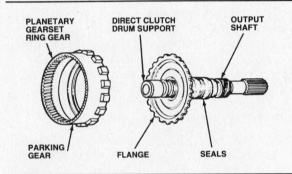

Figure 12-13. The AOD ring gear and output shaft assembly.

The AOD Ravigneaux planetary gearset, figure 12-12, has two sets of pinions that turn on pinion shafts that are part of the carrier. The primary (forward) sun gear meshes with the short pinions. The short pinions mesh with the long pinions. The long pinions mesh with both the secondary (reverse) sun gear and the ring gear. As already mentioned, the primary sun gear is splined to the forward clutch hub. The secondary sun gear is splined through an input shell to the reverse clutch drum.

The ring gear, figure 12-13, fits over the long pinions and is splined to a flange on the output shaft. The outside of the ring gear has external teeth that are engaged by the parking pawl to lock the output shaft to the case in Park. The output from the AOD transmission gearset is always through the ring gear.

The forward end of the output shaft supports the direct clutch drum. The output shaft itself is supported by a bushing in the case, and by the slip yoke and a bushing in the extension housing. The governor assembly is mounted on the output shaft and rotates with it.

AOD POWER FLOW

Now that we have taken a look at the general way in which the AOD transmission is assembled, we can trace the power flow through the apply devices and the planetary gearset in each of the gear ranges.

Park and Neutral

When the gear selector is in Neutral, no clutches or bands are applied. Therefore, no member of the planetary gearset is held or driven, and there can be no output motion. The low-reverse band is applied in Park, but it does not cause any drive condition. In Park and Neutral, the torque converter cover and impeller, the oil pump, and the direct drive shaft all rotate clockwise at engine speed. The torque converter turbine and the input shaft also rotate clockwise, and because there is no load on the turbine, the converter will be in an "unloaded" coupling phase.

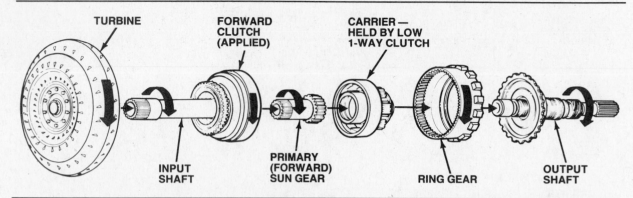

Figure 12-14. AOD power flow in Overdrive and Drive low.

In Neutral, the output shaft is free to rotate in either direction, and the car can be pushed or pulled. In Park, the mechanical parking pawl engages the parking gear on the outside of the ring gear. This locks the output shaft to the transmission case and prevents the car from being moved in either direction.

Overdrive and Drive Low

When the vehicle is stopped and the AOD gear selector is moved to Overdrive (Ⓓ) or Drive (3), the transmission is automatically in low gear, figure 12-14. The forward clutch is applied and turns the primary (forward) sun gear clockwise. The primary sun gear turns the short pinions counterclockwise, and the short pinions then drive the long pinions clockwise.

The long pinions must drive the carrier counterclockwise, or force the ring gear to turn clockwise. Because the low 1-way clutch prevents the carrier from turning counterclockwise, the long pinions drive the ring gear clockwise. The ring gear then drives the output shaft in gear reduction. The long pinions also drive the secondary (reverse) sun gear counterclockwise, but it simply freewheels at this time.

As long as the car is accelerating in Overdrive or Drive low, torque is transferred from the engine through the planetary gearset to the output shaft. If the driver lets up on the throttle, the engine speed drops while the drive wheels keep turning at the same speed. The output shaft and ring gear then drive the long and short pinions, which in turn drive the forward primary (forward) sun gear clockwise at faster than input shaft speed. This removes the counterclockwise-rotating torque load from the low 1-way clutch, so it unlocks and allows the carrier to freewheel. No engine compression braking is available, and this is the major difference between Overdrive or Drive low, and manual low, as you will see later.

In Overdrive and Drive low, the input member of the gearset is the primary (forward) sun gear. The held member is the carrier, and the output member is the ring gear. The input device is the forward clutch, and the holding device is the low 1-way clutch.

Overdrive and Drive Second

As the car continues to accelerate in Overdrive or Drive, the hydraulic system produces an automatic upshift to second gear by applying the intermediate clutch, figure 12-15. The intermediate clutch locks the outer race of the intermediate 1-way clutch to the case. The 1-way clutch prevents the reverse clutch drum, input shell, and secondary (reverse) sun gear from turning counterclockwise. The forward clutch remains applied to transmit engine power from the input shaft to the primary (forward) sun gear.

The primary (forward) sun gear rotates clockwise and turns the short pinions counterclockwise. The short pinions turn the long pinions clockwise. The long pinions walk clockwise around the held secondary (reverse) sun gear, driving the ring gear and output shaft clockwise in gear reduction. The long pinions also turn the carrier clockwise, but the low 1-way clutch overruns so the carrier simply freewheels at this time.

As long as the vehicle is accelerating in Overdrive or Drive second, engine torque is transferred through the planetary gearset to the output shaft. If the driver lets up on the throttle, the intermediate 1-way clutch overruns, allowing the reverse clutch drum, input shell, and secondary (reverse) sun gear to freewheel. As a result, there is no engine compression braking in second gear.

In Overdrive and Drive second, the input member of the gearset is the primary (forward) sun gear. The held member is the secondary

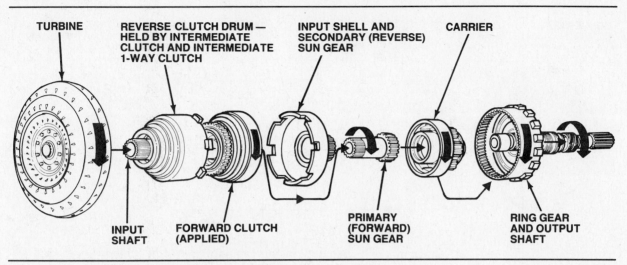

Figure 12-15. AOD power flow in Overdrive and Drive second.

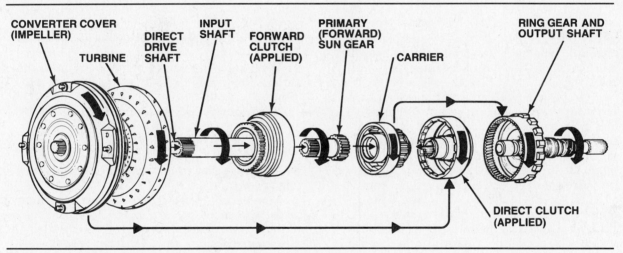

Figure 12-16. AOD power flow in Overdrive and Drive third.

(reverse) sun gear, and the output member is the ring gear. The input device is the forward clutch, and the holding devices are the intermediate clutch and intermediate 1-way clutch.

Overdrive and Drive Third

As the car continues to accelerate in Overdrive or Drive, the transmission automatically up-shifts from second to third gear by applying the direct clutch, figure 12-16. The direct clutch transmits engine power mechanically from the direct drive shaft to the carrier. The forward clutch remains applied and transmits engine power hydraulically from the input shaft to the primary (forward) sun gear. The intermediate clutch remains applied and holds the outer race of the intermediate 1-way clutch.

The carrier is driven clockwise at engine speed by the direct drive shaft. Assuming the

torque converter is in the coupling phase, the primary (forward) sun gear is driven clockwise at nearly engine speed by the input shaft. With two members of the gearset turning at essentially the same speed, the gearset locks up and turns clockwise as a unit. Clockwise rotation of the ring gear then drives the output shaft in direct drive. The secondary (reverse) sun gear, input shell, and reverse clutch drum also rotate clockwise at this time, but the intermediate 1-way clutch overruns so that these parts simply freewheel.

Input torque to the gearset in third gear is split between hydraulic and mechanical sources. The percentage of power supplied by each source varies with the relative speeds of the torque converter impeller and turbine. When the converter is coupled as described above, and impeller and turbine speeds are

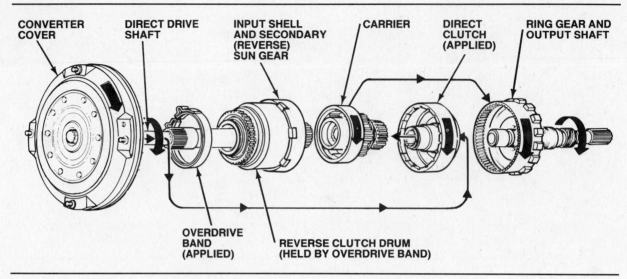

CONVERTER COVER   DIRECT DRIVE SHAFT   INPUT SHELL AND SECONDARY (REVERSE) SUN GEAR   CARRIER   DIRECT CLUTCH (APPLIED)   RING GEAR AND OUTPUT SHAFT

OVERDRIVE BAND (APPLIED)   REVERSE CLUTCH DRUM (HELD BY OVERDRIVE BAND)

Figure 12-17. AOD power flow in Overdrive fourth.

nearly the same, input is over 90 percent mechanical. When the converter is in the torque multiplication phase, and the impeller is turning somewhat faster than the turbine, the carrier and pinions walk around the slower turning primary (forward) sun gear, resulting in a loss of mechanical input. Under these conditions, the hydraulically driven input shaft provides about 40 percent of the input torque, while the mechanically driven direct drive shaft provides about 60 percent of the input torque.

In Overdrive and Drive third, the input members of the gearset are the primary (forward) sun gear and the carrier. The output member is the ring gear. The input devices are the forward and direct clutches.

Overdrive Fourth

As the car continues to accelerate in Overdrive, the transmission automatically upshifts to fourth gear, figure 12-17. To do this, the forward clutch is released, and the overdrive band applied. The overdrive band holds the reverse clutch drum, input shell, and secondary (reverse) sun gear stationary. The direct clutch remains applied and transmits engine power from the direct drive shaft to the carrier. The intermediate clutch, though still applied, is ineffective because it is overridden by the overdrive band.

The carrier rotates clockwise, taking the pinions with it. The long pinions walk clockwise around the stationary secondary (reverse) sun gear, and drive the ring gear and output shaft clockwise in overdrive. The long pinions also

turn the short pinions counterclockwise, which turn the primary (forward) sun gear clockwise, causing the low 1-way clutch to overrun so that all these components simply freewheel at this time.

The input to the gearset in fourth gear is 100 percent mechanical through the direct drive shaft, which is splined to the damper assembly in the converter cover. This eliminates converter slip for maximum fuel economy at cruising speeds.

In Overdrive fourth gear, the input member of the gearset is the carrier. The held member is the secondary (reverse) sun gear, and the output member is the ring gear. The input device is the direct clutch, and the holding device is the overdrive band.

Manual Low

The power flow in manual low is the same as in Overdrive and Drive low, figure 12-14, but the hydraulic system prevents an upshift to second gear. The main difference between Overdrive or Drive low, and manual low, is the use of one apply device.

In Overdrive and Drive low, the low 1-way clutch locks and holds the carrier against counterclockwise rotation. However, the clutch overruns and allows the carrier to freewheel when the vehicle decelerates. In manual low, the low-reverse band is applied to hold the carrier and prevent it from turning in either direction. This provides engine braking during deceleration.

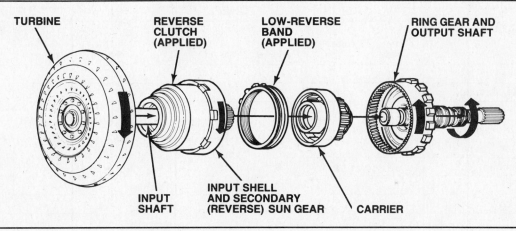

Figure 12-18. AOD power flow in Reverse.

As in Overdrive and Drive low, the input member of the gearset in manual low is the primary (forward) sun gear, the held member is the carrier, and the output member is the ring gear. The input device is the forward clutch, but the holding device is the low-reverse band, replacing the low 1-way clutch. The low 1-way clutch still holds during acceleration, but its action is secondary to that of the low-reverse band.

"Manual second"
The AOD gear selector does not provide a manual second position, but when the selector is placed in manual low at higher speeds, the transmission does shift into "manual second" gear until the vehicle speed drops below 25 mph (40 kph). The power flow in "manual second" gear is the same as in Overdrive or Drive second, figure 12-15, except for the use of one apply device.

In Overdrive and Drive second, the intermediate 1-way clutch locks to hold the reverse clutch drum, input shell, and secondary (reverse) sun gear against counterclockwise rotation. However, the clutch overruns and allows these parts to freewheel when the vehicle decelerates. When the transmission is in "manual second" gear, the overdrive band is applied to prevent the reverse clutch drum, input shell, and secondary (reverse) sun gear carrier from turning in either direction. This provides engine braking during deceleration.

As in Overdrive and Drive second, the input member of the gearset is the primary (forward) sun gear. The held member is the secondary (reverse) sun gear, and the output member is the ring gear. The input device is the forward clutch, but the holding device is the overdrive band, replacing the intermediate clutch and

intermediate 1-way clutch. The intermediate 1-way clutch will still hold if the vehicle is accelerated in "manual second" gear, but its action is secondary to that of the overdrive band.

Reverse

When the gear selector is moved to Reverse, the hydraulic system applies the reverse clutch and the low-reverse band, figure 12-18. The reverse clutch transmits engine power from the input shaft to the secondary (reverse) sun gear. The low-reverse band holds the carrier stationary.

The secondary (reverse) sun gear rotates clockwise and drives the long pinions counterclockwise. The long pinions must drive the carrier clockwise or the ring gear counterclockwise. Since the carrier is held by the low-reverse band, the long pinions drive the ring gear and output shaft counterclockwise for reverse gear. The long pinions also drive the short pinions clockwise, and the short pinions drive the primary (forward) sun gear counterclockwise. However, because the forward clutch is not applied, these components simply freewheel in Reverse.

In Reverse, the input member of the gearset is the secondary (reverse) sun gear. The held member is the carrier, and the output member is the ring gear. The input device is the reverse clutch, and the holding device is the low-reverse band.

Transmission Clutch and Band Applications

Before we move on to look at the AOD hydraulic system, we should review the apply devices

GEAR RANGES	CLUTCHES						BANDS	
	Low 1-Way	Inter-mediate 1-Way	Forward	Direct	Inter-mediate	Reverse	Low-Reverse	Over-drive
Overdrive Low	●		●					
Overdrive Second		●	●		●			
Overdrive Third			●	●	†			
Overdrive Fourth				●	†			●
Drive Low	●		●					
Drive Second		●	●		●			
Drive Third			●	●	†			
Manual Low	*		●				●	
Reverse						●	●	

* The low 1-way clutch will hold in manual low if the low-reverse band fails, but it will not provide engine braking.
† The intermediate clutch is applied in these gear ranges, but it is ineffective because the intermediate 1-way clutch overruns.

Figure 12-19. Ford AOD clutch and band application chart.

used in each gear range. Figure 12-19 is a clutch and band application chart. Also, remember these facts about the apply devices in the AOD transmission:
● The forward, direct, and reverse clutches are input devices.
● The intermediate clutch, both 1-way clutches, and the low-reverse and overdrive bands are holding devices.
● The forward clutch is applied in low, second, and third gears.
● The direct clutch is applied in third and fourth gears.
● The intermediate clutch is applied in second, third, and fourth gears, but is effective only in second because the intermediate 1-way clutch overruns in third and fourth.
● The reverse clutch is applied only in Reverse.
● The low-reverse band is applied in manual low and Reverse. It is also applied in Park, but does not cause any drive conditions.
● The overdrive band is applied in fourth gear. It is also applied in second gear to provide engine compression braking when a downshift to manual low is made at high speeds.
● The low 1-way clutch locks during acceleration in Drive low.
● The intermediate 1-way clutch locks during acceleration in second gear.

THE AOD HYDRAULIC SYSTEM

The hydraulic system of the AOD transmission controls shifting under varying vehicle loads and speeds, as well as in response to manual gear selection by the driver. Figure 12-20 is a diagram of the complete AOD hydraulic system. We will begin our study of this system with the pump.

Hydraulic Pump

The AOD uses a gear-type oil pump, figure 12-21, located at the front of the transmission. The pump gears are installed in the pump body, which bolts to the transmission case. The stator support seals the backside of the pump body, and bolts to it.

Whenever the engine is running, the converter hub turns the pump drive gear. The rotation of the pump gears creates suction that draws fluid through a screen from the sump at the bottom of the transmission case. Oil from the pump flows to the pressure regulator valve, the 3-4 accumulator, the manual valve, the 2-3 shift valve, and the throttle valve.

Hydraulic Valves

The AOD hydraulic system contains the following valves:
1. Pressure regulator valve
2. Converter relief valve
3. Manual valve
4. Throttle valve
5. Throttle limit valve
6. Throttle relief valve
7. Governor valve
8. 1-2 shift valve
9. 1-2 capacity modulator valve
10. 1-2 accumulator valve
11. 2-1 scheduling valve
12. Low servo modulator valve

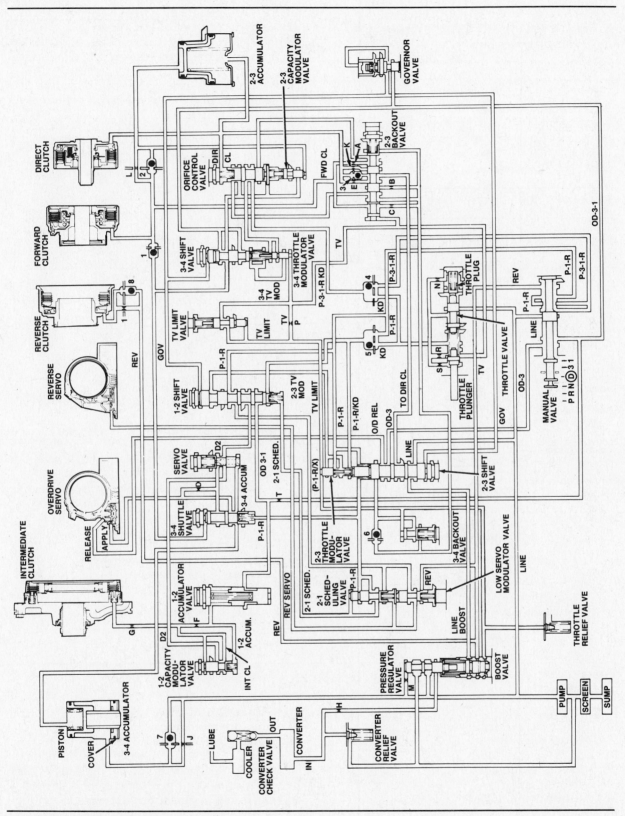

Figure 12-20. The AOD hydraulic system.

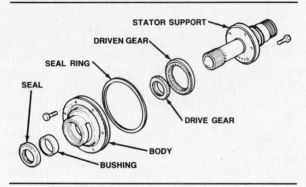

Figure 12-21. The AOD gear-type oil pump.

Valve body and ball-check valves

The AOD valve body, figure 12-22, is located at the bottom of the transmission above the sump. Bores in the valve body contain all the hydraulic system valves except the governor valve. Passages cast in the valve body allow fluid to flow to and between the various valves.

There are eight ball-check valves in the AOD valve body, figure 12-23. They are interchangeable except for one that is orange. The orange check ball must always be returned to its original location.

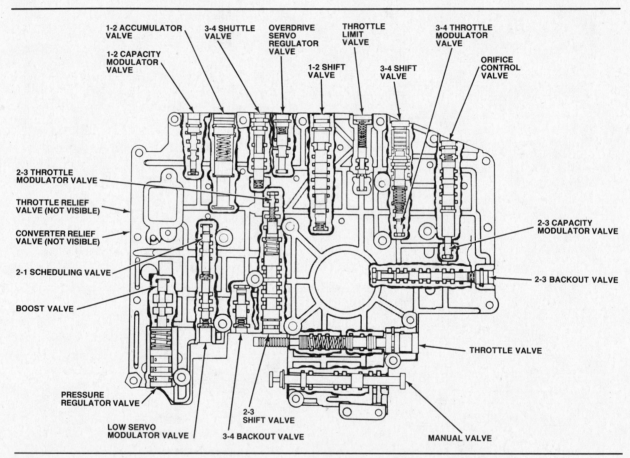

Figure 12-22. The AOD valve body contains all of the hydraulic system valves except the governor.

13. 2-3 shift valve
14. 2-3 throttle modulator valve
15. 2-3 backout valve
16. Orifice control valve
17. 2-3 capacity modulator valve
18. 3-4 shift valve
19. 3-4 throttle modulator valve
20. Overdrive servo regulator valve
21. 3-4 shuttle valve
22. 3-4 backout valve.

Pressure regulator valve

The pressure regulator valve, figure 12-24, regulates mainline pressure by balancing pump pressure against spring force. When the engine is stopped, the spring holds the pressure regulator valve closed. When the engine is started and the pump begins to supply oil, and pump pressure flows to the pressure regulator valve through two passages.

The passage at the end of the valve is the control passage. When pressure builds up here,

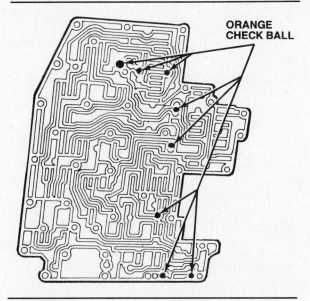

Figure 12-23. AOD valve body check valve locations.

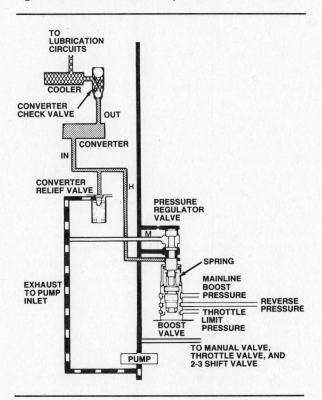

Figure 12-24. AOD pressure regulator valve operation and torque converter oil flow.

it moves the valve against spring force. As the valve moves, it opens a vent to the sump that bleeds off excess pressure. Oil pressure and spring force then balance one another to provide a constant mainline pressure.

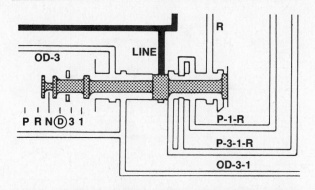

Figure 12-25. The AOD manual valve in Neutral.

The passage at the center of the pressure regulator valve routes oil to the converter circuit. When the converter is full, the converter check valve opens and oil flows to the transmission cooler. Cooler return fluid is routed through the lubrication circuits and then returned to the sump. When the engine is shut off, the converter check valve seats to keep the converter primed full of fluid.

The boost valve is located at the spring end of the pressure regulator valve. Three passages lead to the boost valve, and pressure in any one of them moves the boost valve to increase the tension of the regulator valve spring, and thus raise mainline pressure.

One passage routes throttle limit pressure to the boost valve to increase mainline pressure at larger throttle openings. Mainline pressure is also increased in all gear ranges except third and fourth because pressure is routed to the boost valve through the downshifted 2-3 shift valve. In Reverse, the manual valve sends pressure to the boost valve, increasing mainline pressure to compensate for the higher torque loads. Mainline pressure is increased most in Reverse when all three passages send oil to the boost valve.

Converter relief valve

The converter relief valve, figure 12-24, prevents excessive pressure build up in the torque converter, oil cooler, and lubrication circuits. The converter relief valve is a check ball that is held seated by a spring. If converter pressure rises above 125 psi (862 kPa), the relief valve unseats and exhausts excess oil to the pump inlet. When converter pressure drops, the valve seats again.

Manual valve

The manual valve, figure 12-25, is a directional valve that is moved by a mechanical linkage from the gear selector lever. The manual valve

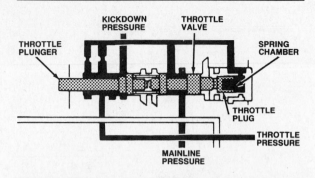

Figure 12-26. AOD throttle valve operation at wide-open throttle.

receives fluid from the pump and directs it to apply devices and other valves in the hydraulic system in order to provide both automatic and manual upshifts and downshifts.

Five passages may be charged with mainline pressure from the manual valve. In Neutral, the lands on the manual valve block oil from any of the passages. In Park, figure 12-25, one land on the manual valve blocks oil from entering the valve altogether. Ford refers to mainline pressure from the manual valve by the name of the passage it flows through.

The OD-3 passage is charged in Overdrive or Drive; it routes pressure to the:

● 2-3 shift valve.

The OD-3-1 passage is charged in all forward gear ranges; it routes pressure to the:

● 1-2 shift valve
● 3-4 shift valve
● Governor
● 1-2 accumulator valve.

The P-3-1-R passage is charged in all gear ranges except Neutral and Overdrive; it routes pressure to the:

● 2-3 capacity modulator valve.

The P-1-R passage is charged in Park, Reverse, and manual low; it routes pressure to the:

● 1-2 shift valve
● 2-3 shift valve.

The R passage is charged in Reverse only; it routes pressure to the:

● Apply area of the reverse clutch
● Low servo modulator valve
● Boost valve.

Throttle valve

The AOD transmission develops throttle pressure using a throttle valve operated by a mechanical linkage, figure 12-26. Throttle pressure indicates engine torque and helps time the transmission upshifts and downshifts. Without throttle pressure, the transmission would upshift and downshift at exactly the same speed for different throttle openings and load conditions.

The throttle valve is controlled by opposition between hydraulic pressure and spring force. The hydraulic pressure is mainline pressure that is routed to the end of the valve opposite the spring. The amount of spring force acting on the opposite end of the valve is controlled by the throttle plunger. The plunger is linked to the accelerator so that its position is determined by throttle opening.

At closed throttle, the throttle plunger exerts no force on the throttle valve spring, and the throttle valve is positioned to block the entry of mainline pressure. As the accelerator is depressed, the linkage moves the throttle plunger to compress the throttle valve spring. This moves the throttle valve and allows mainline pressure into the valley between the two lands of the throttle valve where it becomes throttle pressure.

The throttle pressure that flows out of the valley and into the throttle pressure passages is regulated by the position of the throttle valve. A portion of this pressure is sent through a passage to a chamber behind the throttle plug, and through one or two other passages (depending on the position of the plunger) to a chamber at the far end of the throttle plunger. Pressure in these two chambers helps regulate the throttle valve.

At wide-open throttle, figure 12-26, the throttle plunger is pushed into the throttle valve as far as it can be. The plunger completely compresses the spring, contacts the throttle valve spool, and bottoms the valve to allow mainline pressure into the throttle pressure passage. At the same time, the plunger opens the kickdown passage to route kickdown pressure to the shift valves to force a downshift.

If the link between the accelerator and the throttle plunger is broken or disconnected, a failsafe spring on the valve body moves the plunger to the wide-open-throttle position to create maximum throttle pressure. Although the transmission will not upshift and downshift at the appropriate times, this protects the clutches and bands from slipping and burning.

Throttle pressure is routed from the throttle valve to the:

● 2-3 backout valve
● 3-4 throttle modulator valve
● Throttle limit valve.

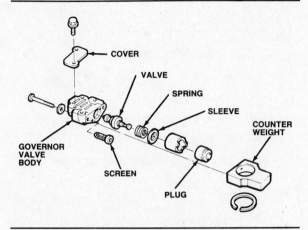

Figure 12-27. The AOD governor assembly.

Throttle limit valve

The throttle limit valve, figure 12-20, is held closed by spring force, and limits throttle pressure to a maximum of 85 psi (586 kPa). As long as throttle pressure is below that level, fluid flows through the valve and into the throttle limit passages.

When throttle pressure rises above 85 psi (586 kPa), the throttle limit valve moves against spring force to restrict the flow of fluid and regulate the throttle pressure. When the throttle limit valve is regulating, the pressure in the throttle valve circuits is called throttle limit pressure. To help decrease throttle pressure, some of the throttle limit pressure is applied to the end of the throttle limit valve opposite spring force.

Throttle limit pressure, which is never greater than 85 psi (586 kPa), is routed from the throttle limit valve to the:
- 2-3 throttle modulator valve
- 3-4 backout valve
- Boost valve in the pressure regulator valve
- Throttle relief valve.

Throttle relief valve

The throttle relief valve prevents excess throttle limit pressure if the throttle limit valve fails to regulate properly. The throttle relief valve consists of a valve and spring located in the throttle limit pressure circuit near the boost valve. If throttle limit pressure rises above 125 psi (862 kPa), the throttle relief valve unseats and exhausts excess fluid to the sump. When the pressure drops below this level, spring force reseats the valve.

Governor valve

We have been discussing the development and control of throttle pressure, which is one of the two auxiliary pressures that work on the shift valves. Before we get to the shift valve operation, we must explain the governor valve (or governor), and how it creates governor pressure. Governor pressure provides a road speed signal to the hydraulic system that causes automatic upshifts to occur as road speed increases, and permits automatic downshifts with decreased road speed.

The AOD governor, figure 12-27, is bolted to a counterweight that is mounted on, and turns with, the transmission output shaft. When the vehicle is stopped, the spring holds the governor valve closed, and there is no governor pressure.

As the car starts to move and the output shaft rotates, centrifugal force moves the governor valve spool outward against spring force to open the OD-3-1 oil inlet port. Mainline pressure in the OD-3-1 circuit then passes through the governor valve where it becomes governor pressure. For any given road speed, the governor valve will move a certain distance to balance spring force. As output shaft speed and centrifugal force increase, the governor valve opens farther, and governor pressure increases. Thus governor pressure is proportional to road speed. At a certain road speed, the governor valve opens all the way, and governor pressure equals mainline pressure.

Governor pressure is produced in all forward gear ranges when the vehicle is moving. Governor pressure is applied to three valves:
- 1-2 shift valve
- 2-3 shift valve
- 3-4 shift valve.

1-2 shift valve

The 1-2 shift valve, figure 12-20, controls upshifts from low to second, and downshifts from second to low. The valve is controlled by opposition between governor pressure and spring force. Other pressures can also affect the valve.

In low gear, OD-3-1 oil from the manual valve flows to the 1-2 shift valve where it is blocked by the downshifted valve. When vehicle speed increases enough for governor pressure to overcome spring force, the valve upshifts. Mainline pressure then flows through the 1-2 shift valve to the overdrive servo regulator valve. From there the pressure is routed through the 1-2 accumulator valve and 1-2 capacity modulator valve, which work together to control the intermediate clutch apply rate and shift the transmission into second gear.

There are also two ports at the 1-2 shift valve for P-1-R oil, which helps to keep the valve downshifted. P-1-R oil flows through the 1-2 shift valve to the low servo modulator valve

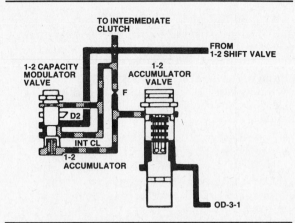

Figure 12-28. The AOD 1-2 capacity modulator valve and 1-2 accumulator valve regulate intermediate clutch apply pressure during a 1-2 shift.

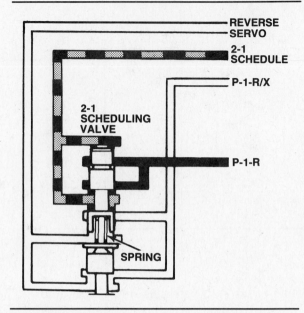

Figure 12-29. The AOD 2-1 scheduling valve regulates pressure to force a 2-1 downshift.

and 2-1 scheduling valve. The 2-1 scheduling valve sends a pressure back to the shift valve spring chamber to lock out second gear. Kickdown pressure can also flow through one of the P-1-R passages to force a downshift. Finally, there is a port for modulated throttle pressure that is sometimes sent from the 2-3 throttle modulator valve to oppose governor pressure.

1-2 capacity modulator valve and 1-2 accumulator valve

When the 1-2 shift valve upshifts, intermediate clutch apply oil (mainline pressure) is routed through the overdrive servo regulator valve to the 1-2 capacity modulator valve, figure 12-28. Mainline pressure moves the valve against spring force to restrict the valve inlet port, thus producing a regulated clutch apply pressure.

Regulated clutch apply pressure is sent to the intermediate clutch, the 1-2 accumulator valve, and two areas of the 1-2 capacity modulator valve to assist in pressure regulation. As pressure rises in the clutch apply circuit, it combines with spring force to move the 1-2 capacity modulator valve back toward the fully open position.

Regulated intermediate clutch apply oil that flows to the 1-2 accumulator valve is routed to the spring chamber. Pressure in this chamber combines with spring force to move the accumulator valve against OD-3-1 oil from the manual valve. The time it takes to bottom the accumulator slows the rate of pressure increase in the intermediate clutch circuit.

When the 1-2 accumulator valve is bottomed, and the 1-2 capacity modulator valve is fully open, the pressure in the intermediate clutch apply circuit reaches mainline pressure. Together, these two valves ensure that clutch

apply pressure builds up gradually so the clutch applies smoothly.

2-1 scheduling valve

The 2-1 scheduling valve, figure 12-29, is located in the same bore as the low servo modulator valve, and determines the 2-1 downshift speed when the gear selector is moved to manual low from the Overdrive or Drive range. It does this by producing a regulated 2-1 scheduling pressure that is sent to the spring chamber of the 1-2 shift valve. After the 2-1 downshift takes place, the 2-1 scheduling works to lock out second gear.

When the gear selector is in manual low and the transmission is in second gear, the 2-1 scheduling valve receives P-1-R oil from the 2-3 shift valve. The P-1-R pressure varies with throttle opening, and moves the 2-1 scheduling valve against spring tension to restrict the valve inlet port, thus producing a regulated 2-1 scheduling pressure. This pressure is routed to the spring end of the 1-2 shift valve, and some is applied to the end of the 2-1 scheduling valve opposite spring force to assist in pressure regulation.

When the 2-1 scheduling valve has been moved far enough against spring force, 2-1 scheduling pressure overcomes governor pressure and the 1-2 shift valve downshifts. P-1-R/X oil then flows from the 1-2 shift valve, through the 1-2 scheduling valve, and into the 2-1 scheduling passage. The P-1-R/X oil bottoms the 2-1 scheduling valve against spring

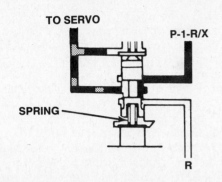

Figure 12-30. The AOD low servo modulator valve regulates low-reverse servo apply pressure in Park or manual low.

force, blocking the flow of P-1-R oil from the 2-3 shift valve. P-1-R/X oil is then applied to the spring end of the 1-2 shift valve to keep it downshifted and lock out second gear.

Low servo modulator valve

The low servo modulator valve, figure 12-30, is located in the same bore as the 2-1 scheduling valve, and regulates low-reverse servo apply pressure for a smooth band apply. In Park, manual low, and Reverse, the low servo modulator valve receives P-1-R/X oil from the downshifted 1-2 shift valve.

In Park and manual low, P-1-R/X pressure moves through the low servo modulator valve and is routed to the end of the valve opposite the spring. Hydraulic pressure moves the valve against spring force to restrict the valve inlet port and thus regulate the oil pressure as it flows to the low-reverse servo.

In Reverse, R oil from the manual valve acts with P-1-R/X oil to bottom the valve against spring force. This stops the valve from regulating, and allows R oil and P-1-R/X oil (both at boosted mainline pressure) to flow directly through the valve to apply the low-reverse band with the extra firmness needed in Reverse.

2-3 shift valve and 2-3 throttle modulator valve

The 2-3 shift valve assembly, figure 12-31, controls upshifts from second to third, and downshifts from third to second. The assembly includes the 2-3 shift valve and the 2-3 throttle modulator valve. Governor pressure is applied to the 2-3 shift valve in all forward gear ranges, and acts on the valve to produce an upshift. The 2-3 throttle modulator valve works with spring force to oppose upshifts, and can also produce a modulated throttle pressure that is applied to the 2-3 and 1-2 shift valves.

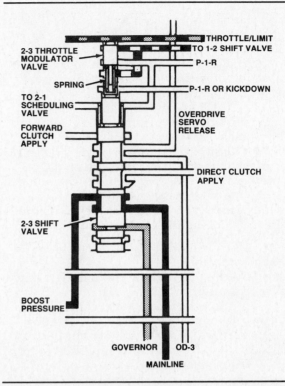

Figure 12-31. The AOD 2-3 shift valve assembly in the second gear position.

Throttle limit pressure is applied to the end of the 2-3 throttle modulator valve to oppose upshifts. If throttle limit pressure moves the valve far enough, a passage to the 2-3 throttle modulator circuit opens. This circuit routes modulated throttle pressure around the 2-3 throttle modulator valve to the spring chamber of the 2-3 shift valve where it helps oppose an upshift. At the same time, modulated throttle pressure is routed to the 1-2 shift valve where it works to help force a further downshift if conditions warrant.

In the forward gear ranges, OD-3 oil is sent to the 2-3 shift valve from the manual valve. When the valve is downshifted, it routes this pressure to the release side of the overdrive servo. When the valve upshifts, it reroutes the OD-3 oil to apply the direct clutch, and pressure from the forward clutch apply circuit then flows into the overdrive servo release passage.

Pump pressure is also routed to the 2-3 shift valve. When the valve is downshifted, pump pressure flows through it to the boost valve in the pressure regulator assembly where it acts to increase mainline pressure in first and second gears. When the valve is upshifted, pump pressure is blocked, and mainline pressure is reduced in third and fourth gears.

In manual low, Park, or Reverse, P-1-R oil flows to three ports in the 2-3 shift valve assembly. Two of the ports allow P-1-R oil to fill the spring chamber of the 2-3 throttle modulator valve; this downshifts the 2-3 shift valve and helps keep it downshifted. From the spring chamber, P-1-R oil flows into the 2-3 throttle modulator circuit, and from there to the 1-2 shift valve where it works to also keep that valve downshifted. At the third port, the P-1-R oil adds to the downshift pressure on the 2-3 shift valve and flows on to the 2-1 scheduling valve. One of the passages into the spring chamber can also be filled with kickdown pressure from the throttle valve anytime the accelerator is completely depressed.

2-3 backout valve

Direct clutch apply oil flows from the 2-3 shift valve to the 2-3 backout valve, figure 12-20, which controls the rate of direct clutch application in relation to throttle pressure. The position of the backout valve is determined by a balance between throttle pressure and spring force. The position of the valve routes direct clutch apply oil through one of two orifices (E or K) based on the valve position. From the backout valve, clutch apply oil flows to the direct clutch apply area, the spring side of the 2-3 accumulator, and the orifice control valve.

The 2-3 backout valve is also part of the forward clutch apply circuit. OD-3-1 pressure from the orifice control valve flows through the 2-3 backout valve to the forward clutch apply area and the 2-3 accumulator. In second and third gears, the 2-3 backout valve moves to allow some of the forward clutch apply oil to flow to the 3-4 backout valve.

Orifice control valve

The orifice control valve, figure 12-20, is part of the forward clutch apply circuit. In all forward gear ranges, the orifice control valve receives OD-3-1 oil from the downshifted 3-4 shift valve, and routes it to the 2-3 backout valve and the forward clutch.

In first and second gears, OD-3-1 pressure bottoms the orifice control valve against spring force. When the 2-3 shift takes place, direct clutch apply pressure combines with spring force to move the orifice control valve and open a passage that allows oil flow to the 3-4 throttle modulator valve and the spring chamber of the 3-4 shift valve. If the gear selector is in Drive, P-3-1-R oil is sent through this passage to lock out a fourth gear upshift. If the gear selector is in Overdrive, P-3-1-R oil is not sent to this passage, so a 3-4 upshift is possible.

2-3 accumulator and 2-3 capacity modulator valve

When the vehicle is shifted into a forward range at a standstill, it starts in low gear, and forward clutch apply pressure pushes down the 2-3 accumulator piston against spring force. The time it takes to bottom the piston slows the rate of pressure buildup in the circuit and smoothes forward clutch application.

The spring chamber of the 2-3 accumulator is part of the direct clutch apply circuit. When the 2-3 shift occurs, direct clutch apply pressure and spring force work together to push the piston back against the forward clutch apply pressure acting on the opposite side of the piston. The time it takes to bottom the piston slows the rate of pressure buildup in the circuit and smoothes direct clutch application.

Moving the accumulator piston to cushion direct clutch apply forces forward clutch apply pressure out of the upper chamber and creates a backpressure that is higher than mainline pressure. This pressure seats a check ball, and although a small amount of the backpressure escapes through a restricting orifice (L), most of the pressure is routed through a passage to the end of the 2-3 capacity modulator valve.

Backpressure moves the 2-3 capacity modulator valve to open a passage leading to the 3-4 shift valve. The rate at which the backpressure escapes through this passage is regulated by the 2-3 capacity modulator valve. Some of the escaping backpressure is applied to the upper end of the valve to help regulate the rate of pressure exhaust, and thus direct clutch apply.

Once the direct clutch is fully applied and backpressure decreases, spring force returns the 2-3 capacity modulator valve to its original position.

3-4 shift valve and 3-4 throttle modulator valve

The 3-4 shift valve assembly, figure 12-32, controls upshifts from third to fourth, and downshifts from fourth to third. The assembly includes the 3-4 shift valve and the 3-4 throttle modulator valve. Governor pressure is applied to the 3-4 shift valve in all forward gear ranges, and acts on the valve to produce an upshift. The 3-4 throttle modulator valve works with spring force to oppose upshifts, and can also produce a modulated throttle pressure that is applied to the 3-4 shift valve.

Throttle limit pressure is applied to the end of the 3-4 throttle modulator valve to oppose upshifts. If throttle limit pressure moves the valve far enough, a passage to the 3-4 throttle modulator circuit opens. This circuit routes modulated throttle pressure around the 3-4

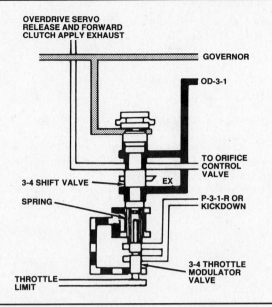

Figure 12-32. The AOD 3-4 shift valve assembly in the third gear position.

throttle modulator valve to the spring chamber of the 3-4 shift valve where it helps oppose an upshift.

In the forward gear ranges, OD-3-1 oil is sent to the 3-4 shift valve from the manual valve. When the valve is downshifted, it routes this pressure to the orifice control valve. When the valve upshifts, OD-3-1 oil is blocked, and forward clutch apply and overdrive servo release oil are allowed to exhaust.

There are two ports at the 3-4 throttle modulator valve for P-3-1-R oil or kickdown pressure. P-3-1-R oil flows to the valve in Park, Drive, manual low, and Reverse to prevent an upshift to fourth at any speed. Whenever the throttle is wide-open, the throttle valve sends kickdown pressure to the 3-4 throttle modulator valve to force a 4-3 downshift.

Overdrive servo regulator valve
The overdrive servo regulator valve, figure 12-20, receives mainline pressure from the up-shifted 1-2 shift valve, and sends the pressure out through two ports. One port is unregulated and routes mainline pressure to the 1-2 capacity modulator valve, where it becomes intermediate clutch apply oil. The other port is controlled by the overdrive servo regulator valve position.

In second and third gear, spring force determines the overdrive servo regulator valve position, and causes the valve to regulate the oil at less than mainline pressure. Therefore, even though the oil flows through the 3-4 shuttle valve to the apply side of the overdrive servo,

the servo cannot apply because the release pressure is mainline pressure plus spring force.

In fourth gear, 3-4 accumulator pressure from the 3-4 shuttle valve acts on the end of the servo regulator valve opposite spring force, and moves the valve so that full mainline pressure can charge the servo apply passage. Because the servo release circuit exhausts at this time, mainline pressure then overcomes spring force to apply the servo and overdrive band.

3-4 shuttle valve and 3-4 accumulator
The top of the 3-4 shuttle valve opposite spring force, figure 12-20, receives forward clutch apply oil. The spring chamber receives P-1-R oil. The center of the valve receives regulated overdrive servo apply oil from the overdrive servo regulator valve. Depending on the 3-4 shuttle valve position, P-1-R oil is either kept in the overdrive servo apply circuit or directed into the 3-4 accumulator circuit.

In Drive and Overdrive low, second, and third gears, P-1-R oil is not sent to the 3-4 shuttle valve, so forward clutch apply oil bottoms the valve against spring force. In this position, regulated overdrive servo apply oil flows freely through the valve to the apply side of the servo. However, it cannot overcome the combined mainline pressure and spring force on the release side of the servo, so the overdrive band does not apply.

In Overdrive fourth gear, forward clutch apply oil exhausts, so spring force moves the 3-4 shuttle valve up. In this position, servo apply oil flows to the apply side of the servo, the end of the overdrive servo regulator valve opposite the spring, and the 3-4 accumulator circuit.

At the overdrive servo regulator valve, accumulator pressure moves the regulator valve up so that full mainline pressure can apply the servo. At the 3-4 accumulator, the accumulator pressure combines with spring force to move the accumulator piston against mainline pressure. The time it takes to bottom the piston slows the rate of pressure buildup in the circuit and smoothes overdrive servo application.

3-4 backout valve
The 3-4 backout valve, figure 12-20, regulates the rate of forward clutch release in relation to throttle pressure during a 3-4 upshift. The 3-4 backout valve receives forward clutch apply oil from the 2-3 backout valve in second and third gears. It routes this pressure to a port at the 2-3 shift valve. The pressure from the 2-3 backout valve also flows to this port through a passage containing a ball-check valve.

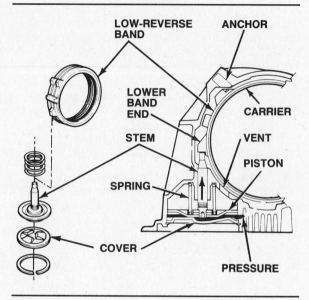

Figure 12-33. The AOD low-reverse band and servo assembly.

When the 2-3 shift valve is downshifted, it routes OD-3 oil to the release side of the overdrive servo, and blocks forward clutch apply pressure. When the 2-3 shift valve upshifts, the forward clutch apply oil flows through the ball-check valve and 2-3 shift valve to fill the overdrive servo release circuit. When the 3-4 shift occurs, forward clutch apply oil exhausts. This seats the ball-check valve and forces the oil to exhaust through the 3-4 backout valve to the 2-3 backout valve.

The position of the 3-4 backout valve is controlled by opposition between throttle limit pressure and spring force. If throttle pressure is high, it pushes the 3-4 backout valve down further to allow faster exhausting of pressure. This releases the forward clutch rapidly to match the timing of the overdrive band application. If throttle pressure is low, the 3-4 backout valve does not move as far, and pressure exhaust occurs more slowly to match the timing of a more gradual upshift.

MULTIPLE-DISC CLUTCHES AND SERVOS

The preceding paragraphs examined the individual valves in the hydraulic system. Before we trace the complete hydraulic system operation in each driving range, we will take a quick look at the multiple-disc clutches and servos.

Forward Clutch

The forward clutch, figure 12-6, is an input device applied in first, second, and third gears. To apply the forward clutch, hydraulic pressure is routed behind the clutch piston through a passage in the stator support. The clutch is released by a single large coil spring.

Direct Clutch

The direct clutch, figure 12-11, is an input device applied in third and fourth gears. The direct clutch handles a minimum of 60 percent of the total input torque in third gear. In fourth gear, it handles 100 percent of the torque input. The direct clutch is applied hydraulically and released by several small coil springs.

Intermediate Clutch

The intermediate clutch, figure 12-8, is a holding device applied in second gear. It is also applied in third and fourth gear, but the intermediate 1-way clutch that it holds overruns in these gears. The intermediate clutch piston is located in a bore machined into the back of the oil pump housing. The clutch plates are installed in the transmission case. The clutch is applied hydraulically and released by several small coil springs.

Reverse Clutch

The reverse clutch, figure 12-7, is applied only in Reverse gear. The reverse clutch piston is applied by hydraulic pressure that is routed through a passage in the stator support. The clutch piston moves against a Belleville spring that forces the clutch plates together. The Belleville spring operates as a lever to increase clutch application force. The clutch is also released by the Belleville spring.

Low-Reverse Servo

The low-reverse servo, figure 12-33, is located rearward of the case center support, in a bore located in the bottom of the case. The servo is hydraulically applied and released, with the help of a single, large, coil release spring. The band is adjusted during transmission assembly by selecting between different length apply stems.

Overdrive Servo

The overdrive servo, figure 12-34, is located forward of the case center support, in a bore in the bottom of the case. The overdrive servo is located on the same side of the case as the low-reverse servo. The servo is hydraulically applied and released, with the help of a single, large, coil release spring. The band is adjusted

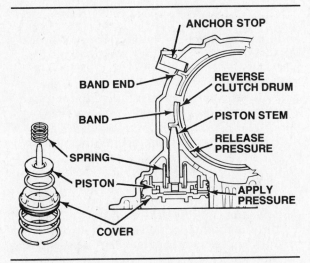

Figure 12-34. The AOD overdrive servo assembly.

during transmission assembly by selecting between different length apply stems.

HYDRAULIC SYSTEM SUMMARY

We have examined the individual valves, clutches, and servos in the hydraulic system and discussed their functions. The following paragraphs summarize their combined operations in the various gear ranges. Refer to figure 12-20 to trace the hydraulic system operation.

Park

In Park, pump oil flows to: the pressure regulator valve, the 3-4 accumulator, the 2-3 shift valve, the throttle valve, and the manual valve. The pressure regulator valve regulates mainline pressure and routes oil to the torque converter, oil cooler, and lubrication circuits. The 3-4 accumulator bottoms against spring force. The 2-3 shift valve routes pressure to the boost valve in the pressure regulator valve assembly. Mainline pressure routed to the throttle valve is blocked by the valve spool.

The manual valve routes P-3-1-R oil to the orifice control valve, where it passes through to the 3-4 throttle modulator valve and the spring chamber of the 3-4 shift valve. The manual valve also routes P-1-R oil to the 1-2 shift valve and the 2-3 throttle modulator valve. P-1-R pressure from the 1-2 shift valve flows to the 2-1 scheduling valve and the low pressure modulator valve. P-1-R pressure from the 2-3 throttle modulator valve flows to the 3-4 shuttle valve and 2-1 scheduling valve. The low servo modulator valve sends pressure to the apply side of the low-reverse servo. The low-reverse band is applied in Park, but does not cause any driving effects.

Neutral

In Neutral, pump oil flows to: the pressure regulator valve, the 3-4 accumulator, the 2-3 shift valve, the throttle valve, and the manual valve. The pressure regulator valve regulates mainline pressure and routes oil to the torque converter, oil cooler, and lubrication circuit. The 3-4 accumulator bottoms against spring force. The 2-3 shift valve routes pressure to the boost valve in the pressure regulator valve assembly. Mainline pressure routed to the throttle valve is blocked by the valve spool.

All of the above oil routings are the same as in Park. The only difference in Neutral is that mainline pressure is blocked at the manual valve by the valve spool. As a result, no pressure applications are made by the manual valve. In Neutral, all clutches and bands are released.

Overdrive and Drive Low

When the gear selector is moved to Drive or Overdrive, the oil flow described for Neutral is still effective, except that the manual valve opens the OD-3 and OD-3-1 ports. OD-3 oil flows to the 2-3 shift valve, and from there to the release side of the overdrive servo.

OD-3-1 oil flows to the 1-2 accumulator valve, 1-2 shift valve, governor, and 3-4 shift valve. OD-3-1 oil flows through the 3-4 shift valve to the 3-4 shuttle valve, the forward clutch apply area, the 2-3 accumulator, the orifice control valve, and the 2-3 capacity modulator valve. The 3-4 shuttle valve bottoms against spring force. The 2-3 accumulator bottoms against spring force to cushion forward clutch application. The orifice control valve bottoms against spring force to open a passage to the 2-3 backout valve. The 2-3 capacity valve moves against spring force to regulate fluid flow through the orifice control valve to the 2-3 backout valve. The 2-3 backout valve directs apply oil through a restricting orifice (A) to cushion forward clutch application.

If the transmission gear selector is in Drive, the manual valve also routes P-3-1-R oil to the orifice control valve. The P-3-1-R oil then flows to the 3-4 throttle modulator valve and the 3-4 shift valve spring chamber to lock out a fourth gear upshift.

As the accelerator is depressed, the throttle valve produces throttle pressure that is routed to the 2-3 backout valve, the 3-4 throttle modulator valve, and the throttle limit valve. The 3-4 throttle modulator valve can produce a modulated throttle pressure that is applied to the 3-4 shift valve spring chamber. The throttle limit valve regulates pressure in the throttle limit circuit to a maximum of 85 psi (586 kPa). Throttle limit pressure is applied to the 2-3 throttle modulator valve, the 3-4 backout valve, the throttle relief valve, and the boost valve in the pressure regulator valve assembly. The 2-3 throttle modulator valve can produce a modulated throttle pressure that is applied to the 1-2 shift valve spring chamber. These throttle pressure applications are made in proportion to throttle opening in all gear ranges.

The forward clutch is the only hydraulic device applied in Overdrive and Drive low.

Overdrive and Drive Second

As vehicle speed and governor pressure increase, governor pressure upshifts the 1-2 shift valve. OD-3-1 oil that was blocked then flows to the overdrive servo regulator valve. From the overdrive servo regulator valve, mainline oil is sent to the 1-2 capacity modulator valve and the 3-4 shuttle valve.

The 1-2 capacity modulator valve routes pressure to the 1-2 accumulator valve, and to the apply area of the intermediate clutch. Mainline pressure and spring force move the 1-2 accumulator valve against mainline pressure alone to cushion the intermediate clutch application.

The overdrive servo regulator valve works with the 3-4 shuttle valve to produce a lower-than-mainline apply pressure that is sent to the overdrive servo. The servo does not apply, however, because the release passage is charged with mainline pressure from the 2-3 shift valve.

Throttle pressure moves the 2-3 backout valve so that some forward clutch apply oil is routed to the 3-4 backout valve. This oil flows through the 2-3 backout valve and a passage fitted with a check ball to the 2-3 shift valve where it is blocked.

In Overdrive and Drive second, the forward clutch remains applied, and works together with the intermediate clutch to provide second gear.

Drive and Overdrive Third

As vehicle speed and governor pressure continue to increase, governor pressure overcomes combined throttle limit pressure and spring force to upshift the 2-3 shift valve. The upshifted 2-3 shift valve blocks mainline pressure from the boost valve in the pressure regulator valve assembly, and vents the passage that leads to the boost valve. OD-3 oil that formerly filled the overdrive servo release passage is now routed into the direct clutch apply circuit, and forward clutch apply oil from the 3-4 backout valve flows into the overdrive servo release passage.

Direct clutch apply oil flows through the 2-3 backout valve, where it is diverted through one of two restricting orifices on its way to the clutch. A portion of the apply oil is also diverted to the spring chamber of the orifice control valve, and to the bottom of the 2-3 accumulator. The 2-3 accumulator cushions the direct clutch application.

If the transmission is in Drive range, P-3-1-R oil flows through the orifice control valve to the 3-4 throttle modulator valve, the 3-4 throttle modulator pressure passage, and the spring chamber of the 3-4 shift valve, to prevent an upshift to fourth.

In Overdrive and Drive third, the forward clutch remains applied, and works together with the direct clutch to provide third gear. The intermediate clutch also remains applied, but the 1-way clutch it controls overruns and is ineffective.

Overdrive Fourth

As vehicle speed and governor pressure continue to increase, governor pressure overcomes combined throttle pressure and spring force to upshift the 3-4 shift valve. The upshifted 3-4 shift valve blocks OD-3-1 oil from the forward clutch apply circuit, which includes the overdrive servo release passage and the spring end of the 3-4 shuttle valve. Pressure in these passages is exhausted through the 3-4 shift valve.

Once the 3-4 shift valve upshifts, the 3-4 shuttle valve works with the overdrive servo regulator valve to regulate servo apply pressure for smooth band application. The 3-4 accumulator also cushions the overdrive servo application.

In fourth gear, the direct clutch remains applied, and works together with the overdrive servo to provide fourth gear. The forward clutch is released, and although the intermediate clutch remains applied, it is ineffective because the 1-way clutch it controls overruns.

Drive and Overdrive Range Forced Downshifts

A 4-3 or 3-2 part-throttle downshift can be forced at moderate speeds. As the driver presses down on the accelerator, the throttle plunger moves to increase throttle pressure. Throttle pressure applied to 3-4 throttle modulator valve, and throttle limit pressure applied to the 2-3 throttle modulator valve, create modulated throttle pressures that are applied to the spring chambers of those shift valves. If the combination of throttle pressure (or modulated throttle pressure) and spring force is great enough, it overcomes governor pressure to downshift one or both of the shift valves. A 2-1 part-throttle downshift is not possible with the AOD transmission.

Full-throttle kickdown downshifts can be forced at higher speeds if the driver depresses the accelerator completely. When this happens, throttle pressure equals mainline pressure, and the throttle plunger moves far enough to open the kickdown circuit. Mainline pressure is then routed through certain branches of the P-1-R and P-3-1-R circuits to the spring ends of all three shift valves.

When the accelerator is floored, the 3-4 shift valve will downshift immediately; the transmission cannot remain in fourth gear at wide open throttle. The 2-3 and 1-2 shift valves will downshift, in that order, when combined kickdown pressure and spring force overcome governor pressure acting on the opposite ends of the valves. The transmission may downshift either one or two additional gears depending on the vehicle speed.

Manual Low

In manual low, the manual valve opens the OD-3-1, P-3-1-R, and P-1-R ports. OD-3-1 applies the forward clutch as described for Overdrive and Drive low. P-3-1-R oil flows through the orifice control valve to the 3-4 throttle modulator valve and modulated throttle pressure circuit, to the spring chamber of the 3-4 shift valve where it forces a downshift from fourth if necessary.

P-1-R oil flows to the 2-3 throttle modulator valve and modulated throttle pressure circuit, and to the spring chamber of the 2-3 shift valve where it forces a downshift from third if necessary. From the 2-3 shift valve, P-1-R oil flows to the 2-1 scheduling valve, which routes it to the

spring chamber of the 1-2 shift valve to help force a 2-1 downshift if necessary, and then to prevent a 1-2 upshift. From the 1-2 shift valve, P-1-R oil flows to the low servo modulator valve which regulates the pressure and sends it to the low-reverse servo.

If the driver shifts to manual low at a high road speed, governor pressure keeps the 1-2 shift valve upshifted, and OD-3-1 oil continues to flow through it to the overdrive servo regulator valve. The overdrive servo regulator valve works with the 3-4 shuttle valve to produce a mainline-strength servo apply pressure that is routed to the 1-2 capacity modulator valve. The 1-2 capacity modulator valve keeps the intermediate clutch applied, and the transmission in second gear, until road speed drops below 25 mph (40 kph). At that point, governor pressure has fallen far enough that the 1-2 shift valve can be downshifted.

The application of the overdrive servo provides engine braking in "manual second" gear. When the 1-2 shift valve downshifts, OD-3-1 oil is blocked, and the intermediate clutch and overdrive servo are both released.

In manual low, both the forward clutch and the low-reverse band are applied.

Reverse

In Reverse, the manual valve opens its P-1-R, P-3-1-R, and R ports. P-1-R oil applies the low-reverse servo as described for manual low and Reverse. P-3-1-R oil flows through the orifice control valve to the 3-4 shift valve assembly as described for Park, Drive third, and manual low. R oil flows to the boost valve in the pressure regulator valve assembly, and to the apply area of the reverse clutch. Because the boost valve receives pressure from both the 2-3 shift valve and the manual valve, as well as a variable pressure from the throttle limit valve, mainline pressure is higher in Reverse than in any other gear range.

The low-reverse band and the reverse clutch are applied in Reverse.

Review Questions

Select the single most correct answer
Compare your answers to the correct answers on page 603

1. Mechanic A says the AOD planetary gearset has basically the same power flow as the Ford FMX in the lower three gears. Mechanic B says the AOD case is similar in some ways to that of the C3.
Who is right?
   a. Mechanic A
   b. Mechanic B
   c. Both mechanic A and B
   d. Neither mechanic A nor B

2. The AOD transmission was introduced in:
   a. 1979
   b. 1980
   c. 1981
   d. 1982

3. Mechanic A says the AOD secondary (reverse) sun gear is integral with the input shell. Mechanic B says the AOD secondary (reverse) sun gear is driven by the forward clutch.
Who is right?
   a. Mechanic A
   b. Mechanic B
   c. Both mechanic A and B
   d. Neither mechanic A nor B

4. Mechanic A says the AOD gearset short pinions are meshed with the primary (forward) sun gear. Mechanic B says the AOD gearset short pinions are also meshed with the ring gear.
Who is right?
   a. Mechanic A
   b. Mechanic B
   c. Both mechanic A and B
   d. Neither mechanic A nor B

5. Mechanic A says the AOD gearset primary (forward) sun gear is driven whenever the forward clutch is applied. Mechanic B says the AOD gearset ring gear drives the output shaft.
Who is right?
   a. Mechanic A
   b. Mechanic B
   c. Both mechanic A and B
   d. Neither mechanic A nor B

6. Mechanic A says the AOD torque converter uses a damper assembly to turn the input shaft. Mechanic B says the AOD torque converter has a clutch that locks up for 100 percent mechanical drive in fourth gear.
Who is right?
   a. Mechanic A
   b. Mechanic B
   c. Both mechanic A and B
   d. Neither mechanic A nor B

7. Mechanic A says the AOD intermediate clutch locks the secondary (reverse) sun gear to the case.
Mechanic B says the AOD intermediate one-way clutch holds the secondary (reverse) sun gear in second gear.
Who is right?
   a. Mechanic A
   b. Mechanic B
   c. Both mechanic A and B
   d. Neither mechanic A nor B

8. Mechanic A says the AOD low-reverse band holds the planet carrier stationary in Drive low. Mechanic B says the AOD low-reverse band holds the planet carrier stationary in Reverse.
Who is right?
   a. Mechanic A
   b. Mechanic B
   c. Both mechanic A and B
   d. Neither mechanic A nor B

9. Mechanic A says torque is delivered to the gearset by both hydraulic and mechanical means when the AOD transmission is in third gear.
Mechanic B says the AOD overdrive band holds the planet carrier stationary in fourth gear.
Who is right?
   a. Mechanic A
   b. Mechanic B
   c. Both mechanic A and B
   d. Neither mechanic A nor B

10. Mechanic A says the direct drive shaft transmits torque to the forward clutch when the AOD transmission is in fourth gear. Mechanic B says fourth gear is locked out hydraulically when the AOD gear selector is placed in Drive (3).
Who is right?
   a. Mechanic A
   b. Mechanic B
   c. Both mechanic A and B
   d. Neither mechanic A nor B

11. Mechanic A says the AOD low-reverse band is applied for engine compression braking in manual low. Mechanic B says the AOD low-reverse band is also applied in Park.
Who is right?
   a. Mechanic A
   b. Mechanic B
   c. Both mechanic A and B
   d. Neither mechanic A nor B

12. Mechanic A says the AOD transmission can be shifted from Ⓓ to D, or D to Ⓓ, at any vehicle speed.
Mechanic B says placing the gear selector in manual low (1) above 25 mph or 40 kph will cause the AOD transmission to shift into Drive second.
Who is right?
   a. Mechanic A
   b. Mechanic B
   c. Both mechanic A and B
   d. Neither mechanic A nor B

13. Mechanic A says the AOD direct clutch is located in front of the center support that is splined to the case.
Mechanic B says the AOD low 1-way clutch is supported by the center support.
Who is right?
   a. Mechanic A
   b. Mechanic B
   c. Both mechanic A and B
   d. Neither mechanic A nor B

14. Mechanic A says the AOD overdrive servo is installed in a bore from outside the transmission case.
Mechanic B says the AOD overdrive band is installed around the direct clutch drum.
Who is right?
   a. Mechanic A
   b. Mechanic B
   c. Both mechanic A and B
   d. Neither mechanic A nor B

15. Mechanic A says the AOD reverse clutch hub is integral with the forward clutch drum.
Mechanic B says the AOD reverse clutch uses a Belleville spring to increase clutch application pressure, and to release the clutch piston.
Who is right?
   a. Mechanic A
   b. Mechanic B
   c. Both mechanic A and B
   d. Neither mechanic A nor B

16. Mechanic A says the intermediate band holds the reverse clutch drum, stationary when the AOD transmission is in Drive second gear.
Mechanic B says the AOD intermediate clutch and intermediate 1-way clutch prevent counter-clockwise rotation of the reverse clutch drum.
Who is right?
a. Mechanic A
b. Mechanic B
c. Both mechanic A and B
d. Neither mechanic A nor B

17. Mechanic A says two members of the planetary gear train are driven at nearly the same speed when the AOD transmission is in third gear.
Mechanic B says the direct clutch drives the planet carrier when the AOD transmission is in third gear.
Who is right?
a. Mechanic A
b. Mechanic B
c. Both mechanic A and B
d. Neither mechanic A nor B

18. Mechanic A says the overdrive band holds the secondary (reverse) sun gear stationary when the AOD transmission is in fourth gear.
Mechanic B says torque delivery is 100 percent mechanical through the input shaft when the AOD transmission is in fourth gear.
Who is right?
a. Mechanic A
b. Mechanic B
c. Both mechanic A and B
d. Neither mechanic A nor B

19. Mechanic A says the AOD 1-2 shift valve controls upshifts and downshifts between first and second gears.
Mechanic B says the AOD 2-3 shift valve is used to reduce mainline pressure in third and fourth gears.
Who is right?
a. Mechanic A
b. Mechanic B
c. Both mechanic A and B
d. Neither mechanic A nor B

20. Mechanic A says the 2-3 accumulator cushions forward clutch apply when the AOD transmission is first placed in gear.
Mechanic B says the AOD throttle plunger opens the kickdown circuit at wide open throttle.
Who is right?
a. Mechanic A
b. Mechanic B
c. Both mechanic A and B
d. Neither mechanic A nor B

13

The Ford AXOD Transaxle

HISTORY AND MODEL VARIATIONS

The Ford AXOD 4-speed automatic transaxle, figure 13-1, was introduced in 1986 on Taurus/Sable models with the 3.0-liter V-6 engine. In 1988, use of the AXOD was expanded to include the FWD Continental with the 3.8-liter V-6 engine — a powerteam that was also made optional in the Taurus/Sable.

The AXOD has several distinct differences from Ford's 3-speed automatic transaxle, the ATX, and from other Ford automatics in general. In fact, the AXOD is similar in many ways to General Motors THM 125 and 440-T4 automatics. The AXOD uses neither a Simpson nor a Ravigneaux compound planetary geartrain. It has four multiple-disc clutches, a 1-way roller clutch, a 1-way sprag clutch, and two bands that provide its various gear ratios. For maximum fuel efficiency, the AXOD torque converter contains a hydraulically applied converter clutch.

The AXOD torque converter does not drive the transaxle geartrain directly. Instead, the converter drives a chain that transfers engine power to the geartrain positioned behind and below the crankshaft centerline, figure 13-2. The chain is located in a separate case that bolts to the end of the transaxle case on the driver's side of the car.

Another unusual feature of the AXOD is that the valve body also mounts on the driver's side of the transaxle, rather than at the bottom in the sump. The valve body bolts to the chain case, and the oil pump is contained in a casting that bolts to the valve body. The AXOD oil pump is a variable-displacement vane-type pump driven by a shaft splined to the torque converter cover. Both the oil pump and valve body are housed under a stamped steel cover that bolts to the chain case.

Two other major distinctions between the AXOD and other Ford automatics are its governor and final drive assembly. The AXOD uses a check-ball governor of the same basic design as that used in General Motors THM 125 transaxles and THM 200 transmissions. And in place of final drive spur gears like those in the ATX, the AXOD uses a planetary gearset for its final drive.

Fluid Recommendations

Ford initially recommended Type H fluid for the AXOD transaxle. When that fluid became obsolete, they changed the recommendation to MERCON® ATF. The use of MERCON® is a warranty *requirement* for all 1988 and later AXOD transaxles.

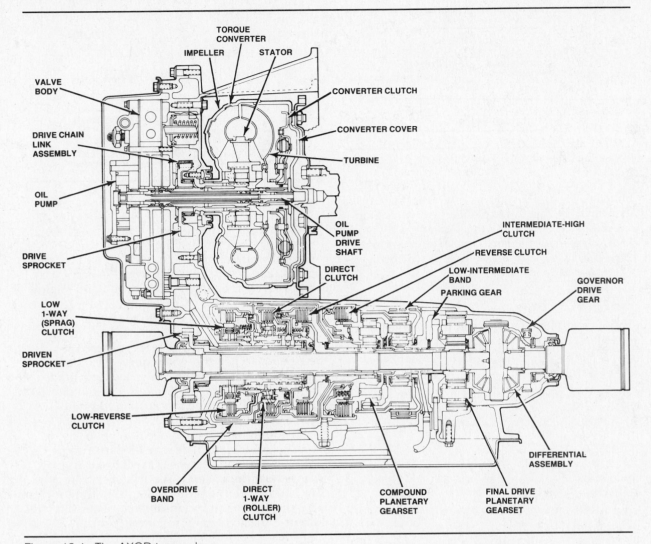

Figure 13-1. The AXOD transaxle.

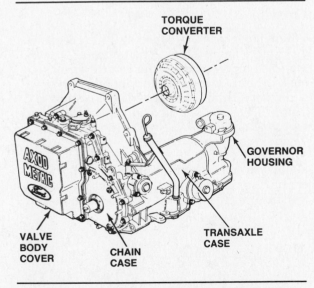

Figure 13-2. The AXOD geartrain is located behind and below the engine crankshaft centerline.

Transaxle Identification

Identifying a Ford automatic transaxle is a 2-step process. First, you determine the model of transaxle. All Ford vehicles have an identification plate on the rear of the left front door. In addition to the vehicle identification number and other data, this plate has a code letter that identifies the automatic transaxle model. Like all other Ford 4-speed automatics, the AXOD is identified by the letter T. This transaxle can also be identified by the terms "AXOD" and "METRIC" stamped into the valve body cover.

In addition to the transaxle model, you must know the exact transaxle assembly part number and/or other identifying characters to obtain repair parts. This information is stamped on a tag attached to the transaxle. On the AXOD, the tag is located on top of the torque converter

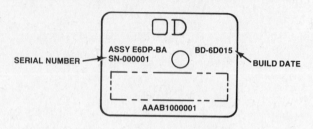

Figure 13-3. The AXOD identification tag.

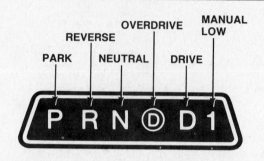

Figure 13-4. AXOD gear selector positions.

housing. Figure 13-3 shows a typical AXOD identification tag and explains the meaning of the data stamped on the tag. As required by Federal regulations, Ford transaxles also carry the serial number of the vehicle in which they were originally installed.

Major Changes

Through the 1988 model year, there have been no major changes to the AXOD transaxle.

GEAR RANGES

The AXOD gear selector has six positions, figure 13-4:
P — Park
R — Reverse
N — Neutral
Ⓓ — Overdrive
D — Drive
1 — Manual low.
The AXOD gear ratios are:
- First (low) — 2.77:1
- Second (intermediate) — 1.54:1
- Third (direct drive) — 1.00:1
- Fourth (overdrive) — 0.694:1
- Reverse — 2.263:1.

Park and Neutral

The engine can be started only when the transaxle is in Park or Neutral. No clutches or bands are applied in these positions, so there is no power flow through the transaxle. In Neutral, the output shafts are free to turn, and the car can be pushed or pulled. In Park, a mechanical pawl engages a parking gear splined to the final drive sun gear. This locks the final drive assembly to the case, and prevents the output shafts from turning so that the vehicle cannot be moved.

Overdrive

Overdrive (Ⓓ) is the normal driving position for the AXOD transaxle. When the vehicle is stopped and the gear selector is placed in Ⓓ,

the transaxle is automatically in low gear. As the vehicle accelerates, the transaxle automatically upshifts to second, third, and fourth gears. The speeds at which upshifts occur vary, and are controlled by vehicle speed and engine load.

Depending on the vehicle operating conditions, the torque converter clutch may be applied in third and fourth gears to increase efficiency. When the driver depresses the throttle to accelerate or maintain speed up a grade or into headwinds, the converter clutch disengages to allow torque multiplication in the converter.

If additional power is needed when the transaxle is in fourth gear, a 4-3 part-throttle (torque-demand) downshift will occur when the accelerator is depressed past a certain point. At speeds below 55 mph (88 kph) in third gear, a 3-2 kickdown (through-detent) downshift will occur if the accelerator is completely depressed. At speeds below 25 mph (40 kph), a 3-1 or 2-1 kickdown downshift is also possible.

Drive

When the vehicle is stopped and the gear selector is placed in D (Drive), the transaxle is automatically in low gear. As the vehicle accelerates, the transaxle automatically upshifts to second and third gears. It does not upshift to fourth. The speeds at which upshifts occur vary, and are controlled by vehicle speed and engine load.

At speeds below 55 mph (88 kph), a 3-2 kickdown downshift will take place when the accelerator is completely depressed. At speeds below 25 mph (40 kph) a 3-1 or 2-1 kickdown downshift is also possible.

FORD TERM	TEXTBOOK TERM
Clutch cylinder	Clutch drum
Converter pump	Impeller
Reactor	Stator
Internal gear	Ring gear
Line pressure	Mainline pressure
Main regulator valve	Pressure regulator valve
T.V. pressure	Throttle pressure
T.V. limit valve	Throttle limit valve
2-3 T.V. modulator valve	2-3 throttle modulator valve
3-4 modulator valve	3-4 throttle modulator valve
Forward clutch	Low-reverse clutch
Intermediate clutch	Intermediate-high clutch
Planetary	Carrier and pinions

Figure 13-5. Ford AXOD transaxle nomenclature table.

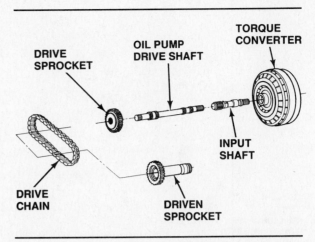

Figure 13-6. In the AXOD, engine power passes from the torque converter, through drive and driven sprockets joined by a drive chain, to the geartrain.

gear. It does not upshift to second. Because different apply devices are used in manual low than are used in Drive low, this gear range can be used to provide engine compression braking to slow the vehicle.

If the driver shifts to manual low at high speed, the transaxle will not shift directly to low gear. To avoid engine or driveline damage, the transaxle first downshifts to second, and then automatically downshifts to low when the speed drops below approximately 28 mph (45 kph). The transaxle then remains in low gear until the gear selector is moved to another position.

Reverse

When the gear selector is moved to R, the transaxle shifts into Reverse. The transaxle should be shifted to Reverse only when the vehicle is standing still. The lockout feature of the shift gate should prevent accidental shifting into Reverse while the vehicle is moving forward. However, if the driver intentionally moves the selector to R while moving forward, the transaxle *will* shift into Reverse, and damage may result.

FORD TRANSAXLE NOMENCLATURE

For the sake of clarity, this text uses consistent terms for transaxle parts and functions. However, the various manufacturers may use different terms. Therefore, before we examine the buildup of the AXOD, along with the operation of its geartrain and hydraulic system, you should be aware of the unique names that Ford uses for some transaxle parts and operations. This will make it easier in the future if you

should look in a Ford manual and see a reference to the "reactor". You will realize that this is simply the stator, and is similar to the same part in any other transaxle. Figure 13-5 lists Ford nomenclature and the textbook terms for some transaxle parts and operations.

AXOD BUILDUP

The AXOD case, figure 13-2, is a 1-piece aluminum casting with the chain cover assembly, valve body cover, and oil pan bolted to it. The bellhousing is cast integral with the transaxle case.

The AXOD torque converter bolts to the engine flexplate, which has the starter ring gear around its outer diameter. The torque converter is a welded assembly and cannot be disassembled. It is installed into the transaxle by sliding it onto the stationary stator support. The splines in the stator 1-way clutch engage the splines on the stator support. The splines in the turbine engage the input shaft. The splines in the converter cover engage the oil pump shaft, which passes through the inside of the input shaft to drive the oil pump. The hub at the rear of the torque converter housing rides on a needle bearing and seal assembly.

The AXOD uses a variable-displacement vane-type oil pump mounted on the valve body. The valve body bolts to the outboard side of the chain cover assembly. Because the converter cover is bolted to the engine flexplate, the oil pump shaft turns and the oil pump delivers fluid whenever the engine is running. The pump draws fluid from the sump through a screen, and routes the fluid through passages in the valve body, chain cover, and transaxle case to several valves in the hydraulic system.

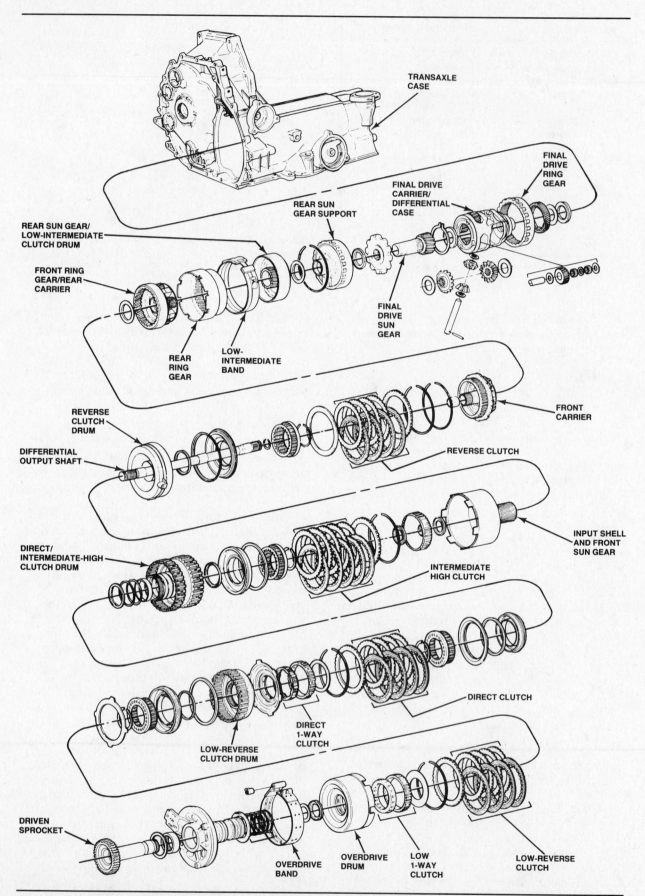

Figure 13-7. An exploded view of the AXOD clutches, 1-way clutches, bands, compound planetary geartrain, and final drive/differential assemblies.

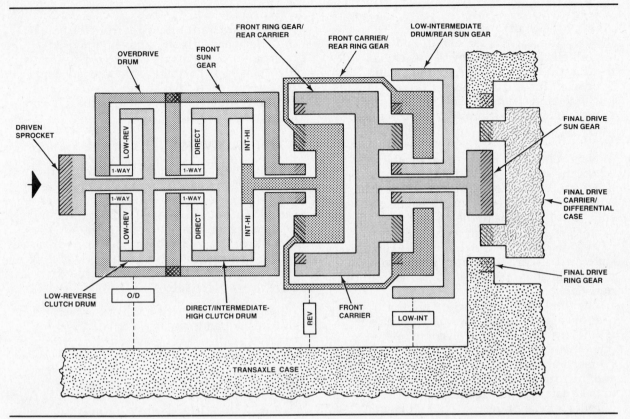

Figure 13-8. A schematic drawing of the AXOD clutches, 1-way clutches, bands, compound planetary geartrain, and final drive/differential assemblies.

The input shaft turns the drive sprocket, which turns the drive chain. The drive chain then turns the driven sprocket that provides power to the planetary gearset, figure 13-6. The driven sprocket assembly includes a shaft that is splined to the low-reverse and direct/intermediate-high clutch drums. As we continue with the remainder of the AXOD buildup, refer to figures 13-7 and 13-8 to identify the components being discussed.

The outside of each low-reverse clutch steel disc is splined to the inside of the low-reverse clutch drum and turns with it. The inside of each friction disc is splined to the clutch hub, which is splined to the outer race of the low 1-way (sprag) clutch. The inner race of the low 1-way clutch is splined to the overdrive drum. The overdrive drum has lugs on it that engage the input shell and front sun gear, which are a single assembly. When the low-reverse clutch is applied, it drives the outer race of the low 1-way clutch counterclockwise. This causes the 1-way clutch to lock up so that the inner race drives the overdrive drum, input shell, and front sun gear counterclockwise.

The overdrive drum is surrounded by the overdrive band. When the overdrive band is applied, it prevents the overdrive drum from turning in either direction. Therefore, the input shell and front sun gear are held stationary as well.

The front sun gear can also be driven by the inner race of the direct 1-way (roller) clutch, which is part of a plate that has lugs on its outer edge that engage the overdrive drum, input shell, and front sun gear. The outer race of the direct 1-way clutch is splined to the inside of each direct clutch friction disc. The outside of each direct clutch steel disc is splined to the inside of the direct clutch drum. When the direct clutch is applied, it drives the outer race of the direct 1-way clutch counterclockwise. This causes the 1-way clutch to lock up so that the inner race drives the overdrive drum, input shell, and front sun gear counterclockwise.

The intermediate-high clutch drum, is part of the same assembly as the direct clutch drum. The outside of each intermediate-high clutch steel disc is splined to the inside of the clutch drum. The inside of each friction disc is splined to the outside of the clutch hub. The intermediate-high clutch hub is splined to the front carrier. When the intermediate-high clutch is applied, it transmits engine torque from the driven sprocket to the front carrier.

All of the geartrain components discussed thus far (the low-reverse, direct, and intermediate-high clutches, the low 1-way clutch, and the direct 1-way clutch) are contained in a "housing" made up of the overdrive drum, input shell, and the front sun gear. The driven sprocket is the single input to this assembly; the front sun gear and front carrier are the two outputs.

Next behind the parts described above comes the reverse clutch. The reverse clutch drum is anchored to the transaxle case. The outside of each reverse clutch steel disc is splined to the inside of the reverse clutch drum. The inside of each friction disc is splined to the outside of the clutch hub. The reverse clutch hub is part of the front carrier. When the reverse clutch is applied, it holds the front carrier stationary.

The front carrier is attached to the rear ring gear by lugs so that the two parts act as one unit. They are said to be "in common". Two other members of the compound planetary gearset are also in common, the front ring gear and the rear carrier. The rear carrier is splined to the final drive sun gear, and is always the output member of the compound planetary gearset. The rear carrier turns on a support that is splined to the case.

The rear sun gear is attached to the low-intermediate drum, which is surrounded by the low-intermediate band. When the low-intermediate band is applied, it holds the rear sun gear and prevents it from turning in either direction.

The members of the compound planetary gearset interrelate in the normal fashion. Each sun gear meshes with its pinion gears. Each set of pinion gears meshes with its ring gear, and turns on pinion shafts attached to its carrier.

The final drive and differential assembly transmits power from the rear carrier to the halfshafts of the vehicle. In place of conventional final drive gears, the AXOD uses a simple planetary gearset with one sun gear, one carrier, four pinions and one ring gear. The sun gear is driven by the rear carrier of the transaxle gearset. The sun gear meshes with the pinions, which turn on pinion shafts in the final drive carrier, which is built into the differential case. The pinions mesh with the final drive ring gear, which is splined to the transaxle case. A single gear attached to the carrier/differential case drives both the governor and the speedometer cable.

The differential case contains four bevel gears: two pinion gears and two side gears. The pinion gears are installed on a pinion shaft that is retained by a pin in the differential case. The pinion gears transmit power from the case to the side gears. A male spline on the right-side halfshaft enters the transaxle to engage one of the side gears. The other side gear is splined to the differential output shaft which delivers power, through a male spline, to the left-side halfshaft.

AXOD POWER FLOW

Now that we have taken a quick look at the general way in which the AXOD transaxle is assembled, we can trace the flow of power through the apply devices and the compound planetary gearset in each of the gear ranges. Refer to figure 13-8 and the specific figure for each gear range to help trace the power flow.

Park and Neutral

When the gear selector is in Park or Neutral, no clutches or bands are applied, and the torque converter clutch is released. Since no members of the planetary gearsets are held, there can be no output motion. The torque converter impeller/cover and oil pump drive shaft rotate clockwise at engine speed. The oil pump drive shaft turns the oil pump. The converter turbine, the drive chain, and the drive and driven sprockets also rotate.

In Neutral, the planetary gearsets idle and the final drive sun gear is free to rotate in either direction, so the vehicle can be pushed or pulled. In Park, a mechanical pawl engages the parking gear splined to the sun gear. Because both the ring gear and sun gear are then held, the final drive planetary gearset locks up and prevents the vehicle from moving.

Overdrive and Drive Low

When the car is stopped and the AXOD gear selector is moved to Overdrive (Ⓓ) or Drive (D), the transaxle is automatically in low gear, figure 13-9. The low-reverse clutch is applied and transmits engine power from the driven sprocket through the low 1-way clutch to the front sun gear. The low-intermediate band is also applied to hold the rear sun gear stationary.

The front sun gear rotates counterclockwise and turns the front pinions clockwise. For reasons to be explained in a moment, the front ring gear is also turning counterclockwise, but at a slower speed than the sun gear. As a result, the front pinions turn the front carrier counterclockwise. The front carrier and rear ring gear are joined by lugs, so the rear ring gear also turns counterclockwise. The rear ring gear

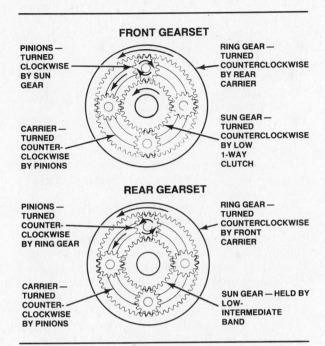

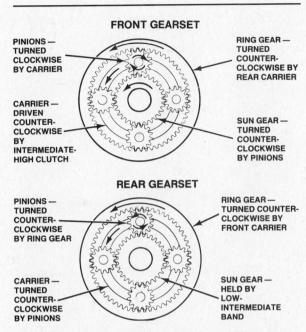

Figure 13-9. AXOD power flow in Overdrive and Drive low.

Figure 13-10. AXOD power flow in Overdrive and Drive second.

turns the rear pinions counterclockwise, and the pinions walk around the stationary rear sun gear, turning the rear carrier and front ring gear assembly counterclockwise in gear reduction.

The rear carrier is splined to the final drive sun gear, which turns the final drive pinions clockwise. Because the final drive ring gear is splined to the transaxle case, it is always held from rotating. Therefore, the pinions walk around the inside of the ring gear and drive the carrier/differential case counterclockwise in gear reduction. Power is then passed through the differential pinions and side gears to the differential output shaft (left side) and both halfshafts. This power flow through the final drive gearset and differential assembly is the same in all forward gears.

As long as the car is accelerating in Overdrive or Drive low, torque is transferred from the engine through the planetary gearset to the halfshafts. If the driver lets up on the throttle, the engine speed drops while the drive wheels keep turning at the same speed. This removes the counterclockwise-rotating torque load from the low 1-way clutch, so it unlocks and allows the front sun gear to freewheel. No engine compression braking is available, and this is the major difference between Overdrive or Drive low, and manual low, as we will see later.

Gear reduction in low gear is produced by both the front and rear gearsets of the transaxle. The input member of the compound planetary gearset is the front sun gear. The held

member is the rear sun gear, and the output member is the rear carrier. The input device is the low-reverse clutch together with the low 1-way clutch. The holding device is the low-intermediate band. In the final drive gearset, the input member is always the sun gear, the held member is always the ring gear, and the output member is always the carrier.

Overdrive and Drive Second

As the vehicle continues to accelerate in Overdrive or Drive, the hydraulic system produces an automatic upshift to second gear by applying the intermediate-high clutch to transmit engine power from the driven sprocket to the front carrier and rear ring gear, figure 13-10. The low-intermediate band remains applied to hold the rear sun gear.

The front carrier rotates counterclockwise at turbine speed. For reasons to be explained in a moment, the front ring gear is also turning counterclockwise, but at a slower speed than the carrier. As a result, the front pinions turn clockwise and drive the front sun gear and inner race of the low 1-way clutch counterclockwise faster than the outer race of the low 1-way clutch. This causes the low 1-way clutch to overrun, making the low-reverse clutch ineffective.

The rear ring gear, attached through lugs to the front carrier, also turns counterclockwise.

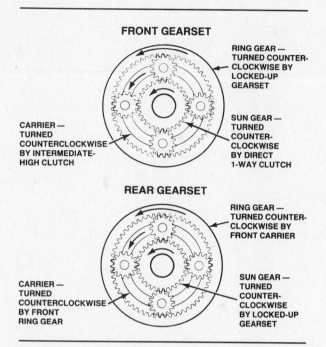

FRONT GEARSET

RING GEAR —
TURNED COUNTER-
CLOCKWISE BY
LOCKED-UP
GEARSET

CARRIER —
TURNED
COUNTERCLOCKWISE
BY INTERMEDIATE-
HIGH CLUTCH

SUN GEAR —
TURNED
COUNTER-
CLOCKWISE
BY DIRECT
1-WAY CLUTCH

REAR GEARSET

RING GEAR —
TURNED COUNTER-
CLOCKWISE BY
FRONT CARRIER

CARRIER —
TURNED
COUNTERCLOCKWISE
BY FRONT
RING GEAR

SUN GEAR —
TURNED
COUNTER-
CLOCKWISE
BY LOCKED-UP
GEARSET

Figure 13-11. AXOD power flow in Overdrive and Drive third.

The rear ring gear drives the rear pinions counterclockwise around the stationary rear sun gear. The pinions turn the rear carrier and front ring gear assembly counterclockwise in gear reduction. The rear carrier transmits power to the final drive gearset and differential assembly as described earlier.

Gear reduction in second gear is produced by the rear gearset of the transaxle. The input member is the front carrier and rear ring gear assembly. The held member is the rear sun gear, and the output member is the rear carrier. The input device is the intermediate-high clutch, and the holding device is the low-intermediate band.

Drive and Overdrive Third

As the vehicle continues to accelerate in Overdrive or Drive, the hydraulic system produces an automatic upshift to third gear by applying the direct clutch and releasing the low-intermediate band, figure 13-11. The direct clutch transmits engine power from the driven sprocket through the direct 1-way clutch to the front sun gear. The intermediate-high clutch remains applied to transmit engine power from the driven sprocket to the front carrier. The low-reverse clutch also remains applied, but is ineffective because the low 1-way clutch overruns.

Because both the front sun gear and the front carrier are turning counterclockwise at the same speed, the front planetary gearset locks

up and turns as a unit in direct drive. Since the front ring gear is in common with the rear carrier, the rear carrier turns counterclockwise and transmits power to the final drive gearset and differential assembly as described earlier.

Remember that the rear ring gear is in common with the front carrier, which means that two elements of the rear gearset (the carrier and ring gear) are also turning at the same speed. As a result the rear gearset is locked up in third gear, although only the carrier is transferring power to the final drive.

As long as the car is accelerating in Overdrive or Drive third, torque is transferred from the engine through the planetary gearset to the halfshafts. If the driver lets up on the throttle, the engine speed drops while the drive wheels keep turning at the same speed. The rear carrier and front ring gear assembly then attempts to walk counterclockwise around front carrier, attempting to turn the front pinions counterclockwise. The front pinions then try to turn the front sun gear, input shell, and overdrive drum clockwise, but the low 1-way clutch locks up to provide engine compression braking.

Direct drive in third gear is produced by the front gearset of the transaxle. The front sun gear and the front carrier are input and held members equally. The output member is the front ring gear and rear carrier assembly. The input devices are the intermediate-high clutch, and the direct clutch together with the direct 1-way clutch. No holding device is necessary because the gearset locks up when any two members turn at the same speed.

In third gear, it is possible to have the torque converter clutch lock up. The solenoid that controls the converter clutch is switched on and off by the Electronic Engine Control (EEC-IV) system. The electronic elements of this system are discussed in Chapter 8. The system hydraulic controls are covered later in this chapter.

Third gear power flow is the same whether the gear selector is in the Overdrive or Drive range. The only difference is that in the Drive range, the hydraulic system prevents an upshift to fourth gear.

Overdrive Fourth

As the vehicle continues to accelerate in Overdrive, the hydraulic system produces an automatic upshift to fourth gear by releasing the low-reverse clutch and applying the overdrive band, figure 13-12. The overdrive band holds the front sun gear, and thus the inner race of the direct 1-way clutch, stationary. The intermediate-high clutch remains applied, and transmits engine power from the driven

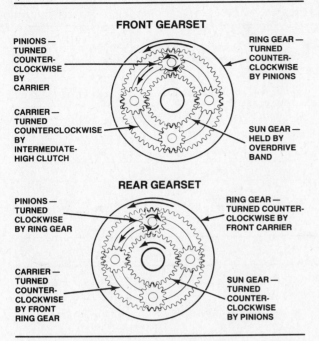

Figure 13-12. AXOD power flow in Overdrive fourth.

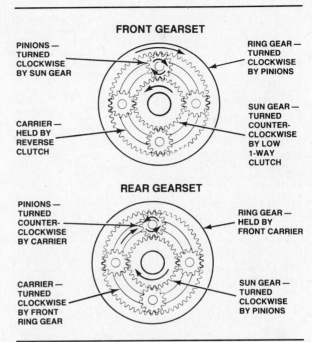

Figure 13-13. AXOD power flow in Reverse.

sprocket to the front carrier. The direct clutch also remains applied, driving the outer race of the direct 1-way clutch, but the 1-way clutch overruns so the direct clutch is ineffective.

The front carrier rotates counterclockwise and drives the front pinions counterclockwise around the stationary front sun gear. The pinions drive the front ring gear and rear carrier assembly counterclockwise in overdrive. The rear carrier transmits power to the final drive gearset and differential assembly as described earlier.

The front carrier turns the rear ring gear counterclockwise, but at a slower speed than the rear carrier is turning. As a result, the rear pinions walk clockwise around the inside of the rear ring gear, and drive the rear sun gear counterclockwise. All of these components simply freewheel at this time.

Overdrive in fourth gear is produced by the front gearset. The input member is the front carrier, the held member is the front sun gear, and the output member is the front ring gear and rear carrier assembly. As in third gear, the converter clutch may or may not be applied, depending on driving conditions.

Manual Low

The power flow in manual low is the same as in Overdrive and Drive low, figure 13-9, but the hydraulic system prevents an upshift to second

gear. The main difference between Overdrive or Drive low, and manual low, is the use of one apply device.

In Overdrive and Drive low, the low-reverse clutch is applied and transmits power to the front sun gear through the low 1-way clutch. However, the low 1-way clutch overruns when the vehicle coasts, allowing the front sun gear to freewheel. In manual low, the direct clutch is applied to activate the direct 1-way clutch. The direct 1-way clutch holds the front sun gear when the vehicle coasts to provide engine compression braking.

Reverse

When the gear selector is moved to Reverse, the low-reverse and reverse clutches are applied. The reverse clutch holds the front carrier and rear ring gear stationary, figure 13-13. The low-reverse clutch drives the front sun gear counterclockwise through the low 1-way clutch.

As the front sun gear rotates counterclockwise, it turns the front pinions clockwise. Because the front carrier is stationary, the front pinions drive the front ring gear clockwise. The front ring gear and rear carrier are in common, so both turn clockwise. The rear carrier transmits power to the final drive gearset and differential assembly as described earlier, the only difference being that the rotation of the planetary sun gear, pinion gears, and carrier/differential case is reversed.

GEAR RANGES	CLUTCHES						BANDS	
	Low 1-Way (Sprag)	Direct 1-Way (Roller)	Inter-mediate High	Direct	Low-Reverse (Forward)	Reverse	Low-Inter-mediate	Over-drive
Overdrive Low	•				•		•	
Overdrive Second			•		*		•	
Overdrive Third		•	•	•	*			
Overdrive Fourth			•	*				•
Drive Low	•				•		•	
Drive Second			•		*		•	
Drive Third		•	•	•	*			
Manual Low	•	•		•	•		•	
Reverse	•				•	•		

★ Applied, but ineffective due to an overrunning 1-way clutch.

Figure 13-14. AXOD clutch and band application chart.

The rear ring gear is held by the front carrier. Therefore, as the rear carrier turns clockwise, the rear pinions walk around the inside of the rear ring gear and turn counterclockwise. The pinions then drive the rear sun gear clockwise. All of these components simply freewheel at this time.

Gear reduction in Reverse is produced by the front planetary gearset. The input member is the front sun gear, the held member is the front carrier, and the output member is the front ring gear and rear carrier assembly. The input device is the low-reverse clutch together with the low 1-way clutch. The holding device is the reverse clutch.

AXOD Clutch and Band Applications

Before we move on to a close look at the AXOD hydraulic system, we should review the apply devices that are used in each of the gear ranges. Figure 13-14 is a clutch and band application chart for the AXOD. Also remember these facts about the apply devices in this transaxle:

• The low-reverse clutch, low 1-way clutch, direct clutch, direct 1-way clutch, and intermediate-high clutch are input devices.
• The low 1-way clutch is applied by the low-reverse clutch in low, second, third, and Reverse gears. It is effective only in Reverse and low; it overruns in second and third.
• The direct 1-way clutch is applied by the direct clutch in third and fourth gears, and in manual low. It is effective only in third and manual low; it overruns in fourth.

• The intermediate-high clutch is applied in second, third, and fourth gears.
• The reverse clutch is applied in Reverse.
• The low-intermediate band is applied in low and second gears.
• The overdrive band is applied in fourth gear.

AXOD HYDRAULIC SYSTEM

The hydraulic system of the AXOD transaxle controls shifting under varying vehicle loads and speeds, as well as in response to manual gear selection by the driver. Figure 13-15 is a diagram of the complete AXOD hydraulic system. We will begin our study of this system with the pump.

Hydraulic Pump

In the AXOD, fluid is stored in the sump at the bottom of the transaxle, and in an upper fluid reservoir located in the lower section of the valve body cover. The upper reservoir controls the oil level in the sump based on fluid temperature. As the fluid heats up and expands, a thermostatic element increases the amount of fluid retained in the upper reservoir. The oil pump draws fluid from the sump at the bottom of the transaxle through a filter attached to the case.

The AXOD uses a variable-displacement vane-type pump, figure 13-16, located on the driver's side of the transaxle. The pump bore ring, pivot pin, rotor, and vanes are installed in the pump housing, which is bolted to the valve body. The bore ring pivots on a pin, and when

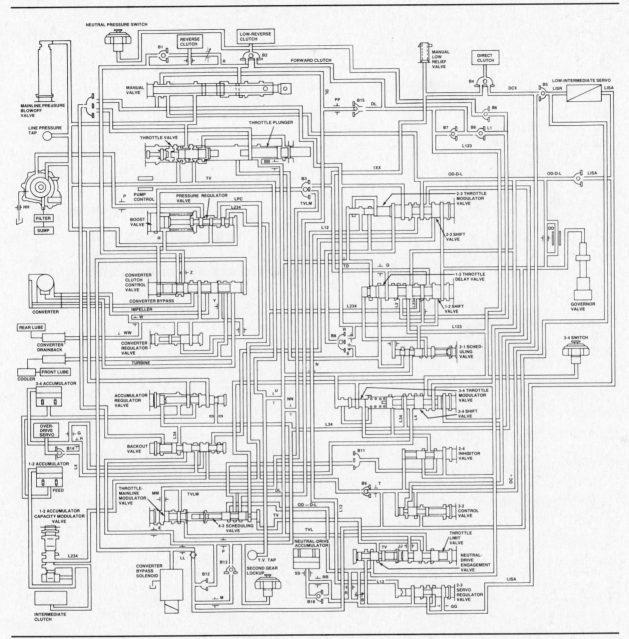

Figure 13-15. The AXOD hydraulic system.

pressure is low in the hydraulic system, the pump priming spring tends to push the ring toward its maximum output position. When pressure in the hydraulic system rises, the pressure regulator valve sends oil through the decrease passage to move the bore ring toward its minimum output position. In this manner, the oil pump output is controlled so that it is proportional to demand.

Whenever the engine is running, the converter cover turns the pump drive shaft and rotor. The rotation of the pump rotor and vanes creates suction that draws fluid through a filter from the sump at the bottom of the transaxle case. Oil from the pump flows to the pressure regulator valve, throttle valve, mainline pressure blow-off valve, converter clutch control valve, and accumulator regulator valve.

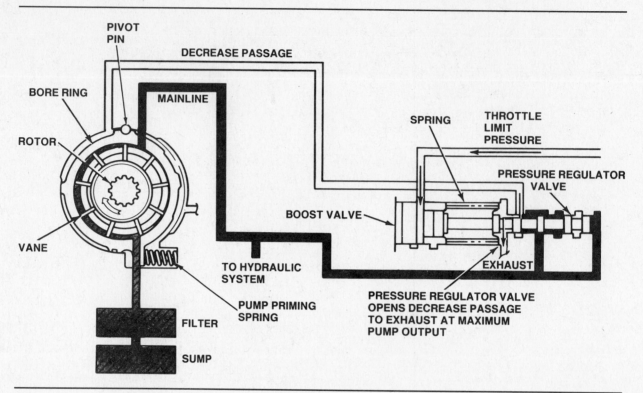

Figure 13-16. The AXOD oil pump and pressure regulator valve.

Hydraulic Valves

The AXOD hydraulic system contains the following valves:

1. Pressure regulator valve
2. Mainline pressure blow off valve
3. Manual valve
4. Throttle valve
5. Throttle limit valve
6. Throttle-mainline modulator valve
7. Governor valve
8. Neutral-Drive engagement valve
9. 1-2 shift valve
10. 1-2 throttle delay valve
11. 2-1 scheduling valve
12. 1-2 accumulator capacity modulator valve
13. 2-3 shift valve
14. 2-3 throttle modulator valve
15. 2-3 servo regulator valve
16. 3-2 control valve
17. Backout valve
18. 3-4 shift valve
19. 3-4 throttle modulator valve
20. 4-3 scheduling valve
21. 2-4 inhibitor valve
22. Accumulator regulator valve
23. Converter regulator valve
24. Converter clutch control valve
25. Manual low relief valve.

Valve body and ball-check valves

In the AXOD, all of the valves except the governor valve, the 1-2 accumulator capacity modulator valve, the mainline pressure blow off valve, and the manual low relief valve are located in the valve body, figure 13-17. The governor valve installs in an opening on top of the transaxle above the differential case. The remaining three valves are located in the oil pump housing that bolts to the valve body, figures 13-18 and 13-19. The pump housing also has three pressure switches on it that are used by the EEC-IV system in controlling the converter clutch.

The passages that carry oil between the valves are cast into the main valve body, the oil pump housing, and the chain cover. Figure 13-19 shows the location of the check balls in both the oil pump housing and the valve body. There are nine check balls in the pump housing, and six in the valve body.

A pair of converter drainback valves, located in the valve body, prevent fluid from draining out of the converter when the engine is not running. Once the engine is started, these valves direct oil from the converter to the rear lubrication circuit, and to the transaxle cooler and front lubrication circuit.

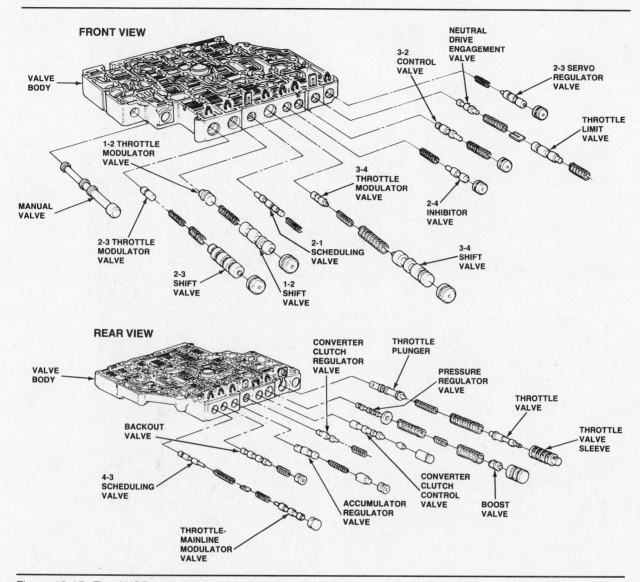

Figure 13-17. The AXOD valve body and hydraulic control valves.

Pressure regulator valve

The pump delivers fluid through two passages to the pressure regulator valve, figure 13-16. One passage is at the end of the valve, the other is at the center of the valve. Hydraulic pressure at the end of the valve acts to move the valve against spring force. Hydraulic pressure at the center of the valve enters between two lands of equal diameter, and does not move the valve in either direction.

When the engine is started and mainline pressure begins to build, pressure at the end of the pressure regulator valve pushes the valve against spring force. When the valve moves far enough, it opens a port that leads from the center chamber into the decrease passage. Mainline pressure then runs into the decrease passage where it moves the pump bore ring against pump priming spring force to reduce pump output. The pressure regulator valve continues to move until hydraulic pressure and spring force reach a balance and provide a constant mainline pressure.

When the transaxle shifts into gear, or from one gear to another, increased fluid flow demands cause mainline pressure to drop slightly. This drop in mainline pressure allows the spring to move the pressure regulator valve so that mainline pressure feed is cut off and the pump decrease circuit exhausts. The pump priming spring then moves the bore ring to increase pump output and bring mainline pressure back to the correct level.

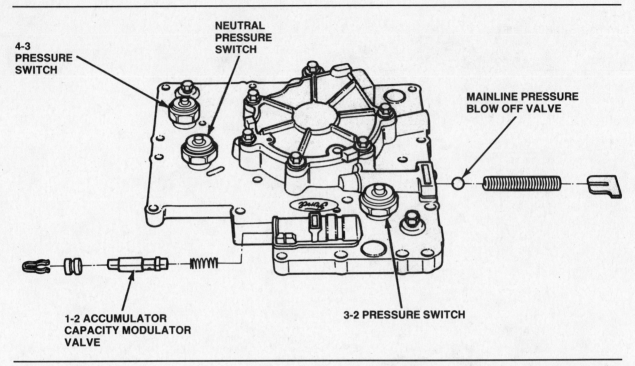

Figure 13-18. The AXOD oil pump housing bolts to the valve body and contains several valves and switches.

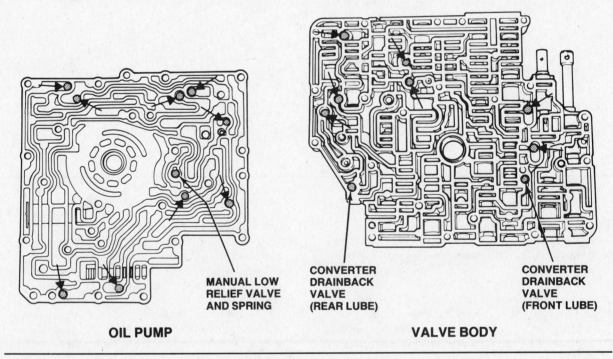

Figure 13-19. AXOD oil pump housing and valve body check ball locations.

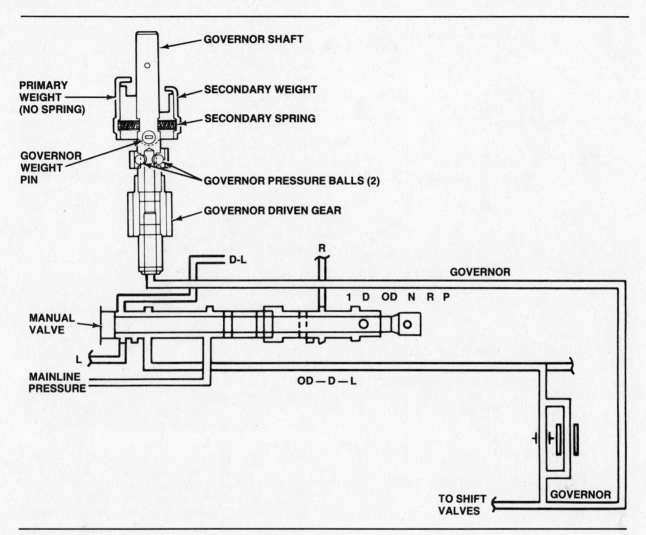

Figure 13-20. The AXOD manual valve and governor valve.

In second, third, and fourth gears, when less mainline pressure is required to hold the apply devices securely, the manual valve routes pressure to the end of the pressure regulator valve opposite spring force. This moves the valve to reduce mainline pressure, and is similar in function to the cutback valve in earlier Ford automatics.

The boost valve at the spring end of the pressure regulator valve raises mainline pressure by increasing the force of the spring acting on the pressure regulator valve. Throttle limit pressure acts on the boost valve, which moves to compress the spring and increase its force. The spring then moves the pressure regulator valve to increase mainline pressure.

Mainline pressure blow off valve

The mainline pressure blow off valve limits mainline pressure to a maximum of 285 to 315 psi (1965 to 2172 kPa). If mainline pressure exceeds this value, the mainline pressure blow off valve unseats to exhaust excess oil to the sump. Once mainline pressure drops below this level, spring force reseats the valve.

Manual valve

The manual valve, figure 13-20, is a directional valve that is moved by a mechanical linkage from the gear selector lever. The manual valve receives fluid from the pump and directs it to apply devices and other valves in the hydraulic system in order to provide both automatic and manual upshifts and downshifts.

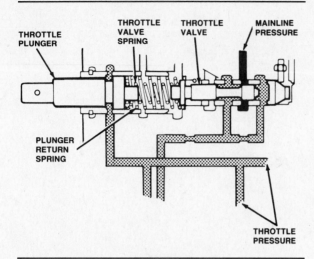

Figure 13-21. A typical throttle valve at closed throttle.

Four passages may be charged with mainline pressure from the manual valve. When the gear selector is in Park or Neutral, the manual valve does not direct pressure to any passages. Ford refers to mainline pressure from the manual valve by the name of the passage it flows through.

The OD-D-L passage is charged in Overdrive, Drive, and manual low; it routes pressure to the:

- 1-2 shift valve
- 2-3 shift valve
- 3-4 shift valve
- 2-4 inhibitor valve
- 3-2 control valve
- 2-3 servo regulator valve
- 4-3 scheduling valve
- Neutral-Drive engagement valve
- Governor valve
- Accumulator regulator valve.

The D-L passage is charged in Drive and manual low; it routes pressure to the:

- Low-reverse clutch
- 1-2 shift valve
- 3-4 shift valve
- 3-4 throttle modulator valve
- 4-3 scheduling valve.

The L passage is charged in manual low; it routes pressure to the:

- Backout valve
- 2-3 shift valve
- 2-3 throttle modulator valve
- 1-2 shift valve
- 2-1 scheduling valve
- downshifted 1-2 shift valve.

The R passage is charged in Reverse; it routes pressure to the:

- Low-reverse clutch
- Reverse clutch
- Boost valve
- Converter clutch control valve.

Throttle valve

The AXOD transaxle develops throttle pressure using a throttle valve operated by a mechanical linkage. Figure 13-21 shows a typical throttle valve of this type. Throttle pressure indicates engine torque and helps time transaxle upshifts and downshifts. Without throttle pressure, the transaxle would upshift and downshift at exactly the same speed for very different throttle openings and load conditions.

The throttle valve is controlled by opposition between hydraulic pressure and spring force. The hydraulic pressure is mainline pressure that is routed to the end of the valve opposite the spring. The amount of spring force acting on the opposite end of the valve is controlled by the throttle plunger. The plunger is linked to the accelerator by a cable so that its position is determined by throttle opening.

At closed throttle, figure 13-21, the throttle plunger exerts no force on the throttle valve spring, and the throttle valve is positioned to block the entry of mainline pressure. As the accelerator is depressed, the linkage moves the throttle plunger to compress the throttle valve spring. This moves the throttle valve and allows mainline pressure into the valley between the two lands of the throttle valve where it becomes throttle pressure. The throttle pressure that flows out of the valley and into the throttle pressure passages is regulated by the position of the throttle valve.

At wide-open throttle, figure 13-22, the throttle plunger is pushed into the throttle valve as far as it can be. The plunger completely compresses the spring, contacts the throttle valve spool, and bottoms the valve to allow mainline pressure into the throttle pressure passage. At the same time, the plunger opens the kickdown passage to route kickdown pressure to the shift valves to force a downshift.

If the link between the accelerator and the throttle plunger is broken or disconnected, a failsafe spring moves the plunger to the wide-open-throttle position to create maximum throttle pressure. Although the transaxle will not upshift and downshift at the appropriate times, this protects the clutches and bands from slipping and burning.

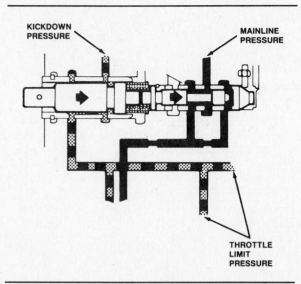

Figure 13-22. A typical throttle valve at wide open throttle.

Throttle pressure is routed from the throttle valve to the:
- Throttle-mainline modulator valve
- 1-2 throttle delay valve
- 2-3 throttle modulator valve
- 3-4 throttle modulator valve
- 2-1 scheduling valve
- 4-3 scheduling valve
- Throttle limit valve.

Throttle limit valve
The throttle limit valve in the AXOD is located in the same bore as the Neutral-Drive engagement valve, figure 13-15. The throttle limit valve restricts throttle pressure to a maximum of 82 to 87 psi (565 to 600 kPa). As long as throttle pressure is below that level, fluid flows through the valve and into the throttle limit passages. Throttle pressure that has passed through the throttle limit valve is called throttle limit pressure.

Throttle limit pressure is routed to the pressure regulator valve, the shift valves, and other valves in the hydraulic system unless throttle pressure falls below 9 psi (62 kPa). Below this pressure, spring force closes the throttle limit valve and cuts off throttle pressure flow to the various valves.

Throttle-mainline modulator valve
The throttle-mainline modulator valve, figure 13-15, modifies throttle limit pressure so that mainline pressure can closely match engine torque and transaxle capacity requirements. The throttle-mainline modulator valve is controlled by spring force at one end and throttle limit pressure at the other end. It sends modified throttle limit pressure to the pressure regulator valve and boost valve to raise and lower mainline pressure based on throttle opening.

Governor valve
We have been discussing the development and control of throttle pressure, which is one of the two auxiliary pressures that work on the shift valves. Before we get to the shift valve operation, we must explain the governor valve (or governor), and how it creates governor pressure. Governor pressure provides a road speed signal to the hydraulic system that causes automatic upshifts to occur as road speed increases, and permits automatic downshifts with decreased road speed.

The AXOD governor, figure 13-20, mounts in an opening on the top of the transaxle, and is driven by a gear on the differential case. The AXOD governor uses check balls to regulate governor pressure. The check-ball governor consists of two check balls, a governor shaft, a primary weight, a secondary weight, a governor weight pin, and a secondary spring. Mainline pressure enters the governor through a fluid passage in the center of the shaft. Two exhaust ports are drilled into the fluid passage, and governor pressure is determined by the amount of fluid allowed to flow out of these ports.

The exhaust port openings are controlled by check balls that seat in pockets formed at the ends of the ports. As the governor rotates, centrifugal force causes the primary weight to act on one check ball, and the secondary weight to act on the other. Each weight acts on the check ball located on the opposite side of the governor shaft. The secondary weight is assisted in its job by the secondary spring, which helps regulate pressure more smoothly as vehicle speed increases.

When the vehicle is stopped and the governor is not turning, there is no centrifugal force and all of the oil is exhausted from the governor valve. Governor pressure is zero at this time. As the governor begins to turn, centrifugal force moves the weights outward. This force is relayed to the check balls, partially seating them in their pockets to restrict fluid flow from the exhaust ports. As governor speed increases, the forces on the check balls become greater and less fluid is allowed to escape; this causes governor pressure to increase. When the governor speed remains constant, the forces acting on the check balls are also constant, and governor pressure stabilizes.

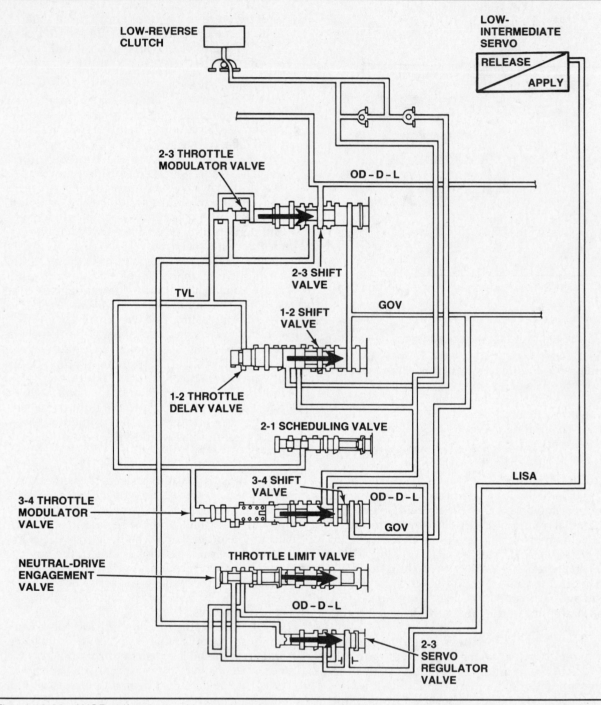

Figure 13-23. AXOD shift valve positions in first gear.

At higher vehicle speeds, centrifugal force becomes great enough to hold both check balls fully seated so that no fluid escapes from the governor. At this point, governor pressure equals mainline pressure.

Governor pressure is sent to the shift valves to oppose throttle limit pressure and spring force, and upshift the valves at the appropriate road speeds. Governor pressure is also routed to the 2-3 control valve.

Neutral-Drive engagement valve
The Neutral-Drive engagement valve, figure 13-15, works with the Neutral-Drive accumulator to control the flow of mainline pressure that applies the low-intermediate band when a shift is made from Neutral to Drive. By routing some

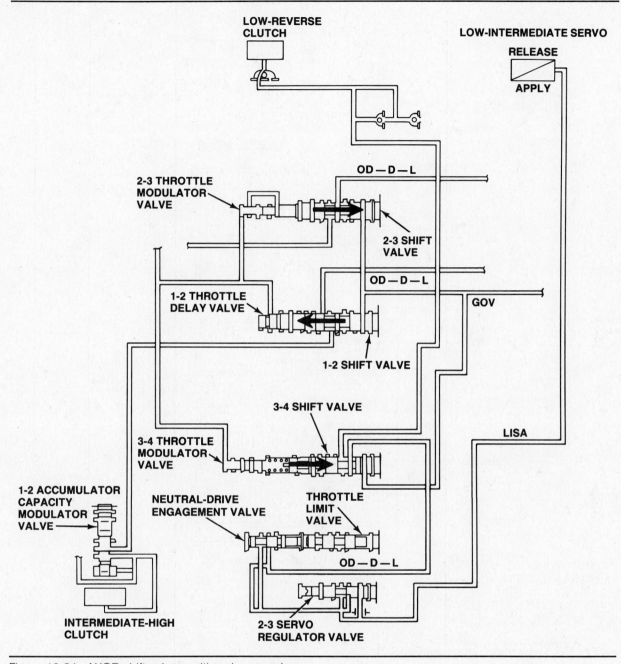

Figure 13-24. AXOD shift valve positions in second gear.

of this pressure through restricting orifices, the Neutral-Drive engagement valve smoothes application of the low-intermediate band.

The Neutral-Drive engagement valve is controlled by spring force opposing throttle pressure. The higher the throttle pressure, the more firmly the valve causes the band to apply.

1-2 shift valve and 1-2 throttle delay valve
The 1-2 shift valve, figure 13-23, controls upshifts from low to second, and downshifts from

second to low. The assembly includes the 1-2 shift valve and the 1-2 throttle delay valve. The valve is controlled by opposition between governor pressure on one end, and combined throttle limit pressure and spring force on the other end. Throttle limit pressure acts on the 1-2 throttle delay valve to increase spring force and oppose an upshift.

In low gear, OD-D-L oil from the manual valve flows to the 1-2 shift valve where it is

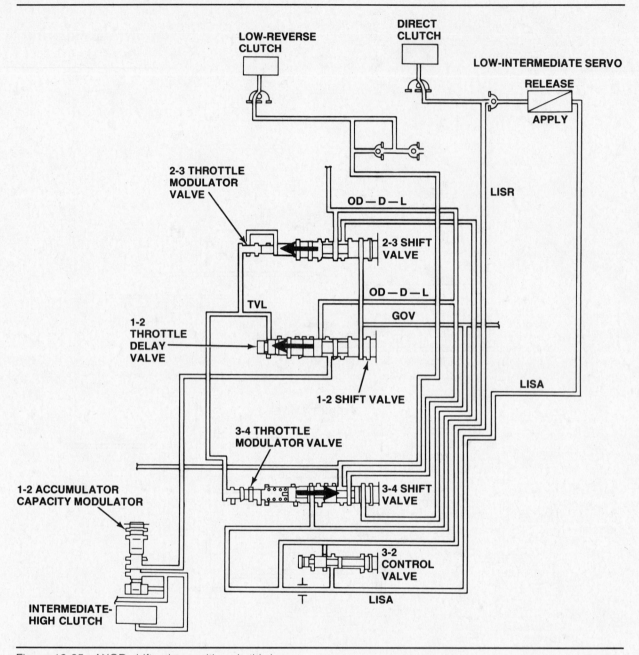

Figure 13-25. AXOD shift valve positions in third gear.

blocked by the downshifted valve. When vehicle speed increases enough for governor pressure to overcome throttle limit pressure and spring force, the valve upshifts. Mainline pressure then flows through the upshifted valve to the intermediate-high clutch apply area, and second gear is engaged, figure 13-24.

When the accelerator pedal is depressed far enough at vehicle speeds below 25 mph (40 kph), the throttle valve plunger directs kickdown pressure into the throttle limit passages. This acts on the throttle delay valve to increase

spring force and move the 1-2 shift valve back to its downshifted position. Intermediate-high clutch apply pressure is then exhausted through an orifice, causing a 2-1 downshift.

2-1 scheduling valve
The 2-1 scheduling valve, figure 13-23, times the 2-1 downshift when the gear selector is moved to manual low from a higher gear. In manual low, mainline pressure (L oil) is directed to the 2-1 scheduling valve, which regulates it to create a scheduling pressure that is routed to the 1-2 throttle delay valve. This pressure acts on

the throttle delay valve to help throttle limit pressure and spring force overcome governor and mainline pressure to downshift the 1-2 shift valve.

1-2 accumulator capacity modulator valve

The 1-2 accumulator capacity modulator valve, figure 13-24, works together with the 1-2 accumulator to smooth application of the intermediate-high clutch. The valve is normally held closed by spring force. When the 1-2 shift occurs, mainline pressure from the 1-2 shift valve opens the 1-2 accumulator capacity modulator valve. The valve then regulates the rate of pressure increase in the circuit, and routes the pressure to the intermediate-high clutch and the 1-2 accumulator. After the 1-2 shift is completed, the 1-2 accumulator capacity modulator valve remains open until the transaxle shifts to either low or Reverse.

2-3 shift valve and 2-3 throttle modulator valve

The 2-3 shift valve assembly, figure 13-24, controls upshifts from second to third, and downshifts from third to second. The assembly includes the 2-3 shift valve and the 2-3 throttle modulator valve. Governor pressure is applied to the 2-3 shift valve in all forward gear ranges, and acts on the valve to produce an upshift. The 2-3 throttle modulator valve works with spring force to oppose upshifts, and can also produce a modulated throttle pressure that is applied to the 2-3 and 1-2 shift valves.

Throttle limit pressure is applied to the end of the 2-3 throttle modulator valve to oppose upshifts. If throttle limit pressure moves the valve far enough, a passage to the 2-3 throttle modulator circuit opens. This circuit routes modulated throttle pressure around the 2-3 throttle modulator valve to the spring chamber of the 2-3 shift valve where it helps oppose an upshift. At the same time, modulated throttle pressure is routed to the 1-2 shift valve where it works to help force a further downshift if conditions warrant.

In the forward gear ranges, OD-D-L oil is sent from the manual valve to the 2-3 shift valve where it is blocked by the downshifted valve. When governor pressure overcomes spring force and throttle limit pressure, the 2-3 shift valve upshifts. OD-D-L oil then flows through the upshifted valve to the direct clutch apply area and the release side of the low-intermediate servo, shifting the transaxle into third gear, figure 13-25. Mainline pressure that flows through the 2-3 shift valve is also routed to the downshifted 3-4 shift valve, the accumulator regulator valve, and the backout valve.

When the accelerator pedal is depressed far enough at vehicle speeds below 58 mph (93

kph), the throttle valve plunger directs kickdown pressure into the throttle limit passages. This moves the 2-3 shift valve back to its downshifted position and puts the transaxle in second gear. The kickdown pressure also flows to the 1-2 throttle delay valve where it will cause a further downshift if conditions warrant.

2-3 servo regulator valve

The 2-3 servo regulator valve, figure 13-15, regulates low-intermediate servo apply pressure, and affects the low-intermediate servo release. In low and second gears, OD-D-L oil from the downshifted 2-3 shift valve flows through the Neutral-Drive engagement valve to the 2-3 servo regulator valve. This moves the 2-3 servo regulator valve to the right, and the pressure flows through to the apply side of the servo.

3-2 control valve

The 3-2 control valve, figure 13-25, is located in the direct clutch apply circuit, and regulates the rate at which 2-3 upshifts and 3-2 downshifts occur in relation to road speed. Spring force holds the 3-2 control valve closed when the car is stopped, and governor pressure gradually opens it as road speed increases.

Backout valve

The backout valve, figure 13-15, is in the intermediate-high clutch, direct clutch, and overdrive servo apply circuits where it controls the orifices used for the 2-3 and 3-4 upshifts. The backout valve also delays 3-2 closed-throttle downshifts. The backout valve is controlled by opposition between spring force and mainline pressure at one end, and throttle limit pressure at the other.

During closed-throttle 2-3 and 3-4 upshifts, spring force and mainline pressure position the backout valve so that it blocks direct clutch or overdrive servo apply pressure from one of the two orifices in the apply circuit. Therefore, the apply pressure must flow slowly through a single orifice, delaying the clutch or band application. During upshifts at larger throttle openings, throttle limit pressure moves the backout valve so both orifices are open and the upshift can take place more quickly.

During closed-throttle coastdown, the backout valve inhibits the 3-2 downshift so that a 3-1 downshift is made instead. With the throttle closed, the throttle plunger opens a passage that exhausts throttle limit pressure in the backout circuit. Spring force and mainline pressure then position the backout valve so that intermediate-high clutch apply oil from the 1-2 shift valve is routed into the direct clutch apply circuit.

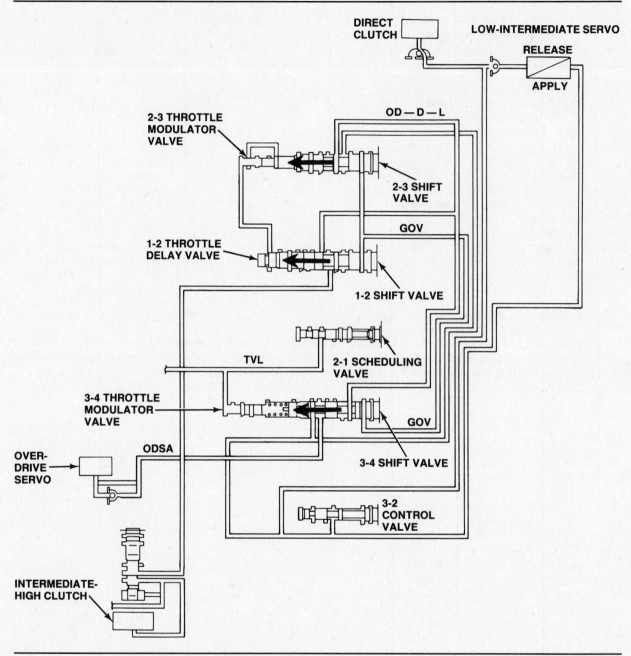

Figure 13-26. AXOD shift valve positions in fourth gear.

When the vehicle speed drops below 15 mph (24 kph), 2-3 shift valve spring force overcomes governor pressure to downshift the valve. However, the pressure routed through the backout valve keeps the direct clutch applied, and the transaxle in third gear. When the vehicle speed drops below 10 mph (16 kph), the 1-2 shift valve downshifts, intermediate-high clutch apply oil exhausts, the pressure keeping the transaxle in third is removed, and the transaxle shifts from third to first gear.

3-4 shift valve and 3-4 throttle modulator valve

The 3-4 shift valve assembly, figure 13-25, controls upshifts from third to fourth, and downshifts from fourth to third. The assembly includes the 3-4 shift valve and the 3-4 throttle modulator valve. Governor pressure is applied to the 3-4 shift valve in all forward gear ranges, and acts on the valve to produce an upshift. The 3-4 throttle modulator valve works with spring force to oppose upshifts, and can also produce a modulated throttle pressure that is applied to the 3-4 shift valve.

Throttle limit pressure is applied to the end of the 3-4 throttle modulator valve to oppose upshifts. If throttle limit pressure moves the valve far enough, a passage to the 3-4 throttle modulator circuit opens. This circuit routes modulated throttle pressure around the 3-4 throttle modulator valve to the spring chamber of the 3-4 shift valve where it helps oppose an upshift. In Drive, pressure from the 2-4 inhibitor valve combines with spring force and throttle limit pressure to keep the 3-4 shift valve downshifted at any speed.

In the forward gear ranges, OD-D-L oil is sent from the manual valve to the 3-4 shift valve where it is blocked by the downshifted valve. When governor pressure overcomes spring force and throttle limit pressure, the 3-4 shift valve upshifts, figure 13-26. When this happens, low-reverse clutch apply pressure is exhausted, and OD-D-L pressure flows through the 3-4 shift valve to the apply side of the overdrive servo.

The 3-4 accumulator cushions application of the overdrive band by absorbing some of the overdrive servo apply oil. The firmness of the shift is controlled by the accumulator regulator valve.

4-3 scheduling valve

If throttle pressure becomes greater than 62 psi (427 kPa) when the transaxle is in fourth gear, throttle limit pressure acting on the 4-3 scheduling valve, figure 13-15, will move the valve to allow OD-D-L to flow through the valve to the 3-4 shift valve. The OD-D-L pressure combines with throttle limit pressure and spring force to downshift the 3-4 shift valve and put the transaxle into third gear.

If the driver manually downshifts the transaxle from Ⓓ to D in fourth gear, D-L oil is directed through the 4-3 scheduling valve and the 2-4 inhibitor valve to force a 4-3 downshift. As long as the transaxle remains in Drive, D-L oil will prevent the 3-4 shift valve from upshifting.

2-4 inhibitor valve

During a normal sequence of transaxle upshifts, the 2-4 inhibitor valve prevents a 2-4 upshift by directing line pressure to hold the 3-4 shift valve in the downshifted position whenever the direct clutch is not applied. As mentioned above, the 2-4 inhibitor valve also works with the 4-3 scheduling valve to force a 4-3 downshift when the driver manually downshifts the transaxle from Ⓓ to D in fourth gear.

Accumulator regulator valve

The 1-2 and 3-4 accumulators, figure 13-15, are controlled by spring force and accumulator pressure opposing band or clutch apply pressure. Accumulator pressure is controlled by the accumulator regulator valve, which is controlled by a balance between spring force and throttle limit pressure.

When mainline pressure is fed to the accumulator regulator valve, the valve regulates the pressure and sends it to the 1-2 and 3-4 accumulators to help the accumulator springs cushion the shifts. The 1-2 accumulator cushions the application of the intermediate-high clutch, and the 3-4 accumulator cushions the application of the overdrive band.

Converter regulator valve

Fluid from the pressure regulator valve flows to the converter regulator valve, figure 13-27, which limits converter pressure to a maximum of 100 psi (689 kPa). The converter regulator valve is a check ball that is held seated by a spring. Whenever mainline pressure exceeds the limit, the valve unseats and vents excess pressure to the sump. From the converter regulator valve, the fluid flows to the converter clutch control valve.

Converter clutch control valve

The converter clutch control valve receives fluid from the converter regulator valve, and directs it into either the release or apply passage of the converter clutch. The valve is controlled by opposition between pressure from the pump at one end, and pressure from the converter clutch solenoid at the other.

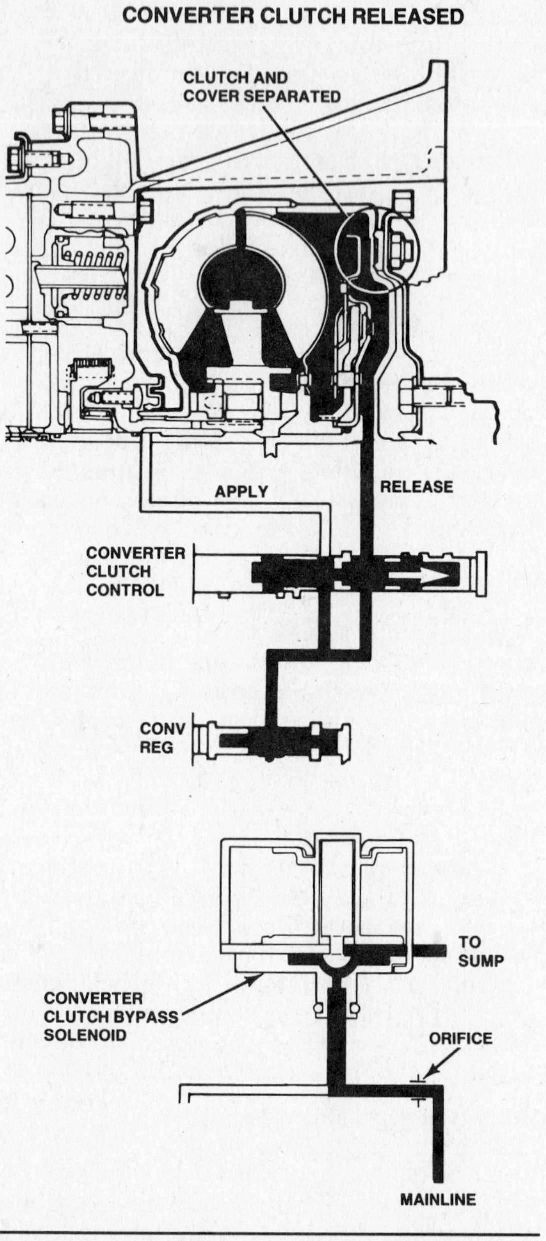

CONVERTER CLUTCH RELEASED

CONVERTER CLUTCH APPLIED

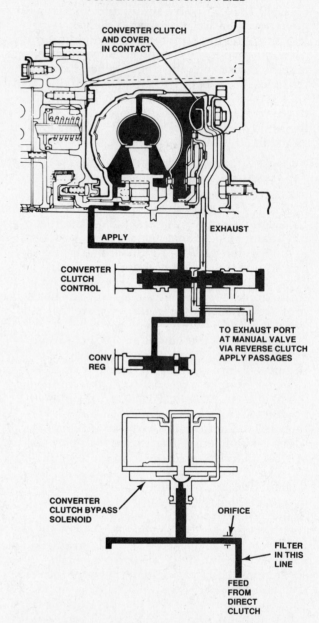

Figure 13-27. The AXOD converter regulator valve limits converter pressure to a maximum of 100 psi (689 kPa).

Figure 13-28. The converter clutch control valve directs fluid to apply and release the torque converter clutch.

When the converter clutch solenoid is not energized, figure 13-27, it exhausts the mainline pressure routed to it. Pump pressure then moves the converter control valve so that the fluid flows through the release passage between the converter clutch and the converter cover, keeping them separated.

When the converter clutch solenoid is energized, figure 13-28, it blocks the escape of mainline pressure. This pressure then moves the

converter clutch control valve against pump pressure. This allows regulated converter pressure to pass through the apply passage to force the converter clutch against the torque converter cover.

The mainline oil routed to the converter clutch solenoid comes from the upshifted 2-3 and 3-4 shift valves. As a result, the converter clutch can only apply in third and fourth gears.

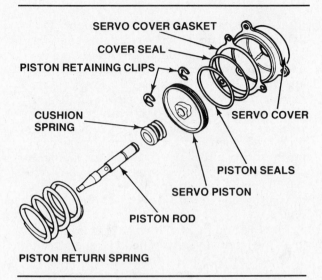

Figure 13-29. The AXOD low-intermediate servo.

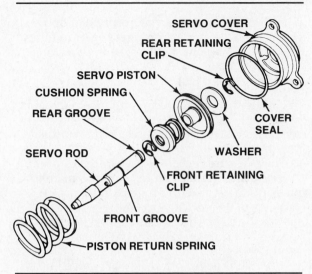

Figure 13-30. The AXOD overdrive servo.

The electronic controls that energize and de-energize the converter clutch solenoid are described in Chapter 8.

MULTIPLE-DISC CLUTCHES AND SERVOS

The preceding paragraphs examined the individual valves in the hydraulic system. Before we trace the complete hydraulic system operation in each driving range, we will take a quick look at the multiple-disc clutches and servos.

Multiple-Disc Clutches

The AXOD transaxle has four multiple-disc clutches. Each clutch is applied hydraulically, and released by several small coil springs when oil is exhausted from the clutch piston cavity.

The low-reverse clutch is applied in low, second, and third gears by oil from the downshifted 1-2, and 3-4 shift valves. In Reverse, the low-reverse clutch is applied by pressure in the manual valve R passage. A check ball prevents pressure from the shift valves from exhausting through the R passage in the forward gears, and R oil from exhausting to the shift valves in Reverse.

The direct clutch is applied in third and fourth gears by mainline pressure from the upshifted 2-3 shift valve. The direct clutch is applied in manual low by L oil through the downshifted 1-2 shift valve. A check ball prevents 2-3 shift valve pressure from exhausting through the 1-2 shift valve in third and fourth, and L oil from exhausting through the 2-3 shift valve in manual low.

The intermediate-high clutch is applied in second, third, and fourth gears by mainline pressure from the upshifted 1-2 shift valve.

The reverse clutch is applied in Reverse by mainline pressure in the manual valve R passage.

Servos

The AXOD has two servos, one to apply the low-intermediate band, figure 13-29, and one to apply the overdrive band, figure 13-30. Both servo assemblies include a servo cover, a piston, an apply rod, and a return spring. Hydraulic pressure moves the piston against spring force, and the apply rod mechanically applies the band. When there is no hydraulic pressure in the piston cavity, the spring releases the servo and band.

In the low-intermediate servo, spring release force is assisted by hydraulic pressure. This servo also acts as an accumulator for the 2-3 upshift. Direct clutch apply oil is routed into the release side of the low-intermediate servo, and the time it takes this pressure and spring force to overcome servo apply pressure and release the servo allows a delay of the direct clutch application.

Both AXOD bands can be adjusted externally by substituting different length apply rods.

HYDRAULIC SYSTEM SUMMARY

We have examined the individual valves, clutches, and servos in the hydraulic system and discussed their functions. The following paragraphs summarize their combined operations in the various gear ranges. Refer to figure 13-15 to trace the hydraulic system operation.

Park and Neutral

In Park and Neutral, pump pressure is directed to the pressure regulator valve, throttle valve, converter clutch control valve, and accumulator regulator valve.

The pressure regulator valve receives pump pressure and regulates it to provide a mainline pressure that varies in proportion to throttle pressure. The regulated mainline pressure is sent to the mainline pressure blow-off valve, converter regulator valve, and converter clutch control valve.

The mainline pressure blow off valve limits pump pressure to a maximum of 285 to 315 psi (1965 to 2172 kPa). The converter regulator valve limits converter pressure to a maximum of 100 psi (689 kPa). Converter pressure feeds the transaxle cooler and lubrication circuits. Pump pressure moves the converter clutch control valve so that converter pressure releases the converter clutch.

The throttle valve produces throttle pressure proportional to throttle position, and routes that pressure to the throttle limit valve. The throttle limit valve limits throttle pressure to a maximum of 82 to 87 psi (565 to 600 kPa), and the throttle-mainline modulator valve modifies the throttle limit pressure. Throttle limit pressure flows to the spring ends of the shift valves. Pressure from the throttle line modulator valve flows to the pressure regulator valve and converter regulator valve.

In Park and Neutral, no clutches or bands are applied, and the converter clutch is released. In Park, a mechanical pawl engages the parking gear attached to the final drive sun gear. This locks the final drive planetary gearset and prevents the car from being moved.

Overdrive and Drive Low

When the transaxle is in Overdrive, Drive, or manual low, mainline pressure is routed through the OD-D-L circuit to the: 1-2 shift valve, 2-3 shift valve, 3-4 shift valve, 2-4 inhibitor valve, 3-2 control valve, 2-3 servo regulator valve, 4-3 scheduling valve, Neutral-Drive engagement valve, governor valve, and accumulator regulator valve.

All of the shift valves are downshifted in low. Pressure flows through the 1-2 and 3-4 shift valves to apply the low-reverse clutch. The oil that flows through the 2-3 shift valve forces the 2-3 servo regulator valve to the right. The 2-3 servo regulator valve then routes oil to the Neutral-Drive accumulator, Neutral-Drive engagement valve, and low-intermediate servo apply circuit where it applies the low-intermediate band. The oil that flows through the 3-4 shift valve goes to the 4-3 scheduling valve, the 1-2 shift valve, and the governor. Governor pressure is then applied to each shift valve opposite spring force. The accumulator regulator valve modifies mainline pressure in relation to throttle opening, and applies that pressure to the spring side of the 1-2 and 3-4 accumulator pistons.

In Drive and manual low, D-L oil is routed from the manual valve to the low-reverse clutch circuit to provide quick low-reverse clutch apply if the selector is moved from Overdrive to Drive in fourth gear. In Drive and manual low, both D-L and OD-D-L oils are routed to the 4-3 scheduling valve so that, regardless of throttle pressure, mainline pressure is directed through the 4-3 scheduling valve and 2-4 inhibit valve to the 3-4 shift valve. This forces the 3-4 shift valve to remain downshifted.

In Overdrive or drive low, the low-reverse clutch and the low-intermediate band are applied.

Overdrive and Drive Second

As governor pressure increases with vehicle speed, it overcomes throttle pressure and spring force, and upshifts the 1-2 shift valve. Mainline pressure flows through the upshifted valve to the: pressure regulator valve, 1-2 accumulator, 1-2 accumulator capacity modulator valve, intermediate-high clutch apply passage.

The application to the pressure regulator valve reduces mainline pressure in second, third, and fourth gears by pushing the valve against spring and boost pressure. The 1-2 accumulator capacity modulator valve controls the 1-2 accumulator to cushion application of the intermediate-high clutch in relation to throttle pressure. Pressure from the shift valve is modulated by the 1-2 accumulator capacity modulator valve, and then directed to apply the intermediate-high clutch. When the intermediate-high clutch is applied, the transaxle is in second gear.

The low-reverse clutch is still applied in second gear but is ineffective because the low 1-way clutch overruns. The intermediate-high clutch is now the input device. The low-intermediate band is still applied.

Overdrive and Drive Third

As governor pressure continues to increase, it overcomes throttle pressure and spring force to upshift the 2-3 shift valve. Low-intermediate servo apply pressure exhausts, and mainline pressure flows through the 2-3 shift valve to the: low-intermediate servo release passage, 3-4 shift valve, accumulator regulator valve, direct clutch apply passage, and converter clutch solenoid.

The 2-3 servo regulator valve regulates the rate at which the low-intermediate servo release. The oil that flows to the 3-4 shift valve is blocked until the next upshift. The pressure at the accumulator regulator valve causes the regulator to change pressure in preparation for the 3-4 shift.

The direct clutch apply pressure flows through the backout valve and two orifices into the direct clutch circuit. The direct clutch circuit applies the direct clutch and also feeds the solenoid circuit, allowing converter clutch application when the solenoid is activated.

With the release of the low-intermediate band and the application of the direct clutch, the transaxle is in third. The intermediate-high clutch is still applied. The application or release of the converter clutch is controlled by the converter clutch solenoid, which is controlled by the EEC-IV system.

Overdrive Fourth

As governor pressure continues to increase, it moves the 3-4 shift valve against throttle pressure and spring force for a 3-4 upshift. The upshifted valve exhausts low-reverse clutch apply oil so the clutch releases, and allows OD-D-L oil to flow to the overdrive servo apply passage. The 3-4 accumulator cushions the application of the overdrive band by absorbing some of the overdrive servo apply oil. The accumulator regulator valve controls the firmness of the shift.

When the overdrive band is applied and the low-reverse clutch is released, the transaxle is in fourth gear. The direct clutch is still applied. The converter clutch may or may not be applied, depending on solenoid operation.

Drive and Overdrive Range Forced Downshifts

When the throttle is opened far enough, the throttle plunger feeds throttle limit pressure into the kickdown circuit. This kickdown pressure is directed to the 1-2 and 2-3 shift valves to force downshifts at higher speeds. Kickdown pressure is not needed at the 4-3 shift valve because the 4-3 scheduling valve will force a 4-3 downshift before the throttle plunger opens the kickdown circuit.

Kickdown pressure and throttle limit pressure act on the 2-3 throttle modulator valve to move the 2-3 shift valve and cause a wide-open throttle downshift at speeds below 58 mph (93 kph). Direct clutch and low-intermediate servo release exhaust are controlled by the 3-2 control valve. Pressure from the 1-2 shift valve flows through the 2-3 servo regulator valve to apply the low-intermediate band.

Kickdown pressure and throttle limit pressure act on the 1-2 throttle delay valve to move the 1-2 shift valve and cause a wide-open throttle downshift at speeds below 25 mph (40 kph). Intermediate-high clutch apply pressure then exhausts.

A kickdown downshift can shift the transaxle from third to second, second to first, or from third to first.

Manual Low

In manual low, the manual valve directs mainline pressure to the OD-D-L, D-L, and L circuits. Mainline pressure in the L circuit flows to the: spring end of the 2-3 shift valve, backout valve, 2-1 scheduling valve, and 1-2 shift valve.

L oil downshifts the 2-3 shift valve, which exhausts oil through the backout valve and causes the transaxle to downshift to second. L oil also is directed to the 2-1 scheduling valve, which directs pressure to the 1-2 shift valve and causes a downshift to low at approximately 28 mph (45 kph).

When the 1-2 shift valve downshifts, L oil flows through it to the manual low relief valve and boost valve in the pressure regulator valve. The manual low relief valve limits the pressure to 55 psi (379 kPa) and directs it to the direct clutch to provide engine compression braking. The boost valve increases mainline pressure to hold the apply devices more securely.

Reverse

In Reverse, the manual valve opens the R passage and directs mainline pressure to the: low-reverse clutch, reverse clutch, and regulator boost valve. R oil applies the low-reverse and reverse clutches to put the transaxle in Reverse. R oil at the regulator boost valve increases mainline pressure to hold the apply devices more securely.

Review Questions

Select the single most correct answer
Compare your answers to the correct answers on page 603

1. Mechanic A says the AXOD transaxle is a 4-speed unit that uses a Ravigneaux compound planetary gearset.
   Mechanic B says the AXOD transaxle has a lockup torque converter.
   Who is right?
   a. Mechanic A
   b. Mechanic B
   c. Both mechanic A and B
   d. Neither mechanic A nor B

2. The AXOD transaxle was introduced in:
   a. 1984
   b. 1985
   c. 1986
   d. 1987

3. Mechanic A says the AXOD has four multiple-disc clutches, two 1-way clutches, and two bands.
   Mechanic B says the low 1-way clutch is a sprag clutch.
   Who is right?
   a. Mechanic A
   b. Mechanic B
   c. Both mechanic A and B
   d. Neither mechanic A nor B

4. Mechanic A says the AXOD input shaft is splined to the converter cover and turns the drive sprocket.
   Mechanic B says the AXOD drive chain turns the driven sprocket counterclockwise, as viewed from the driver's side of the car.
   Who is right?
   a. Mechanic A
   b. Mechanic B
   c. Both mechanic A and B
   d. Neither mechanic A nor B

5. Mechanic A says the AXOD valve body bolts to the transaxle case.
   Mechanic B says the AXOD oil pump housing bolts to the transaxle case.
   Who is right?
   a. Mechanic A
   b. Mechanic B
   c. Both mechanic A and B
   d. Neither mechanic A nor B

6. Mechanic A says the word "METRIC" stamped on the bottom pan identifies the AXOD transaxle.
   Mechanic B says the AXOD gear selector has seven positions.
   Who is right?
   a. Mechanic A
   b. Mechanic B
   c. Both mechanic A and B
   d. Neither mechanic A nor B

7. The fourth gear overdrive ratio of the AXOD geartrain is:
   a. 0.764:1
   b. 0.852:1
   c. 0.694:1
   d. None of the above

8. Mechanic A says MERCON® is the only ATF now recommended for the AXOD transaxle.
   Mechanic B says the AXOD transaxle can provide a 3-2 part-throttle downshift at speeds below 55 mph or 88 kph.
   Who is right?
   a. Mechanic A
   b. Mechanic B
   c. Both mechanic A and B
   d. Neither mechanic A nor B

9. Mechanic A says the AXOD differential operates differently than those in RWD vehicles.
   Mechanic B says the AXOD final drive planetary gearset does the same job as the ring and pinion gears in a RWD axle.
   Who is right?
   a. Mechanic A
   b. Mechanic B
   c. Both mechanic A and B
   d. Neither mechanic A nor B

10. Mechanic A says the AXOD final drive ring gear is splined to the transaxle case.
    Mechanic B says the AXOD parking gear is splined to the final drive sun gear.
    Who is right?
    a. Mechanic A
    b. Mechanic B
    c. Both mechanic A and B
    d. Neither mechanic A nor B

11. Mechanic A says the AXOD final drive gearset operates in gear reduction at all times.
    Mechanic B says the AXOD final drive sun gear is driven by the rear ring gear of the compound planetary gearset.
    Who is right?
    a. Mechanic A
    b. Mechanic B
    c. Both mechanic A and B
    d. Neither mechanic A nor B

12. Mechanic A says the AXOD oil pump is driven by the torque converter hub.
    Mechanic B says the AXOD oil pump is a variable-vane pump.
    Who is right?
    a. Mechanic A
    b. Mechanic B
    c. Both mechanic A and B
    d. Neither mechanic A nor B

13. Mechanic A says the AXOD low-reverse clutch is applied in all gear ranges except fourth.
    Mechanic B says the AXOD low 1-way clutch is effective in manual low, Drive low, and Reverse.
    Who is right?
    a. Mechanic A
    b. Mechanic B
    c. Both mechanic A and B
    d. Neither mechanic A nor B

14. Mechanic A says the AXOD direct clutch is applied in third and fourth gears only.
    Mechanic B says the AXOD direct 1-way clutch is effective in third gear and manual low only.
    Who is right?
    a. Mechanic A
    b. Mechanic B
    c. Both mechanic A and B
    d. Neither mechanic A nor B

15. Mechanic A says the AXOD intermediate-high clutch is applied in second, third, and fourth gears.
    Mechanic B says when the AXOD intermediate-high clutch is applied, it locks the driven sprocket to the front carrier of the compound planetary gearset.
    Who is right?
    a. Mechanic A
    b. Mechanic B
    c. Both mechanic A and B
    d. Neither mechanic A nor B

16. Mechanic A says the AXOD reverse clutch is applied in Reverse only.
    Mechanic B says the AXOD low-intermediate band surrounds the direct/intermediate-high clutch drum.
    Who is right?
    a. Mechanic A
    b. Mechanic B
    c. Both mechanic A and B
    d. Neither mechanic A nor B

17. Mechanic A says the AXOD overdrive band holds the forward planetary sun gear stationary when applied.
    Mechanic B says the AXOD overdrive band is applied in fourth gear only.
    Who is right?
    a. Mechanic A
    b. Mechanic B
    c. Both mechanic A and B
    d. Neither mechanic A nor B

18. Mechanic A says the AXOD mainline pressure blow off valve keeps mainline pressure below 285 to 315 psi (1965 to 2172 kPa).
    Mechanic B says the AXOD pressure regulator valve controls line pressure in relation to governor pressure.
    Who is right?
    a. Mechanic A
    b. Mechanic B
    c. Both mechanic A and B
    d. Neither mechanic A nor B

19. Mechanic A says the AXOD manual valve routes mainline pressure to six passages, depending on its position.
    Mechanic B says the AXOD manual valve is connected by a cable to the throttle linkage.
    Who is right?
    a. Mechanic A
    b. Mechanic B
    c. Both mechanic A and B
    d. Neither mechanic A nor B

20. Mechanic A says the AXOD governor assembly is similar to that in certain General Motors transaxles.
    Mechanic B says that all mainline pressure routed to the AXOD governor escapes when the vehicle is stopped.
    Who is right?
    a. Mechanic A
    b. Mechanic B
    c. Both mechanic A and B
    d. Neither mechanic A nor B

14

Turbo Hydra-matic 125 and 125C Transaxles

HISTORY AND MODEL VARIATIONS

The General Motors (GM) Turbo Hydra-matic (THM) 125 transaxle is designed for use with small-displacement, transverse-mounted engines in FWD applications. The THM 125 was introduced in the 1980 GM X-body cars, and the THM 125C, figure 14-1, was added in 1981. Some GM service manuals and parts catalogs refer to the THM 125 as the M-34 transaxle, and the THM 125C as the MD-9 transaxle.

The THM 125 and 125C transaxles are essentially identical, except that the 125C has a torque converter clutch that locks up to improve fuel economy. In this chapter, we will treat the THM 125/125C as a single transaxle except where specific differences need to be called out. Since its introduction, the THM 125/125C has been used in virtually all of GM's domestically built compact and intermediate size FWD passenger cars.

The THM 125/125C is a 3-speed automatic transaxle that uses the Simpson compound planetary gearset. Three multiple-disc clutches, one band, and a single 1-way roller clutch provide the various gear ratios. In place of final drive spur gears, like those in the Chrysler Torqueflite transaxle or the Ford ATX, the THM 125/125C uses a simple planetary gearset for its final drive.

The THM 125/125C torque converter does not drive the transaxle geartrain directly. Instead, the converter drives two sprockets and a chain that transfers engine power to the geartrain positioned behind and below the crankshaft centerline. The chain is located at the end of the transaxle case on the driver's side of the car, and is enclosed by the case cover.

The THM 125/125C valve body also mounts on the driver's side of the transaxle, rather than at the bottom in the sump. The valve body and oil pump share a common casting that bolts to the case cover. On THM 125C transaxles, an auxiliary valve body containing the converter clutch control valves is mounted on the end of the pump/valve body casting. The THM 125/125C oil pump is a variable-displacement, vane-type pump that is driven by a shaft splined to the torque converter cover. The oil pump, valve body, and auxiliary valve body (THM 125C) are housed under a stamped steel cover that bolts to the transaxle case cover.

Fluid Recommendations

Like all GM automatic transmissions and transaxles, the THM 125/125C requires DEXRON®-IID fluid.

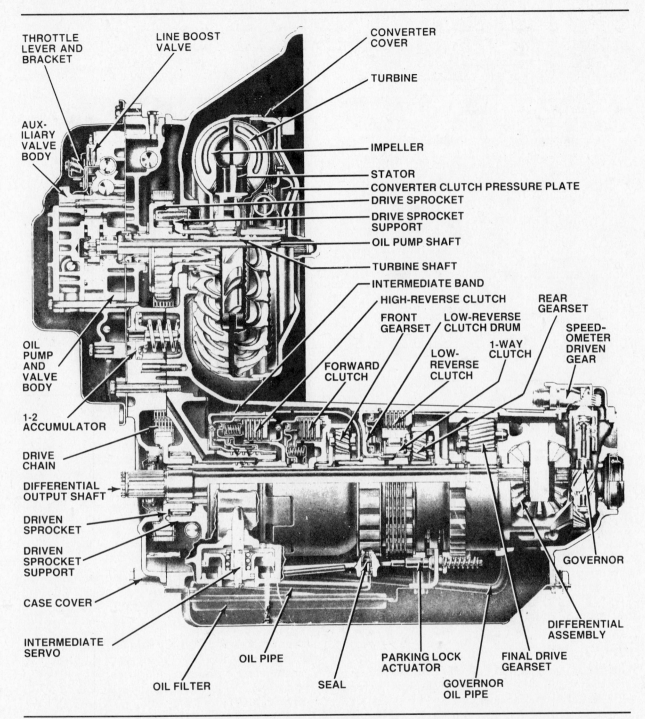

Figure 14-1. The THM 125C transaxle; the THM 125 is basically the same except that it does not have a lockup torque converter.

Transaxle Identification

Identifying the THM 125/125C transaxle is relatively easy because, as of the 1988 model year, it is the only 3-speed automatic used in FWD GM cars with transverse engines. The THM 125/125C can also be identified by the distinctive shape of its valve body cover, and by the words "HYDRAMATIC" and "METRIC" stamped on the valve body cover. As required by Federal regulations, GM transaxles also carry the serial number of the vehicle in which they were originally installed.

In addition to the transaxle model, you must know the exact assembly part number and

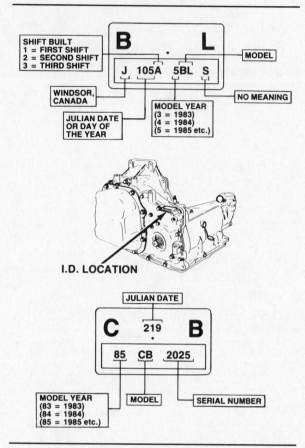

Figure 14-2. THM 125/125C identification tags.

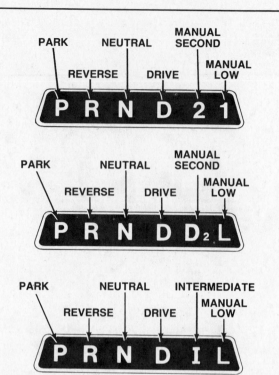

Figure 14-3. THM 125/125C gear selector positions.

other identifying characters in order to obtain proper repair parts. On the THM 125/125C, this information is stamped on a tag riveted to the top of the case cover above the left-side halfshaft. The coding that appears on the tag varies with the plant in which the transaxle was manufactured. Figure 14-2 shows two typical THM 125/125C identification tags, and explains the meaning of the data stamped on them.

Major Changes

The major changes that GM has made in the THM 125/125C transaxle are as follows:

1982 • Valve body spacer plate and gaskets changed; new parts cannot be used in earlier models.
• High-reverse clutch discs redesigned; new parts may be used in earlier models.
• Intermediate band and high-reverse clutch drum revised. New parts available in a service package that may be installed in earlier models.
1983 • Auxiliary valve body redesigned; new parts cannot be used in earlier models.
1984 • Transaxle case modified. If this case is used as a replacement for an earlier trans-

axle, all of the running changes may be incorporated into the new case.
• New integral bearing and seal introduced.
• New thrust bearing assembly/sun gear introduced.
1985 • Forward and high-reverse clutch assembly revised; new parts may be used in earlier models.
1986 • Intermediate servo orifice eliminated.
• Transaxle case modified.

GEAR RANGES

Depending on the GM division and the particular application, there are minor variations in the exact labeling of the gear selector used with the THM 125/125C transaxle. However, all of the gear selectors have six positions, figure 14-3:
P — Park
R — Reverse
N — Neutral
D — Drive
I, S, or L2 — Manual second
L or L1 — Manual low.
Gear ratios for the THM 125/125C are:
First (low) — 2.84:1
Second (intermediate) — 1.60:1
Third (direct drive) — 1.00:1
Reverse — 2.07:1.

GENERAL MOTORS TERM	TEXTBOOK TERM
Clutch housing	Clutch drum
Converter pump	Impeller
Detent downshift	Full-throttle downshift
Internal gear	Ring gear
Input drum	Input shell
Input planetary gearset	Front planetary gearset
Reaction planetary gearset	Rear planetary gearset
Control valve assembly	Valve body
Line pressure	Mainline pressure
T.V. pressure	Throttle pressure
T.V. boost valve	Boost valve
S.T.V. pressure	Shift throttle pressure
T.V. plunger	Throttle plunger
1-2, 2-3 T.V. valves	1-2, 2-3, 3-4 throttle valves
Direct clutch	High-reverse clutch

Figure 14-4. THM 125/125C transaxle nomenclature table.

Park and Neutral

The engine can be started only when the transaxle is in Park or Neutral. No clutches or bands are applied in these positions, and there is no flow of power through the transaxle. In Neutral, the output shafts are free to turn, and the car can be pushed. In Park, a mechanical pawl engages the parking gear on the outside of the rear ring gear. This locks the final drive planetary gearset and prevents the vehicle from being moved.

Drive

When the vehicle is stopped and the gear selector is placed in Drive (D), the transaxle is automatically in low gear. As the vehicle accelerates, the transaxle automatically upshifts to second, and then to third gear. The speeds at which upshifts occur vary, and are controlled by vehicle speed and throttle opening.

The driver can force downshifts in Drive by opening the throttle partially or completely. The speeds at which downshifts occur vary for different engine, axle ratio, tire size, and vehicle weight combinations. Below 50 mph (80 kph), the driver can force a part-throttle 3-2 downshift by opening the throttle beyond a certain point. Below 65 mph (105 kph), the driver can force a 3-2 detent (kickdown) downshift by fully opening the throttle. A 3-1 or 2-1 detent (kickdown) downshift can be forced below 30 mph (48 kph).

When the car decelerates in Drive to a stop, the transaxle makes two downshifts automatically, first a 3-2 downshift, then a 2-1 downshift.

Manual Second

When the vehicle is stopped and the gear selector is placed in manual second (I, S, or L2), the transaxle is automatically in low gear. As the vehicle accelerates, the transaxle automatically upshifts to second; it does not upshift to third. The speed at which the upshift occurs will vary, and is controlled by vehicle speed and throttle opening.

A 2-1 detent (kickdown) downshift can be forced below 30 mph (48 kph) if the driver depresses the accelerator completely. The transaxle automatically downshifts to low as the vehicle decelerates to a stop.

The driver can manually downshift from third to second by moving the selector from D to I, S, or L2. In this case, the transaxle shifts to second gear immediately, regardless of road speed or throttle position. Manual downshifting can provide engine compression braking, or gear reduction for hill climbing, pulling heavy loads, or operating in congested traffic.

Manual Low

When the vehicle is stopped and the gear selector is placed in manual low (L or L1), the car starts and stays in first gear; it does not upshift to second or third. Because different apply devices are used in manual low than are used in Drive low, this gear range can be used to provide engine compression braking to slow the vehicle.

If the driver shifts to manual low at high speed, the transaxle will not shift directly to low gear. To avoid engine or driveline damage, the transaxle first downshifts to second, and then automatically downshifts to low when the speed drops below approximately 40 mph (64 kph). The transaxle then remains in low gear until the gear selector is moved to another position.

Reverse

When the gear selector is moved to R, the transaxle shifts into Reverse. The transaxle should be shifted to Reverse only when the vehicle is standing still. The lockout feature of the shift gate should prevent accidental shifting into Reverse while the vehicle is moving forward. However, if the driver intentionally moves the selector to R while moving forward, the transaxle *will* shift into Reverse, and damage may result.

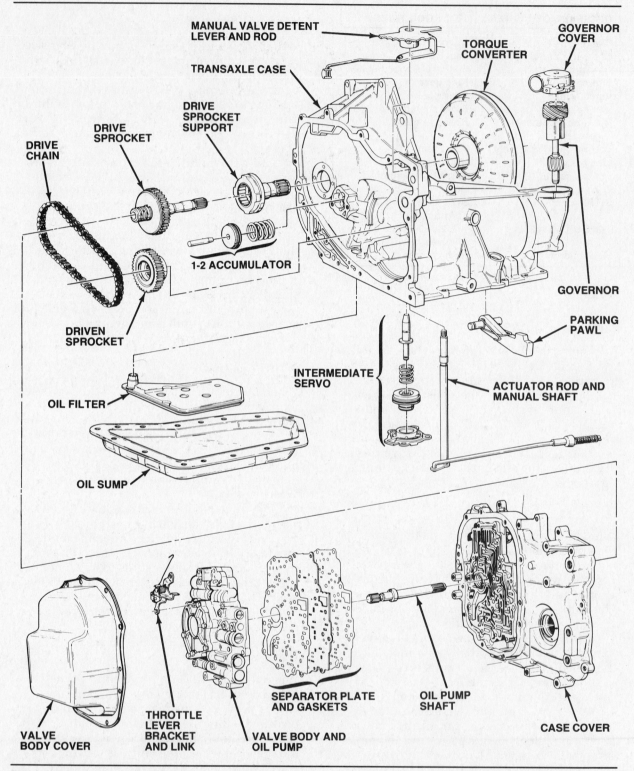

Figure 14-5. An exploded view of the THM 125/125C case, including the drive and driven sprockets, and the drive chain.

GM TRANSAXLE NOMENCLATURE

For the sake of clarity, this text uses consistent terms for transaxle parts and functions. How-ever, the various manufacturers may use differ-ent terms. Therefore, before we examine the buildup of the THM 125/125C, and the opera-

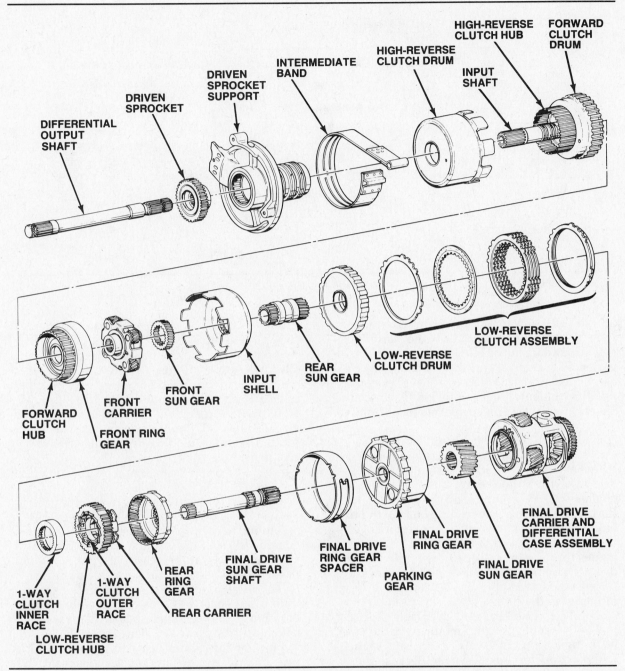

Figure 14-6. An exploded view of the THM 125/125C clutches, 1-way clutch, band, compound planetary geartrain, and final drive/differential assemblies.

tion of its geartrain and hydraulic system, you should be aware of the unique names that GM uses for some transaxle parts and operations. This will make it easier in the future if you should look in a GM manual and see a reference to the ''direct clutch''. You will realize that this is simply the high-reverse clutch and is similar to the same part in any other transaxle that uses a similar gear train and apply devices. Figure 14-4 lists GM nomenclature and the

textbook terms for some transaxle parts and operations.

THM 125/125C BUILDUP

The THM 125/125C case, figure 14-5, is a 1-piece aluminum casting with the case cover and oil pan bolted to it. The bellhousing is integral with the transaxle case. The governor mounts in an opening on the top right-hand side of the case.

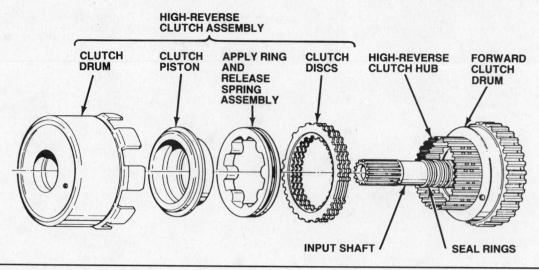

**HIGH-REVERSE
CLUTCH ASSEMBLY**

**CLUTCH
DRUM**   **CLUTCH
PISTON**   **APPLY RING
AND
RELEASE
SPRING
ASSEMBLY**   **CLUTCH
DISCS**   **HIGH-REVERSE
CLUTCH HUB**   **FORWARD
CLUTCH
DRUM**

INPUT SHAFT   **SEAL RINGS**

Figure 14-7. The THM 125/125C input shaft and high-reverse clutch assembly.

The THM 125/125C torque converter bolts to the engine flexplate, which has the starter ring gear around its outer diameter. The torque converter is a welded assembly and cannot be disassembled. The THM 125 converter is a conventional 3-element design (impeller, turbine, and stator). The THM 125C converter also contains a hydraulically applied clutch.

The torque converter is installed into the transaxle by sliding it onto the stationary drive sprocket support, which is bolted to the transaxle case. The splines in the stator 1-way clutch engage the splines on the drive sprocket support. The splines in the turbine engage the turbine shaft. The splines in the converter cover engage the oil pump shaft, figure 14-5, which passes through the inside of the turbine shaft to drive the oil pump. The hub at the rear of the torque converter housing rides on a bushing mounted in the case.

The THM 125/125C uses a variable-displacement vane-type oil pump mounted in the valve body, figure 14-5. The valve body bolts to the outboard side of the case cover. Because the converter cover is bolted to the engine flexplate, the oil pump shaft turns and the oil pump delivers fluid whenever the engine is running. The pump draws fluid from the sump through a filter, and routes the fluid through passages in the valve body, case cover, and transaxle case to several valves in the hydraulic system.

The turbine shaft turns the drive sprocket, which turns the drive chain. The drive chain then turns the driven sprocket that provides power to the transaxle geartrain, figure 14-5. The drive and driven sprockets have different

numbers of teeth, depending on the application. For example, early THM 125s come in two models. In the CV version, used with Chevrolet-built 2.8-liter V-6 engines, both the drive and driven sprockets have 35-teeth, giving a 1:1 ratio in the chain drive. The PZ version, used with Pontiac-built 2.5-liter 4-cylinder engines, has a 37-tooth drive sprocket and a 33 tooth driven sprocket that provide a .891:1 overdrive ratio to the geartrain.

Figure 14-6 shows a general overall view of the THM 125/125C geartrain. As we complete the buildup of this transaxle, refer to figure 14-6, and the detailed figures for each specific assembly, to identify the components being discussed. All references to rotation are made as if you are looking at the geartrain from the front of the engine, that is, the right side of the car.

The driven sprocket is splined to the geartrain input shaft, which is one assembly with the high-reverse clutch hub and the forward clutch drum, figure 14-7. This assembly is held in place by the driven sprocket support. Seal rings on the input shaft seal oil passages to the various clutches. Whenever the vehicle is moving, the drive chain turns the driven sprocket, input shaft, high-reverse clutch hub, and forward clutch drum clockwise.

The inside of each high-reverse clutch friction disc is splined to the outside of the high-reverse clutch hub and turned by it. The outside of each steel disc is splined to the inside of the high-reverse clutch drum. When the high-reverse clutch is applied, the drum is rotated clockwise by the input shaft.

The high-reverse clutch drum is splined to the input shell, figure 14-8, which is splined to the rear sun gear. The front sun gear is also

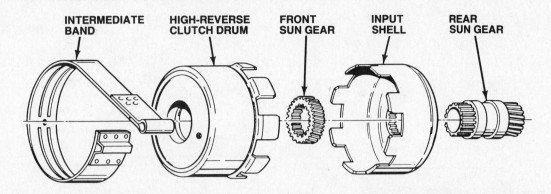

Figure 14-8. The THM 125/125C intermediate band, high-reverse clutch drum, input shell, and sun gear assembly.

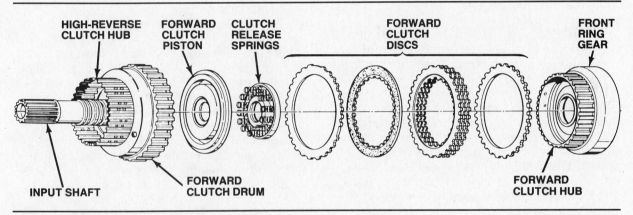

Figure 14-9. The THM 125/125C input shaft and forward clutch assembly.

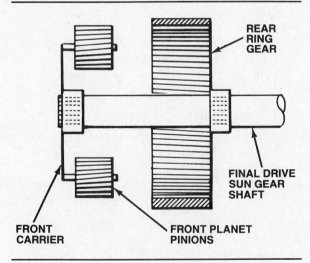

Figure 14-10. The THM 125/125C front carrier and rear ring gear are splined to the final drive sun gear shaft.

splined to the rear sun gear, and both gears rotate as a unit with the input shell at all times. The sun gear assembly turns on bushings located on the final drive sun gear shaft, but the assembly is not splined to the shaft. When the high-reverse clutch is applied, the clutch drum, input shell, and sun gear assembly are turned clockwise by the input shaft.

The intermediate band is wrapped around the outside of the high-reverse clutch drum. When the intermediate band is applied, the clutch drum, input shell, and sun gear assembly are all held stationary. This occurs only in second gear.

The outside of each forward clutch steel disc is splined to the inside of the forward clutch drum, figure 14-9, which is part of the input shaft. The inside of each friction disc is splined to the outside of the front ring gear in the planetary gearset. Whenever the forward clutch is applied, the front ring gear is connected to the input shaft. If the transaxle is in gear and the vehicle's brakes are applied, a load is placed on the drive train and the turbine will not be able to turn; therefore, the input shaft cannot turn.

None of the members of the compound planetary gearset are attached directly to the input shaft. However, two members of the gearset *are* splined to the final drive sun gear shaft, the front carrier and the rear ring gear, figure 14-10.

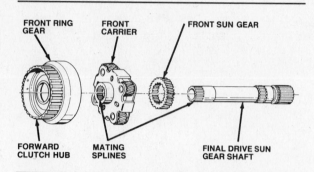

Figure 14-11. The front gearset of the THM 125/125C compound planetary geartrain.

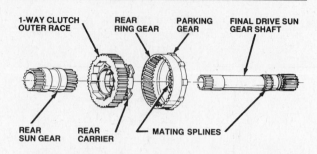

Figure 14-12. The rear gearset of the THM 125/125C compound planetary geartrain.

This means that one or the other of these two units is always the final driving member of the planetary gearset.

The front ring gear meshes with the front pinions. The front pinions turn in the front carrier and mesh with the front sun gear. The front carrier is splined to the final drive sun gear shaft, figure 14-11.

The rear sun gear meshes with the rear pinions. The rear pinions turn in the rear carrier, and mesh with the rear ring gear, figure 14-12. The rear ring gear, like the front carrier, is splined to the final drive sun gear shaft. The outside of the rear ring gear forms the parking gear, which is locked to the transaxle case by the parking pawl when the gear selector is placed in Park.

The outer race of the 1-way clutch is located inside the rear carrier, figure 14-13. The inner race of the 1-way clutch is splined to the low-reverse clutch drum, which is splined to the transaxle case. The 1-way clutch holds the rear carrier if it attempts to turn counterclockwise, but the clutch overruns when the rear carrier turns clockwise.

The inside of each low-reverse clutch friction disc is splined to the outside of the rear carrier, figure 14-13. The outside of each steel disc is splined to the inside of the transaxle case. When the low-reverse clutch is applied, it over-rides the 1-way clutch, and holds the rear carrier from turning in *either* direction.

The final drive and differential assembly, figure 14-14, transmits power from the trans-axle gearset to the halfshafts of the vehicle. In place of conventional final drive gears, the THM 125/125C uses a simple planetary gearset with one sun gear, one carrier, four pinions and one ring gear. The sun gear is driven by the final drive sun gear shaft, which is splined to the front carrier and rear ring gear of the trans-axle gearset. The sun gear meshes with the final drive pinions, which turn on shafts in the final drive carrier. The carrier is built into the

differential case. The pinions mesh with the final drive ring gear, which is splined to the transaxle case.

The differential case contains four bevel gears: two pinion gears and two side gears. The pinion gears are installed on a pinion shaft that is retained by a pin in the differential case. The pinion gears transmit power from the case to the side gears. A male spline on the right-side halfshaft enters the transaxle to engage the right side gears. The left side gear is splined to the differential output shaft which delivers power, through a male spline, to the left-side halfshaft. A bushing and seal for the halfshaft is installed at each side of the transaxle case.

THM 125/125C POWER FLOW

Now that we have taken a quick look at the general way in which the THM 125/125C is assembled, we can trace the flow of power through the apply devices and the compound planetary gearset in each of the gear ranges. Refer to figure 14-6 and the specific figure for each gear range to help trace the power flow.

Park and Neutral

When the gear selector is in Park or Neutral, no clutches or bands are applied, and the torque converter clutch (THM 125C) is released. Since no member of the planetary gearset is held, there can be no output motion. The torque converter impeller/cover and oil pump drive shaft rotate clockwise at engine speed. The oil pump drive shaft turns the oil pump. The converter turbine, turbine shaft, drive chain, drive and driven sprockets, and input shaft also rotate clockwise at engine speed, because the converter is in an unloaded coupling phase.

In Neutral, the compound planetary gearset idles, and the final drive gearset is free to rotate in either direction; thus the vehicle can be pushed or pulled. In Park, a mechanical pawl engages the parking gear on the outside of the

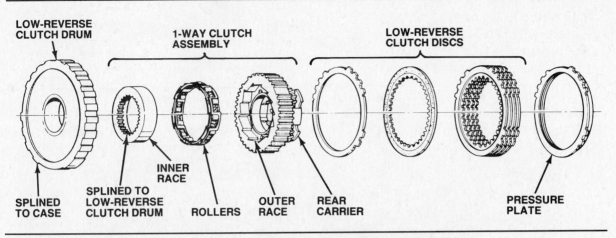

Figure 14-13. The THM 125/125C low-reverse clutch and 1-way roller clutch assemblies.

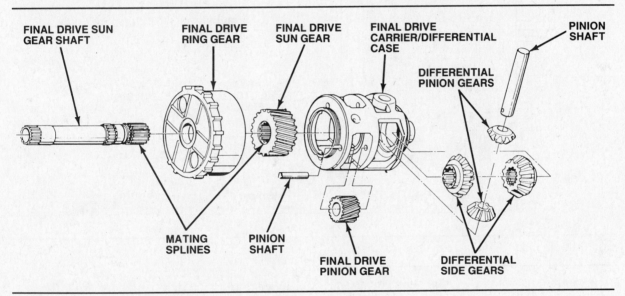

Figure 14-14. The THM 125/125C final drive and differential assembly.

rear ring gear of the compound planetary gearset. The rear ring gear is splined to the final drive sun gear shaft. Because the final drive ring gear and sun gear are both held, the final drive gearset locks up and prevents the vehicle from moving.

Drive Low

When the car is stopped and the gear selector is moved to Drive (D), the transaxle is automatically in low gear, figure 14-15. The forward clutch is applied and transmits engine power from the driven sprocket to the front ring gear which turns clockwise.

Before the car begins to move, the final drive carrier is held stationary by the differential case, halfshafts, and wheels. This also holds

the final drive sun gear shaft, front carrier, and rear ring gear stationary. This means the clockwise rotation of the front ring gear turns the front pinions clockwise on the stationary carrier. Front pinion rotation then causes the sun gear to turn counterclockwise.

Counterclockwise rotation of the sun gear turns the rear pinions clockwise. The rear pinions then attempt to walk counterclockwise around the inside of the rear ring gear, which is splined to the final drive sun gear shaft. This action causes the rear carrier to try to rotate counterclockwise, but the 1-way clutch locks up and prevents rotation. Therefore, the torque from the rear pinions is transferred to the rear ring gear and the final drive sun gear shaft. Because the 1-way clutch holds the carrier more securely than the drive wheels can hold the

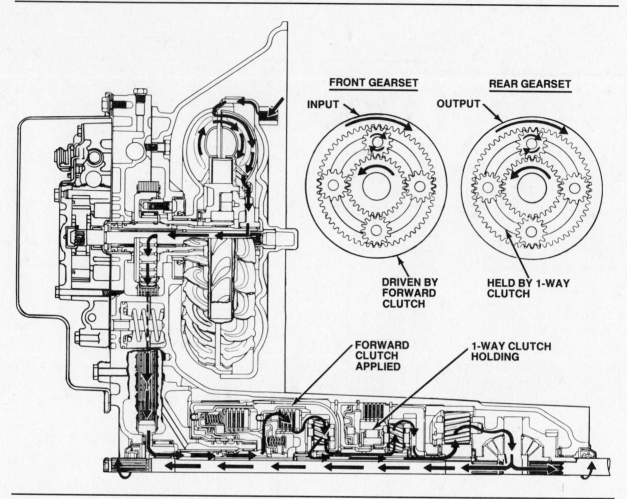

Figure 14-15. THM 125/125C power flow in Drive low.

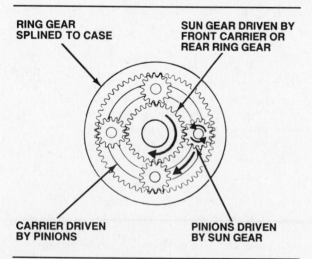

Figure 14-16. THM 125/125C final drive gearset power flow in all forward gears. The sun gear, pinions, and carrier rotate in the opposite directions in Reverse.

ring gear and final drive sun gear shaft, the ring gear and sun gear shaft rotate clockwise in gear reduction.

The final drive sun gear turns the final drive pinions counterclockwise, figure 14-16. The final drive ring gear is splined to the transaxle case, and is always held from rotating. Therefore, the pinions walk around the inside of the ring gear and drive the carrier/differential case clockwise in gear reduction. Power is then passed through the differential pinion and side gears to the differential output shaft (left side) and both halfshafts. This power flow through the final drive gearset and differential assembly is the same in all forward gears.

As long as the car is accelerating in Drive low, torque is transferred from the engine, through the planetary gearset, to the final drive sun gear shaft. If the driver lets up on the throttle, engine speed drops while the drive wheels try to keep turning at the same speed. This removes the counterclockwise-rotating torque

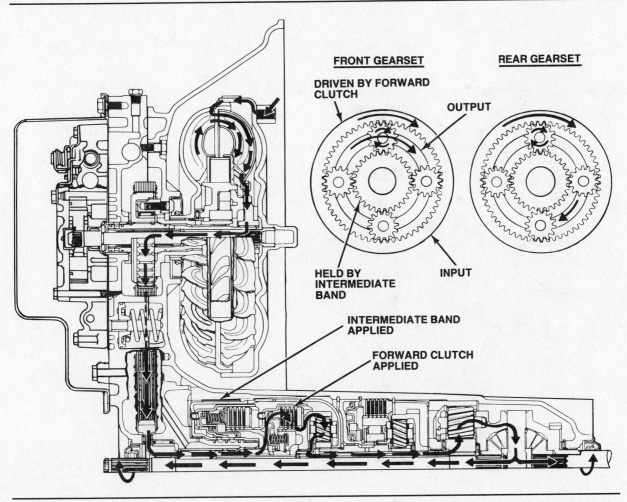

Figure 14-17. THM 125/125C power flow in Drive second.

load from the 1-way clutch, and it unlocks. This allows the rear gearset to freewheel, so no engine compression braking is available. This is a major difference between Drive low and manual low, as we will see later.

Gear reduction in low gear is produced by both the front and the rear gearsets of the transaxle. The input member is the front ring gear, the held member is the rear carrier, and the output member is the rear ring gear. The input device is the forward clutch, and the holding device is the 1-way clutch. In the final drive gearset, the input member is always the sun gear, the held member is always the ring gear, and the output member is always the carrier.

Drive Second

As the car continues to accelerate in Drive, the transaxle hydraulic system produces an automatic upshift to second gear by applying the intermediate band, figure 14-17. The forward clutch remains applied to transmit engine power to the front ring gear. The intermediate band locks the high-reverse clutch drum, input shell, and sun gear assembly to the transaxle case.

As the front ring gear turns clockwise, it turns the front pinions clockwise. The front pinions walk around the stationary sun gear and drive the front carrier, and the final drive sun gear shaft splined to it, clockwise in gear reduction. The final drive sun gear shaft transmits power to the final drive gearset and differential assembly as described earlier.

The ring gear of the rear gearset also rotates clockwise, and drives the rear pinions clockwise. The rear pinions walk around the stationary sun gear assembly and drive the rear carrier clockwise. However, because the 1-way clutch overruns, the rear gearset simply freewheels at this time.

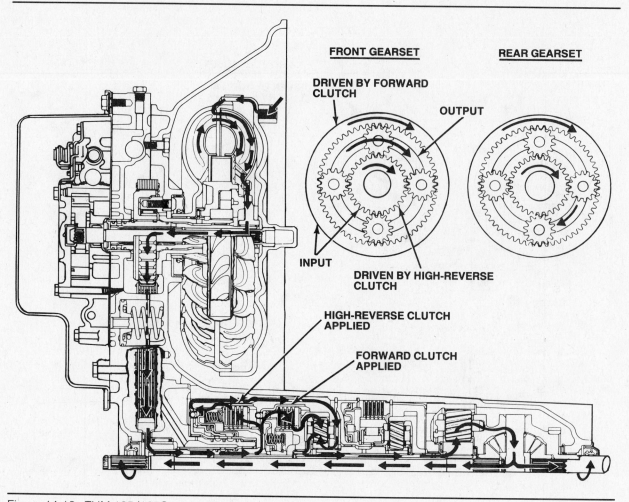

Figure 14-18. THM 125/125C power flow in Drive high.

All of the gear reduction for second gear is produced by the front planetary gearset. The input member is the ring gear, the held member is the sun gear, and the output member is the carrier. The input device is the forward clutch, and the holding device is the intermediate band.

Drive High

As the car continues to accelerate in Drive, the transaxle hydraulic system produces an automatic upshift from second to third gear, figure 14-18. To make this happen, the forward clutch remains applied to transmit engine power to the front ring gear. The intermediate band is released, and the high-reverse clutch is applied to transmit engine power to the sun gear assembly.

When the high-reverse clutch is applied, the clutch drum, input shell, and sun gear assembly are locked to the input shaft and rotate clockwise at input shaft speed. At the same

time, the forward clutch locks the front ring gear to the input shaft. With two members locked together through the input shaft, the front gearset turns as a unit in direct drive. Input torque then passes through the front carrier to the final drive sun gear shaft, which transmits power to the final drive gearset and differential assembly as described earlier.

The front carrier and sun gear shaft also drive the rear ring gear at input shaft speed, and the rear sun gear turns with the front sun gear at input shaft speed. This means that the rear gearset is also locked and turns as a unit, although it does not transmit any engine power at this time.

In third gear, there is a direct 1:1 gear ratio between input shaft speed and final drive sun gear shaft speed. There is no reduction or increase in either speed or torque through the compound planetary gearset. However, the final drive planetary gearset does still provide gear reduction to the differential and halfshafts.

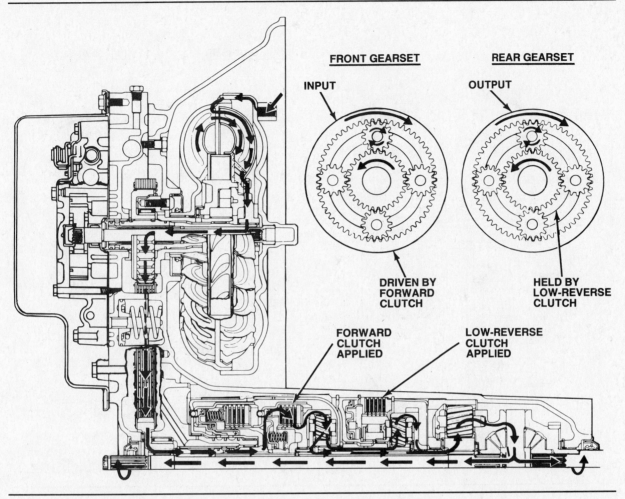

Figure 14-19. THM 125/125C power flow in manual low.

All of the power in third gear is transmitted through the front gearset. The ring gear and sun gear are equally input members. The carrier is the output member. The input devices are the forward clutch and the high-reverse clutch. No holding device is needed because the planetary gearset is locked.

Manual Second

The only difference between manual second (I, S, or L2 on the gear selector) and Drive is that the hydraulic system prevents the transaxle from upshifting into high gear. The flow of power through the planetary gearsets in first and second gears is identical, and the same apply devices are used.

When the gear selector is in manual second and the car accelerates from a standstill, the transaxle starts in Drive low, figure 14-15. The forward clutch drives the front ring gear, and the 1-way clutch holds the rear carrier. As the vehicle continues to accelerate, the transaxle

automatically shifts to second by applying the intermediate band to hold the sun gear assembly, figure 14-17.

If the driver moves the gear selector to I, S, or L2 while the car is moving in Drive third, the hydraulic system will downshift the transaxle to second gear by releasing the high-reverse clutch and applying the intermediate band. This capability provides the driver with a controlled downshift for engine compression braking or improved acceleration at any speed or throttle position.

Manual Low

In manual low (L1 or L on the gear selector), the hydraulic system keeps the transaxle in low gear by preventing the intermediate band from being applied. The flow of power through the planetary gearset is the same as it is in Drive low, with the exception of one apply device.

In manual low, figure 14-19, the low-reverse clutch is applied to hold the rear carrier. The

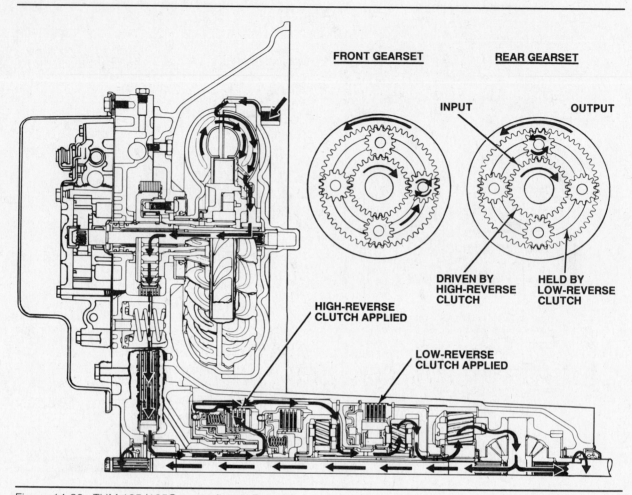

FRONT GEARSET

REAR GEARSET

INPUT

OUTPUT

DRIVEN BY
HIGH-REVERSE
CLUTCH

HELD BY
LOW-REVERSE
CLUTCH

HIGH-REVERSE
CLUTCH APPLIED

LOW-REVERSE
CLUTCH APPLIED

Figure 14-20. THM 125/125C power flow in Reverse.

1-way clutch does this job in Drive low, but it holds the carrier only during acceleration. The 1-way clutch overruns and allows the carrier to freewheel during deceleration. In manual low, the low-reverse clutch locks the rear carrier to the transaxle case during acceleration, and during deceleration to provide engine compression braking. The 1-way clutch continues to hold the rear carrier during acceleration in manual low, but its action is secondary to that of the low-reverse clutch.

The driver can move the gear selector from D to L1 or L while the car is moving. At speeds below approximately 40 mph (64 kph), the transaxle downshifts directly from high to low gear by releasing the high-reverse clutch and applying the low-reverse clutch. At higher speeds, the transaxle first shifts into second gear by releasing the high-reverse clutch and applying the intermediate band. As speed drops below approximately 40 mph (64 kph), the transaxle automatically downshifts from

second to manual low by releasing the intermediate band and applying the low-reverse clutch.

Gear reduction in manual low gear is produced by both the front and the rear gearsets of the transaxle. The input member is the front ring gear, the held member is the rear carrier, and the output member is the rear ring gear. The input device is the forward clutch, and the holding device is the low-reverse clutch.

Reverse

When the THM 125/125C gear selector is moved to Reverse, the input shaft still is turned clockwise by the engine. The transaxle must reverse this rotation to a counterclockwise direction. This is accomplished by releasing the forward clutch, and applying the high-reverse and low-reverse clutches, figure 14-20.

When the high-reverse clutch is applied, the clutch drum, input shell, and sun gear assembly are locked to the input shaft and rotate

GEAR RANGES	CLUTCHES				BANDS
	1-Way Roller (Overrunning)	Forward (Rear)	High-Reverse (Front, or Direct)	Low-Reverse (Rear)	Intermediate (Kickdown, or Front)
Drive Low	●	●			
Drive Second		●			●
Drive High		●	●		
Manual Low	*	●		●	
Manual Second		●			●
Reverse			●	●	

* The 1-way roller clutch will hold in manual low if the low-reverse clutch fails, but will not provide engine braking.

Figure 14-21. THM 125/125C clutch and band application chart.

clockwise. The sun gear then drives the rear pinions counterclockwise. Because the low-reverse clutch holds the rear carrier stationary, the rear pinions drive the rear ring gear counterclockwise in gear reduction. The rear ring gear is splined to the final drive sun gear shaft, which transmits power to the final drive gearset and differential assembly as described earlier.

In Reverse, the front sun gear also rotates clockwise and the front carrier rotates counterclockwise. Together, they drive the front pinions counterclockwise. The front pinions then drive the front ring gear counterclockwise. However, because the forward clutch is not applied, the front gearset freewheels at this time.

All of the gear reduction in Reverse is produced by the rear planetary gearset. The input member is the sun gear, the held member is the carrier, and the output member is the ring gear. The input device is the high-reverse clutch, and the holding device is the low-reverse clutch.

THM 125/125C Clutch and Band Applications

Before we move on to a closer look at the hydraulic system, we should review the apply devices that are used in each of the gear ranges in this transaxle. Figure 14-21 is a clutch and band application chart for the THM 125/125C. Also, remember these facts about the apply devices in this transaxle:

● The forward and high-reverse clutches are input devices.
● The forward clutch is applied in all forward gears.
● The high-reverse clutch is applied in third and Reverse gears.
● The 1-way clutch, intermediate band, and low-reverse clutch are holding devices.
● The 1-way clutch holds on acceleration in Drive low.

● The intermediate band is applied in second gear.
● The low-reverse clutch is applied in Reverse and manual low.

THM 125/125C HYDRAULIC SYSTEMS

The THM 125/125C hydraulic system controls shifting under varying vehicle loads and speeds, as well as in response to manual gear selection by the driver. Figure 14-22 is a diagram of the complete THM 125 hydraulic system. The THM 125C hydraulic system is similar, but adds two valves in an auxiliary valve body to control the converter clutch. We will begin our study of the hydraulic system with the pump.

Hydraulic Pump

In the THM 125/125C, fluid is stored in the sump at the bottom of the transaxle, and in an upper fluid reservoir located in the lower section of the valve body cover, figure 14-23. The upper reservoir controls the oil level in the sump based on fluid temperature. As the fluid heats up and expands, a thermostatic element increases the amount of fluid retained in the upper reservoir.

The THM 125/125C uses a variable-displacement, vane-type oil pump, figure 14-24, that is located on the driver's side of the transaxle. The pump slide, pivot pin, rotor, vanes, and vane ring are installed in the pump housing, which is part of the valve body casting. The slide pivots on a pin, and when pressure is low in the hydraulic system, the pump priming spring tends to push the ring toward its maximum output position. When pressure in the hydraulic system rises, the pressure regulator valve sends oil through the decrease passage to move the slide toward its minimum output position.

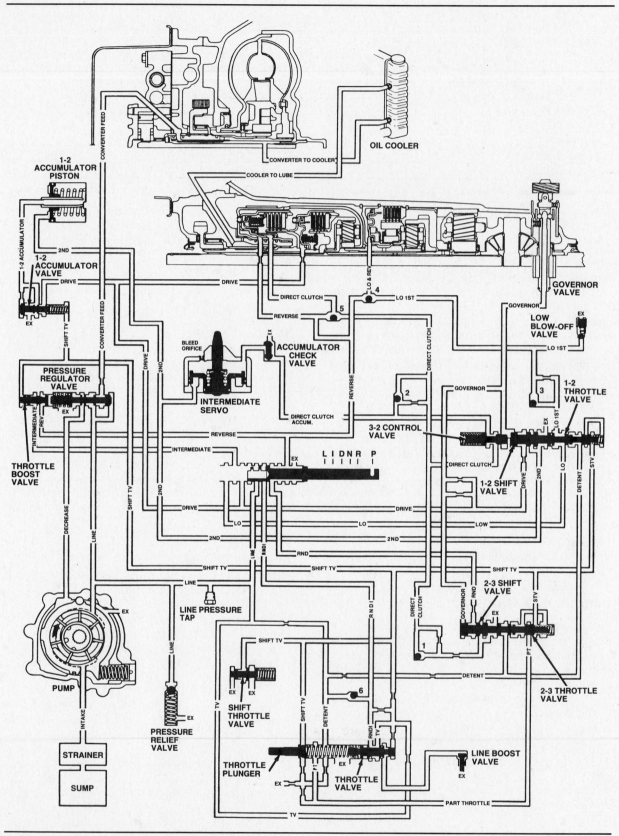

Figure 14-22. The THM 125/125C hydraulic system.

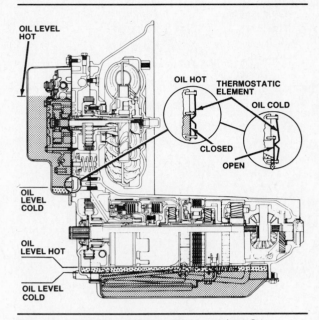

Figure 14-23. Oil level in the THM 125/125C is controlled by a thermostatic element.

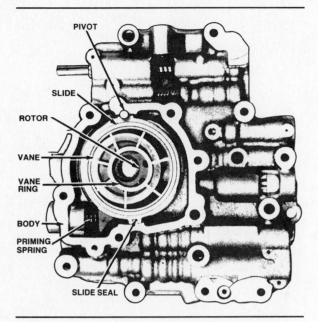

Figure 14-24. The THM 125/125C oil pump assembly.

In this manner, oil pump output is controlled so that it is proportional to demand.

Whenever the engine is running, the converter cover turns the pump drive shaft and rotor. The rotation of the pump rotor and vanes creates suction that draws fluid through a filter from the sump at the bottom of the transaxle case. Oil from the pump flows to the pressure regulator and pressure relief valves in the valve body, and to the manual valve in the case.

Hydraulic Valves

The THM 125/125C hydraulic system contains the following valves:
1. Pressure regulator valve
2. Pressure relief valve
3. Manual valve
4. Throttle valve
5. Shift throttle valve
6. Line boost valve
7. Governor valve
8. 1-2 shift valve
9. 1-2 throttle valve
10. 1-2 accumulator valve
11. Low blow-off valve
12. 2-3 shift valve
13. 2-3 throttle valve
14. 3-2 control valve
15. Converter clutch control valve (125C only)
16. Converter clutch regulator valve (125C only).

Valve body and ball-check valves
Except for the manual valve, governor valve, converter clutch control valve, and converter clutch regulator valve, all of the hydraulic valves in the THM 125/125C are housed in the valve body, figure 14-25. The manual valve is located in the transaxle case. The governor valve installs in an opening on top of the transaxle above the final drive carrier/differential case.

A separator plate and two gaskets are installed between the valve body and the case cover, figure 14-5. The valve body has passages to route fluid between various valves, and additional passages are cast into the case cover where the valve body bolts to it. These cast passages form the back half of the valve body. The THM 125/125C uses six check balls. One is located in the valve body, while the other five are located in the case cover, along with the low blow-off valve plug.

The THM 125C has an auxiliary valve body that houses the converter clutch control valve and the converter clutch regulator valve, figure 14-26. Depending on the application, the auxiliary valve body may also contain one or more pressure switches that are used in controlling the torque converter clutch (TCC). The electrical and electronic elements of General Motors TCC systems are discussed in Chapter 8. The hydraulic parts of the THM 125C system are covered later in this chapter.

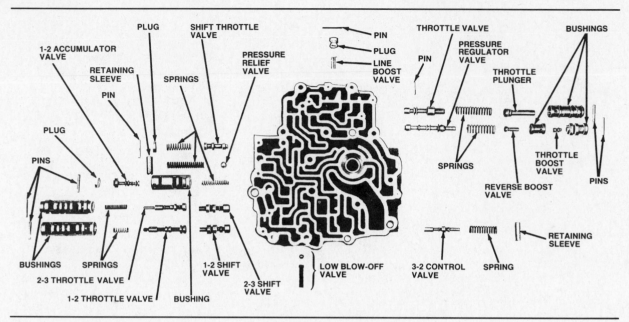

Figure 14-25. The THM 125/125C valve body and hydraulic control valves.

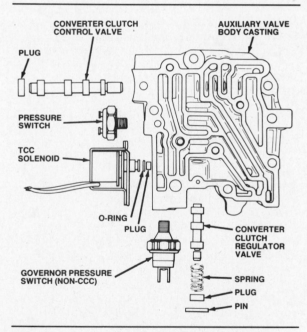

Figure 14-26. The THM 125C auxiliary valve body bolts to the main valve body and contains the valves and switches used to control the torque converter clutch.

Pressure regulator valve

The pump delivers fluid through two passages to the pressure regulator valve, figure 14-27. One passage is at the end of the valve opposite the boost valves, the other is at the center of the valve. Hydraulic pressure at the end of the valve acts to move the valve against spring force. Hydraulic pressure at the center of the

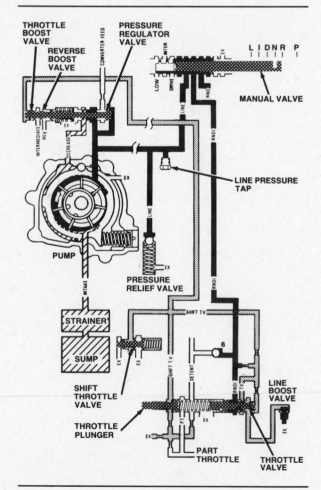

Figure 14-27. The THM 125/125C oil pump, pressure regulator valve, manual valve, and throttle valve (transaxle in Neutral).

valve enters between two lands of equal diameter, and does not move the valve in either direction.

When the engine is started and mainline pressure begins to build, pressure at the end of the regulator valve moves the valve against spring force. This opens the converter feed passage and sends oil to the torque converter. Converter return oil is routed to the transaxle cooler, and cooler return oil is directed to the transaxle lubrication circuits.

After the converter is filled, pressure buildup at the regulator valve moves it farther against spring force. This opens a port in the center chamber of the valve that leads into the decrease passage. Pressure then flows through the passage to the pump where it moves the slide against priming spring force to reduce pump output. The pressure regulator valve continues to move until hydraulic pressure and spring force reach a balance that provides a constant mainline pressure.

When the transaxle shifts into gear, or from one gear to another, increased fluid flow demands cause mainline pressure to drop slightly. This drop in mainline pressure allows the spring to move the pressure regulator valve so that mainline pressure feed is cut off and the pump decrease circuit exhausts. The pump priming spring then moves the slide to increase pump output and bring mainline pressure back to the correct level.

The throttle boost and reverse boost valves at the spring end of the pressure regulator valve, figure 14-27, are used to raise mainline pressure under certain driving conditions. This is done by routing hydraulic pressure to the boost valves, which then move to increase the spring force acting on the pressure regulator valve.

Shift throttle pressure acts on the throttle boost valve to increase mainline pressure in response to higher torque loads on the drivetrain. Mainline pressure acts on the reverse boost valve in Reverse, manual low, and manual second to increase mainline pressure and hold the apply devices more securely under the higher torque loads that are normal in these gear positions.

Pressure relief valve

The pressure relief valve, figure 14-27, limits mainline pressure to a maximum of 400 psi (2,758 kPa). If mainline pressure exceeds this value, the pressure relief valve unseats to exhaust excess oil to the sump. Once mainline pressure drops below 400 psi (2,758 kPa), spring force reseats the valve.

Manual valve

The manual valve, figure 14-27, is a directional valve that is moved by a mechanical linkage from the gear selector lever. The manual valve receives fluid from the pump and directs it to apply devices and other valves in the hydraulic system in order to provide both automatic and manual upshifts and downshifts. The manual valve is held in each gear position by a spring-loaded lever.

Six passages may be charged with mainline pressure from the manual valve. When the gear selector is in Park, mainline pressure is blocked from entering the manual valve, and the valve does not direct pressure to any passages. GM refers to mainline pressure from the manual valve by the name of the passage it flows through.

The RNDI passage is charged in Reverse, Neutral, Drive, and manual second ranges; it routes pressure to the:

- Throttle valve
- Converter clutch regulator valve (125C).

The RND passage is charged in Reverse, Neutral, and Drive ranges; it routes pressure to the:

- 2-3 shift valve.

The DRIVE passage is charged in Drive, manual second, and manual low ranges; it routes pressure to the:

- Apply area of the forward clutch
- Governor valve
- 1-2 shift valve
- 1-2 accumulator valve.

The INTERMEDIATE passage is charged in manual second and manual low; it routes pressure to the:

- Reverse boost valve.

The LOW passage is charged in manual low; it routes pressure to the:

- 1-2 shift valve
- Low blow-off valve
- Apply area of the low-reverse clutch.

The REVERSE passage is charged in Reverse; it routes pressure to the:

- Reverse boost valve
- Apply areas of the high-reverse clutch
- Apply area of the low-reverse clutch.

Throttle valve and shift throttle valve

The THM 125/125C transaxle develops throttle pressure using a throttle valve operated by a mechanical linkage, figure 14-28. Throttle pressure indicates engine torque and helps time the transaxle upshifts and downshifts. Without throttle pressure, the transaxle would upshift

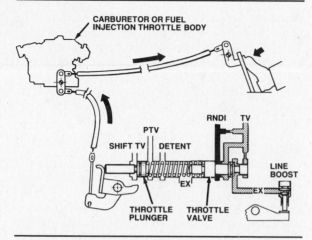

Figure 14-28. The THM 125/125C throttle valve is operated by a cable linkage from the throttle plate in the carburetor or fuel injection throttle body.

and downshift at exactly the same speed for very different throttle openings and load conditions.

The throttle valve is controlled by opposition between hydraulic pressure and spring force. The hydraulic pressure is mainline pressure (RNDI oil) that is routed to the end of the valve opposite the spring. The amount of spring force acting on the opposite end of the valve is controlled by the throttle plunger. The plunger is linked to the accelerator by a cable so that its position is determined by throttle opening.

At closed throttle, the throttle plunger exerts no force on the throttle valve spring. RNDI oil entering the throttle valve acts on the differential area of the valve, and moves the valve to the left to restrict the throttle pressure outlet port. Thus, there is little or no throttle pressure at this time.

As the accelerator is depressed, the cable linkage moves the throttle plunger to compress the throttle valve spring. This moves the throttle valve to the right, opens the outlet port, and allows regulated mainline pressure to exit the valve as throttle pressure. As the throttle is opened wider, spring force rises, the throttle valve moves farther, and throttle pressure continues to increase. However, because throttle pressure is developed from mainline pressure, it never exceeds mainline pressure.

To help balance the throttle valve, some throttle pressure is directed to the end of the valve opposite spring force. This acts to move the throttle valve to the left against spring force, and also uncovers another exhaust port that helps limit the pressure increase. At any given throttle opening, the throttle valve,

figure 14-28, reaches a balanced position, and can regulate pressure between 0 and 105 psi.

Throttle pressure is routed to the shift throttle valve, figure 14-27, which limits the pressure to a maximum of 90 psi (620 kPa). The shift throttle valve is a differential spool valve. As throttle pressure enters the center of the valve, it acts with greater force on the larger land to move the valve against spring force. If throttle pressure rises above 90 psi (620 kPa), the shift throttle valve moves far enough to open an exhaust passage that vents excess pressure to the sump. Pressure in the throttle pressure circuit that is limited by the shift throttle valve is called shift throttle pressure.

Shift throttle pressure is routed to the throttle plunger, the throttle boost valve, the 1-2 accumulator valve, and the 1-2 and 2-3 throttle valves. Shift throttle pressure to the throttle plunger helps regulate throttle pressure by increasing the spring force acting on the throttle valve; it also reduces the effort needed to hold the throttle valve plunger in place. Shift throttle pressure to the throttle boost valve raises mainline pressure in proportion to throttle opening by increasing the force acting on the spring end of the pressure regulator valve. Shift throttle pressure to the 1-2 accumulator valve combines with spring force to help cushion application of the intermediate band. Shift throttle pressure to the 1-2 and 2-3 throttle valves combines with spring force to help time the upshifts. In the THM 125C, shift throttle pressure is also routed to the spring end of the converter clutch regulator valve where it combines with spring force to help cushion application of the TCC.

When the accelerator is depressed far enough, the throttle plunger opens a port that allows shift throttle pressure to flow through a passage to the 2-3 throttle valve. Oil in this passage is called part-throttle pressure. Part-throttle pressure on the 2-3 throttle valve combines with spring force and moves the 2-3 shift valve to produce a part-throttle 3-2 downshift when the throttle is opened past a certain point while cruising in high gear.

When the accelerator is completely depressed, the throttle plunger opens a second port that allows shift throttle pressure to flow through a passage to the 1-2 and 2-3 throttle valves. Oil in this passage is called detent pressure. Detent pressure on the 1-2 and 2-3 throttle valves combines with spring force to move the 1-2 and/or 2-3 shift valve (depending on vehicle speed and gear position) to produce a wide-open throttle 3-2, 3-1, or 2-1 downshift.

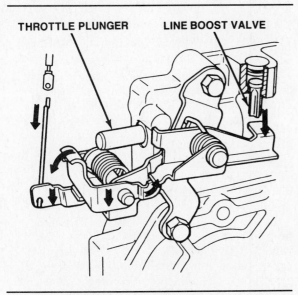

Figure 14-29. The throttle plunger linkage in the THM 125/125C also controls the line boost valve.

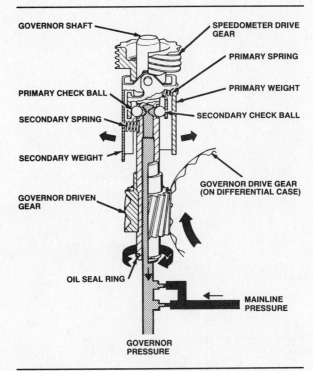

Figure 14-30. The THM 125/125C check-ball governor.

Line boost valve

The line boost valve, figure 14-28, is controlled by an extension of the throttle plunger cable linkage. In normal operation, the line boost valve is held off its seat to exhaust excess throttle pressure. However, if the throttle plunger cable breaks, is disconnected, or is badly adjusted, the linkage moves downward and allows the line boost valve to seat, figure 14-29. With no exhaust port to relieve it, throttle pressure then remains high regardless of actual throttle position. Although the transaxle will not upshift and downshift at the appropriate times, this protects the clutches and bands from slipping and burning.

Governor system

We have been discussing the development and control of throttle pressure, which is one of the two auxiliary pressures that work on the shift valves. Before we get to the shift valve operation, we must explain the governor valve (or governor), and how it creates governor pressure. Governor pressure provides a road speed signal to the hydraulic system that causes automatic upshifts to occur as road speed increases, and permits automatic downshifts with decreased road speed.

The THM 125/125C governor, figure 14-30, mounts in an opening on the top of the transaxle, and is driven by a gear on the differential case. The THM 125/125C governor uses check balls to regulate governor pressure, and consists of a governor shaft, a governor weight pin, primary and secondary weights, primary

and secondary springs, and two check balls. A gear at the top of the governor shaft turns the speedometer cable driven gear, which fits into an opening in the governor cover.

Whenever the engine is running and the gear selector is in a position other than Park, mainline pressure (DRIVE oil) passes through two restricting orifices and enters a fluid passage in the center of the governor shaft. Two exhaust ports are drilled through the shaft into the fluid passage. Governor pressure is determined by the amount of fluid that is allowed to flow out of these ports.

The exhaust port openings are controlled by check balls that seat in pockets formed at the ends of the ports. As the governor rotates, centrifugal force causes the primary weight to act on one check ball, and the secondary weight to act on the other. Each weight acts on the check ball located on the opposite side of the governor shaft. The weights are assisted in their jobs by springs that help to regulate pressure more smoothly as vehicle speed increases.

When the vehicle is stopped and the governor is not turning, there is no centrifugal force. At this time, DRIVE oil pressure overcomes the force of the primary and secondary springs, and unseats both of the check balls. All of the oil is exhausted, and governor pressure is zero at this time.

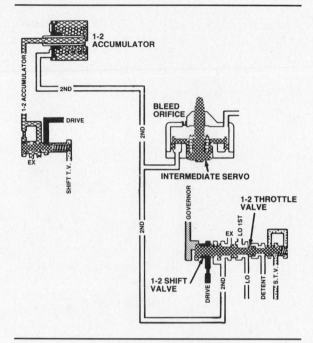

Figure 14-31. The 1-2 shift valve assembly, intermediate servo, 1-2 accumulator, and 1-2 accumulator valve in their positions just before a 1-2 upshift of the THM 125/125C.

As the vehicle accelerates and the gear on the differential case begins to turn the governor, centrifugal force moves the weights outward. This force is relayed through the springs to the check balls, partially seating them in their pockets to restrict fluid flow from the exhaust ports; this causes governor pressure to increase. As long as governor speed increases, the forces on the check balls become greater and less fluid is allowed to escape; thus governor pressure continues to rise. When governor speed remains constant, the forces acting on the check balls are also constant, and governor pressure stabilizes.

The heavier primary weight and spring are more sensitive to changes in governor rpm than are their secondary counterparts. Therefore, the primary weight and spring work to control governor pressure at lower vehicle speeds. At higher speeds, centrifugal force seats the primary check ball, and the lighter secondary weight and spring take over governor pressure regulation. Above a certain speed, centrifugal force becomes great enough to seat both check balls so that no fluid escapes from the governor. At this point, governor pressure equals mainline pressure.

Governor pressure is sent to the shift valves to oppose shift throttle pressure and spring force, and upshift the valves at the appropriate road speeds. Governor pressure is also routed to the 3-2 control valve.

1-2 shift valve
The 1-2 shift valve controls upshifts from low to second, and downshifts from second to low. The assembly includes the 1-2 shift valve, the 1-2 throttle valve, and a spring, figure 14-31. The valve is controlled by opposition between governor pressure on one end, and combined shift throttle pressure and spring force on the other end. Shift throttle pressure and spring force act on the 1-2 throttle valve to oppose an upshift.

In low gear, mainline pressure (DRIVE oil) from the manual valve flows to the 1-2 shift valve where it is blocked by the downshifted valve. When vehicle speed increases enough for governor pressure to overcome shift throttle pressure and spring force acting on the 1-2 throttle valve, the 1-2 shift valve upshifts. DRIVE oil then flows through the upshifted valve, and into the second gear passage where it travels to the apply side of the intermediate servo. Pressure in the second gear circuit is also routed to the 1-2 accumulator to cushion intermediate band engagement.

A closed-throttle 2-1 downshift occurs when governor pressure drops to a very low level (road speed decreases), and shift throttle pressure and spring force act on the 1-2 throttle valve to move the 1-2 shift valve against governor pressure.

A full-throttle 2-1 detent downshift occurs at speeds below approximately 30 mph (48 kph) when detent pressure from the throttle valve flows to the 1-2 throttle valve. The combination of detent pressure, shift throttle pressure, and spring force moves the 1-2 shift valve against governor pressure and causes a 2-1 downshift.

In manual low, mainline pressure (LOW oil) from the manual valve fills a chamber between the 1-2 shift valve and the 1-2 throttle valve. This downshifts the 1-2 shift valve and keeps it downshifted regardless of throttle opening or road speed. LOW oil is also routed to the low blow-off valve.

1-2 accumulator valve
The 1-2 accumulator valve, figure 14-31, cushions the application of the intermediate band in relation to throttle opening. It does this by creating a 1-2 accumulator pressure that combines with spring force in the accumulator to control the rate at which the accumulator piston moves through its range of travel.

When the transaxle is in any forward gear range, mainline pressure (DRIVE oil) is routed

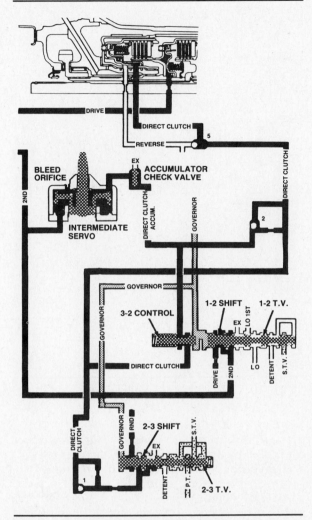

Figure 14-32. The 1-2 and 2-3 shift valve assemblies, intermediate servo, and 3-2 control valve in their positions during a part-throttle 3-2 downshift of the THM 125/125C.

to the 1-2 accumulator valve. The DRIVE pressure moves the valve against spring force and shift throttle pressure to create a variable 1-2 accumulator pressure. When shift throttle pressure is low, accumulator pressure is also low. When shift throttle pressure is high, accumulator pressure is also high. Accumulator pressure flows into the spring chamber of the accumulator.

When the 1-2 shift valve upshifts, mainline pressure flows through the second gear (servo apply) circuit to the accumulator piston where it opposes spring force and accumulator pressure. The amount of time it takes the servo apply pressure to bottom the accumulator piston determines the amount of force with which the servo and band are applied.

When accumulator pressure is low, indicating a light load on the drivetrain, servo apply oil rapidly bottoms the piston. This absorbs much of the pressure in the circuit and provides an easy band application. When accumulator pressure is high, indicating a heavy load on the drivetrain, servo apply oil takes much longer to bottom the piston. As a result, most of the pressure in the apply circuit is immediately available at the servo to apply the band much more firmly. In both cases, once the accumulator piston is bottomed, full mainline pressure is present in the circuit to hold the band securely applied.

Low blow-off valve

In manual low, mainline pressure (LOW oil) is routed from the manual valve to the 1-2 shift valve assembly, where it downshifts the 1-2 shift valve. From there, LOW oil flows to the apply area of the low-reverse clutch, and to the low blow-off valve, figure 14-22. The low blow-off valve limits direct clutch apply pressure to a maximum of 35 psi (241 kPa) for a smooth coasting downshift.

2-3 shift valve and 2-3 throttle valve

The 2-3 shift valve controls upshifts from second to third, and downshifts from third to second. The assembly includes the 2-3 shift valve, the 2-3 throttle valve, and a spring, figure 14-32. The valve is controlled by opposition between governor pressure on one end, and combined shift throttle pressure and spring force on the other. Shift throttle pressure and spring force act on the 2-3 throttle valve to oppose an upshift.

In second gear, mainline pressure (RND oil) from the manual valve flows to the 2-3 shift valve where it is blocked by the downshifted valve. When vehicle speed increases enough for governor pressure to overcome shift throttle pressure and spring force acting on the 2-3 throttle valve, the 2-3 shift valve upshifts. RND oil then flows through the upshifted valve, and into the high-reverse (direct) clutch passage where it applies the high-reverse clutch piston.

Some pressure in the high-reverse clutch circuit is routed to the release side of the intermediate servo, where the release action of the servo acts as an accumulator for high-reverse clutch application. The time that it takes the clutch apply pressure to release the intermediate servo provides a delay in high-reverse clutch application. Pressure in the high-reverse clutch circuit closes the accumulator check valve to seal the exhaust port in the passage at this time.

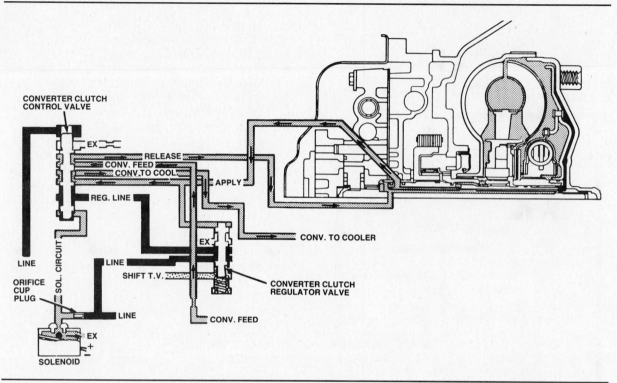

Figure 14-33. The THM 125C converter clutch regulator valve and control valve positions with the clutch released.

A closed-throttle 3-2 downshift occurs when governor pressure drops (road speed decreases) to the point where shift throttle pressure and spring force act on the 2-3 throttle valve to move the 2-3 shift valve against governor pressure.

A part-throttle 3-2 downshift occurs at speeds below 50 mph (80 kph) when part-throttle pressure from the throttle valve flows to the 2-3 throttle valve. The combination of part-throttle pressure, shift throttle pressure, and spring force moves the 2-3 shift valve against governor pressure to cause a 3-2 downshift.

A full-throttle 3-2 detent downshift occurs at speeds below 65 mph (105 kph) when detent pressure from the throttle valve flows to the 2-3 throttle valve. The combination of detent pressure, part-throttle pressure, shift throttle pressure, and spring force moves the 2-3 shift valve against governor pressure and causes a 3-2 downshift, figure 14-32. If the vehicle speed is below approximately 30 mph (48 kph) detent pressure will also downshift the 1-2 shift valve, resulting in a 3-1 downshift.

3-2 control valve

The 3-2 control valve, figure 14-32, regulates the rate of high-reverse clutch release and intermediate band application during part- and full-

throttle (detent) 3-2 downshifts. The valve is controlled by opposition between spring force and governor pressure. When the vehicle is stopped, spring force holds the 3-2 control valve wide open. As speed rises, governor pressure moves the valve against spring force to restrict the valve opening. Above approximately 50 mph (80 kph), governor pressure closes the valve entirely.

During part-throttle 3-2 downshifts below 50 mph (80 kph), the 3-2 control valve opening regulates the downshift in relation to vehicle speed. Some of the exhausting high-reverse (direct) clutch apply oil and accumulator oil flows through the 3-2 control valve. From the valve, clutch exhaust oil seats a check ball (1) and flows through two orifices to the 2-3 shift valve where it finally exhausts. This allows the high-reverse clutch to release, and the intermediate band to apply, at a rate that provides the best shift feel for a given vehicle speed.

During full-throttle 3-2 detent downshifts above 50 mph (80 kph), the 3-2 control valve is closed. As a result, exhausting accumulator oil from the intermediate servo seats a second check ball (2) and passes through an additional orifice before joining the oil exhausting from the high-reverse (direct) clutch. All of the oil then flows to seat the first check ball (1), pass

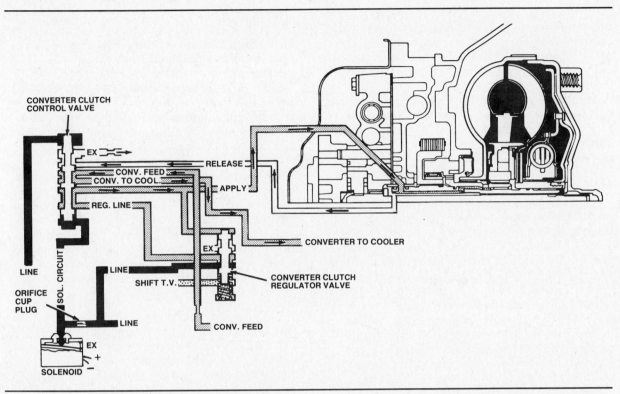

Figure 14-34. The THM 125C converter clutch regulator valve and control valve positions with the clutch applied.

through the two orifices, and exhaust at the 2-3 shift valve as already described. This causes band application to be delayed even more, allowing engine speed to increase and better match the higher gear ratio of second gear.

Converter clutch control valve (THM 125C)
The converter clutch control valve, figure 14-33, receives converter feed pressure from the pressure regulator valve, and regulated mainline pressure from the converter clutch regulator valve. It then directs these pressures to the torque converter and transaxle cooler. The position of the valve and the passages through which it directs the oil determine whether the converter clutch is released or applied. Converter clutch control valve position is controlled by opposition between mainline pressure at one end, and pressure from the TCC solenoid circuit at the other.

The electrical and electronic controls used to control the TCC solenoid are described in Chapter 8. The paragraphs below describe the hydraulic operation of the converter clutch system.

When the TCC solenoid is de-energized, figure 14-33, oil is exhausted from the solenoid circuit, and mainline pressure holds the converter clutch control valve in the release position. In this position, pressure from the converter clutch regulator valve is blocked, and

converter feed oil is routed to the converter through the clutch release passage. Oil enters the converter between the clutch pressure plate and the converter cover; this prevents the clutch from applying. Oil leaving the converter returns to the converter clutch control valve through the clutch apply passage, and is then routed to the transaxle cooler.

When the TCC solenoid is energized, figure 14-34, its exhaust port is closed and mainline pressure travels through the solenoid circuit to move the converter clutch control valve to the apply position. In this position, pressure from the converter clutch regulator valve is routed to the converter through the clutch apply passage. Oil enters the converter between the clutch pressure plate and the turbine; this forces the pressure plate into contact with the converter cover, locking up the converter. Oil leaving the converter returns to the converter clutch control valve through the clutch release passage, and is then exhausted. Converter feed oil routed to the converter clutch control valve, while the valve is in the apply position, is sent directly to the transaxle cooler.

Converter clutch regulator valve (THM 125C)
The converter clutch regulator valve controls converter clutch apply pressure in relation to throttle position. The valve is controlled by opposition between spring force and shift throttle

pressure on one end, and clutch apply pressure on the other. The converter clutch regulator valve receives mainline pressure from the manual valve in every gear range except Park and manual low.

When the converter clutch is released, figure 14-33, spring force and shift throttle pressure fully open the regulator valve. Mainline pressure then flows through the regulator valve to the converter clutch control valve where it is blocked because the control valve is in the release position.

When the converter clutch control valve moves to the apply position, figure 14-34, regulated mainline pressure from the regulator valve is routed to the converter, and also to the end of the regulator valve opposite spring force and shift throttle pressure. This moves the valve to restrict the flow of mainline pressure to the control valve and into the apply circuit.

When shift throttle pressure is low, indicating a light load on the driveline, clutch apply pressure moves the regulator valve farther to create a greater restriction; this slows the buildup of pressure in the apply circuit to smooth TCC application. When shift throttle pressure is high, indicating a heavy load on the driveline, apply pressure cannot move the regulator valve as far so a smaller restriction is created; this allows a faster buildup of pressure in the apply circuit, and firmer application of the converter clutch.

MULTIPLE-DISC CLUTCHES AND SERVOS

The preceding paragraphs examined the individual valves in the hydraulic system. Before we trace the complete hydraulic system operation in each driving range, we will take a quick look at the multiple-disc clutches and servos.

Clutches

The THM 125/125C transaxle has three multiple-disc clutches. Each clutch is applied hydraulically, and released by several small coil springs when oil is exhausted from the clutch piston cavity. The forward clutch is applied in all forward gear ranges by DRIVE oil from the manual valve.

The low-reverse clutch is applied in manual low and Reverse. In manual low, it is applied by LOW oil from the downshifted 1-2 shift valve. In Reverse, it is applied by REVERSE oil from the manual valve. A check ball prevents LOW oil from exhausting through the manual valve in manual low, and REVERSE oil from exhausting through the 1-2 shift valve in Reverse.

The high-reverse (direct) clutch is applied in Drive high and Reverse. In Drive high, it is applied by direct clutch oil from the upshifted 2-3 shift valve. In Reverse, it is applied by REVERSE oil from the manual valve. A check ball prevents direct clutch oil from exhausting through the manual valve in Drive high, and REVERSE oil from exhausting through the 2-3 shift valve in Reverse.

The high-reverse clutch is designed so that the piston has two separate surfaces on which hydraulic pressure may act. In high gear, when torque loads are low, pressure is applied to the smaller, inner area of the piston. In Reverse, when torque loads are greater, pressure is applied to the inner area of the piston, and to the larger, outer area. This allows the clutch to develop greater holding force in Reverse.

Servo

The THM 125/125C has a single servo that is used to apply the intermediate band. The servo is located at the bottom of the transaxle case, figure 14-5, and consists of a piston, an apply pin, a cushion spring, and a cover held in place by four bolts.

When a 1-2 upshift occurs, hydraulic pressure moves the piston against spring force, and the apply pin mechanically applies the band. When there is no hydraulic pressure in the piston cavity, the spring releases the servo and band.

During a 2-3 upshift, high-reverse clutch apply pressure is routed to the spring side of the intermediate servo piston. The servo then acts as an accumulator to smooth application of the high-reverse clutch. The time it takes hydraulic pressure and spring force to release the servo provides a delay in application of the high-reverse clutch.

The THM 125/125C intermediate band cannot be adjusted from outside the transaxle. Band adjustment is controlled by the length of the servo apply pin, which is selected at the time of transaxle assembly.

HYDRAULIC SYSTEM SUMMARY

We have examined the individual valves, clutches, and servos in the hydraulic system and discussed their functions. The following paragraphs summarize their combined operations in the various gear ranges. Refer to figure 14-22 to trace the hydraulic system operation.

Park

In Park, oil flows from the pump through the pressure regulator valve to the manual valve where it is blocked. All clutches and the band are released in Park, and valve positions are determined by spring force alone. A mechanical linkage engages the parking pawl with the parking gear on the outside of the rear ring gear to lock the final drive sun gear shaft. This locks the final drive planetary gearset to prevent the vehicle from being moved.

Neutral

In Neutral, oil flows from the pump through the pressure regulator valve to the manual valve. The manual valve routes RND oil to the 2-3 shift valve where it is blocked. This RND pressure application is made in all gear ranges except Park, manual second, and manual low.

The manual valve also routes RNDI oil to the throttle valve, which develops a throttle pressure that increases in proportion to throttle opening. Throttle pressure is sent to the end of the throttle valve to help balance the valve, and to the shift throttle valve. These RNDI pressure applications are made in all gear ranges except Park and manual low.

The shift throttle valve limits throttle pressure to a maximum of 90 psi (620 kPa), and routes shift throttle pressure to the: throttle plunger, throttle boost valve, 1-2 accumulator valve, 1-2 and 2-3 throttle valves, and converter clutch regulator valve (THM 125C).

Shift throttle pressure to the throttle plunger helps regulate throttle pressure and reduce the effort needed to hold the plunger in place. Shift throttle pressure to the throttle boost valve raises mainline pressure in proportion to throttle opening. Shift throttle pressure to the 1-2 accumulator valve helps cushion application of the intermediate band. Shift throttle pressure to the 1-2 and 2-3 throttle valves tends to force a downshift. Shift throttle pressure to the converter clutch regulator valve helps cushion application of the TCC.

Drive Low

When the vehicle is stopped and the gear selector is moved to Drive, the manual valve routes DRIVE oil to the: apply area of the forward clutch, 1-2 shift valve, 1-2 accumulator valve, and governor valve. These DRIVE pressure applications are made in all forward gear ranges.

The forward clutch applies to place the transaxle in first gear. Drive oil is blocked by the downshifted 1-2 shift valve. The 1-2 accumulator valve regulates DRIVE pressure in relation to throttle opening, then applies that pressure to the spring side of the 1-2 accumulator piston. The governor valve regulates mainline pressure in relation to vehicle speed, then applies that

pressure to the: 1-2 shift valve, 2-3 shift valve, 3-2 control valve, and governor pressure switch (THM 125C, non CCC only).

Governor pressure acting on the 1-2 and 2-3 shift valves tends to force an upshift. Governor pressure acting on the 3-2 control valve moves the valve against spring force to regulate the rate of intermediate band application in relation to vehicle speed during a 3-2 downshift.

In Drive low gear, the forward clutch is applied, and the 1-way clutch is holding.

Drive Second

As vehicle speed and governor pressure increase, governor pressure acting on the 1-2 shift valve overcomes shift throttle pressure and spring force, and upshifts the 1-2 shift valve. This allows DRIVE oil to flow through the 1-2 shift valve to the apply side of the intermediate servo and the 1-2 accumulator. The intermediate servo applies the intermediate band to place the transaxle in second gear. DRIVE oil in the second gear circuit moves the 1-2 accumulator piston through its range of travel to smooth band application.

In Drive second gear, the forward clutch and intermediate band are applied.

Drive High

As vehicle speed and governor pressure continue to increase, governor pressure overcomes shift throttle pressure and spring force, and upshifts the 2-3 shift valve. This allows RND pressure to flow through the shift valve to the: inner apply area of the high-reverse (direct) clutch, release side of the intermediate servo, accumulator check valve, and 3-2 control valve.

RND pressure applies the high-reverse (direct) clutch to place the transaxle in third gear. RND oil to the intermediate servo moves the servo piston through its range of travel to help release the intermediate band and cushion engagement of the high-reverse clutch. The accumulator check valve seats at this time, and RND oil simply passes through the 3-2 control valve.

When a vehicle with a 125C transaxle reaches converter clutch engagement speed in third gear, the TCC solenoid is energized to close the exhaust port. Solenoid circuit oil then moves the converter clutch control valve against mainline pressure to its apply position, and regulated mainline pressure from the converter clutch regulator valve passes into the converter apply passage, applying the converter clutch.

In Drive high gear, the forward clutch and high-reverse clutch are applied. The converter clutch in the 125C may or may not be applied, depending on vehicle operating conditions.

Drive Range Forced Downshifts

A part-throttle 3-2 downshift can be forced at speeds below approximately 50 mph (80 kph). When the accelerator is depressed beyond a certain point, the throttle plunger opens a port that allows shift throttle pressure into the part-throttle passages. Part-throttle pressure combines with shift throttle pressure and spring force acting on the 2-3 throttle valve to move the 2-3 shift valve against governor pressure and force a downshift.

A full-throttle 3-2 downshift can be forced at speeds below approximately 65 mph (105 kph). When the throttle is completely depressed, the throttle plunger opens a port that allows shift throttle pressure into the detent passages. Detent pressure combines with part-throttle pressure, shift throttle pressure, and spring force acting on the 2-3 throttle valve to move the 2-3 shift valve against governor pressure and force a downshift.

A full-throttle 2-1 downshift can be forced at speeds below approximately 30 mph (48 kph). When the throttle is completely depressed, the throttle plunger opens a port that allows shift throttle pressure into the detent passages. Detent pressure combines with part-throttle pressure, shift throttle pressure, and spring force acting on the 1-2 throttle valve to move the 1-2 shift valve against governor pressure and force a downshift.

Manual Second

When the gear selector is placed in manual second, the manual valve sends INTERMEDIATE oil to the reverse boost valve in the pressure regulator valve to increase mainline pressure. This pressure application is also made in manual low.

The manual valve also opens an exhaust port for the RND passage, which is the feed (through the 2-3 shift valve) for the high-reverse clutch and the release side of the intermediate servo. With RND oil exhausted, the high-reverse clutch is released and the intermediate servo applies the band, regardless of vehicle speed.

In manual second gear, the forward clutch and intermediate band are applied. If the vehicle speed drops low enough, shift throttle pressure and spring force acting on the 1-2 throttle valve will move the 1-2 shift valve against governor pressure and cause a coasting downshift to Drive low.

Manual Low

When the gear selector is placed in manual low, the manual valve routes LOW oil to the: 1-2 shift valve, apply area of the low-reverse clutch, and low blow-off valve. LOW oil to the 1-2 shift valve moves and holds the valve against governor pressure. LOW oil then flows through the downshifted valve to apply the low-reverse clutch and lock the transaxle in first gear. Intermediate servo oil exhausts through the 1-2 shift valve.

In manual low, the manual valve also blocks RNDI oil to the throttle valve. As a result, there are no shift throttle pressure applications to oppose governor pressure at the 1-2 shift valve. Once the valve is in its downshifted position, LOW oil and spring force hold it there regardless of vehicle speed and governor pressure.

If the gear selector is placed in manual low at a vehicle speed above approximately 40 mph (64 kph), governor pressure acting on the 1-2 shift valve will be high enough to prevent a downshift. If the transaxle is in high gear at the time the gear selector is moved, the transaxle will downshift to second gear. The 1-2 shift valve blocks LOW oil until governor pressure drops below 40 mph (64 kph) and the valve downshifts to allow LOW oil to flow to the low-reverse clutch.

In manual low gear, the forward clutch and the low-reverse clutch are applied.

Reverse

When the gear selector is placed in Reverse, the manual valve routes REVERSE oil through the RND circuit, the RNDI circuit, and the reverse circuit. REVERSE pressure in the RND circuit is blocked at the 2-3 shift valve. REVERSE pressure in the RNDI circuit is routed through the throttle valve and into the shift throttle circuit where it acts on the throttle boost valve in the pressure regulator valve to increase mainline pressure.

REVERSE oil in the reverse circuit is routed to the: inner and outer apply areas of the high-reverse clutch, apply area of the low-reverse clutch, and reverse boost valve. The high-reverse and low-reverse clutches apply to place the transaxle in reverse gear. The reverse boost valve acts on the spring end of the pressure regulator valve to increase mainline pressure and hold the apply devices more securely.

Both the DRIVE and INTERMEDIATE circuits are exhausted in Reverse, so the forward clutch and intermediate band are released. The high-reverse and low-reverse clutches are applied.

Review Questions

Select the single most correct answer
Compare your answers to the correct answers on page 603

1. Mechanic A says the THM 125/125C was engineered for use with small-displacement transverse-mounted engines in FWD cars.
   Mechanic B says the THM 125 was introduced in the 1980 model year.
   Who is right?
   a. Mechanic A
   b. Mechanic B
   c. Both mechanic A and B
   d. Neither mechanic A nor B

2. Mechanic A says the THM 125/125C uses spur gears for its final drive.
   Mechanic B says the THM 125/125C differential case and final drive carrier are a single assembly.
   Who is right?
   a. Mechanic A
   b. Mechanic B
   c. Both mechanic A and B
   d. Neither mechanic A nor B

3. Mechanic A says the THM 125/125C uses three multiple-disc clutches, one band, and a single 1-way roller clutch.
   Mechanic B says the THM 125/125C uses the Simpson compound planetary gearset.
   Who is right?
   a. Mechanic A
   b. Mechanic B
   c. Both mechanic A and B
   d. Neither mechanic A nor B

4. Mechanic A says that (through the 1988 model year) the THM 125/125C is the only 3-speed automatic transaxle used in GM FWD vehicles with transverse engines.
   Mechanic B says the THM 125/125C can be identified by the words "HYDRAMATIC" and "METRIC" stamped on the valve body cover.
   Who is right?
   a. Mechanic A
   b. Mechanic B
   c. Both mechanic A and B
   d. Neither mechanic A nor B

5. Mechanic A says the THM 125/125C requires DEXRON®-IID ATF.
   Mechanic B says THM 125 and 125C transaxles use different cases that cannot be interchanged.
   Who is right?
   a. Mechanic A
   b. Mechanic B
   c. Both mechanic A and B
   d. Neither mechanic A nor B

6. Mechanic A says the manual second range on the THM 125/125C gear selector may be marked I, S, or L2.
   Mechanic B says the manual low range on the THM 125/125C gear selector may be marked either L or L1.
   Who is right?
   a. Mechanic A
   b. Mechanic B
   c. Both mechanic A and B
   d. Neither mechanic A nor B

7. Mechanic A says the intermediate band is applied when the THM 125/125C is in Neutral.
   Mechanic B says a pawl locks the differential case to the transaxle case when the THM 125/125C is in Park.
   Who is right?
   a. Mechanic A
   b. Mechanic B
   c. Both mechanic A and B
   d. Neither mechanic A nor B

8. Mechanic A says a part-throttle 3-2 downshift is possible below 50 mph (80 kph) with the THM 125/125C.
   Mechanic B says a part-throttle 2-1 downshift is possible below 30 mph (48 kph) with the THM 125/125C.
   Who is right?
   a. Mechanic A
   b. Mechanic B
   c. Both mechanic A and B
   d. Neither mechanic A nor B

9. Mechanic A says the THM 125/125C forward clutch is also called the low-reverse clutch.
   Mechanic B says the THM 125/125C high-reverse clutch is also called the direct clutch.
   Who is right?
   a. Mechanic A
   b. Mechanic B
   c. Both mechanic A and B
   d. Neither mechanic A nor B

10. Mechanic A says the buildup of the THM 125 and 125C is essentially the same.
    Mechanic B says the THM 125C uses a 3-element torque converter with a hydraulically applied clutch.
    Who is right?
    a. Mechanic A
    b. Mechanic B
    c. Both mechanic A and B
    d. Neither mechanic A nor B

11. Mechanic A says the THM 125/125C uses a variable-displacement vane-type oil pump.
    Mechanic B says the THM 125/125C oil pump is driven by the torque converter hub.
    Who is right?
    a. Mechanic A
    b. Mechanic B
    c. Both mechanic A and B
    d. Neither mechanic A nor B

12. Mechanic A says one difference between various models of the THM 125/125C is the ratio between the drive and driven sprockets.
    Mechanic B says the THM 125/125C driven sprocket drives the input shaft, the high-reverse clutch drum, and the forward clutch hub.
    Who is right?
    a. Mechanic A
    b. Mechanic B
    c. Both mechanic A and B
    d. Neither mechanic A nor B

13. Mechanic A says no member of the planetary gearset is attached directly to the input shaft on the THM 125/125C.
    Mechanic B says the THM 125/125C front and rear sun gears are splined together.
    Who is right?
    a. Mechanic A
    b. Mechanic B
    c. Both mechanic A and B
    d. Neither mechanic A nor B

14. Mechanic A says the THM 125/125C forward clutch drives the input shell and sun gear assembly in high gear.
    Mechanic B says the THM 125/125C high-reverse clutch drives the front ring gear in all forward gears.
    Who is right?
    a. Mechanic A
    b. Mechanic B
    c. Both mechanic A and B
    d. Neither mechanic A nor B

15. Mechanic A says the THM 125/125C 1-way roller clutch prevents the front carrier from rotating counterclockwise.
    Mechanic B says the inner race of the THM 125/125C 1-way roller clutch is splined to the low-reverse clutch drum.
    Who is right?
    a. Mechanic A
    b. Mechanic B
    c. Both mechanic A and B
    d. Neither mechanic A nor B

16. Mechanic A says the THM 125/125C final drive pinion gears rotate counterclockwise when the vehicle is moving forward. Mechanic B says the THM 125/125C final drive ring gear rotates counterclockwise when the vehicle is moving in Reverse. Who is right?
    a. Mechanic A
    b. Mechanic B
    c. Both mechanic A and B
    d. Neither mechanic A nor B

17. Mechanic A says first gear reduction in the THM 125/125C is provided by the front planetary gearset only. Mechanic B says second gear reduction in the THM 125/125C is provided by the rear planetary gearset only. Who is right?
    a. Mechanic A
    b. Mechanic B
    c. Both mechanic A and B
    d. Neither mechanic A nor B

18. Mechanic A says the THM 125/125C forward and high-reverse clutches are input devices. Mechanic B says the THM 125/125C intermediate band is applied in second gear only. Who is right?
    a. Mechanic A
    b. Mechanic B
    c. Both mechanic A and B
    d. Neither mechanic A nor B

19. Mechanic A says the THM 125/125C throttle valve is cable actuated. Mechanic B says that, if the THM 125/125C throttle valve linkage is disconnected, the line boost valve will increase mainline pressure to its maximum level. Who is right?
    a. Mechanic A
    b. Mechanic B
    c. Both mechanic A and B
    d. Neither mechanic A nor B

20. Mechanic A says the THM 125/125C intermediate band is adjusted by the length of the servo rod, or apply pin. Mechanic B says the THM 125/125C intermediate servo piston also serves as an accumulator for the low-reverse clutch. Who is right?
    a. Mechanic A
    b. Mechanic B
    c. Both mechanic A and B
    d. Neither mechanic A nor B

15

Turbo Hydra-matic 180 and 180C Transmissions

HISTORY AND MODEL VARIATIONS

The General Motors (GM) Turbo Hydra-matic (THM) 180 transmission is designed for use in RWD automobiles with small-displacement engines. The THM 180 was used from 1969-75 in the imported Opel models sold by Buick. The THM 180 was also used in the Chevrolet Chevette and the Pontiac T1000 (Acadian in Canada) from 1977-81. The THM 180C, figure 15-1, replaced the THM 180 in 1982-86 models of these same cars. Some GM shop manuals and parts catalogs refer to the THM 180 as the MD-3 transmission, and the THM 180C as the MD-2 transmission.

The THM 180 and 180C transmissions are basically identical, except that the 180C has a torque converter clutch that locks up to improve fuel economy. In this chapter, we will treat the THM 180/180C as a single transmission, except where it is necessary to call out specific differences.

The THM 180/180C is a 3-speed automatic transmission that uses the Ravigneaux compound planetary gearset. Three multiple-disc clutches, one band, and a single 1-way (sprag) clutch provide the various gear ratios. All of these transmissions are built in Strasbourg, France.

Fluid Recommendations

Like all GM automatic transmissions and transaxles, the THM 180/180C requires DEXRON®-IID fluid.

Transmission Identification

Identifying the THM 180/180C transmission is relatively easy because it is the only automatic used in the GM car models listed at the beginning of this chapter. The THM 180/180C can also be identified by the fact that its bellhousing is a separate casting from the case, and that it has a vacuum modulator threaded into the right rear of the case. As required by Federal regulations, all GM transmissions carry the serial number of the vehicle in which they were originally installed.

In addition to the transmission model, you must know the exact assembly part number and other identifying characters in order to obtain proper repair parts. On the THM 180/180C, this information is stamped on a tag riveted to the left forward side of the case above the oil pan. Figure 15-2 shows a typical THM 180/180C identification tag, and explains the meaning of

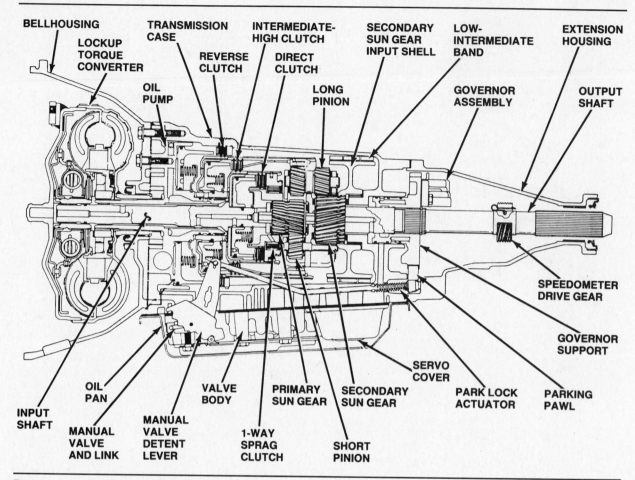

Figure 15-1. The THM 180C transmission; the THM 180 is basically the same except that it does not have a lockup torque converter.

the data stamped on it. If the case is ever replaced, the complete set of identification numbers should be transferred to the new case.

Major Changes

The major changes that GM has made in the THM 180/180C transmission are as follows:

1982 ● THM 180C with torque converter clutch introduced.

1984 ● Oil pump body, gears, and thrust plate redesigned to eliminate a whining noise.

GEAR RANGES

The THM 180/180C gear selector has six positions, figure 15-3:

P — Park
R — Reverse
N — Neutral
D — Drive
L2 — Manual second
L1 — Manual low.

Gear ratios for the THM 180/180C are:
● First (low) — 2.40:1
● Second (intermediate) — 1.48:1
● Third (direct drive) — 1.00:1
● Reverse — 1.91:1.

Park and Neutral

The engine can be started only when the transmission is in Park or Neutral. No clutches or bands are applied in these positions, and there is no flow of power through the transmission. In Neutral, the output shaft is free to turn, and the car can be pushed or pulled. In Park, a mechanical pawl engages the parking gear which is part of the governor support that is splined to the output shaft. This locks the shaft to the transmission case to prevent the vehicle from being moved.

Drive

When the vehicle is stopped and the gear selector is placed in Drive (D), the transmission is

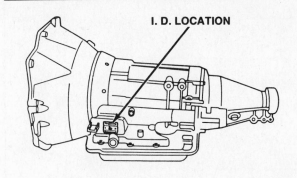

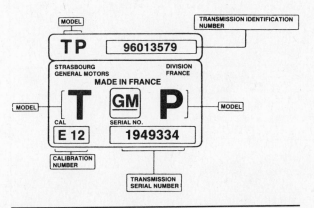

Figure 15-2. The THM 180/180C identification tag.

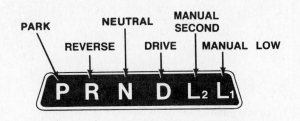

Figure 15-3. THM 180/180C gear selector positions.

automatically in low gear. As the vehicle accelerates, the transmission automatically upshifts to second, and then to third gear. The speeds at which upshifts occur vary, and are controlled by vehicle speed and engine load.

The driver can force downshifts in Drive by opening the throttle partially or completely. The exact speeds at which these downshifts occur vary for different engine, axle ratio, tire size, and vehicle weight combinations. Below approximately 40 mph (64 kph), the driver can force a part-throttle 3-2 downshift by opening the throttle beyond a certain point. Below approximately 65 mph (105 kph), the driver can force a 3-2 detent (kickdown) downshift by fully opening the throttle. A 3-1 or 2-1 detent downshift can be forced below approximately 25 mph (40 kph).

When the car decelerates in Drive to a stop, the transmission makes two downshifts automatically, first a 3-2 downshift, then a 2-1 downshift.

Manual Second

When the vehicle is stopped and the gear selector is placed in manual second (L2), the transmission is automatically in low gear. As the vehicle accelerates, the transmission automatically upshifts to second; it does not upshift to third. The speed at which the upshift occurs will vary, and is controlled by vehicle speed and engine load.

A 2-1 detent (kickdown) downshift can be forced below approximately 25 mph (40 kph) if the driver depresses the accelerator completely. The transmission automatically downshifts to low as the vehicle decelerates to a stop.

The driver can manually downshift from third to second by moving the selector from D to L2. In this case, the transmission shifts to second gear immediately, regardless of road speed or throttle position. Manual downshifting can provide engine compression braking, or gear reduction for hill climbing, pulling heavy loads, or operating in congested traffic.

Manual Low

When the vehicle is stopped and the gear selector is placed in manual low (L1), the car starts and stays in first gear; it does not upshift to second or third. Because different apply devices are used in manual low than are used in Drive low, this gear range can be used to provide engine compression braking to slow the vehicle.

If the driver shifts to manual low at high speed, the transmission will not shift directly to low gear. To avoid engine or driveline damage, the transmission first downshifts to second, and then automatically downshifts to low when the speed drops below approximately 25 mph (40 kph). The transmission then remains in low gear until the gear selector is moved to another position.

Reverse

When the gear selector is moved to R, the transmission shifts into Reverse. The transmission should be shifted to Reverse only when the vehicle is standing still. The lockout feature of the shift gate should prevent accidental shifting into Reverse while the vehicle is moving forward. However, if the driver intentionally moves the selector to R while moving forward,

GENERAL MOTORS TERM	TEXTBOOK TERM
Clutch housing	Clutch drum
Converter pump	Impeller
Detent downshift	Full-throttle downshift
Turbine shaft	Input shaft
Internal gear	Ring gear
Sun gear drum	Input shell
Input sun gear	Primary sun gear
Reaction sun gear	Secondary sun gear
Control valve assembly	Valve body
Line pressure	Mainline pressure
Modulator valve	Throttle valve
Modulator pressure	Throttle pressure
Third clutch	Direct clutch
Sprag assembly	1-way sprag clutch
Second clutch	Intermediate-high clutch
Low band	Low-intermediate band
Low servo	Low-intermediate servo

Figure 15-4. THM 180/180C transmission nomenclature table.

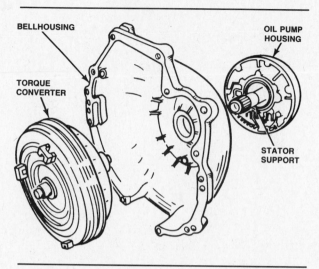

Figure 15-5. The THM 180/180C torque converter, bellhousing, and oil pump housing.

the transmission *will* shift into Reverse, and damage may result.

GM TRANSMISSION NOMENCLATURE

For the sake of clarity, this text uses consistent terms for transmission parts and functions. However, the various manufacturers may use different terms. Therefore, before we examine the buildup of the THM 180/180C, and the operation of its geartrain and hydraulic system, you should be aware of the unique names that GM uses for some transmission parts and operations. This will make it easier in the future if you should look in a GM manual and see a reference to the "input sun gear". You will realize that this is simply the primary sun gear and is similar to the same part in any other transmission that uses a similar geartrain and apply devices. Figure 15-4 lists GM nomenclature and the textbook terms for some transmission parts and operations.

THM 180/180C BUILDUP

The THM 180/180C bellhousing and extension housing are separate castings that bolt to the main case of the transmission. The torque converter fits inside the bellhousing, and is a welded unit that cannot be disassembled. The converter bolts to the engine flexplate, which has the starter ring gear around its outer diameter. The THM 180 converter is a conventional 3-element design (impeller, turbine, and stator). The THM 180C converter also contains a hydraulically applied clutch.

The torque converter is installed in the transmission by sliding it onto the stationary stator support, figure 15-5, which is part of the oil pump housing. The splines in the stator 1-way clutch engage the splines on the stator support, and the splines in the turbine engage the transmission input shaft. The torque converter hub at the rear of the converter housing engages the lugs on the oil pump inner gear to drive the pump. The converter housing and impeller are bolted to the engine flexplate so the oil pump turns and pumps fluid whenever the engine is running.

The THM 180/180C uses a gear-type oil pump mounted at the front of the transmission case. The bellhousing casting serves as the pump cover. The pump's inner gear is driven by the torque converter hub. The pump draws fluid from the sump through a filter and delivers it to the pressure regulator valve, the manual valve, and other valves in the hydraulic system.

Figure 15-6 shows an exploded view of the THM 180/180C transmission and geartrain. As we complete the buildup of this transmission, refer to figure 15-6, and the detailed figures for each assembly, to identify the components being discussed.

The transmission input shaft, figure 15-7, passes through the stator support, where it is supported by bushings, and into the transmission case. The direct clutch drum/intermediate-high clutch hub is welded to the end of the input shaft, and the entire assembly rotates clockwise whenever the torque converter turbine is driven by the impeller.

The outside of each direct clutch steel disc is splined to the inside of the direct clutch drum

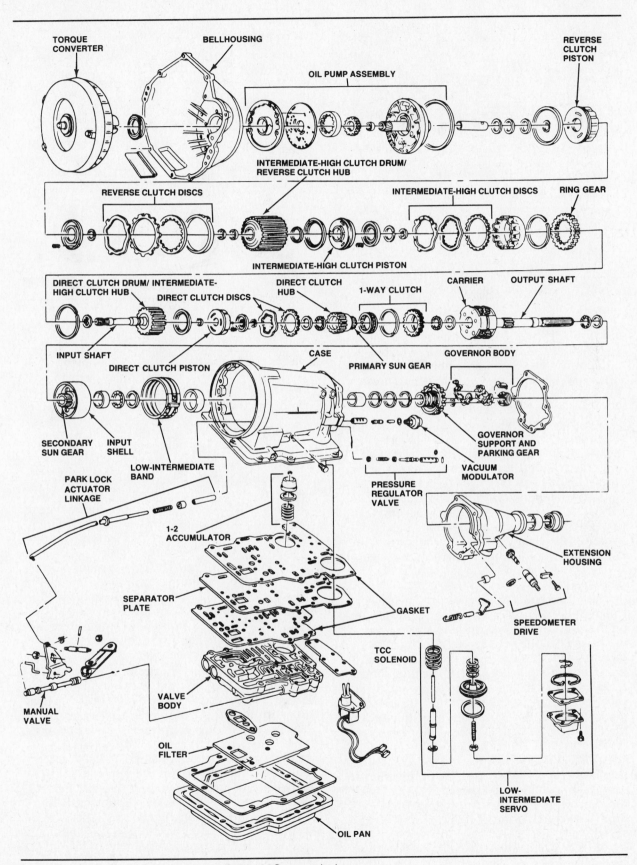

Figure 15-6. An exploded view of the THM 180C transmission.

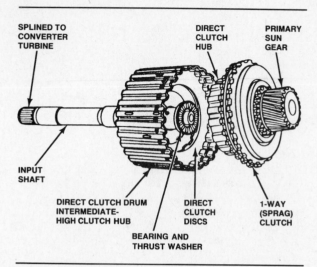

Figure 15-7. The THM 180/180C input shaft, direct clutch assembly, 1-way clutch, and primary sun gear.

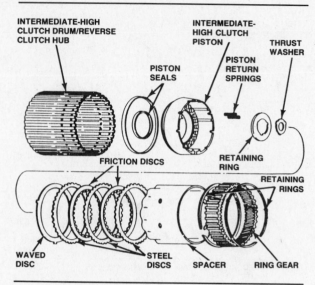

Figure 15-8. The THM 180/180C intermediate-high clutch assembly and planetary ring gear.

and turned by it. The inside of each friction disc is splined to the direct clutch hub, which is part of the primary sun gear assembly. When the direct clutch is applied, the input shaft and direct clutch drum drive the primary sun gear clockwise at turbine speed.

The outside of the 1-way (sprag) clutch outer race, figure 15-7, is also splined to the inside of the direct clutch drum and turned by it. The inner race of the 1-way clutch is part of the primary sun gear assembly. Therefore, whenever the input shaft and direct clutch drum turn, they drive the primary sun gear clockwise at turbine speed through the 1-way clutch. However, if the primary sun gear turns faster than the input shaft and direct clutch drum, the 1-way clutch overruns and the sun gear freewheels.

The intermediate-high clutch drum/reverse clutch hub, figure 15-8, is supported by a hub on the rear of the oil pump housing. Seal rings on the hub seal the oil passages to the intermediate-high clutch piston. The outside of each intermediate-high clutch steel disc is splined to the inside of the clutch drum. The inside of each friction disc is splined to the outside of the direct clutch drum, which serves as the hub for the intermediate-high clutch. Remember that the direct clutch drum turns with the input shaft. The outside of the planetary gearset ring gear is also splined to the inside of the intermediate-high clutch drum. Therefore, whenever the intermediate-high clutch is applied, it drives the clutch drum and ring gear at turbine speed.

The reverse clutch piston is housed in a bore on the back side of the oil pump body, figure 15-6. The outside of each reverse clutch steel

disc is splined to the inside of the transmission case, figure 15-9. The inside of each friction disc is splined to the outside of the intermediate-high clutch drum, which serves as the hub for the reverse clutch. When the reverse clutch is applied, it holds the intermediate-high clutch drum, and thus the planetary ring gear, stationary.

The ring gear surrounds the planetary gearset carrier and meshes with the front portions of the long pinions, figure 15-10. The ring gear may be driven by the intermediate-high clutch, or held by the reverse clutch. The long pinions mesh with the short pinions, which mesh with the primary sun gear. The primary sun gear may be driven by the 1-way clutch or the direct clutch. The rear portions of the long pinions mesh with the secondary sun gear, which is attached to a sun gear shell surrounded by the low-intermediate band. When the band is applied, the secondary sun gear is held.

Both the long and short pinions turn on pinion shafts in the carrier. The carrier is welded to the transmission output shaft, and is always the output member of the planetary gearset. The output shaft is supported by two bushings, one at the rear of the case, and another at the rear of the extension housing. Thrust washers and bearings are used: at the rear of the case, between the carrier and secondary sun gear, and between the carrier and input shaft/direct clutch drum assembly, figure 15-9. These washers and bearings take up slack, reduce friction, and keep the various components separated from one another.

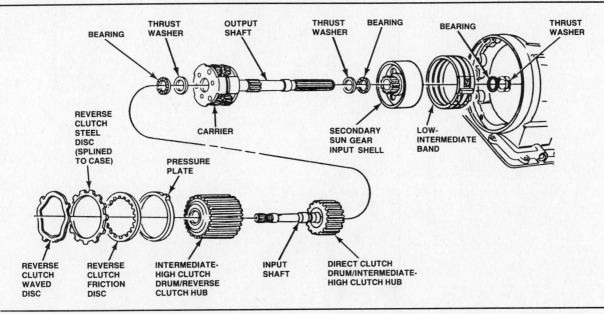

Figure 15-9. The THM 180/180C input and output shafts, including the reverse clutch, low-intermediate band, and the thrust washers and bushings used to control endplay.

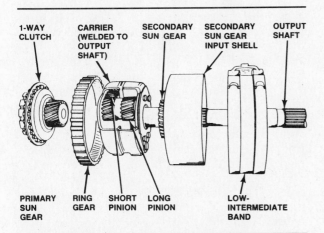

Figure 15-10. The THM 180/180C Ravigneaux planetary gearset.

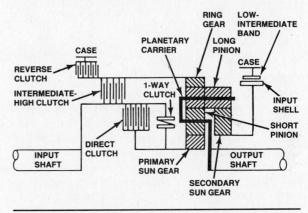

Figure 15-12. A schematic of the THM 180/180C geartrain and apply devices.

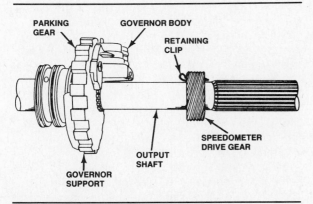

Figure 15-11. The governor assembly and speedometer drive gear are mounted on the THM 180/180C output shaft.

The output shaft passes through the extension housing and out the rear of the transmission. The governor support is splined to the output shaft, and has the parking gear cast into its outer edge, figure 15-11. The governor body bolts to the support. The speedometer drive gear is also splined to the output shaft, and is secured by a retaining clip.

THM 180/180C POWER FLOW

Now that we have taken a quick look at the general way in which the THM 180/180C is assembled, we can trace the flow of power through the apply devices and the compound planetary gearset in each of the gear ranges. Figure 15-12 shows a schematic of the THM

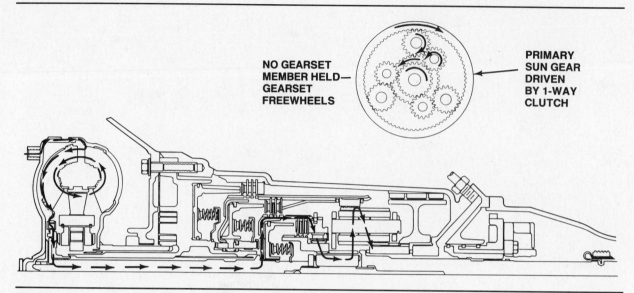

NO GEARSET MEMBER HELD— GEARSET FREEWHEELS

PRIMARY SUN GEAR DRIVEN BY 1-WAY CLUTCH

Figure 15-13. THM 180/180C power flow in Park and Neutral.

180/180C geartrain and apply devices. Refer to this figure and the specific figure for each gear range to help trace the power flow.

Park and Neutral

When the gear selector is in Park or Neutral, figure 15-13, no clutches or bands are applied, and the torque converter clutch (THM 180C) is released. The torque converter cover/impeller and the oil pump rotate clockwise at engine speed. The torque converter turbine and the input shaft also rotate at engine speed.

The input shaft drives the direct clutch drum clockwise. This turns the 1-way clutch outer race clockwise, causing the clutch to lock up and drive the primary sun gear clockwise. The primary sun gear turns the short pinions counterclockwise. The short pinions turn the long pinions clockwise. The long pinions turn the secondary sun gear counterclockwise, and the ring gear clockwise. However, because no member of the planetary gearset is held, there is no output motion; the entire gearset simply freewheels at this time.

In Neutral, the output shaft is free to rotate in either direction, so the vehicle can be pushed or pulled. In Park, a mechanical pawl engages the parking gear on the outside of the governor support and locks the output shaft to the transmission case to prevent the vehicle from moving in either direction.

Drive Low

When the car is stopped and the gear selector is moved to Drive (D), the transmission is automatically in low gear, figure 15-14. The low-intermediate band is applied to hold the shell and secondary sun gear stationary.

The input shaft drives the direct clutch drum clockwise. This turns the 1-way clutch outer race clockwise, causing the clutch to lock up and drive the primary sun gear clockwise. The primary sun gear turns the short pinions counterclockwise. The short pinions turn the long pinions clockwise. The long pinions walk around the held secondary sun gear and drive the carrier and output shaft clockwise in gear reduction for first gear. The long pinions also turn the ring gear clockwise, but it simply freewheels at this time.

As long as the car is accelerating in Drive low, torque is transferred from the engine, through the planetary gearset, to the output shaft. If the driver lets up on the throttle, engine speed drops while the drive wheels try to keep turning at the same speed. This removes the clockwise-rotating torque load from the 1-way clutch, and it unlocks. This allows the primary sun gear to freewheel, so no engine compression braking is available. This is a major difference between Drive low and manual low, as we will see later.

In low gear, the input member of the gearset is the primary sun gear. The held member is the secondary sun gear, and the output member is the carrier. The 1-way clutch is the input device, and the low-intermediate band is the holding device.

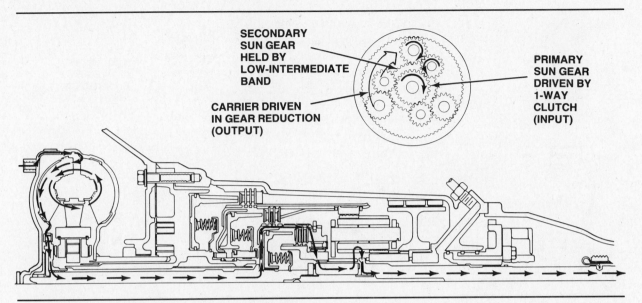

Figure 15-14. THM 180/180C power flow in Drive low.

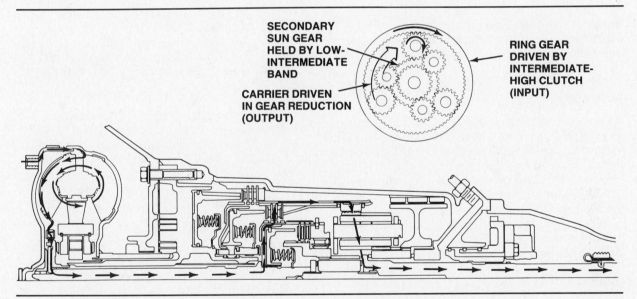

Figure 15-15. THM 180/180C power flow in Drive second.

Drive Second

As the car continues to accelerate in Drive, the transmission hydraulic system produces an automatic upshift to second gear by applying the intermediate-high clutch, figure 15-15. The low-intermediate band remains applied to hold the secondary sun gear.

The input shaft drives the direct clutch drum/intermediate-high clutch hub clockwise. The locked intermediate-high clutch drives the clutch drum and planetary ring gear clockwise. The ring gear drives the long pinions clockwise.

The long pinions walk around the held secondary sun gear and drive the carrier and output shaft clockwise in gear reduction for second gear.

The long pinions also drive the short pinions counterclockwise, and the short pinons drive the primary sun gear clockwise. However, the primary sun gear turns faster than input shaft speed, so the 1-way clutch overruns. As a result, the short pinions and primary sun gear simply freewheel at this time.

In second gear, the input member of the gearset is the ring gear. The held member is the

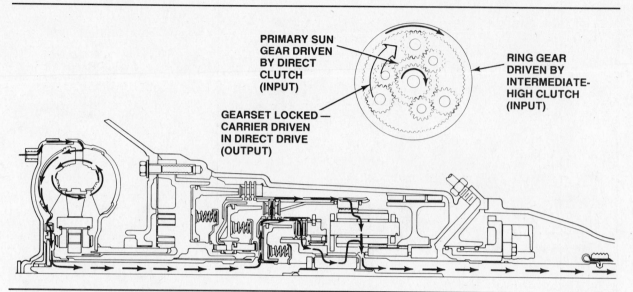

PRIMARY SUN
GEAR DRIVEN
BY DIRECT
CLUTCH
(INPUT)

RING GEAR
DRIVEN BY
INTERMEDIATE-
HIGH CLUTCH
(INPUT)

GEARSET LOCKED —
CARRIER DRIVEN
IN DIRECT DRIVE
(OUTPUT)

Figure 15-16. THM 180/180C power flow in Drive high.

secondary sun gear, and the output member is the carrier. The intermediate-high clutch is the input device, and the low-intermediate band is the holding device.

Drive High

As the car continues to accelerate in Drive, the transmission hydraulic system produces an automatic upshift from second to third gear, figure 15-16. To make this happen, the low-intermediate band is released and the direct clutch is applied. The intermediate-high clutch remains applied to drive the ring gear.

When the direct clutch is applied, it transfers power from the input shaft to the primary sun gear. Thus, both the primary sun gear and the ring gear are driven clockwise at turbine speed. As a result, the gearset locks up and rotates clockwise as a unit, turning the carrier and output shaft in direct drive. There is no reduction or increase in either speed or torque through the transmission. The 1-way clutch also acts to drive the primary sun gear in Drive high, but its action is secondary to that of the direct clutch.

In third gear, the ring gear and primary sun gear are input members equally. The carrier is the output member. The two input devices are the direct clutch and the intermediate-high clutch.

Manual Second

The only difference between manual second (L2 on the gear selector) and Drive second is

that the hydraulic system prevents the transmission from upshifting into high gear. The flow of power through the planetary gearset in first and second gears is identical, and the same apply devices are used.

When the gear selector is in manual second and the car accelerates from a standstill, the transmission starts in Drive low, figure 15-14. The 1-way clutch drives the primary sun gear, and the low-intermediate band holds the secondary sun gear. As the vehicle accelerates, the transmission automatically shifts to second by applying the intermediate-high clutch to drive the ring gear, figure 15-15.

If the driver moves the gear selector to L2 while the car is moving in Drive third, the hydraulic system will downshift the transmission to second gear by releasing the direct clutch and applying the low-intermediate band. This capability provides the driver with a controlled downshift for engine compression braking or improved acceleration at any speed or throttle position.

Manual Low

In manual low (L1 on the gear selector), the hydraulic system prevents the intermediate-high clutch from being applied, thus keeping the transmission in low gear. The flow of power through the planetary gearset is the same as in Drive low, with the exception of one apply device.

In manual low, figure 15-17, the direct clutch is applied to drive the primary sun gear. The 1-way clutch does this job in Drive low, but it

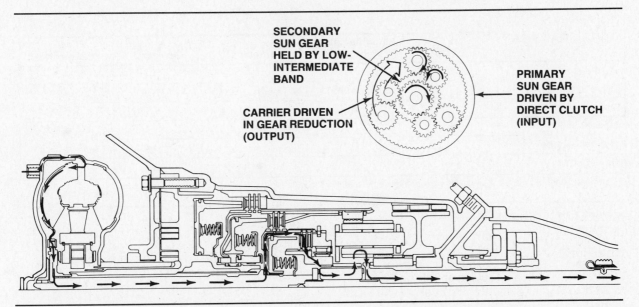

Figure 15-17. THM 180/180C power flow in manual low.

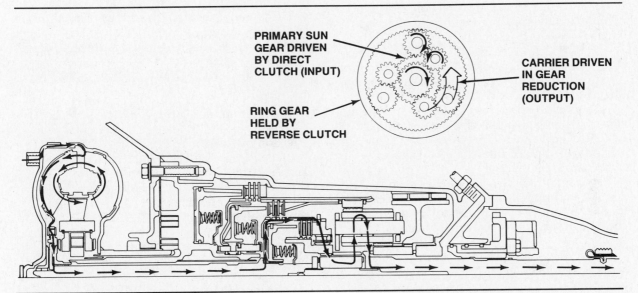

Figure 15-18. THM 180/180C power flow in Reverse.

drives the primary sun gear only during acceleration. The 1-way clutch overruns and allows the sun gear to freewheel during deceleration. In manual low, the direct clutch locks the primary sun gear to the input shaft during acceleration, and during deceleration to provide engine compression braking. The 1-way clutch continues to drive the primary sun gear during acceleration in manual low, but its action is secondary to that of the direct clutch.

The driver can move the gear selector from D to L1 while the car is moving. At speeds below approximately 25 mph (40 kph), the transmission downshifts directly from high to low gear by releasing the direct and intermediate-high clutches, and applying the low-intermediate band. At higher speeds, the transmission first shifts into second gear by releasing the direct clutch and applying the low-intermediate band. As speed drops below approximately 25 mph (40 kph), the transmission automatically downshifts from second to manual low by releasing the intermediate-high clutch.

In manual low, as in Drive low, the primary sun gear is the input member of the gearset, while the secondary sun gear is the held member and the carrier is the output member. The direct clutch is the input device, replacing the 1-way clutch, and the low-intermediate band is the holding device.

GEAR RANGES	CLUTCHES				BAND
	1-Way	Intermediate-High	Direct	Reverse	Low-Intermediate
Drive Low	●				●
Drive Second		●			●
Drive High		●	●		
Manual Low	*		●		●
Manual Second		●			●
Reverse	*		●	●	

★ The 1-way clutch will provide input in these gear ranges if the direct clutch fails, but it will not provide engine braking.

Figure 15-19. THM 180/180C clutch and band application chart.

Reverse

When the THM 180/180C gear selector is moved to Reverse (R), the input shaft still is turned clockwise by the torque converter turbine. The transmission must reverse this rotation to a counterclockwise direction. This is accomplished by applying the direct and reverse clutches, figure 15-18.

The direct clutch drives the primary sun gear clockwise. The primary sun gear turns the short pinions counterclockwise, and the short pinions turn the long pinions clockwise. The long pinions then walk counterclockwise around the inside of the ring gear, which is held stationary by the reverse clutch. This drives the carrier and output shaft counterclockwise in gear reduction for reverse gear.

The long pinions also drive the secondary sun gear counterclockwise. However, because the low-intermediate band is not applied, the secondary sun gear simply freewheels at this time.

In Reverse, the input member of the gearset is the primary sun gear. The held member is the ring gear, and the output member is the carrier. The direct clutch is the input device, and the reverse clutch is the holding device.

THM 180/180C Clutch and Band Applications

Before we move on to a closer look at the hydraulic system, we should review the apply devices that are used in each of the gear ranges in this transmission. Figure 15-19 is a clutch and band application chart for the THM 180/180C. Also, remember these facts about the apply devices in this transmission:

● The 1-way, intermediate-high, and direct clutches are input devices.
● The 1-way clutch is applied in low gear.

● The intermediate-high clutch is applied in second and third gears.
● The direct clutch is applied in third gear, manual low, and Reverse.
● The low-intermediate band and reverse clutch are holding devices.
● The low-intermediate band is applied in first and second gears.
● The reverse clutch is applied in Reverse only.

THM 180/180C HYDRAULIC SYSTEM

The THM 180/180C hydraulic system controls shifting under varying vehicle loads and speeds, as well as in response to manual gear selection by the driver. Figure 15-20 is a diagram of a complete THM 180 hydraulic system. The THM 180C system shown in figure 15-21 is similar, but includes a converter clutch control valve, TCC solenoid, and revisions to some control valves and oil passages. Where necessary, these will be called out in the sections that follow. We will begin our study of the THM 180/180C hydraulic system with the pump.

Hydraulic Pump

Fluid is stored in the oil pan at the bottom of the transmission. The oil pump draws fluid through a filter attached to the bottom of the valve body. Fluid flows through a passage in the valve body and then through a passage in the front of the transmission case and bellhousing to the pump. Remember that the bellhousing is the cover for the oil pump.

The THM 180/180C uses a gear-type oil pump that is driven by the torque converter hub and delivers fluid directly to the valve or valves in the pump housing. The THM 180 pump housing contains only the pressure regulator valve assembly. The THM 180C pump

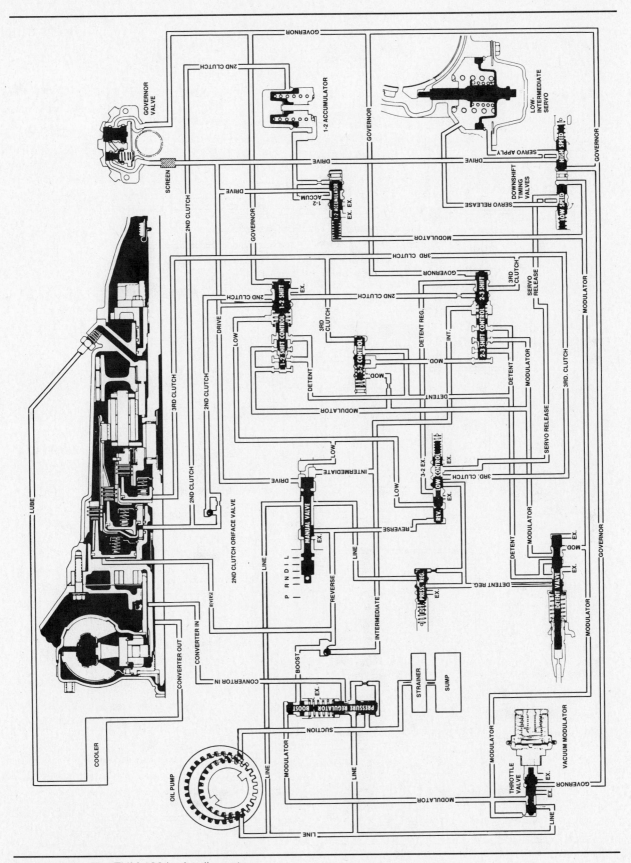

Figure 15-20. The THM 180 hydraulic system.

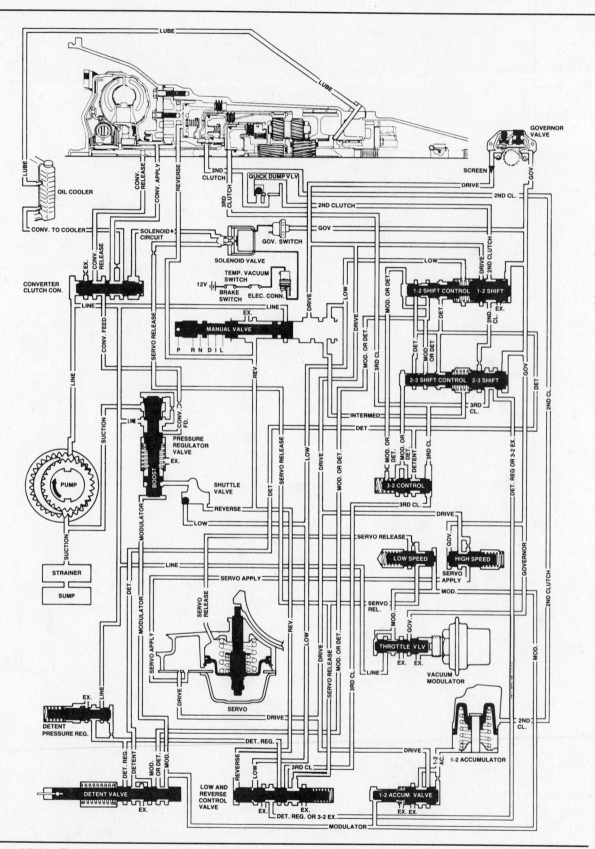

Figure 15-21. The THM 180C hydraulic system.

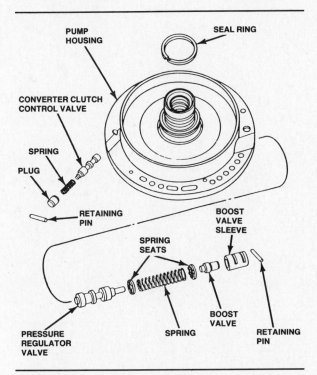

Figure 15-22. The THM 180C oil pump housing; the THM 180 housing does not contain a converter clutch control valve.

housing, figure 15-22, also contains the converter clutch control valve. For this reason, the two housings are not interchangeable.

Hydraulic Valves

The THM 180/180C hydraulic system contains the following valves:

1. Pressure regulator valve assembly
2. Manual valve
3. Throttle valve
4. Detent pressure regulator valve
5. Governor valve
6. Detent valve
7. 1-2 shift valve assembly
8. Second clutch orifice valve/quick dump valve
9. 1-2 accumulator valve
10. 2-3 shift valve assembly
11. 3-2 control valve
12. Manual low and reverse control valves
13. Shuttle valve
14. Low-speed timing valve
15. High-speed timing valve
16. Converter clutch control valve (180C only)

Valve body and ball-check valves
Except for the pressure regulator valve, governor valve, throttle valve, and converter clutch control valve, all of the hydraulic valves in the

THM 180/180C are housed in the valve body, figure 15-23. The pressure regulator and converter clutch control valves are located in the oil pump housing as already mentioned. The throttle valve is located at the right rear of the case where it is acted on by the vacuum modulator. The governor valve mounts on the output shaft in the extension housing.

The valve body bolts to the bottom of the transmission case. A separator plate and two gaskets are installed between the valve body and the case, figure 15-6. The valve body has passages to route fluid between various valves, and it forms the bottom of the 1-2 accumulator bore. Additional fluid passages, cast into the bottom of the transmission case, form the upper half of the valve body. The valve bodies for the 180/180C are different enough that they cannot be interchanged.

The THM 180 uses a single ball-check valve; the 180C uses two. In both transmissions, the check ball(s) are located in the case half of the valve body, figure 15-24.

Pressure regulator valve assembly
The pump delivers fluid through two passages to the pressure regulator valve, figure 15-25. One passage is at the end of the valve opposite the boost valves, the other is at the center of the valve. Hydraulic pressure at the end of the valve acts to move the valve against spring force. Hydraulic pressure at the center of the valve enters between two lands of equal diameter, and does not move the valve in either direction.

When the engine is started and mainline pressure begins to build, pressure at the end of the regulator valve moves the valve against spring force to open the converter inlet passage. In the THM 180, oil is routed to the torque converter, figure 15-20. Return oil from the converter flows to the transmission cooler, and from there to the lubrication circuit. In the THM 180C, figure 15-21, oil flows to the converter clutch control valve. This valve routes oil to the converter, transmission cooler, and lubrication circuit through different passages depending on whether the converter clutch is applied or released. This will be explained later in the chapter.

After the converter is filled, pressure buildup at the regulator valve moves it farther against spring force. This opens a port to the pump suction passage which vents excess pressure. Hydraulic pressure and spring force then reach a balance that provides a constant mainline pressure.

The boost valve at the spring end of the pressure regulator valve is used to raise mainline

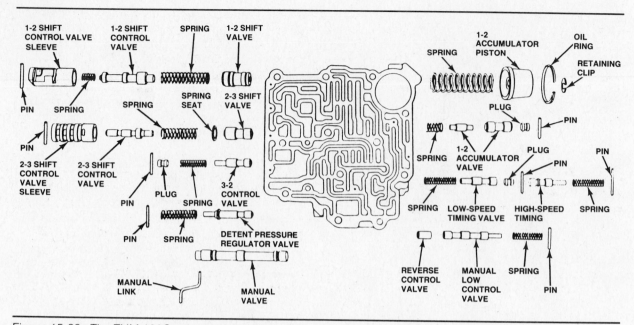

Figure 15-23. The THM 180C valve body and hydraulic control valves; the THM 180 valve body is similar.

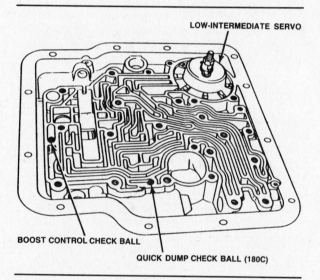

Figure 15-24. THM 180/180C check ball locations.

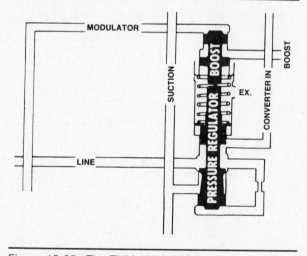

Figure 15-25. The THM 180/180C pressure regulator valve and boost valve.

pressure under certain driving conditions. In all gear positions, throttle (modulator) pressure from the throttle valve is routed to the end of the boost valve. This moves the valve to increase the spring force acting on the pressure regulator valve, thus increasing mainline pressure in proportion to throttle opening.

In Reverse, manual low, and manual second (THM 180 only) boost pressure from the manual valve is routed between the two lands of the boost valve. This pressure acts with greater force on the larger land, and again moves the valve to increase the spring force acting on the

pressure regulator valve. This increases mainline pressure up to a maximum of 160 psi to hold the apply devices more securely under the increased torque loads that can be expected in these gears.

Regulated mainline pressure is routed to the:
• Manual valve
• Throttle valve
• Detent pressure regulator valve (180C only)
• Converter clutch control valve (180C only).

Manual valve
The manual valve, figure 15-26, is a directional valve that is moved by a mechanical linkage

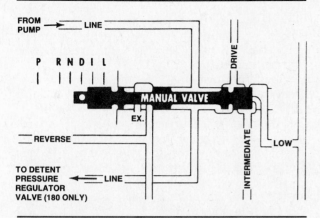

Figure 15-26. The THM 180 manual valve in Drive; the THM 180C valve is similar, but does not have the passage that routes mainline pressure to the detent pressure regulator valve.

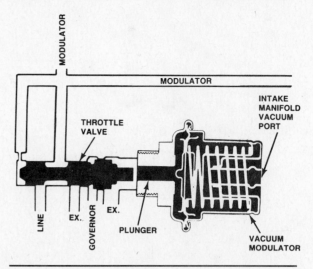

Figure 15-27. The THM 180/180C throttle valve is operated by a vacuum modulator connected to engine intake manifold vacuum.

from the gear selector lever. The manual valve receives regulated mainline pressure and directs it to apply devices and other valves in the hydraulic system to provide both automatic and manual upshifts and downshifts. The manual valve is held in each gear position by a spring-loaded lever.

Four (THM 180C) or five (THM 180) passages may be charged with mainline pressure from the manual valve. When the gear selector is in Park, mainline pressure is blocked from entering the manual valve, and the valve does not direct pressure to any passages. When the THM 180C gear selector is in Neutral, mainline pressure enters the manual valve, but does not flow into any passages. When the THM 180 gear selector is in Neutral, mainline pressure flows through the manual valve to the detent pressure regulator valve, but it does not pass into any of the other four passages. This port remains open in all gear selector positions. GM refers to mainline pressure in the other four passages from the manual valve by the name of the passage.

The DRIVE passage is charged in Drive, manual second, and manual low; it routes pressure to the:

- Governor valve
- 1-2 shift valve
- 1-2 accumulator valve
- Apply side of the low-intermediate servo
- High-speed timing valve.

The INTERMEDIATE passage is charged in manual second and manual low; it routes pressure to the:

- Spring end of the 2-3 shift valve
- Boost valve (THM 180 only).

The LOW passage is charged in manual low; it routes pressure to the:

- Spring end of the 1-2 shift valve
- Area between the reverse and manual low control valves
- Boost valve (THM 180C only).

The REVERSE passage is charged in Reverse; it routes pressure to the:

- Apply area of the reverse clutch
- Boost valve
- Reverse control valve.

Throttle valve

The THM 180/180C transmission develops throttle pressure using a throttle valve operated by a vacuum modulator, figure 15-27. For this reason, GM sometimes refers to this type of throttle valve as a modulator valve, and uses the term modulator pressure in place of throttle pressure. Throttle pressure indicates engine torque and helps time the transmission upshifts and downshifts. Without throttle pressure, the transmission would upshift and downshift at exactly the same speed for very different throttle openings and load conditions.

The vacuum modulator contains two chambers separated by a flexible rubber diaphragm. The chamber nearest the throttle valve spool is open to atmospheric pressure. The chamber farthest from the valve spool is sealed and receives engine intake manifold vacuum. The vacuum chamber also contains a spring that acts on the back side of the diaphragm to move a plunger and force the valve spool into its bore.

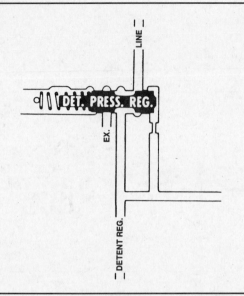

Figure 15-28. The THM 180/180C detent pressure regulator valve provides a constant regulated detent pressure.

When the engine is started, manifold vacuum is applied to the vacuum chamber of the modulator. If manifold vacuum is high enough, such as when the engine is idling or under light load, atmospheric pressure against the modulator diaphragm will oppose spring pressure with enough force to move the valve spool all the way to the right, closing the mainline pressure inlet port. When engine vacuum reaches approximately 16 inches (405 mm) of mercury, throttle pressure drops to zero.

Under most vehicle operating conditions, manifold vacuum will be insufficient to allow atmospheric pressure to cut off throttle pressure altogether. Instead, the valve spool will only partially obstruct the mainline pressure inlet port, thus regulating throttle pressure at the outlet port, figure 15-27.

When the engine comes under heavy load, manifold vacuum drops almost to zero. At these times, pressure is nearly equal in both chambers of the vacuum modulator, and spring pressure moves the valve spool all the way into its bore. Fluid then passes straight through the valve, and throttle pressure equals mainline pressure.

Besides providing a signal to other valves, throttle pressure is sent through an orifice to the end of the throttle valve opposite the vacuum modulator. This pressure acts against spring force to help balance the position of the valve and regulate throttle pressure. If manifold pressure increases (such as during deceleration), spring tension on the throttle valve decreases, so any pressure in the throttle circuit

is able to push the throttle valve to the right. Thereby cutting off line pressure and opening an exhaust to allow throttle valve pressure to return to zero.

Governor pressure is also routed to the throttle valve to help regulate throttle pressure in relation to vehicle speed. This is done because less throttle pressure is needed as speed increases. Governor pressure enters between two lands of the throttle valve and acts on the larger land to move the valve against spring force. This decreases throttle pressure as the governor pressure and vehicle speed increase.

Throttle (modulator) pressure from the throttle valve is routed to the:

- Boost valve
- 1-2 shift throttle valve
- 2-3 shift control valve
- 3-2 control valve
- Detent valve
- 1-2 accumulator valve
- Low-speed timing valve.

Detent pressure regulator valve

The detent pressure regulator valve, figure 15-28, receives mainline pressure from the manual valve (THM 180), or directly from the pump (THM 180C). This pressure opposes spring force to produce a constant regulated detent pressure. A portion of the pressure is routed through an orifice to one end of the detent pressure regulator valve to help oppose spring force and regulate the pressure.

Regulated detent pressure is routed to the:

- Detent valve
- Manual low control valve.

Governor Valve

We have been discussing the development and control of throttle pressure, which is one of the two auxiliary pressures that work on the shift valves. Before we get to the shift valve operation, we must explain the governor valve (or governor), and how it creates governor pressure. Governor pressure provides a road speed signal to the hydraulic system that causes automatic upshifts to occur as road speed increases, and permits automatic downshifts with decreased road speed.

The THM 180/180C governor body, figure 15-29, mounts on the governor support which is splined to the output shaft in the extension housing. The THM 180/180C governor contains primary and secondary spool valves, and a secondary spring.

The manual valve sends mainline pressure through a screen to the inlet port of the secondary valve, which is held open by spring force.

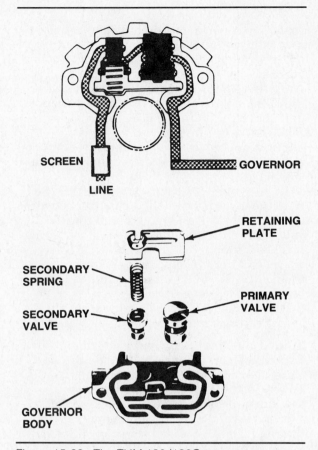

Figure 15-29. The THM 180/180C governor.

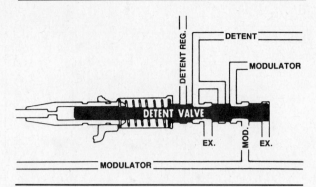

Figure 15-30. The THM 180/180C detent valve.

The oil enters between the two lands of the valve, and acts with greater force on the larger land to move the valve inward against spring force until the smaller land of the valve blocks the mainline inlet port. If the pressure between the lands is still greater than spring force, the valve moves further until an exhaust port opens. This balance between spring force and mainline pressure produces a fixed pressure that is sent to the inlet port primary valve.

Pressure enters the primary valve, which has no spring, between the two lands of the valve. When the vehicle is stopped, the pressure acts with greater force on the larger land to move the primary valve inward. This closes the inlet port and opens an exhaust port that vents the pressure coming from the secondary valve. As a result, governor pressure is zero.

As the vehicle starts to move and the output shaft and governor begin to rotate, centrifugal force acts with spring force on the secondary valve to increase the pressure sent to the primary valve. At the same time, centrifugal force acts on the primary valve to open the inlet port, close the exhaust port, and direct pressure into the governor pressure circuit.

At lower speeds, governor pressure is controlled by combined movement of the primary and secondary valves. Once vehicle speed reaches the point where the primary valve is fully open, further pressure increases are controlled by the secondary valve alone. When both the primary and secondary valve are fully open, governor pressure equals mainline pressure.

Governor pressure is directed to the:

- Throttle valve
- 1-2 shift valve
- 2-3 shift valve
- High-speed timing valve.

Detent valve

The detent valve, figure 15-30, is controlled by a mechanical linkage connected to the accelerator. The linkage moves the valve against a spring to force downshifts at full throttle. Both regulated detent pressure and throttle (modulator) pressure are routed to the detent valve.

At part-throttle, throttle pressure flows through the detent valve to the 1-2 shift control valve, 2-3 shift control valve, and 3-2 control valve. Regulated detent pressure is blocked from entering the valve at this time, and any residual pressure in the detent circuit is exhausted.

At full-throttle, the detent valve blocks throttle pressure from entering the passages to the 1-2 shift valve, 2-3 shift valve, and 3-2 control valve. Regulated detent pressure is sent through the detent passages *and* the throttle pressure passages to act on these valves and force a 3-2, 2-1, or 3-1 downshift depending on the vehicle speed.

1-2 shift valve assembly

The 1-2 shift valve controls upshifts from low to second, and downshifts from second to low. The assembly includes the 1-2 shift valve, the

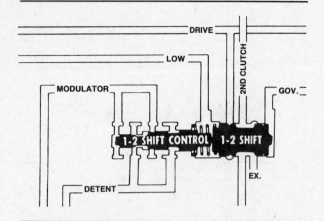

Figure 15-31. The THM 180/180C 1-2 shift valve assembly.

1-2 shift control valve, and a spring, figure 15-31. The valve is controlled by opposition between governor pressure on one end, and combined throttle (modulator) pressure and spring force on the other end. Spring force, and throttle pressure acting on two lands of the 1-2 shift control valve, oppose an upshift.

In low gear, mainline pressure (DRIVE oil) from the manual valve flows to the 1-2 shift valve where it is blocked by the downshifted valve. The downshifted valve also opens an exhaust port in the second gear passage. When vehicle speed increases enough for governor pressure to overcome throttle pressure and spring force, the 1-2 shift valve upshifts. DRIVE oil then flows through the upshifted valve, and into the second gear passage to apply the intermediate-high clutch and place the transmission in second gear. DRIVE oil is also routed to the 2-3 shift valve where it is blocked.

To help prevent hunting between first and second gears, DRIVE oil acts on a larger-diameter land of the upshifted 1-2 shift valve so that, even if governor pressure does not continue to increase, a higher throttle pressure is needed to downshift the 1-2 shift valve. In addition, throttle pressure acting on the inner land of the 1-2 shift control valve is exhausted through the detent passage. This reduces the amount of pressure acting on the 1-2 shift control valve to force a downshift.

A closed-throttle 2-1 downshift occurs when governor pressure drops to a very low level (road speed decreases), and throttle pressure and spring force move the 1-2 shift valve against governor pressure.

A part-throttle 2-1 downshift occurs at low speeds when the throttle is opened past a certain point and combined throttle pressure and spring force move the 1-2 shift valve against governor pressure.

A full-throttle 2-1 detent downshift occurs at speeds below approximately 25 mph (40 kph) when pressure from the detent valve flows to the 1-2 shift control valve. The combination of detent pressure, throttle pressure, and spring force moves the 1-2 shift valve against governor pressure and forces a downshift. A full-throttle 2-1 downshift can be forced at a higher speed than a part-throttle 2-1 downshift.

In manual low, mainline pressure (LOW oil) from the manual valve fills a chamber between the 1-2 shift valve and the 1-2 shift control valve. The combination of LOW oil and spring force keeps the 1-2 shift valve downshifted regardless of throttle opening or road speed. In the THM 180C, LOW oil is also routed to the boost valve to increase mainline pressure. In the THM 180, INTERMEDIATE oil is routed to the boost valve in manual low.

Second clutch orifice valve/quick dump valve
For correct shift timing, the intermediate-high clutch should apply slowly but release quickly. To control the application and release rates of the clutch, the second gear passage contains two orifices and a check ball. In the THM 180, figure 15-20, this is called the second clutch orifice valve. In the THM 180C, figure 15-21, it is called the quick dump valve. Both valves work exactly the same.

When pressure flows *to* the clutch, the check ball seats, closing off the main passage and forcing fluid to pass through the smaller restricting orifice. When the clutch releases, the check ball unseats and allows unrestricted fluid flow *from* the clutch through both orifices. The 1-2 accumulator valve also helps cushion intermediate-high clutch application as described in the next section.

1-2 accumulator valve
The 1-2 accumulator valve, figure 15-32, cushions the application of the intermediate band in relation to throttle opening. It does this by creating a 1-2 accumulator pressure that combines with spring force to control the rate at which the accumulator piston moves through its range of travel.

When the transmission is in any forward gear range, mainline pressure (DRIVE oil) is routed to the 1-2 accumulator valve. The DRIVE pressure moves the valve against spring force and throttle (modulator) pressure to create a variable 1-2 accumulator pressure. When throttle pressure is low, accumulator pressure is also low. When throttle pressure is high, accumulator pressure is also high. The 1-2 accumulator

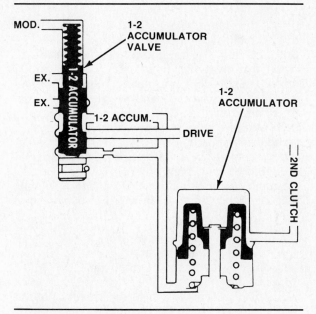

Figure 15-32. The THM 180/180C 1-2 accumulator valve and accumulator.

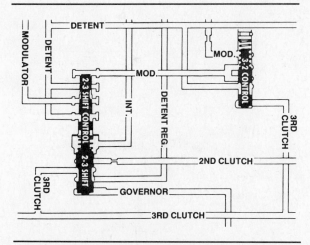

Figure 15-33. The THM 180/180C 2-3 shift valve assembly and 3-2 control valve.

pressure is routed to the spring chamber of the 1-2 accumulator.

When the 1-2 shift valve upshifts, intermediate-high clutch apply pressure (mainline pressure) flows through the second gear circuit to the accumulator piston where it opposes spring force and 1-2 accumulator pressure. The amount of time it takes second gear pressure to bottom the accumulator piston determines the amount of force with which the clutch is applied.

When the 1-2 accumulator pressure is low, indicating a light load on the drivetrain, clutch apply oil rapidly bottoms the piston. This absorbs much of the pressure in the circuit and provides an easy clutch application. When accumulator pressure is high, indicating a heavy load on the drivetrain, clutch apply oil takes longer to bottom the piston. As a result, most of the pressure in the second gear circuit is immediately available to apply the clutch much more firmly. In both cases, once the accumulator piston is bottomed, full mainline pressure is present in the circuit to hold the clutch securely applied.

2-3 shift valve assembly

The 2-3 shift valve controls upshifts from second to third, and downshifts from third to second. The assembly includes the 2-3 shift valve, the 2-3 shift control valve, and a spring, figure 15-33. The valve is controlled by opposition between governor pressure on one end, and combined throttle (modulator) pressure and spring force on the other. Spring force, and throttle

pressure acting on two lands of the 2-3 shift control valve, oppose an upshift.

When the transmission is in low and second gears, throttle pressure and spring force keep the 2-3 shift valve downshifted. In second gear, mainline pressure flows from the 1-2 shift valve to the 2-3 shift valve where it is blocked by the downshifted valve. When vehicle speed increases enough for governor pressure to overcome throttle pressure and spring force, the 2-3 shift valve upshifts. Mainline pressure then flows through the valve and into the third gear circuit where it applies the direct clutch to place the transmission in high gear. Pressure in the third gear circuit is also routed to the 3-2 control valve (see below), and to the manual low control valve which releases the low-intermediate band.

To help prevent hunting between second and third gears, mainline pressure acts on a larger-diameter land of the upshifted 2-3 shift valve so that, even if governor pressure does not continue to increase, a higher throttle pressure is needed to downshift the 2-3 shift valve. In addition, the 3-2 control valve reduces the amount of throttle pressure acting on the 2-3 shift control valve as described in the next section.

A closed-throttle 3-2 downshift occurs when governor pressure drops (road speed decreases) to the point where throttle pressure and spring force move the 2-3 shift valve against governor pressure.

A part-throttle 3-2 downshift occurs at speeds below approximately 40 mph (64 kph) when the throttle is opened past a certain point and combined throttle pressure and spring force move the 2-3 shift valve against governor pressure.

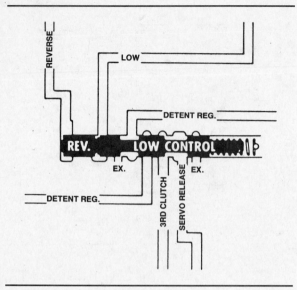

Figure 15-34. The THM 180/180C manual low and reverse control valves.

A full-throttle 3-2 detent downshift occurs at speeds below approximately 65 mph (105 kph) when pressure from the detent valve flows to the 2-3 shift control valve. The combination of detent pressure, throttle pressure, and spring force moves the 2-3 shift valve against governor pressure and forces a downshift. If the vehicle speed is below approximately 25 mph (40 kph), detent pressure will also downshift the 1-2 shift valve to cause a 3-1 downshift.

In manual second, mainline pressure (INTERMEDIATE oil) from the manual valve fills a chamber between the 2-3 shift valve and the 2-3 shift control valve. The combination of INTERMEDIATE oil and spring force keeps the 2-3 shift valve downshifted regardless of throttle opening or road speed. In the THM 180, INTERMEDIATE oil also flows to the boost valve to increase mainline pressure.

3-2 control valve

The 3-2 control valve, figure 15-33, helps time 3-2 downshifts. The valve is regulated by opposition between spring force and throttle (modulator) pressure on one end, and third gear oil at the other. When the transmission is in second gear, spring force and throttle pressure bottom the 3-2 control valve, and throttle pressure is routed through the valve to the end of the 2-3 shift control valve where it helps oppose an upshift.

When the 2-3 shift valve upshifts, direct clutch apply oil is routed to the end of the 3-2 control valve opposite spring force and throttle pressure. At smaller throttle openings when throttle pressure is low, direct clutch oil moves the 3-2 control valve to exhaust the throttle pressure that was applied to the end of the 2-3 shift control valve. This reduces the pressure opposing governor pressure at the 2-3 shift valve, and locks out a part-throttle downshift.

At larger throttle openings, increased throttle pressure combines with spring force to overcome direct clutch apply pressure. This moves the 3-2 control valve and reapplies throttle pressure to the end 2-3 shift control valve, producing a part-throttle 3-2 downshift so long as the vehicle speed is below approximately 40 mph (64 kph).

Manual low and reverse control valves

The manual low and reverse control valves, figure 15-34, are located in the same bore and do a number of jobs. When the transmission shifts from second to third gear, direct clutch apply oil flows from the 2-3 shift valve to the manual low control valve. From there, it flows to the low-speed timing valve and the release side of the low-intermediate servo. This releases the low-intermediate band as the direct clutch is being applied. In the THM 180C, servo release oil is also routed to the TCC solenoid.

In manual low, the low-intermediate band remains applied even though the direct clutch is applied to override the 1-way clutch and provide engine compression braking. To do this, the manual valve sends LOW oil to the area between the manual low and reverse control valves. This bottoms the manual low control valve against spring force to exhaust the servo release passage and prevent release of the band. Because the transmission is in low gear, direct clutch apply oil cannot come from the 2-3 shift valve as it does in third gear. Instead, the movement of the manual low control valve allows regulated detent pressure to flow into the direct clutch apply circuit. Thus, both the direct clutch and the low-intermediate band are applied.

When the transmission is placed in Reverse, the manual valve directs REVERSE oil to the reverse clutch, the boost valve, and the end of the reverse control valve. The clutch applies, the boost valve increases mainline pressure, and both the reverse control valve and the manual low control valve bottom against spring force. Because the low control valve is then in the same position as in manual low, regulated detent pressure applies the direct clutch. The low-intermediate band does not apply in Reverse because the manual valve exhausts the servo apply oil.

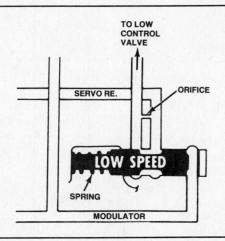

Figure 15-35. The THM 180/180C low-speed timing valve.

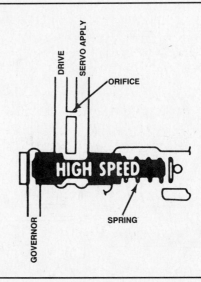

Figure 15-36. The THM 180/180C high-speed timing valve.

Shuttle valve

The shuttle valve, figures 15-20 and 15-21, sometimes called the boost control valve, is a simple 2-way check ball used to control the flow of oil to the boost valve in manual low and reverse. In manual low, the manual valve sends LOW oil to the shuttle valve. The LOW oil seats the ball against the reverse passage and flows through the boost passage to the boost valve in the pressure regulator valve assembly. In Reverse, REVERSE oil from the manual valve seats the ball against the low passage and flows to the boost valve.

Low-speed timing valve

The low-speed and high-speed timing valves work together to allow vehicle speed and throttle position to control the rate of low-intermediate band application during a 3-2 downshift. The low-speed timing valve is described below; the high-speed timing valve is covered in the next section.

The low-speed timing valve, figure 15-35, delays low-intermediate band application during a coasting 3-2 downshift until after the direct clutch is released. It does this by regulating the rate at which servo release oil flows to the low control valve where it exhausts. The low-speed timing valve is controlled by opposition between spring force at one end and throttle (modulator) pressure at the other.

When the car coasts, throttle pressure is zero and spring force closes the valve so that all of the servo release pressure must exhaust through a restricting orifice. This slows application of the low-intermediate servo and band, giving the direct clutch time to release before the band applies.

During part- and full-throttle downshifts, throttle pressure moves the low-speed timing valve against spring force. This allows part of the servo release pressure to exhaust through the valve and into the passage leading to the low control valve. The higher the throttle pressure, the farther the valve moves, the faster the servo release pressure exhausts, and the quicker the low-intermediate band applies.

High-speed timing valve

The high-speed timing valve, figure 15-36, delays low-intermediate band application during a high-speed, full-throttle forced 3-2 downshift until after the direct clutch is released. It does this by regulating the rate at which servo apply (DRIVE) oil flows to the servo. The high-speed timing valve is controlled by opposition between spring force at one end and governor pressure at the other.

During a low-speed 3-2 downshift, governor pressure is low and spring force positions the high-speed timing valve so that most of the DRIVE oil flows quickly through the valve to apply the low-intermediate servo and band.

During a high-speed 3-2 downshift, governor pressure moves the high-speed timing valve against spring force. This restricts the flow of apply oil to the low intermediate servo, and slows band application. The higher the governor pressure, the farther the valve moves, the slower the servo apply pressure flows to the servo, and the longer the low-intermediate band takes to apply. When governor pressure is very high, the high-speed timing valve is completely closed, and all of the servo apply oil must flow through a restricting orifice.

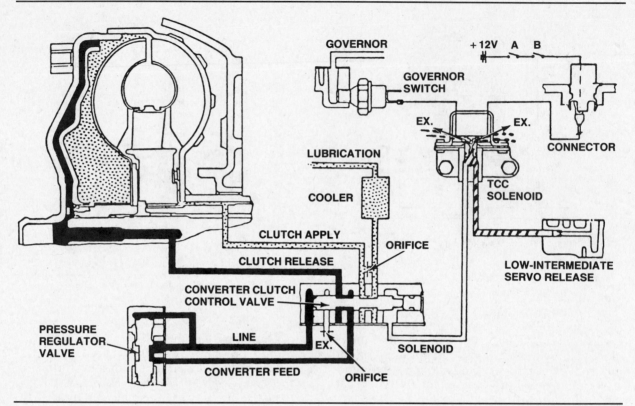

Figure 15-37. The THM 180C TCC circuit with the converter clutch control valve in the release position.

Converter clutch control valve (THM 180C)

The converter clutch control valve, figure 15-37, receives converter feed oil from the pressure regulator valve, and directs this pressure to the torque converter and transmission cooler. The position of the valve, and the passages through which it directs the oil, determine whether the converter clutch is released or applied.

The converter clutch control valve position is controlled by opposition between mainline pressure at one end, and TCC solenoid pressure at the other. Mainline pressure acts to keep the control valve in the release position; TCC solenoid pressure acts to move the control valve to the apply position. TCC solenoid pressure is actually low-intermediate servo release oil. Because this servo does not release until the transmission is in third gear, the converter clutch cannot apply in first or second gears.

The electrical and electronic controls used to control the TCC solenoid are described in Chapter 8. The paragraphs below describe the hydraulic operation of the converter clutch system.

When the TCC solenoid is de-energized, figure 15-37, oil is exhausted from the solenoid circuit, and mainline pressure holds the con-

verter clutch control valve in the release position. In this position, converter feed oil is routed to the converter through the clutch release passage. Oil enters the converter between the clutch pressure plate and the converter cover; this prevents the clutch from applying. Oil leaving the converter returns to the converter clutch control valve through the clutch apply passage, and is then routed to the transmission cooler and lubrication circuit.

When the TCC solenoid is energized, figure 15-38, its exhaust port closes and low-intermediate servo release pressure travels through the solenoid circuit to move the converter clutch control valve to the apply position. In this position, converter feed oil is routed to the converter through the clutch apply passage. Oil enters the converter between the clutch pressure plate and the turbine; this forces the pressure plate into contact with the converter cover, locking up the converter. Oil leaving the converter returns to the converter clutch control valve through the clutch release passage, and is then exhausted through a restricting orifice. A portion of the converter feed oil in the clutch apply passage is diverted through another restricting orifice to the transmission cooler and lubrication circuit.

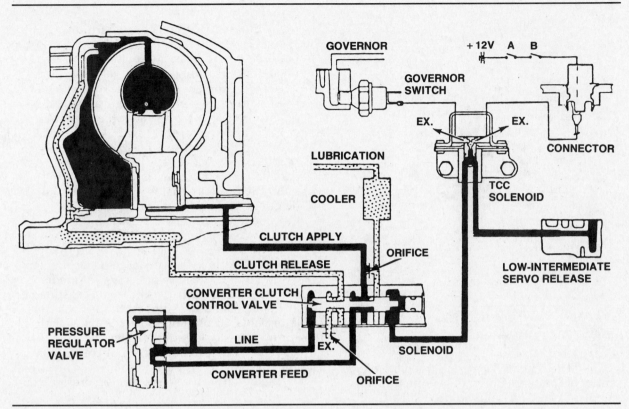

Figure 15-38. The THM 180C TCC circuit with the converter clutch control valve in the apply position.

MULTIPLE-DISC CLUTCHES AND SERVO

The preceding paragraphs examined the individual valves in the THM 180/180C hydraulic system. Before we trace the complete hydraulic system operation in each driving range, we will take a quick look at the multiple-disc clutches and servos.

Clutches

The THM 180/180C transmission has three multiple-disc clutches. Each clutch is applied hydraulically, and released by several small coil springs when oil is exhausted from the clutch piston cavity.

The reverse clutch is applied in Reverse gear only to hold the ring gear of the planetary gearset. The reverse clutch is applied by REVERSE oil from the manual valve.

The intermediate-high clutch is applied in second and third gears to drive the ring gear of the planetary gearset. It is applied by DRIVE oil routed through the 1-2 shift valve into the second gear circuit.

The direct clutch is applied in third gear, manual low, and Reverse to drive the primary

sun gear of the planetary gearset. In third gear, the direct clutch is applied by DRIVE oil routed through the 1-2 and 2-3 shift valves into the third gear circuit. In manual low and Reverse, the direct clutch is applied by regulated detent pressure from the manual low control valve.

Servo

The THM 180/180C has a single servo that is located at the bottom of the transmission case, figure 15-6. The servo performs two functions. It applies the low-intermediate band, and it acts as an accumulator for the direct clutch during a 2-3 upshift.

In first and second gears, servo apply pressure acting on the apply side of the low-intermediate servo moves the piston against spring force, and the apply pin mechanically applies the band, figure 15-39.

During a 2-3 upshift, direct clutch apply oil is routed through the manual low control valve into the servo release passage. This pressure combines with spring force to bottom the servo piston and release the low-intermediate band. The time it takes to release the servo delays direct clutch application to smooth the upshift.

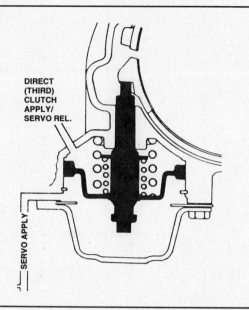

DIRECT
(THIRD)
CLUTCH
APPLY/
SERVO REL.

SERVO APPLY

Figure 15-39. The THM 180/180C low-intermediate servo.

The THM 180/180C low-intermediate band cannot be adjusted from outside the transmission. Band adjustment is controlled by the length of the servo apply pin, which is adjusted at the time of transmission assembly.

HYDRAULIC SYSTEM SUMMARY

We have examined the individual valves, clutches, and servos in the hydraulic system and discussed their functions. The following paragraphs summarize their combined operations in the various gear ranges. Refer to figures 15-20 and 15-21 to trace the hydraulic system operation.

Park and Neutral

In Park and Neutral, the clutches and band are released, and oil flows from the pump to the: pressure regulator valve, manual valve, throttle valve, detent pressure regulator valve (THM 180C), and converter clutch control valve (THM 180C).

The pressure regulator valve regulates pump oil to create mainline pressure. From the THM 180 pressure regulator valve, oil flows to the torque converter, transmission cooler, and lubrication circuit. From the THM 180C pressure regulator valve, oil flows to the converter clutch control valve, and then to the torque converter, transmission cooler, and lubrication circuit.

In Park, mainline pressure is blocked from entering the manual valve. When the THM 180C is in Neutral, mainline pressure enters

the manual valve but is blocked from leaving. When the THM 180 is in Neutral, mainline pressure passes through the manual valve to the detent pressure regulator valve.

The detent pressure regulator valve regulates mainline pressure to create a regulated detent pressure that is sent to the detent valve and manual low control valve. At both valves, regulated detent pressure is blocked.

The throttle valve regulates mainline pressure based on intake manifold vacuum to create a throttle pressure that is proportional to throttle opening. Throttle pressure is sent to the: low-speed timing valve, 1-2 accumulator valve, pressure regulator valve, and detent valve.

At the low-speed timing valve, throttle pressure acts to move the valve against spring force to control the rate at which low-intermediate servo release oil exhausts during a coasting or low-speed 3-2 downshift. At the 1-2 accumulator valve, throttle pressure combines with spring force to bottom the valve. At the boost valve, throttle pressure acts to increase mainline pressure in proportion to throttle opening.

The detent valve sends throttle pressure to the: 1-2 shift control valve, 2-3 shift control valve, and 3-2 control valve. Throttle pressure at the 1-2 shift control valve acts on two lands to help spring force keep the 1-2 shift valve downshifted. Throttle pressure at the 2-3 shift control valve acts on one land to help spring force keep the 2-3 shift valve downshifted. Throttle pressure at the 3-2 control valve passes through the valve to act on a second land of the 2-3 shift control valve to help spring force keep the 2-3 shift valve downshifted.

In the THM 180C, mainline pressure directed to the converter clutch control valve moves the valve to the release position so that the torque converter is not applied.

Drive Low

When the vehicle is stopped and the gear selector is moved to Drive, the manual valve routes DRIVE oil to the: 1-2 shift valve, high-speed timing valve, 1-2 accumulator valve, and governor valve. These DRIVE pressure applications are made in all forward gear ranges.

DRIVE oil routed to the 1-2 shift valve is blocked by the downshifted valve. DRIVE oil to the high-speed timing valve passes through the valve (and a restricting orifice) to apply the low-intermediate band. DRIVE oil to the 1-2 accumulator valve acts to oppose spring force and governor pressure, creating a 1-2 accumulator pressure that is applied to the spring side of the 1-2 accumulator piston.

The governor regulates DRIVE pressure based on output shaft rotation to create a governor pressure that is proportional to vehicle speed. Governor pressure is applied to the: 1-2 shift valve, 2-3 shift valve, throttle valve, and high-speed timing valve. In the THM 180C, governor pressure is also applied to a governor pressure switch used in controlling the TCC solenoid. These governor pressure applications are made in all forward gear ranges.

Governor pressure at the 1-2 and 2-3 shift valves acts to move the valves against throttle pressure and spring force to cause an upshift. Governor pressure at the throttle valve acts on the valve to reduce throttle pressure at higher vehicle speeds. Governor pressure at the high-speed timing valve moves the valve against spring force to delay application of the low-intermediate band during a high-speed 3-2 downshift.

In Drive low, the low-intermediate band is applied, and the 1-way clutch is holding.

Drive Second

As vehicle speed increases, governor pressure acting on the 1-2 shift valve overcomes throttle pressure and spring force to upshift the valve. This allows mainline pressure (DRIVE) oil to flow through the 1-2 shift valve into the second gear circuit where it is routed to the: 2-3 shift valve, intermediate-high clutch apply area, and 1-2 accumulator. Mainline pressure to the 2-3 shift valve is blocked by the downshifted valve. Mainline pressure applies the intermediate-high clutch to place the transmission in second gear. Mainline pressure to the 1-2 accumulator moves the accumulator piston through its range of travel to smooth intermediate-high clutch application.

In Drive second gear, the intermediate-high clutch and the low-intermediate band are applied.

Drive High

As vehicle speed continues to increase, governor pressure acting on the 2-3 shift valve overcomes throttle pressure and spring force to upshift the valve. This allows mainline pressure (DRIVE oil from the 1-2 shift valve) to flow through the 2-3 shift valve into the third gear circuit where it is routed to the: direct clutch apply area, 3-2 control valve, manual low control valve, and the release side of the low-intermediate servo.

Mainline pressure applies the direct clutch to place the transmission in high gear. Mainline pressure to the 3-2 control valve moves the valve against throttle pressure and spring force to cut off throttle pressure from one land of the 2-3 shift control valve, locking out a part-throttle downshift. Mainline pressure to the manual low control valve flows through the valve to the low-speed timing valve and the TCC solenoid (THM 180C).

Mainline pressure to the low-speed timing valve is regulated by the valve and sent to the release side of the low-intermediate servo to release the band. Direct clutch engagement is cushioned by the stroking motion of the servo piston moving to the released position. Low-intermediate servo release pressure to the TCC solenoid is vented when the solenoid is not energized. When the solenoid is energized, the pressure moves the converter clutch control valve to the apply position against mainline pressure, thus engaging the converter clutch.

In Drive high, the intermediate-high and direct clutches are applied. The THM 180C torque converter clutch may lock up if the system controls energize the TCC solenoid.

Drive Range Forced Downshifts

A part-throttle 3-2 downshift can be forced at speeds below approximately 40 mph (64 kph). When the accelerator is depressed beyond a certain point, throttle pressure combines with spring force to move the 3-2 control valve against mainline pressure, and route throttle pressure to the 2-3 shift control valve. Throttle pressure then combines with spring force to move the 2-3 shift valve against governor pressure and force a downshift.

A full-throttle 3-2 downshift can be forced at speeds below approximately 65 mph (105 kph). When the throttle is completely depressed, a mechanical linkage moves the detent valve to allow detent pressure into the detent and throttle pressure passages. Detent oil is routed through these passages and the 3-2 control valve to act on the 2-3 shift control valve. Detent pressure then combines with spring force to move the 2-3 shift valve against governor pressure and force a downshift.

A full-throttle 2-1 downshift can be forced at speeds below approximately 25 mph (40 kph). When the throttle is completely depressed, a mechanical linkage moves the detent valve to allow detent pressure into the detent and throttle pressure passages. Detent pressure acting on the 1-2 shift control valve then combines with spring force to move the 1-2 shift valve against governor pressure and force a downshift.

Manual Second

When the gear selector is placed in manual second, the manual valve routes mainline pressure (INTERMEDIATE oil) to the area between the 2-3 shift control valve and the 2-3 shift valve. INTERMEDIATE pressure then combines with spring force to hold the 2-3 shift valve in the downshifted position.

When the 2-3 shift valve is downshifted, direct clutch apply pressure and low-intermediate servo release oil are exhausted through the manual low control valve. This releases the direct clutch and reapplies the low-intermediate band to place the transmission in second gear regardless of vehicle speed.

When the THM 180 is placed in manual second, the manual valve also sends INTERMEDIATE oil to the boost valve in the pressure regulator valve to increase mainline pressure. This pressure application is made in manual low as well.

In manual second, as in Drive second, the intermediate-high clutch and the low-intermediate band are applied. If the vehicle speed drops low enough, throttle pressure and spring force will move the 1-2 shift valve against governor pressure and cause a coasting downshift to Drive low.

Manual Low

When the gear selector is placed in manual low, the manual valve routes LOW oil to the: area between the 1-2 shift control valve and the 1-2 shift valve, and the area between the manual low control valve and the reverse control valve.

LOW pressure to the area between the 1-2 shift control valve and the 1-2 shift valve combines with spring force to hold the 1-2 shift valve in the downshifted position. Pressure

in the second gear circuit is then exhausted through the 1-2 shift valve.

LOW pressure to the area between the manual low control valve and the reverse control valve bottoms the manual low control valve against spring force. This allows regulated detent pressure to flow through the manual low control valve into the third gear circuit to apply the direct clutch. The direct clutch overrides the 1-way clutch to provide engine compression braking.

When the THM 180C is placed in manual low, the manual valve also sends LOW oil to the boost valve in the pressure regulator valve to increase mainline pressure. Remember that this pressure application is made by INTERMEDIATE oil in the THM 180.

In manual low, the direct clutch and the low-intermediate band are applied.

Reverse

In Reverse, the manual valve routes mainline pressure (REVERSE oil) to the: reverse clutch, boost valve, and reverse control valve. REVERSE oil to the reverse clutch applies the clutch to place the transmission in reverse gear. REVERSE oil to the boost valve increases mainline pressure to hold the apply devices more securely. REVERSE oil to the reverse control valve bottoms that valve and the manual low control valve against spring force. With the manual low control valve bottomed, regulated detent pressure passes through it into the third gear circuit to apply the direct clutch.

The DRIVE, INTERMEDIATE, and LOW circuits are all exhausted in Reverse, so the low-intermediate band and intermediate-high clutch are released. The reverse and direct clutches are applied.

Review Questions

Select the single most correct answer
Compare your answers to the correct answers on page 603

1. Mechanic A says the THM 180 was introduced in the early 1970s. Mechanic B says the THM 180C replaced the THM 180 in 1982. Who is right?
   a. Mechanic A
   b. Mechanic B
   c. Both mechanic A and B
   d. Neither mechanic A nor B

2. Mechanic A says the THM 180/180C identification tag is attached to the left side of the case. Mechanic B says THM 180/180C transmissions are designed to use DEXRON®-IIC fluid. Who is right?
   a. Mechanic A
   b. Mechanic B
   c. Both mechanic A and B
   d. Neither mechanic A nor B

3. Mechanic A says the THM 180/180C has a 1-piece case. Mechanic B says the THM 180/180C stator support is part of the oil pump housing. Who is right?
   a. Mechanic A
   b. Mechanic B
   c. Both mechanic A and B
   d. Neither mechanic A nor B

4. Mechanic A says the THM 180 uses a 3-element torque converter. Mechanic B says the THM 180C contains a centrifugally applied clutch. Who is right?
   a. Mechanic A
   b. Mechanic B
   c. Both mechanic A and B
   d. Neither mechanic A nor B

5. Mechanic A says the THM 180/180C low-intermediate band is applied in Park. Mechanic B says the THM 180/180C can provide a part-throttle 3-2 downshift at any vehicle speed below approximately 65 mph (105 kph). Who is right?
   a. Mechanic A
   b. Mechanic B
   c. Both mechanic A and B
   d. Neither mechanic A nor B

6. Mechanic A says the THM 180/180C uses a 1-way roller clutch to drive the primary sun gear. Mechanic B says the THM 180/180C direct clutch drives the primary sun gear. Who is right?
   a. Mechanic A
   b. Mechanic B
   c. Both mechanic A and B
   d. Neither mechanic A nor B

7. Mechanic A says the THM 180/180C can provide a full-throttle 2-1 downshift at any vehicle speed below approximately 25 mph (40 kph). Mechanic B says that when the THM 180/180C gear selector is moved from Drive to manual low at 40 mph (64 kph), the transmission immediately shifts into low gear. Who is right?
   a. Mechanic A
   b. Mechanic B
   c. Both mechanic A and B
   d. Neither mechanic A nor B

8. Mechanic A says that when the THM 180/180C is in Reverse, the reverse clutch holds the secondary sun gear stationary. Mechanic B says that when the THM 180/180C is in Reverse, the intermediate-high clutch drives the ring gear. Who is right?
   a. Mechanic A
   b. Mechanic B
   c. Both mechanic A and B
   d. Neither mechanic A nor B

9. Mechanic A says the THM 180/180C direct clutch drum is welded to the input shaft. Mechanic B says the THM 180/180C intermediate-high clutch hub is welded to the input shaft. Who is right?
   a. Mechanic A
   b. Mechanic B
   c. Both mechanic A and B
   d. Neither mechanic A nor B

10. Mechanic A says the THM 180/180C 1-way clutch outer race is splined to the direct clutch drum. Mechanic B says the inner race of the THM 180/180C 1-way clutch is part of the secondary sun gear. Who is right?
    a. Mechanic A
    b. Mechanic B
    c. Both mechanic A and B
    d. Neither mechanic A nor B

11. Mechanic A says the THM 180/180C reverse clutch piston is located in the back side of the oil pump housing. Mechanic B says that no member of the THM 180/180C planetary gearset is attached directly to the input shaft. Who is right?
    a. Mechanic A
    b. Mechanic B
    c. Both mechanic A and B
    d. Neither mechanic A nor B

12. Mechanic A says the THM 180/180C carrier is driven by the low-intermediate band. Mechanic B says the THM 180/180C carrier is always the output member of the gearset. Who is right?
    a. Mechanic A
    b. Mechanic B
    c. Both mechanic A and B
    d. Neither mechanic A nor B

13. Mechanic A says the THM 180/180C 1-way clutch overruns during deceleration in Drive low gear. Mechanic B says the THM 180/180C 1-way clutch overruns during both acceleration and deceleration in Drive second gear. Who is right?
    a. Mechanic A
    b. Mechanic B
    c. Both mechanic A and B
    d. Neither mechanic A nor B

14. Mechanic A says THM 180/180C throttle (modulator) pressure is high when engine manifold vacuum is high. Mechanic B says THM 180/180C governor pressure is used to reduce throttle (modulator) pressure at higher vehicle speeds. Who is right?
    a. Mechanic A
    b. Mechanic B
    c. Both mechanic A and B
    d. Neither mechanic A nor B

15. Mechanic A says the THM 180/180C direct clutch is applied in manual low to prevent the 1-way clutch from overrunning. Mechanic B says the THM 180/180C direct clutch is applied in manual low to drive the planetary ring gear. Who is right?
    a. Mechanic A
    b. Mechanic B
    c. Both mechanic A and B
    d. Neither mechanic A nor B

16. Mechanic A says the THM 180/180C reverse clutch locks the secondary sun gear to the transmission case. Mechanic B says the THM 180/180C reverse control valve routes apply pressure to the reverse clutch. Who is right?
    a. Mechanic A
    b. Mechanic B
    c. Both mechanic A and B
    d. Neither mechanic A nor B

17. Mechanic A says both the reverse clutch and the direct clutch are applied when the THM 180/180C is in Reverse gear.
    Mechanic B says the THM 180/180C reverse clutch is applied to drive the ring gear.
    Who is right?
    a. Mechanic A
    b. Mechanic B
    c. Both mechanic A and B
    d. Neither mechanic A nor B

18. Mechanic A says the THM 180C oil pump housing contains the pressure regulator valve and converter clutch control valve.
    Mechanic B says the case half of the THM 180 valve body contains two ball-check valves.
    Who is right?
    a. Mechanic A
    b. Mechanic B
    c. Both mechanic A and B
    d. Neither mechanic A nor B.

19. Mechanic A says that the boost valve in the THM 180 is used to raise mainline pressure in manual second and manual low.
    Mechanic B says that the boost valve in the THM 180C is used to raise mainline pressure in proportion to throttle opening.
    Who is right?
    a. Mechanic A
    b. Mechanic B
    c. Both mechanic A and B
    d. Neither mechanic A nor B

20. Mechanic A says the THM 180 manual valve directs fluid to only four passages, while the THM 180C manual valve directs fluid to five passages.
    Mechanic B says the THM 180/180C detent valve is mechanically controlled to provide full-throttle downshifts.
    Who is right?
    a. Mechanic A
    b. Mechanic B
    c. Both mechanic A and B
    d. Neither mechanic A nor B

16

Turbo
Hydra-matic
200, 200C,
200-4R, 325
and 325-4L
Transmissions

HISTORY AND MODEL VARIATIONS

The General Motors (GM) Turbo Hydra-matic (THM) 200, 200C, 200-4R, 325, and 325-4L transmissions are light- and medium-duty gearboxes. The 200 series units are used in RWD applications. The 325 transmissions are used in FWD vehicles with longitudinally mounted engines. The 325 series gearboxes are not considered transaxles because the differential and final drive assembly is not built into the transmission case, but is a separate unit that bolts to the gearbox. All of these transmissions are based on the Simpson geartrain and built to metric specifications.

The 3-speed THM 200 was introduced in 1976 on some compact and subcompact cars. In 1977, use of the THM 200 was expanded to include several full-size cars with smaller engines. All GM car divisions used the THM 200, and some shop manuals and parts catalogs refer to this gearbox as the M-29 transmission.

In 1979, the THM 200 was replaced by the THM 200C, figure 16-1. The THM 200 and 200C transmissions are essentially identical, except that the 200C has a torque converter clutch that locks up to improve fuel economy. In this chapter, we will treat the THM 200/200C as a single transmission except where specific differences need to be called out. Like the THM 200, all GM car divisions use the THM 200C.

The THM 200-4R, figure 16-2, was introduced in 1982. This transmission adds an extra gearset to the THM 200/200C design to provide an overdrive fourth gear. Also, like the THM 200C, the THM 200-4R is equipped with a lockup torque converter. The THM 200-4R is used primarily on larger models produced by all of the GM car divisions.

In 1979, the 3-speed THM 325 was introduced on the FWD Buick Riviera, Cadillac Eldorado, and Oldsmobile Toronado. In 1982-85 models of these cars, the THM 325 was replaced by the THM 325-4L, which adds an extra gearset to the THM 325 design to provide an overdrive fourth gear. The THM 325-4L is also equipped with a lockup torque converter. The geartrains and hydraulic systems of the THM 325 and 325-4L are basically the same as those in the THM 200/200C and 200-4L; however, the FWD transmissions are modified so they can be mounted to one side of, and slightly below, the engine, figure 16-3.

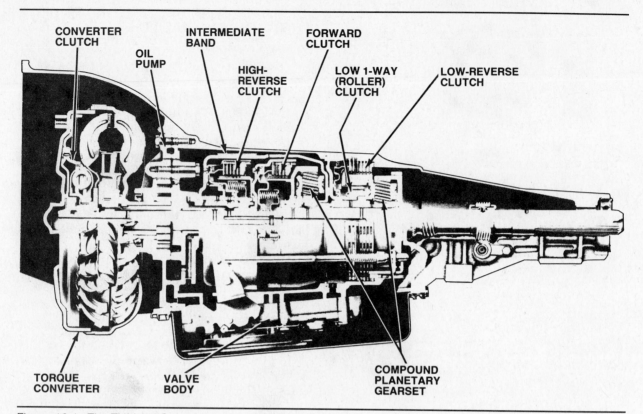

Figure 16-1. The THM 200C transmission; the THM 200 is basically the same except that it does not have a lockup torque converter.

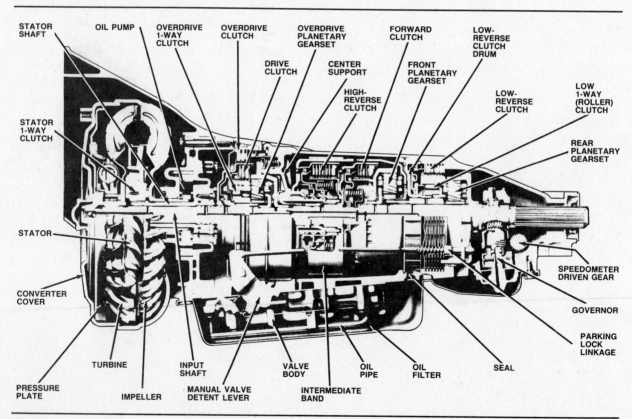

Figure 16-2. The THM 200-4R transmission; note the overdrive gearset and clutches added in front of the Simpson geartrain.

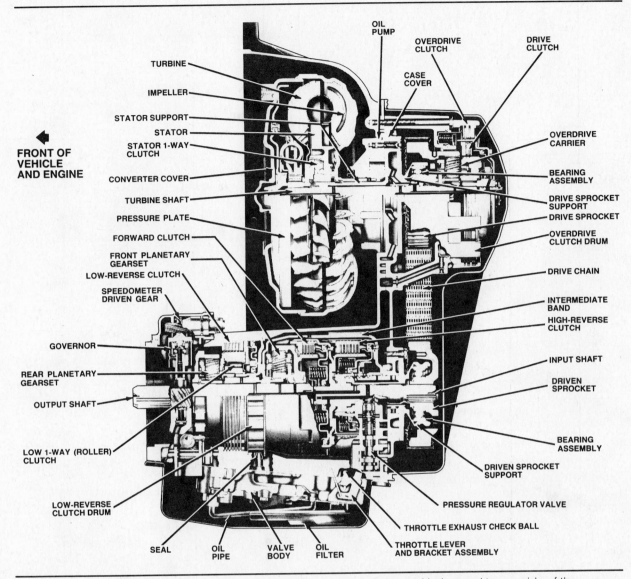

Figure 16-3. The THM 325-4L transmission; note that the geartrain is located below and to one side of the crankshaft centerline.

Fluid Recommendations

Like all GM automatic transmissions and trans-axles, the THM 200/200C, 200-4R, 325, and 325-4L transmissions require DEXRON®-IID fluid.

Transmission Identification

THM 200/200C and 200-4L transmissions can be identified by the word "METRIC" stamped on the bottom of the oil pan. The THM 200-4R is further distinguished by the Overdrive position on its gear selector. Also, as required by Federal law, the serial number of the vehicle in

which the transmission was originally installed is stamped in one of three places:

● On the lower left side of the pan below the parking pawl
● On the lower left side of the pan above the parking pawl
● On the lower right side of the bellhousing.

THM 325 and 325-4L transmissions are simple to identify because they are the only 3- and 4-speed automatic gearboxes used in the GM car models listed at the beginning of this chapter. Once again, as required by Federal regulations, these transmissions carry the serial number of the vehicle in which they were originally installed.

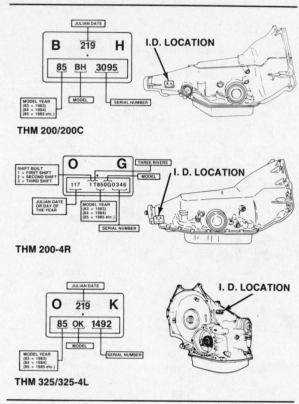

THM 200/200C

THM 200-4R

THM 325/325-4L

Figure 16-4. The identification tags for THM 200 and 325 series transmissions.

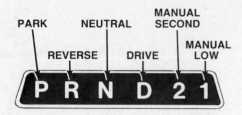

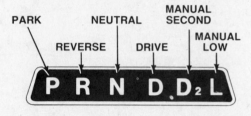

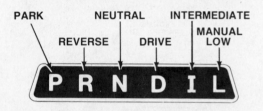

Figure 16-5. THM 200/200C and 325 gear selector positions.

In addition to the transmission model, you must know the exact assembly part number and other identifying characters in order to obtain proper repair parts. On the THM 200/200C and 200-4L, this information is stamped on a tag riveted to the right rear of the transmission case. On the THM 325 and 325-4L, a similar tag is riveted to the left side of the bellhousing. The coding that appears on the tags varies with the transmission model and the plant in which it was manufactured. Figure 16-4 shows three typical identification tags for the transmissions covered in this chapter, and explains the meaning of the data stamped on them. If the transmission case is ever replaced, the complete set of identification numbers should be transferred to the new part.

Major Changes

The type and extent of changes made to the transmissions covered in this chapter varies with the transmission model.

THM 200/200C transmissions
1982 ● Rear ring gear and output shaft modified.
     ● Input shaft design revised.

1983 ● Pump body redesigned.
     ● New third gear oil pressure switch used.
1985 ● Torque converter oil pump drive lugs redesigned.

THM 200-4R transmissions
1982 ● New output shaft snap ring used.
     ● Oil pump redesigned to eliminate a buzzing noise.
1983 ● Torque converter clutch valve assembly eliminated.
     ● 1-piece rod assembly used for Park lock.
     ● Oil pan, gasket, and bolts revised for better sealing.
1985 ● Torque converter oil pump drive lugs redesigned.

THM 325-4L transmissions
1982 ● Sprocket cover and case revised.
     ● Bottom oil pan redesigned.
1983 ● Oil pan, gasket, and bolts revised for better sealing.
1985 ● New servo cover retaining ring.
     ● Torque converter oil pump drive lugs redesigned.

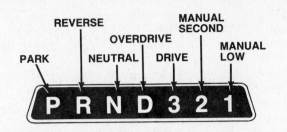

Figure 16-6. THM 200-4R and 325-4L gear selector positions.

GEAR RANGES

Depending on the GM division and the particular application, there are minor variations in the exact labeling of the gear selectors used with the 3-speed THM 200/200C and THM 325 transmissions. However, all of the gear selectors have six positions, figure 16-5:

P — Park
R — Reverse
N — Neutral
D — Drive
I, S, or L2 — Manual second
L or L1 — Manual low.

The gear selectors used with the 4-speed THM 200-4R and THM 325-4L transmissions have seven positions, figure 16-6:

P — Park
R — Reverse
N — Neutral
D — Overdrive
3 — Drive
2 — Manual second
1 — Manual low.

Gear ratios for these THM transmissions are:

- First (low) — 2.74:1
- Second (intermediate) — 1.57:1
- Third (direct drive) — 1.00:1
- Fourth (overdrive) — 0.67:1
- Reverse — 2.07:1

Park and Neutral

The engine can be started only when the transmission is in Park or Neutral. No clutches or bands are applied in these positions, and there is no flow of power through the transmission. In Neutral, the output shaft is free to turn, and the car can be pushed or pulled. In Park, a mechanical pawl engages the parking gear on the output shaft. This locks the output shaft to the transmission case and prevents the vehicle from being moved.

Overdrive

Overdrive (D) is the normal driving range for the 200-4R and 325-4L transmissions. When the car is stopped and the gear selector is placed in Overdrive, the transmission is automatically in low gear. As the vehicle accelerates, the transmission automatically upshifts to second, third, and then to fourth gear. The speeds at which upshifts occur vary, and are controlled by vehicle speed and throttle position. In the THM 200-4R and THM 325-4L, the torque converter clutch may be applied in second gear in some models, and in third and fourth gears in all models.

When a 200-4R or 325-4L transmission is in Overdrive, the driver can force downshifts by opening the throttle partially or completely. However, not all models of these transmissions are capable of providing part-throttle downshifts. The exact speeds at which downshifts occur vary for different engine, axle ratio, tire size, and vehicle weight combinations.

On some transmissions, the driver can force a part-throttle 4-3 downshift by opening the throttle beyond a certain point. On all transmissions, a 4-3 detent downshift occurs at speeds above approximately 70 mph (113 kph) when the throttle is fully opened. A full-throttle 4-2 or 3-2 downshift can be forced at speeds below approximately 70 mph (113 kph). A full-throttle 3-1 or 2-1 downshift can take place below approximately 30 mph (48 kph).

When the car decelerates in Overdrive to a stop, the transmission makes three downshifts automatically, first a 4-3 downshift, then a 3-2 downshift, then a 2-1 downshift.

Drive

Drive (D on the gear selector) is the normal driving range for 3-speed THM 200/200C and 325 transmissions. With the 4-speed THM 200-4R and 325-4L transmissions, Drive (3 on the gear selector) can be used to lock out fourth gear and provide engine compression braking, or gear reduction for hill climbing, pulling heavy loads, or operating in congested traffic.

When the vehicle is stopped and the gear selector is placed in Drive, the transmission is automatically in low gear. As the vehicle accelerates, the transmission automatically upshifts to second, and then to third gear. The 4-speed THM 200-4R and 300-4L transmissions do not upshift to fourth gear. The speeds at which upshifts occur vary, and are controlled by vehicle speed and engine load. In the THM 200C, the torque converter clutch may be applied in third gear only.

GENERAL MOTORS TERM	TEXTBOOK TERM
Clutch housing	Clutch drum
Converter pump	Impeller
Detent downshift	Full-throttle downshift
Internal gear	Ring gear
Input drum	Input shell
Input planetary gearset	Front planetary gearset
Reaction planetary gearset	Rear planetary gearset
Control valve assembly	Valve body
Line pressure	Mainline pressure
T.V. valve	Throttle valve
T.V. pressure	Throttle pressure
T.V. limit valve	Throttle limit valve
T.V. plunger	Throttle plunger
M.T.V. up and down valves	Modulated throttle upshift and downshift valves
Shift T.V. valve	Shift throttle valve
Fourth clutch	Overdrive clutch
Overrun clutch	Drive clutch
Direct clutch	High-reverse clutch

Figure 16-7. GM transmission nomenclature table.

The driver can force downshifts in Drive by opening the throttle partially or completely. The speeds at which downshifts occur vary for different engine, axle ratio, tire size, and vehicle weight combinations. The downshift speeds for the THM 200-4R and 325-4L transmissions are given in the preceding section. The downshift speeds for the THM 200/200C and 325 transmission are as follows.

Below approximately 45 mph (72 kph) in the THM 200/200C, or 50 mph (80 kph) in the THM 325, the driver can force a part-throttle 3-2 downshift by opening the throttle beyond a certain point. Below approximately 65 mph (105 kph), the driver can force a 3-2 detent downshift by fully opening the throttle. A full-throttle 3-1 or 2-1 downshift can be forced below approximately 25 mph (40 kph) in the THM 200/200C, or 40 mph (64 kph) in the THM 325.

When the car decelerates in Drive to a stop, the transmission makes two downshifts automatically, first a 3-2 downshift, then a 2-1 downshift.

Manual Second

When the vehicle is stopped and the gear selector is placed in manual second (I, S, 2, or L2), the transmission is automatically in low gear. As the vehicle accelerates, the transmission automatically upshifts to second gear; it does not upshift to third or fourth. The speed at which the upshift occurs will vary, and is controlled by vehicle speed and engine load.

A 2-1 detent downshift can be forced below the speeds given in the preceding two sections

if the driver depresses the accelerator completely. The transmission automatically downshifts to low as the vehicle decelerates to a stop.

The driver can manually downshift from fourth or third to second by moving the selector from D or 3 to I, S, 2, or L2. In this case, the transmission shifts to second gear immediately, regardless of road speed or throttle position. Second gear can then be used to provide engine compression braking, or gear reduction for hill climbing, pulling heavy loads, or operating in congested traffic.

Manual Low

When the vehicle is stopped and the gear selector is placed in manual low (L, 1, or L1), The car starts and stays in first gear; it does not upshift to second, third, or fourth. Because different apply devices are used in manual low than are used in Overdrive or Drive low, this gear range can be used to provide engine compression braking to slow the vehicle.

If the driver shifts to manual low at high speed, the transmission will not shift directly to low gear. To avoid engine or driveline damage, the transmission first downshifts to second, and then automatically downshifts to low when the vehicle speed drops to between 40 and 25 mph (64 and 40 kph), depending on the transmission model. The transmission then remains in low gear until the gear selector is moved to another position.

Reverse

When the gear selector is moved to R, the transmission shifts into Reverse. The transmission should be shifted to Reverse only when the vehicle is standing still. The lockout feature of the shift gate should prevent accidental shifting into Reverse while the vehicle is moving forward. However, if the driver intentionally moves the selector to R while moving forward, the transmission *will* shift into Reverse, and damage may result.

GM TRANSMISSION NOMENCLATURE

For the sake of clarity, this text uses consistent terms for transmission parts and functions. However, the various manufacturers may use different terms. Therefore, before we examine the buildup of the transmissions covered in this chapter, and the operation of their geartrains and hydraulic systems, you should be aware of the unique names that GM uses for some transmission parts and operations. This will make it

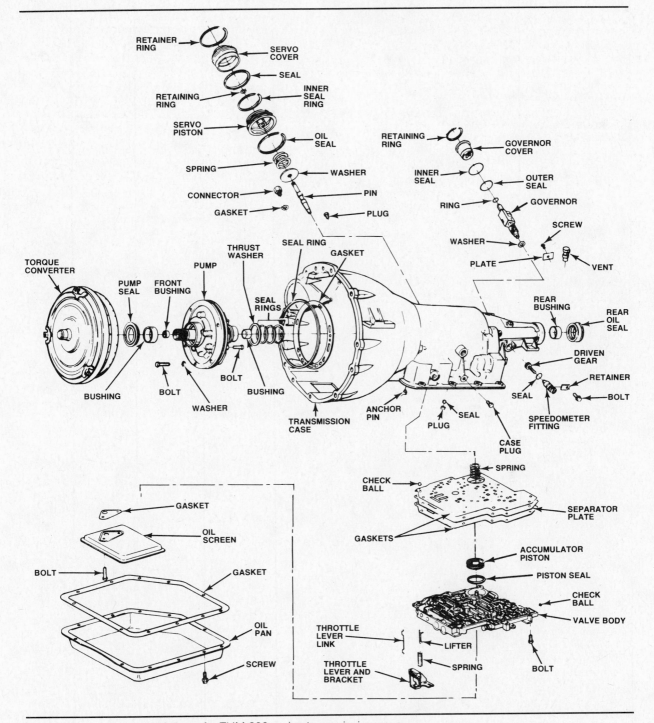

Figure 16-8. An exploded view of a THM 200 series transmission case.

easier in the future if you should look in a GM manual and see a reference to an "internal gear". You will realize that this is a ring gear, and is similar to the same part in any other transmission that uses a similar geartrain and apply devices. Figure 16-7 lists GM nomenclature and hte textbook terms for some transmission parts and operations.

THM 200/200C, 200-4R, 325, AND 325-4L BUILDUP

Although the transmissions covered in this chapter have a great deal in common, they also have some significant differences. For this reason, the buildup is broken down into four sec-

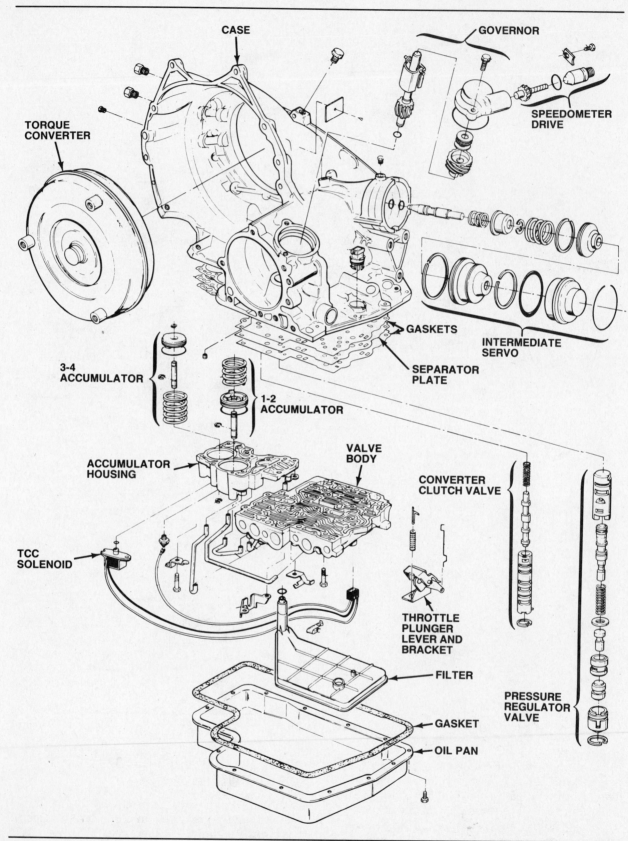

Figure 16-9. An exploded view of a THM 325 series transmission case.

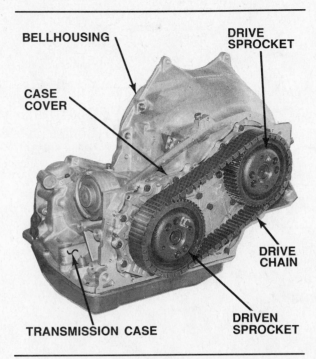

Figure 16-10. The THM 325 drive chain assembly.

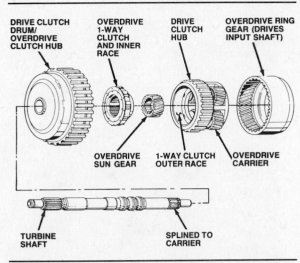

Figure 16-11. The THM 200-4R overdrive gearset assembly.

tions to help clarify the ways in which these gearboxes are assembled:

- Initial buildup
- THM 325 series drive chain assembly
- THM 200-4R and 325-4L overdrive assembly
- Final buildup.

Initial Buildup

All the transmissions covered in this chapter use a 1-piece case casting that incorporates the bellhousing and extension housing. Figure 16-8 shows an exploded view of a typical THM 200 series transmission case; figure 16-9 shows a similar view for the THM 325 series transmissions.

In the THM 200 series, the oil pump assembly bolts to the front of the case, and the oil pan bolts to the bottom. In the THM 325 series, the oil pump assembly bolts to a case cover that bolts to the back side of the main case. The drive chain cover also bolts to the back of the transmission case, and the oil pan bolts to the bottom of the case.

The torque converter fits inside the bellhousing, and is a welded unit that cannot be disassembled. The converter bolts to the engine flexplate, which has the starter ring gear around its outer diameter. The THM 200 and 325 converters are conventional 3-element designs (impeller, turbine, and stator). The THM 200C, 200-4L, and 325-4R converters also contain a hydraulically applied clutch.

The torque converter is installed in the transmission by sliding it onto the stationary stator support. The stator support is part of the oil pump housing. The splines in the stator 1-way clutch engage the splines on the stator support. In the THM 200/200C, the splines in the turbine engage the transmission input shaft. In the THM 200-4R and 325 series, the splines in the turbine engage the turbine shaft.

The hub at the rear of the torque converter housing engages the lugs on the oil pump inner gear or rotor to drive the pump. The converter housing and impeller are bolted to the engine flexplate so the oil pump turns and pumps fluid whenever the engine is running.

The THM 200/200C, 325, and 325-4L transmissions have gear-type oil pumps. The THM 200/200C pump mounts in the transmission case directly behind the torque converter. The THM 325 series oil pump, also positioned directly behind the torque converter, bolts to the case cover. The THM 200-4R has a variable-displacement vane-type pump mounted in the same position as the gear-type pump in the THM 200/200C. In all these transmissions, the oil pump draws fluid from the sump through the filter, and delivers it to the pressure regulator valve, the manual valve, and other valves in the hydraulic system.

THM 325 Series Intermediate Buildup

In the THM 325, the turbine shaft is splined to and turns the drive sprocket, which turns the drive chain, figure 16-10. The drive chain then turns the driven sprocket, which turns the transmission input shaft. From this point on,

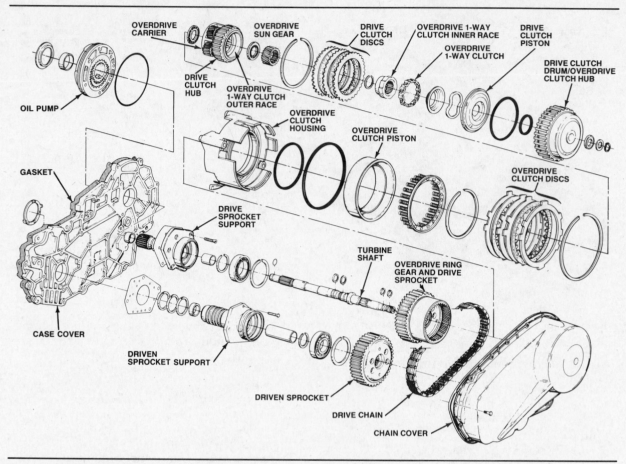

Figure 16-12. An exploded view of the THM 325-4L overdrive gearset assembly, including the drive and overdrive clutches, and the drive chain assembly.

the buildup is as described in the Final Buildup section later in the chapter. The THM 325-4L also uses sprockets and a drive chain to get power to the transmission geartrain, but the drive sprocket is built into the overdrive assembly described in the next section.

Power flow to and through the geartrain of the THM 325 series transmissions can be looked at in two ways. Viewed from the *front* of the engine, the torque converter and turbine shaft rotate clockwise and turn the drive sprocket, drive chain, driven sprocket, and input shaft clockwise as well. However, when viewed from the *rear* of the engine, looking toward the "front" or input end of the transmission, these components turn counterclockwise. To simplify the power flow descriptions for the transmissions covered in this chapter, all references to rotation in the THM 325 series gearboxes will be made as if you are looking at the transmission from the *front* of the engine.

200-4R and 325-4L Overdrive Assembly

Both the THM 200-4R and the 325-4L are equipped with an overdrive assembly that is added to the Simpson geartrain to provide fourth gear. The overdrive assembly mounts on the turbine shaft, figures 16-11 and 16-12, and consists of an overdrive planetary gearset, overdrive 1-way clutch, overdrive clutch, and drive clutch. The paragraphs below describe the overdrive assembly buildup. Its operation is described in the Power Flow section.

The overdrive gearset carrier is splined to the turbine shaft and turns with it. The overdrive pinions turn on pinion shafts in the carrier, and mesh with both the overdrive sun gear and the overdrive ring gear. The ring gear is always the output member of the overdrive gearset.

In the THM 200-4R, figure 16-11 the overdrive ring gear is attached to the input shaft that provides power to the Simpson geartrain. In the THM 325-4L, figure 16-12, the overdrive ring gear has the drive sprocket around its outer edge. As in the THM 325, the drive sprocket turns the drive chain, the drive chain

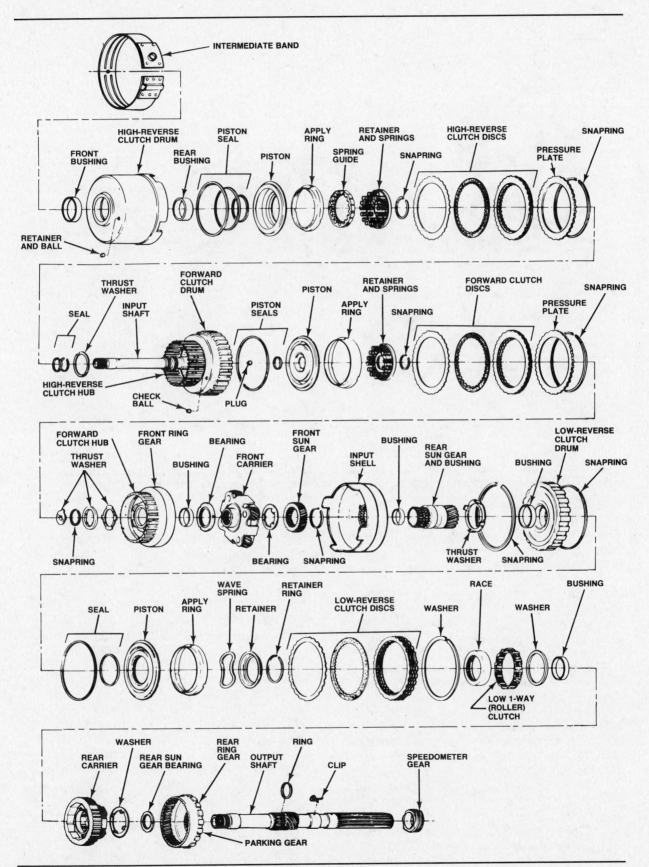

Figure 16-13. An exploded view of the 3-speed Simpson geartrain that is common to all the transmission discussed in this chapter.

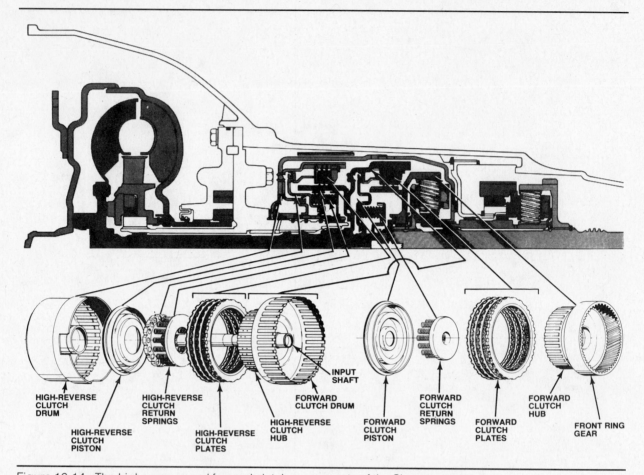

Figure 16-14. The high-reverse and forward clutch components of the Simpson geartrain.

turns the driven sprocket, and the driven sprocket turns the transmission input shaft.

The outer race of the overdrive 1-way clutch is splined to the carrier. The inner race, together with the overdrive sun gear, is splined to the drive clutch drum/overdrive clutch hub. The overdrive 1-way clutch locks the overdrive sun gear to the overdrive carrier whenever the sun gear attempts to rotate clockwise faster than turbine (carrier) speed.

The drive clutch drum/overdrive clutch hub is a single assembly. The inside of each drive clutch friction disc is splined to the outside of the carrier. The outside of each steel disc is splined to the inside of the drive clutch drum. When the drive clutch is applied, it overrides the 1-way clutch and locks the overdrive sun gear to the overdrive carrier. In the 200-4R, the outside of each overdrive steel disc is splined to the inside of the transmission case and held in place with a snapring.

The inside of each overdrive clutch friction disc is splined to the outside of the overdrive clutch hub. In the 325-4L, the outside of each

overdrive clutch steel disc is splined to the overdrive clutch housing, which bolts to the transmission case. When the overdrive clutch is applied, it locks the sun gear to the case.

Final Buildup

The input shaft is turned clockwise by the torque converter turbine (THM 200/200C), the overdrive ring gear (THM 200-4R), or the driven sprocket (THM 325 and 325-4L). The input shaft is supported on bushings in the stator support (THM 200/200C), center support (THM 200-4R), or driven sprocket support (THM 325 and 325-4L), and passes through it into the transmission case to power the Simpson geartrain.

From this point on, all the transmissions covered in this chapter are essentially the same. We will use a basic THM 200 transmission to illustrate the remaining buildup. Figure 16-13 shows a general overall view of the Simpson geartrain. As we complete the buildup, refer to figure 16-13 and the detailed figures for each specific assembly, to identify the components being discussed.

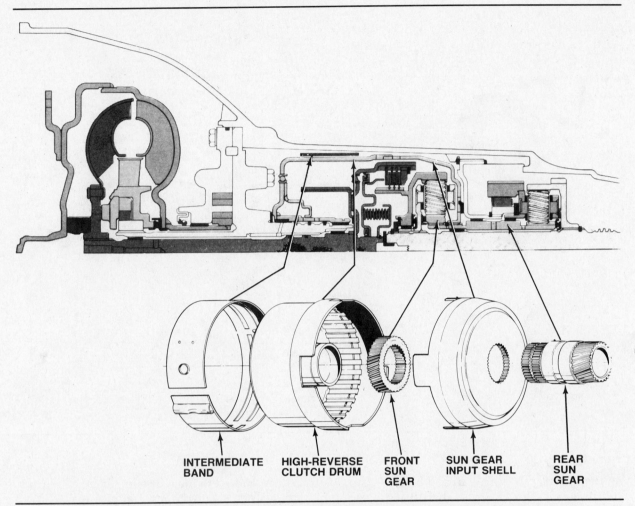

INTERMEDIATE BAND   HIGH-REVERSE CLUTCH DRUM   FRONT SUN GEAR   SUN GEAR INPUT SHELL   REAR SUN GEAR

Figure 16-15. The sun gear assembly of the Simpson geartrain, and its input and holding devices.

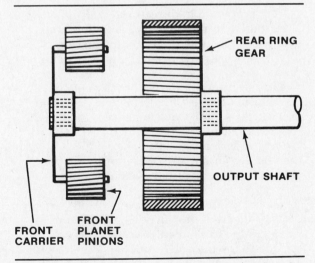

REAR RING GEAR

OUTPUT SHAFT

FRONT CARRIER   FRONT PLANET PINIONS

Figure 16-16. The front carrier and rear ring gear of the Simpson geartrain are splined to the output shaft.

The input shaft is one assembly with the high-reverse clutch hub and the forward clutch drum, figure 16-14. The inside of each high-reverse clutch friction disc is splined to the outside of the high-reverse clutch hub and turned by it. The outside of each steel disc is splined to the inside of the high-reverse clutch drum. When the high-reverse clutch is applied, the drum is rotated clockwise by the input shaft.

The outside of each forward clutch steel disc is splined to the inside of the forward clutch drum, figure 16-14, which is part of the input shaft. The inside of each friction disc is splined to the outside of the front ring gear in the planetary gearset. Whenever the forward clutch is applied, the front ring gear is turned clockwise by the input shaft. The forward clutch is applied in all forward gears.

The high-reverse clutch drum is splined to the input shell, figure 16-15, which is splined to the rear sun gear. The front sun gear is also

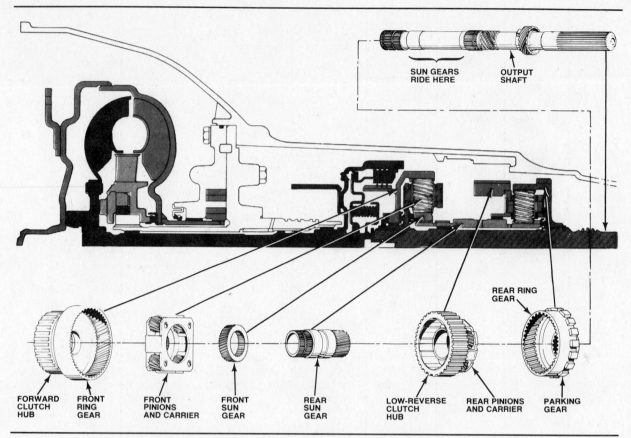

Figure 16-17. The complete Simpson gearset and the output shaft.

splined to the rear sun gear, and both gears rotate as a unit with the input shell at all times. The sun gear assembly turns on bushings located on the final drive sun gear shaft, but the assembly is not splined to the shaft. When the high-reverse clutch is applied, the clutch drum, input shell, and sun gear assembly are turned clockwise by the input shaft.

The intermediate band is wrapped around the outside of the high-reverse clutch drum, figure 16-15. When the intermediate band is applied, the clutch drum, input shell, and sun gear assembly are all held stationary. This occurs only in second gear.

None of the members of the compound planetary gearset are attached directly to the input shaft. However, two members of the gearset *are* splined to the output shaft, the front carrier and the rear ring gear, figure 16-16. This means that one or the other of these two units is always the final driving member of the Simpson planetary gearset.

The front ring gear meshes with the front pinions, figure 16-17. The front pinions turn in the front carrier and mesh with the front sun gear. As already mentioned, the front carrier is splined to the output shaft, and the front sun

gear is splined to the rear sun gear. The rear sun gear meshes with the rear pinions, which turn in the rear carrier and mesh with the rear ring gear. The rear ring gear, like the front carrier, is splined to the output shaft. The outside of the rear ring gear forms the parking gear, which is locked to the transmission case by the parking pawl when the gear selector is placed in Park.

The outer race of the low 1-way clutch is located inside the rear carrier, figure 16-18. The inner race of the 1-way clutch is splined to the low-reverse clutch drum, which is splined to the transmission case. The 1-way clutch holds the rear carrier if it attempts to turn counterclockwise, but the clutch overruns when the rear carrier turns clockwise.

The inside of each low-reverse clutch friction disc is splined to the outside of the rear carrier, figure 16-18. The outside of each steel disc is splined to the transmission case. When the low-reverse clutch is applied, it overrides the low 1-way clutch, and holds the rear carrier from turning in *either* direction.

The output shaft passes through the rear of the transmission case. In the THM 200 series, the governor drive gear is cut into the output

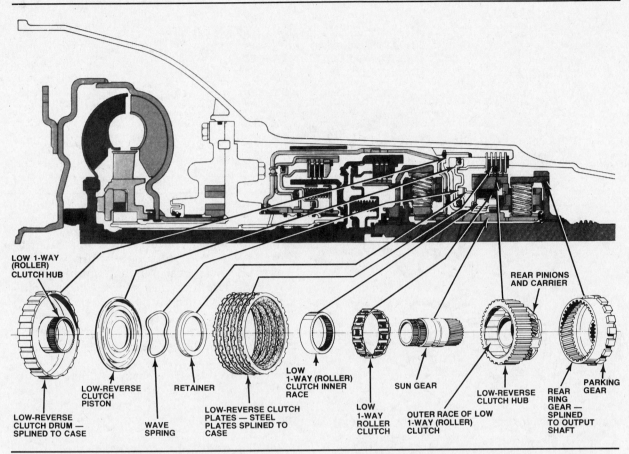

LOW 1-WAY (ROLLER) CLUTCH HUB

REAR PINIONS AND CARRIER

LOW-REVERSE CLUTCH PISTON

RETAINER

LOW 1-WAY (ROLLER) CLUTCH INNER RACE

SUN GEAR

REAR RING GEAR — SPLINED TO OUTPUT SHAFT

PARKING GEAR

LOW-REVERSE CLUTCH DRUM — SPLINED TO CASE

WAVE SPRING

LOW-REVERSE CLUTCH PLATES — STEEL PLATES SPLINED TO CASE

LOW 1-WAY ROLLER CLUTCH

OUTER RACE OF LOW 1-WAY (ROLLER) CLUTCH

LOW-REVERSE CLUTCH HUB

Figure 16-18. The low-reverse clutch, low 1-way clutch, and the rear planetary gearset of the Simpson geartrain.

shaft behind the rear ring gear, and the speedometer drive gear is clipped to the output shaft between the governor drive gear and the output yoke splines. The bushing and seal for the output shaft are installed at the rear of the transmission case. In the THM 325 series, the governor is driven by a separate gear splined to the output shaft, and the speedometer drive gear is attached to the top of the governor assembly. The rear bushing and seal are installed in the final drive/differential housing, which bolts to the transmission case.

THM 200/200C, 200-4R, 325, AND 325-4L POWER FLOW

Now that we have examined how these transmissions are assembled, we can trace the flow of power through the apply devices, the overdrive gearset (THM 200-4R and 325-4L), and the compound planetary gearset in each of the gear ranges. We will again use the THM 200 to illustrate power flow for the three forward speeds of the Simpson geartrain.

Park and Neutral

When the gear selector is in Park or Neutral, figure 16-19, no clutches or bands are applied, and the torque converter clutch (where fitted) is released. Because no member of the Simpson compound planetary gearset is held, there can be no output motion. The torque converter impeller/cover, and the oil pump drive gear or rotor, turn clockwise at engine speed.

In the THM 200/200C, the torque converter turbine and the transmission input shaft rotate clockwise, at turbine speed. In the THM 325, the torque converter turbine, turbine shaft, drive sprocket, drive chain, driven sprocket, and input shaft rotate clockwise at turbine speed. In the THM 200-4R and 325-4L, the overdrive 1-way clutch locks up, and the overdrive gearset turns clockwise at turbine speed. In the THM 200-4R, the overdrive gearset ring gear turns the input shaft. In the THM 325-4R, the drive gear on the outside of the overdrive gearset ring gear turns the drive chain, driven sprocket, and input shaft.

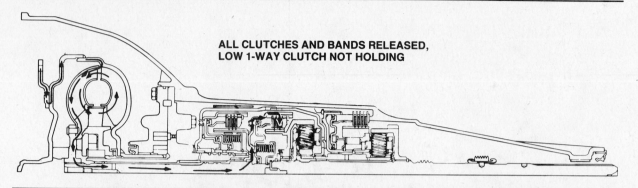

**ALL CLUTCHES AND BANDS RELEASED,
LOW 1-WAY CLUTCH NOT HOLDING**

Figure 16-19. Power flow through the Simpson geartrain in Park and Neutral.

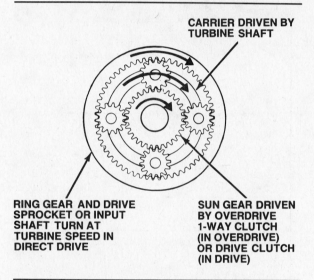

**CARRIER DRIVEN BY
TURBINE SHAFT**

**RING GEAR AND DRIVE
SPROCKET OR INPUT
SHAFT TURN AT
TURBINE SPEED IN
DIRECT DRIVE**

**SUN GEAR DRIVEN
BY OVERDRIVE
1-WAY CLUTCH
(IN OVERDRIVE)
OR DRIVE CLUTCH
(IN DRIVE)**

Figure 16-20. Power flow through the THM 200-4R and 325-4L overdrive gearset assembly in direct drive.

In Neutral, the Simpson compound planetary gearset is not connected to power, and the output shaft is free to rotate in either direction so the vehicle can be pushed or pulled. In Park, a mechanical pawl locks the parking gear on the outside of the rear ring gear to the case. The rear ring gear is splined to the output shaft, so the vehicle is prevented from moving in either direction.

THM 200-4R and 325-4L Overdrive and Drive Ranges

Before we look at the power flow through these transmissions in the forward gear ranges, it is necessary to first examine the operation of the THM 200-4R and 325-4L overdrive assembly. The power flow through the overdrive gearset varies depending on whether the gear selector is in the Overdrive (D) or Drive (3) position.

When the gear selector is in Overdrive, neither the overdrive clutch nor the drive clutch is applied in first gear. When the turbine shaft drives the overdrive carrier clockwise, the pinions attempt to turn the overdrive ring gear and sun gear clockwise. Because the ring gear is held by the vehicle driveline, the pinions attempt to walk around the ring gear and drive the overdrive sun gear clockwise faster than the carrier. This causes the 1-way clutch to lock up so that both the sun gear and the carrier rotate at turbine speed. With two members driven at the same speed, the overdrive gearset locks up and drives the ring gear in direct drive, figure 16-20. The ring gear then turns the input shaft (200-4R) or drive sprocket (325-4L) to send power to the Simpson planetary geartrain.

As long as the car is accelerating in Overdrive, torque is transferred from the engine, through the overdrive gearset, to the input shaft. If the driver lets up on the throttle, engine speed drops while the drive wheels, acting through the geartrain and input shaft, try to keep the ring gear turning at the same speed. This removes the clockwise-rotating torque load from the 1-way clutch, and it unlocks, allowing the overdrive gearset to freewheel so that no engine compression braking is available. As you will see, this is a major difference between the Overdrive and Drive gear selector positions.

When the gear selector is placed in Drive (3), manual second (2), or manual low (1), the drive clutch is applied to lock the overdrive sun gear to the overdrive carrier. With two members driven at the same speed, the planetary gearset locks up and drives the ring gear in direct drive. As before, the ring gear turns the input shaft (200-4R) or drive sprocket (325-4L) to send power to the Simpson planetary geartrain. The overdrive 1-way clutch also helps transmit power at this time, but its action is secondary to that of the drive clutch. Locking the overdrive

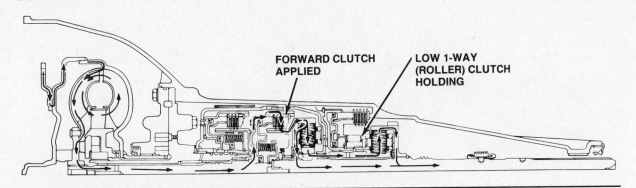

Figure 16-21. Power flow through the Simpson geartrain in Overdrive and Drive Low.

gearset in this manner makes engine braking possible under certain conditions in the lower three forward gears. How this happens is explained below.

Overdrive (THM 200-4R and 325-4L) and Drive Low

When the gear selector is placed in Overdrive (D on 4-speeds) or Drive (D on 3-speeds, 3 on 4-speeds) the car starts to move with the transmission automatically in low gear, figure 16-21. The forward clutch is applied to transmit engine power from the input shaft to the front ring gear. In 3-speed transmissions, the forward clutch is the only hydraulic apply device that operates in Drive low. In 4-speed transmissions, only the forward clutch operates in Overdrive low, but in Drive low the drive clutch is also applied.

Before the car begins to move, the transmission output shaft is held stationary by the driveline and wheels. This also holds the front carrier and rear ring gear stationary. As a result, the clockwise rotation of the front ring gear turns the front pinions clockwise on the stationary carrier. The front pinion rotation then causes the sun gear to turn counterclockwise.

Counterclockwise rotation of the sun gear turns the rear pinions clockwise. The rear pinions then try to walk clockwise around the inside of the rear ring gear, which is splined to the stationary output shaft. This action causes the rear carrier to try to rotate counterclockwise. However, the rear carrier contains the outer race of the low 1-way clutch.

As the rear carrier tries to rotate counterclockwise, the 1-way clutch locks up and prevents rotation. Therefore, the torque from the rear pinions is transferred to the rear ring gear and output shaft instead of the rear carrier. Because the 1-way clutch holds the carrier more

securely than the drive wheels can hold the output shaft and ring gear, the output shaft and rear ring gear begin to rotate clockwise in gear reduction.

As long as the car is accelerating in Overdrive or Drive low, torque is transferred from the input shaft, through the Simpson planetary gearset, to the output shaft. If the driver lets up on the throttle, the engine speed drops while the drive wheels try to keep turning at the same speed. This removes the clockwise-rotating torque load from the low 1-way clutch, which then unlocks and allows the rear planetary gearset to freewheel so that no engine compression braking is available. This is a major difference between Drive or Overdrive low, and manual low, as we will see later.

Note that even when a 200-4R or 325-4L transmission is in Drive low, with the drive clutch applied to lock the overdrive gearset, there is no engine braking. This is because the low 1-way clutch in the Simpson gearset allows the rear planetary gearset to freewheel.

In Overdrive and Drive low, gear reduction is produced by a combination of the front and rear planetary gearsets. The front ring gear is the input member; the rear carrier is the held member; and the rear ring gear is the output member.

In all but the 200-4R and 325-4L transmissions, there are only two apply devices working in Drive low, the forward clutch and the low 1-way clutch. The forward clutch is the input device. The low 1-way clutch is the holding device. In the 200-4R and 325-4L, one or two additional apply devices are used, depending on the gear selector position. In Overdrive low, the overdrive 1-way clutch is an input device. In Drive low, the drive clutch is also used as an input device.

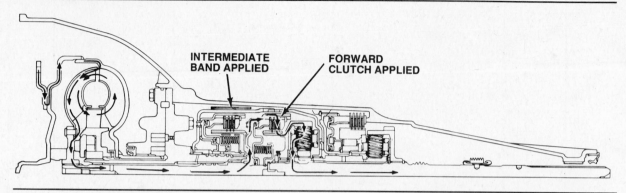

Figure 16-22. Power flow through the Simpson geartrain in Overdrive and Drive Second.

Overdrive (THM 200-4R and 325-4L) and Drive Second

As the car continues to accelerate in Overdrive or Drive, the transmission hydraulic system produces an automatic upshift to second gear by applying the intermediate band, figure 16-22. The forward clutch remains applied, and power from the engine is still delivered to the front ring gear. The front ring gear rotates clockwise and turns the front pinions clockwise.

At the same time, the intermediate band locks the high-reverse clutch drum and sun gear input shell to the transmission case. This holds the sun gear stationary. As the front ring gear turns the front pinions clockwise, they walk around the stationary sun gear. This causes the front carrier to rotate clockwise. The front carrier is splined to the output shaft and drives it in gear reduction.

As long as the car is accelerating in Overdrive or Drive second, torque is transferred from the engine, through the Simpson planetary gearset, to the output shaft. As vehicle speed rises and the driver lets up on the accelerator pedal, the transmission upshifts to third — even when a 200-4R or 325-4L is in Drive second with the drive clutch applied and the overdrive gearset locked. Because of this upshift, there is no engine compression braking. This is a major difference between Overdrive or Drive second, and manual second, as we will see later.

All of the gear reduction for second gear is produced by the front planetary gearset. The rear gearset simply idles. The 1-way clutch that held the rear carrier in Overdrive or Drive low overruns in second because the sun gear is held. The rear ring gear, which is splined to the output shaft, continues to rotate with the shaft, but freewheels because no holding device is applied to the rear gearset.

In all but the 200-4R and 325-4L transmissions, there are again only two apply devices working in Overdrive or Drive second, the forward clutch and the intermediate band. The forward clutch is the input device. The intermediate band is the holding device. In the 200-4R and 325-4L, one or two additional apply devices are used, depending on the gear selector position. In Overdrive second, the overdrive 1-way clutch is an input device. In Drive second, the overdrive clutch is also used as an input device.

Overdrive (THM 200-4R and 325-4L) and Drive Third

As the car continues to accelerate in Overdrive or Drive, the transmission automatically upshifts from second to third gear. For this to happen, the forward clutch remains applied, the intermediate band is released, and the high-reverse clutch is applied, figure 16-23.

When the high-reverse clutch is applied, the high-reverse clutch drum and the sun gear input shell are locked to the transmission input shaft. Because the sun gear input shell is splined to the sun gear, this causes the sun gear to revolve clockwise with the input shaft at turbine speed.

The forward clutch continues to lock the front ring gear to the input shaft. This means that two members of the planetary gearset, the sun gear and the ring gear, are locked to the input shaft, and thus to each other. With two members of the gearset locked together, the entire gearset turns as a unit.

With the front planetary gearset locked, there is no relative motion between the sun gear, ring gear, carrier, or the pinions. As a result, input torque to the sun and ring gears is transmitted through the planet carrier, which is splined to the output shaft, to drive the output shaft. In third gear, there is a direct 1:1 gear ratio between the input or center shaft speed

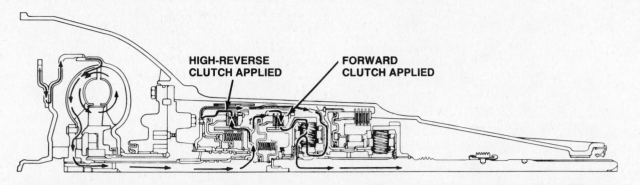

Figure 16-23. Power flow through the Simpson geartrain in Overdrive and Drive third.

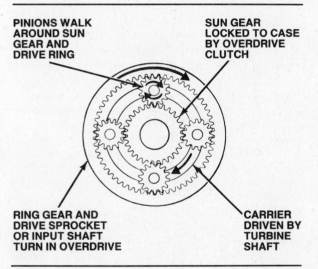

Figure 16-24. Power flow through the THM 200-4R and 325-4L overdrive gearset assembly in overdrive fourth.

and the output shaft speed. There is no reduction or increase in either speed or torque through the transmission.

All the transmissions covered in this chapter provide engine compression braking in Drive third because there is a direct mechanical connection between the input shaft and the output shaft; no 1-way clutches are involved. In the THM 200-4R and 325-4L, the hydraulic system prevents an upshift to fourth, and the drive clutch is applied to override the 1-way clutch and lock the overdrive gearset. However, compression braking is not possible when a THM 200-4R or 325-4L is in Overdrive third because the transmission will upshift to fourth at higher speeds, or the overdrive 1-way clutch will overrun at lower speeds because the drive clutch is not applied.

In third gear, the sun gear and front ring gear of the Simpson planetary gearset are input members equally. The front carrier is the output

member. In all but THM 200-4R and 325-4L transmissions, there are again only two apply devices working in Drive third, the forward clutch and the high-reverse clutch. Both clutches are input devices equally. In the THM 200-4R and 325-4L, one or two additional apply devices are used, depending on the gear selector position. In Overdrive third, the overdrive 1-way clutch is an input device. In Drive third, the overdrive clutch is also used as an input device.

Overdrive (THM 200-4R and 325-4L) Fourth

As the vehicle continues to accelerate in Overdrive, the transmission shifts into fourth gear. To do this, the overdrive clutch is applied. This locks the overdrive clutch hub, and the overdrive sun gear splined to it, to the transmission case. As the turbine shaft continues to turn the overdrive carrier, the pinions walk around the sun gear and overdrive the ring gear at faster-than-turbine speed, figure 16-24.

The overdrive 1-way clutch overruns in fourth gear because the carrier and outer race turn faster than the inner race which is splined to the overdrive clutch hub and locked to the case along with the sun gear. In fourth gear, the overdrive ring gear provides the input to the Simpson compound planetary gearset, which is still in direct drive, figure 16-23. The Simpson gearset then transmits the overdrive speed to the output shaft.

The input member of the overdrive gearset is the overdrive carrier; the held member is the overdrive sun gear; and the output member is the overdrive ring gear. There is no input device because the turbine shaft is splined directly to the overdrive carrier. The holding device is the overdrive clutch. The input and output members and devices of the Simpson gearset are the same as in Overdrive third.

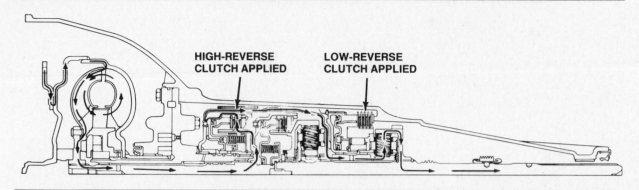

Figure 16-25. Power flow through the Simpson geartrain in Reverse.

Manual Second

When the gear selector is placed in manual second (I, S, L2 or 2) and the vehicle is accelerated from a stop, the transmission starts in Drive low and upshifts to Drive second, figure 16-22. The hydraulic system prevents the transmission from upshifting into third or fourth gear. Other than this, there are no differences between manual second and Drive second. The flow of power through the planetary gearsets is identical, and the same apply devices are used. Note that the drive clutch is applied when the THM 200-4R and 325-4L are in Drive second.

At any time, the driver can manually downshift from third to second by moving the gear selector from D (3-speeds) or 3 (4-speeds) to 2. In this case, the high-reverse clutch is released, and the intermediate band is applied. In the THM 200-4R and 325-4L, the driver can also manually downshift from Overdrive Ⓓ to 2. In this case, the high-reverse clutch and overdrive clutch are released, and the drive clutch and intermediate band are applied. These capabilities provide the driver with a controlled downshift for engine braking or improved acceleration at any speed or throttle position.

If the car is allowed to decelerate with the gear selector in manual second, the transmission automatically downshifts to low gear as speed decreases.

Manual Low

When the gear selector is placed in manual low (L, L1, or 1), the hydraulic system prevents an upshift to second, third, or fourth, and the flow of power through the planetary gearset is basically the same as in Drive low. Note that in the THM 200-4R and 325-4L transmissions the drive clutch is applied in Drive low. The only difference between Drive low and manual low is the use of one apply device.

In manual low, a low-reverse clutch is applied to hold the rear carrier. In Drive low, the 1-way clutch does this job, but it holds the rear carrier only during acceleration; the 1-way clutch overruns during deceleration and allows the rear gearset to freewheel. In manual low, the low-reverse clutch holds the rear carrier during acceleration *and* deceleration, which provides engine braking on deceleration. The 1-way clutch still helps hold the rear carrier during acceleration in manual low, but its action is secondary as long as the low-reverse clutch is applied.

The driver can move the gear selector from D to L, L1, or 1 while the car is moving. At lower speeds, the transmission will downshift directly to low gear by releasing the appropriate apply devices and applying the low-reverse clutch, and the drive clutch in the THM 200-4R and 325-4L. At higher speeds, the transmission first shifts into second by releasing the appropriate apply devices and applying the intermediate band, and the drive clutch in the THM 200-4R and 325-4L. Then, once vehicle speed drops below a certain level, the transmission automatically downshifts to low by releasing the intermediate band and applying the low-reverse clutch.

Reverse

When the gear selector is moved to Reverse, the input shaft is still turned clockwise by the engine. The transmission must reverse this rotation to a counterclockwise direction.

In Reverse, the forward clutch is released and the high-reverse clutch is applied, figure 16-25. This connects the high-reverse clutch drum and the sun gear input shell to the input shaft so that the input shaft drives the sun gear clockwise. The sun gear drives the rear pinions counterclockwise.

GEAR RANGES	CLUTCHES							BAND
	Low 1-way Roller	Forward	High-Reverse (Direct)	Low-Reverse	Over-drive (Fourth)	Drive (Over-run)	Over-drive 1-way	Inter-mediate
Overdrive Low	●	●					●	
Overdrive Second		●					●	●
Overdrive Third		●	●				●	
Overdrive Fourth		●	●		●		●	
Drive Low	●	●				●		
Drive Second		●				●		●
Drive Third		●	●			●		
Manual Low	*	●		●		●		
Manual Second		●				●		●
Reverse			●	●			●	

* The low 1-way roller clutch holds in manual low if the low-reverse clutch fails, but will not provide engine braking.

Figure 16-26. THM 200/200C, 200-4R, 325, and 325-4L clutch and band application chart.

The other apply device used in Reverse is the low-reverse clutch which holds the rear carrier. Because the rear carrier is held stationary, the counterclockwise rotation of the rear pinions drives the rear ring gear counterclockwise. The rear ring gear is splined to the output shaft and turns it counterclockwise in gear reduction to provide Reverse.

All gear reduction in Reverse is produced by the rear planetary gearset. Although there is input to the front gearset through the sun gear and the front pinions, no member of the gearset is held so it simply freewheels.

In Reverse, the input member of the Simpson planetary gearset is the sun gear; the held member is the rear carrier; and the output member is the rear ring gear. Two apply devices are used: the high-reverse clutch is the input device, and the low-reverse multiple-disc clutch is the holding device. In the THM 200-4R and 325-4L transmissions, the overdrive 1-way clutch is also holding in Reverse.

THM 200/200C, 325, 200-4R, and 325-4L Clutch and Band Applications

Before we move on to a closer look at the hydraulic system, we should review the apply devices that are used in each of the gear ranges. Figure 16-26 is a clutch and band application chart for the transmissions covered in this chapter. Also, remember these facts about the apply devices in these transmissions:
● The forward, high-reverse, drive, and overdrive 1-way clutches are input devices.
● The forward clutch is applied in all forward gear ranges.

● The high-reverse clutch is applied in third, fourth, and Reverse gears.
● The drive clutch is applied in the Drive, manual second and manual low gear selector positions.
● The overdrive 1-way clutch locks during acceleration in the Overdrive, Neutral, and Park gear selector positions; it freewheels during deceleration.
● The low-reverse, low 1-way, and overdrive clutches, along with the intermediate band, are holding devices.
● The low-reverse clutch is applied in manual low and Reverse.
● The low 1-way clutch locks during acceleration in Overdrive and Drive low; it freewheels during deceleration.
● The overdrive clutch is applied in fourth gear.
● The intermediate band is applied in second gear.

THM 200/200C, 200-4R, 325, and 325-4L HYDRAULIC SYSTEM

The hydraulic systems of the transmissions covered in this chapter control shifting under varying vehicle loads and speeds, as well as in response to manual gear selection by the driver. Figure 16-27 is a diagram of the complete THM 200C hydraulic system; the THM 200 and 325 systems are similar. Figure 16-28 is a diagram of the complete THM 325-4L hydraulic system; the THM 200-4R system is similar. You will find it helpful to refer back to these diagrams as we discuss the various components that make up the hydraulic system. We will begin our study with the pump.

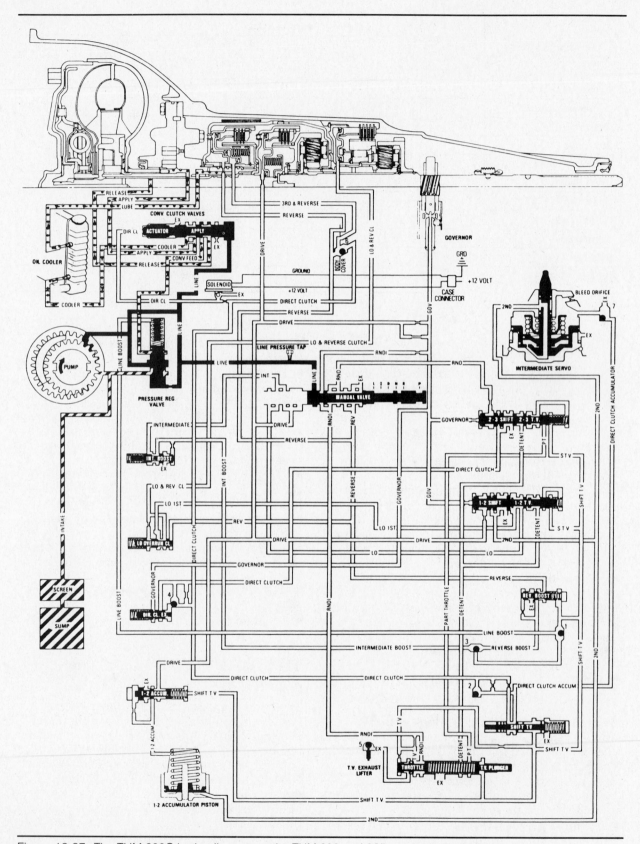

Figure 16-27. The THM 200C hydraulic system; the THM 200 and 325 systems are similar.

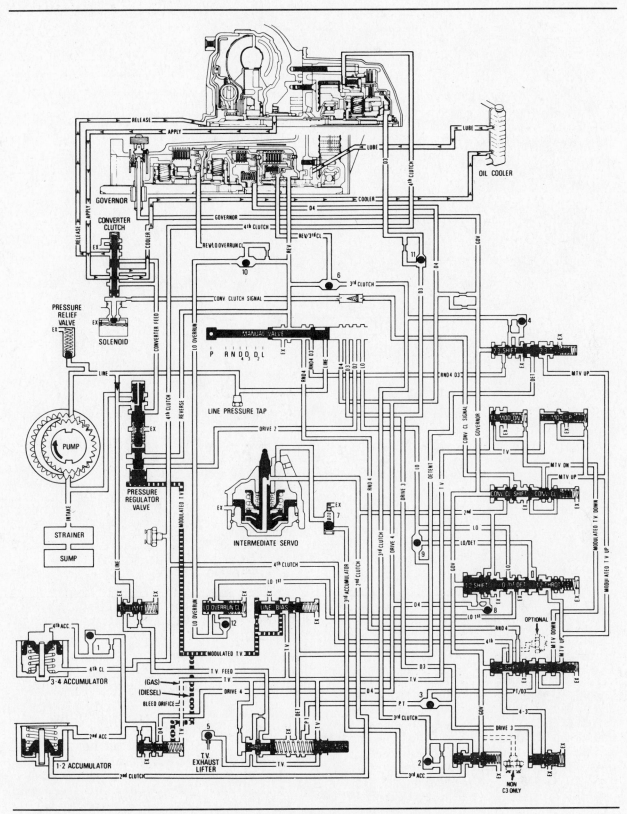

Figure 16-28. The THM 325-4L hydraulic system; the THM 200-4R system is similar.

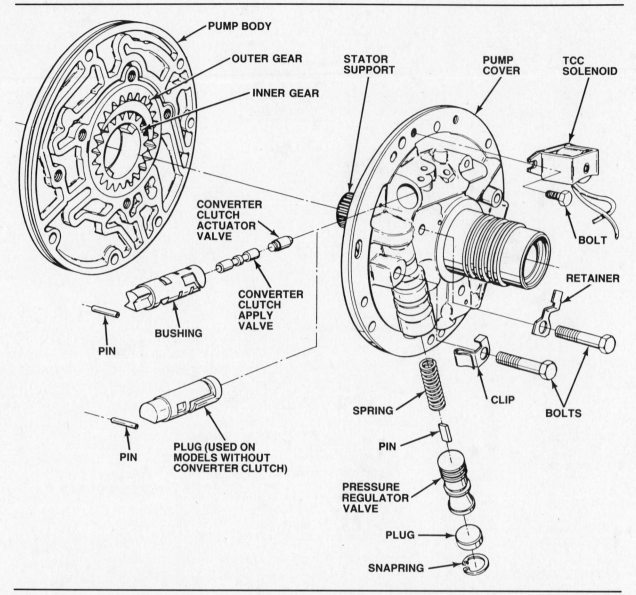

Figure 16-29. The THM 200C oil pump assembly. The THM 200 pump cover does not contain a converter clutch valve assembly or TCC solenoid. The THM 325 series pump cover does not contain any hydraulic valves.

Hydraulic Pump

In all of these transmissions, fluid is stored in the oil pan at the bottom of the transmission. The oil pump draws fluid through the filter attached to the bottom of the valve body. Fluid then flows through a passage in the valve body, and through another passage in the front of the transmission case (THM 200 series) or case cover (THM 325 series), to the pump.

The THM 200/200C, 325, and 325-4L use a gear-type oil pump, figure 16-29. In the THM 200/200C, the pump delivers fluid directly to the pressure regulator valve located in the pump cover. The THM 200C pump cover also contains the converter clutch valve assembly, and has the TCC solenoid attached to its back side. Later model THM 200 transmissions use the THM 200C pump cover, but have a solid plug in place of the converter clutch valve assembly. The THM 325 series pumps have no valves in the pump cover; all the valves are located in the valve body or case.

In mid-1976, a new passage and check ball were added to the THM 200 pump cover as part of the Reverse gear circuit. Early pumps without a check ball can be used *only* with early valve bodies. The later check-ball pumps can be used with any THM 200 valve body.

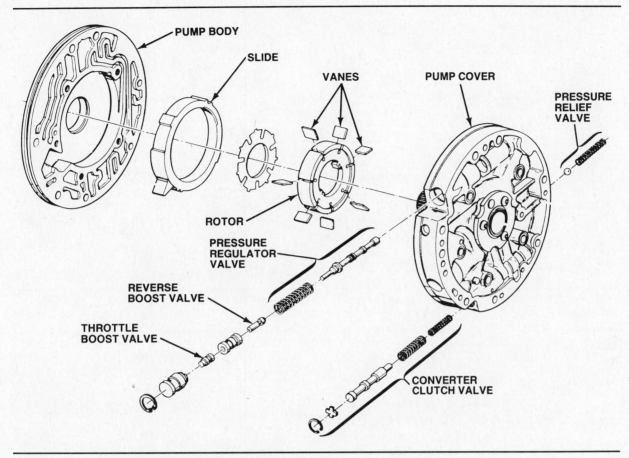

Figure 16-30. The THM 200-4R oil pump assembly.

The THM 200-4R uses a variable-displacement vane-type pump, figure 16-30. The pump delivers fluid directly to the pressure regulator valve assembly in the pump cover. The converter clutch apply valve and the pressure relief valve are also located in the pump cover.

Hydraulic Valves

The hydraulic systems of the transmissions discussed in this chapter contain the following valves:
1. Pressure regulator valve assembly
2. Pressure relief valve (THM 200-4R and 325-4L)
3. Manual valve
4. Throttle valve assembly
5. Throttle limit valve (THM 200-4R and 325-4L)
6. Shift throttle valve (THM 200/200C and 325)
7. Line bias valve (THM 200-4R and 325-4L)
8. Modulated throttle upshift valve (THM 200-4R and 325-4L)
9. Modulated throttle downshift valve (THM 200-4R and 325-4L)
10. Governor valve
11. 1-2 shift valve assembly
12. Accumulator valve
13. Low overrun clutch valve
14. 2-3 shift valve assembly
15. Direct clutch exhaust valve (THM 200/200C and 325)
16. 3-2 control valve (THM 200-4R and 325-4L)
17. Intermediate boost valve (THM 200/200C and 325)
18. Reverse boost valve (THM 200/200C and 325)
19. 3-4 shift valve assembly (THM 200-4R and 325-4L)
20. 4-3 control valve (THM 200-4R and 325-4L)
21. Converter clutch actuator valve (200C only)
22. Converter clutch apply valve (200C only)
23. Converter clutch valve (THM 200-4R and 325-4L)
24. Converter clutch throttle valve (THM 200-4R and 325-4L)
25. Converter clutch shift valve (THM 200-4R and 325-4L).

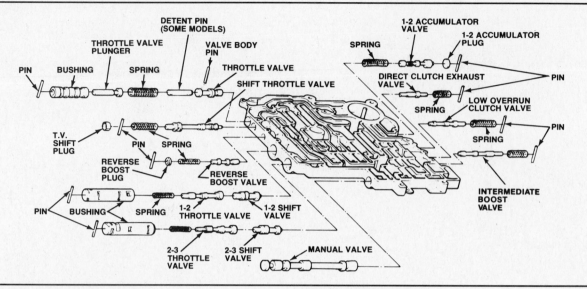

Figure 16-31. The THM 200 valve body and hydraulic control valves; the THM 200C and 325 valve bodies are similar.

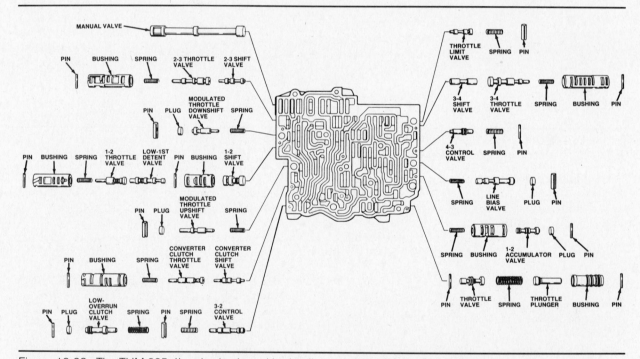

Figure 16-32. The THM 325-4L valve body and hydraulic control valves; the THM 200-4R valve body is similar.

Valve body and ball-check valves

The valve bodies of the transmissions covered in this chapter are bolted to the bottom of the transmission case. A separator plate and two gaskets are installed between the valve body and the case. The valve body has hydraulic passages that route fluid between valves. Additional hydraulic fluid passages cast into the bottom of the transmission case form the upper half of the valve body. The oil filter attaches to the bottom of the valve body.

Because of the oil pump change described earlier, the THM 200 has had three valve bodies: the first was used with an early-model spacer plate and a non-check-ball pump; the second was used with a late-model spacer plate and either pump; the third was used with a late-model spacer plate and a check-ball pump. The three valve bodies are not interchangeable.

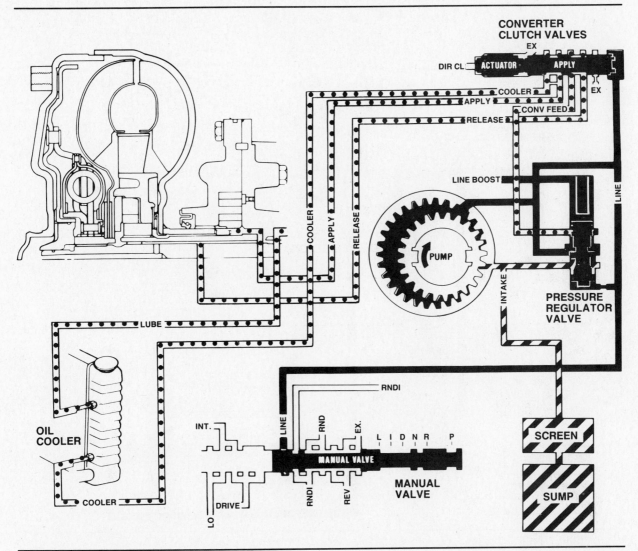

Figure 16-33. The THM 200C pressure regulator valve and manual valve; the THM 200 and 325 valves are similar.

In the THM 200/200C, all of the valves except the pressure regulator valve, converter clutch apply and actuator valves (THM 200C), and governor valve are housed in the valve body, figure 16-31. The pressure regulator valve and the two converter clutch valves (THM 200C) are installed in the oil pump cover, figure 16-29. The governor valve installs in the extension housing of the transmission case. The valve body casting also contains the bottom of the intermediate servo bore. The THM 200/200C has five check balls, four in the valve body and one in the case.

The THM 325 has the same hydraulic valves as the THM 200. All the valves except the pressure regulator valve and governor valve are housed in the valve body. The pressure regulator valve is in the transmission case, figure 16-9. The governor valve installs in the top of the case near the output end. The valve body casting also contains the bottom of the intermediate servo bore. The THM 325 has seven check balls, three in the valve body and four in the passages between the separator plate and the case.

The THM 200-4R and 325-4L have basically the same valves and hydraulic circuitry. All the valves except the pressure regulator valve, converter clutch valve, and governor valve are housed in the valve body, figure 16-32. In the THM 200-4R, the pressure regulator valve and converter clutch valve are installed in the oil pump cover, figure 16-30; the governor valve installs in the rear of the transmission case. In the THM 325-4L, the pressure regulator valve and converter clutch valve are installed in the transmission case, figure 16-9; the governor

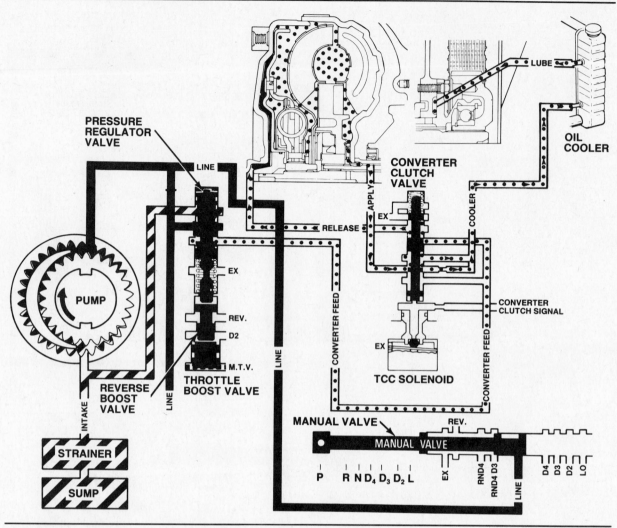

Figure 16-34. The THM 325-4L pressure regulator valve assembly and manual valve.

valve installs in the top of the case near the output end. Both the THM 200-4R and the 325-4L have twelve check balls, three in the valve body and nine in the passages between the separator plate and the case.

Pressure regulator valve assembly
The exact design of the pressure regulator valve varies with the type of transmission and oil pump. In the THM 200/200C and 325, figure 16-33, pressure from the pump is applied through two passages to the center and one end of the pressure regulator valve. In the THM 325-4L, figure 16-34, pressure from the pump is applied through a single passage to the center of the pressure regulator valve; an internal passage in the valve spool routes pressure to the end of the valve. In the THM 200-4R, figure 16-35A, pressure from the pump

is routed directly to the end of the valve, and through a screen to the center of the valve.

When the engine is started and mainline pressure begins to build, pressure at the end of the regulator valve moves the valve against spring force to open the converter feed passage. In the transmissions without lockup torque converters, oil flows through the feed passage directly to the converter. Converter return oil is routed to the transmission cooler, and the cooled oil is then directed to the transmission lubrication circuits.

In transmissions with lockup torque converters, oil flows through the feed passage to the converter clutch apply valve (200C), figure 16-33, or the converter clutch valve (200-4R and 325-4L), figure 16-34. These valves route oil to the converter, the transmission cooler, and the lubrication circuit through different passages

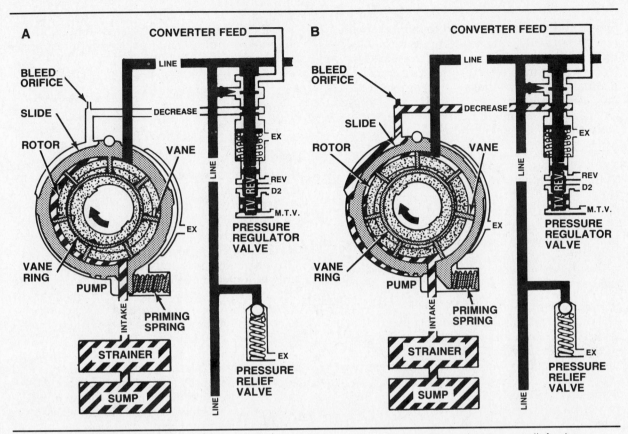

Figure 16-35. The THM 200-4R pressure regulator valve assembly, manual valve, and pressure relief valve.

depending on whether the torque converter clutch is applied or released. This will be explained later when we cover the operation of the converter clutch.

After the converter is filled, pressure buildup at the regulator valve moves it farther against spring force. In all but the THM 200-4R this opens a port to the pump suction passage to vent excess pressure. Hydraulic pressure and spring force then reach a balance that provides a constant mainline pressure.

In the THM 200-4R, movement of the pressure regulator valve opens a port to the decrease passage. Pressure then flows through the passage to the pump where it moves the slide against priming spring force to reduce pump output, figure 16-35B. The pressure regulator valve continues to move until hydraulic pressure and spring force reach a balance that provides a constant mainline pressure.

When the THM 200-4R shifts into gear, or from one gear to another, increased fluid flow demands cause mainline pressure to drop slightly. This drop in pressure allows the spring to move the pressure regulator valve so that the mainline pressure feed is cut off and the pump decrease circuit exhausts. The pump priming

spring then moves the slide to increase pump output and bring mainline pressure back to the correct level.

In the THM 200-4R and 325-4L, figures 16-34 and 16-35, the throttle boost and reverse boost valves at the spring end of the pressure regulator valve are used to increase mainline pressure under certain driving conditions. To do this, hydraulic pressure is routed to the boost valves, which then move to increase the spring force acting on the pressure regulator valve. Modulated throttle pressure acts on the throttle boost valve to increase mainline pressure in response to higher torque loads on the drivetrain. Mainline pressure acts on the reverse boost valve in Reverse, manual low, and manual second to increase mainline pressure and hold the apply devices more securely under the higher torque loads that are normal in these gear positions.

In the THM 200/200C and 325, figure 16-33, mainline pressure is increased under certain driving conditions by other pressures that are routed through the line boost passage to the spring end of the pressure regulator valve. These pressures combine with spring force to regulate mainline pressure at a higher level.

Shift throttle pressure is used to increase mainline pressure in response to higher torque loads on the drivetrain. Reverse boost pressure (in Reverse) and intermediate boost pressure (in manual low and manual second) are used to increase mainline pressure and hold the apply devices more securely under the higher torque loads that are normal in these gear positions.

Pressure relief valve — THM 200-4R and 325-4L

In the THM 200-4R and 325-4L, a pressure relief valve, figure 14-35, limits mainline pressure to a maximum of 320 to 360 psi (2,240 to 2,520 kPa). If mainline pressure exceeds this value, the pressure relief valve unseats to exhaust excess oil to the sump. Once mainline pressure drops below the regulated level, spring force reseats the valve.

Manual valve

The manual valve is a directional valve that is moved by a mechanical linkage from the gear selector lever. The THM 200/200C and 325 manual valves receive fluid from the pressure regulator valve. The THM 200-4R and 325-4L manual valves receive fluid directly from the pump via the pressure regulator. Both valves direct mainline pressure to apply devices and other valves in the hydraulic system in order to provide both automatic and manual upshifts and downshifts. The manual valve is held in each gear position by a spring-loaded lever.

In the THM 200/200C and 325, figure 16-33, six passages may be charged with mainline pressure from the manual valve. When the gear selector is in Park, mainline pressure is blocked from entering the manual valve, and the valve does not direct pressure to any passages. GM refers to mainline pressure from the manual valve by the name of the passage it flows through.

The RNDI passage is charged in Reverse, Neutral, Drive, and manual second ranges; it routes pressure to the:

- Throttle valve
- Governor valve.

The RND passage is charged in Reverse, Neutral, and Drive ranges; it routes pressure to the:

- 2-3 shift valve.

The DRIVE passage is charged in Drive, manual second, and manual low ranges; it routes pressure to the:

- Apply area of the forward clutch
- Governor valve
- 1-2 shift valve
- 1-2 accumulator valve.

The INTERMEDIATE passage is charged in manual second and manual low; it routes pressure to the:

- Intermediate boost valve.

The LOW passage is charged in manual low; it routes pressure to the:

- 1-2 shift valve.

The REVERSE passage is charged in Reverse; it routes pressure to the:

- Reverse boost valve
- Apply areas of the high-reverse clutch
- Apply area of the low-reverse clutch.

The THM 200 hydraulic system was changed slightly in mid-1976. In early models, the high-reverse clutch inner apply area (used in both Drive and Reverse) and outer apply area (used only in Reverse) were supplied by two separate circuits. The reverse circuit contained a branch through the 2-3 shift valve and direct clutch exhaust valve to the high-reverse clutch inner apply area.

In later models, the oil pump cover contains a passage controlled by a check ball that connects the two apply areas. The check ball *prevents* fluid in the inner apply area from entering the outer area. However, the check ball *allows* fluid from the outer apply area to enter the inner area. The reverse circuit in later models does not include the 2-3 shift valve and direct clutch exhaust valve because fluid in the high-reverse clutch outer apply area circuit can flow through the pump passage and past the check ball into the inner apply chamber.

In the THM 200-4R and 325-4L, figure 16-34, seven passages may be charged with mainline pressure from the manual valve. When the gear selector is in Park, mainline pressure is blocked from entering the manual valve, and the valve does not direct pressure to any passages. Once again, GM refers to mainline pressure from the manual valve by the name of the passage it flows through.

The RND4 passage is charged in Reverse, Neutral, and Overdrive; it routes pressure to the:

- 3-4 shift valve.

The RND4-D3 passage is charged in Reverse, Neutral, Overdrive, and Drive; it routes pressure to the:

- 2-3 shift valve.

The D4 passage is charged in Overdrive, Drive, manual second, and manual low; it routes pressure to the:

- Apply area of the forward clutch
- Governor valve
- 1-2 shift valve
- 1-2 accumulator valve.

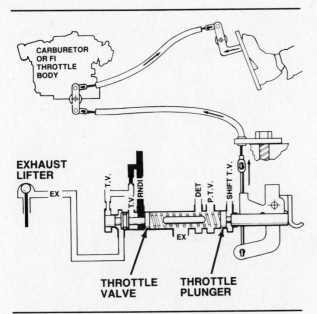

Figure 16-36. This throttle valve, typical of those in the THM 200/200C and 325, is operated by a cable connected to the accelerator linkage.

The D3 passage is in Drive, manual second, and manual low; it routes pressure to the:

- Apply area of the drive clutch
- 4-3 control valve
- 3-4 shift valve.

The D2 passage is charged in manual second and manual low; it routes pressure to the:

- Pressure regulator valve.

The LOW passage is charged in manual low; it routes pressure to the:

- 1-2 shift valve
- Low overrun clutch valve
- Apply area of the low-reverse clutch.

The REVERSE passage is charged in Reverse; it routes oil to the:

- Pressure regulator valve
- Apply area of the low-reverse clutch
- Inner and outer apply areas of the high-reverse clutch.

Throttle valve assembly
Throttle limit valve — THM 200-4R and 325-4L
Shift throttle valve — THM 200/200C and 325

All of the transmissions covered in this chapter develop throttle pressure using a throttle valve operated by a mechanical linkage, figure 16-36. Throttle pressure indicates engine torque and helps time the transmission upshifts and downshifts. Without throttle pressure, the transmission would upshift and downshift at exactly the same speed for very different throttle openings and load conditions.

The throttle valve is controlled by opposition between hydraulic pressure and spring force. In the THM 200/200C and 325, figure 16-36, the hydraulic pressure is mainline pressure (RNDI oil) from the manual valve. In the THM 200-4R and 325-4L, figure 16-37, the hydraulic pressure is throttle feed pressure from the throttle limit valve. The amount of spring force acting on the opposite end of the throttle valve is controlled by the throttle plunger. The plunger is linked to the accelerator by a cable so that its position is determined by throttle opening.

The THM 200-4R and 325-4L throttle limit valve, figure 16-37, restricts throttle feed pressure to a maximum of 90 psi (620 kPa). Mainline pressure from the pump passes through the throttle limit valve where it acts on two lands to move the valve against spring force. If throttle feed pressure exceeds 90 psi (620 kPa) the valve moves far enough to block the entry of mainline pressure and exhaust excess throttle feed pressure.

At closed throttle, the throttle plunger exerts no force on the throttle valve spring. RNDI oil or throttle feed pressure entering the throttle valve acts on the valve and moves it to the right to restrict the throttle pressure outlet port. Thus, there is little or no throttle pressure at this time.

As the accelerator is depressed, the cable linkage moves the throttle plunger to compress the throttle valve spring. This moves the throttle valve to the left, opens the outlet port, and allows regulated mainline or converter feed pressure to exit the valve as throttle pressure. As the throttle is opened wider, spring force rises, the throttle valve moves farther, and the throttle pressure continues to increase. However, because throttle pressure is developed from mainline pressure, it never exceeds mainline pressure.

To help balance the throttle valve, some throttle pressure is directed to the end of the valve opposite spring force. This acts to move the throttle valve to the right against spring force, and also uncovers a check-ball-regulated exhaust port that helps limit the pressure increase. At any given throttle opening, the throttle valve reaches a balanced position and provides a constant level of throttle pressure.

The check-ball-regulated exhaust port is controlled by an extension of the throttle plunger cable linkage, figure 16-38. In normal operation, the check ball is held off its seat to exhaust excess throttle pressure. However, if the throttle plunger cable breaks, is disconnected, or is badly adjusted, the linkage moves downward and allows the check ball to seat. With no exhaust port to relieve it, throttle pressure then

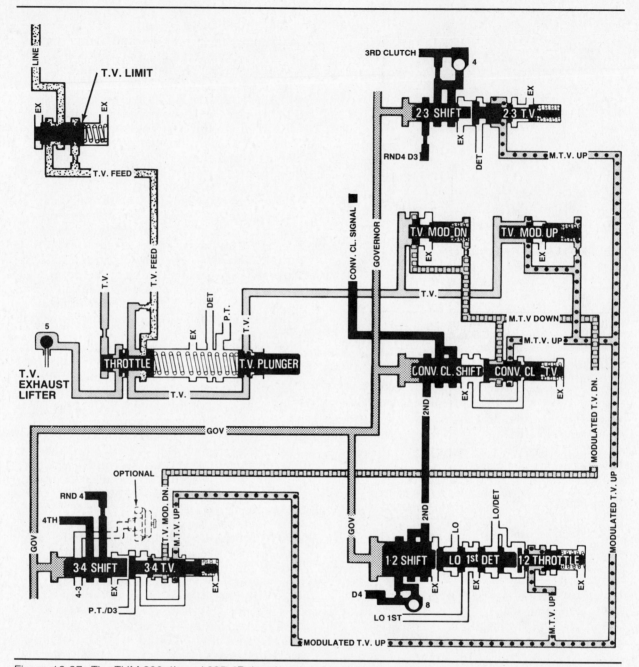

Figure 16-37. The THM 200-4L and 325-4R throttle valve, throttle limit valve, modulated throttle upshift and downshift valves, and the 1-2, 2-3, and 3-4 shift valve assemblies.

remains high regardless of actual throttle position. Although the transmission will not upshift and downshift at the appropriate times, this protects the clutches and band from slipping and burning.

In the THM 200/200C and 325, throttle pressure is routed to the shift throttle valve, figure 16-39, which limits the pressure to a maximum of 90 psi (620 kPa). The shift throttle valve is a differential spool valve. As throttle pressure

enters the center of the valve, it acts with greater force on the larger land to move the valve against spring force. If throttle pressure rises above 90 psi (620 kPa), the shift throttle valve moves far enough to open an exhaust passage that vents excess pressure to the sump. Pressure in the throttle pressure circuit that is limited by the shift throttle valve is called shift throttle pressure. The shift throttle valve also works with the direct clutch exhaust valve to

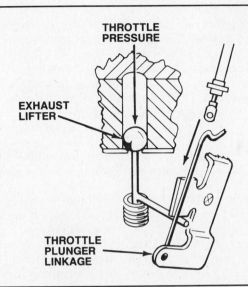

Figure 16-38. The throttle pressure exhaust port is blocked by a check ball if the throttle plunger cable breaks, is disconnected, or is badly adjusted.

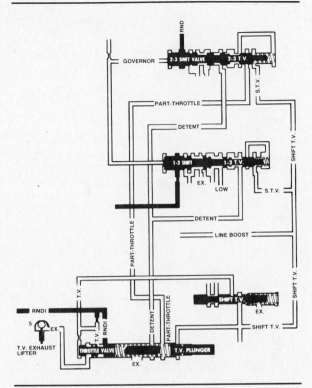

Figure 16-39. The THM 200/200C and 325 throttle valve, shift throttle valve, and 1-2 and 2-3 shift valve assemblies.

control 3-2 downshifts. This will be explained later in the chapter.

Shift throttle pressure is routed to the throttle plunger, line boost circuit, reverse boost valve, 1-2 accumulator valve, and the 1-2 and 2-3 throttle valves, figure 16-27. Shift throttle pressure to the throttle plunger helps regulate throttle pressure by increasing the spring force acting on the throttle valve; it also reduces the effort needed to hold the throttle valve plunger in place. Shift throttle pressure in the line boost circuit acts on the pressure regulator valve to increase mainline pressure in proportion to throttle opening. The actions of shift throttle pressure on the other valves are explained later in the chapter.

In the THM 200-4R and 325-4L, throttle pressure is routed to the throttle plunger, line bias valve, accumulator valve (some models), and the modulated throttle upshift and downshift valves, figure 16-28. Throttle pressure to the throttle plunger helps regulate throttle pressure by increasing the spring force acting on the throttle valve; it also reduces the effort needed to hold the throttle valve plunger in place. The actions of throttle pressure on the other valves will be explained later in the chapter.

When the accelerator is depressed far enough, the THM 200/200C and 325 throttle plunger opens a port that allows shift throttle pressure to flow through a passage to the 2-3 throttle valve, figure 16-27. The THM 200-4R and 325-4L throttle plunger opens a similar port that allows throttle pressure to flow through a passage to the 3-4 throttle valve, figure 16-28.

Oil in these passages is called part-throttle pressure. Part-throttle pressure combines with spring force acting on the 2-3 or 3-4 throttle valve to move the 2-3 or 3-4 shift valve and produce a part-throttle downshift from high gear.

When the accelerator is completely depressed, the THM 200/200C and 325 throttle plunger opens a port that allows shift throttle pressure to flow through a passage to the 1-2 and 2-3 throttle valves, figure 16-27. The THM 200-4R and 325-4L throttle plunger opens a similar port that allows throttle pressure to flow through a passage to the 1-2 and 2-3 throttle valves, and the line bias valve, figure 16-28. Oil in these passages is called detent pressure. Detent pressure on the 1-2 and 2-3 throttle valves combines with spring force to move the 1-2 and/or 2-3 shift valve (depending on vehicle speed and gear position) to produce a wide-open throttle 3-2, 3-1, or 2-1 downshift. The action of detent pressure on the line bias valve is explained in the next section.

Line bias valve — THM 200-4R and 325-4L

The line bias valve, figure 16-28, receives throttle pressure and regulates it to create a modulated throttle pressure whose rate of increase more closely matches engine torque rather than just the degree of throttle opening. Throttle

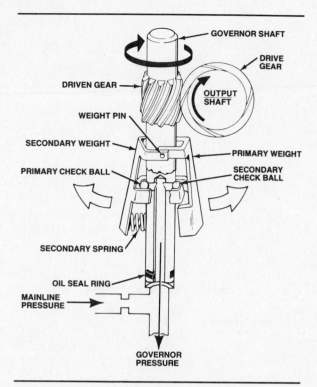

Figure 16-40. This THM 200 series governor is typical of those used in the THM 325 series as well.

pressure enters the line bias valve where it becomes modulated throttle pressure. This pressure is then routed to the end of the valve to oppose spring force. As throttle pressure increases, modulated throttle pressure moves the line bias valve to restrict the inlet port and control the rate of pressure increase in the modulated throttle pressure circuit.

Modulated throttle pressure is routed to the throttle boost valve in the pressure regulator valve assembly to raise mainline pressure as throttle opening and engine torque increase. In some models, modulated throttle pressure is also directed to the spring end of the accumulator valve. When the throttle is wide open, detent pressure is sent to the line bias valve where it combines with spring force to bottom the valve. This allows unmodified throttle pressure to flow into the modulated throttle pressure circuit.

Modulated throttle upshift valve — THM 200-4R and 325-4L
Modulated throttle downshift valve — THM 200-4R and 325-4L

The THM 200-4R and 325-4L do not use a shift throttle valve, so shift throttle pressure is not available to control shift points. Instead, modulated throttle upshift and downshift valves, figure 16-37, are used to create pressures that

perform this job. Both valves are controlled by opposition between throttle pressure on one end and spring force on the other.

The modulated throttle upshift valve is closed by spring force, and begins to open at approximately 10 psi (69 kPa) of throttle pressure. This valve delays upshifts at small throttle openings by directing modulated throttle pressure to the 1-2, 2-3, and 3-4 throttle valves where it combines with spring force to hold the shift valves in the downshifted position against governor pressure.

The modulated throttle downshift valve is closed by spring force and does not open until there is approximately 40 psi (276 kPa) of throttle pressure. This valve helps force a 4-3 downshift at larger throttle openings by directing modulated throttle downshift pressure to the 3-4 throttle valve where it combines with spring force to downshift the shift valve against governor pressure.

Modulated throttle upshift and downshift pressures are also routed to the converter clutch throttle valve. The actions of these pressures on this valve is explained later in the chapter.

Governor valve

We have been discussing the development and control of throttle pressure, which is one of the two auxiliary pressures that work on the shift valves. Before we get to the shift valve operation, we must explain the governor valve (or governor), and how it creates governor pressure. Governor pressure provides a road speed signal to the hydraulic system that causes automatic upshifts to occur as road speed increases, and permits automatic downshifts with decreased road speed.

All of the transmissions covered in this chapter use a check-ball governor, figure 16-40, that is driven by a gear on the output shaft. However, although all of the governors work in basically the same way, they are not interchangeable. The check-ball governor design consists of a governor shaft, a governor weight pin, primary and secondary weights, primary (some models) and secondary springs, and two check balls. On the THM 325 series transmissions, a gear at the top of the governor shaft turns the speedometer cable driven gear, which fits into an opening in the governor cover.

In the THM 200/200C and 325, mainline pressure (DRIVE oil) passes through two restricting orifices and enters the governor circuit whenever the engine is running and the gear selector is in a position other than Park. RNDI oil is also routed into the governor circuit through a third restricting orifice when the gear selector is in any position except Park and manual low. In the THM 200-4R and 325-4L, mainline pressure (D4 oil) enters the governor circuit through two restricting orifices whenever the gear selector is in a forward gear range.

Pressure enters the governor through a passage drilled down the center of the governor shaft. Two exhaust ports are drilled through the shaft into this passage. Governor pressure is determined by the amount of fluid that is allowed to flow out of these exhaust ports.

The exhaust port openings are controlled by check balls that seat in pockets formed at the ends of the ports. As the governor rotates, centrifugal force causes the primary weight to act on one check ball, and the secondary weight to act on the other. Each weight acts on the check ball located on the opposite side of the governor shaft. The secondary weight, and sometimes the primary weight, is assisted in its job by a spring that helps to regulate pressure more smoothly as vehicle speed increases.

When the vehicle is stopped and the governor is not turning, there is no centrifugal force. At this time, DRIVE/RNDI or D4 oil pressure overcomes any spring force and unseats both of the check balls. All of the oil is exhausted, and governor pressure is zero at this time.

As the vehicle accelerates and the gear on the output shaft begins to turn the governor, centrifugal force moves the weights outward. This force is relayed directly or through springs to the check balls, partially seating them in their pockets to restrict fluid flow from the exhaust ports; this causes governor pressure to increase. As long as governor speed increases, the forces on the check balls become greater and less fluid is allowed to escape; thus governor pressure continues to rise. When governor speed remains constant, the forces acting on the check balls are also constant, and governor pressure stabilizes.

The heavier primary weight and spring (where fitted) are more sensitive to changes in governor rpm than are the secondary weight and spring. Therefore, the primary weight and spring work to control governor pressure at lower vehicle speeds. At higher speeds, centrifugal force seats the primary check ball, and the lighter secondary weight and spring take over governor pressure regulation. Above a certain speed, centrifugal force becomes great enough to seat both check balls so that no fluid escapes from the governor. At this point, governor pressure equals mainline pressure.

Governor pressure sent to the shift valves opposes spring force and shift throttle pressure or modulated throttle pressure to upshift the valves at the appropriate road speeds. In the THM 200/200C and 325, governor pressure is also routed to the direct clutch exhaust valve. In the THM 200-4R and 325-4L, governor pressure is also sent to the converter clutch shift valve and 3-2 control valve.

1-2 shift valve assembly

The 1-2 shift valve controls upshifts from low to second, and downshifts from second to low. In the THM 200/200C and 325, the assembly includes the 1-2 shift valve, the 1-2 throttle valve, and a spring, figure 16-39. The THM 200-4R and 325-4L assembly, figure 16-41, also includes a low-first detent valve. The 1-2 shift valve is controlled by opposition between governor pressure on one end, and combined spring force and shift throttle pressure or modulated throttle upshift pressure on the other end.

In low gear, mainline pressure (DRIVE or D4 oil) from the manual valve flows to the 1-2 shift valve where it is blocked by the downshifted valve. When vehicle speed increases enough for governor pressure to overcome spring force and shift throttle pressure or modulated throttle upshift pressure acting on the 1-2 throttle valve, the 1-2 shift valve upshifts. DRIVE or D4 oil then flows through the upshifted valve and into the second gear circuit where it travels to the apply side of the intermediate servo, and to the 1-2 accumulator, which cushions intermediate band engagement. In the THM 200-4R and 325-4L, second gear oil also flows to the converter clutch shift valve.

To prevent hunting between gears after the 1-2 upshift, shift throttle pressure is blocked from entering the 1-2 shift valve assembly of the THM 200/200C and 325. In the THM 200-4R and 325-4L, modulated throttle upshift pressure acts on a smaller land of the 1-2 throttle valve after the upshift.

A closed-throttle 2-1 downshift occurs when governor pressure drops to a very low level (road speed decreases), and spring force and shift throttle pressure or modulated throttle upshift pressure act on the 1-2 throttle valve to move the 1-2 shift valve against governor pressure. The transmissions covered in this chapter do not offer a part-throttle 2-1 downshift.

A full-throttle 2-1 detent downshift occurs below approximately 25 mph (40 kph) in the THM 200/200C, 40 mph (64 kph) in the THM 325, or 30 mph (48 kph) in the THM 200-4R and 325-4L, when detent pressure from the throttle valve flows to the 1-2 throttle valve or low-first detent valve. The combination of detent pressure, spring force, and shift throttle pressure or modulated throttle upshift pressure moves the

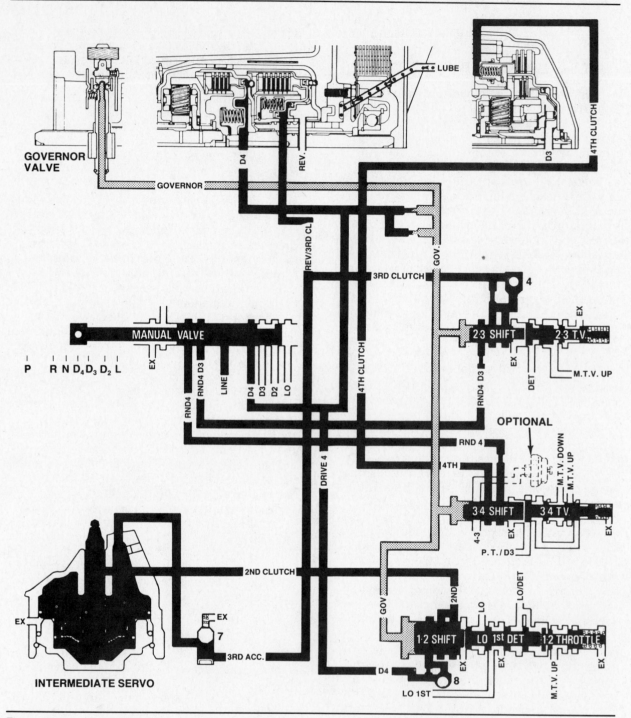

Figure 16-41. The THM 325-4L shift valves; the THM 200-4R valves are similar.

1-2 shift valve against governor pressure and causes a 2-1 detent downshift.

In manual low, mainline pressure (LOW oil) from the manual valve fills a chamber between the 1-2 shift valve and the 1-2 throttle valve (THM 200/200C and 325), or the chamber between the low-first detent valve and the 1-2 throttle valve. This downshifts the 1-2 shift valve against governor pressure, and keeps it downshifted regardless of throttle opening or road speed. In the THM 200-R4 and 325-4L, LOW oil is also routed to the low overrun clutch valve.

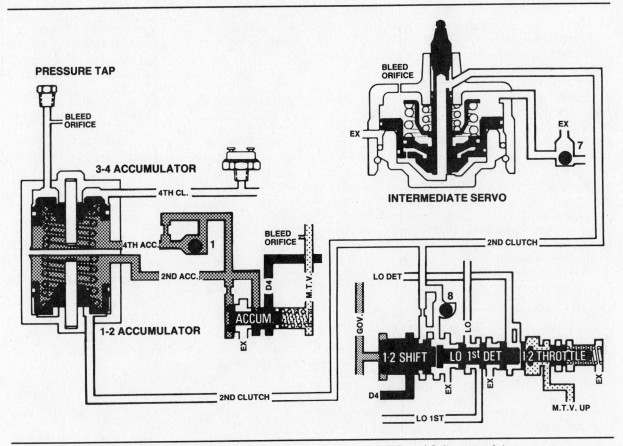

Figure 16-42. The THM 200-4R and 325-4L accumulator valve and the 1-2 and 3-4 accumulators.

Accumulator valve

The accumulator valve, figure 16-42, cushions application of the intermediate band during 1-2 upshifts, and the overdrive clutch (THM 200-4R and 325-4L) during 3-4 upshifts. It does this in relation to throttle opening by creating an accumulator pressure that combines with spring force in the 1-2 and 3-4 accumulators to control the rate at which the accumulator pistons move through their ranges of travel. Because it operates only during the 1-2 upshift, the THM 200/200C and 325 accumulator valve is sometimes called the 1-2 accumulator valve.

When the transmission is in any forward gear range, mainline pressure (DRIVE or D4 oil) is routed to the accumulator valve. The DRIVE or D4 pressure moves the valve against spring force and shift throttle or modulated throttle pressure to create a variable accumulator pressure. When shift throttle or modulated throttle pressure is low, accumulator pressure is also low. When shift throttle or modulated throttle pressure is high, accumulator pressure is also high. Accumulator pressure flows into the spring chambers of the 1-2 and 3-4 accumulators.

When the 1-2 or 3-4 shift valve upshifts, mainline pressure flows through the second or fourth gear (servo or clutch apply) circuit to the accumulator piston where it opposes spring force and accumulator pressure. The amount of time it takes the servo apply pressure to bottom the accumulator piston determines the amount of force with which the band or clutch is applied.

When accumulator pressure is low, indicating a light load on the drivetrain, apply oil rapidly bottoms the piston. This absorbs much of the pressure in the circuit and provides an easy band or clutch application. When accumulator pressure is high, indicating a heavy load on the drivetrain, apply oil takes much longer to bottom the piston. As a result, most of the pressure in the apply circuit is immediately available to apply the band or clutch much more firmly. In both cases, once the accumulator piston is bottomed, full mainline pressure is present in the circuit to hold the band or clutch securely applied.

Low overrun clutch valve

The low overrun clutch valve, figures 16-27 and 16-28, is a pressure-reducing valve that works only in the manual low range. Mainline pressure (LOW oil) is routed from the 1-2 shift valve assembly to the low overrun clutch valve, which regulates the pressure down to approximately 30 psi (205 kPa). This reduced pressure is then routed to the apply area of the low-reverse clutch where it provides a smooth manual 2-1 downshift.

When a THM 200/200C or 325 transmission is placed in Reverse, mainline (REVERSE) pressure from the manual valve is applied to the end of the low overrun clutch valve opposite spring force. This bottoms the valve so that apply pressure to the low-reverse clutch is *not* regulated to a lower level.

2-3 shift valve assembly

The 2-3 shift valve controls upshifts from second to third, and downshifts from third to second. In the THM 200/200C and 325, the assembly includes the 2-3 shift valve, the 2-3 throttle valve, and a spring, figure 16-39. The THM 200-4R and 325-4L assembly, figure 16-41, is essentially the same. The 2-3 shift valve is controlled by opposition between governor pressure on one end, and combined spring force and shift throttle pressure or modulated throttle upshift pressure on the other end.

Mainline pressure (RND or RND4-D3 oil) is routed from the manual valve to the 2-3 shift valve where it is blocked by the downshifted valve. When vehicle speed increases enough for governor pressure to overcome spring force and shift throttle pressure or modulated throttle upshift pressure acting on the 2-3 throttle valve, the 2-3 shift valve upshifts. In the THM 200/200C and 325, RND oil flows through the upshifted valve to the high-reverse clutch, release side of the intermediate servo, direct clutch exhaust valve, shift throttle valve, and TCC solenoid (200C). In the THM 200-4R and 325-4L, RND4-D3 oil flows through the upshifted valve to the high-reverse clutch, release side of the intermediate servo, and 3-2 control valve.

Third gear circuit pressure to the high-reverse clutch applies the clutch to place the transmission in third gear. The pressure routed to the release side of the intermediate servo combines with spring force to release the servo and intermediate band. The stroking action of the servo piston acts as an accumulator for high-reverse clutch application. The time it takes third gear oil to release the intermediate servo provides a delay in high-reverse clutch application. Pressure in the high-reverse clutch

circuit closes the accumulator check valve to seal the exhaust port in the passage at this time.

To prevent hunting between gears after the 2-3 upshift, shift throttle pressure is blocked from entering the 2-3 shift valve assembly of the THM 200/200C and 325. In the THM 200-4R and 325-4L, modulated throttle upshift pressure acts on a second land of the 2-3 throttle valve to reduce the force applied to the 2-3 shift valve after the upshift.

A closed-throttle 3-2 downshift occurs when governor pressure drops (road speed decreases) to the point where spring force and shift throttle pressure or modulated throttle upshift pressure act on the 2-3 throttle valve to move the 2-3 shift valve against governor pressure.

A part-throttle 3-2 downshift occurs at speeds below approximately 45 mph (72 kph) in the THM 200/200C, or 50 mph (80 kph) in the THM 325, when part-throttle pressure from the throttle valve flows to the 2-3 throttle valve. The combination of part-throttle pressure, shift throttle pressure, and spring force moves the 2-3 shift valve against governor pressure to cause a 3-2 downshift. The THM 200-4R and 325-4L do not offer a part-throttle 3-2 downshift.

A full-throttle 3-2 detent downshift occurs at speeds below approximately 65 mph (105 kph) in the THM 200/200C and 325, or 70 mph (113 kph) in the THM 200-4R and 325-4L, when detent pressure from the throttle valve flows to the 2-3 throttle valve. The combination of detent pressure, spring force, part-throttle pressure, and shift throttle pressure or modulated throttle upshift pressure moves the 2-3 shift valve against governor pressure and causes a 3-2 downshift. If the vehicle speed is below approximately 25 mph (40 kph) in the THM 200/200C, 40 mph (64 kph) in the 325, or 30 mph (48 kph) in the THM 200-4R and 325-4L, detent pressure will also downshift the 1-2 shift valve, resulting in a 3-1 downshift.

Direct clutch exhaust valve — THM 200/200C and 325

The direct clutch exhaust valve, figure 16-43, works with the shift throttle valve to regulate the rate of high-reverse (direct) clutch release and intermediate band application during part- and full-throttle (detent) 3-2 downshifts.

The shift throttle valve is controlled by opposition between throttle pressure and spring force. When the accelerator is fully released, fluid in the intermediate servo release (high-reverse clutch accumulator) passage is free to flow through the shift throttle valve to the direct clutch exhaust valve. As throttle pressure

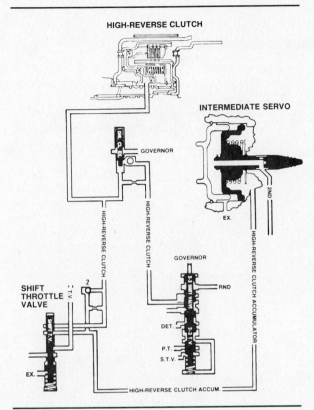

Figure 16-43. The THM 200/200C and 325 direct clutch exhaust valve.

rises, movement of the shift throttle valve gradually restricts the passage until, at approximately 45 mph (72 kph), the passage is blocked.

The direct clutch exhaust valve is controlled by opposition between spring force and governor pressure. When the vehicle is stopped, spring force holds the direct clutch exhaust valve closed. As vehicle speed rises, governor pressure gradually opens the valve against spring force until at approximately 45 mph (72 kph) the valve is completely open.

During coasting and part-throttle 3-2 downshifts below approximately 45 mph (72 kph), intermediate servo release (high-reverse clutch accumulator) oil flows unrestricted through the shift throttle valve to the direct clutch exhaust valve. High-reverse clutch oil also flows to the direct clutch exhaust valve. All the exhausting fluid in the third gear circuit then flows through a restricting orifice and the direct clutch exhaust valve opening to the 2-3 shift valve where it exhausts. The direct clutch exhaust valve position varies in relation to governor pressure, which allows the high-reverse clutch to release, and the intermediate band to apply, at rates that provide the best shift feel for a given vehicle speed.

During full-throttle 3-2 detent downshifts above approximately 45 mph (72 kph), the direct clutch exhaust valve is fully open, which allows high-reverse clutch oil to flow unrestricted to the 2-3 shift valve where it exhausts. At the same time, the shift throttle valve closes the unrestricted intermediate servo release (high-reverse clutch accumulator) passage. As a result, exhausting accumulator oil seats a check ball (2) and passes through a restricting orifice before joining the oil exhausting from the high-reverse clutch. This causes band application to be delayed, allowing engine speed to increase and better match the higher gear ratio of second gear.

3-2 control valve — THM 200-4R and 325-4L

The 3-2 control valve, figure 16-44, regulates the rate of high-reverse clutch release and intermediate band application during part- and full-throttle (detent) 3-2 downshifts. The valve is controlled by opposition between spring force and governor pressure. When the vehicle is stopped, spring force holds the 3-2 control valve wide open. As speed rises, governor pressure moves the valve against spring force to restrict the valve opening. Above approximately 50 mph (80 kph), governor pressure closes the valve entirely.

During part-throttle 3-2 downshifts below approximately 50 mph (80 kph), the 3-2 control valve opening regulates the downshift in relation to vehicle speed. Oil exhausting from the high-reverse clutch flows through a passage to the 2-3 shift valve where it seats a check ball (4), The oil then passes through a restricting orifice and into the shift valve where it exhausts. Oil exhausting from the release side of the intermediate servo seats two additional check balls (7 and 2) and flows to the 3-2 control valve. The oil then passes through a restricting orifice and the 3-2 control valve. The size of the opening in the 3-2 control valve allows the high-reverse clutch to release, and the intermediate band to apply, at a rate that provides the best shift feel for a given vehicle speed. From the valve, accumulator release oil flows to the 2-3 shift valve where it exhausts.

During full-throttle 3-2 detent downshifts above approximately 50 mph (80 kph), the 3-2 control valve is closed. As a result, all of the exhausting accumulator oil from the intermediate servo must pass through the restricting orifice before joining the oil exhausting from the high-reverse clutch. This causes band application to be delayed even more, allowing engine speed to increase and better match the higher gear ratio of second gear.

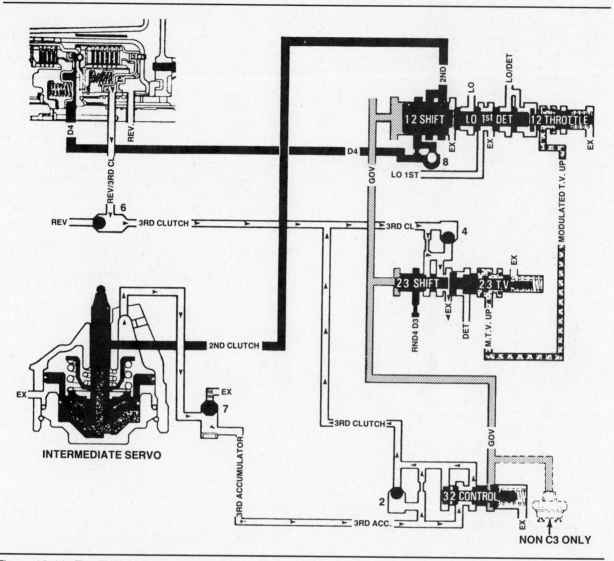

Figure 16-44. The THM 200-4R and 325-4L 3-2 control valve.

**Intermediate boost valve —
THM 200/200C and 325**

In manual second and manual low, the intermediate boost valve, figure 16-27, receives mainline pressure (INTERMEDIATE oil) from the manual valve and regulates it to create intermediate boost pressure. This pressure is routed through the line boost circuit to the spring end of the pressure regulator valve where it acts on the valve to increase mainline pressure.

The boost pressure passages are controlled by two ball check valves (1 and 3). These valves allow reverse boost pressure into the line boost circuit when the gear selector is in Reverse. They also allow shift throttle pressure to replace intermediate boost pressure in the line boost circuit if the throttle is opened far

enough. This allows mainline pressure to be boosted during full-throttle operation regardless of gear selector position.

Reverse boost valve — THM 200/200C and 325

When the gear selector is placed in Reverse, the reverse boost valve, figure 16-27, receives mainline pressure (REVERSE oil) from the manual valve and regulates it to create reverse boost pressure. The reverse boost valve is controlled by opposition between mainline pressure on one end, and spring force and shift throttle pressure on the other. As a result, reverse boost pressure increases with throttle opening.

In Reverse, reverse boost pressure is routed through the line boost circuit to the spring end of the pressure regulator valve where it acts on the valve to increase mainline pressure. This holds the apply devices more securely under the increased torque loads that can be expected in Reverse.

3-4 shift valve assembly — THM 200-4R and 325-4L

The 3-4 shift valve controls upshifts from third to fourth, and downshifts from fourth to third. The assembly consists of the 3-4 shift valve, 3-4 throttle valve, and a spring, figure 16-41. The 3-4 shift valve is controlled by opposition between governor pressure on one end, and combined spring force and modulated throttle upshift pressure on the other end.

In Reverse, Neutral, and Overdrive, mainline pressure (RND4 oil) is routed from the manual valve to the 3-4 shift valve where it is blocked by the downshifted valve. When vehicle speed increases enough for governor pressure to overcome spring force and modulated throttle upshift pressure acting on the 3-4 throttle valve, the 3-4 shift valve upshifts. RND4 oil then flows through the upshifted valve to apply the overdrive clutch and place the transmission in fourth gear. Some of the fourth gear circuit pressure flows to the 3-4 accumulator, figure 16-42, which cushions overdrive clutch engagement as described earlier. To prevent hunting between gears, modulated throttle upshift pressure is blocked from entering the 3-4 shift valve assembly after the 3-4 upshift.

A closed-throttle 4-3 downshift occurs when governor pressure drops (road speed decreases) to the point where spring force acts on the 3-4 throttle valve to move the 3-4 shift valve against governor pressure.

In all THM 200-4R and 325-4L transmissions, throttle pressure acting on the modulated throttle downshift valve can cause a part-throttle 4-3 downshift. As explained earlier, when throttle pressure rises above 40 psi (276 kPa), the modulated throttle downshift valve opens and sends modulated throttle downshift pressure to the 3-4 throttle valve. The combination of spring force and modulated throttle downshift pressure acting on the 3-4 throttle valve then moves the 3-4 shift valve against governor pressure to cause a 4-3 downshift.

In some THM 200-4R and 325-4L transmissions, a part-throttle 4-3 downshift is also possible when the accelerator linkage moves the throttle plunger far enough to uncover the part-throttle port in the throttle valve. This routes part-throttle pressure through the part-throttle/ D3 passage to the 3-4 throttle valve. The combination of part-throttle pressure and spring force acting on the 3-4 throttle valve then moves the 3-4 shift valve against governor pressure to

cause a 4-3 downshift. In the THM 200-4R and 325-4L models that do not offer this part-throttle downshift, the throttle valve bushing blocks the part-throttle/D3 passage until the detent port is opened for a full-throttle downshift.

A full-throttle 4-3 detent downshift occurs at speeds above approximately 70 mph (113 kph) when the accelerator is completely depressed and detent pressure from the throttle valve flows through the part-throttle/D3 passage to the 3-4 throttle valve. The combination of detent pressure and spring force acting on the 3-4 throttle valve moves the 3-4 shift valve against governor pressure and causes a 4-3 downshift. If the vehicle speed is below approximately 70 mph (113 kph) detent pressure will also downshift the 2-3 shift valve, resulting in a 4-2 downshift.

When a THM 200-4R or 325-4L transmission is placed in Drive (3 on the gear selector), mainline pressure (D3 oil) from the manual valve is routed through the part-throttle/D3 passage to the 4-3 throttle valve. The mainline pressure downshifts the 3-4 shift valve against governor pressure, and keeps the valve downshifted regardless of vehicle speed or throttle position.

4-3 control valve — THM 200-4R and 325-4L

The 4-3 control valve, figure 16-28, controls overdrive clutch release during a manual downshift from Overdrive to Drive (D to 3 on the gear selector). The valve is controlled by opposition between spring force on one end and D3 oil on the other.

When the gear selector is moved from Overdrive to Drive, the manual valve cuts off RND4 oil to the 3-4 shift valve, and sends mainline pressure into the D3 circuit. D3 oil applies the drive clutch, closes the 3-4 shift valve as described in the previous section, and flows to the 4-3 control valve. Overdrive clutch apply oil flows through the downshifted 3-4 shift valve to the 4-3 control valve where it is blocked. As pressure in the D3 circuit rises, it opens the 4-3 control valve and allows the overdrive clutch apply oil to exhaust.

Converter clutch actuator valve — THM 200C
Converter clutch apply valve — THM 200C

In the THM 200C, the torque converter clutch is controlled by a TCC solenoid and a converter clutch valve assembly that contains an actuator valve and an apply valve, figure 16-45. When the transmission shifts into third gear, high-reverse (direct) clutch apply oil flows to the TCC solenoid. This allows converter clutch application when the TCC solenoid is energized.

The electrical and electronic controls used to control the TCC solenoid are described in

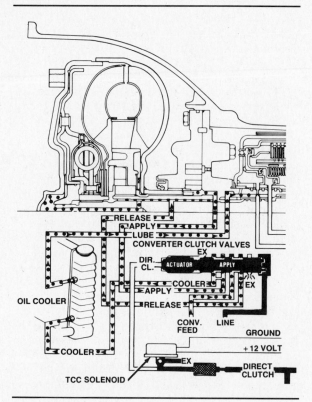

Figure 16-45. The THM 200C converter clutch circuit in the unapplied position.

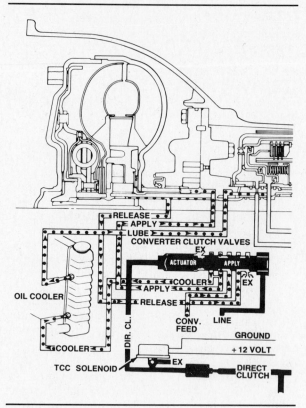

Figure 16-46. The THM 200C converter clutch circuit in the applied position.

Chapter 8. The paragraphs below describe the hydraulic operation of the converter clutch system.

When the TCC solenoid is de-energized, figure 16-45, oil is exhausted from the solenoid circuit, and mainline pressure holds the converter clutch actuator and apply valves in the release position. In this position, converter feed oil is routed through the converter clutch apply valve to the torque converter through the clutch release passage. Oil enters the converter between the clutch pressure plate and the converter cover, which prevents the clutch from applying. Oil leaving the converter returns to the converter clutch apply valve through the clutch apply passage, and is then routed to the transmission cooler and the lubrication circuit.

When the TCC solenoid is energized, figure 16-46, its exhaust port is closed and mainline pressure travels through the solenoid circuit to move the converter clutch actuator and apply valves to the apply position. In this position, converter feed pressure is routed to the converter through the clutch apply passage. Oil enters the converter between the clutch pressure plate and the turbine, which forces the pressure plate into contact with the converter cover, locking up the converter. Oil leaving the converter returns to the converter clutch control

valve through the clutch release passage, and is then exhausted. A portion of the converter feed oil in the clutch apply passage is diverted through a restricting orifice to the transmission cooler and lubrication circuit.

Converter clutch valve — THM 200-4R and 325-4L

In the THM 200-4R and 325-4L, the torque converter clutch is controlled by the TCC solenoid, converter clutch valve, converter clutch shift valve, and converter clutch throttle valve, figure 16-47. Under certain conditions in second, third, and fourth gears, converter clutch signal pressure is routed to the TCC solenoid from the converter clutch shift and throttle valves. This allows converter clutch application when the TCC solenoid is energized.

The electrical and electronic controls used to control the TCC solenoid are described in Chapter 8. The paragraphs below describe the hydraulic operation of the converter clutch system. The operation of the converter clutch shift and throttle valves is described in the next section.

When the TCC solenoid is de-energized, figure 16-47, oil is exhausted from the solenoid circuit, and spring force holds the converter

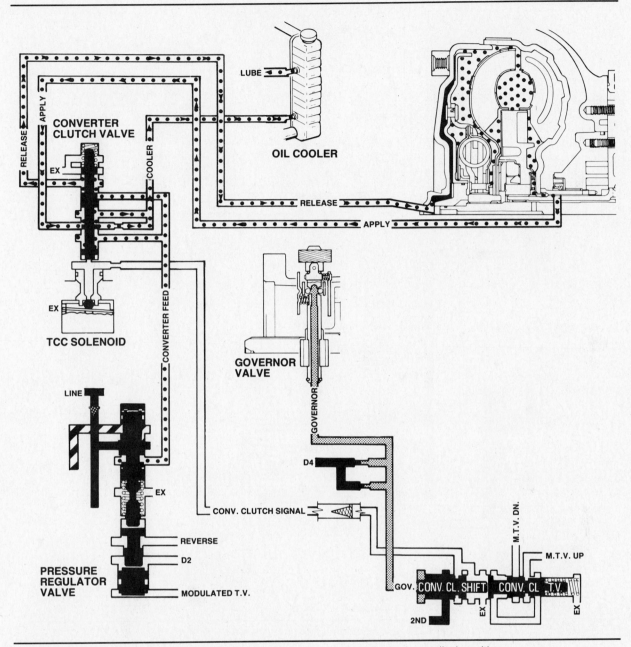

Figure 16-47. The THM 200-4R and 325-4L converter clutch circuit in the unapplied position.

clutch valve in the release position. In this position, converter feed oil is routed through the converter clutch apply valve to the torque converter through the clutch release passage. Oil enters the converter between the clutch pressure plate and the converter cover, which prevents the clutch from applying. Oil leaving the converter returns to the converter clutch apply valve through the clutch apply passage, and is then routed to the transmission cooler and the lubrication circuit.

When the TCC solenoid is energized, figure 16-48, its exhaust port is closed and converter clutch signal pressure in the solenoid circuit moves the converter clutch valve to the apply position. In this position, converter feed pressure is routed to the converter through the clutch apply passage. Oil enters the converter between the clutch pressure plate and the turbine, which forces the pressure plate into contact with the converter cover, locking up the converter. Oil in the clutch release passage is

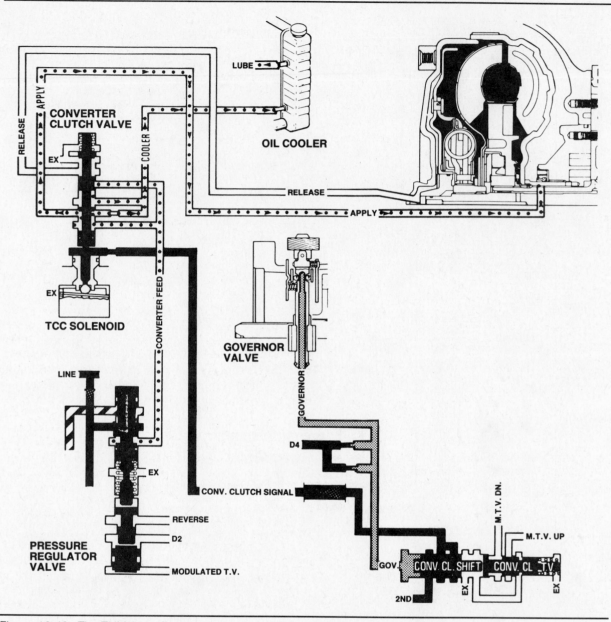

Figure 16-48. The THM 200-4R and 325-4L converter clutch circuit in the applied position.

exhausted through the converter clutch valve. A portion of the converter feed oil in the clutch apply passage is diverted through a restricting orifice to the transmission cooler and lubrication circuit.

Converter clutch shift valve —
THM 200-4R and 325-4L
Converter clutch throttle valve —
THM 200-4R and 325-4L
The converter clutch shift valve and throttle valve determine when converter clutch signal pressure is sent to the TCC solenoid to enable clutch lockup. The two valves are located in the same bore, and are controlled by opposition between governor pressure on one end, and combined spring force and modulated throttle upshift pressure on the other.

When the transmission upshifts to second gear, mainline pressure (second gear oil) is routed from the 1-2 shift valve to the converter clutch shift valve. At lower speeds, figure 16-47, second gear oil is blocked because spring force and modulated throttle upshift pressure acting on the converter clutch throttle valve keep the converter clutch shift valve downshifted against governor pressure.

When vehicle speed increases to the point where governor pressure overcomes spring force and modulated throttle upshift pressure, the converter clutch throttle valve upshifts the shift valve and directs converter clutch signal pressure to the TCC solenoid, figure 16-48. This allows converter clutch application when the TCC solenoid is energized.

To prevent repeated locking and unlocking of the converter clutch, modulated throttle upshift pressure is blocked from entering the converter clutch throttle valve after the converter clutch shift valve upshifts. This reduces the pressure acting to downshift the converter clutch shift valve.

If throttle pressure exceeds 40 psi (276 kPa), the modulated throttle downshift valve will route modulated throttle downshift pressure to the converter clutch throttle valve. This pressure will downshift the converter clutch shift valve and unlock the converter clutch to allow torque multiplication in the converter for better acceleration or pulling power.

MULTIPLE-DISC CLUTCHES AND SERVOS

The preceding paragraphs examined the individual valves in the THM 200/200C, 200-4R, 325, and 325-4L hydraulic systems. Before we trace the complete hydraulic system operation in each driving range, we will take a quick look at the multiple-disc clutches and servo used in these transmissions.

Clutches

The THM 200/200C and 325 transmissions have three multiple-disc clutches: the forward clutch, the low-reverse clutch, and the high-reverse clutch. Each clutch is applied hydraulically, and released by a waved spring when oil is exhausted from the clutch piston cavity. The forward clutch is applied in all forward gear ranges by DRIVE oil from the manual valve.

The low-reverse clutch is applied in manual low and Reverse. In manual low, it is applied by LOW oil from the downshifted 1-2 shift valve. In Reverse, it is applied by REVERSE oil from the manual valve. The low overrun clutch valve prevents LOW oil from exhausting through the manual valve in manual low, and REVERSE oil from exhausting through the 1-2 shift valve in Reverse.

The high-reverse (direct) clutch is applied in Drive high and Reverse. In Drive high, it is applied by direct clutch oil from the upshifted 2-3 shift valve. In Reverse, it is applied by REVERSE oil from the manual valve. A check ball

prevents direct clutch oil from exhausting through the manual valve in Drive high, and REVERSE oil from exhausting through the 2-3 shift valve in Reverse.

The THM 200-4R and 325-4L have the same three multiple-disc clutches described above. In addition, they have two additional multiple-disc clutches, the drive clutch and overdrive clutches. Each clutch is applied hydraulically, but the forward, high-reverse, and overdrive clutches are released by coil-type return springs, while the drive and low-reverse clutches are released by wave plate springs that are part of the clutch packs. The forward clutch is applied in all forward gear ranges by D4 oil from the manual valve.

The low-reverse clutch is applied in manual low and Reverse. In manual low, it is applied by LOW oil from the downshifted 1-2 shift valve. In Reverse, it is applied by REVERSE oil from the manual valve. A check ball prevents LOW oil from exhausting through the manual valve in manual low, and REVERSE oil from exhausting through the 1-2 shift valve in Reverse.

The high-reverse (direct) clutch is applied in Overdrive high, Drive high, and Reverse. In Overdrive and Drive high, it is applied by RND4-D3 oil from the upshifted 2-3 shift valve. In Reverse, it is applied by REVERSE oil from the manual valve. A check ball prevents RND4-D3 oil from exhausting through the manual valve in Drive high, and REVERSE oil from exhausting through the 2-3 shift valve in Reverse.

The drive clutch is applied in Drive, manual second, and manual low by D3 oil from the manual valve. The overdrive clutch is applied in fourth gear by RND4 oil from the upshifted 3-4 shift valve.

The high-reverse clutch in all the transmissions covered in this chapter is designed so that the piston has two separate surfaces on which hydraulic pressure may act. In high gear, when torque loads are low, pressure is applied to the smaller, inner area of the piston. In Reverse, when torque loads are greater, pressure is applied to the inner area of the piston, and to the larger, outer area. This allows the clutch to develop greater holding force in Reverse.

Servo

All the transmissions covered in this chapter have a single servo that is used to apply the intermediate band. In THM 200 series transmissions, the servo is located on the right side of the case, figure 16-8. In THM 325 series transmissions, the servo is located on the top left rear side of the transmission case, figure 16-9.

The servo assembly basically consists of a piston, an apply pin, one or two springs, and a cover.

When a 1-2 upshift occurs, hydraulic pressure moves the piston against spring force, and the apply pin mechanically applies the band. When there is no hydraulic pressure in the piston cavity, the spring releases the servo and band.

During a 2-3 upshift, high-reverse clutch apply pressure is routed to the spring side of the intermediate servo piston. The servo acts as an accumulator to smooth application of the high-reverse clutch. The time it takes hydraulic pressure and spring force to release the servo provides a delay in application of the high-reverse clutch.

The intermediate band in the transmissions covered in this chapter cannot be adjusted from outside the transmission. Band adjustment is controlled by the length of the servo apply pin, which is selected at the time of transmission assembly.

HYDRAULIC SYSTEM SUMMARY — THM 200/200C AND 325

We have examined the individual valves, clutches, and servos in the hydraulic system, and discussed their various functions. The following paragraphs summarize their combined operation in the various gear ranges for the THM 200/200C and 325 transmissions. Refer to figure 16-27 to trace the hydraulic system operation.

Park

In Park, pump oil enters the mainline circuit and flows to the: pressure regulator valve, manual valve, and converter clutch apply valve (THM 200C). The pressure regulator valve regulates the pump oil to establish mainline pressure. Mainline pressure at the manual valve is blocked. Mainline pressure at the converter clutch apply valve holds the valve in the release position so that the converter clutch is not applied. All clutches and the band are released, and the valve positions are determined by spring force alone. A mechanical linkage engages the parking pawl with the parking gear on the outside of the rear ring gear to lock the output shaft and prevent the vehicle from being moved.

Neutral

In Neutral, the manual valve routes mainline pressure through the RND and RNDI passages.

Mainline pressure in the RND passage flows to the 2-3 shift valve where it is blocked. Mainline pressure in the RNDI passage flows to the governor and throttle valves. Because the vehicle is not moving, any pressure in the governor circuit is exhausted through the governor valve at this time.

The throttle valve regulates mainline pressure to create throttle pressure which is routed to the shift throttle valve. The shift throttle valve regulates throttle pressure to create shift throttle pressure which is limited to a maximum of 90 psi (620 kPa) and applied to the: throttle plunger, pressure regulator valve, accumulator valve, reverse boost valve, and 1-2 and 2-3 throttle valves.

Shift throttle pressure to the throttle plunger helps regulate throttle pressure and reduce the effort needed to hold the plunger in place. Shift throttle pressure at the pressure regulator valve acts to increase mainline pressure in proportion to throttle opening. Shift throttle pressure at the accumulator valve helps determine the accumulator pressure sent to the 1-2 accumulator to smooth application of the intermediate band. Shift throttle pressure at the reverse boost valve helps determine the reverse boost pressure sent to the pressure regulator valve when the transmission is in Reverse. Shift throttle pressure at the 1-2 and 2-3 throttle valves acts with spring force to delay upshifts and cause downshifts.

Drive Low

When the vehicle is stopped and the gear selector is placed in Drive, the manual valve routes DRIVE oil to the: apply area of the forward clutch, 1-2 shift valve, accumulator valve, and governor valve. DRIVE oil applies the forward clutch to place the transmission in first gear. DRIVE oil at the 1-2 shift valve is blocked by the downshifted valve. The accumulator valve regulates DRIVE pressure in relation to throttle opening, then applies it to the spring side of the 1-2 accumulator piston.

The governor valve regulates mainline pressure to create governor pressure which is sent to the 1-2 shift valve, 2-3 shift valve, and direct clutch overrun valve. Governor pressure at the 1-2 and 2-3 shift valves opposes spring force and other hydraulic pressures to cause upshifts. Governor pressure at the direct clutch overrun valve opposes spring force to help control the rate of direct clutch apply during a 2-3 upshift.

In Drive low gear, the forward clutch is applied, and the 1-way clutch is holding.

Drive Second

As vehicle speed and governor pressure increase, governor pressure acting on the 1-2 shift valve overcomes shift throttle pressure and spring force, and upshifts the 1-2 shift valve. This allows DRIVE oil to flow through the 1-2 shift valve to the apply side of the intermediate servo and the 1-2 accumulator. The intermediate servo applies the intermediate band to place the transmission in second gear. DRIVE oil in the second gear circuit moves the 1-2 accumulator piston through its range of travel to smooth band application.

In Drive second gear, the forward clutch and intermediate band are applied.

Drive Third

As vehicle speed and governor pressure continue to increase, governor pressure overcomes shift throttle pressure and spring force, and upshifts the 2-3 shift valve. This allows RND pressure to flow through the shift valve to the: inner apply area of the high-reverse clutch, release side of the intermediate servo, and TCC solenoid (THM 200C).

RND pressure applies the high-reverse clutch to place the transmission in third gear. RND oil to the intermediate servo moves the servo piston through its range of travel to release the intermediate band and help cushion engagement of the high-reverse clutch.

When a THM 200C reaches converter clutch engagement speed in third gear, the TCC solenoid is energized to close the exhaust port. Solenoid circuit oil then moves the converter clutch actuator and apply valve against mainline pressure to the apply position. The apply valve then reroutes converter feed pressure through the apply passage to lock the converter clutch.

In Drive high gear, the forward clutch and high-reverse clutch are applied. The converter clutch in the THM 200C may or may not be applied, depending on vehicle operating conditions.

Drive Range Forced Downshifts

A part-throttle 3-2 downshift can be forced at speeds below approximately 45 mph (72 kph) in the THM 200/200C, or 50 mph (80 kph) in the THM 325. When the accelerator is depressed beyond a certain point, the throttle plunger opens a port that allows shift throttle pressure into the part-throttle passages. Part-throttle pressure combines with shift throttle pressure and spring force acting on the 2-3 throttle valve to move the 2-3 shift valve against governor pressure and force a downshift.

A full-throttle 3-2 downshift can be forced at speeds below approximately 65 mph (105 kph). When the throttle is completely depressed, the throttle plunger opens a port that allows shift throttle pressure into the detent passages. Detent pressure combines with part-throttle pressure, shift throttle pressure, and spring force acting on the 2-3 throttle valve to move the 2-3 shift valve against governor pressure and force a downshift.

A full-throttle 2-1 downshift can be forced at speeds below approximately 25 mph (40 kph) in the THM 200/200C, or 40 mph (64 kph) in the THM 325. When the throttle is completely depressed, the throttle plunger opens a port that allows shift throttle pressure into the detent passages. Detent pressure combines with part-throttle pressure, shift throttle pressure, and spring force acting on the 1-2 throttle valve to move the 1-2 shift valve against governor pressure and force a downshift.

Manual Second

When the gear selector is placed in manual second, the manual valve sends INTERMEDIATE oil to the intermediate boost valve which modulates the pressure and applies it to the pressure regulator valve to increase mainline pressure. This pressure application is also made in manual low.

The manual valve also opens an exhaust port for the RND passage, which is the feed (through the 2-3 shift valve) for the high-reverse clutch and the release side of the intermediate servo. With RND oil exhausted, the high-reverse clutch is released, and the intermediate servo applies the band, regardless of vehicle speed.

In manual second gear, the forward clutch and intermediate band are applied. If the vehicle speed drops low enough, shift throttle pressure and spring force acting on the 1-2 throttle valve will move the 1-2 shift valve against governor pressure and cause a coasting downshift to Drive low.

Manual Low

When the gear selector is placed in manual low, the manual valve routes LOW oil to the 1-2 shift valve where it moves and holds the valve against governor pressure. LOW oil then flows through the downshifted valve to the low overrun clutch valve, which regulates the pressure down to approximately 35 psi (240 kPa) and routes it to the apply area of the low-reverse clutch, placing the transmission in first gear. Intermediate servo oil exhausts through the 1-2 shift valve.

In manual low, the manual valve also blocks RNDI oil to the throttle valve. As a result, there are no shift throttle pressure applications to oppose governor pressure at the 1-2 shift valve. Once the valve is downshifted, LOW oil and spring force hold it there regardless of vehicle speed and governor pressure.

If the gear selector is placed in manual low at a vehicle speed above approximately 25 mph (40 kph) in the THM 200/200C, or 40 mph (64 kph) in the THM 325, governor pressure acting on the 1-2 shift valve will be high enough to prevent a downshift. If the transmission is in high gear at the time the gear selector is moved, the transmission will downshift to second gear. The 1-2 shift valve will block LOW oil until the vehicle speed drops below the limits given above. The valve then downshifts to allow LOW oil to flow to the low-reverse clutch.

In manual low gear, the forward clutch and the low-reverse clutch are applied.

Reverse

When the gear selector is placed in Reverse, the manual valve routes mainline pressure through RND circuit, the RNDI circuit, and the REVERSE circuit. In early THM 200 transmissions, mainline pressure in the RND circuit is routed through the 2-3 shift valve to the inner apply area of the high-reverse clutch. In later THM 200 models, RND pressure is blocked at the 2-3 shift valve, and the inner apply area of the high-reverse clutch is fed by a check-ball controlled passage from the REVERSE circuit. Mainline pressure in the RNDI circuit is routed through the throttle valve and into the shift throttle circuit where it acts on the reverse boost valve.

Mainline pressure in the REVERSE circuit is routed to the: inner (except early THM 200) and outer apply areas of the high-reverse clutch, apply area of the low-reverse clutch, and reverse boost valve. The high-reverse and low-reverse clutches apply to place the transmission in reverse gear. The reverse boost valve moves against spring force and shift throttle pressure to route reverse boost pressure through the line boost passage to the pressure regulator valve. This increases mainline pressure to hold the apply devices more securely.

Both the DRIVE and INTERMEDIATE circuits are exhausted in Reverse, so the forward clutch and intermediate band are released. The high-reverse and low-reverse clutches are applied.

HYDRAULIC SYSTEM SUMMARY — THM 200-4R AND 325-4L

We have examined the individual valves, clutches, and servos in the hydraulic system, and discussed their various functions. The following paragraphs summarize their combined operation in the various gear ranges for the THM 200-4R and 325-4L transmissions. Refer to figure 16-28 to trace the hydraulic system operation.

Park

When the transmission is in Park, pump oil enters the mainline circuit and flows to the: pressure regulator valve, manual valve, pressure relief valve, and throttle limit valve. The pressure regulator valve regulates the pump oil to establish mainline pressure. Mainline pressure at the manual valve is blocked. The pressure relief valve limits mainline pressure to a maximum of 320 to 360 psi (2240 to 2520 kPa). The throttle limit valve regulates mainline pressure to a maximum of 90 psi (620 kPa) and directs it to the throttle valve as throttle feed pressure.

The throttle valve regulates mainline pressure to create throttle pressure which is routed to the: throttle plunger, line bias valve, accumulator valve (some models), and modulated throttle upshift and downshift valves. Throttle pressure to the throttle plunger helps regulate throttle pressure and reduce the effort needed to hold the plunger in place. The line bias valve regulates throttle pressure to create modulated throttle pressure that is sent to the throttle boost valve and accumulator valve (some models). Modulated throttle pressure at the throttle boost valve acts on the pressure regulator valve to increase mainline pressure in proportion to throttle opening and engine torque. Throttle pressure or modulated throttle pressure at the accumulator valve helps determine the accumulator pressure that is sent to the 1-2 and 3-4 accumulators to smooth application of the intermediate band and overdrive clutch.

Throttle pressure to the modulated throttle upshift valve opposes spring force and opens the valve above approximately 10 psi (69 kPa) to create a modulated throttle upshift pressure that is sent to the converter clutch throttle valve, and the 1-2, 2-3, and 3-4 throttle valves. Modulated throttle upshift pressure at the converter clutch throttle valve combines with spring force to hold the valves in the release position so that no pressure is directed into the converter clutch signal circuit. Modulated throttle upshift pressure at the 1-2, 2-3, and 3-4 throttle valve acts with spring force to delay upshifts and cause downshifts.

Throttle pressure to the modulated throttle downshift valve opposes spring force and opens the valve above approximately 40 psi (276 kPa) to create a modulated throttle downshift pressure that is sent to the converter clutch throttle valve and the 3-4 throttle valve. Modulated throttle downshift pressure at the converter clutch throttle valve combines with spring force to move the converter clutch shift valve to the release position. Modulated throttle downshift pressure at the 3-4 throttle valve acts with spring force to force a part-throttle downshift.

In Park, all clutches and the band are released, and valve positions are determined by spring force alone. A mechanical linkage engages the parking pawl with the parking gear on the outside of the rear ring gear to lock the output shaft and prevent the vehicle from being moved.

Neutral

In Neutral, the manual valve routes mainline pressure through the RND4-D3 and RND4 passages. Mainline pressure in the RND4-D3 passage flows to the 2-3 shift valve where it is blocked. This pressure application is also made in Reverse, Overdrive, and Drive. Mainline pressure in the RND4 passage flows to the 3-4 shift valve where it is also blocked. This pressure application is also made in Reverse and Overdrive.

Overdrive and Drive Low

When the vehicle is stopped and the gear selector is placed in Overdrive, the manual valve routes D4 oil to the: apply area of the forward clutch, 1-2 shift valve, accumulator valve, and governor valve. D4 oil applies the forward clutch to place the transmission in first gear. D4 oil at the 1-2 shift valve is blocked by the downshifted valve. The accumulator valve regulates D4 pressure in relation to throttle opening, then applies it to the spring side of the 1-2 accumulator piston, and the non-spring side of the 3-4 accumulator piston.

The governor valve regulates mainline pressure to create governor pressure which is sent to the 1-2, 2-3, and 3-4 shift valves, converter clutch shift valve, and 3-2 control valve. Governor pressure at the 1-2, 2-3, and 3-4 shift valves opposes spring force and other hydraulic pressures to cause upshifts. Governor pressure at the converter clutch shift valve opposes spring force and other hydraulic pressures that hold the valve in the release position. Governor pressure to the 3-2 control valve helps control the rate of direct clutch apply during a 2-3 upshift.

When the vehicle is stopped and the gear selector is placed in Drive, the same mainline pressure applications are made as in Overdrive. In addition, the manual valve routes D3 oil to the apply area of the drive clutch, the 3-4 throttle valve, and the 4-3 control valve. D3 oil applies the drive clutch to lock the overdrive gearset and make engine braking possible in the lower two gears. D3 oil to the 3-4 throttle valve combines with spring force to downshift the 3-4 shift valve and/or hold it in the downshifted position. D3 oil to the 4-3 control valve acts against spring force to control the rate of overdrive clutch oil exhaust during a manual 4-3 downshift.

In Overdrive low gear, the forward clutch is applied, the low 1-way clutch is holding, and the overdrive 1-way clutch is holding. In Drive low gear, the forward clutch is applied, the low 1-way clutch is holding, and the drive clutch is applied to override the overdrive 1-way clutch.

Overdrive and Drive Second

As vehicle speed and governor pressure increase, governor pressure acting on the 1-2 shift valve overcomes modulated throttle upshift pressure and spring force, and upshifts the 1-2 shift valve. This allows D4 oil to flow through the 1-2 shift valve to the apply side of the intermediate servo, the 1-2 accumulator, and the converter clutch shift valve. The intermediate servo applies the intermediate band to place the transmission in second gear. Second gear oil moves the 1-2 accumulator piston through its range of travel to smooth band application.

Second gear oil is blocked at the converter clutch shift valve until governor pressure moves the valve to the apply position by overcoming spring force and modulated throttle upshift pressure acting on the converter clutch throttle valve. This will happen in either second or third gear, depending on the transmission application. When the converter clutch shift valve moves to the apply position, second gear oil flows through the converter clutch signal circuit to the TCC solenoid. The clutch will then apply when the TCC solenoid is energized.

In Overdrive second gear, the forward clutch and intermediate band are applied, and the overdrive 1-way clutch is holding. In Drive second gear, the drive clutch is applied to override the overdrive 1-way clutch. The converter clutch may or may not be applied in second gear, depending on vehicle operating conditions and the transmission application.

Overdrive and Drive Third

As vehicle speed and governor pressure continue to increase, governor pressure overcomes modulated throttle upshift pressure and spring force, and upshifts the 2-3 shift valve. This allows RND4-D3 pressure to flow through the shift valve to the inner apply area of the high-reverse clutch, release side of the intermediate servo, and the 3-2 control valve.

Third gear oil applies the high-reverse clutch to place the transmission in third gear. Third gear oil to the intermediate servo moves the servo piston through its range of travel to release the intermediate band and help cushion engagement of the high-reverse clutch.

In Overdrive third gear, the forward and high-reverse clutches are applied, and the overdrive 1-way clutch is holding. In Drive third gear, the drive clutch is applied to override the overdrive 1-way clutch. The converter clutch may or may not be applied in third gear, depending on vehicle operating conditions.

Overdrive Fourth

As vehicle speed and governor pressure continue to increase, governor pressure overcomes modulated throttle upshift pressure and spring force, and upshifts the 3-4 shift valve. This allows RND4 pressure to flow through the shift valve into the fourth gear circuit where it flows to the apply area of the overdrive clutch and the 3-4 accumulator.

Fourth gear oil applies the overdrive clutch to place the transmission in fourth gear. Fourth gear oil also moves the 3-4 accumulator piston through its range of travel to smooth clutch application. In Overdrive fourth, the forward, high-reverse, and overdrive clutches are applied.

Overdrive and Drive Range Forced Downshifts

A part throttle 4-3 downshift can be forced when throttle pressure acting on the modulated throttle downshift valve rises above 40 psi (276 kPa). This opens the valve and sends modulated throttle downshift pressure to the 3-4 throttle valve. The combination of spring force and modulated throttle downshift pressure acting on the 3-4 throttle valve then moves the 3-4 shift valve against governor pressure to cause a 4-3 downshift.

A part-throttle 4-3 downshift is also possible in some THM 200-4R and 325-4L transmissions when part-throttle pressure from the throttle valve is routed to the 3-4 throttle valve. The combination of spring force, part-throttle pressure, and modulated throttle downshift pressure acting on the 3-4 throttle valve then moves the 3-4 shift valve against governor pressure to cause a 4-3 downshift.

A full-throttle 4-3 detent downshift can be forced at speeds above approximately 70 mph (113 kph) when detent pressure from the throttle valve flows to the 3-4 throttle valve. The combination of spring force, detent pressure, and modulated throttle downshift pressure acting on the 3-4 throttle valve moves the 3-4 shift valve against governor pressure to cause a 4-3 downshift.

In all types of 4-3 downshifts, the downshifted 3-4 shift valve blocks RND4 oil from entering the fourth gear circuit. Exhausting overdrive clutch and 3-4 accumulator oil passes through the 3-4 shift valve to the 4-3 control valve which controls the venting of the oil for a smooth downshift.

A full-throttle 4-2 or 3-2 detent downshift can be forced at speeds below approximately 70 mph (113 kph) when detent pressure from the throttle valve flows to the 3-2 throttle valve. The combination of spring force and detent pressure acting on the 2-3 throttle valve moves the 2-3 shift valve against governor pressure to cause a 3-2 downshift.

A full-throttle 2-1 downshift can be forced at speeds below approximately 30 mph (48 kph) when detent pressure from the throttle valve is routed to the low-first detent valve. Detent pressure acting on the low-first detent valve combines with spring force and modulated throttle upshift pressure acting on the 1-2 throttle valve to move the 1-2 shift valve against governor pressure and force a downshift.

Manual Second

When the gear selector is placed in manual second, the manual valve sends D2 oil to the area between the throttle boost valve and reverse boost valve in the pressure regulator valve. D2 oil acting on the reverse boost valve combines with spring force to increase mainline pressure and hold the apply devices more securely. This pressure application is also made in manual low.

The manual valve also opens an exhaust port for the RND4-D3 passage, which is the feed (through the 2-3 shift valve) for the high-reverse clutch and the release side of the intermediate servo. With RND4-D3 oil exhausted, the high-reverse clutch is released, and the intermediate servo applies the band to place the transmission in second gear regardless of vehicle speed.

In manual second gear, the forward clutch, drive clutch, and intermediate band are applied. If the vehicle speed drops low enough, spring force acting on the 1-2 throttle valve will move the 1-2 shift valve against governor pressure and cause a coasting downshift to Drive low.

Manual Low

When the gear selector is placed in manual low, the manual valve routes LOW oil to the area between the 1-2 throttle valve and the low-first detent valve. This pressure application downshifts the 1-2 shift valve and/or holds it downshifted against governor pressure regardless of vehicle speed. LOW oil then flows through the downshifted valve to the low overrun clutch valve, which regulates the pressure down to approximately 35 psi (240 kPa) and routes it to the apply area of the low-reverse clutch, placing the transmission in first gear. Intermediate servo oil exhausts through the 1-2 shift valve.

If the gear selector is placed in manual low at a vehicle speed above approximately 30 mph (48 kph), governor pressure acting on the 1-2 shift valve will be high enough to prevent a downshift. If the transmission is in third or fourth gear at the time the gear selector is

moved, the transmission will downshift to second gear. The 1-2 shift valve then blocks LOW oil until the vehicle speed drops below 30 mph (48 kph). The valve then downshifts to allow LOW oil to flow to the low-reverse clutch.

In manual low gear, the forward, low-reverse, and drive clutches are applied.

Reverse

When the gear selector is placed in Reverse, the manual valve routes mainline pressure through the RND4-D3 circuit, the RND4 circuit, and the REVERSE circuit. Mainline pressure in the RND4-D3 passage flows to the 2-3 shift valve where it is blocked. Mainline pressure in the RND4 passage flows to the 3-4 shift valve where it is also blocked.

Mainline pressure in the REVERSE circuit is routed to the inner and outer apply areas of the high-reverse clutch, the apply area of the low-reverse clutch, and the reverse boost valve. The high-reverse and low-reverse clutches apply to place the transmission in reverse gear. The reverse boost valve acts on the pressure regulator valve to increase mainline pressure and hold the apply devices more securely.

The D4, D3, D2, and LOW circuits are all exhausted in Reverse, so the forward clutch, drive clutch, overdrive clutch, and intermediate band are all released. The high-reverse and low-reverse clutches are applied.

Review Questions

Select the single most correct answer
Compare your answers to the correct answers on page 603

1. Mechanic A says the THM 200 transmission was introduced in 1978, and is sometimes called the M-29 transmission.
   Mechanic B says the THM 200-4R transmission was introduced in 1982, and has been used mainly in larger cars.
   Who is right?
   a. Mechanic A
   b. Mechanic B
   c. Both mechanic A and B
   d. Neither mechanic A nor B

2. The THM 325 transmission is:
   a. Based on the THM 200 gear train
   b. A 3-speed gearbox
   c. Used with longitudinally mounted engines.
   d. All of the above

3. Mechanic A says THM 325 series transmissions are used only in FWD applications.
   Mechanic B says the THM 200C uses the Simpson compound planetary gearset and has a lockup torque converter.
   Who is right?
   a. Mechanic A
   b. Mechanic B
   c. Both mechanic A and B
   d. Neither mechanic A nor B

4. Mechanic A says THM 200 series transmissions may be identified by a tag riveted to the bellhousing.
   Mechanic B says the D on a THM 325-4L gear selector identifies the Drive range.
   Who is right?
   a. Mechanic A
   b. Mechanic B
   c. Both mechanic A and B
   d. Neither mechanic A nor B

5. Mechanic A says the various gear ratios for the THM 200 and 325 series transmissions are identical.
   Mechanic B says the THM 325 is actually a transaxle because the differential and final drive assembly bolts to its case.
   Who is right?
   a. Mechanic A
   b. Mechanic B
   c. Both mechanic A and B
   d. Neither mechanic A nor B

6. Mechanic A says the direct clutch is also called the high-reverse clutch.
   Mechanic B says the direct clutch drives the front ring gear of the Simpson gearset.
   Who is right?
   a. Mechanic A
   b. Mechanic B
   c. Both mechanic A and B
   d. Neither mechanic A nor B

7. Mechanic A says THM 200 series transmissions have the pressure regulator valve built into the oil pump.
   Mechanic B says the THM 200-4R uses a variable displacement vane-type oil pump.
   Who is right?
   a. Mechanic A
   b. Mechanic B
   c. Both mechanic A and B
   d. Neither mechanic A nor B

8. "Drive oil" is the GM name for one form of:
   a. Throttle pressure
   b. Governor pressure
   c. Modulator pressure
   d. Mainline pressure

9. Mechanic A says the THM 200-4R and 325-4L have seven positions on the gear selector.
   Mechanic B says the THM 200-4R and 325-4L offer engine braking in Drive first, second, and third.
   Who is right?
   a. Mechanic A
   b. Mechanic B
   c. Both mechanic A and B
   d. Neither mechanic A nor B

10. Mechanic A says the sun gear in the THM 200/200C is made of two gears splined together.
    Mechanic B says the sun gear in the THM 200/200C is held by an input shell in third gear.
    Who is right?
    a. Mechanic A
    b. Mechanic B
    c. Both mechanic A and B
    d. Neither mechanic A nor B

11. Mechanic A says all of the transmissions covered in this chapter use an intermediate band.
    Mechanic B says application of the intermediate band is cushioned by the 1-2 accumulator.
    Who is right?
    a. Mechanic A
    b. Mechanic B
    c. Both mechanic A and B
    d. Neither mechanic A nor B

12. Mechanic A says the forward clutch is applied in all forward gears except overdrive fourth.
    Mechanic B says the forward clutch, when applied, drives the front ring gear of the Simpson geartrain.
    Who is right?
    a. Mechanic A
    b. Mechanic B
    c. Both mechanic A and B
    d. Neither mechanic A nor B

13. Mechanic A says overdrive fourth gear in the THM 200-4R is obtained by applying the overdrive clutch to lock the overdrive sun gear to the case.
    Mechanic B says the overdrive carrier is driven by the turbine shaft in overdrive fourth gear.
    Who is right?
    a. Mechanic A
    b. Mechanic B
    c. Both mechanic A and B
    d. Neither mechanic A nor B

14. The throttle valve in all of the transmissions covered in this chapter is controlled by a:
    a. Vacuum modulator
    b. Solenoid
    c. Servo
    d. Cable

15. Mechanic A says the THM 200C torque converter clutch may or may not be applied in second gear, depending on the transmission model and vehicle operating conditions.
    Mechanic B says the torque converter clutch in the THM 200-4R and 325-4L applies when the TCC solenoid stops exhausting high-reverse (direct) clutch oil.
    Who is right?
    a. Mechanic A
    b. Mechanic B
    c. Both mechanic A and B
    d. Neither mechanic A nor B

16. When decelerating to a stop from third gear, all of the transmissions in this chapter:
    a. Downshift twice, 3-2 and 2-1
    b. Downshift once, 3-1
    c. Do not downshift until the vehicle has stopped
    d. None of the above

17. Mechanic A says intermediate band adjustment for the transmissions covered in this chapter is controlled by an adjusting screw on the side of the case.
    Mechanic B says intermediate band adjustment for the transmissions covered in this chapter is controlled by a self-adjusting mechanism inside the case.
    Who is right?
    a. Mechanic A
    b. Mechanic B
    c. Both mechanic A and B
    d. Neither mechanic A nor B

18. Mechanic A says the THM 200-4R and 325-4L drive clutch is used to override the overdrive 1-way clutch when the gear selector is in Drive only.
    Mechanic B says the overdrive ring gear is always the output member of the overdrive gearset.
    Who is right?
    a. Mechanic A
    b. Mechanic B
    c. Both mechanic A and B
    d. Neither mechanic A nor B

19. Mechanic A says engine compression braking is not possible at higher speeds in Overdrive third because the transmission will up-shift to fourth gear.
    Mechanic B says engine compression braking is not possible at lower speeds in Overdrive third because the overdrive 1-way clutch will overrun.
    Who is right?
    a. Mechanic A
    b. Mechanic B
    c. Both mechanic A and B
    d. Neither mechanic A nor B

20. Mechanic A says the throttle limit valve in the THM 200/200C and 325 limits throttle pressure to a maximum of 90 psi (620 kPa).
    Mechanic B says the shift throttle valve in the THM 200-4R and 325-4L limits shift throttle pressure to a maximum of 90 psi (620 kPa).
    Who is right?
    a. Mechanic A
    b. Mechanic B
    c. Both mechanic A and B
    d. Neither mechanic A nor B

17

Turbo Hydra-matic 250, 250C, 350, 350C, and 375B Transmissions

HISTORY AND MODEL VARIATIONS

The General Motors (GM) Turbo Hydra-matic (THM) 250, 250C, 350, 350C, and 375B transmissions are all 3-speed units built around the Simpson compound planetary geartrain. The THM 250 and 250C, figure 17-1, are light-duty transmissions used primarily in 1974-84 compact and intermediate GM car models, although they have also seen limited service in selected full-size cars. The THM 350 and 350C transmissions, figure 17-2, are medium-duty units that are used in 1969-86 intermediate and full-size cars and light trucks from all GM divisions except Cadillac.

The THM 375B is a heavy-duty version of the THM 350 that is identical to the THM 350, except that it has additional high-reverse clutch plates for increased torque capacity. Some models also have a longer output shaft and extension housing. The THM 375B will not be called out separately in this chapter, but all statements regarding the THM 350 and 350C apply equally to the THM 375B.

The THM 250 and 250C transmissions are basically identical except that the 250C, figure 17-3, has a torque converter clutch that locks up to improve fuel economy. The THM 350 and 350C transmissions are related in the same way. In this chapter, we will treat the THM 250/250C and 350/350C transmissions as single designs, except where it is necessary to call out specific differences.

The transmissions covered in this chapter differ primarily in the type of holding devices used for second gear. In the THM 250/250C, a band is the only second gear holding device. In the THM 350/350C, the second gear holding devices include a multiple-disc clutch, a 1-way roller clutch, and a band for manual second.

Fluid Recommendations

Like all GM automatic transmissions and transaxles, the THM 250/250C and 350/350C require DEXRON®-IID fluid.

Transmission Identification

In addition to the transmission model, you must know the exact assembly part number and other identifying characters in order to obtain proper repair parts. The identification numbers on 1969-72 THM 350 transmissions are stamped on the 1-2 accumulator cover located on the center right side of the transmission

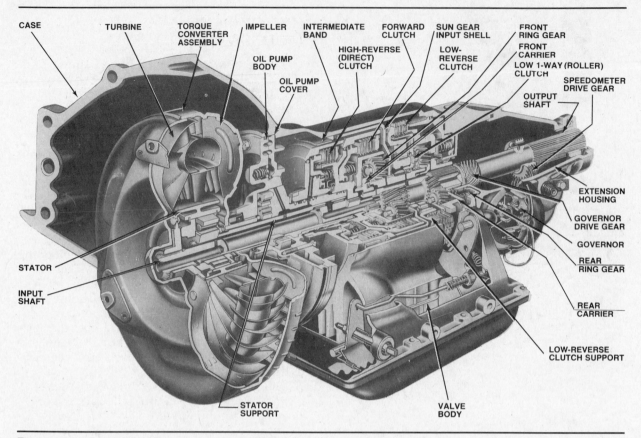

Figure 17-1. The THM 250 transmission; the THM 250C is basically the same with the addition of a lockup torque converter.

case, figure 17-4. On 1973 and later THM 350/350C transmissions, and on all THM 250/250C gearboxes, the identification numbers are stamped on the governor cover, figure 17-5.

The coding that appears on the accumulator or governor cover varies with the transmission. Figures 17-4 and 17-5 show typical identification codes and explain their meanings. If the accumulator or governor cover is replaced on a transmission, the complete set of identification numbers should be stamped on the new part.

As required by Federal regulations, GM transmissions also carry the serial number of the vehicle in which they were originally installed. On the transmissions covered in this chapter, the serial number is stamped on the lower left side of the case, next to the manual valve shaft.

Major Changes

The major changes that GM has made to the THM 250/250C and 350/350C transmissions are as follows:

1972 ● Steel valve body check balls replaced by viton parts.
1977 ● Number of roller thrust bearings increased from two to five.
1980 ● TCC introduced.
1982 ● Sun gear input shell drilled with holes to reduce weight.
1985 ● THM 350C torque converter lugs redesigned.

GEAR RANGES

Depending on the GM division and the particular application, there are minor variations in the exact labeling of the gear selectors used with

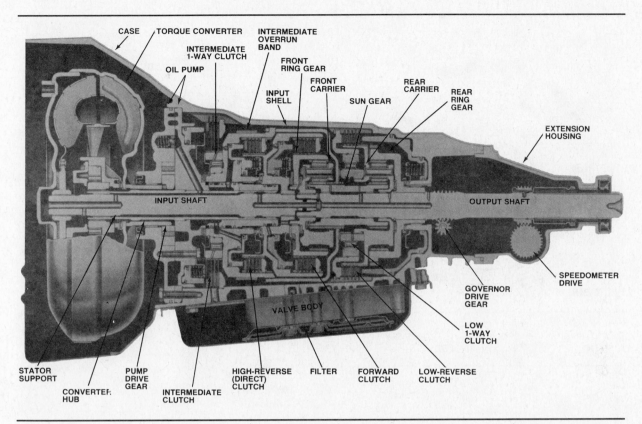

Figure 17-2. The THM 350 transmission; the THM 350C is basically the same with the addition of a lockup torque converter.

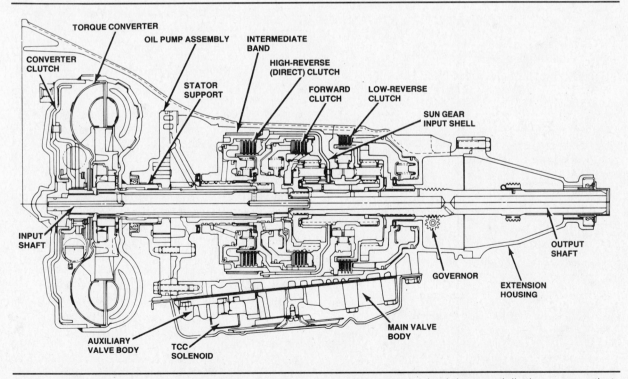

Figure 17-3. A cutaway of the THM 250C transmission; the torque converter clutch is essentially the same as that in the THM 350C.

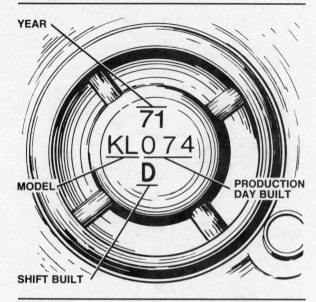

Figure 17-4. On 1972 and earlier THM 350 transmissions, the identification numbers are stamped on the 1-2 accumulator cover.

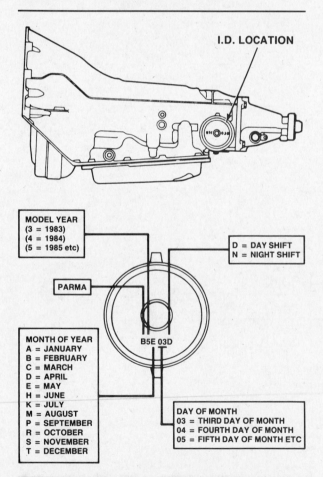

Figure 17-5. On 1973 and newer THM 350/350C transmissions, and all THM 250/250C gearboxes, the identification numbers are stamped on the governor cover.

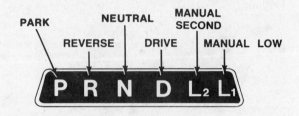

Figure 17-6. Typical THM 250/250C and 350/350C gear selector positions.

the THM 250/250C and 350/350C transmissions. However, all of the selectors have six positions, figure 17-6:

P — Park
R — Reverse
N — Neutral
D or 3 — Drive
I, S, 2, or L2 — Manual second
L, 1, or L1 — Manual low.

Gear ratios for the THM 250/250C and 350/350C are:

- First (low) — 2.52:1
- Second (intermediate) — 1.52:1
- Third (direct drive) — 1.00:1
- Reverse — 1.93:1.

Park and Neutral

The engine can be started only when the transmission is in Park or Neutral. No clutches or bands are applied in these positions, and there is no flow of power through the transmission. In Neutral, the output shaft is free to turn, and the car can be pushed or pulled. In Park, a mechanical pawl engages the parking gear on the outside of the rear ring gear. This locks the output shaft to the case and prevents the vehicle from being moved.

Drive

When the vehicle is stopped and the gear selector is placed in Drive (D or 3), the transmission is automatically in low gear. As the vehicle accelerates, the transmission automatically upshifts to second, and then to third gear. The speeds at which upshifts occur vary, and are controlled by vehicle speed and engine load. In the THM 250C and 350C, the torque converter clutch can apply in second gear in some applications, and in third gear in all applications.

The driver can force downshifts in Drive by opening the throttle partially or completely. The exact speeds at which these downshifts occur vary for different engine, axle ratio, tire size, and vehicle weight combinations. Below

GENERAL MOTORS TERM	TEXTBOOK TERM
Clutch housing	Clutch drum
Converter pump	Impeller
Detent downshift	Full-throttle downshift
Turbine shaft	Input shaft
Internal gear	Ring gear
Reaction carrier	Rear carrier
Output internal gear	Rear ring gear
Output carrier	Front carrier
Input internal gear	Front ring gear
Line pressure	Mainline pressure
Modulator valve	Throttle valve
Modulator pressure	Throttle pressure
Control valve assembly	Valve body
Direct, or front, clutch	High-reverse clutch
Low and reverse roller clutch	Low 1-way (roller) clutch
Intermediate overrun roller clutch (350/350C)	Intermediate 1-way (roller) clutch
Kickdown, or front, band	Intermediate band
Front servo	Intermediate servo
Faced clutch plate	Clutch friction disc
Separator plate	Clutch steel disc

Figure 17-7. General Motors transmission nomenclature table.

approximately 65 mph (105 kph) with a THM 250/250C, or 50 mph (80 kph) with a THM 350/350C, the driver can force a part-throttle 3-2 downshift by opening the throttle beyond a certain point. Below approximately 65 mph (105 kph) with a THM 250/250C, or 75 mph (121 kph) with a THM 350/350C, the driver can force a full-throttle 3-2 detent downshift by flooring the accelerator pedal. A 3-1 or 2-1 detent downshift can be forced below approximately 35 mph (56 kph) with a THM 250/250C, or 40 mph (64 kph) with a THM 350/350C.

When the car decelerates in Drive to a stop, the transmission makes two coasting downshifts automatically, first a 3-2 downshift, then a 2-1 downshift.

Manual Second

When the vehicle is stopped and the gear selector is placed in manual second (I, S, 2, or L2), the transmission is automatically in low gear. As the vehicle accelerates, the transmission automatically upshifts to second; it does not upshift to third. The speed at which the upshift occurs will vary, and is controlled by vehicle speed and engine load.

A 2-1 detent downshift can be forced below approximately 35 mph (56 kph) with a THM 250/250C, or 40 mph (64 kph) with a THM 350/350C, if the driver depresses the accelerator completely. The transmission automatically downshifts to low as the vehicle decelerates to a stop.

The driver can manually downshift from third to second by moving the selector from D or 3 to I, S, 2, or L2. In this case, the transmission shifts to second gear immediately, regardless of road speed or throttle position. Manual downshifting can provide engine compression braking, or gear reduction for hill climbing, pulling heavy loads, or operating in congested traffic. To obtain engine compression braking, the THM 350/350C uses an additional apply device in manual second that is not used in Drive second.

Manual Low

When the vehicle is stopped and the gear selector is placed in manual low (L, 1, or L1), the car starts and stays in first gear; it does not upshift to second or third. Because different apply devices are used in manual low than are used in Drive low, this gear range can be used to provide engine compression braking to slow the vehicle.

If the driver shifts to manual low at high speed, the transmission will not shift directly to low gear. To avoid engine or driveline damage, the transmission first downshifts to second, and then automatically downshifts to low when the speed drops below approximately 42 mph (68 kph) for a THM 250/250C, or 50 mph (80 kph) for a THM 350/350C. The transmission then remains in low gear until the gear selector is moved to another position.

Reverse

When the gear selector is moved to R, the transmission shifts into Reverse. The transmission should be shifted to Reverse only when the vehicle is standing still. The lockout feature of the shift gate should prevent accidental shifting into Reverse while the vehicle is moving forward. However, if the driver intentionally moves the selector to R while moving forward, the transmission *will* shift into Reverse, and damage may result.

GM TRANSMISSION NOMENCLATURE

For the sake of clarity, this text uses consistent terms for transmission parts and functions. However, the various manufacturers may use different terms. Therefore, before we examine the buildup of the THM 250/250C and 350/350C, and the operation of their geartrains and hydraulic systems, you should be aware of the unique names that GM uses for some transmission parts and operations. This will make it

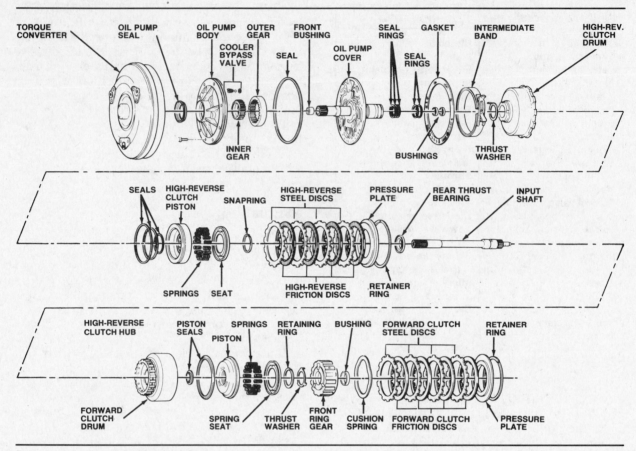

Figure 17-8. The THM 250/250C torque converter, oil pump, intermediate band, high-reverse clutch, and forward clutch.

easier in the future if you should look in a GM manual and see a reference to the "direct clutch." You will realize that this is simply the high-reverse clutch and is similar to the same part in any other transmission that uses a similar geartrain and apply devices. Figure 17-7 lists GM nomenclature and the textbook terms for some transmission parts and operations.

THM 250/250C AND 350/350C BUILDUP

All the transmissions covered in this chapter use a 1-piece case casting that incorporates the bellhousing. The extension housing is a separate casting that bolts to the back of the case. Figures 17-8 and 17-9 show exploded views of a THM 250/250C transmission; figures 17-10 and 17-11 show similar views for the THM 350/350C gearbox. Refer to these figures to identify the components being discussed as we describe the buildup of the THM 250/250C and 350/350C transmissions.

The torque converter fits inside the bellhousing, and is a welded unit that cannot be disassembled. The converter bolts to the engine flexplate, which has the starter ring gear around its outer diameter. The THM 250 and 350 converters are conventional 3-element designs (impeller, turbine, and stator). The THM 250C, and 350C converters also contain a hydraulically applied clutch.

The torque converter is installed in the transmission by sliding it onto the stationary stator support. The stator support is part of the oil pump housing. The splines in the stator 1-way clutch engage the splines on the stator support, and the splines in the turbine engage the transmission input shaft. The hub at the rear of the torque converter housing engages the lugs on the oil pump inner gear to drive the pump. The converter housing and impeller are bolted to the engine flexplate so the oil pump turns and pumps fluid whenever the engine is running.

All of the transmissions covered in this chapter have gear-type oil pumps that bolt to the front of the transmission case directly behind the torque converter. The pump draws fluid from the sump through the filter, and delivers it to the pressure regulator valve, the manual valve, and other valves in the hydraulic system.

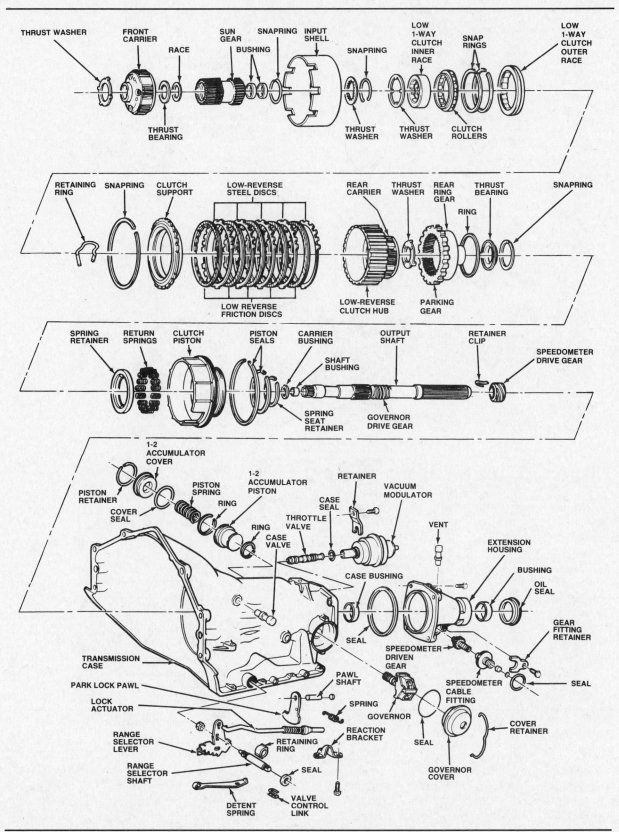

Figure 17-9. The THM 250/250C planetary gearset, low 1-way clutch, low-reverse clutch, case, 1-2 accumulator, throttle valve and vacuum modulator, governor, and extension housing.

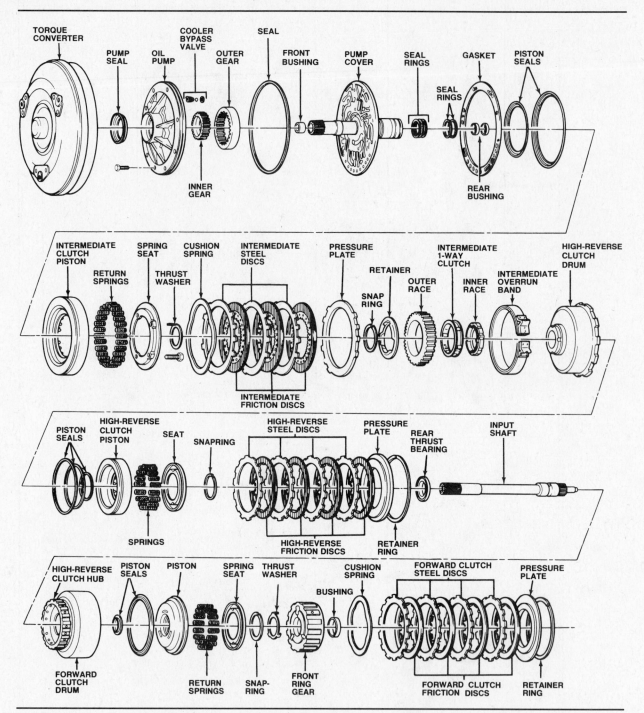

Figure 17-10. The THM 350/350C torque converter, oil pump, intermediate clutch, intermediate 1-way clutch, intermediate overrun band, high-reverse clutch, and forward clutch.

The transmission input shaft is turned clockwise by the torque converter turbine. The shaft passes through the stator support, where it is supported by bushings, and into the transmission case. The output shaft is supported by a bushing in the rear of the transmission case and by a bushing in the rear of the extension housing. The rear of the input shaft fits within the front of the output shaft, but the two shafts do not touch.

The 1-piece forward clutch drum and high-reverse clutch hub assembly is splined to the end of the input shaft and rotates clockwise

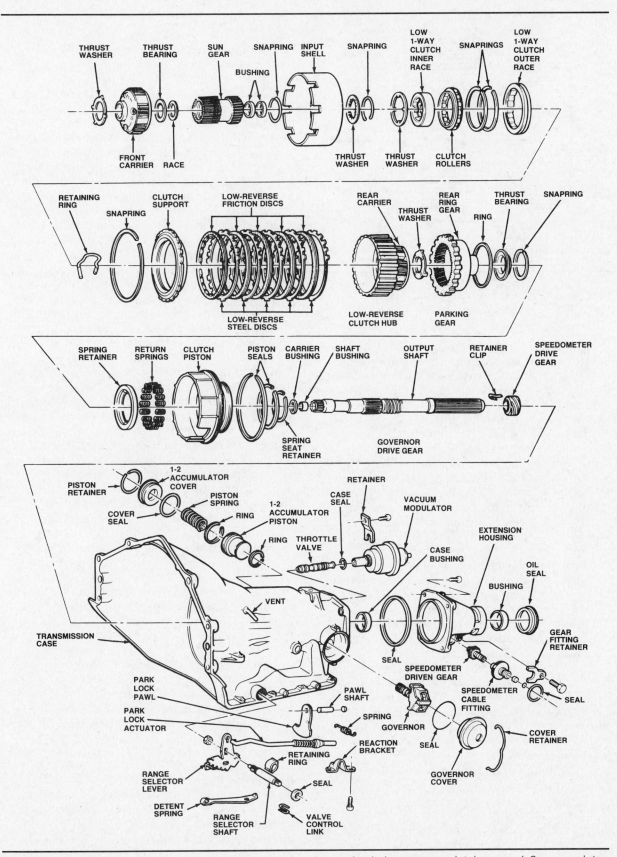

Figure 17-11. The THM 350/350C planetary gearset, low 1-way clutch, low-reverse clutch, case, 1-2 accumulator, throttle valve and vacuum modulator, governor, and extension housing.

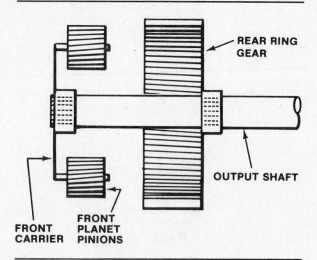

REAR RING GEAR

OUTPUT SHAFT

FRONT CARRIER

FRONT PLANET PINIONS

Figure 17-12. The THM 250/250C and 350/350C front carrier and rear ring gear are splined to the output shaft.

with it whenever the torque converter turbine is being driven by the impeller. The forward clutch drum and high-reverse clutch hub assembly is piloted on the outside of the stator support hub. Seal rings are installed on the hub to seal oil passages to the clutches.

The outside of each forward clutch steel disc is splined to the inside of the forward clutch drum, which is splined to the input shaft. The inside of each friction disc is splined to the outside of the front ring gear in the planetary gearset. Whenever the forward clutch is applied, the front ring gear is turned clockwise by the input shaft. The forward clutch is applied in all forward gears.

The inside of each high-reverse clutch friction disc is splined to the outside of the high-reverse clutch hub and turned by it. The outside of each steel disc is splined to the inside of the high-reverse clutch drum. When the high-reverse clutch is applied, the drum is rotated clockwise by the input shaft.

The high-reverse clutch drum is splined to the input shell, which is splined to the sun gear. The high-reverse clutch drum, input shell, and sun gear rotate as a unit at all times. The sun gear turns on bushings on the output shaft, but it is not splined to the shaft. When the high-reverse clutch is applied, the clutch drum, input shell, and sun gear are turned clockwise by the input shaft.

The major differences between the THM 250/250C and 350/350C involve the apply devices used for second gear. In the THM 250/250C, figure 17-8, an intermediate band is wrapped around the outside of the high-reverse clutch drum. When the intermediate

band is applied, the clutch drum, input shell, and sun gear are all held stationary. This occurs in both Drive second and manual second gears.

In the THM 350/350C, figure 17-10, an intermediate overrun band is wrapped around the outside of the high-reverse clutch drum. In addition, the inner race of an intermediate 1-way (roller) clutch is splined to the high-reverse clutch drum. The outer race of the 1-way clutch forms the hub of the intermediate multiple-disc clutch. The inside of each intermediate clutch friction disc is splined to the outside of the clutch hub. The outside of each intermediate clutch steel disc is splined to the transmission case.

When the THM 350/350C is in Drive second, the intermediate clutch locks the outer race of the intermediate 1-way clutch to the case. The 1-way clutch then allows clockwise rotation of the high-reverse clutch drum, input shell, and the sun gear; but it prevents counterclockwise rotation of these parts. In manual second, the intermediate overrun band is applied to prevent the high-reverse clutch drum, input shell, and sun gear from rotating in *either* direction. We will examine these actions in more detail when we study the second gear power flow.

None of the members of the compound planetary gearset are attached directly to the input shaft. However, two members of the gearset *are* splined to the output shaft, the front carrier and the rear ring gear, figure 17-12. This means that one or the other of these two units is always the final driving member of the Simpson planetary gearset.

The front ring gear meshes with the front pinions. The front pinions turn in the front carrier and mesh with the sun gear. As just mentioned, the front carrier is splined to the output shaft. The sun gear also meshes with the rear pinions, which turn in the rear carrier and mesh with the rear ring gear. The rear ring gear, like the front carrier, is splined to the output shaft. The outside of the rear ring gear forms the parking gear, which is locked to the transmission case by the parking pawl when the gear selector is placed in Park.

The outer race of the low 1-way (roller) clutch, figure 17-13, is located inside the low-reverse clutch support. The support is splined to the transmission case, and also serves as the pressure plate for the low-reverse clutch. The inner race of the low 1-way clutch is splined to the rear carrier. The 1-way clutch holds the rear carrier if it attempts to turn counterclockwise, but the clutch overruns when the rear carrier turns clockwise.

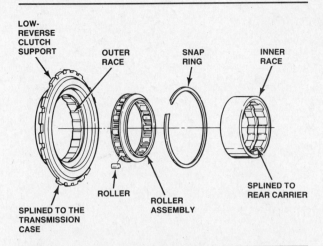

LOW-REVERSE CLUTCH SUPPORT

OUTER RACE

SNAP RING

INNER RACE

SPLINED TO THE TRANSMISSION CASE

ROLLER

ROLLER ASSEMBLY

SPLINED TO REAR CARRIER

Figure 17-13. The THM 250/250C and 350/350C low 1-way (roller) clutch assembly.

The inside of each low-reverse clutch friction disc is splined to the outside of the rear carrier. The outside of each steel disc is splined to the transmission case. The clutch piston operates in a bore at the rear of the case. When the low-reverse clutch is applied, it overrides the low 1-way clutch and holds the rear carrier from turning in *either* direction.

The output shaft passes through the rear of the transmission case, into the extension housing. The governor drive gear is cut into the output shaft behind the splines where the rear ring gear mounts. The speedometer drive gear is clipped to the output shaft between the governor drive gear and the output yoke splines. The governor is mounted in the left rear of the main case and driven by a nylon pinion gear that engages the gear on the output shaft. A bushing and seal for the output yoke are installed at the rear of the extension housing.

THM 250/250C AND 350/350C POWER FLOW

Now that we have taken a quick look at the general way in which the THM 250/250C and 350/350C are assembled, we can trace the flow of power through the apply devices and the Simpson compound planetary gearset in each of the gear ranges. We will use the THM 350/350C to illustrate most gear ranges. The power flow in the THM 250/250C is similar, but the second gear apply devices differ as explained in the appropriate sections below.

Park and Neutral

When the gear selector is in Park or Neutral, no clutches or bands are applied, and the torque converter clutch (where fitted) is released. Because no member of the Simpson compound planetary gearset is held, there can be no output motion. The torque converter impeller/cover and oil pump drive gear turn clockwise at engine speed. The torque converter turbine and transmission input shaft also rotate clockwise at engine speed.

In Neutral, the Simpson compound planetary gearset idles, and the output shaft is free to rotate in either direction so the vehicle can be pushed or pulled. In Park, a mechanical pawl locks the parking gear on the outside of the rear ring gear to the case. The rear ring gear is splined to the output shaft, so the vehicle is prevented from moving in either direction.

Drive Low

When the gear selector is placed in Drive (D or 3) the car starts to move with the transmission automatically in low gear, figure 17-14. The forward clutch is applied to transmit engine power from the input shaft to the front ring gear.

Before the car begins to move, the transmission output shaft is held stationary by the driveline and wheels. This also holds the front carrier and rear ring gear stationary. As a result, the clockwise rotation of the front ring gear turns the front pinions clockwise on the stationary carrier. The front pinion rotation then causes the sun gear to turn counterclockwise.

Counterclockwise rotation of the sun gear turns the rear pinions clockwise. The rear pinions then try to walk clockwise around the inside of the rear ring gear, which is splined to the stationary output shaft. This action causes the rear carrier to try to rotate counterclockwise, but the low 1-way clutch locks up to prevent such rotation. Therefore, the torque from the rear pinions is transferred to the rear ring gear and output shaft instead of the rear carrier. Because the 1-way clutch holds the carrier more securely than the driveline can hold the ring gear and output shaft, the rear ring gear and output shaft begin to rotate clockwise in gear reduction.

As long as the car is accelerating in Drive low, torque is transferred from the input shaft, through the Simpson planetary gearset, to the output shaft. If the driver lets up on the throttle, the engine speed drops while the drive wheels try to keep turning at the same speed. This removes the clockwise-rotating torque load from the low 1-way clutch, which unlocks and allows the rear planetary gearset to freewheel so that no engine compression braking is available. This is a major difference between Drive low, and manual low, as we will see later.

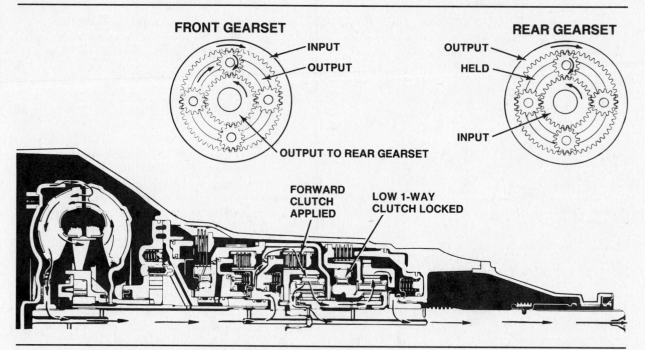

Figure 17-14. THM 250/250C and 350/350C power flow in Drive low.

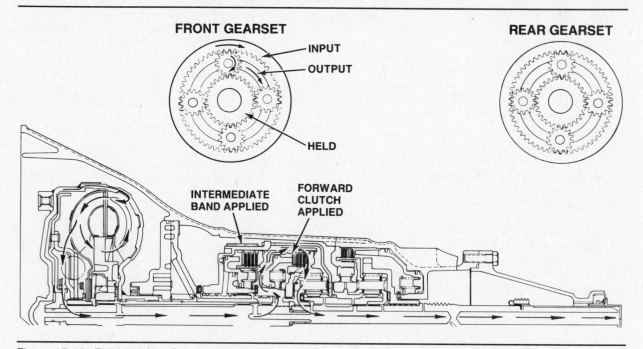

Figure 17-15. THM 250/250C power flow in Drive second.

In Drive low, gear reduction is produced by a combination of the front and rear planetary gearsets. The front ring gear is the input member; the rear carrier is the held member; and the rear ring gear is the output member. The forward clutch is the input device, and the low 1-way clutch is the holding device.

Drive Second — THM 250/250C

As the car continues to accelerate in Drive, the transmission hydraulic system produces an automatic upshift to second gear by applying the intermediate band, figure 17-15. The forward clutch remains applied, and power from the engine is still delivered to the front ring gear. The

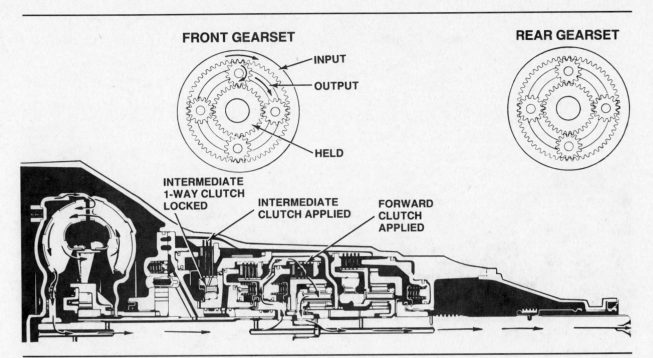

Figure 17-16. THM 350/350C power flow in Drive second.

front ring gear rotates clockwise and turns the front pinions clockwise.

At the same time, the intermediate band locks the high-reverse clutch drum and sun gear input shell to the transmission case. This holds the sun gear stationary. As the front ring gear turns the front pinions clockwise, they walk around the stationary sun gear. This causes the front carrier to rotate clockwise. The front carrier is splined to the output shaft and drives it in gear reduction.

As long as the car is accelerating in Drive second, torque is transferred from the engine, through the Simpson planetary gearset, to the output shaft. As the vehicle speed rises above the minimum shift point and the driver eases up on the throttle, the transmission upshifts to third. Because of this upshift, there is usually no engine compression braking in Drive second. This is a major difference from manual second, as we will see later.

If the vehicle speed is below the minimum 2-3 upshift point when the throttle is released, there *will* be engine compression braking because the intermediate band holds the sun gear from turning in either direction. This is one difference between the THM 250/250C and the 350/350C as explained in the next section.

All of the gear reduction for second gear is produced by the front planetary gearset. The rear gearset simply idles. The low 1-way clutch that held the rear carrier in Drive low overruns in second because the pinions in the rear carrier

are meshed with the rear portion of the stationary sun gear. The rear ring gear is splined to the output shaft and as the ring gear turns, the rear pinions are forced to walk around the stationary sun gear, causing the carrier to turn clockwise and the 1-way clutch to freewheel. The rear ring gear continues to rotate with the output shaft, but freewheels because no holding device is applied to the rear gearset.

In Drive second gear, the front ring gear is the input member; the sun gear is the held member; and the front carrier is the output member. The forward clutch is the input device, and the intermediate band is the holding device.

Drive Second — THM 350/350C

As the car continues to accelerate in Drive, the transmission hydraulic system produces an automatic upshift to second gear by applying the intermediate clutch, figure 17-16. The forward clutch remains applied, and power from the engine is still delivered to the front ring gear. The front ring gear rotates clockwise and turns the front pinions clockwise.

At the same time, the intermediate clutch locks the outer race of the intermediate 1-way clutch to the transmission case. Therefore, when the front pinions attempt to drive the sun gear counterclockwise, the intermediate 1-way clutch locks up and holds the sun gear, input shell, and high-reverse clutch drum stationary.

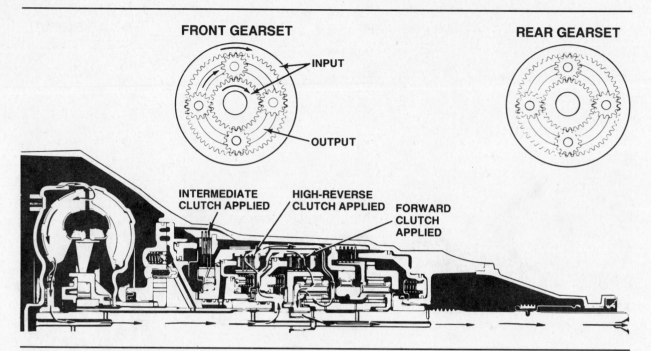

Figure 17-17. THM 250/250C and 350/350C power flow in Drive high.

With the sun gear held stationary, the flow of power through the planetary gearset is exactly the same as in the THM 250/250C. As the front ring gear turns the front pinions clockwise, they walk around the stationary sun gear. This causes the front carrier to rotate clockwise and drive the output shaft in gear reduction.

As long as the car is accelerating in Drive second, torque is transferred from the engine, through the Simpson planetary gearset, to the output shaft. As the vehicle speed rises above the minimum shift point and the driver eases up on the throttle, the transmission upshifts to third. Because of this upshift, there is usually no engine compression braking in Drive second. This is a major difference from manual second, as we will see later.

If the vehicle speed is below the minimum 2-3 upshift point when the throttle is released, there still *will not* be any engine compression braking because the intermediate 1-way clutch overruns and allows the sun gear to freewheel. This is one difference between the THM 350/350C and the 250/250C described in the previous section.

All of the gear reduction for second gear is produced by the front planetary gearset. The rear gearset simply idles. The low 1-way clutch that held the rear carrier in Drive low overruns in second because the pinions in the rear carrier are meshed with the rear portion of the stationary sun gear. The rear ring gear is splined to the

output shaft and as the ring gear turns, the rear pinions are forced to walk around the stationary sun gear, causing the carrier to turn clockwise and the 1-way clutch to freewheel. The rear ring gear continues to rotate with the output shaft, but freewheels because no holding device is applied to the rear gearset.

In Drive second gear, the front ring gear is the input member; the sun gear is the held member; and the front carrier is the output member. The forward clutch is the input device, and the intermediate clutch and intermediate 1-way clutch are the holding devices.

Drive Third

As the car continues to accelerate in Drive, the transmission automatically upshifts from second to third gear. For this to happen, the forward clutch and intermediate clutch (THM 350/350C) remain applied, the intermediate band (THM 250/250C) is released, and the high-reverse clutch is applied, figure 17-17.

When the high-reverse clutch is applied, the high-reverse clutch drum, input shell, and sun gear are all locked to the transmission input shaft and rotate clockwise at turbine speed. The forward clutch continues to lock the front ring gear to the input shaft. This means that two members of the planetary gearset, the sun gear and the ring gear, are locked to the input shaft, and thus to each other. With two members of the gearset locked together, the entire gearset turns as a unit.

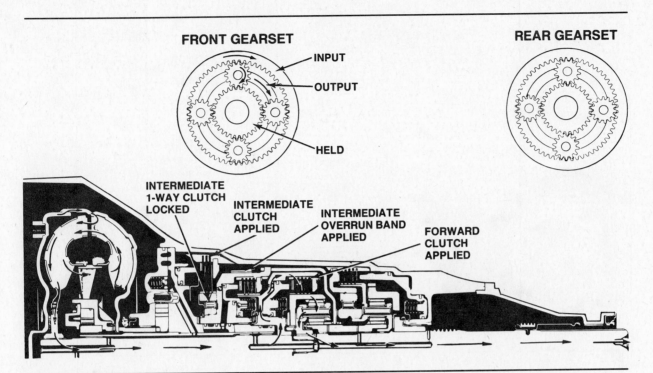

FRONT GEARSET

— INPUT

— OUTPUT

— HELD

REAR GEARSET

INTERMEDIATE
1-WAY CLUTCH
LOCKED

INTERMEDIATE
CLUTCH
APPLIED

INTERMEDIATE
OVERRUN BAND
APPLIED

FORWARD
CLUTCH
APPLIED

Figure 17-18. THM 350/350C power flow in manual second.

With the front planetary gearset locked, there is no relative motion between the sun gear, ring gear, carrier, or the pinions. As a result, input torque to the sun and ring gears is transmitted through the planet carrier, which is splined to the output shaft, to drive the output shaft. In third gear, there is a direct 1:1 gear ratio between input shaft speed and output shaft speed. There is no reduction or increase in either speed or torque through the transmission.

It is not necessary to release the THM 350/350C intermediate clutch when the high-reverse is applied because the intermediate 1-way clutch overruns and allows the sun gear to rotate clockwise as already described. This method of engaging third gear without releasing the intermediate clutch allows the THM 350/350C to provide a very smooth 2-3 upshift.

All of the transmissions covered in this chapter provide engine compression braking in Drive third because there is a direct mechanical connection between the input shaft and the output shaft; no 1-way clutches are involved.

In third gear, the sun gear and front ring gear of the Simpson planetary gearset are input equally. The front carrier is the output member. The forward clutch and the high-reverse clutch are input devices equally. The THM 350/350C intermediate clutch remains applied, but is ineffective in third gear.

Manual Second — THM 250/250C

When the THM 250/250C gear selector is placed in manual second (I, S, 2, or L2) and the vehicle is accelerated from a stop, the transmission starts in Drive low and upshifts to Drive second, figure 17-15. The hydraulic system prevents the transmission from upshifting into third gear. Other than this, there are no differences between manual second and Drive second. The flow of power through the planetary gearsets is identical, and the same apply devices are used.

At any time, the driver can manually downshift from third to second by moving the gear selector from D or 3 to I, S, 2, or L2. In the THM 250/250C, the high-reverse clutch is released, and the intermediate band is applied. This provides the driver with a controlled downshift for engine braking or improved acceleration at any speed or throttle position.

If the car is allowed to decelerate with the gear selector in manual second, the transmission automatically downshifts to low gear as speed decreases.

Manual Second — THM 350/350C

When the THM 350/350C gear selector is placed in manual second (I, S, 2, or L2) and the vehicle is accelerated from a stop, the transmission starts in Drive low and upshifts to manual second, figure 17-18. The hydraulic system pre-

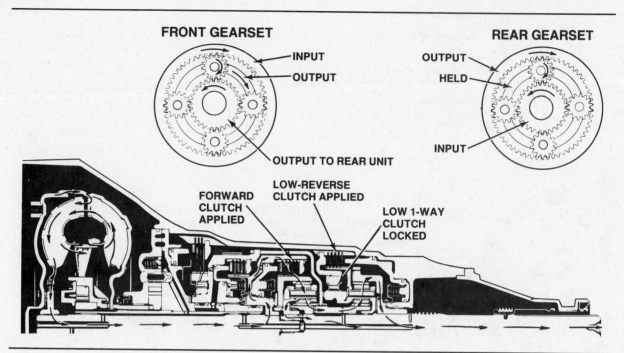

Figure 17-19. THM 250/250C and 350/350C power flow in manual low.

vents an upshift into third gear, and applies the intermediate overrun band in addition to the intermediate clutch. Other than as described below, there are no differences between manual second and Drive second. The flow of power through the planetary gearsets is identical, and the same apply devices are used.

The THM 350/350C intermediate overrun band locks the high-reverse clutch drum, input shell, and sun gear to the case, and prevents them from turning in either direction. The intermediate clutch and intermediate 1-way clutch do this job in Drive second, but they hold the sun gear only during acceleration; the 1-way clutch overruns during deceleration and allows the front gearset to freewheel. In manual second, the intermediate overrun band holds the sun gear during deceleration to provide engine compression braking.

The THM 350/350C intermediate clutch remains applied in manual second, so the intermediate 1-way clutch still holds the sun gear during acceleration. The intermediate overrun band only has to hold the sun gear during deceleration; it does not have to absorb the full engine torque of acceleration. As a result, the intermediate overrun band can be smaller than the intermediate bands in other transmissions.

At any time, the driver can manually downshift from third to second by moving the gear selector from D or 3 to I, S, 2, or L2. In the THM 350/350C, the high-reverse clutch is released, and the intermediate overrun band is applied. This provides the driver with a controlled downshift for engine braking or improved acceleration at any speed or throttle position.

If the car is allowed to decelerate with the gear selector in manual second, the transmission automatically downshifts to low gear as speed decreases.

Manual Low

When the gear selector is placed in manual low (L, 1, or L1), the hydraulic system prevents an upshift to second or third gear, and the flow of power through the planetary gearset is basically the same as in Drive low. The only difference between Drive low and manual low is the use of one apply device.

In manual low, figure 17-19, the low-reverse clutch is applied to hold the rear carrier. In Drive low, the 1-way clutch does this job, but it holds the rear carrier only during acceleration; the 1-way clutch overruns during deceleration and allows the rear gearset to freewheel. In manual low, the low-reverse clutch holds the rear carrier during acceleration *and* deceleration, which provides engine braking on deceleration. The 1-way clutch still helps hold the rear carrier during acceleration in manual low, but its action is secondary as long as the low-reverse clutch is applied.

The driver can move the gear selector from D or 3 to L, 1, or L1 while the car is moving. At

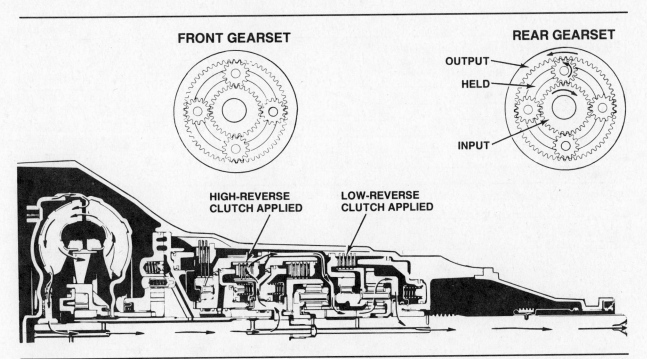

Figure 17-20. THM 250/250C and 350/350C power flow in Reverse.

GEAR RANGES	CLUTCHES				BANDS
	1-Way Roller (Overrunning)	Forward (Rear)	High-Reverse (Front, or Direct)	Low-Reverse (Rear)	Intermediate (Kickdown, or Front)
Drive Low	●	●			
Drive Second		●			●
Drive High		●	●		
Manual Low	*	●		●	
Manual Second		●			●
Reverse			●	●	

* The 1-way roller clutch will hold in manual low if the low-reverse clutch fails, but will not provide engine braking.

Figure 17-21. THM 250/250C clutch and band application chart.

lower speeds, the transmission will downshift directly to low gear by releasing the appropriate apply devices and applying the low-reverse clutch. At higher speeds, the transmission first shifts into second by releasing the appropriate apply devices and applying the intermediate band or clutch. Then, once vehicle speed drops below a certain level, the transmission automatically downshifts to low by releasing the intermediate band or clutch and applying the low-reverse clutch.

Reverse

When the gear selector is moved to Reverse, the input shaft is still turned clockwise by the engine. The transmission must reverse this rotation to a counterclockwise direction.

In Reverse, the forward clutch is released and the high-reverse clutch is applied, figure 17-20. This connects the high-reverse clutch drum and the sun gear input shell to the input shaft so that the input shaft drives the sun gear clockwise. The sun gear drives the rear pinions counterclockwise.

The other apply device used in Reverse is the low-reverse clutch which holds the rear carrier. Because the rear carrier is held stationary, the counterclockwise rotation of the rear pinions drives the rear ring gear counterclockwise. The rear ring gear is splined to the output shaft and turns it counterclockwise in gear reduction.

GEAR RANGES	CLUTCHES						BAND
	Low 1-Way (Roller)	Intermediate 1-Way (Roller)	Forward	High-Reverse (Direct)	Intermediate	Low-Reverse	Intermediate (Overrun)
Drive Low	●		●				
Drive Second		●	●		●		
Drive High			●	●	●		
Manual Low	*		●			●	
Manual Second		●	●		●		●
Reverse				●		●	

*The low 1-way clutch will hold in manual low if the low-reverse clutch fails, but it will not provide engine braking.

Figure 17-22. THM 350/350C clutch and band application chart.

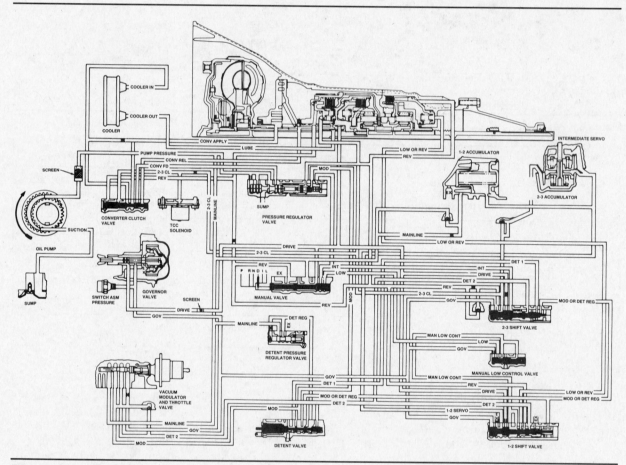

Figure 17-23. The THM 250C hydraulic system.

All gear reduction in Reverse is produced by the rear planetary gearset. Although there is input to the front gearset through the sun gear and the front pinions, no member of the gearset is held so it simply freewheels.

In Reverse, the input member of the Simpson planetary gearset is the sun gear; the held member is the rear carrier; and the output member is the rear ring gear. Two apply devices are used: the high-reverse clutch is the input device, and the low-reverse clutch is the holding device.

THM 250/250C and 350/350C Clutch and Band Applications

Before we move on to a closer look at the hydraulic system, we should review the apply de-

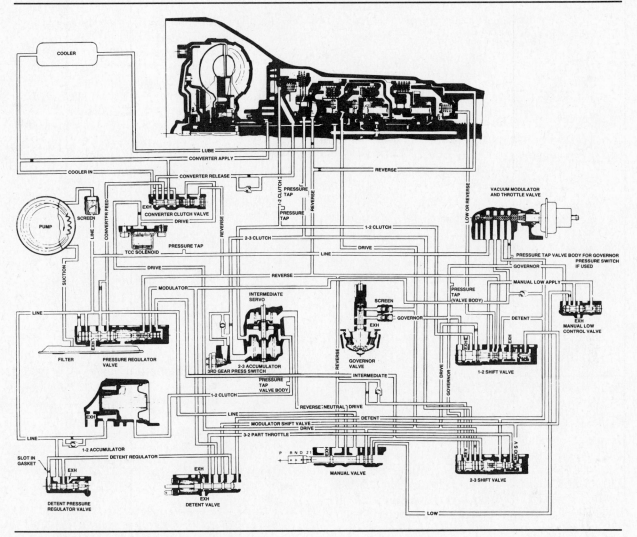

Figure 17-24. The THM 350C hydraulic system.

vices that are used in each of the gear ranges. Figure 17-21 is a clutch and band application chart for the THM 250/250C transmission, and figure 17-22 is a similar chart for the THM 350/350C transmission. Also, remember these facts about the apply devices used in the transmissions covered in this chapter:

- The forward and high-reverse clutches are input devices.
- The forward clutch is applied in all forward gear ranges.
- The high-reverse clutch is applied in third and Reverse gears.
- The low-reverse, low 1-way, intermediate, and intermediate 1-way clutches, along with the intermediate band and intermediate overrun band, are holding devices.
- The low-reverse clutch is applied in manual low and Reverse.

- The low 1-way clutch locks during acceleration in Drive low; it overruns during deceleration.
- The THM 350/350C intermediate clutch is applied in second and third gears, but it is ineffective in third because the intermediate 1-way clutch overruns.
- The THM 350/350C intermediate 1-way clutch locks during acceleration in Drive second; it overruns during deceleration.
- The THM 250/250C intermediate band is applied in second gear.
- The THM 350/350C intermediate overrun band is applied in manual second.

THM 250/250C AND 350/350C HYDRAULIC SYSTEMS

The hydraulic systems of the transmissions covered in this chapter control shifting under

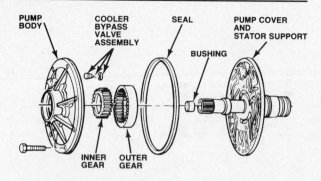

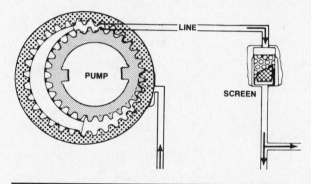

Figure 17-25. The THM 250/250C and 350/350C oil pump assembly.

varying vehicle loads and speeds, as well as in response to manual gear selection by the driver. Figure 17-23 is a diagram of the complete THM 250C hydraulic system; the THM 250 system is similar. Figure 17-24 is a diagram of the complete THM 350C hydraulic system; the THM 350 system is similar. You will find it helpful to refer back to these diagrams as we discuss the various components that make up the hydraulic system. The systems of both transmissions are basically the same, with slight variations for the different apply devices. We will begin our study with the pump.

Hydraulic Pump

All of the transmissions covered in this chapter use a gear-type oil pump, figure 17-25, that is driven by the torque converter hub. Fluid is stored in the oil pan (sump) at the bottom of the transmission, and the pump draws fluid through a filter attached to the bottom of the valve body. Fluid flows through a passage in the valve body and then through a passage in the front of the transmission case and bellhousing to the pump.

In early THM 350 transmissions, the outlet passage from the pump contains a priming valve that is located in the pump body. This cylindrical valve provides a vent for any air that

may be trapped in the pump. A spring holds the priming valve open until trapped air is vented. When outlet pressure reaches approximately 5 psi (34 kPa), the valve is forced closed.

A screen in the pump outlet passage in the transmission case prevents larger pieces of debris from entering the hydraulic system. Fluid from the pump is routed through this screen to the:

- Pressure regulator valve
- Detent pressure regulator valve
- Manual valve
- Throttle valve
- 1-2 accumulator
- Intermediate overrun servo (THM 350/350C).

Hydraulic Valves

The THM 250/250C and 350/350C contain the following valves:

1. Pressure regulator valve assembly
2. Manual valve
3. Throttle valve
4. Governor valve
5. Detent pressure regulator valve
6. Detent valve
7. 1-2 shift valve assembly
8. 1-2 accumulator
9. 2-3 shift valve assembly
10. 2-3 accumulator
11. Manual low control valve
12. Converter clutch valve (THM 250C and 350C).

Valve body and ball-check valves

Except for the governor valve, throttle valve, and converter clutch control valve, all of the hydraulic valves in the THM 250/250C and 350/350C are housed in the valve body, figures 17-26 and 17-27. The governor valve is located in the left rear of the transmission case where it is driven by the output shaft. The throttle valve is located at the right rear of the case where it is acted on by the vacuum modulator. The converter clutch valve in the THM 250C and 350C is located in an auxiliary valve body, figure 17-28, that bolts to the transmission case at the front of the separator plate.

The valve body bolts to the bottom of the transmission case. A separator plate and two gaskets are installed between the valve body and the case. The valve body has passages that route fluid between the various valves. The valve body also forms the bottom of the intermediate servo bore in the THM 250/250C, and the 2-3 accumulator bore in the THM 350/350C. Additional fluid passages, cast into the bottom of the transmission case, form the upper half of the valve body.

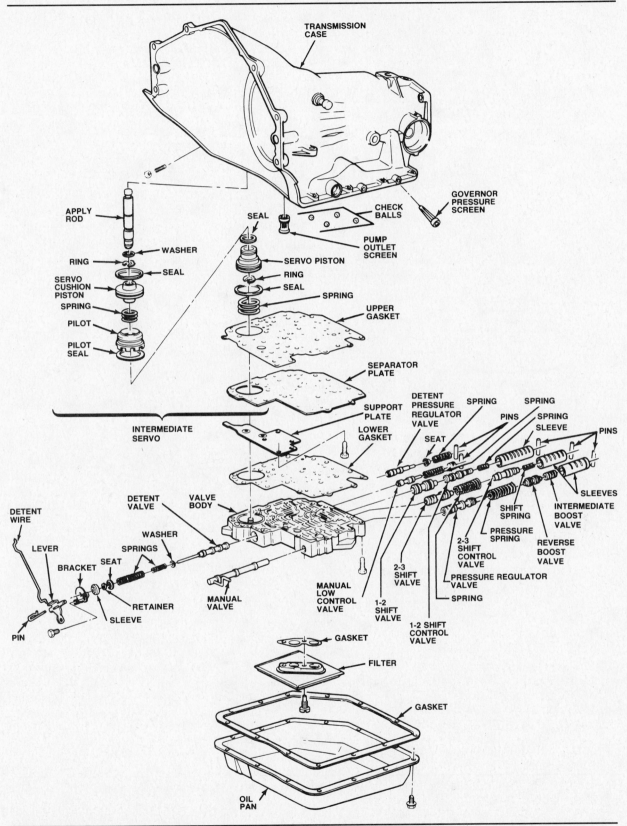

Figure 17-26. The THM 250 intermediate servo, 2-3 accumulator, valve body, and hydraulic valves; the THM 250C is similar.

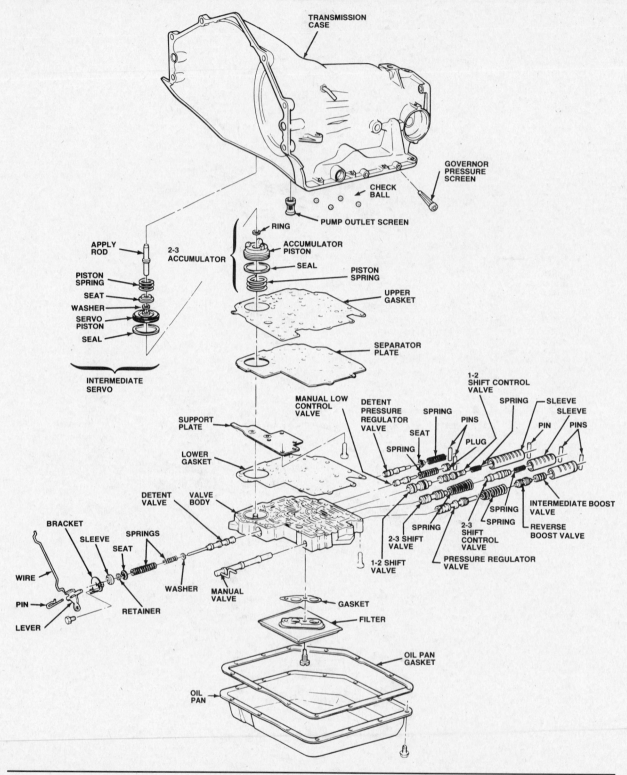

Figure 17-27. The THM 350 intermediate servo, 2-3 accumulator, valve body, and hydraulic valves; the THM 350C is similar.

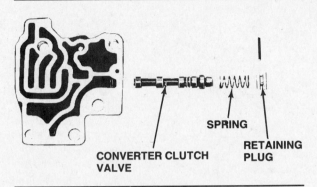

Figure 17-28. The THM 250C and 350C auxiliary valve body.

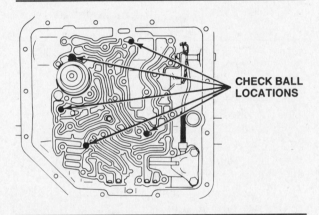

Figure 17-29. THM 350C check ball locations; the THM 250/250C and 350 check balls are in similar locations.

The fluid passages in the case contain the steel or teflon check balls of the hydraulic system. The THM 250/250C and 350 have four check balls. The THM 350C, figure 17-29, has five. In addition to the pump outlet screen already mentioned, all of the transmissions covered in this chapter also have a governor inlet screen that is located in the transmission case.

Pressure regulator valve assembly

The pressure regulator valve assembly, figure 17-30, consists of the pressure regulator valve spool, a spring, the reverse boost valve, and the intermediate boost valve. The pressure regulator valve receives mainline pressure from the oil pump through three passages. One passage is at the end of the valve opposite the boost valves, one is blocked by a land on the valve, and the other leads to a valley between two lands on the valve. Hydraulic pressure at the end of the valve acts to move the valve against spring force. Hydraulic pressure in the valley of the valve acts on two lands of equal diameter, and does not move the valve in either direction.

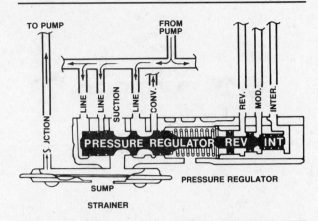

Figure 17-30. The THM 250/250C and 350/350C pressure regulator valve, and reverse and intermediate boost valves.

When the engine is started and mainline pressure begins to build, pressure at the end of the regulator valve moves the valve against spring force to open the converter inlet passage. In the THM 250 and 350, oil is routed to the torque converter. Return oil from the converter flows to the transmission cooler, and from there to the lubrication circuit. A cooler bypass valve located in the cooler feed circuit permits oil to flow directly from the converter to the lubrication circuit if the cooler is restricted.

In the THM 250C and 350C, converter feed oil from the pressure regulator valve flows to the converter clutch valve. This valve routes oil to the converter, transmission cooler, and lubrication circuit through different passages depending on whether the converter clutch is applied or released. This will be explained later in the chapter.

After the converter is filled, pressure build up at the regulator valve moves it farther against spring force. This opens a port to the transmission sump or pump suction passage to vent excess pressure. Hydraulic pressure and spring force then reach a balance that provides a constant mainline pressure.

The reverse and intermediate boost valves at the spring end of the pressure regulator valve are used to increase mainline pressure under certain driving conditions. To do this, hydraulic pressure is routed to the boost valves, which then move to increase the spring force acting on the pressure regulator valve.

Modulated throttle pressure acts on the small land of the reverse boost valve to increase mainline pressure in response to higher torque loads on the drivetrain. Mainline pressure (REVERSE oil) acts on the large land of the reverse boost valve in Reverse to increase mainline

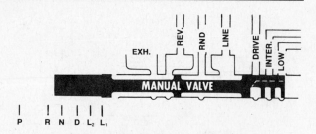

Figure 17-31. The THM 350/350C manual valve in Neutral; the THM 250/250C valve is similar, but does not have the RND passage that routes mainline pressure to the release side of the intermediate overrun servo.

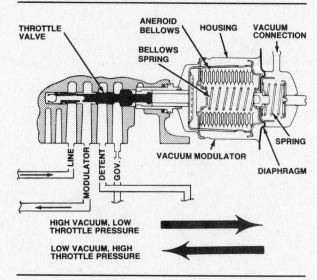

Figure 17-32. The THM 250/250C and 350/350C throttle valve is operated by a vacuum modulator connected to engine intake manifold vacuum.

pressure and hold the apply devices more securely under the higher torque loads that are normal in that gear position. Mainline pressure (INTERMEDIATE oil) acts on the intermediate boost valve in manual second and manual low to again increase mainline pressure and hold apply devices more securely.

Manual valve

The manual valve is a directional valve moved by a mechanical linkage from the gear selector lever. The manual valve receives fluid from the pump, and directs mainline pressure to apply devices and other valves in the hydraulic system in order to provide both automatic and manual upshifts and downshifts. The manual valve is held in each gear position by a spring-loaded lever.

In the THM 350/350C, figure 17-31, five passages may be charged with mainline pressure from the manual valve. In the THM 250/250C, the RND passage is not required, and only four passages are charged with mainline pressure from the manual valve. GM refers to mainline pressure from the manual valve by the name of the passage it flows through. When the gear selector is in Park, mainline pressure is blocked from entering the manual valve, and the valve does not direct pressure to any passages.

The RND passage is charged in Reverse, Neutral and Drive; it routes pressure to the:

● Release side of the intermediate overrun servo.

The DRIVE passage is charged in Drive, manual second, and manual low; it routes pressure to the:

● 1-2 shift valve
● 2-3 shift valve
● Governor valve
● Apply area of the forward clutch
● TCC solenoid (Buick-built THM 250C and all THM 350C).

The INTERMEDIATE passage is charged in manual second and manual low; it routes pressure to the:

● 2-3 shift valve
● Intermediate boost valve.

The LOW passage is charged in manual low; it routes pressure to the:

● Manual low control valve
● Spring end of the 1-2 shift valve.

The REVERSE passage is charged in Reverse; it routes pressure to the:

● Reverse boost valve
● Outer apply area of the low-reverse clutch
● Inner apply area of the low-reverse clutch, through the 1-2 shift valve
● Outer apply area of the high-reverse clutch
● Inner apply area of the high-reverse clutch, through the 2-3 shift valve
● Converter clutch valve (THM 250C and 350C).

Throttle valve

The THM 250/250C and 350/350C transmissions develop throttle pressure using a throttle valve operated by a vacuum modulator, figure 17-32. For this reason, GM sometimes refers to this type of throttle valve as a modulator valve, and uses the term modulator pressure in place of throttle pressure. Throttle pressure indicates engine torque and helps time shifts. Without throttle pressure, the transmission would upshift and downshift at exactly the same speed for very different throttle openings and load conditions. Throttle pressure also boosts mainline pressure to insure that clutches hold securely under load.

The vacuum modulator contains two chambers separated by a flexible rubber diaphragm. The chamber nearest the throttle valve spool is open to atmospheric pressure. The chamber farthest from the valve spool is sealed and receives engine intake manifold vacuum. The vacuum chamber also contains a spring that acts on the back side of the diaphragm to move a plunger and force the throttle valve spool into its bore.

When the engine is started, manifold vacuum is applied to the vacuum chamber of the modulator. If manifold vacuum is high enough, such as when the engine is idling or under light load, atmospheric pressure against the modulator diaphragm will oppose spring pressure with enough force to move the valve spool all the way to the right, closing the mainline pressure inlet port and reducing the throttle pressure to zero.

Under most vehicle operating conditions, however, manifold vacuum will be insufficient to allow atmospheric pressure to cut off throttle pressure altogether. Instead, the valve spool will only partially obstruct the mainline pressure inlet port, thus regulating throttle pressure at the outlet port. A portion of the throttle pressure is sent through an internal passage to the end of the throttle valve opposite the vacuum modulator. This pressure acts against spring force to help balance the position of the valve. As a result of these opposing forces, throttle pressure increases with reduced intake manifold vacuum, and stabilizes whenever the vacuum level remains constant.

When the engine comes under heavy load, manifold vacuum drops almost to zero. At these times, pressure is nearly equal in both chambers of the vacuum modulator, and spring force moves the valve spool all the way into its bore. Fluid then passes straight through the throttle valve, allowing throttle pressure and mainline pressure to raise to maximum.

Governor pressure is routed to the throttle valve to help regulate throttle pressure in relation to vehicle speed. This is done because less throttle pressure is needed as speed increases. Governor pressure enters between two lands of the throttle valve and acts on the larger land to move the valve against spring force. This decreases throttle pressure as vehicle speed and governor pressure increase.

Detent pressure is also routed to the throttle valve when the detent valve is open during a full-throttle downshift. If governor pressure moves the throttle valve far enough at higher speeds, detent oil enters the throttle pressure circuit to maintain a minimum of 80 psi (552 kPa) of throttle pressure. This is discussed in greater detail in the section dealing with the detent valve.

Some earlier transmissions (up to 1974) use a vacuum modulator that contains an evacuated metal chamber called an aneroid bellows and a spring, figure 17-32. The aneroid bellows expands at low atmospheric pressure (high altitude) and contracts at high atmospheric pressure (low altitude), so this modulator is often referred to as altitude compensating.

At sea level the bellows and spring are compressed and manifold vacuum is at its highest possible reading, so throttle pressure is unaffected. However, at higher altitudes, manifold vacuum is much lower, giving increased throttle pressure with a regular modulator. With the bellows type, as barometric pressure decreases at higher altitudes, decreasing the pressure on the outside of the bellows causes it to expand. The spring inside the bellows pushes on the atmospheric side of the vacuum diaphragm, thereby "compensating" for the lack of vacuum by reducing throttle pressure.

With a non-compensated modulator, the spring on the vacuum side of the diaphragm pushes on the diaphragm, moving the throttle valve in. This action increases throttle pressure at high altitude because of the lack of vacuum at these heights. Increased throttle pressure causes upshifts to be later, and more aggressive (rougher) because of boosted mainline pressure.

The altitude compensated modulator overcomes this problem because as the bellows expands, it pushes the diaphragm toward the vacuum side as if vacuum was normal. Therefore, upshift speeds and mainline pressure will be the same whether driving in Denver, Colorado (altitude 5,280 feet), or near sea level in Los Angeles, California.

Throttle pressure from the throttle valve is routed to the:

- Reverse boost valve
- Detent valve.

Governor valve

We have been discussing the development and control of throttle pressure, which is one of the two auxiliary pressures that work on the shift valves. Before we get to the shift valve operation, we must explain the governor valve (or governor), and how it creates governor pressure. Governor pressure provides a road speed signal to the hydraulic system that causes automatic upshifts to occur as road speed increases, and permits automatic downshifts with decreased road speed.

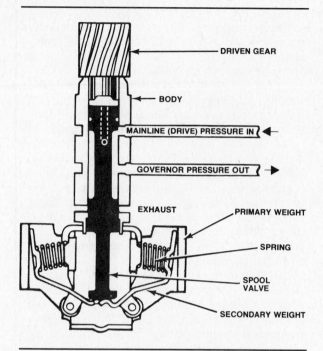

Figure 17-33. The THM 250/250C and 350/350C governor.

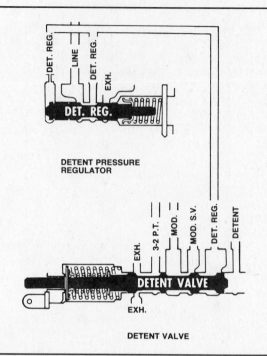

Figure 17-34. The THM 250/250C and 350/350C detent and detent pressure regulator valves.

All of the transmissions covered in this chapter use a gear-driven governor, figure 17-33, that is mounted at the rear of the transmission case. A gear cut into the output shaft turns the driven gear on the end of the governor, which causes the assembly to rotate in the case. The remainder of the governor consists of two sets of weights, two springs, and a spool valve.

When the vehicle is stopped, the mainline pressure inlet is closed or slightly open, and both the governor pressure outlet and the exhaust port are fully open. No governor pressure is developed at this time because any pressure that leaks past the valve spool or enters through the mainline inlet escapes through the exhaust port.

As the vehicle begins to move, the transmission output shaft turns the governor. Centrifugal force throws the weights outward, which levers the valve spool farther into its bore. This begins to close the exhaust port, and opens the mainline pressure inlet port farther, causing governor pressure to build.

Two sets of weights are used to increase the accuracy of pressure regulation at low vehicle speeds. The heavier primary weights move first, and act through springs against the lighter secondary weights that actually move the valve spool. In some transmissions, the spring tensions for the two primary weights differ to provide even smoother regulation of governor pressure. Once governor speed becomes great

enough, the primary weights bottom against their stops, and the secondary weights alone act to move the valve spool. The primary weights control governor pressure up to about 720 governor rpm; above this speed, all regulation is provided by the secondary weights.

During operation, governor pressure is routed through a passage in the center of the spool valve to the end of the valve near the driven gear. This pressure acts against the end of the spool, and helps oppose the lever force applied at the opposite end of the valve. As a result of these opposing forces, governor pressure increases with vehicle speed, and stabilizes whenever vehicle speed is constant.

When the vehicle reaches a certain speed, both sets of weights move out as far as possible. At this time, the secondary weights are bottomed on the primary weights, the exhaust port is fully closed, and both the mainline pressure inlet and the governor pressure outlet are fully open. Therefore, governor pressure equals mainline pressure.

Governor pressure is directed to the:
- 1-2 shift valve
- 2-3 shift valve
- Throttle valve
- Manual low control valve.

Detent pressure regulator valve
The detent pressure regulator valve, figure 17-34, receives mainline pressure directly from

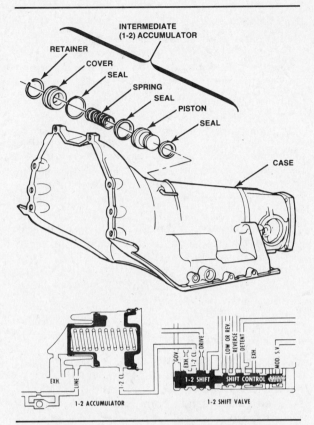

Figure 17-35. The THM 250/250C and 350/350C 1-2 shift valve assembly and 1-2 accumulator.

the pump. This pressure opposes spring force to produce a constant regulated detent pressure. A portion of the pressure is routed through an orifice to one end of the detent pressure regulator valve to help oppose spring force and regulate the pressure. Regulated detent pressure is routed to the detent valve.

Detent valve

The detent valve, figure 17-34, is controlled by a mechanical linkage connected to the accelerator. The linkage moves the valve against a spring to force both part- and full-throttle (detent) downshifts. Both regulated detent pressure and throttle pressure are routed to the detent valve.

At smaller throttle openings, throttle (modulator) pressure flows through the detent valve to the 1-2 and 2-3 shift control valves where it acts on a single land of each valve to help time shifts. Regulated detent pressure is blocked from entering the detent valve at this time, and any residual pressure in the detent circuit is exhausted.

When the accelerator is depressed far enough, the detent valve opens to a position called "detent touch," which allows throttle

pressure to enter the 3-2 part-throttle (or detent 1) passage. Pressure in this passage acts on a second land of the 2-3 shift control valve to cause a part-throttle 3-2 downshift.

At full-throttle, the detent valve reaches the "through-detent" position which allows regulated detent pressure to enter the detent (or detent 2) passage at the end of the valve. This passage routes pressure to additional lands on the 1-2 and 2-3 shift control valves to cause a 3-1 or 2-1 downshift, or delay an upshift, depending on the vehicle speed and the gear engaged at the time the valve is opened.

Remember that throttle pressure is reduced by governor pressure at high speeds. This means that the transmission could reach a point where throttle pressure would drop below 80 psi (552 kPa), even at full throttle. This is prevented by detent pressure unseating a check ball in a passage between the throttle pressure and detent pressure lines. Detent oil then enters the throttle pressure circuit to maintain a constant 80 psi (552 kPa) of throttle pressure. Detent oil can also be routed through the throttle valve and into the throttle pressure circuit as described earlier.

1-2 shift valve assembly

The 1-2 shift valve controls upshifts from low to second, and downshifts from second to low. The assembly includes the 1-2 shift valve, the 1-2 shift control valve, and a spring, figure 17-35. The valve is controlled by opposition between governor pressure on one end, and combined throttle (modulator) pressure and spring force on the other end. Spring force and throttle pressure acting on the 1-2 shift control valve oppose an upshift.

In low gear, mainline pressure (DRIVE oil) from the manual valve flows to the 1-2 shift valve where it is blocked by the downshifted valve. The downshifted valve also opens an exhaust port in the second gear passage. When vehicle speed increases enough for governor pressure to overcome throttle pressure and spring force, the 1-2 shift valve upshifts. DRIVE oil then flows through the upshifted valve, and into the second gear passage to the intermediate servo (THM 250/250C) or clutch (THM 350/350C). This applies the band or clutch to shift the transmission into second gear.

A closed-throttle 2-1 downshift occurs when governor pressure drops to a very low level (road speed decreases), and throttle pressure and spring force acting on the 1-2 shift control valve move the 1-2 shift valve against governor pressure.

A full-throttle 2-1 detent downshift occurs at speeds below approximately 35 mph (56 kph)

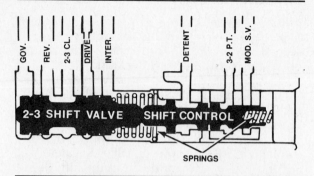

Figure 17-36. The THM 250/250C and 350/350C 2-3 shift valve assembly.

with the THM 250/250C, or 40 mph (64 kph) with the THM 350/350C, when detent pressure from the detent valve flows to the 1-2 shift control valve. The combination of detent pressure, throttle pressure, and spring force moves the 1-2 shift valve against governor pressure and forces a downshift.

In manual low, manual low control or apply pressure is routed from the manual low control valve to fill a chamber between the 1-2 shift valve and the 1-2 shift control valve. This pressure keeps the 1-2 shift valve downshifted regardless of throttle opening or road speed. To increase mainline pressure in manual low, INTERMEDIATE oil from the manual valve is routed to the intermediate boost valve in the pressure regulator valve assembly.

1-2 accumulator
The THM 250/250C and 350/350C have a piston-type accumulator in the intermediate servo or clutch circuit. The 1-2 accumulator, figure 17-35, is installed in the transmission case and consists of a piston, a spring, a cover, a retainer, and seals. Whenever the engine is running, mainline pressure is applied to the inner side of the accumulator piston. This moves the piston outward and compresses the accumulator spring.

When the 1-2 shift valve upshifts, intermediate servo or clutch apply pressure flows to the accumulator where it combines with spring force to oppose mainline pressure. The second gear oil that is diverted to bottom the accumulator piston against mainline pressure smooths application of the intermediate servo or clutch. Once the accumulator piston is bottomed, full mainline pressure exists in the second gear circuit to hold the band or clutch firmly applied.

2-3 shift valve assembly
The 2-3 shift valve controls upshifts from second to third, and downshifts from third to second. The assembly includes the 2-3 shift valve,

the 2-3 shift control valve, and a spring, figure 17-36. The valve is controlled by opposition between governor pressure on one end, and combined throttle (modulator) pressure and spring force on the other. Spring force and throttle pressure acting on the 2-3 shift control valve, oppose an upshift.

In first and second gears, mainline pressure (DRIVE oil) from the manual valve flows to the 2-3 shift valve where it is blocked by the downshifted valve. When vehicle speed increases enough for governor pressure to overcome throttle pressure and spring force, the 2-3 shift valve upshifts. Mainline pressure then flows through the valve and into the third gear circuit where it applies the high-reverse clutch to place the transmission in high gear.

A closed-throttle 3-2 downshift occurs when governor pressure drops (road speed decreases) to the point where throttle pressure and spring force move the 2-3 shift valve against governor pressure.

A part-throttle 3-2 downshift occurs at speeds below approximately 65 mph (105 kph) with the THM 250/250C, or 50 mph (80 kph) with the THM 350/350C, when the throttle is opened past a certain point and 2-3 part throttle pressure from the detent valve combines with throttle pressure and spring force acting on the 2-3 shift control valve to move the 2-3 shift valve against governor pressure.

A full-throttle 3-2 detent downshift occurs at speeds below approximately 65 mph (105 kph) with the THM 250/250C, or 75 mph (121 kph) with the THM 350/350C, when pressure from the detent valve flows to the 2-3 shift control valve. The combination of detent pressure, throttle pressure, and spring force moves the 2-3 shift valve against governor pressure and forces a downshift. If the vehicle speed is below approximately 35 mph (56 kph) with the THM 250/250C, or 40 mph (64 kph) with the THM 350/350C, detent pressure will also downshift the 1-2 shift valve, resulting in a 3-1 downshift.

In manual second, mainline pressure (INTERMEDIATE oil) from the manual valve fills a chamber between the 2-3 shift valve and the 2-3 shift control valve. The combination of INTERMEDIATE oil and spring force keeps the 2-3 shift valve downshifted regardless of throttle opening or road speed. INTERMEDIATE oil also flows to the intermediate boost valve in the pressure regulator valve assembly to increase mainline pressure.

2-3 accumulator — THM 250/250C
In the THM 250/250C, pressure in the high-reverse clutch circuit is also routed to the release side of the intermediate servo where it

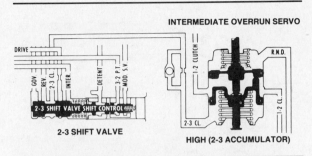

Figure 17-37. The THM 350/350C intermediate overrun servo and 2-3 accumulator.

combines with spring force to release the servo and intermediate band. The stroking action of the servo piston acts as the accumulator for high-reverse clutch application. There is a restriction in the line to the servo to help release the band gradually while the clutch applies.

A check ball is installed in a passage parallel with the restriction. When the servo is released, the check ball is seated. When the servo is applied and pressure on the release side is exhausted, the check ball unseats. This allows the release pressure to vent immediately so that the band does not slip during application.

2-3 accumulator — THM 350/350C

The THM 350/350C has a 2-3 accumulator built into the valve body, figure 17-37. Before the 2-3 upshift, the accumulator piston is held down in its bore by RND oil and spring force above the intermediate overrun servo piston, and by second gear (intermediate clutch) oil in the chamber between the servo and accumulator pistons.

When The 2-3 shift valve upshifts, a portion of the high-reverse clutch apply oil is directed to the bottom of the accumulator piston. This pressure combines with spring force to move the accumulator piston against the pressures above it. This causes pressure to build up gradually in the high-reverse clutch circuit and provide a smooth shift.

Like its THM 250/250C counterpart, the 2-3 accumulator in the THM 350/350C has a restriction and a parallel check ball in the passage leading to it. These provide further modulation of the clutch apply and release pressures, as described in the previous section.

Manual low control valve

The manual low control valve, figure 17-38, prevents engagement of first gear at high speeds. The valve is controlled by opposition between governor pressure at one end, and spring force at the other. When the gear selector is moved to manual low, the manual valve directs mainline pressure (LOW oil) to the manual low control valve.

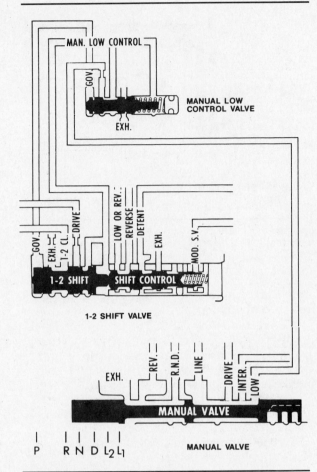

Figure 17-38. The THM 250/250C and 350/350C manual low control valve.

At lower speeds, spring force is stronger than governor pressure and the valve is held open. LOW oil then passes through the valve and is sent as manual low control pressure to the 1-2 shift valve where it downshifts the valve regardless of vehicle speed or throttle position. Some of the pressure is also routed to the end of the manual low control valve to help modulate the manual low control pressure.

At higher speeds, governor pressure overcomes spring force and manual low control pressure. This closes the manual low control valve and blocks LOW oil from the 1-2 shift valve until vehicle speed drops below a certain limit.

Converter clutch valve — 250C and 350C

The converter clutch valve, figure 17-39, receives converter feed oil from the pressure regulator valve, and directs this pressure to the torque converter, transmission cooler, and lubrication circuit. The position of the valve, and the passages through which it directs the oil,

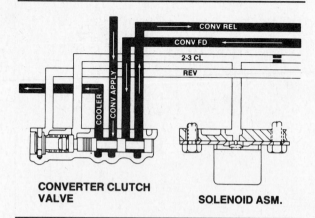

Figure 17-39. The THM 250C and 350C TCC solenoid and converter clutch valve in the release position.

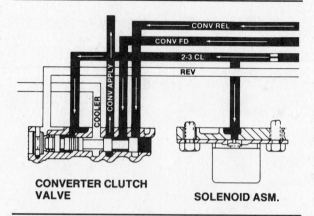

Figure 17-40. The THM 250C and 350C TCC solenoid and converter clutch valve in the apply position.

determine whether the converter clutch is released or applied.

Converter clutch valve position is controlled by opposition between spring force and TCC solenoid pressure. Spring force acts to keep the control valve in the release position; TCC solenoid pressure acts to move the control valve to the apply position. The TCC solenoid pressure may be either high-reverse clutch oil (Chevrolet-built THM 250C), or DRIVE oil (Buick-built THM 250C and all THM 350C). Transmissions that route high-reverse clutch oil to the TCC solenoid can apply the converter clutch in high gear only. Transmissions that route DRIVE oil to the TCC solenoid can apply the converter clutch in lower gears as well.

The electrical and electronic controls used to control the TCC solenoid are described in Chapter 8. The paragraphs below describe the hydraulic operation of the converter clutch system.

When the TCC solenoid is de-energized, figure 17-39, oil is exhausted from the solenoid circuit, and spring force holds the converter clutch valve in the release position. In this position, converter feed oil is routed to the converter through the clutch release passage. Oil enters the converter between the clutch pressure plate and the converter cover; this prevents the clutch from applying. Oil leaving the converter returns to the converter clutch valve through the clutch apply passage, and is then routed to the transmission cooler and lubrication circuit.

When the TCC solenoid is energized, figure 17-40, its exhaust port closes, and high-reverse clutch oil or DRIVE oil travels through the solenoid circuit to move the converter clutch valve to the apply position. In this position, converter

feed oil is routed to the converter through the clutch apply passage. Oil enters the converter between the clutch pressure plate and the turbine; this forces the pressure plate into contact with the converter cover, locking up the converter. Oil leaving the converter returns to the converter clutch valve through the clutch release passage, and is then exhausted through the end of the valve. A portion of the converter feed oil in the clutch apply passage is diverted through a restricting orifice to the transmission cooler and lubrication circuit.

To ensure that the converter clutch will not apply when the gear selector is placed in Reverse, mainline pressure (REVERSE oil) from the manual valve is routed to the spring end of the converter clutch valve. The REVERSE oil combines with spring force to move the converter clutch valve to release position and hold it there.

MULTIPLE-DISC CLUTCHES AND SERVOS

The preceding paragraphs examined the individual valves in the THM 250/250C and 350/350C hydraulic systems. Before we trace the complete hydraulic system operation in each driving range, we will take a quick look at the multiple-disc clutches and servos.

Clutches

The THM 250/250C has three multiple-disc clutches, and the THM 350/350C has four. Each clutch is applied hydraulically, and released by several small coil springs when oil is exhausted from the clutch piston cavity. The low-reverse and high-reverse clutches are designed so that the pistons have two surfaces on which hydraulic pressure acts.

When the low-reverse clutch is engaged in manual low, or the high-reverse clutch is engaged in third gear, pressure is applied only to the smaller, inner area of the clutch piston. When both clutches are applied in Reverse, pressure is applied to the inner areas of the clutch pistons, and also to the larger, outer areas. This allows the clutches to develop greater holding force to compensate for the higher torque loads in Reverse.

Servos

The THM 250/250C and 350/350C each have one servo that applies the intermediate band or intermediate overrun band, figures 17-26 and 17-27. The servo bore is in the case, and the bottom of the bore is formed by the valve body. The intermediate band in the THM 250/250C is externally adjustable. The intermediate overrun band in the THM 350/350C is not adjustable.

HYDRAULIC SYSTEM SUMMARY

We have examined the individual valves, clutches, and servos in the hydraulic system and discussed their functions. The following paragraphs summarize their combined operations in the various gear ranges. Refer to figures 17-23 and 17-24 to trace the hydraulic system operation.

Park and Neutral

In Park and Neutral, the clutches and band are released, and oil flows from the pump to the: priming valve (early THM 350), 1-2 accumulator, pressure regulator valve, manual valve, throttle valve, and detent pressure regulator valve. The priming valve, where fitted, vents any air trapped in the pump circuit. The 1-2 accumulator piston bottoms against spring force.

The pressure regulator valve regulates pump oil to create mainline pressure. From the THM 250 and 350 pressure regulator valve, oil flows to the torque converter, transmission cooler, and lubrication circuit. From the THM 250C and 350C pressure regulator valve, oil flows first to the converter clutch valve, and then to the torque converter, transmission cooler, and lubrication circuit.

In Park, mainline pressure is blocked from entering the manual valve. When the THM 250/250C is in Neutral, mainline pressure enters the manual valve but is blocked from leaving. When the THM 350/350C is in Neutral, mainline pressure passes through the manual valve into the RND passage where it flows to the release side of the intermediate overrun

servo. This pressure application is made in Drive and Reverse also.

The throttle valve regulates mainline pressure based on intake manifold vacuum to create a throttle pressure that is proportional to throttle opening. Throttle pressure is sent to the reverse boost valve and detent valve. At the reverse boost valve, throttle pressure acts to increase mainline pressure in proportion to throttle opening. The detent valve sends throttle pressure to the 1-2 shift control valve and 2-3 shift control valve. Throttle pressure at both shift control valves acts on one land to help spring force keep the 1-2 and 2-3 shift valves downshifted.

The detent pressure regulator valve regulates mainline pressure to create a detent regulator pressure that is limited to a maximum of 80 psi (552 kPa). This pressure is sent to the detent valve where it is blocked.

Drive Low

When the vehicle is stopped and the gear selector is moved to Drive, the manual valve routes DRIVE oil to the: forward clutch, 1-2 shift valve, 2-3 shift valve, governor valve, and TCC solenoid (Buick-built THM 250C and all 350C). These DRIVE pressure applications are made in all forward gear ranges. The forward clutch applies to place the transmission in first gear. DRIVE oil at the 1-2 and 2-3 shift valves is blocked.

As the vehicle begins to move, the governor regulates DRIVE pressure based on output shaft rotation to create a governor pressure that is proportional to vehicle speed. Governor pressure is applied to the: 1-2 shift valve, 2-3 shift valve, manual low control valve, and throttle valve. These governor pressure applications are made in all forward gear ranges.

Governor pressure at the 1-2 and 2-3 shift valves acts to move the valves against throttle pressure and spring force to cause an upshift. Governor pressure at the manual low control valve prevents a manual downshift to low gear above a certain vehicle speed. Governor pressure at the throttle valve acts on the valve to reduce throttle pressure at higher vehicle speeds.

Drive oil to the TCC solenoid (Buick-built THM 250C and all 350C) allows the converter clutch to apply when the solenoid is energized. This does not generally occur until the transmission shifts into second or third gear.

In Drive low, the forward clutch is applied, and the low 1-way clutch is holding.

Drive Second

As vehicle speed increases, governor pressure acting on the 1-2 shift valve overcomes throttle pressure and spring force to upshift the valve. This allows mainline pressure (DRIVE) oil to flow through the 1-2 shift valve into the second gear circuit where it is routed to the: intermediate servo (THM 250/250C) or intermediate clutch (THM 350/350C) apply area, 1-2 accumulator, and 2-3 accumulator (THM 350/350C).

Mainline pressure applies the intermediate band or clutch to place the transmission in second gear. Mainline pressure to the 1-2 accumulator moves the accumulator piston through its range of travel to smooth band or clutch application. Mainline pressure to the 2-3 accumulator (THM 350/350C) bottoms the accumulator piston against spring force.

In Drive second gear, the forward clutch and the intermediate band or clutch are applied.

Drive Third

As vehicle speed continues to increase, governor pressure acting on the 2-3 shift valve overcomes throttle pressure and spring force to upshift the valve. This allows mainline pressure (DRIVE oil) to flow through the 2-3 shift valve into the third gear circuit where it is routed to the: high-reverse clutch apply area, 2-3 accumulator (THM 350/350C) or release side of the low-intermediate servo (THM 250/250C), and TCC solenoid (Chevrolet-built THM 250C).

Mainline pressure applies the high-reverse clutch to place the transmission in high gear. In the THM 350/350C, high-reverse clutch oil applied to the 2-3 accumulator strokes the accumulator piston through its range of travel to cushion clutch engagement. In the THM 250/250C, mainline pressure routed to the release side of the intermediate servo strokes the servo piston through its range of travel to release the intermediate band and cushion the clutch engagement.

In Drive high, the forward, high-reverse, and intermediate (THM 350/350C) clutches are applied. High-reverse clutch oil to the TCC solenoid (Chevrolet-built THM 250C) allows the converter clutch to apply when the solenoid is energized.

Drive Range Detent Downshift

A part-throttle 3-2 downshift can be forced at speeds below approximately 65 mph (105 kph) with the THM 250/250C, or 50 mph (80 kph) with the THM 350/350C. When the accelerator is depressed beyond a certain point, the downshift cable moves the detent valve to direct throttle pressure into the 3-2 part-throttle (or detent 1) passage. This 3-2 part-throttle pressure combines with throttle pressure and spring force acting on the 2-3 shift control valve to move the 2-3 shift valve against governor pressure and force a downshift.

A full-throttle 3-2 downshift can be forced at speeds below approximately 65 mph (105 kph) with the THM 250/250C, or 75 mph (121 kph) with the THM 350/350C. When the throttle is completely depressed, the downshift cable moves the detent valve to allow detent regulator pressure into the detent (or detent 2) passage. Detent pressure is routed to the 2-3 shift control valve where it combines with throttle pressure and spring force to move the 2-3 shift valve against governor pressure and force a downshift. Detent pressure also enters the throttle pressure circuit to maintain at least 80 psi (552 kPa) of throttle pressure.

A full-throttle 2-1 downshift can be forced at speeds below approximately 35 mph (56 kph) with the THM 250/250C, or 40 mph (64 kph) with the THM 350/350C. When the throttle is completely depressed, the downshift cable moves the detent valve to allow detent regulator pressure into the detent (or detent 2) passage. Detent pressure is routed to the 1-2 shift control valve where it combines with throttle pressure and spring force to move the 1-2 shift valve against governor pressure and force a downshift. Detent pressure also enters the throttle pressure circuit to maintain at least 80 psi (552 kPa) of throttle pressure.

During a part- or full-throttle downshift in the THM 250C or 350C, the converter clutch releases just prior to the downshift so the torque converter can multiply torque to help overcome the increased load put on the engine by opening the throttle and downshifting the transmission.

Manual Second

When the gear selector is placed in manual second, the manual valve routes mainline pressure (INTERMEDIATE oil) to the area between the 2-3 shift control valve and the 2-3 shift valve. INTERMEDIATE oil then combines with spring force to hold the 2-3 shift valve in the downshifted position.

The manual valve also routes INTERMEDIATE oil to the intermediate boost valve in the pressure regulator valve assembly to increase mainline pressure. This pressure application is made in manual low as well.

When the 2-3 shift valve in the THM 250/250C is downshifted, the high-reverse clutch releases, and DRIVE oil applies the intermediate servo through the 1-2 shift valve. When the

2-3 shift valve in the THM 350/350C is down-shifted, the high-reverse clutch releases, DRIVE oil is applied to the intermediate clutch through the 1-2 shift valve, and RND oil is vented from the release side of the intermediate overrun servo to apply the intermediate overrun band.

In manual second, the forward clutch, intermediate band (THM 250/250C), intermediate clutch (THM 350/350C), and intermediate overrun band (THM 350/350C) are applied. If vehicle speed drops low enough, throttle pressure and spring force will move the 1-2 shift valve against governor pressure and cause a coasting downshift to Drive low.

Manual Low

When the gear selector is placed in manual low, the manual valve routes LOW oil to the manual low control valve, and from there to the area between the 1-2 shift control valve and the 1-2 shift valve. The manual low control valve prevents a downshift to manual low unless vehicle speed is below a certain limit. Once the manual low control valve is in the downshifted position, mainline pressure and spring force hold it there, regardless of vehicle speed.

LOW pressure to the area between the 1-2 shift control valve and the 1-2 shift valve downshifts the 1-2 shift valve and holds it in the downshifted position. The downshifted 1-2

shift valve directs oil to the low-reverse clutch and vents the intermediate servo or clutch circuit.

In manual low, the forward and low-reverse clutches are applied.

Reverse

In Reverse, the manual valve routes mainline pressure (REVERSE oil) to the: reverse boost valve, outer apply area of the low-reverse clutch, outer apply area of the high-reverse clutch, 1-2 and 2-3 shift valves, and converter clutch valve.

REVERSE oil to the boost valve increases mainline pressure to hold the apply devices more securely. REVERSE oil to the outer apply areas of the low-reverse and high-reverse clutches applies the clutches to place the transmission in reverse gear. REVERSE oil to the 1-2 and 2-3 shift valves is routed through the valves to the inner apply areas of the low-reverse and high-reverse clutches. REVERSE oil to the converter clutch valve moves the valve to the release position.

The DRIVE, INTERMEDIATE, and LOW circuits are all exhausted in Reverse, so the intermediate band or clutch and forward clutch are released. The low-reverse and high-reverse clutches are applied.

Review Questions

Select the single most correct answer
Compare your answers to the correct answers on page 603

1. Mechanic A says GM introduced the THM 350 transmission in 1969.
   Mechanic B says early THM 250 transmissions with the Ravigneaux geartrain did not have a torque converter clutch.
   Who is right?
   a. Mechanic A
   b. Mechanic B
   c. Both mechanic A and B
   d. Neither mechanic A nor B

2. Mechanic A says the THM 350 is identified in some service manuals and parts catalogs as the M-38 transmission.
   Mechanic B says the THM 250 is identified in some service manuals and parts catalogs as the M-31 transmission.
   Who is right?
   a. Mechanic A
   b. Mechanic B
   c. Both mechanic A and B
   d. Neither mechanic A nor B

3. Mechanic A says the THM 375B is a heavy-duty version of the THM 250 transmission.
   Mechanic B says the THM 250C and 350C transmissions are identical except for their torque capacities.
   Who is right?
   a. Mechanic A
   b. Mechanic B
   c. Both mechanic A and B
   d. Neither mechanic A nor B

4. Mechanic A says early-model THM 350 transmissions, have identification numbers stamped on the governor cover.
   Mechanic B says late-model THM 250/250C transmissions have identification numbers stamped on the 1-2 accumulator cover.
   Who is right?
   a. Mechanic A
   b. Mechanic B
   c. Both mechanic A and B
   d. Neither mechanic A nor B

5. Mechanic A says all THM 250/250C and 350/350C transmissions have six gear selector positions.
   Mechanic B says the driver can force a part-throttle 2-1 downshift in a THM 250/250C or 350/350C transmission.
   Who is right?
   a. Mechanic A
   b. Mechanic B
   c. Both mechanic A and B
   d. Neither mechanic A nor B

6. Mechanic A says the THM 350/350C transmission uses the same apply devices in manual second that it does in Drive second.
   Mechanic B says that when the THM 250/250C gear selector is placed in manual second, the transmission starts in Drive low and upshifts to drive second.
   Who is right?
   a. Mechanic A
   b. Mechanic B
   c. Both mechanic A and B
   d. Neither mechanic A nor B

7. Mechanic A says THM 250/250C and 350/350C transmissions make a single 3-1 downshift when decelerating to a stop in Drive. Mechanic B says that when a THM 250/250C or 350/350C is in high gear and the gear selector is moved to manual second, the transmission immediately shifts into second gear regardless of vehicle speed or throttle position. Who is right?
   a. Mechanic A
   b. Mechanic B
   c. Both mechanic A and B
   d. Neither mechanic A nor B

8. Mechanic A says "drive oil" is the GM name for one form of mainline pressure. Mechanic B says the TCC solenoid receives low-reverse clutch oil in some THM 250 transmissions. Who is right?
   a. Mechanic A
   b. Mechanic B
   c. Both mechanic A and B
   d. Neither mechanic A nor B

9. Mechanic A says all THM 250/250C and 350/350C transmissions use a gear-type oil pump with a priming valve. Mechanic B says THM 250/250C and 350/350C transmissions use a detent valve controlled by a cable attached to the throttle linkage. Who is right?
   a. Mechanic A
   b. Mechanic B
   c. Both mechanic A and B
   d. Neither mechanic A nor B

10. Mechanic A says the THM 250/250C and 350/350C manual valve is operated by a mechanical linkage. Mechanic B says the THM 250/250C and 350/350C detent valve is operated by intake manifold vacuum. Who is right?
    a. Mechanic A
    b. Mechanic B
    c. Both mechanic A and B
    d. Neither mechanic A nor B

11. Mechanic A says the THM 250/250C and 350/350C rear carrier is splined to the output shaft. Mechanic B says GM calls the rear carrier of the Simpson geartrain the reaction carrier. Who is right?
    a. Mechanic A
    b. Mechanic B
    c. Both mechanic A and B
    d. Neither mechanic A nor B

12. Mechanic A says the THM 350/350C intermediate 1-way (roller) clutch prevents counterclockwise rotation of the sun gear. Mechanic B says the THM 350/350C intermediate overrun band is always applied at the same time as the intermediate clutch. Who is right?
    a. Mechanic A
    b. Mechanic B
    c. Both mechanic A and B
    d. Neither mechanic A nor B

13. Mechanic A says the THM 250/250C and 350/350C low-reverse clutch locks the rear carrier to the case. Mechanic B says the THM 250/250C and 350/350C high-reverse clutch drum turns as a unit with the sun gear. Who is right?
    a. Mechanic A
    b. Mechanic B
    c. Both mechanic A and B
    d. Neither mechanic A nor B

14. Mechanic A says the THM 250/250C and 350/350C sun gear may be held by the high-reverse clutch. Mechanic B says the THM 250/250C and 350/350C sun gear may be held by the low-reverse clutch. Who is right?
    a. Mechanic A
    b. Mechanic B
    c. Both mechanic A and B
    d. Neither mechanic A nor B

15. Mechanic A says the THM 250/250C and 350/350C governor is mounted on the output shaft. Mechanic B says the THM 250/250C and 350/350C governor sends a road speed signal to the shift valves. Who is right?
    a. Mechanic A
    b. Mechanic B
    c. Both mechanic A and B
    d. Neither mechanic A nor B

16. Mechanic A says the THM 250/250C and 350/350C converter clutch valve is located in an auxiliary valve body at the front of the separator plate. Mechanic B says the THM 250/250C and 350/350C converter clutch valve is controlled by opposition between spring force and mainline pressure. Who is right?
    a. Mechanic A
    b. Mechanic B
    c. Both mechanic A and B
    d. Neither mechanic A nor B

17. Mechanic A says the THM 250/250C transmission has five check balls. Mechanic B says the THM 350C transmission has four check balls. Who is right?
    a. Mechanic A
    b. Mechanic B
    c. Both mechanic A and B
    d. Neither mechanic A nor B

18. Mechanic A says the THM 250/250C and 350/350C reverse boost valve is used to increase mainline pressure in proportion to throttle opening. Mechanic B says the THM 250/250C and 350/350C intermediate boost valve is used to increase mainline pressure in manual low gear. Who is right?
    a. Mechanic A
    b. Mechanic B
    c. Both mechanic A and B
    d. Neither mechanic A nor B

19. Mechanic A says the THM 350/350C manual valve can charge five passages with mainline pressure. Mechanic B says the THM 250/250C manual valve has an additional RND passage leading from the manual valve. Who is right?
    a. Mechanic A
    b. Mechanic B
    c. Both mechanic A and B
    d. Neither mechanic A nor B

20. Mechanic A says the secondary weights in the THM 250/250C and 350/350C governor regulate governor pressure up to approximately 720 governor rpm. Mechanic B says some THM 250/250C and 350/350C throttle valves can alter transmission shift points to compensate for changes in altitude. Who is right?
    a. Mechanic A
    b. Mechanic B
    c. Both mechanic A and B
    d. Neither mechanic A nor B

18

Turbo Hydra-matic 375, 400, 425, and 475 Transmissions

HISTORY AND MODEL VARIATIONS

The General Motors (GM) Turbo Hydra-matic (THM) 375, 400, 425, and 475 transmissions are heavy-duty gearboxes generally used with large engines in full-sized cars and trucks. The THM 375, 400, and 475 are used in RWD applications. The THM 425 is used in FWD vehicles with longitudinally mounted engines. The THM 425 is not considered a transaxle because the differential and final drive assembly is not built into the transmission case, but is a separate unit that bolts to the gearbox.

The THM 400, figure 18-1, was introduced in 1964 and is the oldest of the transmissions covered in this chapter. The THM 400 has been used by all the GM divisions, and by other manufacturers including Jeep, Jaguar, and Rolls-Royce. As automobiles have been downsized over the years, the number of applications for this transmission has declined. As of the 1988 model year, the THM 400 is still used in full-size trucks, and is optional in the RWD Cadillac Brougham. The Buick version of the THM 400 is sometimes called the Super Turbine 400, and some GM parts catalogs and service manuals identify the THM 400 as the M-40 transmission.

The THM 425, figure 18-2, debuted in the 1966 FWD Oldsmobile Toronado. It was also used in the Cadillac Eldorado beginning in 1967. THM 425 transmissions were used in these two cars through the 1978 model year. In 1979, the THM 425 was discontinued in favor of the THM 325 series transmissions covered in Chapter 16. Some GM parts catalogs and service manuals identify the THM 425 as the M-42 transmission.

The THM 375 is a medium-duty version of the THM 400. This transmission should not be confused with the THM 375B, which is a heavy-duty version of the THM 350 transmission. Some GM parts catalogs and service manuals identify the THM 475 as the M-41 transmission. The THM 375 and 475 will not be called out separately in this chapter, but all statements regarding the THM 400 apply equally to these transmissions.

All the transmissions covered in this chapter use the same basic torque converter and 3-speed compound planetary gearset. However, some early models of the THM 400 and 425 have a torque converter with a variable-pitch stator that is electrically and hydraulically controlled to change the converter stall speed in response to throttle position. This feature was discontinued after the 1967 model year, which resulted

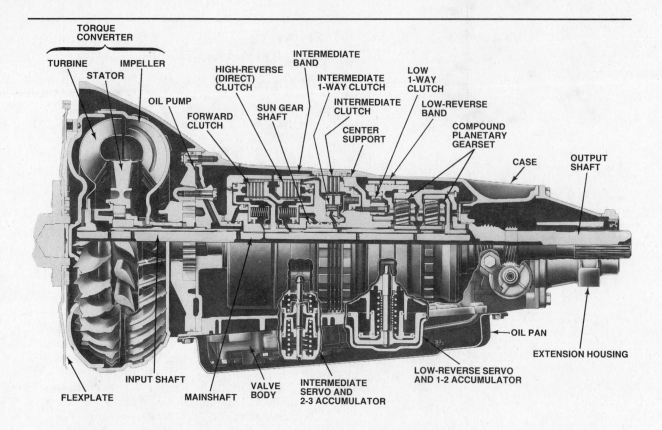

Figure 18-1. The THM 400 transmission; the THM 375 and 475 are essentially the same.

in several changes to the transmission mechanical and hydraulic systems. The variable pitch stator and its related parts differences will be explained later in the chapter.

Other major internal changes to these transmissions center on the oil filter and the two 1-way clutches. Through 1966, the oil filter contained a bypass valve. This valve was eliminated in mid-1967, and a new filter and oil pan were introduced, figure 18-3. All THM 400 and 425 models have two 1-way clutches, but over the years, various combinations of roller and sprag clutches have been used. These changes will be explained in detail during the transmission buildup section later in this chapter.

Fluid Recommendations

Like all GM automatic transmissions and transaxles, the THM 400 and 425 transmissions require DEXRON®-IID fluid.

Transmission Identification

The THM 400 transmission can easily be identified by the shape of its oil pan, and by the governor cover, identification plate, and vacuum modulator, all of which are located on the right side of the transmission case, figure 18-4. The THM 425 transmission is simple to identify because it is the only gearbox used in the GM cars listed at the beginning of this chapter.

In addition to the transmission model, you must know the exact assembly part number and other identifying characters in order to obtain proper repair parts. On the THM 400, figure 18-4, this information is stamped on a tag riveted to the right center of the transmission case. On the THM 425, figure 18-5, a similar tag is riveted to the left side of the bellhousing.

The coding that appears on the identification tag will vary with the transmission model and the plant in which it was manufactured. Figure 18-6 shows a typical identification tag for the THM 400, and explains the meaning of the data stamped on it. If the transmission case is ever replaced, the complete set of identification numbers should be transferred to the new part.

As required by Federal regulations, GM transmissions also carry the serial number of the vehicle in which they were originally installed. On the THM 400 and 425, the serial number is stamped on the lower left side of the case next to the manual valve shaft.

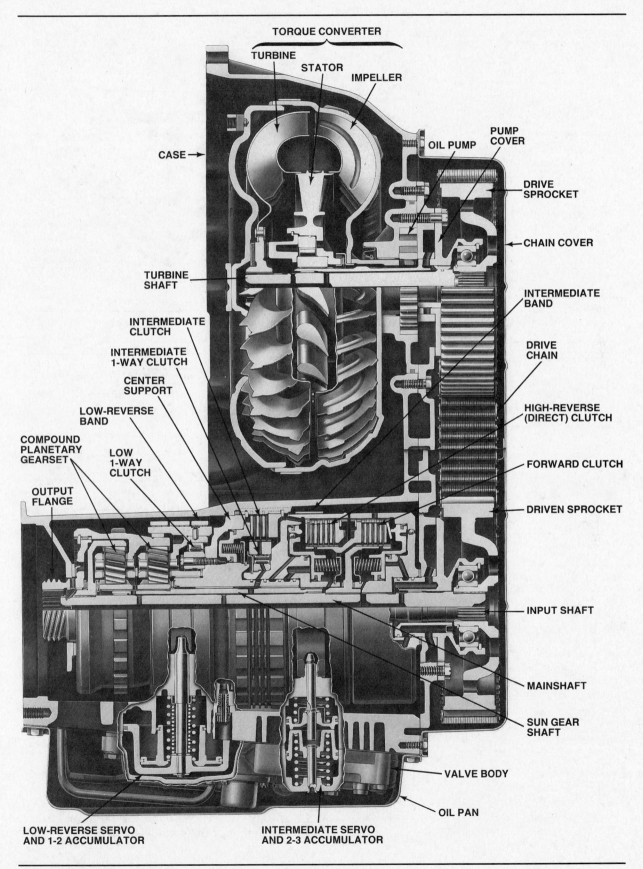

Figure 18-2. The THM 425 transmission.

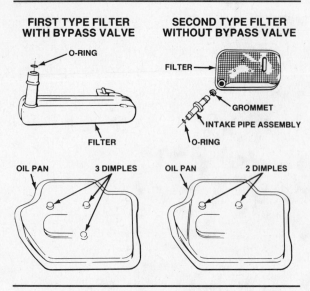

Figure 18-3. THM 400 and 425 transmissions have used two oil filter designs over the years.

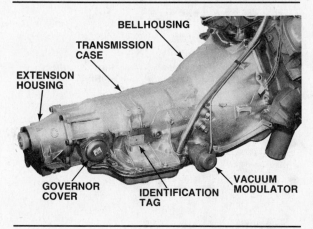

Figure 18-4. The identification tag on the THM 400 is riveted to the right side of the transmission case.

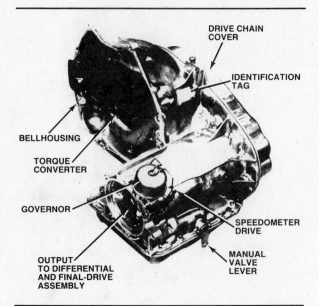

Figure 18-5. The identification tag on the THM 425 is riveted to the left side of the transmission bellhousing.

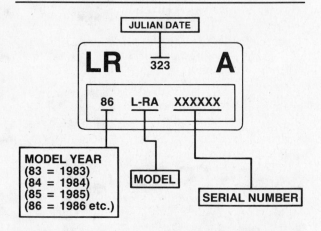

Figure 18-6. Typical THM 400 identification tag markings.

Major Changes

Some of the changes made in early THM 400 and 425 transmissions were explained in the introduction. Because of the dependability of these transmissions, very few other changes have been made. The major changes that have been made in the THM 400 and 425 transmissions are:

1966 ● Low 1-way clutch changed from sprag- to roller-type.
1967 ● Spacer added under the center support.
1971 ● Pressure regulator valve and pump cover changed, the new parts cannot be used in earlier transmissions.

1972 ● Intermediate 1-way clutch changed from sprag- to roller type.
1977 ● Forward clutch redesigned to use bevel-cut O-rings instead of lip seals.
1980 ● Teflon rings added to seal the forward and high-reverse clutches.
1985 ● Four new bolts used on extension housing.

GEAR RANGES

Depending on the GM division and the particular application, there are minor variations in the exact labeling of the gear selectors used with

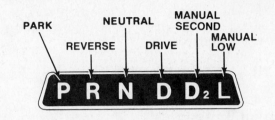

Figure 18-7. Typical THM 400 and 425 gear selector positions.

the THM 400 and 425 transmissions. However, all of the gear selectors have six positions, figure 18-7:

P — Park
R — Reverse
N — Neutral
D or 3 — Drive (also called Drive left)
I, S, 2, or L2 — Manual second (also called Drive right)
L, 1, or L1 — Manual low.

Gear ratios for these THM transmissions are:

- First (low) — 2.48:1
- Second (intermediate) — 1.48:1
- Third (direct drive) — 1.00:1
- Reverse — 2.08:1.

Park and Neutral

The engine can be started only when the transmission is in Park or Neutral. No clutches or bands are applied in these positions, and there is no flow of power through the transmission. In Neutral, the output shaft is free to turn, and the car can be pushed or pulled. In Park, a mechanical pawl engages the parking gear on the output shaft. This locks the output shaft to the transmission case and prevents the vehicle from being moved.

Drive

When the vehicle is stopped and the gear selector is placed in Drive (D or 3), the transmission is automatically in low gear. As the vehicle accelerates, the transmission automatically upshifts to second, and then to third gear. The speeds at which upshifts occur vary, and are controlled by vehicle speed and engine load.

The driver can force downshifts in Drive by opening the throttle partially or completely. The speeds at which downshifts occur vary for different engine, axle ratio, tire size, and vehicle weight combinations.

Below approximately 33 mph (53 kph) in most models, the driver can force a part-throttle 3-2 downshift by opening the throttle beyond a

certain point. Some transmissions, however, do not provide a part-throttle 3-2 downshift. Below approximately 70 mph (113 kph), the driver can force a 3-2 detent downshift by fully opening the throttle. A full-throttle 3-1 or 2-1 downshift can be forced below approximately 20 mph (32 kph).

When the car decelerates in Drive to a stop, the transmission makes two downshifts automatically, first a 3-2 downshift, then a 2-1 downshift.

Manual Second

When the vehicle is stopped and the gear selector is placed in manual second (I, S, 2, or L2), the transmission is automatically in low gear. As the vehicle accelerates, the transmission automatically upshifts to second gear; it does not upshift to third. The speed at which the upshift occurs will vary, and is controlled by vehicle speed and engine load.

A full-throttle 2-1 detent downshift can be forced below approximately 20 mph (32 kph) if the driver depresses the accelerator completely. The transmission automatically downshifts to low as the vehicle decelerates to a stop.

The driver can manually downshift from third to second by moving the selector from D or 3 to I, S, 2, or L2. In this case, the transmission shifts to second gear immediately, regardless of road speed or throttle position. Because different apply devices are used in manual second than are used in Drive second, this gear range can be used to provide engine compression braking, or gear reduction for hill climbing, pulling heavy loads, or operating in congested traffic.

Manual Low

When the vehicle is stopped and the gear selector is placed in manual low (L, 1, or L1), the car starts and stays in first gear; it does not upshift to second or third. Because different apply devices are used in manual low than are used in Drive low, this gear range can be used to provide engine compression braking to slow the vehicle.

If the driver shifts to manual low at high speed, the transmission will not shift directly to low gear. To avoid engine or driveline damage, the transmission first downshifts to second, and then automatically downshifts to low when the vehicle speed drops below approximately 40 mph (64 kph). The transmission then remains in low gear until the gear selector is moved to another position.

GENERAL MOTORS TERM	TEXTBOOK TERM
Clutch housing	Clutch drum
Converter pump	Impeller
Detent downshift	Full-throttle downshift
Internal gear	Ring gear
Control valve assembly	Vavle body
Modulator valve	Throttle valve
Modulator pressure	Throttle pressure
Output carrier	Rear carrier
Reaction carrier	Front carrier
Rear band	Low-reverse band
Rear servo	Low-reverse servo
Front band	Intermediate band
Front servo	Intermediate servo
Low-gear 1-way clutch	Low 1-way clutch
Direct clutch	High-reverse clutch

Figure 18-8. THM 400 and 425 transmission nomenclature table.

Reverse

When the gear selector is moved to R, the transmission shifts into Reverse. The transmission should be shifted to Reverse only when the vehicle is standing still. The lockout feature of the shift gate should prevent accidental shifting into Reverse while the vehicle is moving forward. However, if the driver intentionally moves the selector to R while moving forward, the transmission *will* shift into Reverse, and damage may result.

GM TRANSMISSION NOMENCLATURE

For the sake of clarity, this text uses consistent terms for transmission parts and functions. However, the various manufacturers may use different terms. Therefore, before we examine the buildup of the THM 400 and 425, and the operation of their geartrains and hydraulic systems, you should be aware of the unique names that General Motors uses for some transmission parts and operations. This will make it easier in the future if you should look in a GM manual and see a reference to the "direct clutch." You will realize that this is simply the high-reverse clutch and is similar to the same part in any other transmission that uses a similar gear train and apply devices. Figure 18-8 lists GM nomenclature and the textbook terms for some transmission parts and operations.

THM 400 AND 425 BUILDUP

Although the THM 400 and 425 transmissions have a great deal in common, they also have some significant differences. For this reason,

the buildup is broken down into three sections to help clarify the ways in which these gearboxes are assembled:

- Initial buildup
- THM 425 drive chain assembly
- Final buildup.

Initial Buildup

All the transmissions covered in this chapter use a 1-piece case casting that incorporates the bellhousing. The THM 400 has an extension housing that bolts to the back of the case, figure 18-4. The THM 425 does not have an extension housing because the differential and final drive assembly bolts directly to the output end of the case, figure 18-9.

In the THM 400, the oil pump assembly bolts to the front of the case, and the oil pan bolts to the bottom. In the THM 425, the oil pump assembly bolts to a pump cover that is bolted to the back side of the case. The drive chain cover also bolts to the back of the transmission case, and the oil pan bolts to the bottom of the case.

The torque converter fits inside the bellhousing, and is a welded unit that cannot be disassembled. The converter bolts to the engine flexplate, which has the starter ring gear around its outer diameter. Most THM 400, and all THM 425, torque converters are conventional 3-element designs (impeller, turbine, and stator).

The torque converter is installed in the transmission by sliding it onto the stationary stator support. In the THM 400, figure 18-10, the stator support is part of the oil pump housing. In the THM 425, figure 18-11, the stator support is a separate part bolted to the back side of the pump cover and case. The splines in the stator 1-way clutch engage the splines on the stator support. In the THM 400, the splines in the converter turbine engage the transmission input shaft. In the THM 425, the splines in the turbine engage the turbine shaft. The turbine shaft rides in bushings inside the stator support and passes through the oil pump and pump cover.

The hub at the rear of the torque converter housing engages the lugs on the oil pump inner gear to drive the pump. The converter housing and impeller are bolted to the engine flexplate so the oil pump turns and pumps fluid whenever the engine is running.

All of the transmissions covered in this chapter have gear-type oil pumps. The THM 400 pump mounts to the transmission case directly behind the torque converter. The THM 425 oil pump, also positioned directly behind the

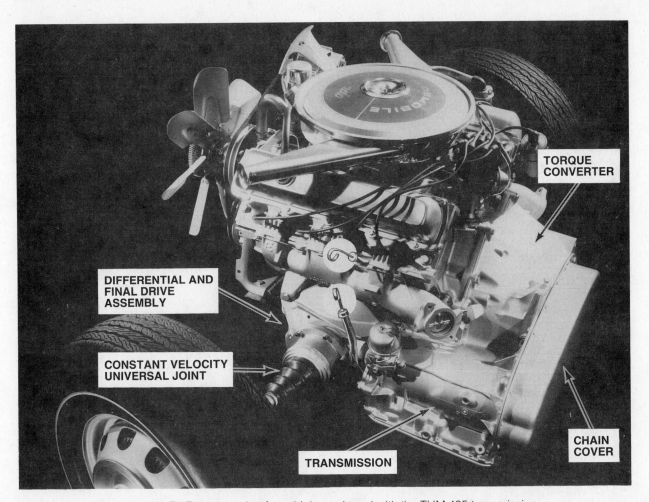

TORQUE CONVERTER

DIFFERENTIAL AND FINAL DRIVE ASSEMBLY

CONSTANT VELOCITY UNIVERSAL JOINT

TRANSMISSION

CHAIN COVER

Figure 18-9. The complete FWD powertrain of a vehicle equipped with the THM 425 transmission.

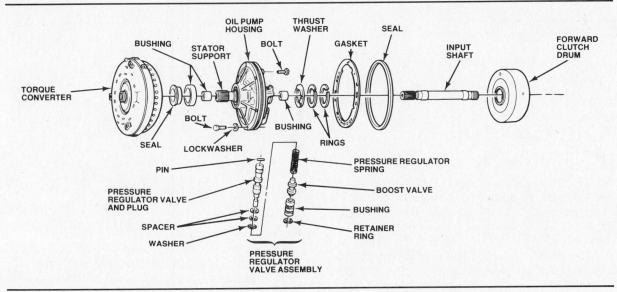

OIL PUMP HOUSING

THRUST WASHER

BUSHING

STATOR SUPPORT

BOLT

GASKET

SEAL

INPUT SHAFT

FORWARD CLUTCH DRUM

TORQUE CONVERTER

BOLT

BUSHING

SEAL

LOCKWASHER

RINGS

PIN

PRESSURE REGULATOR SPRING

PRESSURE REGULATOR VALVE AND PLUG

BOOST VALVE

BUSHING

SPACER

WASHER

RETAINER RING

PRESSURE REGULATOR VALVE ASSEMBLY

Figure 18-10. The THM 400 torque converter, oil pump, input shaft, and forward clutch drum.

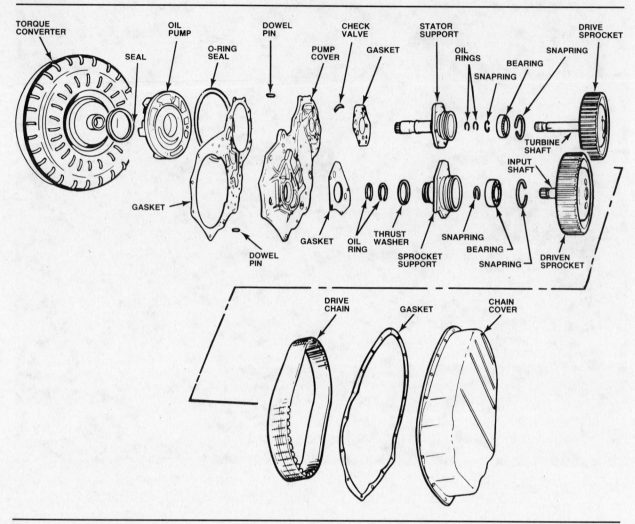

Figure 18-11. The THM 425 torque converter, oil pump, and drive chain assembly.

torque converter, bolts to a pump cover that is bolted to the back of the case. In all of these transmissions, the oil pump draws fluid from the sump through the filter, and delivers it to the pressure regulator valve.

THM 425 Drive Chain Assembly

The THM 425 turbine shaft is splined to and turns the drive sprocket, which rides on ball bearings on the rear of the stator support, figure 18-11. The drive sprocket turns a multiple-link drive chain that connects it to the driven sprocket. The driven sprocket is splined to and turns the transmission input shaft. The driven sprocket rides on ball bearings on a sprocket support that passes through the lower part of the pump cover, and is bolted to the transmission case.

Whenever the torque converter impeller drives the converter turbine, the turbine shaft turns the drive sprocket, drive chain, driven sprocket, and transmission input shaft. Therefore, engine torque makes two 90-degree turns before entering the transmission geartrain. From this point, the THM 425 buildup is as described in the Final Buildup section below.

Power flow to and through the geartrain of the THM 425 transmission can be looked at in two ways. Viewed from the *front* of the engine, the torque converter and turbine shaft rotate clockwise and turn the drive sprocket, drive chain, driven sprocket, and input shaft clockwise as well. However, when viewed from the *rear* of the engine, looking toward the "front" or input end of the transmission geartrain, all these components turn counterclockwise. To simplify the power flow descriptions for the transmissions covered in this chapter, all references to rotation in the THM 425 will be made as if you are looking at the transmission from the *front* of the engine.

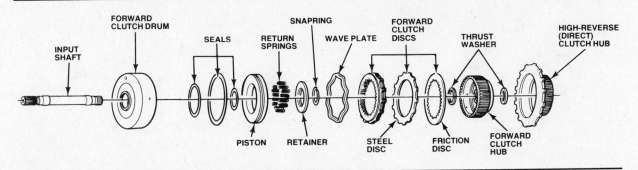

Figure 18-12. The THM 400 and 425 forward clutch assembly.

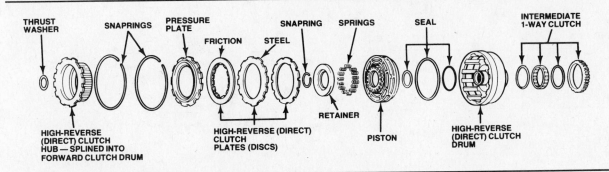

Figure 18-13. The THM 400 and 425 high-reverse clutch assembly.

Final Buildup

The compound planetary gearset in the THM 400 and 425 is similar to the Simpson gearset, but it does not work in quite the same way because these transmissions use a unique 4-shaft layout that includes an input shaft, a mainshaft, a sun gear shaft, and an output shaft. Because of this, the planetary gearset members are attached to different shafts than their counterparts in a Simpson geartrain.

The input shaft is turned clockwise by the torque converter turbine (THM 400) or the driven sprocket (THM 425). The input shaft is supported on bushings in the stator support (THM 400) or the driven sprocket support (THM 425), and passes through it into the transmission case to power the geartrain. The outside of the stator support or driven sprocket support serves as the pilot for the forward clutch drum. From this point on, the THM 400 and 425 transmissions are essentially the same.

The forward clutch drum, figure 18-12, is splined to the end of the input shaft and turns clockwise with it. The outside of each forward clutch steel disc is splined to the inside of the forward clutch drum. The inside of each friction disc is splined to the outside of the forward clutch hub, which is splined to the mainshaft. When the forward clutch is applied, the input

shaft turns the mainshaft clockwise. The mainshaft extends rearward through the center of the hollow sun gear shaft.

The high-reverse clutch hub, figure 18-13, is splined to the forward clutch drum; therefore, it also turns clockwise with the input shaft. However, the high-reverse clutch hub does not contact the forward clutch hub in any way. There is a thrust washer on each side of the forward clutch hub. One washer takes the thrust between the front of the clutch hub and the inside rear of the forward clutch drum, and the other washer takes the thrust between the rear of the forward clutch hub and the inner front of the high-reverse clutch drum. The inside of each high-reverse clutch friction disc is splined to the outside of the high-reverse clutch hub. The outside of each steel disc is splined to the inside of the high-reverse clutch drum, which is splined to the sun gear shaft. When the high-reverse clutch is applied, the input shaft turns the sun gear shaft, and the sun gear splined to it, clockwise.

An extension on the back side of the high-reverse clutch drum forms the inner race of the intermediate 1-way clutch, figure 18-14. A sprag clutch is used on 1964-71 THM 400 transmissions, and *all* THM 425 gearboxes; 1972 and later THM 400s use a roller clutch. The outer

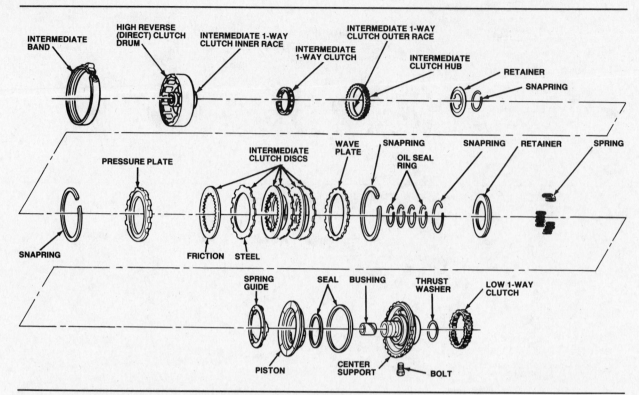

Figure 18-14. The THM 400 and 425 intermediate band, intermediate 1-way clutch, intermediate multiple-disc clutch, and center support.

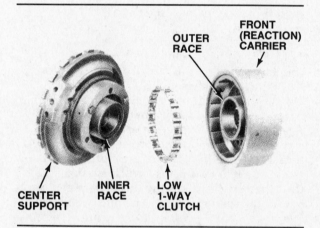

Figure 18-15. The THM 400 and 425 low 1-way clutch assembly.

race of the intermediate 1-way clutch also forms the hub of the intermediate multiple-disc clutch.

The inside of each intermediate clutch friction disc is splined to the outside of the intermediate clutch hub. The outside of each steel disc is splined to the transmission case. The intermediate clutch piston mounts in the front side of the stationary center support, which is splined and bolted to the transmission case. When the intermediate clutch is applied, it

locks the outer race of the intermediate 1-way clutch to the transmission case. The intermediate 1-way clutch then allows the high-reverse clutch drum, sun gear shaft, and sun gear to rotate clockwise, but prevents them from rotating counterclockwise.

The outside of the high-reverse clutch drum forms the apply surface for the intermediate overrun band, figure 18-14, which is wrapped around it. When the intermediate band is applied, the high-reverse clutch drum, sun gear shaft, and sun gear are held from rotating in *either* direction. This occurs in manual second gear.

The inner race of the low 1-way clutch is bolted to the rear of the stationary center support, figure 18-15. A sprag clutch is used on 1964-65 THM 400 transmissions; 1966 and later THM 400s, and *all* THM 425 gearboxes, use a roller clutch. The outer race of the low 1-way clutch is located inside the front carrier of the compound planetary gearset. The low 1-way clutch allows the front carrier to rotate clockwise, but prevents it from turning counterclockwise.

The outside of the front carrier forms the apply surface for the low-reverse band, figure 18-16, which is wrapped around it. When the low-reverse band is applied, the front carrier

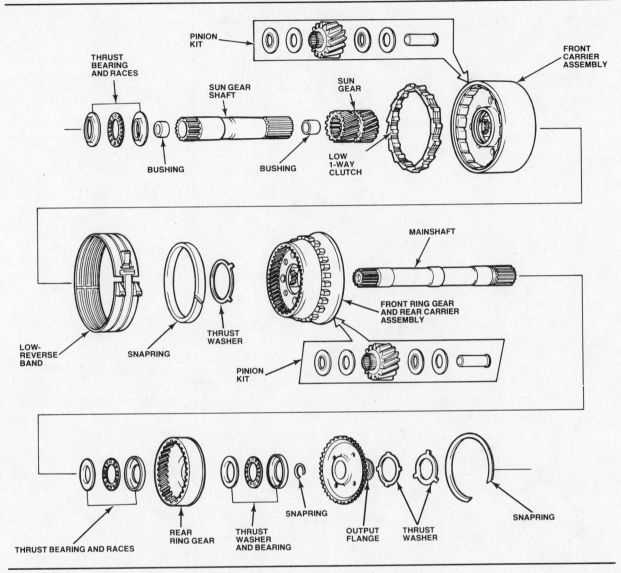

Figure 18-16. The THM 425 low-reverse band and planetary gearset.

is held from rotating in *either* direction. This occurs in manual low and reverse gears.

The mainshaft and sun gear shaft extend rearward through an opening in the center support to drive the geartrain, figures 18-16 and 18-17. As mentioned earlier, the 1-piece sun gear is splined to the hollow sun gear shaft. The forward portion of the sun gear meshes with the front pinions, which turn on pinion shafts in the front carrier. The front pinions mesh with the front ring gear, which is a single assembly with the rear carrier. The rear pinions turn on pinion shafts in the rear carrier, and mesh with the rear portion of the sun gear. The rear pinions also mesh with the rear ring gear, which is splined to the end of the mainshaft. The outside of the front ring gear and rear carrier assembly forms the parking gear, which is

locked to the transmission case by the parking pawl when the gear selector is placed in Park.

In the THM 400, figure 18-17, the front ring gear and rear carrier assembly is splined to the output shaft flange. Just behind this flange, the output shaft is supported by a bushing in the main case. The governor drive gear is cut into the output shaft behind this bushing. The governor mounts in the left rear of the case and is driven by a nylon pinion gear that engages the gear on the output shaft. The speedometer drive gear mounts on the output shaft behind the governor drive gear and is secured by a retaining clip. Splines for the output yoke are cut into the end of the output shaft, and a bushing and a seal are installed in the rear of the extension housing to support the yoke and shaft.

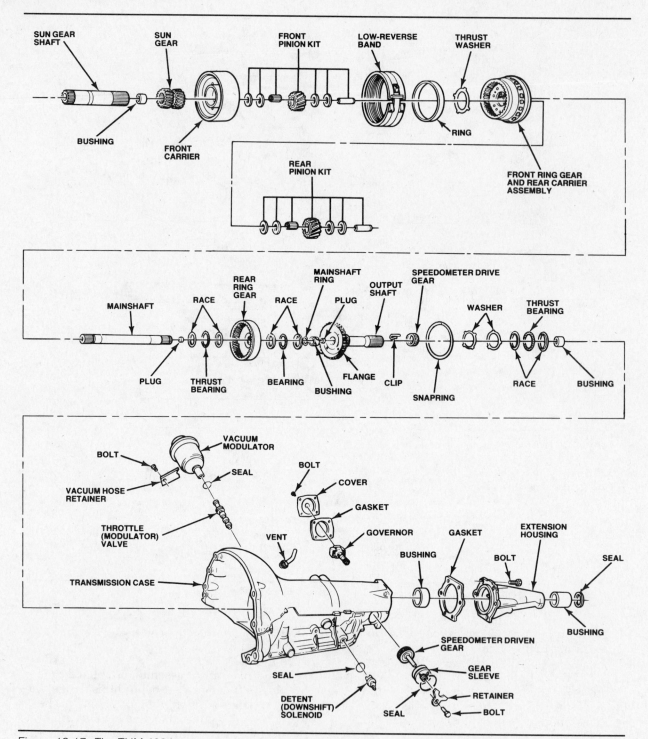

Figure 18-17. The THM 400 low-reverse band and planetary gearset.

In the THM 425, figure 18-16, the front ring gear and rear carrier assembly is splined to the output flange. This is a very short and hollow shaft with internal splines that connect to the differential and final drive assembly. A thrust washer installs between the output flange and the inner face of the transmission case, and governor drive gear teeth are cut into the flange just past the thrust washer. The governor mounts in the top of the case, figure 18-18, and is driven by a nylon pinion gear that engages the output flange. The speedometer drive gear mounts at the top of the governor shaft.

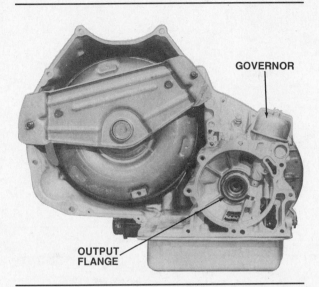

Figure 18-18. The THM 425 output flange and governor drive gear.

THM 400 AND 425 POWER FLOW

Now that we have taken a quick look at the general way in which the THM 400 and 425 are assembled, we can trace the flow of power through the apply devices and the compound planetary gearset in each of the gear ranges. We will use the THM 400 to illustrate each gear range; power flow in the THM 425 is identical.

Park and Neutral

When the gear selector is in Park or Neutral, figure 18-19, no clutches or bands are applied. The torque converter cover/impeller and the oil pump rotate clockwise at engine speed. The torque converter turbine and the input shaft also rotate, but at less than engine speed. The input shaft turns the forward clutch drum and high-reverse clutch hub clockwise. However, because neither clutch is applied, there is no input to the planetary geartrain, and thus, no output motion at this time.

In Neutral, the output shaft is free to rotate in either direction, so the vehicle can be pushed or pulled. In Park, a mechanical pawl engages the parking gear on the outside of the front ring gear and rear carrier assembly. This locks the output shaft to the transmission case to prevent the vehicle from moving in either direction.

Drive Low

When the car is stopped and the gear selector is moved to Drive (D or 3), the transmission is automatically in low gear, figure 18-20. The forward clutch is applied, which causes the input

shaft to turn the mainshaft and the rear ring gear clockwise at turbine speed.

Before the car begins to move, the transmission output shaft is held stationary by the driveline and wheels. This also holds the front ring gear and rear carrier assembly stationary. As a result, the clockwise rotation of the rear ring gear turns the rear pinions clockwise on the stationary carrier. Rear pinion rotation then drives the sun gear counterclockwise.

Counterclockwise rotation of the sun gear turns the front pinions clockwise. The front pinions then try to walk clockwise around the inside of the front ring gear, which is splined to the stationary output shaft. This action causes the front carrier to try to rotate counterclockwise. However, the front carrier contains the outer race of the low 1-way clutch.

As the front carrier tries to rotate counterclockwise, the 1-way clutch locks up and prevents rotation. Therefore, the torque from the front pinions is transferred to the front ring gear and output shaft instead of the front carrier. Because the 1-way clutch holds the carrier more securely than the drive wheels can hold the output shaft and ring gear, the output shaft and front ring gear begin to rotate clockwise in gear reduction.

As long as the car is accelerating in Drive low, torque is transferred from the input shaft, through the planetary gearset, to the output shaft. If the driver lets up on the throttle, the engine speed drops while the drive wheels try to keep turning at the same speed. This removes the counterclockwise-rotating torque load from the low 1-way clutch, which then unlocks and allows the front planetary gearset to freewheel so no engine compression braking is available. This is a major difference between Drive low and manual low as we will see later.

In Drive low, gear reduction is produced by a combination of the front and rear planetary gearsets. The rear ring gear is the input member; the front carrier is the held member; and the front ring gear is the output member. There are only two apply devices working in Drive low. The forward clutch is the input device, and the low 1-way clutch is the holding device.

Drive Second

As the car continues to accelerate in Drive, the transmission hydraulic system produces an automatic upshift to second gear by applying the intermediate clutch, figure 18-21. The forward clutch remains applied, and power from the engine is still delivered to the rear ring gear. The rear ring gear rotates clockwise and turns the

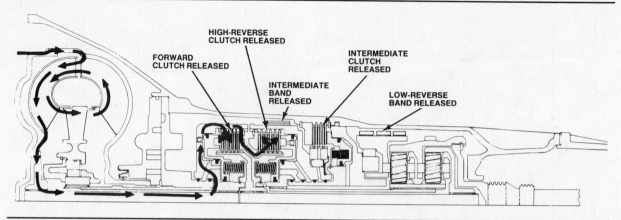

Figure 18-19. THM 400 and 425 power flow in Park and Neutral.

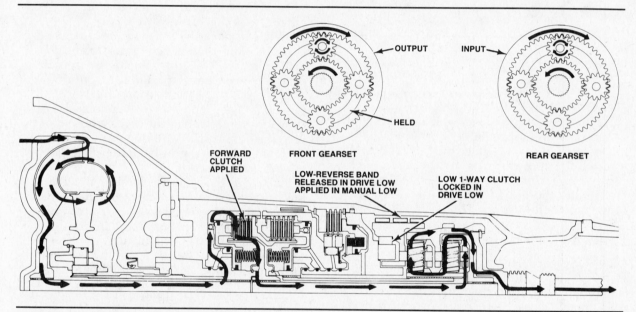

Figure 18-20. THM 400 and 425 power flow in Drive low and manual low.

rear pinions clockwise. The rear pinions then attempt to drive the sun gear counterclockwise.

At the same time, the intermediate clutch locks the outer race of the intermediate 1-way clutch to the transmission case. When the rear pinions attempt to drive the sun gear counterclockwise, the intermediate 1-way clutch locks and holds the sun gear stationary. The rear pinions then walk clockwise around the stationary sun gear, taking the rear carrier with them. The rear carrier is splined to the output shaft and drives it in gear reduction.

As long as the car is accelerating in Drive second, torque is transferred from the engine, through the planetary gearset, to the output shaft. As vehicle speed rises and the driver lets up on the accelerator pedal, the transmission generally upshifts to third. Because of this upshift, there is no engine compression braking. If the vehicle speed is not high enough for an upshift, there still is no engine compression braking because the intermediate 1-way clutch overruns and allows the sun gear to freewheel. This is a major difference between Drive second and manual second, as we will see later.

In Drive second, the front gearset simply idles. The low 1-way clutch that held the front carrier in Drive low overruns in second because the front internal gear, rotating with the output shaft, drives the front pinions clockwise. The pinions are forced to walk around the held sun gear, forcing the carrier to turn clockwise, which unlocks the 1-way clutch and lets it freewheel. The front ring gear, which is splined to

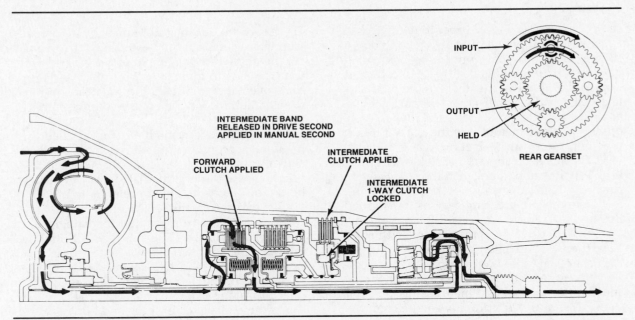

Figure 18-21. THM 400 and 425 power flow in Drive second and manual second.

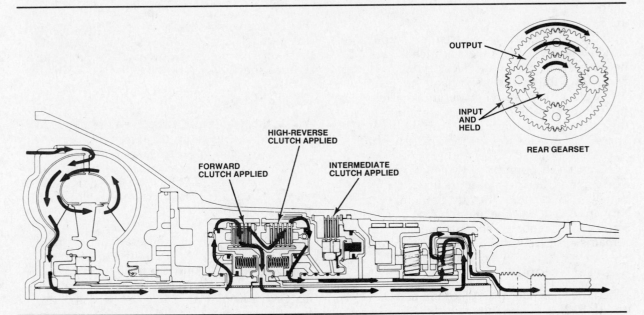

Figure 18-22. THM 400 and 425 power flow in Drive high.

the output shaft, continues to rotate with the shaft, but freewheels because no holding device is applied to the front gearset.

All of the gear reduction for second gear is produced by the rear gearset. The ring gear is the input member, the sun gear is the held member, and the carrier is the output member. There are three apply devices working in Drive second. The forward clutch is the input device, and the intermediate clutch and intermediate 1-way clutch are the holding devices.

Drive Third

As the car continues to accelerate in Drive, the transmission automatically upshifts from second to third gear. For this to happen, the forward and intermediate clutches remain applied, and the high-reverse clutch is also applied, figure 18-22.

When the high-reverse clutch is applied, the high-reverse clutch drum, sun gear shaft, and sun gear are locked to the transmission input shaft. This causes the sun gear to revolve clockwise with the input shaft at turbine speed.

The forward clutch continues to lock the mainshaft and rear ring gear to the input shaft. This means that two members of the rear planetary gearset, the sun gear and the ring gear, are locked to the input shaft, and thus to each other. With two members of the gearset locked together, the entire gearset turns as a unit.

With the rear planetary gearset locked, there is no relative motion between the sun gear, ring gear, carrier, or pinions. As a result, input torque to the sun and ring gears is transmitted through the carrier, which is splined to the output shaft, to drive the output shaft. In third gear, there is a direct 1:1 gear ratio between the input shaft speed and output shaft speed. There is no reduction or increase in either speed or torque through the transmission. Both the THM 400 and 425 provide engine compression braking in Drive third because there is a direct mechanical connection between the input shaft and the output shaft; no 1-way clutches are involved.

It is not necessary to release the intermediate clutch when the high-reverse clutch is applied because the intermediate 1-way clutch overruns and allows the sun gear shaft and sun gear to rotate clockwise as already described. This method of engaging third gear without releasing the intermediate clutch allows the THM 400 and 425 to provide a very smooth 2-3 upshift.

In third gear, all of the engine power is transmitted through the rear planetary gearset. The sun gear and ring gear are input members equally. The carrier is the output member. The forward clutch and the high-reverse clutch are input devices equally. The intermediate clutch remains applied, but is ineffective in third gear.

Manual Second

When the gear selector is placed in manual second (I, S, 2, or L2) and the vehicle is accelerated from a stop, the transmission starts in Drive low and upshifts to manual second, figure 18-21. The hydraulic system prevents an upshift into third gear, and applies the intermediate overrun band in addition to the intermediate clutch. Other than as described below, there are no differences between manual second and Drive second. The flow of power through the planetary gearsets is identical, and the same apply devices are used.

The intermediate band locks the high-reverse clutch drum, sun gear shaft, and sun gear to the case, and prevents them from turning in either direction. The intermediate clutch and intermediate 1-way clutch do this job in

Drive second, but they hold the sun gear only during acceleration; the 1-way clutch overruns during deceleration and allows the rear gearset to freewheel. In manual second, the intermediate overrun band holds the sun gear during deceleration to provide engine compression braking.

The intermediate clutch remains applied in manual second, so the intermediate 1-way clutch still holds the sun gear during acceleration. Because the intermediate band only has to hold the sun gear during deceleration, it does not have to absorb the full engine torque of acceleration. As a result, the intermediate band in the THM 400 and 425 can be smaller than the intermediate bands in other transmissions.

At any time, the driver can manually downshift from third to second by moving the gear selector from D or 3 to I, S, 2, or L2. This releases the high-reverse clutch and applies the intermediate band to provide a controlled downshift for engine braking or improved acceleration at any vehicle speed or throttle position. If the vehicle is allowed to decelerate with the gear selector in manual second, the transmission automatically downshifts to Drive low gear as speed decreases.

Manual Low

When the gear selector is placed in manual low (L, 1, or L1), the hydraulic system prevents an upshift to second or third gear, and the flow of power through the planetary gearset is basically the same as in Drive low. The only difference between Drive low and manual low is the use of one apply device.

In manual low, figure 18-20, the low-reverse band is applied to hold the front carrier. In Drive low, the 1-way clutch does this job, but it holds the carrier only during acceleration; the 1-way clutch overruns during deceleration and allows the front gearset to freewheel. In manual low, the low-reverse band holds the front carrier during acceleration *and* deceleration, which provides engine braking on deceleration. The 1-way clutch still helps hold the front carrier during acceleration in manual low, but its action is secondary as long as the low-reverse band is applied.

The driver can move the gear selector from D or 3 to L, 1, or L1 while the car is moving. At lower speeds, the transmission will downshift directly to low gear by releasing the high-reverse and intermediate clutches, and applying the low-reverse band. At higher speeds, the transmission first shifts into Drive second by releasing only the high-reverse clutch. Then, once vehicle speed drops below approximately

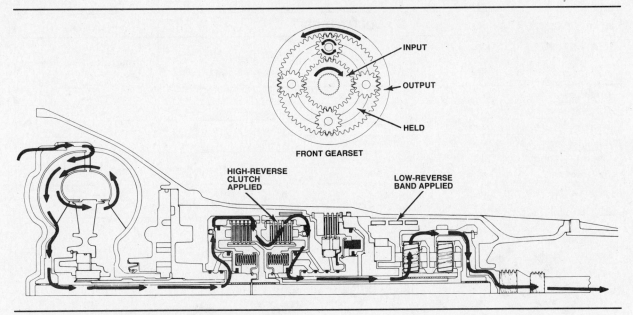

Figure 18-23. THM 400 and 425 power flow in Reverse.

40 mph (64 kph), the transmission automatically downshifts to low by releasing the intermediate clutch and applying the low-reverse band.

Reverse

When the gear selector is moved to Reverse, the input shaft is still turned clockwise by the engine. The transmission must reverse this rotation to a counterclockwise direction.

In Reverse, the forward clutch is released and the high-reverse clutch is applied, figure 18-23. The high-reverse clutch connects the sun gear shaft and sun gear to the input shaft so that the input shaft drives the sun gear clockwise. The sun gear then drives the front pinions counterclockwise.

The other apply device used in Reverse is the low-reverse band, which locks the front carrier to the case. Because the front carrier is held stationary, the counterclockwise rotation of the front pinions drives the front ring gear counterclockwise. The front ring gear is splined to the output shaft and turns it counterclockwise in gear reduction.

All gear reduction in Reverse is produced by the front planetary gearset. Although there is input to the rear gearset through the sun gear and the rear pinions, no member of the gearset is held so it simply freewheels.

In Reverse, the input member of the planetary gearset is the sun gear; the held member is the front carrier; and the output member is the front ring gear. Two apply devices are used: the high-reverse clutch is the input device, and the low-reverse band is the holding device.

THM 400 and 425 Clutch and Band Applications

Before we move on to a closer look at the hydraulic system, we should review the apply devices that are used in each of the gear ranges. Figure 18-24 is a clutch and band application chart for the THM 400 and 425 transmissions. Also, remember these facts about the apply devices used in these transmissions:

● The forward and high-reverse clutches are input devices.
● The forward clutch is applied in all forward gear ranges.
● The high-reverse clutch is applied in third and Reverse gears.
● The low 1-way, intermediate, and intermediate 1-way clutches, along with the low-reverse and intermediate bands, are holding devices.
● The low 1-way clutch locks during acceleration in Drive low; it overruns during deceleration.
● The intermediate clutch is applied in second and third gears, but is ineffective in third because the intermediate 1-way clutch overruns.
● The intermediate 1-way clutch locks during acceleration in Drive second and manual second; it overruns during deceleration.

GEAR RANGES	CLUTCHES					BANDS	
	Low-Gear 1-Way Roller	Intermediate 1-Way Sprag	Forward	High-Reverse (Direct)	Intermediate	Intermediate (Front)	Low-Reverse (Rear)
Drive Low	●		●				
Drive Second		●	●		●		
Drive High			●	●	●		
Manual Low	●		●				●
Manual Second		●	●		●	●	
Reverse				●			●

Figure 18-24. THM 400 and 425 clutch and band application chart.

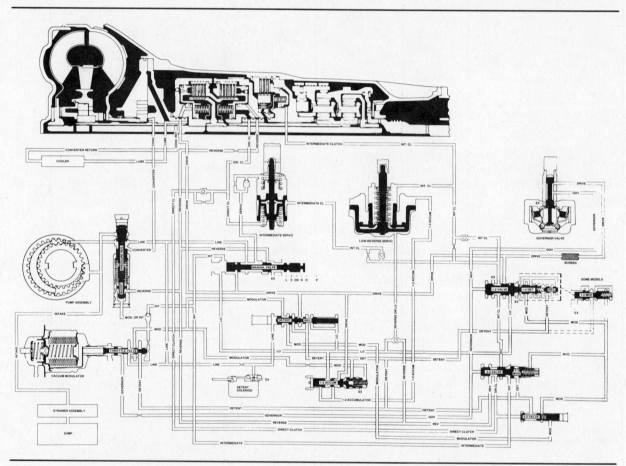

Figure 18-25. The THM 400 hydraulic system; the THM 425 system is similar.

● The low-reverse band is applied in manual low and Reverse gears.
● The intermediate band is applied in manual second gear.

THM 400 AND 425 HYDRAULIC SYSTEMS

The THM 400 and 425 hydraulic systems control shifting under varying vehicle loads and speeds, as well as in response to manual gear selection by the driver. Figure 18-25 is a diagram of the complete THM 400 hydraulic system; the THM 425 system is similar. You will find it helpful to refer back to this diagram as we discuss the various components that make up the hydraulic system. There have been a few changes in the hydraulic system over the years, and these will be detailed in the sections that follow. We will begin our study with the oil pump.

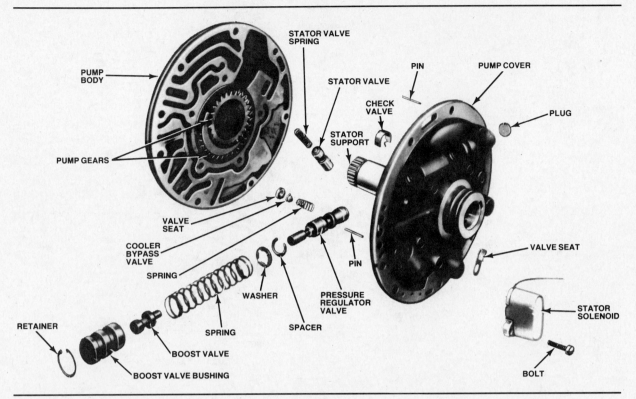

Figure 18-26. An early THM 400 oil pump from a transmission equipped with a torque converter having a variable-pitch stator. The THM 425 pump is similar, but does not contain any hydraulic control valves.

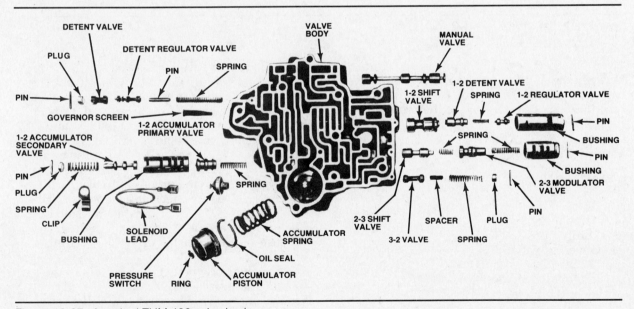

Figure 18-27. A typical THM 400 valve body.

Hydraulic Pump

All of the transmissions covered in this chapter use a gear-type oil pump, figure 18-26, that is driven by the torque converter hub. Fluid is stored in the oil pan (sump) at the bottom of the transmission, and the pump draws fluid through a filter attached to the bottom of the valve body. In the THM 400, fluid flows through a passage in the valve body and then through a passage in the front of the transmission case and bellhousing to the pump. In the

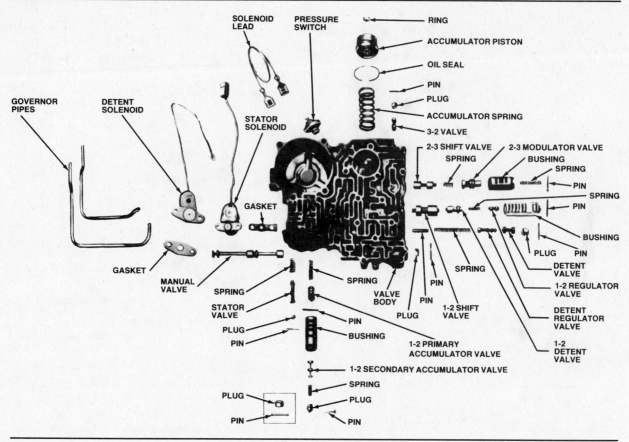

Figure 18-28. A typical THM 425 valve body.

THM 425, fluid flows through passages created between the transmission case and the pump cover to reach the oil pump. The pump in both transmissions delivers fluid to the pressure regulator valve.

Hydraulic Valves

The THM 400 and 425 contain the following valves:
1. Pressure regulator valve assembly
2. Manual valve
3. Throttle valve
4. Governor valve
5. Detent valve assembly
6. 1-2 shift valve assembly
7. 1-2 accumulator valve
8. 2-3 shift valve assembly
9. 3-2 valve
10. Stator valve (models with variable-pitch stator).

In addition, there are one or two electrically controlled solenoids that affect fluid flow:
1. Detent solenoid
2. Stator solenoid (models with variable-pitch stator).

Valve body and ball-check valves

In the THM 400 and 425, all of the valves except the pressure regulator valve, stator valve (THM 400), governor valve, and throttle valve are housed in the valve body, figures 18-27 and 18-28. The pressure regulator valve is located in the oil pump housing (THM 400) or the transmission case (THM 425). The THM 400 stator valve is located in the oil pump housing. The governor valve is located in the left rear (THM 400) or top (THM 425) of the transmission case. The throttle valve is located at the right rear of the case where it is acted on by the vacuum modulator.

The valve body bolts to the bottom of the transmission case. A separator plate and two gaskets are installed between the valve body and the case, figures 18-29 and 18-30. The valve body has passages that route fluid between the various valves. The valve body also forms the bottom of the intermediate servo and 2-3 accumulator bore, and in the THM 425, the low-reverse servo bore. The low-reverse servo in the THM 400 has a separate housing. Additional fluid passages cast into the bottom of the

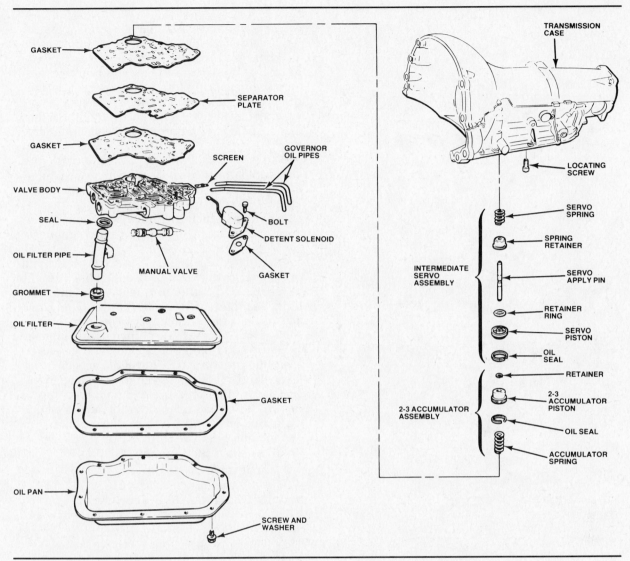

Figure 18-29. THM 400 valve body, intermediate servo, and 2-3 accumulator installation.

transmission case form the upper half of the valve body and contain the hydraulic system check balls.

Because of the large number of running changes made to these transmissions over the years, the springs and valves in some THM 400 or 425 valve bodies may not look exactly like those shown in our illustrations. It is impossible to describe every variation in valve body design and check ball location. For this reason, *it is essential that you refer to the appropriate shop manual and/or service bulletins for the year and model of transmission you are servicing.*

The oil filter attaches to the bottom of the valve body, and some THM 400 and 425 transmissions have one or two small filter screens for the governor inlet or outlet. These screens install in the oil pipes that connect the governor to the valve body.

Pressure regulator valve assembly

The pressure regulator valve assembly, figure 18-31, consists of the pressure regulator valve spool, a spring, and a boost valve. The pressure regulator valve receives mainline pressure from the oil pump through two passages. One passage is at the end of the valve opposite the boost valve, the other leads to a valley between two lands on the valve. Hydraulic pressure at the end of the valve acts to move the valve against spring force. Hydraulic pressure in the valley of the valve acts on two lands of equal diameter, and does not move the valve in either direction.

When the engine is started and mainline pressure begins to build, pressure at the end of the regulator valve moves the valve against spring force to open the converter inlet passage. Oil is then routed to the torque converter.

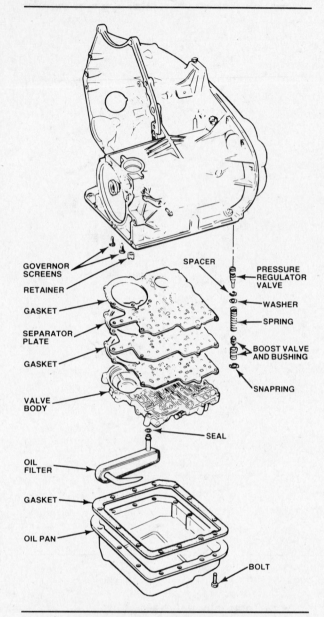

GOVERNOR SCREENS
SPACER
PRESSURE REGULATOR VALVE
RETAINER
WASHER
GASKET
SPRING
SEPARATOR PLATE
BOOST VALVE AND BUSHING
GASKET
SNAPRING
VALVE BODY
SEAL
OIL FILTER
GASKET
OIL PAN
BOLT

Figure 18-30. THM 425 pressure regulator valve and valve body installation.

Converter return oil flows to the transmission cooler, and from there to the lubrication circuit.

Some early transmissions have a cooler bypass circuit and one or two additional valves, figure 18-31. If the cooler becomes clogged and restricts fluid flow, the increasing pressure in the converter return passage opens the cooler bypass valve, allowing fluid to flow directly to the lubrication circuit. A pressure relief valve may also be used in the bypass passage to prevent excessive pressure in the converter and lubrication circuits. If the transmission has a cooler bypass circuit, the bypass and pressure relief valves are installed in the oil pump cover.

After the converter is filled, pressure build up at the regulator valve moves the valve farther against spring force. This opens a port to the pump intake passage to vent excess pressure. Hydraulic pressure and spring force then reach a balance that provides a constant mainline pressure.

The boost valve at the spring end of the pressure regulator valve is used to increase mainline pressure under certain driving conditions. To do this, hydraulic pressure is routed to the boost valve, which then moves to increase the spring force acting on the pressure regulator valve.

Modulated throttle pressure acts on the end of the boost valve to increase mainline pressure in response to higher torque loads on the drivetrain. Mainline pressure (INTERMEDIATE oil) acts on the end of the boost valve in manual second and manual low to increase mainline pressure and hold apply devices more securely. Mainline pressure (REVERSE oil) acts on a larger land of the boost valve in Reverse to increase mainline pressure and hold the apply devices more securely under the higher torque loads that are normal in that gear position.

Mainline pressure from the pressure regulator valve is routed to the:

- Manual valve
- Throttle valve
- Detent valve assembly and solenoid
- Stator valve (models with variable-pitch stator).

Manual valve

The manual valve, figure 18-31, is a directional valve moved by a mechanical linkage from the gear selector lever. The manual valve receives fluid from the pressure regulator valve, and directs mainline pressure to apply devices and other valves in the hydraulic system in order to provide both automatic and manual upshifts and downshifts. The manual valve is held in each gear position by a spring-loaded lever.

In the THM 400 and 425, five passages may be charged with mainline pressure from the manual valve. GM refers to mainline pressure from the manual valve by the name of the passage it flows through. When the gear selector is in Park, mainline pressure is blocked from entering the manual valve, and the valve does not direct pressure to any passages.

The SERVO passage is charged in Neutral, Drive, and Reverse; it routes pressure to the:

- Release side of the intermediate servo.

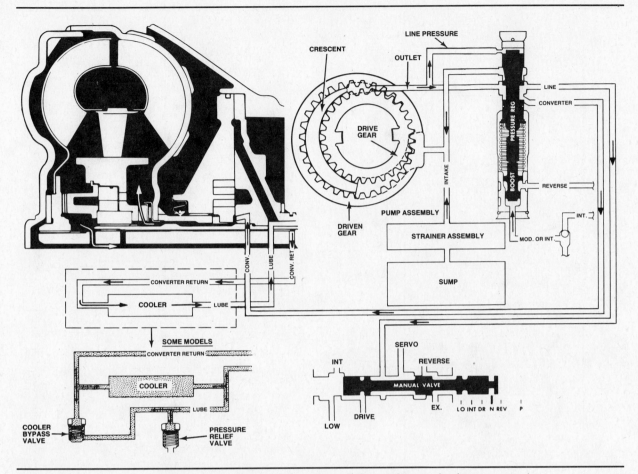

Figure 18-31. THM 400 and 425 pressure regulator valve operation and manual valve passages.

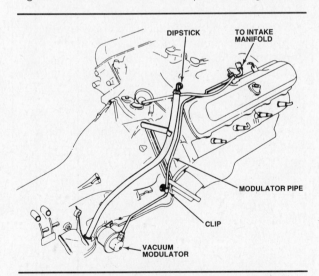

Figure 18-32. The THM 400 and 425 use a throttle valve operated by a vacuum modulator connected to engine intake manifold vacuum.

The DRIVE passage is charged in Drive, manual second, and manual low; it routes pressure to the:

- Apply area of the forward clutch
- Governor valve
- Detent valve assembly
- 1-2 accumulator valve
- 1-2 shift valve.

The INT passage is charged in manual second; it routes pressure to the:

- Boost valve
- Spring end of the 2-3 shift valve.

The LOW passage is charged in manual low; it routes pressure to the:

- Spring end of the detent valve assembly
- 1-2 accumulator valve
- 1-2 shift valve
- Apply area of the low-reverse servo.

The REVERSE passage is charged in Reverse; it routes pressure to the:

- Boost valve
- Large apply area of the high-reverse clutch
- Apply area of the low-reverse servo
- 2-3 shift valve.

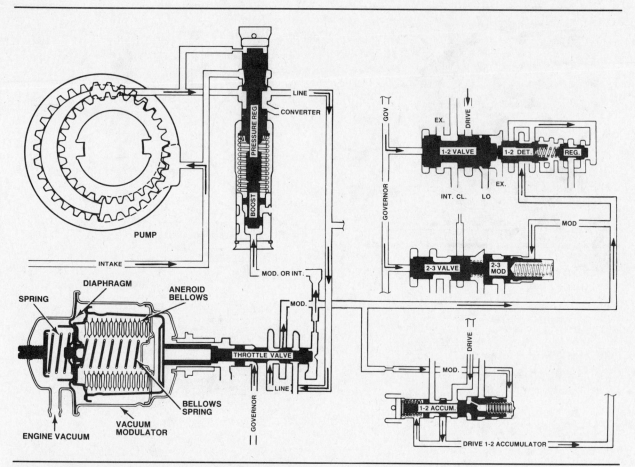

Figure 18-33. THM 400 and 425 throttle valve operation.

Throttle valve

The THM 400 and 425 transmissions develop throttle pressure using a throttle valve operated by a vacuum modulator, figure 18-32. For this reason, GM sometimes refers to this type of throttle valve as a modulator valve, and uses the term modulator pressure in place of throttle pressure. Throttle pressure indicates engine torque and helps time shifts. Without throttle pressure, the transmission would upshift and downshift at exactly the same speed for very different throttle openings and load conditions.

The vacuum modulator contains two chambers separated by a flexible rubber diaphragm, figure 18-33. The chamber nearest the throttle valve spool is open to atmospheric pressure. The chamber farthest from the valve spool is sealed and receives engine intake manifold vacuum. The vacuum chamber also contains a spring that acts on the back side of the diaphragm to move a plunger and force the throttle valve spool into its bore. As intake manifold vacuum changes, the diaphragm moves and the throttle valve travels back and forth in its bore.

When the engine is started, intake manifold vacuum is applied to the vacuum chamber of the modulator. If the vacuum level is high enough, such as when the engine is idling or under light load, atmospheric pressure against the modulator diaphragm will oppose spring pressure with enough force to move the valve spool all the way to the left, closing the mainline pressure inlet port and reducing the throttle pressure to zero.

Under most vehicle operating conditions, however, intake manifold vacuum is low enough that atmospheric pressure cannot move the throttle valve to cut off throttle pressure altogether. Instead, the valve spool only partially obstructs the mainline pressure inlet port, thus regulating throttle pressure at the outlet port. A portion of the throttle pressure is sent through an internal passage to the end of the throttle valve opposite the vacuum modulator. This pressure acts against spring force to help balance the position of the valve. As a result of these opposing forces, throttle pressure increases as the intake manifold vacuum drops, and stabilizes whenever the vacuum level remains constant.

When the engine comes under heavy load, manifold vacuum drops almost to zero. At these times, pressure is nearly equal in both chambers of the vacuum modulator, and spring force moves the valve spool all the way into its bore. Fluid then passes straight through the throttle valve, and throttle pressure equals mainline pressure.

Governor pressure is routed to the throttle valve to help regulate throttle pressure in relation to vehicle speed. This is done because less throttle pressure is needed as speed increases. Governor pressure enters between two lands of the throttle valve and acts on the larger land to move the valve against spring force. This decreases throttle pressure as vehicle speed and governor pressure increase.

Detent pressure is also routed to the throttle valve when the detent valve is open during a full-throttle downshift. If governor pressure moves the throttle valve far enough at higher speeds, detent oil enters the throttle pressure circuit to maintain a minimum of 70 psi (463 kPa) of throttle pressure. This is discussed in greater detail in the section dealing with the detent valve assembly.

Some earlier transmissions (up to 1976) use a vacuum modulator that contains an evacuated metal chamber called an aneroid bellows and a spring. The aneroid bellows expands at low atmospheric pressure (high altitude) and contracts at high atmospheric pressure (low altitude), so this modulator is often referred to as altitude compensating.

At sea level the bellows and spring are compressed and manifold vacuum is at its highest possible reading, so throttle pressure is unaffected. However, at higher altitudes, manifold vacuum is much lower, giving increased throttle pressure with a regular modulator. With the bellows type, as barometric pressure decreases at higher altitudes, decreasing the pressure on the outside of the bellows causes it to expand. The spring inside the bellows pushes on the atmospheric side of the vacuum diaphragm, thereby "compensating" for the lack of vacuum by reducing throttle pressure.

With a non-compensated modulator, the spring on the vacuum side of the diaphragm pushes on the diaphragm, moving the throttle valve in. This action increases throttle pressure at high altitude because of the lack of vacuum at these heights. Increased throttle pressure causes upshifts to be later, and more aggressive (rougher) because of boosted mainline pressure.

The altitude compensated modulator overcomes this problem because as the bellows expands, it pushes the diaphragm toward the

vacuum side as if vacuum was normal. Therefore, upshift speeds and mainline pressure will be the same whether driving in Denver, Colorado (altitude 5,280 feet), or near sea level in Los Angeles, California.

Throttle (modulator) pressure from the throttle valve is routed to the:
- Boost valve
- Spring end of the 1-2 accumulator valve
- Detent valve assembly.

Throttle pressure flows through the detent valve to the:
- 1-2 shift valve assembly
- Spring end of the 2-3 shift valve assembly
- Spring end of the 3-2 valve.

Governor

We have been discussing the development and control of throttle pressure, which is one of the two auxiliary pressures that work on the shift valves. Before we get to the shift valve operation, we must explain the governor valve (or governor), and how it creates governor pressure. Governor pressure provides a road speed signal to the hydraulic system that causes automatic upshifts to occur as road speed increases, and permits automatic downshifts with decreased road speed.

All of the transmissions covered in this chapter use a gear-driven governor, figure 18-34. The THM 400 governor mounts at the left rear of the transmission case. The THM 425 governor mounts on the top of the transmission case. A drive gear cut into the output shaft turns the driven gear on the end of the governor, which causes the assembly to rotate in the case. On the THM 425, the speedometer drive gear is mounted on the top of the governor shaft. The remainder of the governor consists of two sets of weights, two springs, and a spool valve.

When the vehicle is stopped, the mainline pressure inlet is closed or slightly open, and both the governor pressure outlet and the exhaust port are fully open. No governor pressure is developed at this time because any pressure that leaks past the valve spool or enters through the mainline inlet escapes through the exhaust port.

As the vehicle begins to move, the transmission output shaft turns the governor. Centrifugal force throws the weights outward, which levers the valve spool farther into its bore. This begins to close the exhaust port, and opens the mainline pressure inlet port farther, causing governor pressure to build.

Two sets of weights are used to increase the accuracy of pressure regulation at low vehicle speeds. The heavier primary weights move

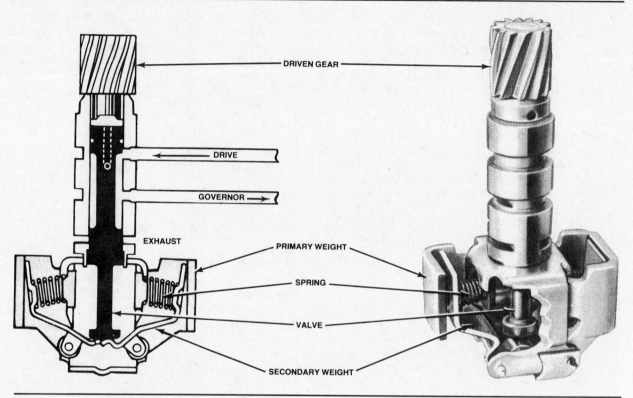

Figure 18-34. The THM 400 and 425 governor.

first, and act through springs against the lighter secondary weights that actually move the valve spool. In some transmissions, the spring tensions for the two primary weights differ to provide even smoother regulation of governor pressure. Once governor speed becomes great enough, the primary weights bottom against their stops, and the secondary weights alone act to move the valve spool. The primary weights control governor pressure up to about 720 governor rpm; above this speed, all regulation is provided by the secondary weights.

During operation, governor pressure is routed through a passage in the center of the spool valve to the end of the valve nearest the mainline pressure inlet. This pressure acts against the end of the spool, and helps oppose the lever force applied at the opposite end of the valve. As a result of these opposing forces, governor pressure increases with vehicle speed, and stabilizes whenever vehicle speed is constant.

When the vehicle reaches a certain speed, both sets of weights move out as far as possible. At this time, the secondary weights are bottomed on the primary weights, the exhaust

port is fully closed, and both the mainline pressure inlet and the governor pressure outlet are fully open. Therefore, governor pressure equals mainline pressure.

Governor pressure is directed to the:
- 1-2 shift valve
- 2-3 shift valve
- Throttle valve.

Detent valve assembly

The detent valve assembly, figure 18-35, is used to control full-throttle downshifts. The assembly consists of the detent valve spool, the detent pressure regulator valve, and a spring. The detent valve assembly is controlled by opposition between mainline pressure on one end, and spring force on the other.

Mainline pressure to the detent valve assembly is regulated by the detent solenoid, which moves a needle valve to open or close an exhaust passage. The solenoid mounts on the transmission case, inside the sump, next to the upper valve body casting. The solenoid is operated by a switch on the throttle linkage.

At any accelerator position other than wide-open throttle, the solenoid is de-energized and the exhaust passage is closed. Mainline pressure then moves the detent valve assembly

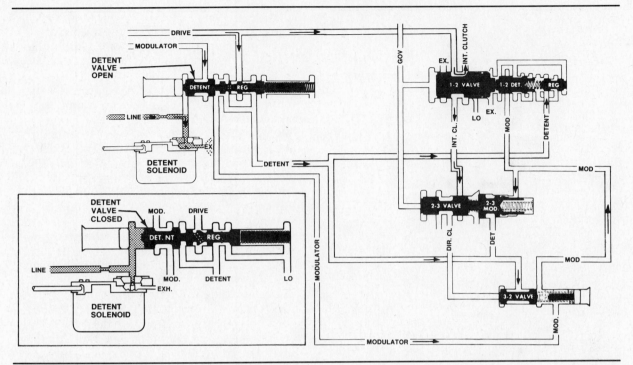

Figure 18-35. The THM 400 and 425 detent valve assembly, detent solenoid, 1-2 shift valve, 2-3 shift valve, and 3-2 valve.

against spring force to the closed position, figure 18-35 (inset). When the detent valve assembly is in this position, throttle pressure passes through the detent valve to the 1-2 shift valve, 2-3 shift valve, and 3-2 valve where it is used to control upshifts and part-throttle 3-2 downshifts.

When the throttle is opened all the way, the solenoid is energized and mainline pressure is vented through the solenoid exhaust port. Spring force then moves the detent valve assembly to the open position, figure 18-35. When the detent valve is in this position, throttle pressure is blocked from flowing through the valve. In addition, DRIVE oil from the manual valve enters the detent pressure regulator valve, where it is regulated to a maximum of 70 psi (463 kPa). Detent pressure is then routed to the:

- Throttle pressure circuit
- 1-2 shift valve assembly
- 1-2 accumulator valve
- 2-3 shift valve assembly
- 3-2 valve
- Throttle valve.

Detent pressure in the throttle pressure circuit and detent circuit opens the 3-2 valve and acts on the 1-2 and 2-3 shift valve assemblies to cause a 3-2 or 3-1 downshift, or delay an upshift, depending on the vehicle speed and the

gear engaged at the time the detent valve assembly is opened. At the same time, detent pressure at the 1-2 accumulator valve increases accumulator pressure in preparation for a smooth 1-2 full-throttle upshift.

Remember that throttle pressure is reduced by governor pressure at high speeds. This means that the transmission could reach a point where throttle pressure would drop below 70 psi (463 kPa), even at full throttle. This is prevented by the detent valve assembly which routes detent pressure into the throttle pressure circuit during a full-throttle downshift to maintain a constant 70 psi (463 kPa) of throttle pressure. Detent oil can also be routed through the throttle valve and into the throttle pressure circuit as described earlier.

1-2 shift valve assembly

The 1-2 shift valve controls upshifts from low to second, and downshifts from second to low. The assembly includes the 1-2 shift valve spool, the 1-2 detent valve, a spring, and the 1-2 regulator valve, figure 18-35. Some transmissions have a 1-piece 1-2 modulator valve that replaces the 1-2 detent valve and 1-2 regulator valve, figure 18-25. The 1-2 shift valve assembly is controlled by opposition between governor pressure on one end, and combined throttle (modulator) pressure and spring force on the other end.

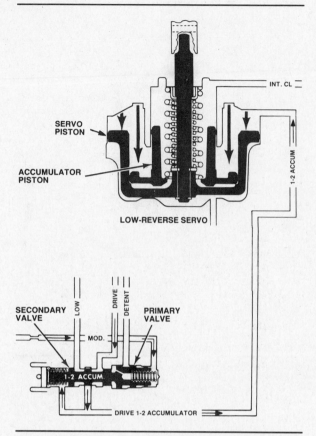

Figure 18-36. The THM 400 and 425 1-2 accumulator and 1-2 accumulator valve with the transmission in first gear.

In low gear, mainline pressure (DRIVE oil) from the manual valve flows to the 1-2 shift valve where it is blocked by the downshifted valve. The downshifted valve also opens an exhaust port in the second gear passage. Throttle pressure is applied to the 1-2 modulator or 1-2 regulator valve, which re-regulates it to a lower pressure that is still proportional to throttle pressure. This regulated throttle pressure is applied to the spring end of the 1-2 detent valve, or a land on the 1-2 modulator valve, to oppose an upshift.

When vehicle speed increases enough for governor pressure to overcome throttle pressure and spring force, the 1-2 shift valve upshifts. DRIVE oil then flows through the upshifted valve to the 2-3 shift valve where it is blocked, and into the intermediate clutch circuit where it applies the intermediate clutch to shift the transmission into second gear.

Mainline pressure in the intermediate clutch circuit is also routed through a restricting orifice to an accumulator in the low-reverse servo assembly to cushion the clutch engagement. How this is done is explained in the 1-2 accumulator valve section.

Mainline pressure in the intermediate clutch circuit is also routed to the center chamber of the intermediate servo. However, in Drive, SERVO oil from the manual valve has filled the release side of the intermediate servo. SERVO pressure combines with spring force to keep the servo released as long as the gear selector remains in Drive.

A closed-throttle 2-1 downshift occurs when governor pressure drops to a very low level (road speed decreases), and throttle pressure and spring force acting on the 1-2 detent valve or 1-2 modulator valve move the 1-2 shift valve against governor pressure.

A full-throttle 2-1 detent downshift occurs at speeds below approximately 20 mph (32 kph) when detent pressure from the detent valve flows to the 1-2 detent valve or 1-2 modulator valve. The combination of detent pressure and spring force moves the 1-2 shift valve against governor pressure and forces a downshift.

In manual low, LOW oil from the manual valve is routed to the 1-2 shift valve. This pressure keeps the 1-2 shift valve downshifted regardless of throttle opening or road speed. To increase mainline pressure in manual low, INTERMEDIATE oil from the manual valve is routed to the boost valve in the pressure regulator valve assembly.

1-2 accumulator valve

The 1-2 accumulator valve, figure 18-36, cushions the application of the intermediate clutch in relation to throttle opening. It does this by creating a 1-2 accumulator pressure that combines with spring force to control the rate at which the accumulator piston moves through its range of travel.

Early 1-2 accumulator valves consisted of a primary valve, a primary spring, a secondary valve, and a secondary spring. Late-model transmissions have a 1-piece accumulator valve in place of the separate primary and secondary valves. Both valve designs are controlled by opposition between spring force and throttle (modulator) pressure.

When the transmision is in any forward gear range, mainline pressure (DRIVE oil) is routed to the 1-2 accumulator valve. When the throttle is closed and there is no throttle pressure, primary and secondary spring force center the 1-2 accumulator valve. At this time, DRIVE pressure is trapped between two lands of the valve and cannot escape.

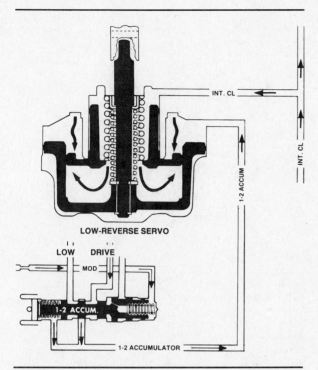

LOW-REVERSE SERVO

1-2 ACCUMULATOR

Figure 18-37. The THM 400 and 425 1-2 accumulator and 1-2 accumulator valve with the transmission in second gear.

As the throttle is opened, throttle pressure is applied to the spring end of the primary valve to move the 1-2 accumulator valve against secondary spring force. This opens the outlet port of the accumulator valve, and regulates the DRIVE pressure to create a variable 1-2 accumulator pressure. When throttle pressure is low, accumulator pressure is also low. When throttle pressure is high, accumulator pressure is also high.

To help regulate accumulator pressure, throttle pressure is routed through a port in the primary valve. This pressure then acts on the primary valve to oppose primary spring force. In addition, accumulator pressure is routed to the spring end of the secondary valve where it acts on the valve to increase secondary spring force. Opposition between the various forces and pressures causes the 1-2 accumulator pressure to increase with the throttle opening, and stabilize when the throttle opening is constant.

Accumulator pressure flows into the area above the 1-2 accumulator piston. The THM 400 and 425 1-2 accumulator is part of the low-reverse servo assembly, figure 18-36. The inside of the low-reverse servo piston forms the bore for the 1-2 accumulator piston. Accumulator pressure combines with spring force to bottom the accumulator piston in its bore.

When the 1-2 upshift occurs, figure 18-37, intermediate clutch oil is sent to the lower side of the accumulator piston where it overcomes spring force and accumulator pressure to move the accumulator piston upward in the servo bore. The amount of time it takes to stroke the accumulator piston to the top of its bore determines the amount of force with which the intermediate clutch is applied.

When accumulator pressure is low, indicating a light load on the drivetrain, intermediate clutch oil rapidly tops out the piston. This absorbs much of the pressure in the circuit and provides an easy clutch application. When accumulator pressure is high, indicating a heavy load on the drivetrain, intermediate clutch oil takes much longer to top out the piston. As a result, most of the pressure in the circuit is immediately available at the clutch to apply it much more firmly. In both cases, once the accumulator piston is topped out, full mainline pressure is present in the circuit to hold the clutch securely applied.

2-3 shift valve assembly

The 2-3 shift valve controls upshifts from second to third, and downshifts from third to second. The assembly includes the 2-3 shift valve spool, the 2-3 modulator valve, and two springs, figure 18-35. The valve is controlled by opposition between governor pressure on one end, and combined throttle (modulator) pressure and spring force on the other. Throttle pressure and spring force act on the 2-3 modulator valve to oppose an upshift.

In second gear, mainline pressure (intermediate clutch oil) from the 1-2 shift valve flows to the 2-3 shift valve where it is blocked by the downshifted valve. When vehicle speed increases enough for governor pressure to overcome throttle pressure and spring force acting on the 2-3 modulator valve, the 2-3 shift valve upshifts. Mainline pressure then flows through the upshifted valve, and into the high-reverse (direct) clutch circuit where it applies the high-reverse clutch to place the transmission in third gear.

Some of the pressure in the high-reverse clutch circuit is routed through a restricting orifice to the 2-3 accumulator, figure 18-38, which is located in the valve body and is part of the intermediate servo assembly. Remember that when the gear selector is placed in Drive, mainline pressure (SERVO oil) from the manual valve is routed to the release side of the intermediate servo. This bottoms both the servo piston and the accumulator piston. In Drive second, intermediate clutch apply oil enters the

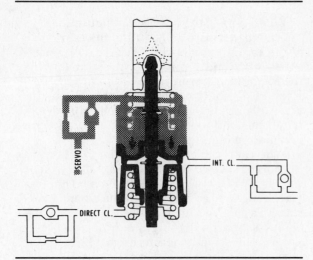

Figure 18-38. The THM 400 and 425 2-3 accumulator with the transmission in second gear.

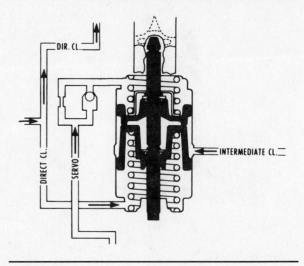

Figure 18-39. The THM 400 and 425 2-3 accumulator with the transmission in third gear.

area between the two pistons, but it cannot apply the servo because of the mainline pressure and spring force present on the release side.

In Drive third, figure 18-39, the high-reverse clutch oil routed to the lower chamber in the servo bore combines with spring force to stroke the accumulator and servo pistons upward, cushioning application of the high reverse clutch. The accumulator and servo pistons move upward until they contact the stop on the apply pin, and the spring in the release chamber. At this point, there is insufficient pressure to apply the servo, so the intermediate band remains released.

A closed-throttle 3-2 downshift occurs when governor pressure drops (road speed decreases) to the point where throttle pressure and spring force act on the 2-3 modulator valve to move the 2-3 shift valve against governor pressure.

A part-throttle 3-2 downshift occurs in some transmissions at speeds below approximately 33 mph (53 kph) when throttle pressure from the 3-2 valve flows to the 2-3 modulator valve. The combination of throttle pressure and spring force moves the 2-3 shift valve against governor pressure to cause a 3-2 downshift.

A full-throttle 3-2 detent downshift occurs at speeds below approximately 70 mph (113 kph) when detent pressure from the detent valve flows to the 2-3 modulator valve. The combination of detent pressure and spring force moves the 2-3 shift valve against governor pressure and causes a 3-2 downshift. If the vehicle speed is below approximately 20 mph (32 kph) detent pressure will also downshift the 1-2 shift valve, resulting in a 3-1 downshift.

3-2 valve
Most THM 400 and 425 transmissions have a 3-2 valve to control part-throttle 3-2 downshifts. The 3-2 valve consists of the valve spool, a spacer, and a spring, figure 18-35. The valve is controlled by opposition between mainline pressure (high-reverse clutch oil) on one end, and throttle (modulator) pressure and spring force on the other.

Throttle pressure must pass through the 3-2 valve to reach the 2-3 shift valve assembly. In Drive second, the 3-2 valve is held in the open position by spring force and throttle pressure. Throttle pressure then passes through the 3-2 valve to the 2-3 modulator valve where it acts to oppose an upshift.

When the 2-3 shift valve does upshift, it routes mainline pressure to the end of the 3-2 valve opposite the spring. This moves the 3-2 valve to the closed position, and cuts off throttle pressure to the 2-3 shift valve assembly. As long as mainline pressure is greater than combined throttle pressure and spring force, the 3-2 valve remains closed, and the driver can depress the throttle for light acceleration in third gear without causing a 3-2 downshift.

When the throttle is depressed past a certain point, the combination of increased throttle pressure and spring force will be great enough to overcome mainline pressure and open the 3-2 valve. Throttle pressure then flows through the 3-2 valve to the 2-3 modulator valve where it acts to move the 2-3 shift valve against governor pressure and force a downshift.

Some THM 400 and 425 transmissions have a 3-2 valve that is controlled by governor pressure applied to keep the valve in the closed

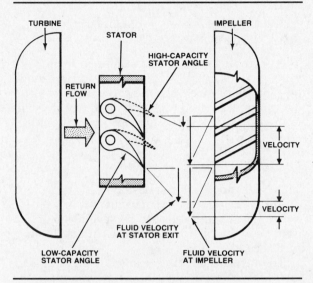

Figure 18-40. The position of variable-pitch stator blades can be changed between high and low angles.

position. At low speeds, part-throttle acceleration can force a 3-2 downshift because governor pressure is low and easily overcome. At high speeds, throttle pressure and spring force are insufficient to overcome governor pressure. Only full-throttle acceleration and the detent valve assembly can force a 3-2 downshift.

A few THM 400 and 425 transmissions do not have a 3-2 valve of any kind. In these cases, throttle pressure is always present at the 2-3 shift valve, and part-throttle acceleration in Drive third always forces a 3-2 downshift.

Stator valve and solenoid

Many pre-1968 THM 400 and 425 transmissions have a torque converter with a variable-pitch stator, figure 18-40. When the stator blades are at a high angle, converter stall speed is increased. This reduces vehicle creep at a stop, and increases torque multiplication for maximum performance. When the stator blades are at a low angle, the converter is more efficient for highway cruising.

The stator blade angle is controlled by a hydraulic piston inside the converter. The piston position is determined by the stator valve and stator solenoid, figure 18-41. In THM 400 transmissions, figure 18-26, the valve is installed in the oil pump and the solenoid bolts to the back of the pump cover. In THM 425 transmissions, figure 18-28, the valve is in the valve body and the solenoid bolts to the outside of the valve body casting.

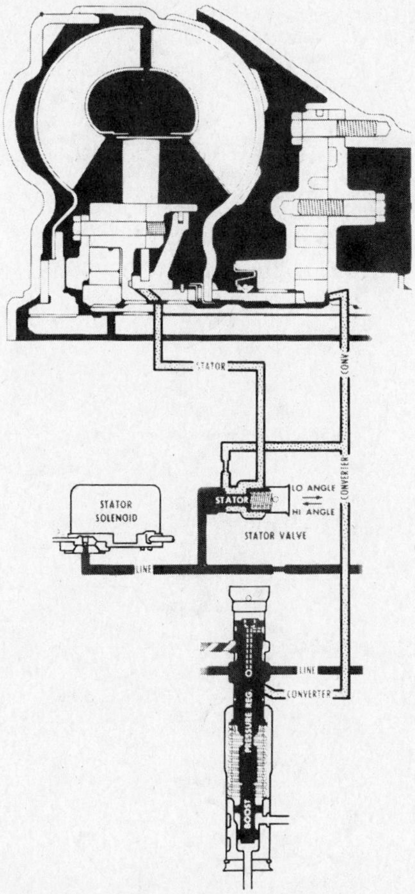

Figure 18-41. The THM 400 and 425 variable-pitch stator control circuit.

The stator solenoid, like the detent solenoid, has a needle valve that controls a mainline pressure exhaust passage. When the solenoid is de-energized, the exhaust passage is closed. Mainline pressure then moves the stator valve against spring force and allows pressure into the stator piston apply area. The piston shifts the stator blades to a low angle for efficiency at cruising speeds.

When the solenoid is energized, the exhaust passage is opened. Mainline pressure exhausts, and spring force closes the stator valve to remove pressure from the stator piston. The force of fluid circulating through the converter then moves the stator blades to a high angle for good performance or less vehicle creep at a stop.

The stator solenoid is controlled by a throttle position switch. When the throttle plates are open at an angle of more than 40 degrees, indicating hard acceleration, the solenoid is energized to shift the stator blades to a high angle

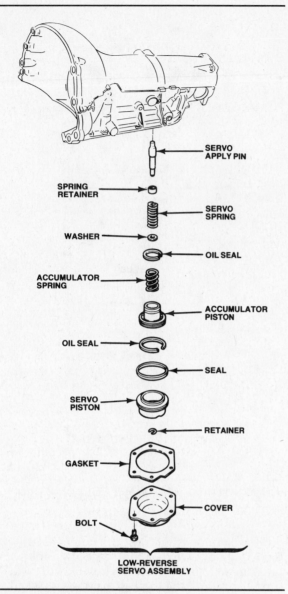

SERVO
APPLY PIN

SPRING
RETAINER

SERVO
SPRING

WASHER

OIL SEAL

ACCUMULATOR
SPRING

ACCUMULATOR
PISTON

OIL SEAL

SEAL

SERVO
PISTON

RETAINER

GASKET

COVER

BOLT

LOW-REVERSE
SERVO ASSEMBLY

Figure 18-42. The THM 400 low-reverse servo and 1-2 accumulator installation.

for maximum performance. Depending on the application, one of three different methods is used to control the stator pitch at idle for reduced vehicle creep:

• The throttle position switch may energize the solenoid when the throttle closes to an angle of 6 degrees or less.
• A switch in the brake light circuit may energize the solenoid while the brakes are applied.
• A switch controlled by the speedometer cable may energize the solenoid below a certain road speed.

MULTIPLE-DISC CLUTCHES AND SERVOS

The preceding paragraphs examined the individual valves in the THM 400 and 425 hydraulic system. Before we trace the complete hydraulic system operation in each driving range, we will take a quick look at the multiple-disc clutches and servos in these transmissions.

Clutches

THM 400 and 425 transmissions have three multiple-disc clutches. Each clutch is applied hydraulically, and released by several small coil springs when oil is exhausted from the clutch piston cavity.

The forward clutch is applied in all forward gear ranges by DRIVE oil from the manual valve. The intermediate clutch is applied in Drive second, Drive high, and manual second by intermediate clutch oil from the 1-2 shift valve. The high-reverse (direct) clutch is applied in Drive high and Reverse by direct clutch oil from the 2-3 shift valve.

The high-reverse clutch is designed so the piston has two surfaces on which hydraulic pressure acts. To engage the high-reverse clutch in high gear, pressure is applied only to the smaller inner area of the piston. When the clutch is applied in Reverse, pressure is routed to the inner area, and also to the larger outer area. This allows the clutch to develop greater holding force in Reverse to compensate for the higher torque loads that are typical in this gear.

Servos

The THM 400 and 425 transmissions have two servos. The low-reverse (rear) servo applies the low-reverse band, and also contains the 1-2 accumulator piston that cushions application of the intermediate clutch. The intermediate (front) servo applies the intermediate band.

The bore for each servo is located in the transmission case. In the THM 400, the bottom of the intermediate servo bore is formed by the valve body, figure 18-29. The low-reverse servo has a separate cover for the bottom of the bore, figure 18-42. In the THM 425, the bottom of the bore for both servos is formed by the valve body.

Band adjustment for the low-reverse band is controlled by the length of the servo rod, or apply pin, which is selected at the time of transmission assembly.

HYDRAULIC SYSTEM SUMMARY

We have examined the individual valves, clutches, and servos in the hydraulic system and discussed their functions. The following paragraphs summarize their combined operations in the various gear ranges. Refer to figure 18-25 to trace the hydraulic system operation.

Park and Neutral

In Park, oil flows from the pump through the pressure regulator valve which regulates it to create converter feed oil and mainline pressure. Converter feed oil is routed to the torque converter, and from there to the transmission cooler and lubrication circuit. Mainline pressure is routed to the: manual valve, detent solenoid, and throttle valve.

Mainline pressure at the manual valve is blocked. The detent solenoid is de-energized, and routes mainline pressure to the detent valve assembly where the pressure moves the valve to the closed position. The throttle valve regulates mainline pressure to create a throttle pressure that is proportional to throttle opening. Throttle pressure is routed to the boost valve, 1-2 accumulator valve, and detent valve.

Throttle pressure at the boost valve increases mainline pressure in proportion to throttle opening. Throttle pressure at the 1-2 accumulator valve combines with spring force to move the valve and create a 1-2 accumulator pressure that is applied to the spring side of the 1-2 accumulator piston. Throttle pressure to the detent valve is routed through the valve to the spring end of the 3-2 valve.

Throttle pressure at the 3-2 valve combines with spring force to bottom the valve. Throttle pressure then passes through the valve to the ends of the 2-3 modulator valve, and 1-2 detent and regulator valves (or 1-2 modulator valve) where it combines with spring force to oppose an upshift.

Oil flow in Neutral is the same as in Park, except that mainline pressure is routed through the manual valve and applied, as SERVO oil, to the release side of the intermediate servo. This pressure application is made in Drive as well.

All clutches and bands are released in Park and Neutral, and valve positions are determined by spring force alone. In Park, a mechanical linkage engages the parking pawl with the parking gear on the outside of the front ring gear and rear carrier assembly to lock the output shaft to the case and prevent the vehicle from being moved.

Drive Low

When the gear selector is moved to Drive, the manual valve directs mainline pressure (DRIVE oil) to the: apply area of the forward clutch, governor, detent valve, 1-2 shift valve, and 1-2 accumulator valve. The forward clutch applies to place the transmission in first gear.

The governor regulates DRIVE oil to create a governor pressure that is proportional to vehicle speed. Governor pressure is routed to the ends of the 1-2 and 2-3 shift valves, and to the throttle valve. Governor pressure at the shift valves acts to cause an upshift. Governor pressure to the throttle valve acts to reduce throttle pressure at higher vehicle speeds.

DRIVE oil at the detent valve and 1-2 shift valve is blocked. DRIVE oil at the 1-2 accumulator valve is modulated by the valve based on throttle pressure to create a 1-2 accumulator pressure that is routed to the spring side of the 1-2 accumulator piston.

In Drive low, the forward clutch is applied, and the low 1-way clutch is holding.

Drive Second

As vehicle speed and governor pressure increase, governor pressure acting on the 1-2 shift valve overcomes throttle pressure and spring force, and upshifts the valve. This allows DRIVE oil to flow through the 1-2 shift valve to the: 2-3 shift valve, apply side of the intermediate clutch, 1-2 accumulator, and the central chamber in the intermediate servo.

Pressure to the 2-3 shift valve is blocked by the downshifted valve. Pressure to the intermediate clutch applies the clutch to place the transmission in second gear. Pressure to the 1-2 accumulator moves the accumulator piston through its range of travel to smooth clutch application. Pressure to the center chamber of the intermediate servo has no effect at this time.

In Drive second, the forward clutch and the intermediate clutch are applied, and the intermediate 1-way clutch is holding.

Drive Third

As vehicle speed and governor pressure continue to increase, governor pressure overcomes throttle pressure and spring force, and upshifts the 2-3 shift valve. This allows mainline pressure to flow through the valve to the: high-reverse (direct) clutch, 2-3 accumulator, and 3-2 valve.

Pressure to the high-reverse clutch fills the smaller inner apply area to apply the clutch and place the transmission in third gear. Pressure to the 2-3 accumulator combines with spring force to move the accumulator piston through its range of travel and cushion high-reverse clutch application. Pressure to the 3-2 valve moves the valve against throttle pressure and spring force to cut off throttle pressure to the 1-2 and 2-3 shift valve assemblies.

In Drive third, the forward, intermediate, and high-reverse clutches are applied.

Drive Range Forced Downshifts

If the transmission has a 3-2 valve, a part-throttle 3-2 downshift can be forced at speeds below approximately 33 mph (53 kph). When the accelerator is depressed beyond a certain point, combined throttle pressure and spring force move the 3-2 valve against mainline pressure. This allows throttle pressure to pass through the valve and act on the 2-3 modulator valve where it combines with spring force to move the 2-3 shift valve against governor pressure and force a downshift.

A full-throttle 3-2 downshift can be forced at speeds below approximately 70 mph (113 kph). When the throttle is completely depressed, the detent solenoid is energized and mainline pressure at the detent valve assembly is exhausted, allowing spring force to move the valve. The detent regulator valve then regulates DRIVE oil to a constant detent pressure and directs it into the throttle and detent pressure circuits. Pressure in these passages then combines with spring force acting on the 2-3 modulator valve to move the 2-3 shift valve against governor pressure and force a downshift.

A full-throttle 2-1 downshift can be forced at speeds below approximately 20 mph (32 kph). When the throttle is completely depressed, the detent solenoid is energized and mainline pressure at the detent valve assembly is exhausted, allowing spring force to move the valve. The detent regulator valve then regulates DRIVE oil to a constant detent pressure and directs it into the throttle and detent pressure circuits. Pressure in these passages then combines with spring force acting on the 1-2 detent or 1-2 modulator valve to move the 1-2 shift valve against governor pressure and force a downshift.

Manual Second

When the gear selector is placed in manual second, the manual valve routes mainline pressure into the DRIVE and INTERMEDIATE circuits. DRIVE oil applies the forward clutch. INTERMEDIATE oil is routed to the: boost valve and

the spring end of the 2-3 shift valve. INTERMEDIATE oil to the boost valve increases mainline pressure. This pressure application is also made in manual low.

INTERMEDIATE oil to the spring end of the 2-3 shift valve downshifts the valve regardless of vehicle speed and governor pressure. In manual second, mainline pressure is applied to the intermediate clutch and intermediate servo through the 1-2 shift valve assembly. Because the intermediate servo release circuit is exhausted at the manual valve, the pressure in the center chamber of the servo can apply the intermediate band to provide engine braking.

In manual second, the forward clutch, intermediate clutch, and intermediate band are applied. If the vehicle speed drops low enough, throttle pressure and spring force acting on the 1-2 detent valve will move the 1-2 shift valve against governor pressure and cause a coasting downshift to Drive low.

Manual Low

When the gear selector is placed in manual low, the manual valve directs mainline pressure to the DRIVE, INTERMEDIATE, and LOW circuits. DRIVE oil applies the forward clutch. INTERMEDIATE oil is routed to the boost valve to increase mainline pressure. LOW oil is routed to the: apply side of the low-reverse servo, 1-2 accumulator valve, and detent valve assembly. LOW oil to the intermediate servo applies the low-reverse band to provide engine braking. LOW oil passes directly through the 1-2 accumulator valve to increase 1-2 accumulator pressure for a smooth band application. LOW oil moves the detent valve assembly to allow mainline pressure into the throttle and detent pressure circuits to downshift the 1-2 and 2-3 shift valves.

Once the 1-2 shift valve is in the downshifted position, spring force combined with mainline pressure in the throttle and detent circuits hold it there regardless of car speed. However, if the transmission is shifted to manual low above approximately 40 mph (64 kph), governor pressure is great enough to hold the 1-2 shift valve in its upshifted position. In this case, only the 2-3 shift valve downshifts immediately, releasing the high reverse clutch. The transmission then remains in Drive second until road speed drops below approximately 40 mph (64 kph) when the 2-1 downshift will occur.

Reverse

When the gear selector is placed in Reverse, the manual valve routes SERVO oil to the release side of the intermediate servo. The manual valve also routes REVERSE oil to the: boost valve, outer apply area of the high reverse-clutch, 2-3 shift valve, and the apply area of the low-reverse servo. REVERSE pressure to the boost valve in the pressure regulator valve increases mainline pressure.

REVERSE oil to the outer apply area of the high-reverse clutch applies the clutch. REVERSE oil to the 2-3 shift valve flows through the valve to the inner apply area of the high-reverse clutch to help apply the clutch. REVERSE oil to the low-reverse servo applies the low-reverse band.

The LOW, INTERMEDIATE, and DRIVE circuits are all exhausted in Reverse, so the forward clutch and intermediate clutch are released. Together, the high-reverse clutch and low-reverse band place the transmission in reverse gear.

Review Questions

Select the single most correct answer
Compare your answers to the correct answers on page 603

1. Mechanic A says GM introduced the THM 400 transmission in 1965.
   Mechanic B says GM introduced the THM 425 transmission in 1966.
   Who is right?
   a. Mechanic A
   b. Mechanic B
   c. Both mechanic A and B
   d. Neither mechanic A nor B

2. Mechanic A says the THM 400 and 425 are heavy-duty transmissions.
   Mechanic B says the THM 375B is a light-duty version of the THM 400 transmission.
   Who is right?
   a. Mechanic A
   b. Mechanic B
   c. Both mechanic A and B
   d. Neither mechanic A nor B

3. Mechanic A says GM refers to the THM 400 high-reverse clutch as the direct clutch.
   Mechanic B says the THM 425 high-reverse clutch drives the sun gear.
   Who is right?
   a. Mechanic A
   b. Mechanic B
   c. Both mechanic A and B
   d. Neither mechanic A nor B

4. Mechanic A says the THM 400 is sometimes called the Super Turbine 400 transmission.
   Mechanic B says the THM 425 is sometimes called the M-42 transmission.
   Who is right?
   a. Mechanic A
   b. Mechanic B
   c. Both mechanic A and B
   d. Neither mechanic A nor B

5. Mechanic A says the THM 400 oil pump is driven by the turbine shaft.
   Mechanic B says the THM 425 oil pump bolts to the transmission case.
   Who is right?
   a. Mechanic A
   b. Mechanic B
   c. Both mechanic A and B
   d. Neither mechanic A nor B

6. Mechanic A says the THM 400 transmission uses four shafts to transmit power from the torque converter to the driveshaft.
   Mechanic B says the THM 425 transmission uses five shafts to transmit power from the torque converter to the differential and final drive assembly.
   Who is right?
   a. Mechanic A
   b. Mechanic B
   c. Both mechanic A and B
   d. Neither mechanic A nor B

7. Mechanic A says the THM 400 transmission makes a single 3-1 downshift when it decelerates to a stop in Drive.
   Mechanic B says the THM 425 transmission makes a 3-2 and a 2-1 downshift when it decelerates to a stop in Drive.
   Who is right?
   a. Mechanic A
   b. Mechanic B
   c. Both mechanic A and B
   d. Neither mechanic A nor B

8. Mechanic A says the THM 400 mainshaft passes through the hollow sun gear shaft.
   Mechanic B says the THM 425 mainshaft is splined to the rear ring gear.
   Who is right?
   a. Mechanic A
   b. Mechanic B
   c. Both mechanic A and B
   d. Neither mechanic A nor B

9. Mechanic A says the THM 400 intermediate clutch is applied in Drive second.
   Mechanic B says the THM 425 intermediate clutch is applied in manual second and Drive third.
   Who is right?
   a. Mechanic A
   b. Mechanic B
   c. Both mechanic A and B
   d. Neither mechanic A nor B

10. Mechanic A says the rear ring gear is the input member of the THM 400 planetary gearset in low and second gears.
    Mechanic B says the third gear input member of the THM 425 planetary gearset turns counter-clockwise when viewed from the front of the engine.
    Who is right?
    a. Mechanic A
    b. Mechanic B
    c. Both mechanic A and B
    d. Neither mechanic A nor B

11. Mechanic A says the forward clutch drives the sun gear when the THM 400 transmission is in Reverse.
    Mechanic B says the sun gear is the input member when the THM 425 transmission is in Reverse.
    Who is right?
    a. Mechanic A
    b. Mechanic B
    c. Both mechanic A and B
    d. Neither mechanic A nor B

12. Mechanic A says GM uses the term "modulator pressure" to refer to throttle pressure in a THM 400 or 425 transmission.
    Mechanic B says throttle pressure in a THM 400 or 425 transmission is controlled by a cable.
    Who is right?
    a. Mechanic A
    b. Mechanic B
    c. Both mechanic A and B
    d. Neither mechanic A nor B

13. Mechanic A says the THM 400 pressure regulator valve is located in the oil pump housing. Mechanic B says the THM 425 pressure regulator valve is located in the transmission case. Who is right?
    a. Mechanic A
    b. Mechanic B
    c. Both mechanic A and B
    d. Neither mechanic A nor B

14. Mechanic A says full-throttle downshifts in the THM 400 are controlled by a cable. Mechanic B says the THM 425 detent valve opens when a detent solenoid exhausts mainline pressure. Who is right?
    a. Mechanic A
    b. Mechanic B
    c. Both mechanic A and B
    d. Neither mechanic A nor B

15. Mechanic A says THM 400 low-reverse band adjustment is controlled by a threaded rod that extends through the transmission case. Mechanic B says THM 425 intermediate band adjustment is controlled by the length of the servo rod or apply pin. Who is right?
    a. Mechanic A
    b. Mechanic B
    c. Both mechanic A and B
    d. Neither mechanic A nor B

16. Mechanic A says the THM 425 transmission never used a torque converter clutch. Mechanic B says some THM 400 transmissions have used a TCC since 1985. Who is right?
    a. Mechanic A
    b. Mechanic B
    c. Both mechanic A and B
    d. Neither mechanic A nor B

17. Mechanic A says the THM 400 applies the high-reverse band and the low-reverse clutch in Reverse. Mechanic B says THM 425 main-line pressure is routed to the large apply area of the high-reverse clutch in Reverse. Who is right?
    a. Mechanic A
    b. Mechanic B
    c. Both mechanic A and B
    d. Neither mechanic A nor B

18. Mechanic A says the THM 400 3-2 valve is used to control full-throttle 3-2 downshifts. Mechanic B says any THM 400 transmissions without a 3-2 valve cannot provide a part throttle 2-3 downshift. Who is right?
    a. Mechanic A
    b. Mechanic B
    c. Both mechanic A and B
    d. Neither mechanic A nor B

19. Mechanic A says THM 400 variable-pitch stator blades are at a high angle when the car is stopped with the engine idling. Mechanic B says the THM 425 torque converter with a variable pitch stator was discontinued after 1967. Who is right?
    a. Mechanic A
    b. Mechanic B
    c. Both mechanic A and B
    d. Neither mechanic A nor B

20. Mechanic A says THM 400 and 425 transmissions have two servos, each containing an accumulator. Mechanic B says THM 400 and 425 transmissions have four multiple-disc clutches. Who is right?
    a. Mechanic A
    b. Mechanic B
    c. Both mechanic A and B
    d. Neither mechanic A nor B

19

The Turbo Hydra-matic 440-T4 Transaxle

HISTORY AND MODEL VARIATIONS

The General Motors (GM) Turbo Hydra-matic (THM) 440-T4 transaxle is designed for use with medium-displacement, transverse-mounted engines in FWD applications. The THM 440-T4, figure 19-1, was introduced in selected 1984 Chevrolet Celebrity and Buick Century models. In 1985, the Oldsmobile Cutlass also began to use the 440-T4, as did the all-new FWD Buick Park Avenue, Cadillac Sedan and Coupe DeVille, and the Oldsmobile Ninety-Eight. Since that time, the THM 440-T4 has been used in most of GM's domestically built intermediate and full-size FWD passenger cars.

The name of the THM 440-T4 indicates several facts about this transaxle. The number "440" means it is intended for use with engines that put out no more than 440 Newton-meters (325 ft-lb) of torque. The "T" signifies that this is a FWD transaxle. The "4" refers to the fact that the THM 440-T4 is a 4-speed gearbox. Some GM service manuals and parts catalogs refer to the THM 440-T4 as the ME-9 transaxle. A heavy-duty version of the THM 440-T4, used in the Cadillac Allanté, is called the F-7.

The THM 440-T4 uses four multiple-disc clutches, a 1-way roller clutch, a 1-way sprag clutch, and two bands to provide its various gear ratios. In place of final drive spur gears, like those in the Chrysler Torqueflite transaxle or the Ford ATX, the THM 440-T4 uses a simple planetary gearset for its final drive. All models of this transaxle are fitted with a torque converter clutch to improve fuel economy. Cadillac versions of the THM 440-T4 have a viscous converter clutch for smoother operation. Like all of GM's newer design transmissions and transaxles, the THM 440-T4 is built to metric specifications.

The THM 440-T4 torque converter does not drive the transaxle geartrain directly. Instead, the converter drives two sprockets and a chain that transfer engine power to the geartrain positioned behind and below the crankshaft centerline. The chain is located at the end of the transaxle case on the driver's side of the car. The THM 440-T4 oil pump and valve body also mount on the driver's side of the transaxle, rather than at the bottom in the sump. The oil pump body and cover bolt to the valve body; the valve body bolts to a channel plate; and the channel plate bolts to the transaxle case. All of these components are housed under a stamped steel chain cover that bolts to the transaxle case and channel plate.

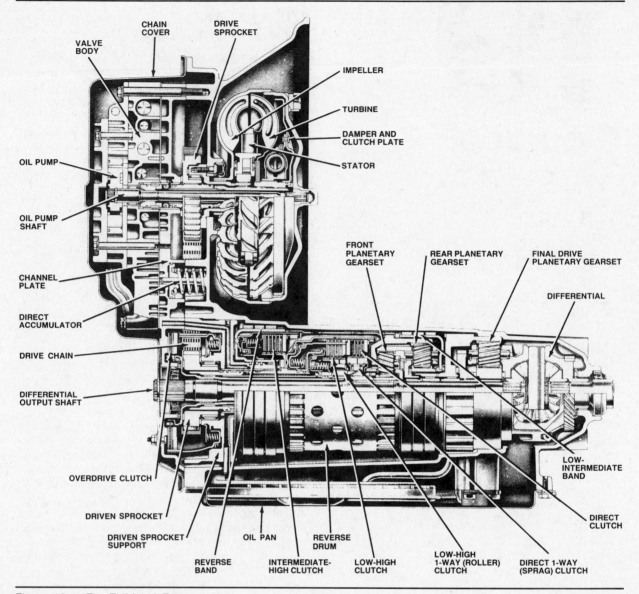

Figure 19-1. The THM 440-T4 transaxle.

Fluid Recommendations

Like all GM automatic transmissions and transaxles, the THM 440-T4 requires DEXRON®-IID fluid.

Transaxle Identification

Identifying the THM 440-T4 transaxle is relatively easy because, as of the 1988 model year, it is the only 4-speed automatic used in FWD GM cars with transverse engines. The THM 440-T4 can also be identified by the distinctive shape of its case. As required by Federal regulations, all GM transaxles carry the serial number of the vehicle in which they were originally installed.

In addition to the transaxle model, you must know the exact assembly part number and other identifying characters in order to obtain proper repair parts. On the THM 440-T4, this information is stamped on a tag riveted to the right rear of the case in line with the right-side halfshaft. The coding that appears on the tag varies; figure 19-2 shows a typical THM 440-T4 identification tag, and explains the meaning of the data stamped on it.

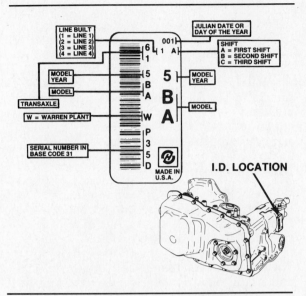

Figure 19-2. The THM 440-T4 identification tag.

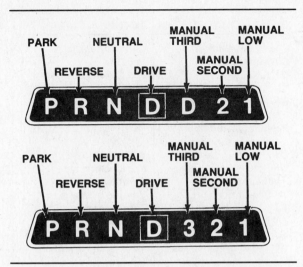

Figure 19-3. THM 440-T4 gear selector positions.

Major Changes

The major changes that GM has made in the THM 440-T4 transaxle are as follows:

1984 • Torque converter clutch screen revised.
1985 • Servo spring retainer revised.

GEAR RANGES

Depending on the GM division and the particular application, there are minor variations in the exact labeling of the gear selector used with the THM 440-T4 transaxle. However, all of the gear selectors have seven positions, figure 19-3:

P — Park
R — Reverse
N — Neutral
Ⓓ — Overdrive
D or 3 — Manual third
2 — Manual second
1 — Manual low.

Gear ratios for the THM 440-T4 are:

• First (low) — 2.92:1
• Second (intermediate) — 1.57:1
• Third (direct drive) — 1.00:1
• Fourth (overdrive) — 0.70:1
• Reverse — 2.38:1.

Park and Neutral

The engine can be started only when the transaxle is in Park or Neutral. In these gear selector positions, the direct clutch is applied and turns the front sun gear. However, because no other member of the planetary gearset is held, there is no flow of power through the transaxle. In Neutral, the differential and final drive assembly is free to turn, and the car can be pushed or pulled. In Park, a mechanical pawl engages the parking gear splined to the final drive sun gear shaft. This locks the final drive planetary gearset and prevents the vehicle from moving.

Overdrive

Overdrive (Ⓓ) is the normal operating range for the THM 440-T4 transaxle. When the vehicle is stopped and the gear selector is placed in Overdrive, the transaxle is automatically in low gear. As the vehicle accelerates, the transaxle automatically upshifts to second, third, and then to fourth gear. The speeds at which the upshifts occur vary, and are controlled by vehicle speed and throttle opening. Depending on vehicle operating conditions, the torque converter clutch may be applied in second, third, and fourth gears.

The driver can force downshifts in Overdrive by opening the throttle partially or completely. The speeds at which downshifts occur vary for different engine, axle ratio, tire size, and vehicle weight combinations. Except at extremely high speeds, the driver can force a part-throttle 4-3 downshift by opening the throttle beyond a certain point. Below approximately 65 mph (105 kph), the driver can force a 3-2 detent downshift by fully opening the throttle. A full-throttle 3-1 or 2-1 detent downshift can be forced below approximately 35 mph (56 kph). When the vehicle decelerates to a stop in Overdrive, the transaxle makes three closed throttle downshifts.

GENERAL MOTORS TERM	TEXTBOOK TERM
Clutch housing	Clutch drum
Converter pump	Impeller
Detent downshift	Full-throttle downshift
Internal gear	Ring gear
Reverse reaction drum	Reverse drum
Input planetary gearset	Front planetary gearset
Reaction planetary gearset	Rear planetary gearset
Control valve assembly	Valve body
Line pressure	Mainline pressure
Modulator valve	Throttle valve
Modulator pressure	Throttle pressure
T.V. limit valve	Throttle limit valve
T.V. plunger	Throttle plunger
M.T.V. valve	Modulated throttle valve
1-2, 2-3, 3-4 T.V. valves	1-2, 2-3, 3-4 throttle valves
Fourth clutch	Overdrive clutch
Third clutch	Low-high clutch
Third roller clutch	Low-high 1-way clutch
Second clutch	Intermediate-high clutch
Input sprag	Direct 1-way clutch
Input clutch	Direct clutch
1-2 band	Low-intermediate band
1-2 servo	Low-intermediate servo

Figure 19-4. THM 440-T4 transaxle nomenclature table.

Manual Third

When the vehicle is stopped and the gear selector is placed in manual third (D or 3), the transaxle is automatically in low gear. As the vehicle accelerates, the transaxle automatically upshifts to second, and then to third gear. It does not upshift to fourth. The speeds at which the upshifts occur vary, and are controlled by vehicle speed and throttle opening. Depending on vehicle operating conditions, the torque converter clutch may be applied in second and third gears.

The driver can force downshifts in manual third by opening the throttle completely. The speeds at which downshifts occur vary for different engine, axle ratio, tire size, and vehicle weight combinations. Below approximately 65 mph (105 kph), the driver can force a 3-2 detent downshift by fully opening the throttle. A full-throttle 3-1 or 2-1 detent downshift can be forced below approximately 35 mph (56 kph).

The driver can manually downshift from fourth to third by moving the selector from Ⓓ to D or 3. In this case, the transaxle shifts to third gear immediately, regardless of road speed or throttle position. Because different apply devices are used in manual third than are used in Overdrive third, this gear range can be used to provide engine compression braking, or gear reduction for hill climbing, pulling heavy loads, or operating in congested traffic.

When the vehicle decelerates to a stop in manual third, the 3-2 coast valve provides a direct, non-orificed exhaust for third clutch oil allowing a 3-2 coast down shift. As the vehicle continues to slow, and governor pressure continues to drop, the transaxle makes a 2-1 shift.

Manual Second

When the vehicle is stopped and the gear selector is placed in manual second (2), the transaxle is automatically in low gear. As the vehicle accelerates, the transaxle automatically upshifts to second; it does not upshift to third or fourth. The speed at which the upshift occurs will vary, and is controlled by vehicle speed and throttle opening. Depending on vehicle operating conditions, the torque converter clutch may be applied in second gear.

A full-throttle 2-1 detent downshift can be forced below approximately 35 mph (56 kph) if the driver depresses the accelerator completely. The transaxle automatically downshifts to first as the vehicle decelerates to a stop.

The driver can manually downshift from fourth or third to second by moving the selector from Ⓓ, D, or 3 to 2. In this case, the transaxle shifts to second gear immediately, regardless of road speed or throttle position. Manually downshifting to second gear provides engine compression braking, or gear reduction for hill climbing, pulling heavy loads, or operating in congested traffic.

Manual Low

When the vehicle is stopped and the gear selector is placed in manual low (1), the car starts and stays in first gear; it does not upshift to second, third, or fourth. Because different apply devices are used in manual low than are used in Overdrive low, this gear range can be used to provide engine compression braking, or gear reduction for hill climbing, pulling heavy loads, or operating in congested traffic.

If the driver shifts to manual low at high speed, the transaxle will not shift directly to first gear. To avoid engine or driveline damage, the transaxle first downshifts to second, and then automatically downshifts to low when vehicle speed drops below approximately 40 mph (64 kph). The transaxle then remains in low gear until the gear selector is moved to another position.

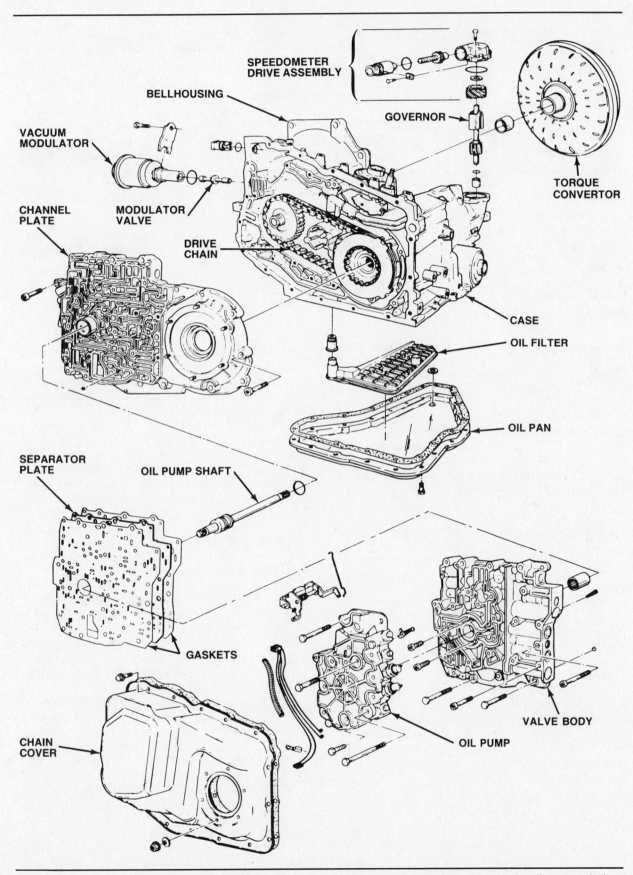

Figure 19-5. An exploded view of the THM 440-T4 case including the channel plate, valve body, oil pump, chain cover, and oil pan.

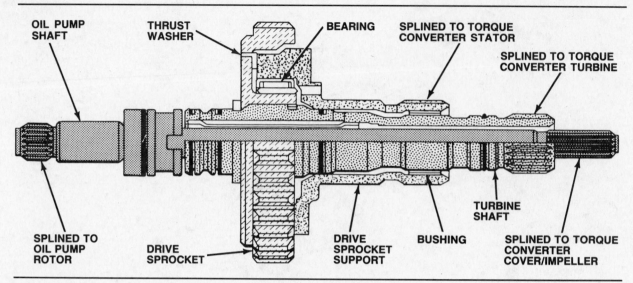

Figure 19-6. A detailed view of the THM 440-T4 oil pump shaft, turbine shaft, drive sprocket support, and drive sprocket.

Reverse

When the gear selector is moved to R, the transaxle shifts into Reverse. The transaxle should be shifted to Reverse only when the vehicle is standing still. The lockout feature of the shift gate should prevent accidental shifting into Reverse while the vehicle is moving forward. However, if the driver intentionally moves the selector to R while moving forward, the transaxle *will* shift into Reverse, and damage may result.

GM TRANSAXLE NOMENCLATURE

For the sake of clarity, this text uses consistent terms for transaxle parts and functions. However, the various manufacturers may use different terms. Therefore, before we examine the buildup of the THM 440-T4, and the operation of its geartrain and hydraulic system, you should be aware of the unique names that GM uses for some transaxle parts and operations. This will make it easier in the future if you should look in a GM manual and see a reference to an "internal gear". You will realize that this is simply a planetary gearset ring gear and is similar to the same part in any other transaxle. Figure 19-4 lists GM nomenclature and the textbook terms for some transaxle parts and operations.

THM 440-T4 BUILDUP

The THM 440-T4 case, figure 19-5, is a 1-piece aluminum casting with the channel plate, valve body, oil pump, chain cover, and oil pan bolted

to it. The bellhousing is integral with the transaxle case. The governor mounts in an opening on the top, back, right-hand side of the case. The vacuum modulator mounts at the upper, front, left-hand side of the case.

The torque converter in the THM 440-T4 bolts to the engine flexplate, which has the starter ring gear around its outer diameter. The torque converter is a welded assembly and cannot be disassembled. The THM 440-T4 converter is a conventional 3-element design (impeller, turbine, and stator) with the addition of a hydraulically applied clutch. Cadillac THM 440-T4 converters contain a viscous converter clutch for smoother operation. This type of clutch is described in detail in Chapter 7.

The torque converter is installed into the transaxle by sliding it onto the stationary drive sprocket support, figure 19-6, which is bolted to the case. The splines in the stator 1-way clutch engage the splines on the drive sprocket support. The splines in the turbine engage the hollow turbine shaft. The splines in the converter cover engage the oil pump shaft, which passes through the inside of the turbine shaft to drive the oil pump. The turbine and oil pump shafts are supported on bushings and needle bearings. The hub at the rear of the torque converter housing rides on a bushing mounted in the transaxle case.

The THM 440-T4 uses a variable-displacement vane-type oil pump mounted on the valve body, figure 19-5. The oil pump and valve body, together with the channel plate, bolt to the left side of the transaxle case. Because the torque converter cover is bolted to the engine

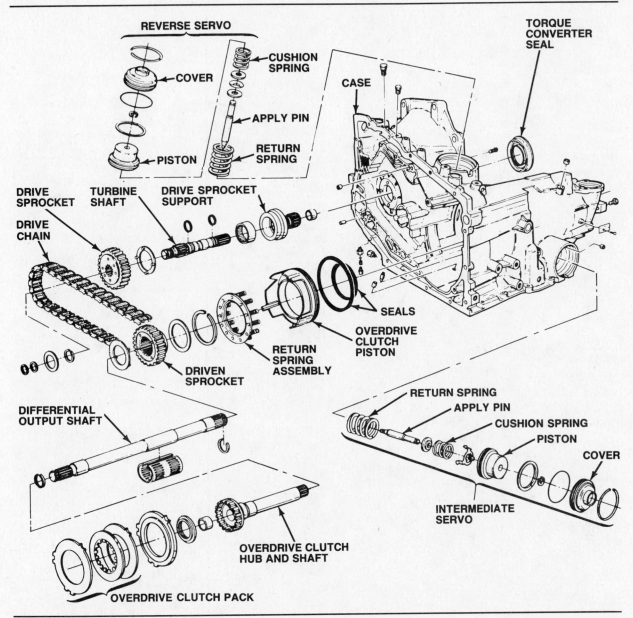

Figure 19-7. An exploded view of the THM 440-T4 drive sprocket, drive chain, driven sprocket, overdrive clutch assembly, reverse servo, and intermediate servo.

flexplate, the oil pump shaft turns and the oil pump delivers fluid whenever the engine is running. The pump draws fluid from the sump through a filter, and routes the fluid through passages in the valve body, channel plate, and transaxle case to various valves in the hydraulic system.

The turbine shaft turns the drive sprocket, figure 19-7, which turns the drive chain. The drive chain then turns the driven sprocket and input shaft to provide power to the transaxle geartrain. In this manner, torque from the engine makes two 90-degree turns before entering the geartrain. To obtain the best balance of fuel economy and performance, the numbers of teeth on the drive and driven sprockets vary for different applications.

The overdrive clutch, figure 19-7, is located outboard of the driven sprocket. The outside of each overdrive clutch steel disc is splined to the channel plate, which bolts to the case. The inside of each friction disc is splined to the outside of the overdrive clutch hub, which is welded to the end of the overdrive shaft. The overdrive shaft extends through the driven sprocket into the transaxle case where it is

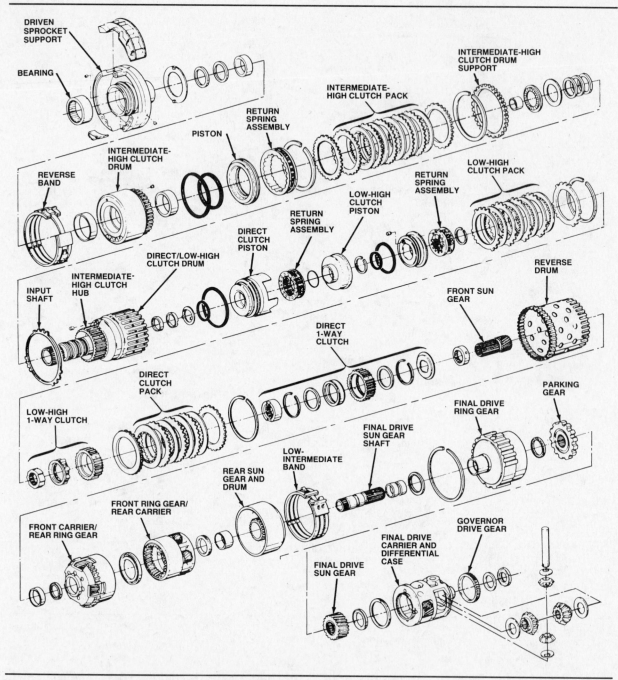

Figure 19-8. An exploded view of the THM 440-T4 clutches, 1-way clutches, bands, compound planetary geartrain, and final drive gearset and differential assembly.

splined to the front sun gear of the compound planetary gearset. When the overdrive clutch is applied in fourth gear, it holds the front sun gear stationary.

Figure 19-8 shows a general overall view of the remainder of the THM 440-T4 geartrain. As we complete the buildup of this transaxle, refer to figure 19-8 and the detailed figures for each specific assembly to identify the components

being discussed. To simplify the buildup and the power flow descriptions later in the chapter, all references to rotation are made as if you are looking at the geartrain from the front of the engine, that is, the right side of the car.

The driven sprocket turns clockwise and is splined to the input shaft, figure 19-9. The input shaft is one assembly with the direct/low-high clutch drum and the intermediate-high

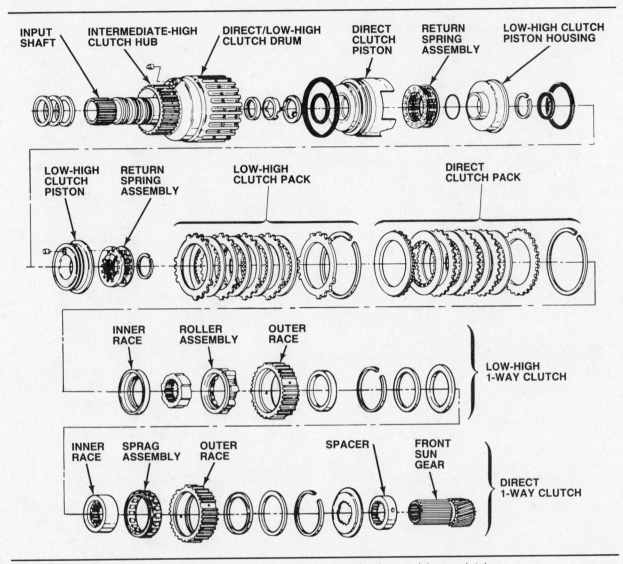

Figure 19-9. The THM 440-T4 direct and low-high multiple-disc clutches and 1-way clutches.

clutch hub. The outside of each direct clutch steel disc is splined to the inside of the direct/low-high clutch drum. The inside of each friction disc is splined to the outer race of the direct 1-way (sprag) clutch. The inner race of the direct 1-way clutch is splined to the front sun gear. When the direct clutch is applied, the input shaft drives the front sun gear clockwise through the direct 1-way clutch. If the front sun gear turns clockwise faster than the input shaft, the direct 1-way clutch overruns.

The outside of each low-high clutch steel disc is splined to the inside of the direct/low-high clutch drum, figure 19-9. The inside of each friction disc is splined to the outer race of the low-high 1-way (roller) clutch. The inner race of the low-high 1-way clutch is splined to the front sun gear. When the low-high clutch

is applied, the low-high 1-way clutch prevents the front sun gear from turning clockwise faster than the input shaft. Figure 19-10 shows how the direct and low-high 1-way clutches combine to lock the front sun gear to the input shaft when both the direct and low-high clutches are applied.

The inside of each intermediate-high clutch friction disc is splined to the outside of the intermediate-high clutch hub, which turns clockwise with the input shaft, figure 19-9. The outside of each steel disc is splined to the inside of the intermediate-high clutch drum, figure 19-11. The intermediate-high clutch drum is splined to the reverse drum, figure 19-12, which is splined to the front carrier. The front carrier is one assembly with the rear ring gear. When

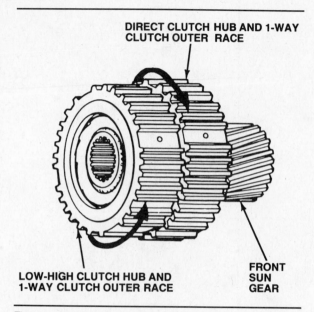

DIRECT CLUTCH HUB AND 1-WAY CLUTCH OUTER RACE

LOW-HIGH CLUTCH HUB AND 1-WAY CLUTCH OUTER RACE

FRONT SUN GEAR

Figure 19-10. The direct and low-high 1-way clutches are both splined to the front sun gear. Each clutch allows the sun gear to freewheel in the direction of the arrow, but locks when the sun gear attempts to turn in the opposite direction.

the intermediate-high clutch is applied in second, third, and fourth gears, the intermediate-high clutch drum, reverse drum, and front carrier/rear ring gear are all driven clockwise by the input shaft.

The reverse band is wrapped around the outside of the intermediate-high clutch drum, figure 19-11. When the reverse band is applied, the intermediate-high clutch drum, reverse drum, and front carrier/rear ring gear are all held stationary. This occurs only in reverse gear.

The front sun gear, which as you will recall is splined to the overdrive shaft and the inner races of the direct and low-high 1-way clutches, meshes with the front pinions, figure 19-12. The front pinions turn on shafts in the front carrier and mesh with the front ring gear, which is one assembly with the rear carrier. The front ring gear/rear carrier is splined to the final drive sun gear shaft, and is always the output member of the compound planetary gearset. The parking gear is also splined to the final drive sun gear shaft.

The rear pinions turn on shafts in the rear carrier, and mesh with the front carrier/rear ring gear, figure 19-12. The rear pinions also mesh with the rear sun gear, which is attached to a drum. The rear sun gear drum is surrounded by the low-intermediate band. When the low-intermediate band is applied in first and second gears, the rear sun gear is held stationary.

The final drive and differential assembly, figure 19-13, transmits power from the transaxle gearset to the halfshafts of the vehicle. In place of conventional final drive gears, the THM 440-T4 uses a simple planetary gearset with one sun gear, one carrier, four pinions and one ring gear. The sun gear is driven by the final drive sun gear shaft, which is splined to the front ring gear/rear carrier of the transaxle gearset. The sun gear meshes with the final drive pinions, which turn on shafts in the final drive carrier. The carrier is part of the differential case, which has the governor drive gear pressed onto it. The final drive pinions mesh with the final drive ring gear, which is splined to the transaxle case.

The differential case contains four bevel gears: two pinion gears and two side gears. The pinion gears are installed on a pinion shaft that is retained by a pin in the differential case. The pinion gears transmit power from the case to the side gears. A male spline on the right-side halfshaft enters the transaxle to engage one of the side gears. The other side gear is splined to the differential output shaft which delivers power, through a male spline, to the left-side halfshaft. A bushing and seal for the halfshaft is installed at each side of the transaxle case.

THM 440-T4 POWER FLOW

Now that we have taken a quick look at the general way in which the THM 440-T4 is assembled, we can trace the flow of power through the apply devices and the compound planetary gearset in each of the gear ranges.

Park and Neutral

When the gear selector is in Park or Neutral, figure 19-14, the torque converter impeller/cover and oil pump shaft rotate clockwise at engine speed. The oil pump shaft drives the oil pump. The converter turbine, turbine shaft, drive chain, drive and driven sprockets, and input shaft also rotate clockwise, but at less than engine speed. In Park and Neutral, the converter clutch is always released.

The input shaft drives the direct clutch drum clockwise. The direct clutch is applied, and the direct 1-way clutch drives the front sun gear clockwise. The front sun gear drives the front pinions counterclockwise, and the pinions attempt to drive the front ring gear counterclockwise. However, the front ring gear/rear carrier is splined to the final drive sun gear shaft and held by the driveline. Therefore, the front pinions walk clockwise around the inside of the front ring gear, taking the front carrier/rear ring gear with them.

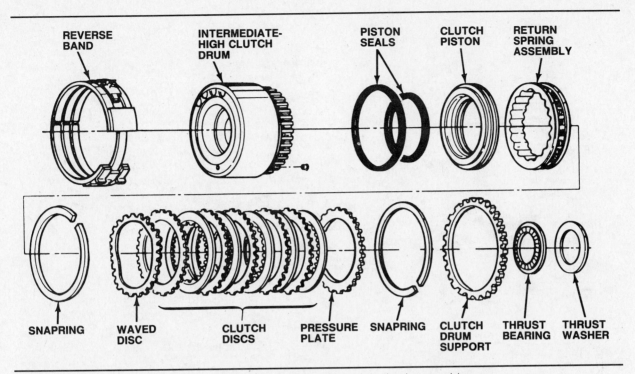

Figure 19-11. The THM 440-T4 reverse band and intermediate-high clutch assembly.

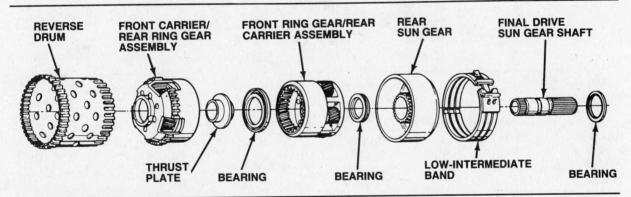

Figure 19-12. The THM 440-T4 reverse drum, compound planetary gearset, low-intermediate band, and final drive sun gear shaft.

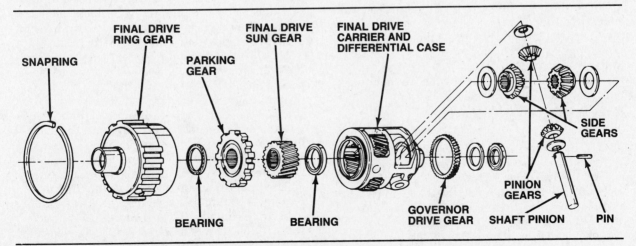

Figure 19-13. The THM 440-T4 final drive and differential assembly.

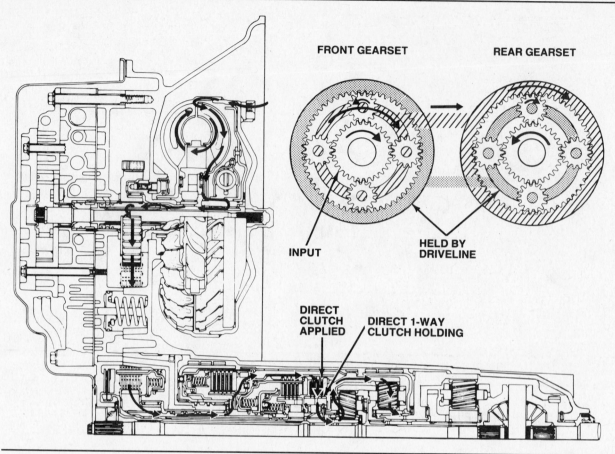

FRONT GEARSET

REAR GEARSET

INPUT

HELD BY
DRIVELINE

DIRECT
CLUTCH
APPLIED

DIRECT 1-WAY
CLUTCH HOLDING

Figure 19-14. THM 440-T4 power flow in Park and Neutral.

The rear ring gear then drives the rear pinions clockwise on the stationary front ring gear/rear carrier, and the rear pinions drive the rear sun gear and its drum counterclockwise. However, because no member of the planetary gearset is held by an apply device, the compound planetary gearset idles, and there is no output motion.

In Neutral, the final drive sun gear is free to rotate in either direction; thus, the vehicle can be pushed or pulled. In Park, a mechanical pawl engages the parking gear, which is splined to the final drive sun gear shaft. Because the final drive ring gear and sun gear are then both held, the final drive gearset locks up and prevents the vehicle from moving.

Overdrive Low

When the car is stopped and the gear selector is moved to Overdrive (Ⓓ), the transaxle is automatically in low gear, figure 19-15. As in Park and Neutral, the direct clutch is applied and transmits engine power from the driven sprocket to the front sun gear which turns

clockwise. The low-intermediate band is also applied to hold the rear sun gear stationary.

Before the car begins to move, the final drive carrier is held stationary by the driveline. This also holds the final drive sun gear, sun gear shaft, and front ring gear/rear carrier stationary. As a result, the clockwise rotation of the front sun gear drives the front pinions counterclockwise, and the pinions walk clockwise around the inside of the front ring gear, taking the front carrier/rear ring gear with them.

Clockwise rotation of the rear ring gear turns the rear pinions clockwise. The rear pinions then walk clockwise around the held rear sun gear, taking the front ring gear/rear carrier with them. Because the low-intermediate band holds the rear sun gear more securely than the driveline can hold the front ring gear/rear carrier, the front ring gear/rear carrier and final drive sun gear shaft rotate clockwise in gear reduction.

The final drive sun gear turns the final drive pinions counterclockwise, figure 19-16. The final drive ring gear is splined to the transaxle case, and is always held from rotating. Therefore, the pinions walk around the inside of the

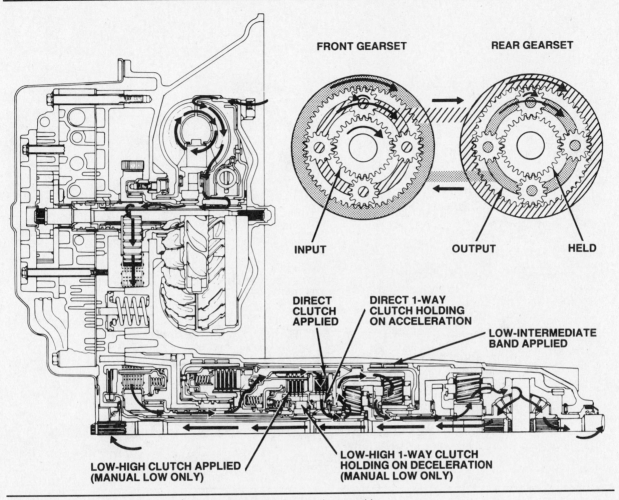

Figure 19-15. THM 440-T4 power flow in Overdrive low and manual low.

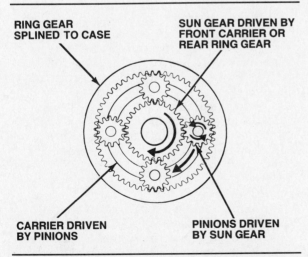

Figure 19-16. THM 440-T4 final drive gearset power flow in all forward gears. The final drive sun gear, pinions, and carrier rotate in the opposite directions in Reverse.

ring gear and drive the carrier/differential case clockwise in gear reduction. Power is then passed through the differential pinion and side gears to the differential output shaft (left side) and both halfshafts. This power flow through the final drive gearset and differential assembly is the same in all forward gears.

As long as the car is accelerating in Overdrive low, torque is transferred from the engine, through the planetary gearset, to the final drive sun gear shaft. If the driver lets up on the throttle, engine speed drops while the drive wheels try to keep turning at the same speed. This removes the counterclockwise-rotating torque load from the direct 1-way clutch, and it unlocks. The front gearset then freewheels, so no engine compression braking is available. This is a major difference between Overdrive low and manual low, as we will see later.

Gear reduction in low gear is produced by both the front and the rear gearsets of the

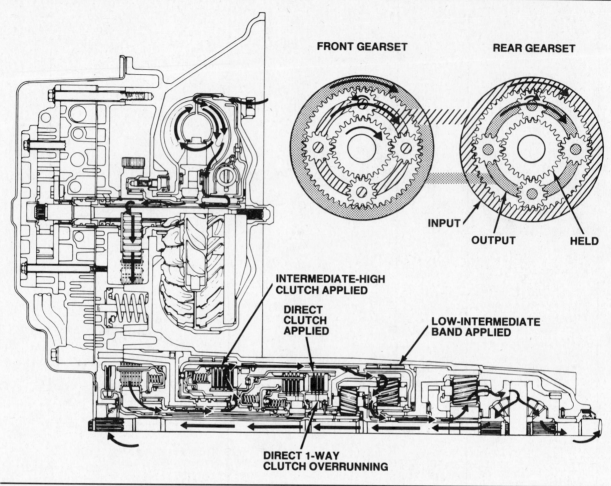

Figure 19-17. THM 440-T4 power flow in Overdrive second and manual second.

transaxle. The input member is the front sun gear, the held member is the rear sun gear, and the output member is the front ring gear/rear carrier. The input device is the direct clutch together with the direct 1-way clutch. The holding device is the low-intermediate band. In the final drive gearset, the input member is always the sun gear, the held member is always the ring gear, and the output member is always the carrier.

Overdrive Second

As the car continues to accelerate in Overdrive, the transaxle hydraulic system produces an automatic upshift to second gear by applying the intermediate-high clutch, figure 19-17. The input shaft then turns the intermediate-high clutch drum, reverse drum, and front carrier/rear ring gear clockwise. The low-intermediate band remains applied to hold the rear sun gear.

The rear ring gear drives the rear pinions clockwise. The rear pinions walk around the held rear sun gear, taking the rear carrier with them. The front ring gear/rear carrier and final drive sun gear shaft then rotate clockwise in gear reduction. The final drive sun gear shaft transmits power to the final drive gearset and differential assembly as described earlier.

In second gear, the front ring gear turns clockwise at the same speed as the final drive sun gear shaft. The front carrier also turns clockwise, but at (higher) turbine speed. As a result, the front pinions turn counterclockwise and drive the front sun gear faster than turbine speed. This causes the direct 1-way clutch to overrun, making the direct clutch ineffective, even though it remains applied in second gear. Thus, the front gearset freewheels at this time.

All of the gear reduction for second gear is produced by the rear planetary gearset. The input member is the ring gear, the held member is the sun gear, and the output member is the carrier. The input device is the intermediate-high clutch, and the holding device is the low-intermediate band.

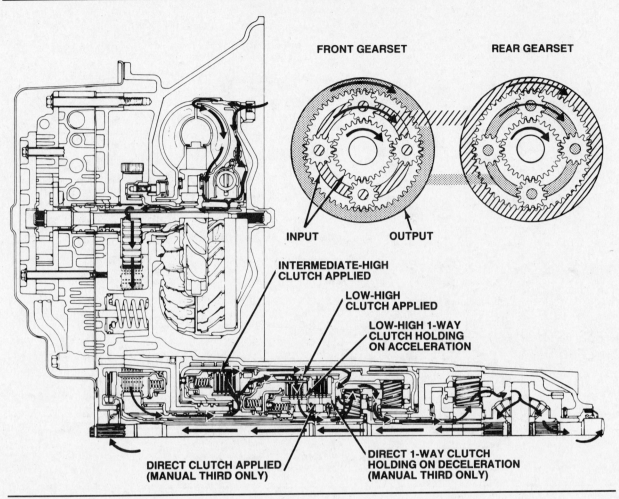

FRONT GEARSET　　　　**REAR GEARSET**

INPUT　　　**OUTPUT**

**INTERMEDIATE-HIGH
CLUTCH APPLIED**

**LOW-HIGH
CLUTCH APPLIED**

**LOW-HIGH 1-WAY
CLUTCH HOLDING
ON ACCELERATION**

**DIRECT CLUTCH APPLIED
(MANUAL THIRD ONLY)**

**DIRECT 1-WAY CLUTCH
HOLDING ON DECELERATION
(MANUAL THIRD ONLY)**

Figure 19-18. THM 440-T4 power flow in Overdrive third and manual third.

Overdrive Third

As the car continues to accelerate in Overdrive, the transaxle hydraulic system produces an automatic upshift from second to third gear, figure 19-18. To make this happen, the direct clutch and low-intermediate band are released, and the low-high clutch is applied. The low-high clutch locks the outer race of the low-high 1-way clutch to the input shaft. The intermediate-high clutch remains applied and drives the front carrier/rear ring gear clockwise at input shaft speed.

As the front carrier is driven, the front pinions attempt to walk counterclockwise around the inside of the front ring gear and drive the front sun gear clockwise. However, the low-high 1-way clutch prevents the front sun gear from turning clockwise faster than input shaft speed. As a result, both the front carrier and the front sun gear rotate clockwise at input shaft speed. With two members turning at the

same speed, the front gearset locks up and turns as a unit in direct drive. Input torque then passes through the front ring gear/rear carrier to the final drive sun gear shaft, which transmits power to the final drive gearset and differential assembly as described earlier.

As long as the car is accelerating in Overdrive third, torque is transferred from the engine, through the planetary gearset, to the final drive sun gear shaft. If the driver lets up on the throttle before the vehicle speed is high enough for a 3-4 upshift, engine speed drops while the drive wheels try to keep turning at the same speed. This removes the clockwise-rotating torque load from the low-high 1-way clutch. The clutch then unlocks and allows the front gearset to freewheel, so no engine compression braking is available. This is a major difference between Overdrive third and manual third, as we will see later.

In third gear, there is a direct 1:1 gear ratio between input shaft speed and final drive sun

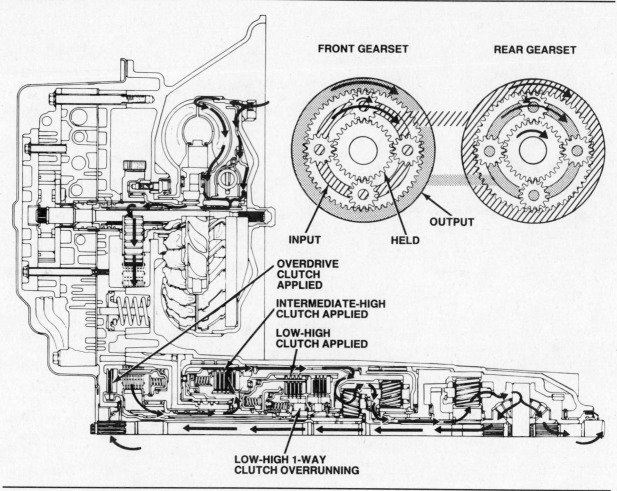

Figure 19-19. THM 440-T4 power flow in Overdrive fourth.

gear shaft speed. There is no reduction or increase in either speed or torque through the compound planetary gearset. However, the final drive planetary gearset does still provide gear reduction to the differential and halfshafts.

All of the power in third gear is transmitted through the front gearset. The carrier and sun gear are equally input members. The ring gear is the output member. The input devices are the intermediate-high clutch, and the low-high clutch together with the low-high 1-way clutch. No holding device is needed because the planetary gearset is locked.

Overdrive Fourth

As the car continues to accelerate in Overdrive, the transaxle hydraulic system produces an automatic upshift from third to fourth gear by applying the overdrive clutch, figure 19-19. The overdrive clutch locks the front sun gear to the

case. The intermediate-high clutch remains applied and drives the front carrier/rear ring gear clockwise at input shaft speed.

As the front carrier is driven, the front pinions walk around the held front sun gear and drive the front ring gear clockwise in overdrive. Input torque then passes through the front ring gear/rear carrier to the final drive sun gear shaft, which transmits power to the final drive gearset and differential assembly as described earlier. Although the low-high clutch remains applied, it is ineffective because the low-high 1-way clutch overruns.

In fourth gear, the rear ring gear turns clockwise at input shaft speed while the rear carrier turns clockwise at (higher) final drive sun gear shaft speed. As a result, the rear pinions turn counterclockwise and drive the rear sun gear clockwise. However, because no member of the rear gearset is held, the gearset simply free-wheels at this time.

All of the power in fourth gear is transmitted through the front gearset. The carrier is the input member, the sun gear is the held member, and the ring gear is the output member. The input device is the intermediate-high clutch. The holding device is the overdrive clutch.

Manual Third

When the vehicle is stopped and the gear selector is moved to manual third (D or 3), the transaxle operates just as in Overdrive except that the hydraulic system prevents an upshift to fourth gear. The flow of power through the compound planetary gearset in the lower two gears is exactly the same as in Overdrive low and Overdrive second, figures 19-15 and 19-17. The flow of power through the gearset in manual third gear is basically the same as in Overdrive third, figure 19-18, with the exception of one apply device.

In manual third, the direct clutch is applied so that the direct 1-way clutch can prevent the front sun gear from turning *slower* than input shaft speed. The low-high 1-way clutch prevents the front sun gear from turning *faster* than input shaft speed during acceleration in Overdrive third, but it overruns and allows the sun gear to slow during deceleration. In manual third, the direct and low-high 1-way clutches combine to lock the front sun gear solidly to the input shaft, and thus provide engine compression braking during deceleration.

If driver moves the gear selector to D or 3 while the car is moving in Overdrive fourth, the hydraulic system immediately downshifts the transaxle to third gear by releasing the overdrive clutch and applying the direct clutch. If the gear selector is moved to D or 3 while the car is moving in Overdrive third, the hydraulic system will apply the direct clutch and lock out an upshift to fourth. This capability provides the driver with a controlled downshift for engine compression braking or improved acceleration at any speed or throttle position.

Manual Second

When the vehicle is stopped and the gear selector is moved to manual second (2), the transaxle operates just as in Overdrive except that the hydraulic system prevents upshifts to third and fourth gears. The flow of power through the compound planetary gearset in first and second gears is identical, and the same apply devices are used.

When the gear selector is in manual second and the car accelerates from a standstill, the transaxle starts in Overdrive low, figure 19-15. The direct clutch and direct 1-way clutch drive the front sun gear, and the low-intermediate band holds the rear sun gear. As the car continues to accelerate, the transaxle automatically shifts to second by applying the intermediate-high clutch to drive the front carrier/rear ring gear, figure 19-17.

If driver moves the gear selector to 2 while the car is moving in Overdrive fourth, Overdrive third, or manual third, the hydraulic system immediately downshifts the transaxle to second gear. If the downshift is from Overdrive fourth, the low-high and overdrive clutches are released, and the low-intermediate band and direct clutch are applied. If the downshift is from Overdrive third, the low-high clutch is released, and the low-intermediate band and direct clutch are applied. The same apply devices are involved in a downshift from manual third, except that the direct clutch is already applied. This capability provides the driver with a controlled downshift for engine compression braking or improved acceleration at any speed or throttle position.

Manual Low

When the vehicle is stopped and the gear selector is moved to manual low (1), the hydraulic system prevents upshifts to second, third, or fourth gears. The flow of power through the compound planetary gearset in first gear is the same as in Overdrive low, figure 19-15, except for the use of one apply device.

In manual low, the low-high clutch is applied so that the low-high 1-way clutch can prevent the front sun gear from turning *faster* than input shaft speed. The direct 1-way clutch prevents the front sun gear from turning *slower* than input shaft speed during acceleration in Overdrive low, but it unlocks and allows the sun gear to overrun during deceleration. In manual low, the low-high and direct 1-way clutches combine to lock the front sun gear solidly to the input shaft, and thus provide engine compression braking during deceleration.

The driver can move the gear selector from Ⓓ, D, 3, or 2 to 1 while the car is moving. At speeds below approximately 40 mph (64 kph) the transaxle downshifts directly to low gear. At higher speeds, the transaxle first downshifts to second, and then automatically downshifts to low when vehicle speed drops below approximately 40 mph (64 kph). This capability provides the driver with a manual downshift for engine compression braking or improved acceleration.

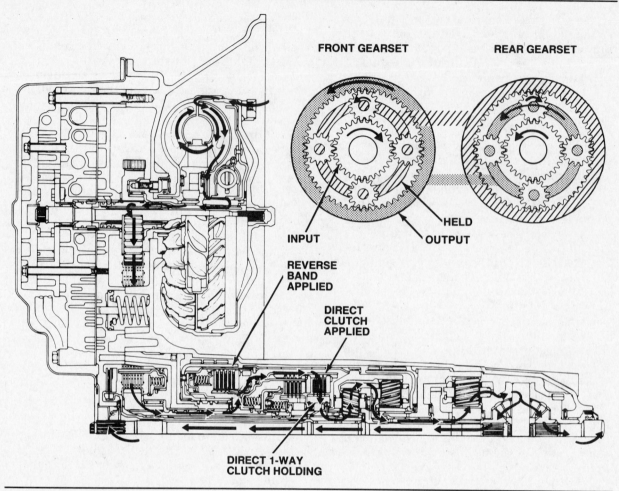

Figure 19-20. THM 440-T4 power flow in Reverse.

Reverse

When the gear selector is moved to Reverse, the input shaft is still turned clockwise by the engine. The transaxle must reverse this rotation to a counterclockwise direction. To do this, the direct clutch and reverse band are applied, figure 19-20.

The input shaft, acting through the direct clutch and the direct 1-way clutch, drives the front sun gear clockwise. The reverse band holds the intermediate-high clutch drum, reverse drum, and front carrier/rear ring gear stationary. As a result, clockwise rotation of the front sun gear turns the front pinions counterclockwise on the stationary front carrier. The front pinions then drive the front ring gear counterclockwise in gear reduction. Input torque passes through the front ring gear/rear carrier to the final drive sun gear shaft, which transmits power to the final drive gearset and differential assembly as described earlier.

As long as the car is accelerating in Reverse, torque is transferred from the engine, through the planetary gearset, to the final drive sun gear shaft. If the driver lets up on the throttle, engine speed drops while the drive wheels try to keep turning at the same speed. This removes the counterclockwise-rotating torque load from the direct 1-way clutch, and it unlocks. The front gearset then freewheels, so that no engine compression braking is available in Reverse.

All of the gear reduction in Reverse is produced by the front gearset of the transaxle. The input member is the sun gear, the held member is the carrier, and the output member is the ring gear. The input device is the direct clutch together with the direct 1-way clutch. The holding device is the reverse band.

GEAR RANGES	CLUTCHES						BANDS	
	1-way Sprag	1-way Roller	Direct (Input)	Intermediate-High (Second)	Low-High (Third)	Over-drive (Fourth)	Low-Intermediate (1-2)	Reverse
Overdrive Low	●		●				●	
Overdrive Second			*	●			●	
Overdrive Third		●		●	●			
Overdrive Fourth				●	*	●		
Manual Third	●	●	●	●	●			
Manual Low	●	●	●		●		●	
Manual Second			*	●			●	
Reverse	●		●					●

★ Applied, but ineffective due to overrunning 1-way clutch.

Figure 19-21. THM 440-T4 clutch and band application chart.

THM 440-T4 Clutch and Band Applications

Before we move on to a closer look at the hydraulic system, we should review the apply devices that are used in each of the gear ranges in this transaxle. Figure 19-21 is a clutch and band application chart for the THM 440-T4. Also, remember these facts about the apply devices in this transaxle:

● The direct, direct 1-way, and intermediate-high clutches are input devices.
● The direct clutch is applied in all gears except Overdrive third and fourth.
● The direct 1-way clutch is locked in first, manual third, and Reverse gears.
● The direct 1-way clutch overruns in second gear.
● The intermediate-high clutch is applied in second, third, and fourth gears.
● The low-high, low-high 1-way, and overdrive clutches, along with the low-intermediate and reverse bands, are holding devices.
● The low-high clutch is applied in third, fourth, and manual low gears.
● The low-high 1-way and manual third clutch is locked in third and manual low gears.
● The low-high 1-way clutch overruns in fourth gear.
● The overdrive clutch is applied in fourth gear.
● The low-intermediate band is applied in first and second gears.
● The reverse band is applied in Reverse.

THE THM 440-T4 HYDRAULIC SYSTEM

The THM 440-T4 hydraulic system controls shifting under varying vehicle loads and speeds, as well as in response to manual gear selection by the driver. Figure 19-22 is a diagram of the complete THM 440-T4 hydraulic system. You will find it helpful to refer back to this diagram as we discuss the various components that make up the hydraulic system. We will begin our study with the oil pump.

Hydraulic Pump

In the THM 440-T4, fluid is stored in the sump at the bottom of the transaxle, and in an upper fluid reservoir located in the chain cover. The upper reservoir controls the oil level in the sump based on fluid temperature. As the fluid heats up and expands, a thermostatic element increases the amount of fluid retained in the upper reservoir.

The THM 440-T4 uses a variable-displacement, vane-type oil pump, figure 19-23, located on the driver's side of the transaxle. The pump slide, pivot pin, rotor, vanes, and vane rings are installed in the pump body, which bolts to the valve body casting. The slide pivots on a pin, and when pressure is low in the hydraulic system, the pump priming spring tends to push the ring toward its maximum output position. When pressure in the hydraulic system rises, the pressure regulator valve sends oil through the decrease passage to move the slide toward its minimum output position. In this manner, oil pump output is controlled so that it is proportional to demand.

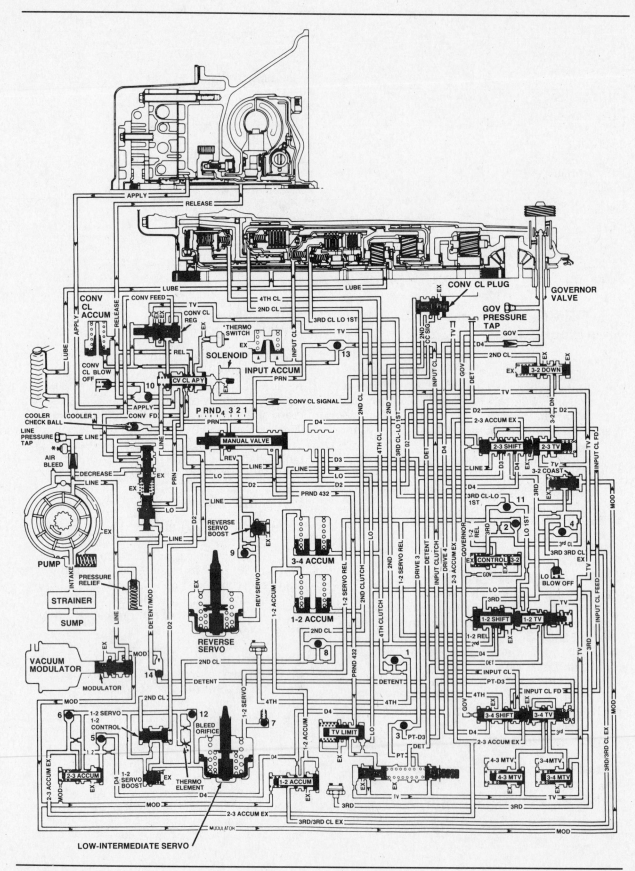

Figure 19-22. The THM 440-T4 hydraulic system.

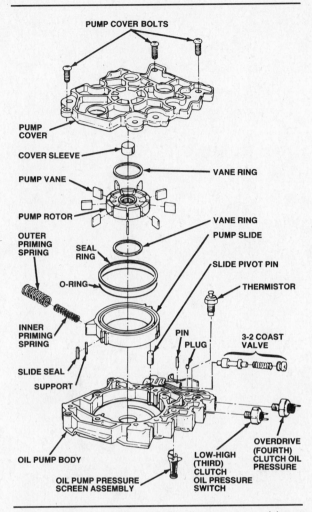

Figure 19-23. The THM 440-T4 oil pump assembly.

Whenever the engine is running, the converter cover turns the oil pump shaft and rotor. The rotation of the pump rotor and vanes creates suction that draws fluid through a filter from the sump at the bottom of the transaxle case. Oil from the pump flows to the: pressure regulator valve, manual valve, converter clutch apply valve, pressure relief valve, modulator valve, and 2-3 shift valve.

THM 440-T4 Hydraulic Valves

The 440-T4 hydraulic system contains the following valves:
1. Pressure regulator valve assembly
2. Pressure relief valve
3. Manual valve
4. Throttle limit valve
5. Throttle valve
6. Modulator valve
7. Governor valve
8. 1-2 servo boost valve
9. 1-2 shift valve assembly
10. 1-2 accumulator valve
11. 2-3 shift valve assembly
12. 2-3 accumulator valve
13. 3-2 downshift valve
14. 3-2 control valve
15. 3-2 coast valve
16. 1-2 servo control valve
17. 3-4 shift valve assembly
18. 3-4 modulated throttle valve
19. 4-3 modulated throttle valve
20. Reverse servo boost valve
21. Converter clutch shift plug (CCC)
22. Converter clutch shift valve (non-CCC)
23. Converter clutch apply valve
24. Converter clutch regulator valve

Valve body and check balls

Except for the manual valve, 3-2 downshift valve, governor valve, modulator valve, and 3-2 coast valve, all of the hydraulic valves in the THM 440-T4 are housed in the valve body, figure 19-24. The manual valve and 3-2 downshift valve are located in the channel plate. The governor valve installs in an opening on top of the transaxle above the final drive carrier and differential case. The modulator valve is located in the front, upper, left side of the case behind the vacuum modulator. The 3-2 coast valve is located in the oil pump body.

A separator plate and two gaskets are installed between the valve body and the channel plate, figure 19-5. The valve body has passages to route fluid between various valves, and additional passages are cast into the channel plate where the valve body bolts to it. These cast passages form the back half of the valve body. The THM 440-T4 uses 14 check balls, figure 19-25. Four are in the valve body, eight are in the channel plate, one is in a capsule in the transaxle case, and one is in the channel plate.

The hydraulic system also includes a converter clutch blow-off check ball, low blow-off check ball, and cooler check ball, figure 19-22. The converter clutch blow-off check ball is located in the converter release circuit where it prevents converter clutch release or apply pressure from exceeding 100 psi (689 kPa). The low blow-off check ball is located in the manual low circuit where it prevents low-high clutch apply pressure from exceeding 70 psi (483 kPa). The cooler check ball is located in the transaxle cooler circuit where it seats to prevent converter drain-back when the engine is shut off.

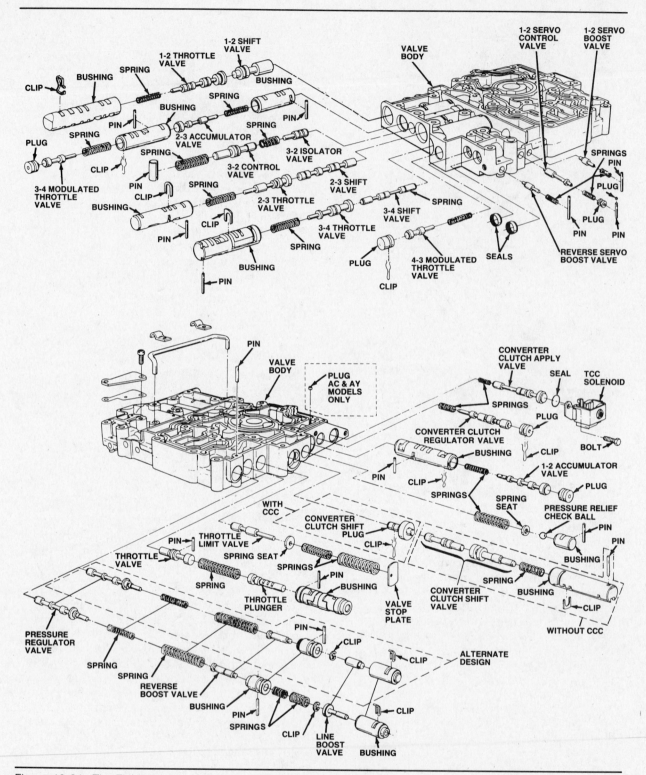

Figure 19-24. The THM 440-T4 valve body and hydraulic control valves.

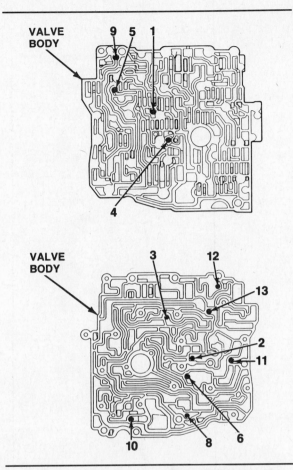

Figure 19-25. THM 440-T4 check ball locations. The #7 check ball is located in a capsule in the transaxle case. The #14 check ball is located in the channel plate.

Pressure regulator valve assembly

The pump delivers fluid through two passages to the pressure regulator valve, figure 19-26. One passage is at the end of the valve opposite the boost valves, the other is at the center of the valve. Hydraulic pressure at the end of the valve acts to move the valve against spring force. Hydraulic pressure at the center of the valve enters between two lands of equal diameter, and does not move the valve in either direction.

When the engine is started and mainline pressure begins to build, pressure at the end of the regulator valve moves the valve against spring force. This opens the converter feed passage and routes oil to the converter clutch apply and regulator valves, figure 19-26A. These valves direct oil to the torque converter, transaxle cooler, and lubrication circuit through different passages depending on whether the converter clutch is applied or released. This is explained in detail later in the chapter.

After the converter is filled, pressure build up at the regulator valve moves it farther against spring force. This opens a port in the center chamber of the valve that leads into the decrease passage, figure 19-26B. Pressure then flows through the passage to the pump where it moves the slide against priming spring force to reduce pump output. The pressure regulator valve continues to move until hydraulic pressure and spring force reach a balance that provides a constant mainline pressure.

When the transaxle shifts into gear, or from one gear to another, increased fluid flow demands cause mainline pressure to drop slightly. This drop in mainline pressure allows the spring to move the pressure regulator valve so that mainline pressure feed is cut off and the pump decrease circuit exhausts. The pump priming spring then moves the slide to increase pump output and bring mainline pressure back to the correct level.

The line boost valve and reverse boost valve at the spring end of the pressure regulator valve are used to raise mainline pressure under certain driving conditions. This is done by routing various hydraulic pressures to the boost valves, which then move to increase the amount of spring force acting on the pressure regulator valve.

Modulator pressure acts on the line boost valve to increase mainline pressure in response to higher drivetrain torque loads. During full throttle operation, detent pressure acts on the line boost valve to increase mainline pressure and hold the apply devices more securely. In Park, Reverse, Neutral, and manual low, oil pressure in the PRN or LO circuit acts on the reverse boost valve to increase mainline pressure and hold the apply devices more securely.

Pressure relief valve

The pressure relief valve, figure 19-24, is a ball-check valve located in the valve body. This valve limits mainline pressure to a maximum somewhere between 245 and 360 psi (1690 and 2480 kPa). If mainline pressure exceeds this value, the pressure relief valve unseats to exhaust excess oil to the sump. Once mainline pressure drops below the maximum, spring force reseats the valve.

Manual valve

The manual valve, figure 19-27, is a directional valve that is moved by a mechanical linkage from the gear selector lever. The manual valve receives fluid from the pump and directs it to apply devices and other valves in the hydraulic system in order to provide both automatic and

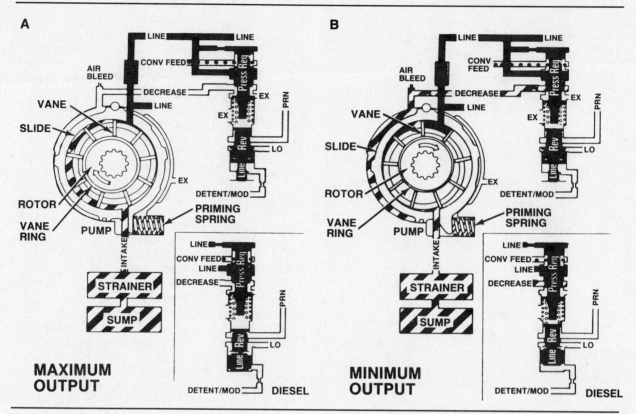

Figure 19-26. The THM 440-T4 pressure regulator valve assembly in the maximum and minimum output positions.

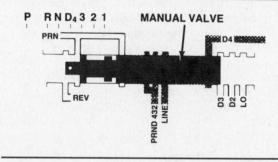

Figure 19-27. The THM 440-T4 manual valve in Overdrive.

manual upshifts and downshifts. The manual valve is held in each gear position by a spring-loaded lever.

Seven passages may be charged with mainline pressure from the manual valve. GM refers to mainline pressure from the manual valve by the name of the passage it flows through.

The PRN passage is charged in Park, Reverse, and Neutral; it routes pressure to the:

● Apply area of the direct (input) clutch
● Direct (input) accumulator
● Reverse boost valve.

The PRND-432 passage is charged in Park, Reverse, Neutral, Overdrive, and manual third; it routes pressure to the:

● Throttle limit valve.

The D4 circuit is charged in Overdrive, manual third, manual second, and manual low; it routes pressure to the:

● Governor valve
● 1-2 shift valve
● 2-3 shift valve
● 3-4 shift valve
● 1-2 accumulator valve
● 2-3 accumulator valve
● 1-2 control valve
● 1-2 servo boost valve
● Apply side of the low-intermediate servo.

The D3 passage is charged in manual third, manual second, and manual low; it routes pressure to the:

● 2-3 shift valve
● 3-4 throttle valve.

The D2 circuit is charged in manual second and manual low; it routes pressure to the:

● 1-2 control valve
● 2-3 throttle valve.

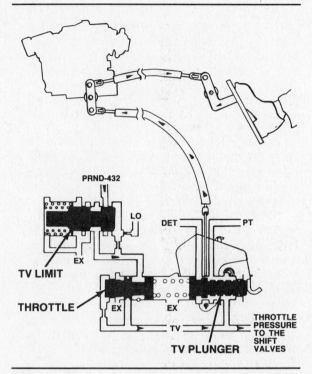

Figure 19-28. The THM 440-T4 throttle limit valve and throttle valve. Throttle pressure is used to regulate shift timing.

The LO circuit is charged in manual low; it routes pressure to the:

- Reverse boost valve
- 1-2 throttle valve
- Apply area of the low-high clutch
- Throttle limit valve.

The REVERSE circuit is charged in Reverse; it routes pressure to the:

- Reverse servo boost valve
- Reverse servo.

Throttle limit valve

The throttle limit valve, figure 19-28 receives mainline pressure (PRND-432 oil) from the manual valve, and limits it to a maximum of 90 psi (621 kPa) before sending it to the throttle valve as throttle feed pressure. The throttle limit valve is controlled by opposition between spring force at one end and mainline pressure at the other.

PRND-432 pressure moves the throttle limit valve against spring force to restrict the flow of fluid through the valve. At 90 psi (621 kPa), the throttle limit valve moves far enough to block the entry of mainline pressure and exhaust any excess pressure in the throttle feed circuit. In manual low, LO oil is routed to the throttle limit valve to bottom the valve and completely cut off throttle feed pressure to the throttle valve.

Throttle valve

The THM 440-T4 transaxle develops throttle pressure using a throttle valve operated by a mechanical linkage, figure 19-28. Throttle pressure indicates engine torque and helps time the transaxle upshifts and downshifts. Without throttle pressure, the transaxle would upshift and downshift at exactly the same speed for very different throttle openings and load conditions.

The throttle valve is controlled by opposition between hydraulic pressure and spring force. The hydraulic pressure is throttle pressure that is routed to the end of the valve opposite the spring. The amount of spring force acting on the other end of the valve is controlled by the throttle plunger. The plunger is linked to the accelerator by a cable so that its position is determined by throttle opening.

At closed throttle, the throttle plunger exerts no force on the throttle valve spring. Throttle feed oil passes through the throttle valve and into the throttle pressure circuit. This pressure acts on the end of the throttle valve, and moves it to the right against spring force to nearly close the outlet port of the throttle valve. Thus, there is little or no throttle pressure at this time.

As the accelerator is depressed, the cable linkage moves the throttle plunger to compress the throttle valve spring. This moves the throttle valve to the left, opens the outlet port, and allows throttle feed pressure to exit the valve as throttle pressure. As the throttle is opened wider, spring force rises, the throttle valve moves farther, and throttle pressure continues to increase. However, because throttle pressure is developed from mainline pressure, it never exceeds mainline pressure.

To help balance the throttle valve, throttle pressure is directed to the end of the valve opposite spring force. This acts to move the throttle valve to the right against spring force. At any given throttle opening, the throttle valve reaches a balanced position and provides a constant level of throttle pressure.

Throttle pressure is routed to the: throttle plunger, 3-4 and 4-3 modulated throttle valves, 1-2 and 2-3 throttle valves, 3-2 downshift valve, converter clutch shift valve (non-CCC), and converter clutch regulator valve. Throttle pressure to the throttle plunger helps regulate throttle pressure by increasing the spring force acting on the throttle valve; it also reduces the effort needed to hold the throttle valve plunger in place. The effect of throttle pressure on the other valves mentioned will be covered in the appropriate sections later in the chapter.

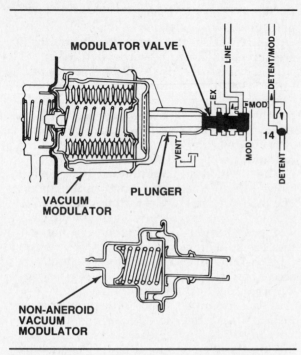

Figure 19-29. The THM 440-T4 vacuum modulator and modulator valve. Modulator pressure is used to regulate shift feel.

When the accelerator is depressed far enough, the throttle plunger opens a port that allows throttle pressure to flow through a passage to the 3-4 throttle valve. Oil in this passage is called part-throttle pressure. Part-throttle pressure on the 3-4 throttle valve combines with spring force and moves the 3-4 shift valve to produce a part-throttle 4-3 downshift when the throttle is opened past a certain point while cruising in fourth gear.

When the accelerator is completely depressed, the throttle plunger opens a second port that allows throttle pressure to flow through a passage to the 1-2 throttle valve, line boost valve, and converter clutch shift valve (non-CCC). Oil in this passage is called detent pressure. Detent pressure on the 1-2 throttle valve combines with spring force to move the 1-2 shift valve to produce a wide-open throttle 2-1 downshift. Detent pressure on the line boost valve increases mainline pressure to hold the apply devices more securely. Detent pressure on the converter clutch shift valve (non-CCC) moves the valve to unlock the torque converter clutch to allow torque multiplication in the converter.

Modulator valve

Most late-model General Motors transaxles use throttle pressure acting on the pressure regulator valve to vary mainline pressure in propor-

tion to engine torque. The THM 440-T4, however, uses modulator pressure for this purpose. The modulator valve that creates this pressure should not be confused with the throttle valve, although in some other transmissions and transaxles GM refers to the throttle valve as a modulator valve. In the THM 440-T4, throttle pressure proportional to throttle *opening* acts on the shift valves to help regulate shift *timing*. Modulator pressure proportional to engine intake manifold vacuum acts on the pressure regulator valve to increase mainline pressure and help regulate shift *feel*.

The THM 440-T4 modulator valve is operated by a vacuum modulator, figure 19-29, that contains two chambers separated by a flexible rubber diaphragm. The chamber nearest the modulator valve spool is open to atmospheric pressure. The chamber farthest from the valve spool is sealed and receives engine intake manifold vacuum. The vacuum chamber also contains a spring that acts on the back side of the diaphragm to move a plunger and force the modulator valve spool into its bore. As intake manifold vacuum changes, the diaphragm and plunger move, and the modulator valve travels back and forth in its bore.

When the engine is started, intake manifold vacuum is applied to the vacuum chamber of the modulator. If the vacuum level is high enough, such as when the engine is idling or under light load, atmospheric pressure against the modulator diaphragm will oppose spring pressure with enough force to move the valve spool all the way to the left, closing the mainline pressure inlet port and reducing modulator pressure to zero.

Under most vehicle operating conditions, however, intake manifold vacuum is low enough that atmospheric pressure cannot move the valve to cut off modulator pressure altogether. Instead, the modulator valve spool only partially obstructs the mainline pressure inlet port, thus regulating modulator pressure at the outlet port.

A portion of the modulator pressure is sent to the end of the valve opposite the vacuum modulator. This pressure acts against spring force to help balance the position of the valve. As a result of these opposing forces, modulator pressure increases as the intake manifold vacuum drops, and stabilizes whenever the vacuum level remains constant.

When the engine comes under heavy load, manifold vacuum drops almost to zero. At these times, pressure is nearly equal in both chambers of the vacuum modulator, and spring force moves the valve spool all the way into its

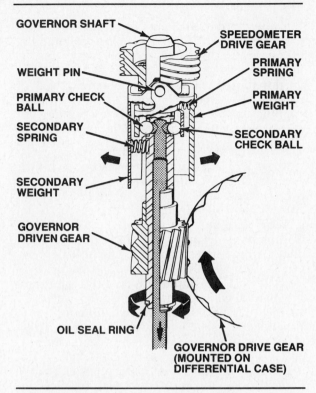

GOVERNOR SHAFT

SPEEDOMETER DRIVE GEAR

WEIGHT PIN

PRIMARY SPRING

PRIMARY CHECK BALL

PRIMARY WEIGHT

SECONDARY SPRING

SECONDARY CHECK BALL

SECONDARY WEIGHT

GOVERNOR DRIVEN GEAR

OIL SEAL RING

GOVERNOR DRIVE GEAR (MOUNTED ON DIFFERENTIAL CASE)

Figure 19-30. The THM 440-T4 check-ball governor.

bore. Fluid then passes straight through the valve so modulator pressure equals mainline pressure.

On some transaxles, the chamber of the vacuum modulator nearest the modulator valve contains an evacuated metal cell called an aneroid bellows. The bellows compensates for reduced intake manifold vacuum and engine torque at higher altitudes. To do this, it expands and contracts with changes in atmospheric pressure, and acts on the vacuum modulator diaphragm to reduce or increase modulator pressure.

At high atmospheric pressures (such as at sea level), the bellows shortens so that the diaphragm and modulator valve move to the right. At low atmospheric pressures (such as at high altitude), the bellows lengthens. A spring inside the bellows ensures that proper expansion takes place. As the bellows expands, the diaphragm and modulator valve move to the left to reduce modulator and mainline pressure. The actions of the aneroid bellows ensure that shift feel remains smooth and uniform at all altitudes.

Governor valve
We have already discussed the development and control of throttle pressure, which is one of the two auxiliary pressures that work on the

shift valves. Before we get to the shift valve operation, we must explain the governor valve (or governor), and how it creates governor pressure. Governor pressure provides a road speed signal to the hydraulic system that causes automatic upshifts to occur as road speed increases, and permits automatic downshifts with decreased road speed.

The THM 440-T4 governor, figure 19-30, mounts in an opening on the top of the transaxle and is driven by a gear on the differential case. This style of governor uses check balls to regulate governor pressure, and consists of a governor shaft, a governor weight pin, primary and secondary weights, primary and secondary springs, and two check balls. A drive gear at the top of the governor shaft turns the speedometer cable driven gear, which fits into an opening in the governor cover.

Whenever the engine is running and the gear selector is in a forward gear range, mainline pressure (D4 oil) passes through a filter screen and a restricting orifice, then enters a fluid passage in the center of the governor shaft. Two exhaust ports are drilled through the shaft into the fluid passage. Governor pressure is determined by the amount of fluid that is allowed to flow out of these ports.

The exhaust port openings are controlled by check balls that seat in pockets formed at the ends of the ports. As the governor rotates, centrifugal force causes the primary weight to act on one check ball, and the secondary weight to act on the other. Each weight acts on the check ball located on the opposite side of the governor shaft. The weights are assisted in their jobs by springs that help to regulate pressure more smoothly as vehicle speed increases.

When the vehicle is stopped and the governor is not turning, there is no centrifugal force. At this time, mainline pressure overcomes the force of the primary and secondary springs, and unseats both of the check balls. All of the oil is exhausted, and governor pressure is zero at this time.

As the vehicle accelerates and the gear on the differential case begins to turn the governor, centrifugal force moves the weights outward. This force is relayed to the check balls, partially seating them in their pockets to restrict fluid flow and cause governor pressure to increase. As long as governor speed increases, the forces on the check balls become greater and less fluid is allowed to escape, thus governor pressure continues to rise. When governor speed remains constant, the forces acting on the check balls are also constant, and governor pressure stabilizes.

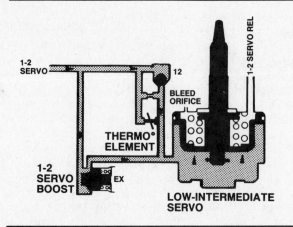

Figure 19-31. The THM 440-T4 1-2 servo boost valve and low-intermediate servo. Note the thermo element used to control fluid flow to the servo when the transaxle is cold.

The heavier primary weight and spring are more sensitive to changes in governor rpm than are their secondary counterparts. Therefore, the primary weight and spring work to control governor pressure at lower vehicle speeds. At higher speeds, centrifugal force seats the primary check ball, and the lighter secondary weight and spring take over governor pressure regulation. Above a certain speed, centrifugal force becomes great enough to seat both check balls so that no fluid escapes from the governor. At this point, governor pressure equals mainline pressure.

Governor pressure is sent to the shift valves to oppose throttle pressure and spring force, and upshift the valves at the appropriate road speeds. Governor pressure is also routed to the 3-2 control valve, and the converter clutch shift plug (CCC) or shift valve (non-CCC).

1-2 servo boost valve

When the vehicle is stopped and the transaxle is placed in any forward gear range, mainline pressure (D4 oil) is routed to the low-intermediate servo to apply the low-intermediate band and place the transaxle in low gear, figure 19-31. Servo apply pressure can reach the servo through two orifices (one controlled by a thermo element) and the 1-2 servo boost valve which is normally held closed by spring force.

Under light acceleration, servo apply pressure is too low to open the 1-2 servo boost valve, and all of the apply oil passes through the orifices. When the transaxle is cold, both orifices are open to allow fluid flow. As the transaxle warms and the fluid thins out, the thermo element gradually blocks one orifice until it is fully closed at normal operating temperature.

Under harder acceleration, the low-intermediate band must hold under increased torque loads. Servo apply pressure increases at this time and opens the 1-2 servo boost valve to quickly route the increased apply pressure directly to the servo. This prevents the band from slipping during sudden throttle applications.

1-2 shift valve assembly

The 1-2 shift valve controls upshifts from low to second, and downshifts from second to low. The assembly includes the 1-2 shift valve, 1-2 throttle valve, and a spring, figure 19-32. The valve is controlled by opposition between governor pressure on one end, and combined throttle pressure and spring force on the other end. Throttle pressure and spring force act on the 1-2 throttle valve to oppose an upshift.

In low gear, mainline pressure (D4 oil) from the manual valve flows to the 1-2 shift valve where it is blocked by the downshifted valve. When vehicle speed increases enough for governor pressure to overcome throttle pressure and spring force acting on the 1-2 throttle valve, the 1-2 shift valve upshifts. D4 oil then flows through the upshifted valve, and into the second gear passage where it travels to the apply area of the intermediate-high (second) clutch, 1-2 accumulator, 1-2 control valve, 3-2 downshift valve, and the converter clutch shift plug (CCC) or shift valve (non-CCC).

A closed-throttle 2-1 downshift occurs when governor pressure drops to a very low level (road speed decreases), and throttle pressure and spring force act on the 1-2 throttle valve to move the 1-2 shift valve against governor pressure.

A full-throttle 2-1 detent downshift occurs at speeds below approximately 35 mph (56 kph) when detent pressure from the throttle valve flows to the 1-2 throttle valve. The combination of detent pressure, throttle pressure, and spring force moves the 1-2 shift valve against governor pressure to cause a 2-1 downshift.

In manual low, mainline pressure (LO oil) from the manual valve fills a chamber between the 1-2 shift valve and the 1-2 throttle valve. This downshifts the 1-2 shift valve and keeps it downshifted. LO oil then passes through the valve to apply the low-high clutch. LO oil is also routed to the low blow-off valve.

1-2 accumulator valve

The 1-2 accumulator valve, figure 19-33, cushions the application of the intermediate-high and overdrive clutches in relation to engine load. It does this by creating a 1-2 accumulator pressure that combines with spring force in the 1-2 and 3-4 accumulators to control the rate at

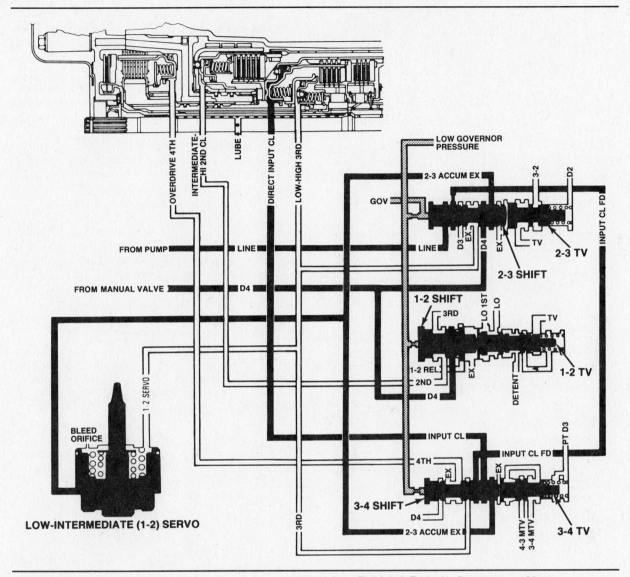

Figure 19-32. The 1-2, 2-3, and 3-4 shift valve assemblies of the THM 440-T4 in the first gear position.

which the accumulator pistons move through their ranges of travel.

When the transaxle is in any forward gear range, mainline pressure (D4 oil) is routed to the 1-2 accumulator valve. The D4 pressure moves the valve against spring force and modulator pressure to create a variable 1-2 accumulator pressure. When modulator pressure is low, accumulator pressure is also low. And when modulator pressure is high, accumulator pressure is also high. Accumulator pressure flows into the spring chambers of the 1-2 and 3-4 accumulators.

When the 1-2 or 3-4 shift valve upshifts, mainline pressure flows through the intermediate-high (second) clutch or overdrive (fourth)

clutch circuit to the appropriate accumulator piston where it opposes spring force and 1-2 accumulator pressure. The amount of time it takes clutch apply pressure to bottom the accumulator piston determines the amount of force with which the clutch is applied.

When accumulator pressure is low, indicating a light load on the drivetrain, clutch apply oil rapidly bottoms the piston. This absorbs much of the pressure in the circuit and provides an easy clutch application. When accumulator pressure is high, indicating a heavy load on the drivetrain, clutch apply oil takes much longer to bottom the piston. As a result, most of the pressure in the apply circuit is immediately available to apply the clutch much

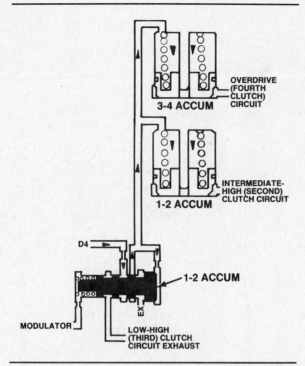

Figure 19-33. The THM 440-T4 1-2 accumulator valve creates a 1-2 accumulator pressure that cushions 1-2 and 3-4 upshifts.

more firmly. In both cases, once the accumulator piston is bottomed, full mainline pressure is present in the circuit to hold the clutch securely applied.

2-3 shift valve assembly

The 2-3 shift valve controls upshifts from second to third, and downshifts from third to second. The assembly includes the 2-3 shift valve, 2-3 throttle valve, and a spring, figure 19-32. The valve is controlled by opposition between governor pressure on one end, and combined throttle pressure and spring force on the other. Throttle pressure and spring force act on the 2-3 throttle valve to oppose an upshift.

Whenever the engine is running, mainline pressure from the pump flows to the 2-3 shift valve. In any gear position except third and fourth, this pressure passes through the downshifted valve and into the direct (input) clutch feed circuit. Pressure in this circuit flows to and through the downshifted 3-4 shift valve, and is then routed to the apply area of the direct clutch.

In second gear, mainline pressure (D4 oil) from the manual valve flows to the 2-3 shift valve where it is blocked by the downshifted valve. When vehicle speed increases enough for governor pressure to overcome throttle pressure and spring force acting on the 2-3

throttle valve, the 2-3 shift valve upshifts. D4 oil then flows through the upshifted valve and into the low-high (third) clutch circuit where it applies the clutch.

Some of the low-high clutch circuit pressure is routed to the release side of the low-intermediate servo. The release action of the servo acts as an accumulator for low-high clutch application. The time it takes clutch apply pressure to release the low-intermediate servo and band provides a delay in low-high clutch application. Clutch apply pressure closes the exhaust check ball in the accumulator passage at this time.

In Overdrive third, pressure in the direct (input) clutch circuit is routed into the D3 passage at the 2-3 shift valve. This pressure then flows to the manual valve where it exhausts to release the direct clutch. In manual third, mainline pressure (D3 oil) flows from the manual valve to the 2-3 shift valve to keep the direct clutch applied and provide engine compression braking.

A closed-throttle 3-2 downshift occurs when governor pressure drops (road speed decreases) to the point where throttle pressure and spring force act on the 2-3 throttle valve to move the 2-3 shift valve against governor pressure.

A full-throttle 3-2 detent downshift occurs at speeds below approximately 65 mph (105 kph) when throttle pressure from the 3-2 downshift valve flows to a second land on the 2-3 throttle valve. The combination of throttle pressure and spring force then moves the 2-3 shift valve against governor pressure and causes a 3-2 downshift. If the vehicle speed is below approximately 35 mph (56 kph) detent pressure will also downshift the 1-2 shift valve, resulting in a 3-1 downshift.

In manual second, mainline pressure (D2 oil) from the manual valve is applied to the spring end of the 2-3 throttle valve. This pressure combines with spring force and throttle pressure to downshift the 2-3 shift valve and keep it downshifted regardless of throttle opening or road speed.

2-3 accumulator valve

As mentioned in the previous section, the release action of the low-intermediate servo serves as an accumulator to cushion application of the low-high (third) clutch. The 2-3 accumulator valve, figure 19-34, controls the rate of low-intermediate servo release in proportion to engine torque in order to provide a consistent shift feel. The 2-3 accumulator valve is controlled by opposition between 2-3 accumulator exhaust pressure on one end, and combined spring force and modulator pressure on the other.

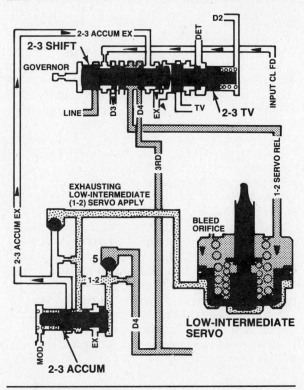

Figure 19-34. The THM 440-T4 2-3 accumulator valve controls the rate of low-intermediate servo release in proportion to engine torque in order to provide a consistent 2-3 shift feel.

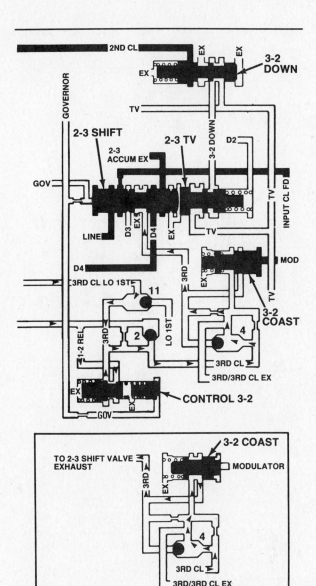

Figure 19-35. THM 440-T4 3-2 control valve and 3-2 coast valve operation.

In first and second gears, mainline pressure (D4 oil) flows through the downshifted 2-3 shift valve into the 2-3 accumulator exhaust circuit. Pressure in this circuit is routed past a check ball to apply the low-intermediate servo. When the 2-3 upshift occurs, low-intermediate servo apply oil, now called 2-3 accumulator oil, exhausts through this same passage.

Before the 2-3 upshift, the 2-3 accumulator valve is held seated by spring force and modulator pressure. When the upshift occurs, 2-3 accumulator exhaust oil flows to the end of the 2-3 accumulator valve opposite the spring. This pressure moves the valve against spring force and modulator pressure to open the exhaust passage to the 2-3 shift valve.

When modulator pressure is high, indicating a heavy load on the drivetrain, 2-3 accumulator exhaust oil moves the valve only a small distance so the low-intermediate servo releases slowly. This limits low-high clutch slippage by ensuring that the clutch applies before the low-intermediate band is fully released. When modulator pressure is low, indicating a light load on the drivetrain, 2-3 accumulator exhaust oil

moves the valve farther so that the low-intermediate servo releases more rapidly. This gives a consistent shift feel because the low-high clutch locks up sooner with less pressure when drivetrain loads are light.

3-2 downshift valve

The 3-2 downshift valve, figure 19-35, regulates the timing of the full-throttle 3-2 detent downshift. The valve is controlled by opposition between spring force on one end, and mainline pressure acting on two lands at the center of the valve.

When the accelerator is fully depressed and mainline pressure reaches approximately 110

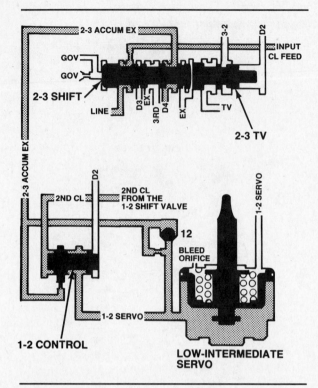

Figure 19-36. Operation of the THM 440-T4 1-2 control valve.

psi (758 kPa), the pressure acts on the larger of the two lands with greater force and moves the 3-2 downshift valve against spring force. This opens a passage that allows throttle pressure to flow through the valve to the 2-3 throttle valve, where it acts with spring force to downshift the 2-3 shift valve.

3-2 control valve

The 3-2 control valve, figure 19-35, regulates the rate of low-high (third) clutch release and low-intermediate band application during 3-2 downshifts. Because of the change in ratio from third to second gear, the engine speed must increase as the low-high clutch releases and the low-intermediate band applies. The faster the vehicle is traveling, the more engine speed must increase, and the longer low-intermediate band application must be delayed.

The 3-2 control valve is controlled by opposition between spring force and governor pressure. When the vehicle is stopped, spring force holds the 3-2 control valve wide open. As speed rises, governor pressure moves the valve against spring force to restrict the valve opening. Above approximately 65 mph (105 kph), governor pressure closes the valve entirely.

During 3-2 downshifts below approximately 65 mph (105 kph), the 3-2 control valve regulates the shift in relation to vehicle speed. Ex-

hausting low-intermediate (1-2) servo release oil seats a check ball (2). Some of the oil then flows through a restricting orifice, while the remainder flows through the 3-2 control valve. All of the exhausting low-intermediate servo release oil then enters the low-high clutch exhaust circuit where it seats a check ball (4) and flows through an orifice to the 2-3 shift valve where it finally exhausts. This allows the low-high clutch to release, and the low-intermediate band to apply, at a rate that provides the best shift feel for a given vehicle speed.

During 3-2 downshifts above approximately 65 mph (105 kph), the 3-2 control valve is closed. As a result, all of the exhausting low-intermediate servo release oil must pass through the restricting orifice before it can enter the low-high clutch exhaust circuit. The oil then seats the first check ball (4), passes through the restricting orifice, and exhaust at the 2-3 shift valve as already described. This causes band application to be delayed even more, allowing engine speed to increase and better match the higher gear ratio of second gear.

3-2 coast valve

The 3-2 coast valve, figure 19-35, allows low-high (third) clutch apply oil to be exhausted more rapidly during a 3-2 coasting downshift. The 3-2 coast valve is controlled by opposition between spring force and modulator pressure.

During part- or full-throttle downshifts, modulator pressure bottoms the 3-2 coast valve against spring force. This closes the valve and forces low-high clutch apply oil to pass through a restricting orifice before it can exhaust at the 2-3 shift valve. When the accelerator is released and there is zero modulator pressure, figure 19-35 (inset) the spring opens the 3-2 coast valve, allowing low-high clutch oil to bypass the orifice and exhaust quickly so that the low-high clutch does not drag.

1-2 servo control valve

The 1-2 servo control valve, figure 19-36, works together with the 3-2 downshift valve to regulate the rate at which the low-intermediate servo applies during a 3-2 downshift. The valve is controlled by opposition between spring force on one end, and mainline pressure on the other end.

When the car is stopped and the transaxle is shifted into any forward gear position, the 1-2 servo control valve is held open by spring force. At this time, 1-2 servo apply oil seats a check ball (12), passes through a restricting orifice, and flows to the low-intermediate servo. Apply oil also flows through a second orifice and the

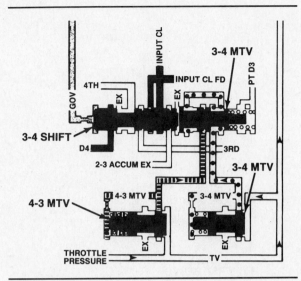

Figure 19-37. The 3-4 and 4-3 modulated throttle valves of the THM 440-T4 re-regulate throttle pressure to help control shifts between third and fourth gears.

1-2 servo control valve on its way to the low-intermediate servo. Oil flow through the two circuits applies the servo without delay.

During a 3-2 downshift, intermediate-high (second) clutch oil holds the 1-2 servo control valve closed against spring force, figure 19-36. This means that all of the low-intermediate servo apply oil must pass through a single orifice. The resulting delay in servo application allows engine speed to increase for a smoother 3-2 downshift.

If the gear selector is moved to manual second gear while the car is in motion, mainline pressure (D2 oil) from the manual valve opens the 1-2 servo control valve immediately. This allows the servo to apply quickly and prevents the low-intermediate band from slipping during strong engine braking or hard acceleration in manual second gear.

3-4 shift valve assembly
The 3-4 shift valve controls upshifts from third to fourth, and downshifts from fourth to third. The assembly consists of the 3-4 shift valve, 3-4 throttle valve, and a spring, figure 19-32. The 3-4 shift valve is controlled by opposition between governor pressure on one end, and combined spring force and 3-4 modulated throttle pressure (see below) on the other end.

In first and second gears, mainline pressure from the 2-3 shift valve flows to and through the downshifted 3-4 shift valve, and is routed to apply the direct (input) clutch. In Overdrive third, the upshifted 2-3 shift valve blocks the flow of mainline pressure, and exhausts the

direct clutch circuit through the D3 passage to the manual valve. In manual third, mainline pressure (D3 oil) flows from the manual valve through the 2-3 shift valve to the 3-4 shift valve in order to keep the direct clutch applied and provide engine compression braking.

In third gear, mainline pressure in the low-high (third) clutch circuit is routed to the 3-4 shift valve where it is blocked by the down-shifted valve. When vehicle speed increases enough for governor pressure to overcome spring force and 3-4 modulated throttle pressure acting on the 3-4 throttle valve, the 3-4 shift valve upshifts. Mainline pressure then flows through the upshifted valve and two orifices to apply the overdrive clutch and place the transaxle in fourth gear.

Some of the fourth gear circuit pressure flows to the 3-4 accumulator, figure 19-33, which cushions overdrive clutch engagement as described earlier. To prevent hunting between gears, 3-4 modulated throttle pressure is blocked from entering the 3-4 shift valve assembly after the 3-4 upshift.

A closed-throttle 4-3 downshift occurs when governor pressure drops (road speed decreases) to the point where spring force acts on the 3-4 throttle valve to move the 3-4 shift valve against governor pressure.

A part-throttle 4-3 downshift can occur at lower speeds when 4-3 modulated throttle pressure (see below) combines with spring force acting on the 3-4 throttle valve to move the 3-4 shift valve against governor pressure. A part-throttle 4-3 downshift also occurs when the accelerator linkage moves the throttle plunger far enough to uncover the part-throttle port in the throttle valve. This routes part-throttle pressure to the spring end of the 3-4 throttle valve. The combination of part-throttle pressure and spring force then moves the 3-4 shift valve against governor pressure to cause a 4-3 downshift.

3-4 and 4-3 modulated throttle valves
The same throttle pressure is not ideal for every upshift and downshift. Therefore, the THM 440-T4 uses two modulated throttle valves, figure 19-37, that re-regulate throttle pressure to lower values before it is sent to the 3-4 throttle valve. Both valves are controlled by opposition between spring force and throttle pressure, and modulated throttle pressure is also routed to the spring end of each valve to help regulate the pressure.

The 3-4 modulated throttle valve determines the modulated throttle pressure that is used to oppose 3-4 upshifts. The 4-3 modulated throttle

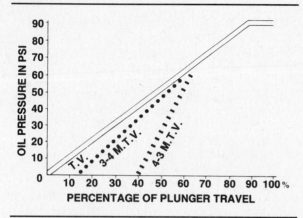

Figure 19-38. This chart shows the relationship between THM 440-T4 throttle pressure and the modulated throttle pressures created by the 3-4 and 4-3 modulated throttle valves.

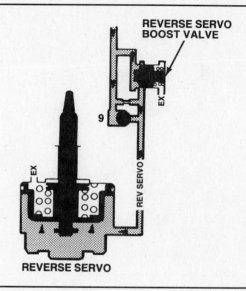

Figure 19-39. Operation of the THM 440-T4 reverse servo boost valve.

valve determines the modulated throttle pressure that is used to force part-throttle 4-3 downshifts. Figure 19-38 shows the relationship between throttle pressure, plunger travel, and the resulting modulated throttle pressures.

Reverse servo boost valve

When the vehicle is stopped and the gear selector is placed in Reverse, mainline pressure (REVERSE oil) is routed to the reverse servo to apply the reverse band and place the transaxle in reverse gear, figure 19-39. Servo apply pressure can reach the servo through an orifice, and also through the reverse servo boost valve which is normally held closed by spring force.

Under light throttle applications, servo apply pressure is too low to open the 1-2 servo boost valve. Instead, the check ball seats and all of the apply oil passes through the orifice to smooth band application. When the servo is released, the check ball unseats to allow servo apply oil to exhaust unrestricted.

During heavier throttle applications, the reverse band must hold under increased torque loads. Servo apply pressure increases at this time and opens the reverse servo boost valve to quickly route the increased apply pressure directly to the servo. This keeps the band from slipping during sudden throttle applications.

Converter clutch shift plug or shift valve

The THM 440-T4 torque converter clutch cannot be applied until a signal pressure is sent to the TCC solenoid and converter clutch apply valve. Intermediate-high (second) clutch apply pressure provides the converter clutch signal pressure in all models of the THM 440-T4 transaxle.

In gasoline-powered vehicles with Computer Command Control (CCC), figure 19-40, intermediate-high (second) clutch apply pressure is routed through the converter clutch shift plug to the TCC solenoid and apply valve as soon as the transaxle shifts into second gear. The shift plug is locked in place and serves only to direct pressure; it does not play any active part in applying or releasing the converter clutch.

In diesel-powered vehicles without CCC, figure 19-41, intermediate-high (second) clutch apply pressure is routed to the converter clutch shift valve on its way to the TCC solenoid and apply valve. The shift valve is controlled by opposition between governor pressure on one end, and spring force on the other end. Throttle pressure also acts on a land at the center of the valve to oppose governor pressure.

When the transaxle upshifts to second gear, intermediate-high clutch apply pressure is routed to the converter clutch shift valve. At lower vehicle speeds and/or larger throttle openings, figure 19-41A, the shift valve is closed by spring force and throttle pressure so the flow of signal pressure to the TCC solenoid and apply valve is blocked. Once the vehicle speed exceeds the minimum value and/or the throttle is closed beyond a certain point, governor pressure opens the shift valve against spring force and throttle pressure, figure 19-41B, and signal pressure is routed to the TCC solenoid and apply valve.

After the shift valve is open, throttle pressure is directed between two lands of the valve and acts with equal force in both directions.

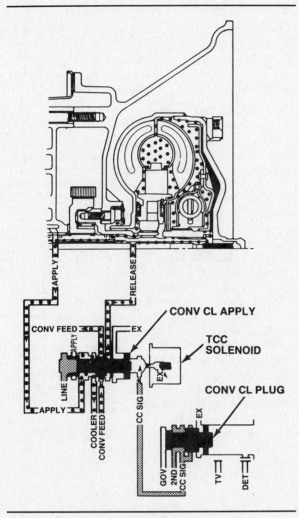

Figure 19-40. The THM 440-T4 torque converter clutch in the released position.

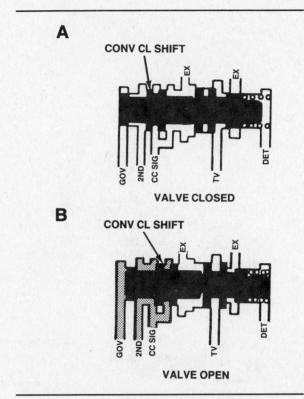

Figure 19-41. Operation of the THM 440-T4 converter clutch shift valve (non-CCC applications).

Therefore, spring force alone acts to close the shift valve and exhaust signal pressure if vehicle speed drops below the minimum allowed. This prevents the shift valve from hunting between the open and closed positions due to minor changes in throttle position.

During full-throttle downshifts, detent pressure is routed to the spring end of the shift valve to close the valve and exhaust converter clutch signal pressure. This unlocks the clutch so that torque multiplication can take place in the converter.

Converter clutch apply valve

The converter clutch apply valve receives converter feed oil from the pressure regulator valve, and converter apply oil from the converter clutch regulator valve. Depending on its position, the apply valve then directs these

pressures to the torque converter and transaxle cooler through various passages that determine whether the converter clutch is released or applied. Converter clutch apply valve position is controlled by opposition between mainline pressure and spring force at one end, and converter clutch signal pressure and the TCC solenoid at the other end.

The electrical and electronic controls used to control the TCC solenoid are described in Chapter 8. The paragraphs below describe the hydraulic operation of the converter clutch system.

When the TCC solenoid is de-energized, figure 19-40, its exhaust port is open and signal oil is exhausted. Thus, mainline pressure holds the converter clutch apply valve in the release position. In this position, apply oil from the converter clutch regulator valve is blocked from entering the apply valve, and converter feed oil is routed to the torque converter through the clutch release passage. Oil enters the converter between the clutch pressure plate and the converter cover; this prevents the clutch from applying. Oil leaving the converter returns to the converter clutch apply valve through the clutch apply passage, and is then routed to the transaxle cooler.

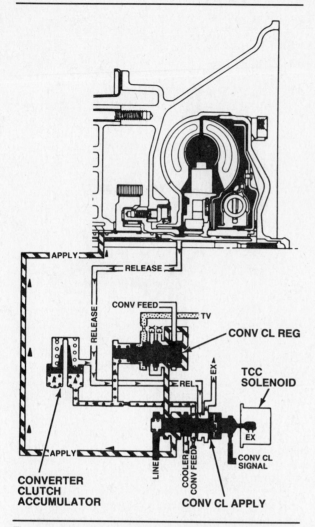

Figure 19-42. Operation of the THM 440-T4 converter clutch apply valve and regulator valve with the clutch applied.

When the TCC solenoid is energized, figure 19-42, its exhaust port is closed and signal pressure moves the converter clutch apply valve to the apply position against spring force and mainline pressure. In this position, mainline pressure is blocked from entering the valve, and apply pressure from the converter clutch regulator valve is routed to the torque converter through the clutch apply passage. Oil enters the converter between the clutch pressure plate and the turbine; this forces the pressure plate into contact with the converter cover, locking up the converter. Oil leaving the converter returns to the converter clutch apply valve through the clutch release passage, and is then exhausted.

After the converter clutch apply valve moves to the apply position, converter feed oil is routed to the converter clutch accumulator and

the converter clutch regulator valve to help regulate the rate of clutch application as described in the next section. A portion of the feed oil is also routed, through an internal passage in the apply valve, to the transaxle cooler.

Converter clutch regulator valve

The converter clutch regulator valve, together with the converter clutch accumulator, controls converter clutch apply feel. The regulator valve, figure 19-42, receives converter clutch feed pressure and regulates it into a converter clutch apply pressure that is proportional to throttle position. The regulator valve is controlled by opposition between apply pressure on one end, and converter feed pressure and spring force (some models) on the other end. Throttle pressure also acts on a land at the center of the valve to oppose apply pressure.

When the converter clutch is released, throttle pressure and spring force act to move the regulator valve to the right, opening the converter feed inlet port. Converter feed pressure then flows through the regulator valve to the converter clutch apply valve where it is blocked because the valve is in the release position. Some of the apply pressure acts on the end of the regulator valve to oppose throttle pressure and spring force. This moves the valve to the left to exhaust a portion of the apply pressure so that apply pressure is always proportional to throttle position.

When the converter clutch apply valve moves to the apply position, apply pressure is routed to the torque converter as already described. In addition, converter feed pressure from the apply valve is routed to the converter clutch accumulator and the end of the regulator valve opposite apply pressure. The accumulator then absorbs the converter feed oil while converter release oil is exhausting and apply oil is engaging the converter clutch. After the accumulator is bottomed, converter feed oil combines with throttle pressure and spring force to bottom the regulator valve against apply pressure, thus allowing full converter feed pressure into the apply circuit to hold the clutch firmly applied.

MULTIPLE-DISC CLUTCHES AND SERVOS

The preceding paragraphs examined the individual valves in the hydraulic system. Before we trace the complete hydraulic system operation in each driving range, we will take a quick look at the multiple-disc clutches and servos.

Clutches

The THM 440-T4 transaxle has four multiple-disc clutches. Each clutch is applied hydraulically, and released by several small coil springs when oil is exhausted from the clutch piston cavity.

The direct (input) clutch, together with the direct 1-way clutch, drives the front sun gear and is applied in all gear ranges except Overdrive third and Overdrive fourth. The intermediate-high (second) clutch drives the front carrier/rear ring gear and is applied in second, third, and fourth gears. The low-high (third) clutch, together with the low-high 1-way clutch, prevents the front sun gear from turning faster than input shaft speed and is applied in third, fourth, and manual low gears to provide engine braking. The overdrive (fourth) clutch is applied to hold the front sun gear in fourth gear only.

Servos

The THM 440-T4 has two piston-type servos that are used to apply the low-intermediate and reverse bands. The servo bores are located in the case, figure 19-7. The reverse servo is at the top left side of the case, and the low-intermediate servo is at the lower right side of the case.

Band adjustment for both the low-intermediate and reverse bands is controlled by the length of the servo rod, or apply pin, which is selected at the time of transaxle assembly. For external adjustment after transaxle assembly, a special General Motors tool may be used to select the correct pin.

HYDRAULIC SYSTEM SUMMARY

We have examined the individual valves, clutches, and servos in the hydraulic system and discussed their functions. The following paragraphs summarize their combined operations in the various gear ranges. Refer to figure 19-21 to trace the hydraulic system operation.

Park and Neutral

In Park and Neutral, oil flows from the pump to the pressure regulator valve, which regulates the pump pressure to create mainline pressure. The pressure regulator valve also creates converter feed pressure which is routed to the converter clutch regulator valve and converter clutch apply valve. The converter clutch regulator valve regulates the feed pressure to create a converter clutch apply pressure that is proportional to throttle opening. This apply pressure is sent to the converter clutch apply valve where it is blocked. Converter feed pressure

sent to the converter clutch apply valve is routed through the valve and into the release passage to the torque converter. Return oil from the converter is routed back through the apply valve, and then to the transaxle cooler and lubrication circuit.

Pressure in the mainline circuit is routed to the pressure relief valve, modulator valve, manual valve, and 2-3 shift valve. The pressure relief valve limits mainline pressure to a preset maximum value between 245 and 360 psi (1690 and 2480 kPa). The modulator valve regulates mainline pressure to create a modulator pressure that is proportional to engine intake manifold vacuum. Modulator pressure is sent to the line boost valve, 1-2 accumulator valve, 2-3 accumulator valve, and 3-2 coast valve.

Modulator pressure at the line boost valve increases mainline pressure in proportion to engine load. Modulator pressure at the 1-2 accumulator valve combines with spring force to help create a 1-2 accumulator pressure that is sent to the spring chambers of the 1-2 and 3-4 accumulators. Modulator pressure at the 2-3 accumulator valve combines with spring force to help control the rate of low-intermediate servo release during a 2-3 upshift. Modulator pressure at the 3-2 control valve opposes spring force to regulate the rate of low-high (third) clutch release and low-intermediate band application during a 3-2 downshift.

Mainline pressure at the manual valve is routed through the PRN port to the direct accumulator and direct clutch to apply the clutch. Mainline pressure also flows from the pump to the direct clutch through the downshifted 2-3 and 3-4 shift valves. The PRN port also routes mainline pressure to the reverse boost valve to increase mainline pressure in Reverse gear.

The manual valve PRND-432 port routes mainline pressure oil to the throttle limit valve, which limits throttle feed oil to a maximum of 90 psi (621 kPa). Throttle feed oil is then sent to the throttle valve, which regulates it to create a throttle pressure that is proportional to throttle opening. Throttle pressure is routed to the throttle plunger, 1-2 and 2-3 throttle valves, 3-4 and 4-3 modulated throttle valves, 3-2 downshift valve, converter clutch shift valve, and converter clutch regulator valve.

Throttle pressure at the throttle plunger helps regulate throttle pressure and makes it easier to hold the plunger in position. Throttle pressure at the 1-2 and 2-3 throttle valves combines with spring force to oppose upshifts. Throttle pressure at the 3-4 and 4-3 modulated throttle valves is re-regulated to produce 3-4 and 4-3 modulated throttle pressures which act on the 3-4 throttle valve to oppose upshifts and force downshifts. Throttle pressure at the 3-2 downshift valve opposes spring force to control the timing of the full-throttle (detent) downshift. Throttle pressure at the converter clutch shift valve opposes governor pressure to prevent the flow of signal pressure to the converter clutch apply valve until the throttle opening is below a certain point. Throttle pressure at the converter clutch regulator valve combines with converter feed oil (and spring force in some models) to create a converter clutch apply pressure that is proportional to throttle opening.

In Park and Neutral, the direct clutch is applied but has no effect because no member of the gearset is held. All of the other clutches and bands are released. All valve positions are determined by spring force alone. If the car is in park, manual linkage engages the parking pawl in the parking gear that is splined to the output shaft, to lock the output shaft.

Overdrive Low

When the gear selector is moved to Overdrive, the PRN port of the manual valve is closed, and the D4 port is opened. Mainline pressure is then routed through the D4 circuit to the 1-2 shift valve, 3-4 shift valve, 1-2 accumulator valve, 2-3 shift valve, and governor.

D4 pressure at the 1-2 and 3-4 shift valve is blocked. D4 pressure at the 1-2 accumulator valve is regulated into a 1-2 accumulator pressure that is sent to the spring chambers of the 1-2 and 3-4 accumulators to cushion the 1-2 and 3-4 upshifts. D4 pressure flows through the 2-3 shift valve to the 3-4 shift valve, 1-2 control valve, 1-2 servo boost valve, 2-3 accumulator valve, and apply side of the low-intermediate servo.

D4 oil in the second passage to the 3-4 shift valve is again blocked. D4 oil at the 1-2 servo control valve helps regulate the rate at which the low-intermediate servo applies during a 3-2 downshift. D4 oil at the 1-2 servo boost valve opens the valve and flows directly to the low-intermediate apply area when mainline pressure rises above a certain value. D4 oil at the low-intermediate servo applies the low-intermediate band to place the transaxle in low gear.

The governor regulates D4 oil to create a governor pressure that is proportional to vehicle speed. Governor pressure is routed to the 1-2, 2-3, and 3-4 shift valves where it opposes spring force and throttle pressure to cause upshifts. Governor pressure is also routed to the 3-2 control valve where it opposes spring force to control the rate of low-high clutch release and low-intermediate band apply during a 3-2 downshift.

In Overdrive low, the direct clutch and low-intermediate band are applied. The direct 1-way clutch holds on acceleration and overruns during deceleration.

Overdrive Second

As vehicle speed increases, governor pressure upshifts the 1-2 shift valve against spring force and throttle pressure. Mainline pressure then flows through the upshifted valve to the apply area of the intermediate-high clutch, 1-2 accumulator, and converter clutch shift plug or shift valve. The intermediate-high clutch applies, cushioned by the 1-2 accumulator, to place the transaxle in second gear.

Mainline pressure to the converter clutch shift plug flows through the plug to the converter clutch apply valve. In models with a converter clutch shift valve, mainline pressure flows through the shift valve to the apply valve once governor pressure is high enough, and throttle pressure is low enough, to open the shift valve.

In Overdrive second gear, the direct and intermediate-high clutches, along with the low-intermediate band, are applied. The direct 1-way clutch overruns. The converter clutch may or may not be applied, depending on vehicle operating conditions.

Overdrive Third

As vehicle speed increases, governor pressure upshifts the 2-3 shift valve against spring force and throttle pressure. Mainline pressure then flows through the upshifted valve to the apply area of the low-high clutch, release side of the low-intermediate servo, and 3-4 shift valve. The low-high clutch applies and the low-intermediate band is released. The release action of the low-intermediate servo cushions application of the low-high clutch. The flow of mainline pressure at the 3-4 shift valve is blocked by the downshifted valve. The upshifted 2-3 shift valve also blocks the entry of mainline pressure to the direct clutch, so the direct clutch releases.

In Overdrive third gear, the intermediate-high and low-high clutches are applied, and the low-high 1-way clutch is holding. The converter clutch may or may not be applied, depending on vehicle operating conditions.

Overdrive Fourth

As governor pressure continues to increase in Overdrive, it upshifts the 3-4 shift valve. This routes mainline pressure to apply the overdrive clutch passage and place the transaxle in fourth gear. The overdrive clutch application is cushioned by the 3-4 accumulator.

In Overdrive fourth gear, the intermediate-high, low-high, and overdrive clutches are applied. The low-high 1-way clutch overruns. The converter clutch may or may not be applied, depending on vehicle operating conditions.

Overdrive Range Forced Downshifts

A part-throttle 4-3 downshift occurs at lower speeds when 4-3 modulated throttle pressure combines with spring force acting on the 3-4 throttle valve to move the 3-4 shift valve against governor pressure. A part-throttle 4-3 downshift also occurs when the accelerator linkage moves the throttle plunger far enough to uncover the part-throttle port in the throttle valve. This routes part-throttle pressure to the spring end of the 3-4 throttle valve. The combination of part-throttle pressure and spring force then moves the 3-4 shift valve against governor pressure to cause a 4-3 downshift.

A full-throttle 3-2 detent downshift occurs at speeds below approximately 65 mph (105 kph) when throttle pressure from the 3-2 downshift valve flows to a second land on the 2-3 throttle valve. The combination of throttle pressure and spring force then moves the 2-3 shift valve against governor pressure and causes a 3-2 downshift. If the vehicle speed is below approximately 35 mph (56 kph) detent pressure will also downshift the 1-2 shift valve, resulting in a 3-1 downshift.

A full-throttle 2-1 detent downshift occurs at speeds below approximately 35 mph (56 kph) when detent pressure from the throttle valve flows to the 1-2 throttle valve. The combination of detent pressure, throttle pressure, and spring force moves the 1-2 shift valve against governor pressure to cause a 2-1 downshift.

Manual Third

When the gear selector is moved to manual third, the hydraulic flow within the transaxle is the same as in Overdrive third, except that the manual valve D3 port is opened. This routes mainline pressure to and through the 2-3 and 3-4 shift valves to apply the direct clutch and provide engine braking. D3 pressure also acts on the 3-4 shift valve to prevent a 3-4 upshift.

In manual third gear, the intermediate-high, low-high, and direct clutches are applied. The low-high and direct 1-way clutches are holding. The converter clutch may or may not be applied, depending on vehicle operating conditions.

Manual Second

When the gear selector is moved to manual second, the hydraulic flow within the transaxle is the same as in Overdrive second, except that the manual valve D2 port is opened. This routes mainline pressure to the spring end of the 2-3 throttle valve where it combines with spring force to downshift the valve. D2 oil also flows to the 1-2 control valve for a quick application of the low-intermediate servo. Low-high clutch and low-intermediate servo release oil exhaust at a rate controlled by the 3-2 control valve and the 3-2 coast valve.

In manual second gear, the direct and intermediate-high clutches, along with the low-intermediate band, are applied. The direct 1-way clutch overruns. The converter clutch may or may not be applied, depending on vehicle operating conditions.

Manual Low

When the gear selector is moved to manual low, LO oil from the manual valve is routed to the reverse boost valve, 1-2 throttle valve, and throttle limit valve. LO pressure at the reverse boost valve increases mainline pressure to hold the apply devices more securely. LO pressure at the 1-2 throttle valve downshifts the 1-2 shift valve and routes LO pressure to the apply area of the low-high clutch. The low-high clutch applies and provides engine braking. LO pressure to the throttle limit valve bottoms the valve against spring force to cut off all throttle pressure.

In manual low, the direct and low-high clutches, along with the low-intermediate band, are applied. The direct and low-high 1-way clutches are holding.

Reverse

When the gear selector is moved to Reverse, REVERSE oil from the manual valve is routed to the apply side of the reverse servo to apply the reverse band. PRN oil is routed to the direct clutch apply area to apply the clutch, and also to the reverse boost valve where it increases mainline pressure to hold the apply devices more securely. PRND-432 oil flows to the throttle limit valve.

In Reverse gear, the direct clutch and reverse band are applied.

Review Questions

Select the single most correct answer
Compare your answers to the correct answers on page 603

1. Mechanic A says the THM 440-T4 transaxle was introduced on cars in the 1984 model year.
   Mechanic B says the THM 440-T4 is sometimes referred to as the ME-9 or F-7 transaxle.
   Who is right?
   a. Mechanic A
   b. Mechanic B
   c. Both mechanic A and B
   d. Neither mechanic A nor B

2. Mechanic A says the "440" in THM 440-T4 means the transaxle cannot handle more than 440 ft-lb of torque.
   Mechanic B says the "T4" in THM 440-T4 means this gearbox is a 4-speed FWD transaxle.
   Who is right?
   a. Mechanic A
   b. Mechanic B
   c. Both mechanic A and B
   d. Neither mechanic A nor B

3. Mechanic A says the THM 440-T4 transaxle is based on the Simpson compound planetary geartrain.
   Mechanic B says the THM 440-T4 transaxle is based on the Ravigneaux compound planetary geartrain.
   Who is right?
   a. Mechanic A
   b. Mechanic B
   c. Both mechanic A and B
   d. Neither mechanic A nor B

4. Mechanic A says the THM 440-T4 transaxle uses four multiple-disc clutches and a single 1-way clutch.
   Mechanic B says the THM 440-T4 transaxle uses two 1-way (roller) clutches.
   Who is right?
   a. Mechanic A
   b. Mechanic B
   c. Both mechanic A and B
   d. Neither mechanic A nor B

5. Mechanic A says the THM 440-T4 gear selector has six positions.
   Mechanic B says the THM 440-T4 direct clutch is applied whenever the engine is running and the car is stopped.
   Who is right?
   a. Mechanic A
   b. Mechanic B
   c. Both mechanic A and B
   d. Neither mechanic A nor B

6. Mechanic A says the THM 440-T4 hydraulic system provides for a part-throttle 4-3 downshift.
   Mechanic B says the THM 440-T4 transaxle makes two downshifts while decelerating to a stop from third gear.
   Who is right?
   a. Mechanic A
   b. Mechanic B
   c. Both mechanic A and B
   d. Neither mechanic A nor B

7. Mechanic A says that when the THM 440-T4 gear selector is in the D or 3 position, engine compression braking is available in third gear.
   Mechanic B says the THM 440-T4 uses different apply devices in manual second than in Overdrive second.
   Who is right?
   a. Mechanic A
   b. Mechanic B
   c. Both mechanic A and B
   d. Neither mechanic A nor B

8. Mechanic A says a mechanical lockout prevents the THM 440-T4 transaxle from being shifted into Reverse while the vehicle is moving forward.
   Mechanic B says if the THM 440-T4 gear selector is shifted to the 1 position above approximately 40 mph (64 kph), the transaxle will shift into second gear.
   Who is right?
   a. Mechanic A
   b. Mechanic B
   c. Both mechanic A and B
   d. Neither mechanic A nor B

9. Mechanic A says the THM 440-T4 transaxle uses a variable-displacement vane-type oil pump.
   Mechanic B says all THM 440-T4 torque converters contain a viscous converter clutch.
   Who is right?
   a. Mechanic A
   b. Mechanic B
   c. Both mechanic A and B
   d. Neither mechanic A nor B

10. Mechanic A says the THM 440-T4 turbine shaft turns the drive sprocket, drive chain, driven sprocket, and input shaft.
    Mechanic B says the THM 440-T4 oil pump shaft passes through the hollow center of the turbine shaft.
    Who is right?
    a. Mechanic A
    b. Mechanic B
    c. Both mechanic A and B
    d. Neither mechanic A nor B

11. Mechanic A says the THM 440-T4 low-high clutch locks the inner race of the low-high 1-way clutch to the input shaft.
    Mechanic B says the THM 440-T4 overdrive clutch shaft is splined to the front sun gear.
    Who is right?
    a. Mechanic A
    b. Mechanic B
    c. Both mechanic A and B
    d. Neither mechanic A nor B

12. Mechanic A says that when the THM 440-T4 direct clutch is applied, the input shaft drives the front sun gear.
    Mechanic B says that when the THM 440-T4 intermediate-high clutch is applied, the input shaft drives the rear ring gear.
    Who is right?
    a. Mechanic A
    b. Mechanic B
    c. Both mechanic A and B
    d. Neither mechanic A nor B

13. Mechanic A says the THM 440-T4 front carrier and rear ring gear are a single assembly.
    Mechanic B says the THM 440-T4 rear carrier and front ring gear are a single assembly.
    Who is right?
    a. Mechanic A
    b. Mechanic B
    c. Both mechanic A and B
    d. Neither mechanic A nor B

14. Mechanic A says the THM 440-T4 front carrier and rear ring gear are always the output members of the planetary gearset.
    Mechanic B says the differential output shaft is locked to the case when the THM 440-T4 transaxle is in Park.
    Who is right?
    a. Mechanic A
    b. Mechanic B
    c. Both mechanic A and B
    d. Neither mechanic A nor B

15. Mechanic A says the THM 440-T4 transaxle uses a compound planetary gearset for its final drive. Mechanic B says the THM 440-T4 final drive ring gear is splined to the case.
Who is right?
a. Mechanic A
b. Mechanic B
c. Both mechanic A and B
d. Neither mechanic A nor B

16. Mechanic A says when the THM 440-T4 direct clutch and low-intermediate band are applied, the transaxle is in low gear. Mechanic B says when the THM 440-T4 direct clutch, low-high clutch, and low-intermediate band are applied, the transaxle is in low gear.
Who is right?
a. Mechanic A
b. Mechanic B
c. Both mechanic A and B
d. Neither mechanic A nor B

17. Mechanic A says the THM 440-T4 transaxle uses modulator pressure to control shift timing. Mechanic B says the THM 440-T4 transaxle uses modulator pressure to help control mainline pressure.
Who is right?
a. Mechanic A
b. Mechanic B
c. Both mechanic A and B
d. Neither mechanic A nor B

18. Mechanic A says the THM 440-T4 reverse band is wrapped around the low-intermediate clutch drum. Mechanic B says the THM 440-T4 low-intermediate band is wrapped around the rear ring gear.
Who is right?
a. Mechanic A
b. Mechanic B
c. Both mechanic A and B
d. Neither mechanic A nor B

19. Mechanic A says the THM 440-T4 transaxle stores fluid in two different reservoirs. Mechanic B says the THM 440-T4 manual valve is located in the channel plate.
Who is right?
a. Mechanic A
b. Mechanic B
c. Both mechanic A and B
d. Neither mechanic A nor B

20. Mechanic A says THM 440-T4 transaxles used with diesel engines have a converter clutch shift plug in place of a converter clutch shift valve. Mechanic B says all THM 440-T4 transaxles use a TCC solenoid to help control torque converter clutch operation.
Who is right?
a. Mechanic A
b. Mechanic B
c. Both mechanic A and B
d. Neither mechanic A nor B

20

The Turbo Hydra-matic 700-R4 Transmission

HISTORY AND MODEL VARIATIONS

The General Motors (GM) THM (THM) 700-R4 transmission, figure 20-1, is a heavy-duty unit designed for use with medium- and large-displacement engines in RWD cars and trucks. The THM 700-R4 was introduced in 1982 Chevrolet Impala, Corvette, and Blazer models, as well as in full-size Chevrolet and GMC pickups. In 1983, the Chevrolet Camaro and Pontiac Firebird and Parisienne also began using this transmission. Since that time, the THM 700-R4 has been used in most full-size GM cars and trucks, and in other selected heavy duty and high-performance applications.

The THM 700-R4 uses five multiple-disc clutches, a low 1-way (roller) clutch, a 1-way sprag clutch, and a band to provide its various gear ratios. All models of this transmission are fitted with a torque converter clutch to improve fuel economy. Like all of GM's newer design transmissions and transaxles, the THM 700-R4 is built to metric specifications. Some GM service manuals and parts catalogs refer to the THM 700-R4 as the MD-8 transmission.

Fluid Recommendations

Like all GM automatic transmissions and transaxles, the THM 700-R4 requires DEXRON®-IID fluid.

Transmission Identification

To obtain proper repair parts, you must know the transmission assembly part number and other identifying characters. Unlike most other late-model GM automatic transmissions, the THM 700-R4 does not have an identification tag riveted to it. Instead, the identification numbers are stamped directly into the case on the right rear pan support rail. Figure 20-2 shows a typical THM 700-R4 identification number and explains the meaning of each character.

As required by Federal regulations, all GM transmissions carry the serial number of the vehicle in which they were originally installed. The THM 700-R4 serial number is also stamped on the lower right side of the transmission case. If the case is ever replaced, the complete set of identification numbers should be transferred to the new case.

Major Changes

The major changes that GM has made in the THM 700-R4 transmission are as follows:

1983 ● Input shaft and drum assembly revised.

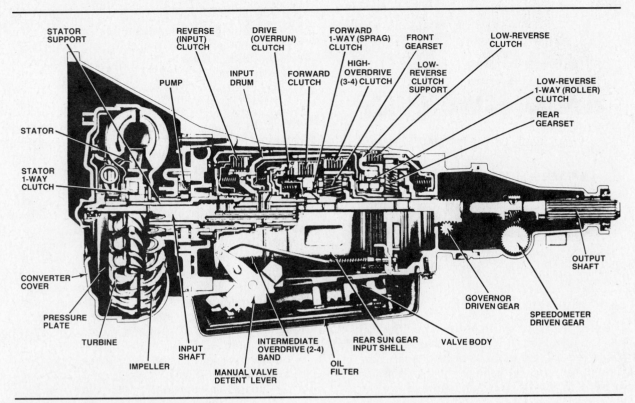

Figure 20-1. The THM 700-R4 transmission.

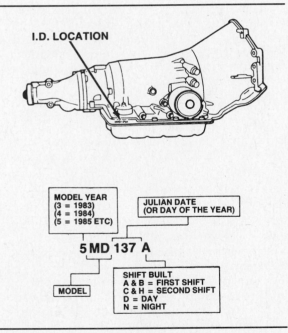

I.D. LOCATION

MODEL YEAR
(3 = 1983)
(4 = 1984)
(5 = 1985 ETC)

JULIAN DATE
(OR DAY OF THE YEAR)

5 MD 137 A

MODEL

SHIFT BUILT
A & B = FIRST SHIFT
C & H = SECOND SHIFT
D = DAY
N = NIGHT

Figure 20-2. The THM 700-R4 identification numbers are stamped on the right rear edge of the oil pan support rail.

1984 ● New torque converter introduced.
　　 ● New rear internal gear.

1985 ● Revised valve body that eliminates the converter clutch shift valve.
　　 ● New pump assembly for transmissions with electronically controlled torque converter clutch.
　　 ● New rear carrier with oil deflector.
1986 ● Wider low 1-way clutch.
1987 ● New ten-vane oil pump.
　　 ● New design reverse clutch.
　　 ● New input shaft.
　　 ● Deeper oil pan.
　　 ● Wider input sprag.

GEAR RANGES

The THM 700-R4 gear selector has seven positions, figure 20-3:

P — Park
R — Reverse
N — Neutral
Ⓓ — Overdrive
D or 3 — Manual third
2 — Manual second
1 — Manual low.

Gear ratios for the THM 700-R4 are:
● First (low) — 3.06:1
● Second (intermediate) — 1.63:1
● Third (direct drive) — 1.00:1
● Fourth (overdrive) — 0.70:1
● Reverse — 2.30:1.

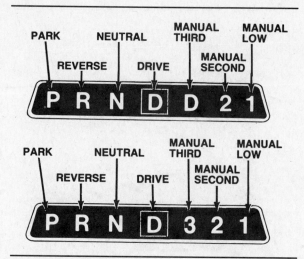

Figure 20-3. THM 700-R4 gear selector positions.

Park and Neutral

The engine can be started only when the transmission is in Park or Neutral. No clutches or bands are applied in these positions, and there is no flow of power through the transmission. In Neutral, the output shaft is free to turn, and the car can be pushed or pulled. In Park, a mechanical pawl engages the parking gear on the output shaft. This locks the output shaft to the transmission case and prevents the vehicle from being moved.

Overdrive

Overdrive (Ⓓ) is the normal driving range for the THM 700-4R transmission. When the car is stopped and the gear selector is placed in Overdrive, the transmission is automatically in low gear. As the vehicle accelerates, the transmission automatically upshifts to second, third, and then to fourth gear. The speeds at which upshifts occur vary, and are controlled by vehicle speed and throttle opening. The torque converter clutch may be applied in second, third, and fourth gears.

In Overdrive, the driver can force downshifts by opening the throttle partially or completely. The exact speeds at which downshifts occur vary for different engine, axle ratio, tire size, and vehicle weight combinations.

The driver can force a part-throttle 4-3 or 3-2 downshift by opening the throttle beyond a certain point. A full-throttle 3-2 detent downshift is possible at speeds below approximately 60 mph (97 kph) when the throttle is fully opened. A full-throttle 3-1 or 2-1 downshift can be forced at speeds below approximately 30 mph (48 kph).

When the car decelerates in Overdrive to a stop, the transmission makes three downshifts automatically, first a 4-3 downshift, then a 3-2 downshift, then a 2-1 downshift.

Manual Third

When the vehicle is stopped and the gear selector is placed in manual third (D or 3), the transmission is automatically in low gear. As the vehicle accelerates, the transmission automatically upshifts to second, and then to third gear; it does not upshift to fourth. The speeds at which upshifts occur vary, and are controlled by vehicle speed and throttle opening. The torque converter clutch may be applied in second and third gears.

In manual third, the driver can force a part-throttle 3-2 downshift by opening the throttle beyond a certain point. A full-throttle 3-2 detent downshift is possible at speeds below approximately 60 mph (97 kph) when the throttle is fully opened. A full-throttle 3-1 or 2-1 downshift can be forced at speeds below approximately 30 mph (48 kph).

The driver can manually downshift from fourth to third by moving the selector lever from Ⓓ to D or 3. In this case, the transmission shifts into third gear immediately, regardless of vehicle speed or throttle position. Because different apply devices are used in manual third than are used in Overdrive third, this gear range can be used to provide engine compression braking, or gear reduction for hill climbing, pulling heavy loads, or operating in congested traffic.

When the car decelerates to a stop in manual third, the transmission makes two downshifts automatically, first a 3-2 downshift, then a 2-1 downshift.

Manual Second

When the vehicle is stopped and the gear selector is placed in manual second (2), the transmission is automatically in low gear. As the vehicle accelerates, the transmission automatically upshifts to second gear; it does not upshift to third or fourth. The speed at which the upshift occurs will vary, and is controlled by vehicle speed and throttle opening.

In manual second, the driver can force a full-throttle 2-1 downshift at speeds below approximately 30 mph (48 kph). The transmission automatically downshifts to low as the vehicle decelerates to a stop.

The driver can manually downshift from fourth or third to second by moving the selector

GENERAL MOTORS TERM	TEXTBOOK TERM
Clutch housing	Clutch drum
Converter pump	Impeller
Detent downshift	Full-throttle downshift
Turbine shaft	Input shaft
Internal gear	Ring gear
Input planetary gearset	Front planetary gearset
Reaction planetary gearset	Rear planetary gearset
Line pressure	Mainline pressure
Control valve assembly	Valve body
T.V. boost valve	Throttle boost valve
T.V. limit valve	Throttle limit valve
T.V. plunger	Throttle plunger
M.T.V. up valve	Modulated throttle upshift valve
M.T.V. down valve	Modulated throttle downshift valve
1-2, 2-3, 3-4 T.V. valves	1-2, 2-3, 3-4 throttle valves
Converter clutch T.V. valve	Converter clutch throttle valve
2-4 band	Intermediate-overdrive band
2-4 servo	Intermediate-overdrive servo
Reverse input clutch	Reverse clutch
Overrun clutch	Drive clutch
3-4, or third, clutch	High-overdrive clutch
Forward sprag clutch	Forward 1-way clutch
Lo roller clutch	Low-reverse 1-way clutch

Figure 20-4. THM 700-R4 transmission nomenclature table.

from Ⓓ, D, or 3 to 2. In this case, the transmission shifts into second gear immediately, regardless of road speed or throttle position. Second gear can then be used to provide engine compression braking, or gear reduction for hill climbing, pulling heavy loads, or operating in congested traffic.

Manual Low

When the vehicle is stopped and the gear selector is placed in manual low (1), the car starts and stays in first gear; it does not upshift to second, third, or fourth. Because different apply devices are used in manual low than are used in Overdrive low, this gear range can be used to provide engine compression braking to slow the vehicle.

If the driver shifts to manual low at high speed, the transmission will not shift directly to low gear. To avoid engine or driveline damage, the transmission first downshifts to second, and then automatically downshifts to low when the vehicle speed drops below approximately

30 mph (48 kph). The transmission then remains in low gear until the gear selector is moved to another position.

Reverse

When the gear selector is moved to R, the transmission shifts into Reverse. The transmission should be shifted to Reverse only when the vehicle is standing still. The lockout feature of the shift gate should prevent accidental shifting into Reverse while the vehicle is moving forward. However, if the driver intentionally moves the selector to R while moving forward, the transmission *will* shift into Reverse, and damage may result.

GM TRANSMISSION NOMENCLATURE

For the sake of clarity, this text uses consistent terms for transmission parts and functions. However, the various manufacturers may use different terms. Therefore, before we examine the buildup of the THM 700-R4 and the operation of its geartrain and hydraulic system, you should be aware of the unique names that General Motors uses for some transmission parts and operations. This will make it easier in the future if you should look in a GM manual and see a reference to an "internal gear". You will realize that this is simply a ring gear and is similar to the same part in any other transmission. Figure 20-4 lists GM nomenclature and the textbook terms for some transmission parts and operations.

THM 700-R4 BUILDUP

The THM 700-R4 has a 1-piece case casting that incorporates the bellhousing, figure 20-5. The extension housing is a separate casting that bolts to the back of the case. The torque converter is a welded unit that fits inside the bellhousing and cannot be disassembled. The converter bolts to the engine flexplate, which has the starter ring gear around its outer diameter. The THM 700-R4 converter is a conventional 3-element design (impeller, turbine, and stator) with the addition of a hydraulically applied clutch.

The torque converter is installed in the transmission by sliding it onto the stationary stator support, which is part of the oil pump housing. The splines in the stator 1-way clutch engage the splines on the stator support, and the splines in the turbine engage the transmission input shaft. The hub at the rear of the torque

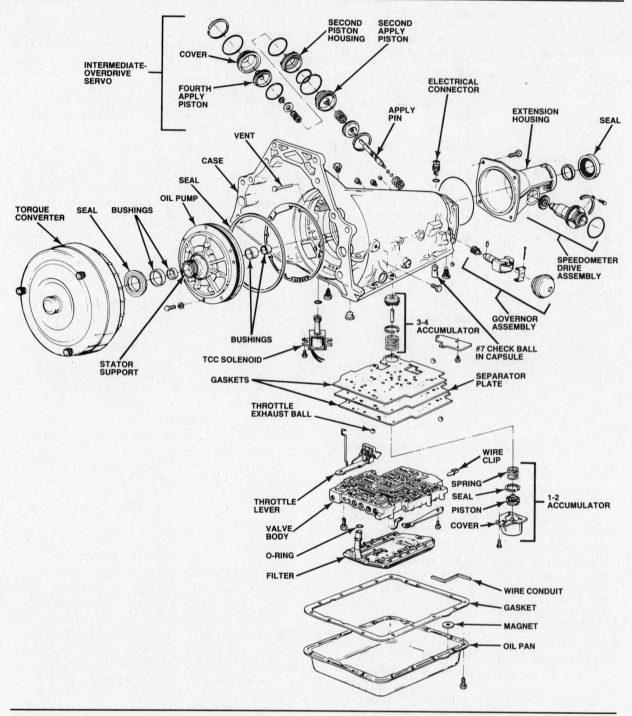

Figure 20-5. An exploded view of the THM 700-R4 case showing the torque converter, oil pump, valve body, oil pan, extension housing, and other components.

converter housing engages the lugs on the oil pump rotor to drive the pump whenever the engine is running.

The THM 700-R4 uses a variable-displacement vane-type oil pump mounted in the front of the transmission case, figure 20-5. The pump draws fluid from the sump through a filter, and routes the fluid through passages in the valve body and transmission case to various valves in the hydraulic system.

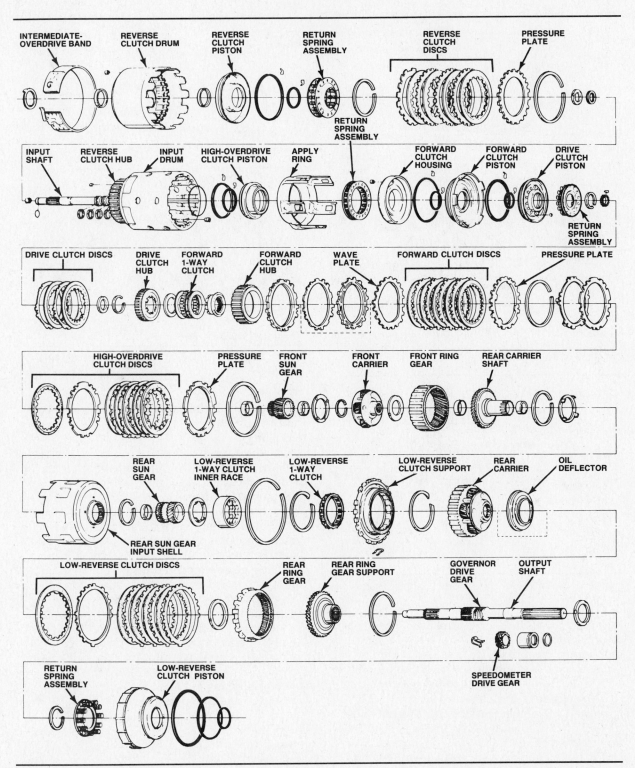

Figure 20-6. An exploded view of the complete THM 700-R4 compound planetary geartrain and all of its apply devices.

Figure 20-6 shows an exploded view of the entire THM 700-R4 geartrain. As we complete the buildup, refer to figure 20-6 and the detailed figures for each specific assembly to identify the components being discussed.

The input shaft is turned clockwise by the torque converter turbine, and passes through the stator support, where it is supported by bushings, into the transmission case. The input shaft is one assembly with the reverse clutch

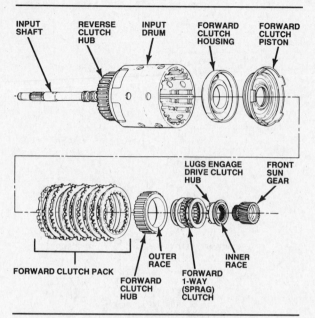

Figure 20-7. The THM 700-R4 input shaft, reverse clutch hub, input drum, forward clutch, forward 1-way clutch, and front sun gear.

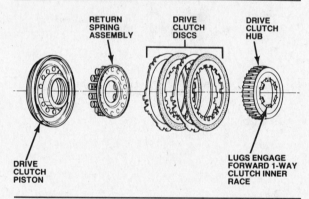

Figure 20-8. The THM 700-R4 drive clutch assembly.

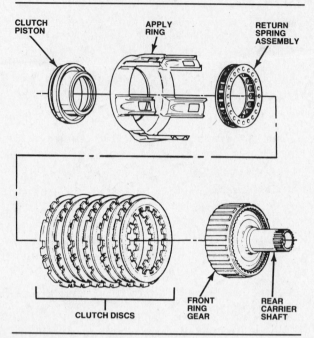

Figure 20-9. The THM 700-R4 high-overdrive clutch assembly.

hub and the input drum, figure 20-7. The input drum is piloted on a hub at the rear of the stator support. Seal rings are installed on the input shaft and the outside of the stator support hub to seal the oil passages to the clutches.

The outside of each forward clutch steel disc is splined to the inside of the input drum and turned by it, figure 20-7. The inside of each friction disc is splined to the outside of the forward clutch hub. The inside of the forward clutch hub forms the outer race of the forward 1-way (sprag) clutch. The inner race of the forward 1-way clutch is splined to the front sun gear. When the forward clutch is applied in all forward gear ranges, the input shaft drives the front sun gear clockwise through the forward

1-way clutch. If the front sun gear turns faster than the input shaft, the forward 1-way clutch overruns.

The outside of each drive (overrun) clutch steel disc is splined to the inside of the input drum and turned by it, figure 20-8. The inside of each friction disc is splined to the drive clutch hub. The drive clutch hub has lugs that engage the inner race of the forward 1-way clutch, and thus, the front sun gear. When the drive clutch is applied in manual third, the front sun gear is locked to the input shaft so that the forward 1-way clutch cannot overrun. This provides engine braking. Because the drive clutch holds the sun gear only during deceleration (the forward clutch and forward 1-way clutch continue to hold during acceleration), it does not have to absorb the full engine torque of acceleration. Therefore, the drive clutch can be made smaller than the other clutches in the transmission.

The outside of each high-overdrive (3-4) clutch steel disc is splined to the inside of the input drum and turned by it, figure 20-9. The inside of each friction disc is splined to the outside of the front ring gear. The front ring gear is splined, through a shaft, to the rear carrier. When the high-overdrive clutch is applied in third and fourth gears, the input shaft drives the front ring gear and rear carrier clockwise.

The inside of each reverse clutch friction disc is splined to the outside of the reverse clutch hub on the input drum and turned by the input

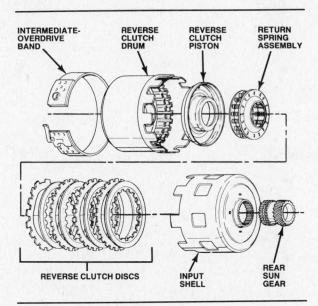

Figure 20-10. The THM 700-R4 intermediate-overdrive band, reverse clutch assembly, input shell, and rear sun gear.

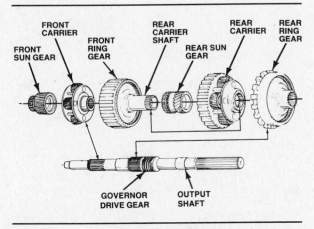

Figure 20-11. The THM 700-R4 front and rear gearsets. The front carrier and ring gear are splined to the output shaft. The front ring gear and rear carrier are splined to one another.

shaft, figure 20-10. The outside of each reverse clutch steel disc is splined to the inside of the reverse clutch drum. The reverse clutch drum has lugs that engage an input shell that is splined to the rear sun gear. When the reverse clutch is applied in reverse gear, the input shaft drives the reverse clutch drum, input shell, and rear sun gear clockwise.

The outside of the reverse clutch drum is wrapped by the intermediate-overdrive (2-4) band, figure 20-10. When the intermediate-overdrive band is applied in second and fourth gears, it holds the reverse clutch drum, input shell, and rear sun gear stationary.

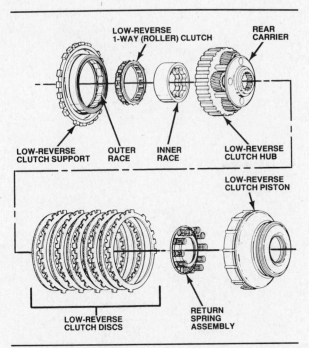

Figure 20-12. The THM 700-R4 low-reverse clutch support, low-reverse 1-way clutch, rear carrier, and low-reverse clutch assembly.

In the THM 700-R4 compound planetary gearset, figure 20-11, the front sun gear is supported by two bushings on the output shaft, but it is not splined to the shaft and does not turn with it. The front sun gear meshes with the front pinions, which turn on shafts in the front carrier. The front carrier is splined to the output shaft. The front pinions also mesh with the front ring gear, which as mentioned earlier is splined through a shaft to the rear carrier.

The rear pinions turn on shafts in the rear carrier, and mesh with the rear sun gear and rear ring gear. The rear sun gear, as mentioned earlier, is splined to the input shell and reverse clutch drum. The rear ring gear is splined to the output shaft, and has the parking gear around its outer edge. Because the front carrier and rear ring gear are splined to the output shaft, one or the other is always the output member of the compound planetary gearset.

The rear carrier is splined to the inner race of the low 1-way (roller) clutch, figure 20-12. The outer race of the Low 1-way clutch is formed by the inside of the low-reverse clutch support, which is splined to the transmission case. The Low 1-way clutch prevents counterclockwise rotation of the rear carrier, but allows the carrier to turn clockwise.

The inside of each low-reverse clutch friction disc is splined to the outside of the rear carrier, figure 20-12. The outside of each low-reverse

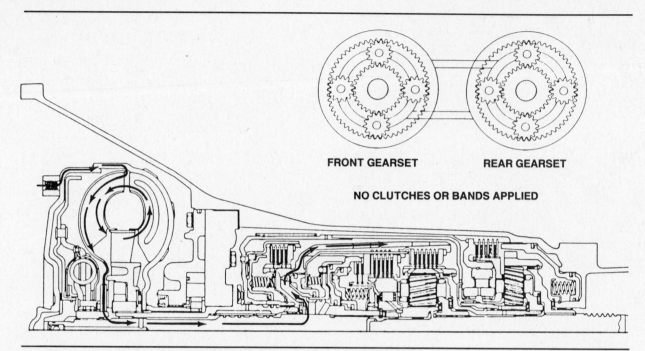

FRONT GEARSET   **REAR GEARSET**

NO CLUTCHES OR BANDS APPLIED

Figure 20-13. THM 700-R4 power flow in Park and Neutral.

clutch steel disc is splined to the transmission case. When the low-reverse clutch is applied, the rear carrier is prevented from turning in either direction.

The output shaft, supported on bushings, passes through the rear of the transmission case and into the extension housing. The governor drive gear is cut into the shaft behind the splines where the rear ring gear mounts, figure 20-6. The speedometer drive gear is clipped to the output shaft between the governor drive gear and the output yoke splines. The governor is mounted in the left rear of the main case and driven by a nylon pinion gear that engages the gear on the output shaft, figure 20-5. A bushing and seal for the output yoke are installed at the rear of the extension housing.

THM 700-R4 POWER FLOW

Now that we have taken a quick look at the general way in which the THM 700-R4 is assembled, we can trace the flow of power through the apply devices and the planetary gearsets in each of the gear ranges.

Park and Neutral

When the gear selector is in Park or Neutral, figure 20-13, no clutches or bands are applied, and the torque converter clutch is released. The torque converter cover/impeller and the oil pump rotate clockwise at engine speed. The torque converter turbine and the input

shaft also rotate clockwise, but at less than engine speed. The input shaft turns the input drum and reverse clutch hub clockwise. However, because none of the clutches engaged with this assembly are applied, there is no input to the planetary geartrain, and thus, no output motion at this time.

In Neutral, the output shaft is free to rotate in either direction, so the vehicle can be pushed or pulled. In Park, a mechanical pawl engages the parking gear on the outside of the rear ring gear. This locks the output shaft to the transmission case to prevent the vehicle from moving in either direction.

Overdrive Low

When the car is stopped and the gear selector is moved to Overdrive (Ⓓ), the transmission is automatically in low gear, figure 20-14. The forward clutch is applied, which causes the input shaft to drive the front sun gear clockwise through the forward 1-way clutch.

Before the car begins to move, the transmission output shaft is held stationary by the driveline and wheels. This also holds the front carrier and rear ring gear stationary. As a result, the clockwise rotation of the front sun gear turns the front pinions counterclockwise on the stationary carrier. The front pinions then attempt to drive the front ring gear, and the rear carrier splined to it, counterclockwise.

However, the rear carrier is splined to the Low 1-way clutch, which locks up to prevent

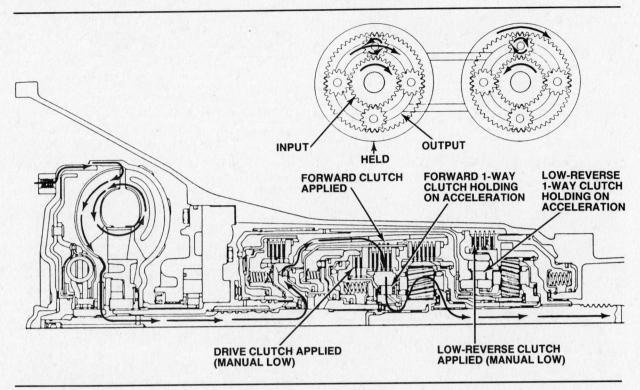

INPUT — HELD — OUTPUT

FORWARD CLUTCH APPLIED — FORWARD 1-WAY CLUTCH HOLDING ON ACCELERATION — LOW-REVERSE 1-WAY CLUTCH HOLDING ON ACCELERATION

DRIVE CLUTCH APPLIED (MANUAL LOW) — LOW-REVERSE CLUTCH APPLIED (MANUAL LOW)

Figure 20-14. THM 700-R4 power flow in Overdrive and manual low.

the rear carrier and front ring gear from turning counterclockwise. Because the Low 1-way clutch holds the front ring gear more securely than the drive wheels can hold the output shaft and front carrier, the front pinions walk clockwise around the inside of the front ring gear, taking the carrier with them. The front carrier then drives the output shaft clockwise in gear reduction.

In Overdrive low, the output shaft drives the rear ring gear clockwise. The rear ring gear drives the rear pinions clockwise on the held rear carrier. And the rear pinions drive the rear sun gear counterclockwise. However, because the rear sun gear is not locked to any other member of the transmission, the rear gearset simply freewheels at this time.

As long as the car is accelerating in Overdrive low, torque is transferred from the input shaft, through the planetary gearset, to the output shaft. However, when the driver lets up on the throttle, the engine speed drops while the drive wheels attempt to keep turning at the same speed. This removes the clockwise-rotating torque load from the forward 1-way clutch, and the counterclockwise-rotating torque load from the Low 1-way clutch. Both clutches then unlock and allow the compound planetary gearset to freewheel so no engine compression

braking is available. This is a major difference between Overdrive low and manual low as we will see later.

In Overdrive low, all of the gear reduction is produced by the front planetary gearset. The front sun gear is the input member; the front ring gear is the held member; and the front carrier is the output member. The forward clutch, together with the forward 1-way clutch, is the input device. The Low 1-way clutch is the holding device.

Overdrive Second

As the vehicle continues to accelerate in Overdrive, the transmission hydraulic system produces an automatic upshift to second gear by applying the intermediate overdrive band, figure 20-15, to hold the reverse drum, input shell, and rear sun gear stationary. The forward clutch remains applied, and the input shaft drives the front sun gear clockwise through the forward 1-way clutch.

As in Overdrive low, the front sun gear rotates clockwise and drives the front pinions counterclockwise. The front pinions then walk around the inside of the front ring gear and drive the front carrier and output shaft clockwise in gear reduction. The output shaft drives the rear ring gear clockwise, and the rear ring

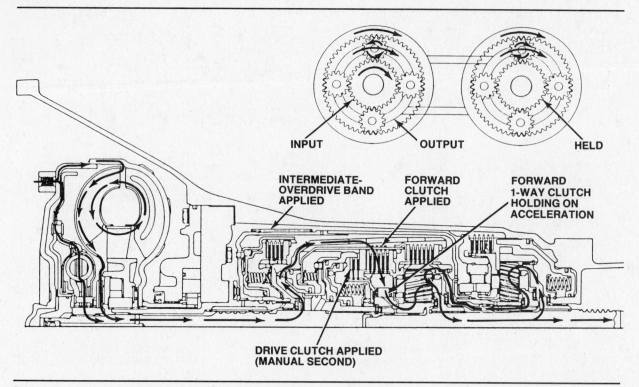

INPUT   OUTPUT   HELD

INTERMEDIATE-
OVERDRIVE BAND
APPLIED

FORWARD
CLUTCH
APPLIED

FORWARD
1-WAY CLUTCH
HOLDING ON
ACCELERATION

DRIVE CLUTCH APPLIED
(MANUAL SECOND)

Figure 20-15. THM 700-R4 power flow in Overdrive and manual second.

gear drives the rear pinions clockwise. The rear pinions then walk clockwise around the stationary rear sun gear, taking the rear carrier with them.

The clockwise rotation of the rear carrier is transferred through the rear carrier shaft to the front ring gear, which was held stationary by the Low 1-way clutch in Overdrive low. Clockwise rotation of the front ring gear causes the front pinions to turn counterclockwise more slowly, while forcing the front carrier to turn clockwise more rapidly. The front carrier then drives the output shaft in gear reduction. The clockwise rotation of the rear carrier removes the counterclockwise-rotating torque from the Low 1-way clutch, causing it to overrun at this time.

As long as the vehicle is accelerating in Overdrive second, torque is transferred from the engine, through the planetary gearset, to the output shaft. As vehicle speed rises and the driver lets up on the accelerator pedal, the transmission generally upshifts to third. Because of this upshift, there is no engine compression braking. If the vehicle speed is not high enough for an upshift, there still is no engine compression braking because the forward 1-way clutch overruns and allows the front sun gear to freewheel. This is a major difference between Overdrive second and manual second, as we will see later.

In Overdrive second, gear reduction is produced by a combination of the front and rear planetary gearsets. The front sun gear is the input member; the rear sun gear is the held member; and the front carrier is the output member. The forward clutch, together with the forward 1-way clutch, is the input device. The intermediate-overdrive band is the holding device.

Overdrive Third

As the vehicle continues to accelerate in Overdrive, the transmission automatically upshifts from second to third gear. For this to happen, the intermediate-overdrive band is released, and the high-overdrive clutch is applied, figure 20-16.

The forward clutch remains applied, and the input shaft drives the front sun gear clockwise through the forward 1-way clutch. The input shaft also drives the front ring gear clockwise through the high-overdrive clutch. This means that two members of the front planetary gearset, the sun gear and the ring gear, are locked to the input shaft, and thus to each other. With two members of the gearset locked together, the entire gearset turns as a unit.

With the front planetary gearset locked, there is no relative motion between the sun gear, ring gear, carrier, or pinions. As a result,

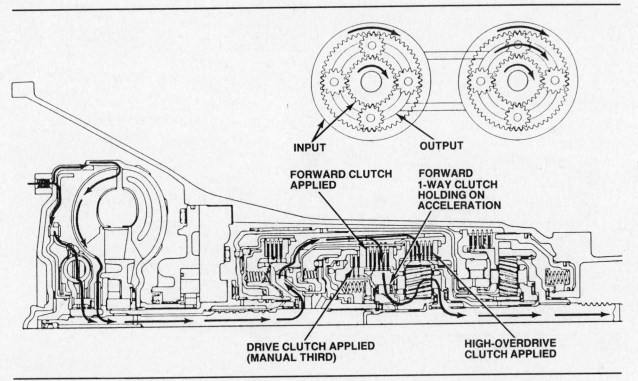

Figure 20-16. THM 700-R4 power flow in Overdrive and manual third.

input torque to the sun and ring gears is transmitted through the carrier to drive the output shaft. In third gear, there is a direct 1:1 gear ratio between the input shaft speed and output shaft speed. There is no reduction or increase in either speed or torque through the transmission.

As long as the vehicle is accelerating in Overdrive third, torque is transferred from the engine, through the planetary gearset, to the output shaft. As vehicle speed rises and the driver lets up on the accelerator pedal, the transmission generally upshifts to fourth. Because of this upshift, there is no engine compression braking. If the vehicle speed is not high enough for an upshift, there still is no engine compression braking because the forward 1-way clutch overruns and allows the front sun gear to freewheel. This is a major difference between Overdrive third and manual third, as we will see later.

In Overdrive third, all of the output is through the front planetary gearset. The front sun gear and front ring gear are input members equally. The front carrier is the output member. The forward clutch, together with the forward 1-way clutch, and the high-overdrive band are the input devices. No holding device is required because the gearset is locked.

Overdrive Fourth

As the vehicle continues to accelerate in Overdrive, the transmission hydraulic system produces an automatic upshift from third to fourth gear by re-applying the intermediate-overdrive band to hold the reverse drum, input shell, and rear sun gear, figure 20-17. The high-overdrive clutch remains applied, and the input shaft drives the front ring gear clockwise. The front ring gear is splined to the rear carrier, and drives it clockwise. As the rear carrier turns, the rear pinions walk clockwise around the held rear sun gear. The rear pinions then drive the rear ring gear and output shaft clockwise in Overdrive.

In fourth gear, the front ring gear turns clockwise at input shaft speed, while the front carrier turns clockwise at the faster output shaft speed. As a result, the front pinions turn counterclockwise and drive the front sun gear clockwise faster than input shaft speed. This causes the forward 1-way clutch to overrun and makes the forward clutch ineffective, even though it remains applied at this time.

Engine compression braking is available in Overdrive fourth gear because there is a solid mechanical connection between the input shaft and the output shaft. No 1-way clutches are involved in transmitting power at this time.

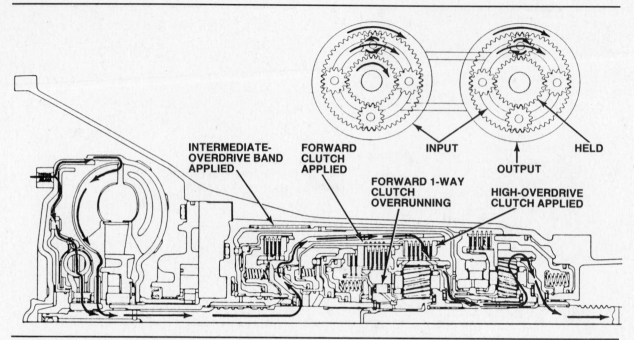

Figure 20-17. THM 700-R4 power flow in Overdrive fourth.

All of the power in fourth gear is transmitted through the rear planetary gearset. The carrier is the input member, the sun gear is the held member, and the ring gear is the output member. The input device is the high-overdrive clutch. The holding device is the intermediate-overdrive band.

Manual Third

When the vehicle is stopped and the gear selector is moved to manual third (D or 3), the transmission operates just as in Overdrive except that the hydraulic system prevents an upshift to fourth gear. The flow of power through the compound planetary gearset in the lower three gears is the same as in Overdrive low, second, and third, except for the use of one apply device.

With the gear selector in the manual third position, the drive clutch is applied to lock the front sun gear to the input shaft. Remember that in Overdrive low, second, and third gears, the forward 1-way clutch drives the front sun gear during acceleration, but overruns and allows the sun gear to freewheel during deceleration. In manual third, figure 20-16, the drive clutch prevents the front sun gear from freewheeling during deceleration, and thus provides engine compression braking in second and third gears. Engine braking is not available in low gear, however, because the Low 1-way clutch overruns.

If the driver moves the gear selector to D or 3 while the car is moving in Overdrive fourth, the hydraulic system will immediately downshift the transmission to third gear by releasing the intermediate-overdrive band and applying the drive clutch. If the gear selector is moved to D or 3 while the car is moving in Overdrive third, the hydraulic system will apply the drive clutch and lock out an upshift to fourth. This capability provides the driver with a controlled downshift for engine compression braking or improved acceleration at any speed or throttle position.

Manual Second

When the vehicle is stopped and the gear selector is moved to manual second (2), the transmission operates just as in manual third except that the hydraulic system prevents an upshift to third gear. The flow of power through the compound planetary gearset in first and second gears is identical, and the same apply devices are used.

When the gear selector is in manual second and the car accelerates from a standstill, the transmission starts in low gear, figure 20-14. The forward clutch and forward 1-way clutch drive the front sun gear, the Low 1-way clutch holds the rear carrier and front ring gear, and the drive clutch prevents the forward 1-way clutch from overrunning. As the car continues to accelerate, the transmission automatically

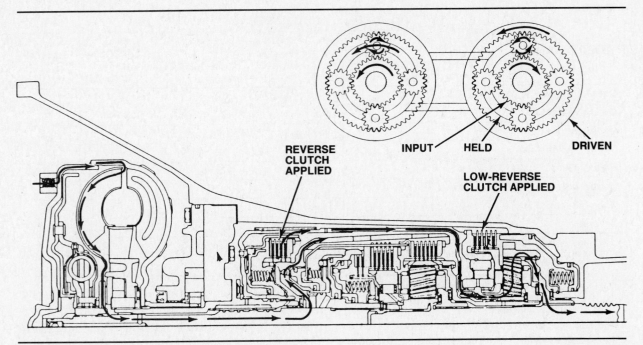

REVERSE CLUTCH APPLIED

INPUT HELD DRIVEN

LOW-REVERSE CLUTCH APPLIED

Figure 20-18. THM 700-R4 power flow in Reverse.

shifts to manual second gear by applying the intermediate-overdrive band to hold the rear sun gear, figure 20-15.

If the driver moves the gear selector to 2 while the car is moving in Overdrive fourth, Overdrive third, or manual third, the hydraulic system immediately downshifts the transmission to second gear. If the downshift is from Overdrive fourth, the high-overdrive clutch is released and the drive clutch is applied. If the downshift is from Overdrive third, the high-overdrive clutch is released and the intermediate-overdrive band and drive clutch are applied. The same apply devices are involved in a downshift from manual third, except that the drive clutch is already applied. This capability provides the driver with a controlled downshift for engine compression braking or improved acceleration at any speed or throttle position.

Manual Low

When the vehicle is stopped and the gear selector is moved to manual low (1), the transmission operates just as in manual third except that the hydraulic system prevents an upshift to second or third gear. The drive clutch is applied to prevent the forward 1-way clutch from overrunning, and first gear power flow through the compound planetary gearset is the same except for the use of one apply device.

In manual low, figure 20-14, the low-reverse clutch is applied to lock the rear carrier and

front ring gear to the case. The low 1-way prevents the front sun gear from turning counterclockwise during acceleration in Overdrive low, but it unlocks and allows the rear carrier and front ring gear to overrun during deceleration. In manual low, the low-reverse clutch locks the rear carrier and front ring gear solidly to the case. This prevents them from turning in either direction and provides engine compression braking during deceleration.

The driver can move the gear selector from Ⓓ, D, 3, or 2 to 1 while the car is moving. At speeds below approximately 30 mph (48 kph) the transmission downshifts directly to low gear. At higher speeds, the transmission first downshifts to second, and then automatically downshifts to low when vehicle speed drops below approximately 30 mph (48 kph). This capability provides the driver with a manual downshift for engine compression braking or improved acceleration.

Reverse

When the gear selector is moved to Reverse, the input shaft is still turned clockwise by the engine. The transmission must reverse this rotation to a counterclockwise direction. To do this, the reverse and low-reverse clutches are applied, figure 20-18.

With the reverse clutch applied, the input shaft drives the reverse drum, input shell, and rear sun gear clockwise. The low-reverse clutch

GEAR RANGES	CLUTCHES							BAND
	1-way Sprag	1-way Roller	Forward	High-Overdrive (3-4)	Low-Reverse	Reverse (Input)	Drive (Overrun)	Intermediate Overdrive (2-4)
Overdrive Low	●	●	●					
Overdrive Second	●		●					●
Overdrive Third	●		●	●				
Overdrive Fourth	†		●	●				●
Manual Third	●		●	●			●	
Manual Low	●	*	●		●		●	
Manual Second	●		●				●	●
Reverse					●	●		

* The 1-way roller clutch holds in manual low if the low-reverse clutch fails, but does not provide engine braking.
† The 1-way sprag clutch is overrunning and ineffective although the forward clutch is applied.

Figure 20-19. THM 700-R4 clutch and band application chart.

locks the rear carrier to the case. As a result the rear sun gear drives the rear pinions counterclockwise on the stationary carrier. The rear pinions then drive the rear ring gear and output shaft counterclockwise in gear reduction.

In Reverse, the front ring gear is also held, while the front carrier is driven counterclockwise. This causes the front pinions to turn clockwise and drive the front sun gear counterclockwise. However, because the forward clutch is released, the forward 1-way clutch is ineffective and the front gearset simply freewheels at this time.

In Reverse, all of the gear reduction is provided by the rear planetary gearset. The rear sun gear is the input member, the rear carrier is the held member, and the rear ring gear is the output member. The reverse clutch is the input device, and the low-reverse clutch is the holding device.

THM 700-R4 Clutch and Band Applications

Before we move on to a closer look at the hydraulic system, we should review the apply devices that are used in each of the gear ranges in this transmission. Figure 20-19 is a clutch and band application chart for the THM 700-R4. Also, remember these facts about the apply devices in this transmission:
● The forward, forward 1-way, high-overdrive, and reverse clutches are input devices.
● The forward clutch is applied in all forward gear ranges.
● The forward 1-way clutch drives the front sun gear in first, second, and third gears, and overruns in fourth.

● The high-overdrive clutch is applied in third and fourth gears.
● The reverse clutch is applied only in Reverse gear.
● The low-reverse, low 1-way, and drive clutches, along with the intermediate-overdrive band, are holding devices.
● The low-reverse clutch is applied in Reverse and manual low gears.
● The Low 1-way clutch holds the rear carrier and front ring gear in Overdrive low gear.
● The drive clutch prevents the front sun gear from freewheeling during deceleration in the manual low, manual second, and manual third gear ranges.
● The intermediate-overdrive band is applied in second and fourth gears.

THM 700-R4 HYDRAULIC SYSTEM

The hydraulic system controls shifting under varying vehicle loads and speeds, as well as in response to manual gear selection by the driver. Figure 20-20 is a diagram of the complete THM 700-R4 hydraulic system. You will find it helpful to refer back to this diagram as we discuss the various components that make up the hydraulic system. We will begin our study with the oil pump.

Hydraulic Pump

The THM 700-R4 uses a variable-displacement, vane-type oil pump, figure 20-21. The pump slide, pivot pin, rotor, vanes, and vane rings are installed in the pump body, which along with the pump cover bolts to the front of the transmission case behind the torque converter.

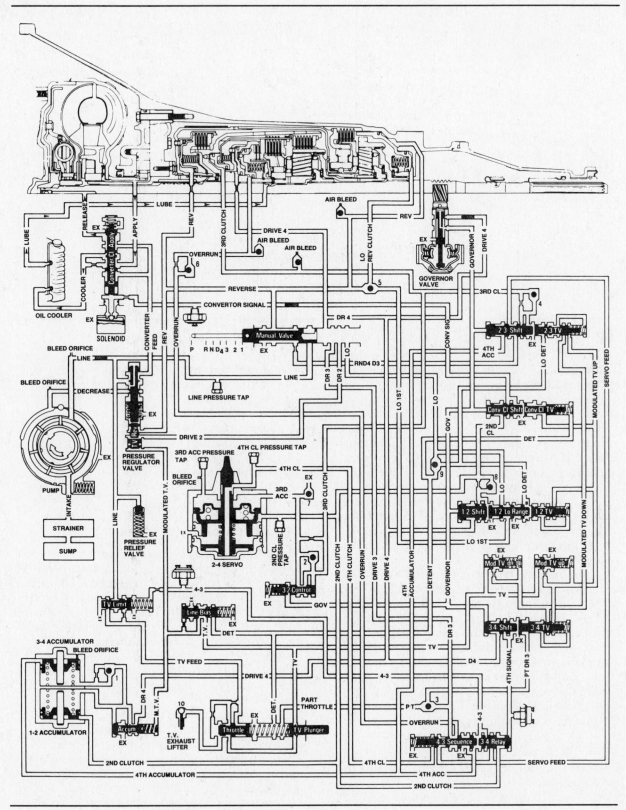

Figure 20-20. The THM 700-R4 hydraulic system.

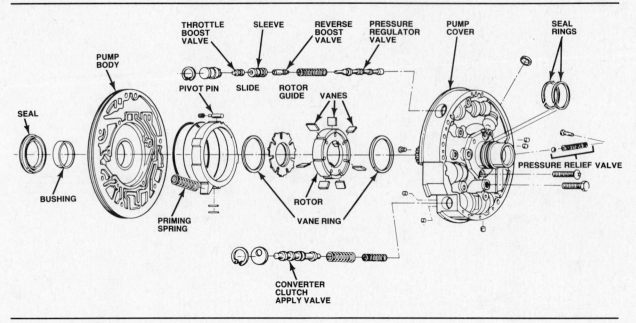

Figure 20-21. The THM 700-R4 oil pump assembly.

The pressure regulator valve assembly, converter clutch apply valve, and pressure relief ball-check valve are all located in the pump cover.

The pump slide pivots on a pin. When pressure in the hydraulic system is low, the pump priming spring tends to push the slide toward its maximum output position. When pressure in the hydraulic system rises, the pressure regulator valve sends oil through the decrease passage to move the slide toward its minimum output position. In this manner, oil pump output is controlled so that it is proportional to demand.

Whenever the engine is running, the converter cover turns the oil pump rotor. The rotation of the pump rotor and vanes creates suction that draws fluid through a filter from the sump at the bottom of the transmission case. The fluid then flows through passages in the valve body and the front of the case to the pump. Oil from the pump flows through a small filter screen to the pressure regulator valve, manual valve, pressure relief valve, and throttle limit valve.

Hydraulic Valves

The THM 700-R4 hydraulic system contains the following valves:
1. Pressure regulator valve assembly
2. Pressure relief valve
3. Manual valve
4. Throttle limit valve
5. Throttle valve
6. Line bias valve
7. Modulated throttle downshift valve
8. Modulated throttle upshift valve
9. Governor valve
10. 1-2 shift valve assembly
11. Accumulator valve
12. 2-3 shift valve assembly
13. 3-2 control valve
14. 3-4 shift valve assembly
15. 3-4 relay valve
16. 4-3 sequence valve
17. Converter clutch shift valve assembly
18. Converter clutch apply valve

Valve body and ball-check valves

Except for the pressure regulator valve, converter clutch apply valve, pressure relief ball-check valve, and the governor valve, all of the hydraulic valves in the THM 700-R4 are housed in the valve body, figure 20-22. The pressure regulator valve, converter clutch apply valve, and pressure relief ball-check valve are located in the oil pump as already described. The governor valve is located at the left rear of the transmission case.

A separator plate and two gaskets are installed between the valve body and the case, figure 20-5. The valve body has passages to route fluid between various valves, and additional passages are cast into the case where the valve body bolts to it. These cast passages form the back half of the valve body. The THM 700-R4 uses 10 check balls, figure 20-23. Four are in the valve body passages and five are in the case passages. The #7 check ball is located in a separate capsule in the case, figure 20-5.

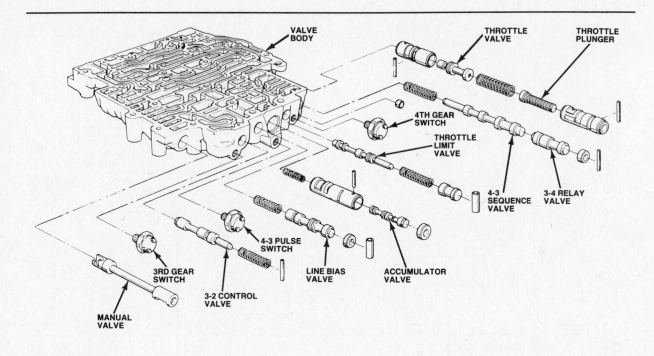

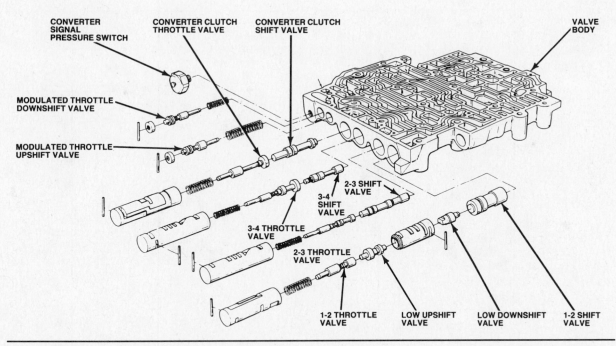

Figure 20-22. The THM 700-R4 valve body and hydraulic control valves.

The oil filter is attached to the bottom of the valve body. The THM 700-R4 also has a pump outlet screen, a governor inlet screen, and a screen in the signal pressure passage to the torque converter clutch apply valve. All three of these small screens are installed in the transmission case, figure 20-23.

Pressure regulator valve assembly
The pump delivers fluid through a single passage to the pressure regulator valve, figure 20-24. Fluid enters at the center of the valve and is routed through an internal passage to the end of the valve spool opposite the boost valves. Hydraulic pressure at the end of the valve acts to move the valve against spring

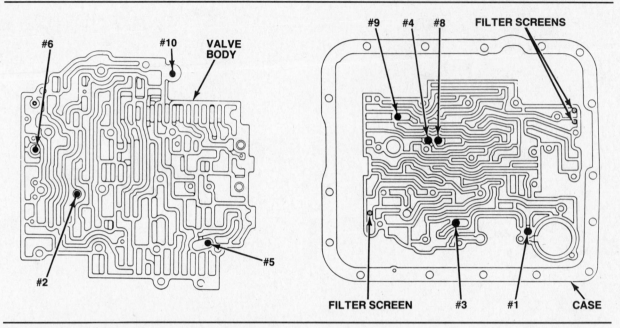

Figure 20-23. THM 700-R4 check ball locations.

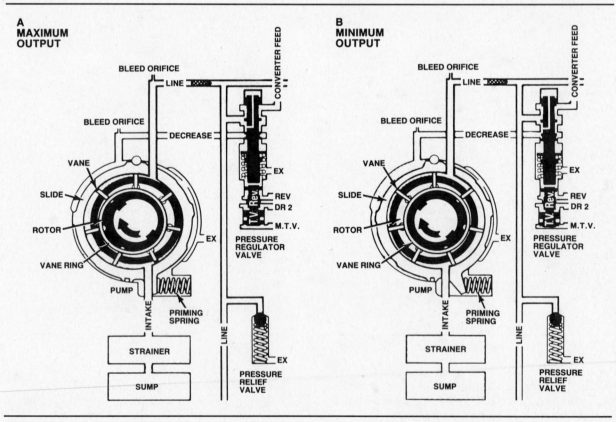

Figure 20-24. The THM 700-R4 pressure regulator valve, reverse and throttle boost valves, and pressure relief valve.

force. Hydraulic pressure at the center of the valve enters between two lands of equal diameter, and does not move the valve in either direction.

When the engine is started, figure 20-24A, pump oil flows through the pressure regulator valve and into the open converter feed passage. Pressure in this passage is routed to the converter clutch apply valve, which directs oil to the torque converter, transmission cooler, and lubrication circuit through different passages depending on whether the converter clutch is applied or released. This is explained in detail later in the chapter.

Once the converter feed circuit is filled, mainline pressure begins to build, and pressure at the end of the regulator valve moves the valve against spring force. This opens a port in the center chamber of the valve that leads into the decrease passage, figure 20-24B. Pressure then flows through the passage to the pump where it moves the slide against priming spring force to reduce pump output. The pressure regulator valve continues to move until hydraulic pressure and spring force reach a balance that provides a constant mainline pressure.

When the transmission shifts into gear, or from one gear to another, increased fluid flow demands cause mainline pressure to drop slightly. This drop in mainline pressure allows the spring to move the pressure regulator valve so that mainline pressure feed is cut off and the pump decrease circuit exhausts. The pump priming spring then moves the slide to increase pump output and bring mainline pressure back to the correct level.

The throttle boost and reverse boost valves at the spring end of the pressure regulator valve are used to raise mainline pressure under certain driving conditions. This is done by routing various hydraulic pressures to the boost valves, which then move to increase the amount of spring force acting on the pressure regulator valve. The increased mainline pressure is used to hold the apply devices more securely.

Modulated throttle pressure acts on the end of the throttle boost valve to increase mainline pressure in response to higher drivetrain torque loads. In Reverse, mainline pressure (REVERSE oil) acts on the *center* of the reverse boost valve to increase mainline pressure. And, in manual low and manual second, mainline pressure (DR2 oil) acts on the *end* of the reverse boost valve to again increase mainline pressure.

Pressure relief valve

The pressure relief valve, figure 20-24, is a ball-check valve located in the pump cover. This valve limits mainline pressure to a maximum of 320 to 360 psi (2206 to 2482 kPa). If mainline pressure exceeds this value, the pressure relief valve unseats to exhaust excess oil to the sump. Once mainline pressure drops below the maximum, spring force reseats the valve.

Manual valve

The manual valve, figure 20-25, is a directional valve that is moved by a mechanical linkage from the gear selector lever. The manual valve receives fluid from the pump and directs it to apply devices and other valves in the hydraulic system in order to provide both automatic and manual upshifts and downshifts. The manual valve is held in each gear position by a spring-loaded lever.

Six passages may be charged with mainline pressure from the manual valve. In Park, the manual valve position blocks mainline pressure from entering the valve, so no passages are charged. GM refers to mainline pressure from the manual valve by the name of the passage it flows through.

The RND4-D3 passage is charged in Reverse, Neutral, Overdrive, and manual third; it routes pressure to the:

- 2-3 shift valve.

The DR4 circuit is charged in Overdrive, manual third, manual second, and manual low; it routes pressure to the:

- Apply area of the forward clutch
- Governor valve
- 1-2 shift valve
- 3-4 shift valve
- Accumulator valve.

The DR3 passage is charged in manual third, manual second, and manual low; it routes pressure to the:

- 4-3 sequence valve
- 3-4 throttle valve.

The DR2 passage is charged in manual second and manual low; it routes pressure to the:

- Reverse boost valve.

The LOW circuit is charged in manual low; it routes pressure to the:

- 1-2 low range or low upshift valve
- 2-3 throttle valve.

The REVERSE circuit is charged in Reverse; it routes pressure to the:

- Inner and outer apply areas of the low-reverse clutch
- Apply area of the reverse clutch
- Reverse boost valve.

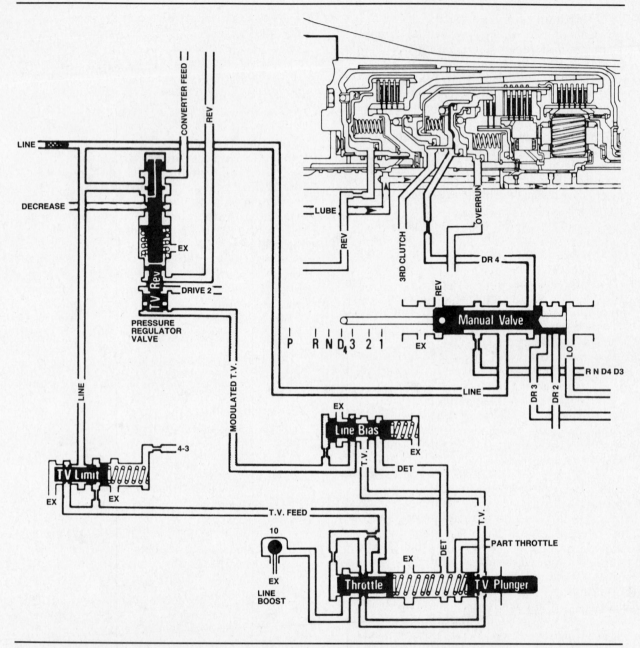

Figure 20-25. The THM 700-R4 manual valve (in Overdrive), throttle limit valve, throttle valve, and line bias valve.

Throttle limit valve

The throttle limit valve, figure 20-25, receives mainline pressure from the pump and limits it to a maximum of 90 psi (621 kPa) before sending it to the throttle valve as throttle feed pressure. The throttle limit valve is controlled by opposition between spring force at one end and mainline pressure at the other.

Mainline pressure moves the throttle limit valve against spring force to restrict the flow of fluid through the valve. At 90 psi (621 kPa), the throttle limit valve moves far enough to block

the entry of mainline pressure and exhaust any excess pressure in the throttle feed circuit. During a 4-3 downshift, fourth gear accumulator oil is routed through a restricting orifice to the spring end of the throttle limit valve where it exhausts.

Throttle valve

The THM 700-R4 transmission develops throttle pressure using a throttle valve operated by a mechanical linkage, figure 20-25. Throttle pressure indicates engine torque and helps time the transmission upshifts and downshifts. Without throttle pressure, the transmission would upshift and downshift at exactly the same speed for very different throttle openings and load conditions.

The throttle valve is controlled by opposition between hydraulic pressure and spring force. The hydraulic pressure is throttle feed pressure that is routed to the end of the valve opposite the spring. The amount of spring force acting on the other end of the valve is controlled by the throttle plunger. The plunger is linked to the accelerator by a cable so that its position is determined by throttle opening.

At closed throttle, the throttle plunger exerts no force on the throttle valve spring. Throttle feed oil passes through a restricting orifice and into the throttle pressure circuit where it acts on the end and one land of the throttle valve. This moves the valve to the right against spring force to nearly close the throttle pressure outlet port. As a result, there is little or no throttle pressure at this time.

As the accelerator is depressed, the cable linkage moves the throttle plunger to compress the throttle valve spring. This moves the throttle valve to the left, opens the outlet port, and allows throttle feed pressure to exit the valve as throttle pressure. As the throttle is opened wider, spring force rises, the throttle valve moves farther, and throttle pressure continues to increase. However, because throttle pressure is developed from mainline pressure, it never exceeds mainline pressure.

To help balance the throttle valve, throttle pressure is directed to the end of the valve opposite spring force. This acts to move the throttle valve to the right against spring force, and also uncovers a check-ball-regulated exhaust port that helps limit the pressure increase. Therefore, at any given throttle opening, the throttle valve reaches a balanced position and provides a constant level of throttle pressure.

The exhaust port check ball, called the line boost throttle exhaust check ball, is controlled by an extension of the throttle plunger cable linkage. In normal operation, the check ball is held off its seat to exhaust excess throttle pressure. However, if the throttle plunger cable breaks, is disconnected, or is badly adjusted, the linkage moves downward and allows the check ball to seat. With no exhaust port to relieve it, throttle pressure then remains high regardless of actual throttle position. Although the transmission will not upshift and downshift at the appropriate times, this protects the clutches and band from slipping and burning.

Throttle pressure is routed to the: throttle plunger, line bias valve, modulated throttle upshift valve, and the modulated throttle downshift valve. Throttle pressure to the throttle plunger helps regulate throttle pressure by increasing the spring force acting on the throttle valve; it also reduces the effort needed to hold the throttle valve plunger in place. The effect of throttle pressure on the other valves mentioned will be covered in the appropriate sections later in the chapter.

When the accelerator is depressed far enough, the throttle plunger opens a port that allows throttle pressure to flow through a passage to the 3-4 throttle valve. Oil in this passage is called part-throttle pressure. Part-throttle pressure on the 3-4 throttle valve combines with spring force to move the 3-4 shift valve and force a part-throttle 4-3 downshift when the throttle is opened past a certain point in fourth gear.

When the accelerator is completely depressed, the throttle plunger opens a second port that allows throttle pressure to flow through a passage to the 2-3 throttle valve, 1-2 low range or low downshift valve, converter clutch throttle valve, and line bias valve. Oil in this passage is called detent pressure. Detent pressure on the 2-3 throttle valve and 1-2 low range or low downshift valve combines with spring force to move the 2-3 and/or 1-2 shift valves to force full-throttle 3-2, 2-1, or 3-1 downshifts. Detent pressure on the converter clutch throttle valve moves the converter clutch shift valve to unlock the converter clutch and allow torque multiplication in the converter. The action of detent pressure on the line bias valve is explained in the next section.

Line bias valve

The line bias valve, figure 20-25, receives throttle pressure and regulates it to create a modulated throttle pressure whose rate of increase more closely matches engine torque rather than just the degree of throttle opening. Throttle pressure enters the line bias valve where it becomes modulated throttle pressure. This pressure is then routed to the end of the valve to oppose spring force. As throttle pressure increases, modulated throttle pressure moves the line bias valve to restrict the outlet port and control the rate of pressure increase in the modulated throttle pressure circuit.

Modulated throttle pressure is routed to the throttle boost valve in the pressure regulator valve assembly to raise mainline pressure as throttle opening and engine torque increase.

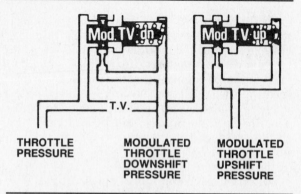

Figure 20-26. The THM 700-R4 modulated throttle upshift and downshift valves.

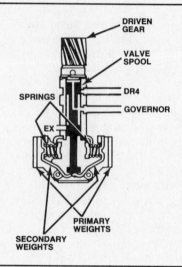

Figure 20-27. The THM 700-R4 governor.

When the throttle is wide open, detent pressure is sent to the line bias valve where it combines with spring force to bottom the valve. This allows unmodified throttle pressure to flow into the modulated throttle pressure circuit.

Modulated throttle upshift valve
Modulated throttle downshift valve

The modulated throttle upshift and downshift valves, figure 20-26, re-regulate throttle pressure to create modulated throttle upshift and downshift pressures that act on the shift valves to help control the shift points. Both valves are controlled by opposition between throttle pressure on one end, and combined spring force and modulated throttle pressure on the other.

The modulated throttle upshift valve is closed by spring force, and begins to open at approximately 10 psi (69 kPa) of throttle pressure. This valve delays upshifts at small throttle openings by directing modulated throttle upshift pressure to the 1-2, 2-3, and 3-4 throttle valves where it combines with spring force to hold the shift valves in their downshifted positions against governor pressure. Modulated throttle upshift pressure is also routed to the converter clutch throttle valve. The action of this pressure on the valve is explained later in the chapter.

The modulated throttle downshift valve is closed by spring force and does not open until there is approximately 40 psi (276 kPa) of throttle pressure. This valve helps force 4-3 and 3-2 downshifts at larger throttle openings by directing modulated throttle downshift pressure to the 3-4 and 2-3 throttle valves where it combines with spring force to downshift the shift valves against governor pressure. This is called a modulated downshift.

Governor valve

We have already discussed the development and control of throttle pressure, which is one of the two auxiliary pressures that work on the shift valves. Before we get to the shift valve operation, we must explain the governor valve (or governor), and how it creates governor pressure. Governor pressure provides a road speed signal to the hydraulic system that causes automatic upshifts to occur as road speed increases, and permits automatic downshifts with decreased road speed.

The THM 700-R4 governor, figure 20-27, mounts in an opening at the left rear of the transmission. A gear cut into the output shaft turns the driven gear on the end of the governor, which causes the assembly to rotate in the case. The remainder of the governor consists of two sets of weights, two springs, and a spool valve.

When the vehicle is stopped, the mainline pressure (DR4 oil) inlet is closed or slightly open, and both the governor pressure outlet and the exhaust port are fully open. No governor pressure is developed at this time because any pressure that leaks past the valve spool or enters through the mainline inlet escapes through the exhaust port.

As the vehicle begins to move, the transmission output shaft turns the governor. Centrifugal force throws the weights outward, which levers the valve spool farther into its bore. This begins to close the exhaust port, and opens the mainline pressure inlet port farther, causing governor pressure to build.

Two sets of weights are used to increase the accuracy of pressure regulation at low vehicle speeds. The heavier primary weights move

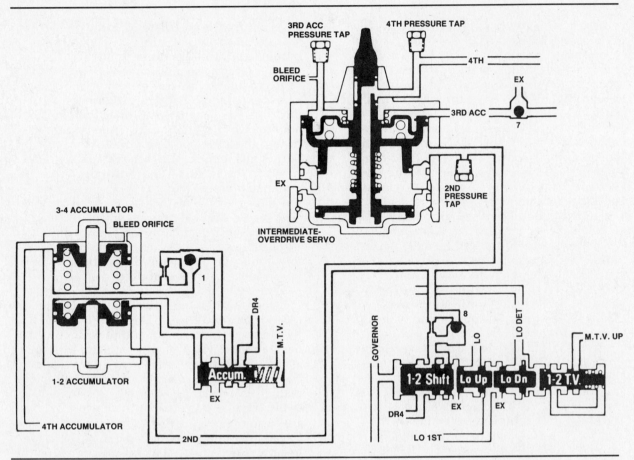

Figure 20-28. The THM 700-R4 1-2 shift valve assembly, accumulator valve, 1-2 and 3-4 accumulators, and intermediate-overdrive servo in their second gear positions.

first, and act through springs against the lighter secondary weights that actually move the valve spool. Once governor speed becomes great enough, the primary weights bottom against their stops, and the secondary weights alone act to move the valve spool.

During operation, governor pressure is routed through a passage in the center of the spool valve to the end of the valve near the driven gear. This pressure acts against the end of the spool, and helps oppose the lever force applied at the opposite end of the valve. As a result of these opposing forces, governor pressure increases with vehicle speed, and stabilizes whenever vehicle speed is constant.

When the vehicle reaches a certain speed, both sets of weights move out as far as possible. At this time, the secondary weights are bottomed on the primary weights, the exhaust port is fully closed, and both the mainline pressure inlet and the governor pressure outlet are fully open. Therefore, governor pressure equals mainline pressure.

Governor pressure is routed to the ends of the 1-2, 2-3, and 3-4 shift valve where it opposes spring force and other hydraulic pressures to cause an upshift. Governor pressure is also routed to the converter clutch shift valve and the 3-2 control valve. The actions of governor pressure on these valves will be discussed in the appropriate sections later in the chapter.

1-2 shift valve assembly

The 1-2 shift valve controls upshifts from low to second, and downshifts from second to low. The assembly includes the 1-2 shift valve, the low upshift and low downshift valves, the 1-2 throttle valve, and a spring, figure 20-28. Some transmissions have a 1-piece low range valve in place of the separate low upshift and low downshift valves, figure 20-20. The 1-2 shift valve is controlled by opposition between governor pressure on one end, and combined spring force and modulated throttle upshift pressure on the other end.

In low gear, mainline pressure (DR4 oil) from the manual valve flows to the 1-2 shift valve where it is blocked by the downshifted valve. When vehicle speed increases enough for governor pressure to overcome spring force and modulated throttle upshift pressure acting on the 1-2 throttle valve, the 1-2 shift valve upshifts. DR4 oil then flows through the upshifted valve into the second gear circuit where it travels to the apply side of the intermediate-overdrive servo, and to the 1-2 accumulator which cushions intermediate-overdrive band engagement. Second gear oil also flows to the converter clutch shift valve and the 3-4 relay valve.

To prevent hunting between gears after the 1-2 upshift, modulated throttle upshift pressure is blocked from entering the 1-2 shift valve assembly. A closed-throttle 2-1 downshift occurs when governor pressure drops to a very low level (road speed decreases), and spring force acts on the 1-2 throttle valve to move the 1-2 shift valve against governor pressure. The THM 700-R4 transmission does not offer a part-throttle 2-1 downshift.

A full-throttle 2-1 downshift occurs below approximately 30 mph (48 kph) when detent pressure from the throttle valve flows to the low downshift valve, or a single land on the low range valve. The combination of detent pressure and spring force then moves the 1-2 shift valve against governor pressure and causes a 2-1 detent downshift.

In manual low, mainline pressure (LOW oil) from the manual valve is routed to the low downshift and the low upshift valves, or to two lands on the low range valve. Together with spring force acting on the 1-2 throttle valve, this pressure downshifts the 1-2 shift valve against governor pressure, and keeps it downshifted regardless of throttle opening or road speed. LOW oil then flows through the downshifted valve to the inner apply area of the low-reverse clutch.

Accumulator valve
The accumulator valve, figure 20-28, cushions application of the intermediate-overdrive band during 1-2 upshifts, and the high-overdrive clutch during 3-4 upshifts. It does this in relation to throttle opening by creating an accumulator pressure that combines with spring force in the 1-2 and 3-4 accumulators to control the rate at which the accumulator pistons move through their ranges of travel.

When the transmission is in any forward gear range, mainline pressure (DR4 oil) is routed to the accumulator valve. The D4 pressure moves through the valve and then acts on its end to move the valve against spring force and modulated throttle pressure. This restricts the valve outlet to create a variable accumulator

pressure. When modulated throttle pressure is low, accumulator pressure is also low. When modulated throttle pressure is high, accumulator pressure is also high.

Accumulator pressure flows into the spring chambers of the 1-2 and 3-4 piston-type accumulators which are stacked on top of one another. The 3-4 accumulator bore is in the case next to the valve body, and the 1-2 accumulator bore is inside the cover of the 3-4 accumulator.

When the 1-2 or 3-4 shift valve upshifts, mainline pressure flows through the second or fourth gear (servo apply) circuit to the accumulator piston where it opposes spring force and accumulator pressure. The amount of time it takes servo apply pressure to bottom the accumulator piston determines the amount of force with which the band is applied.

When accumulator pressure is low, indicating a light load on the drivetrain, apply oil rapidly bottoms the piston. This absorbs much of the pressure in the circuit and provides an easy band application. When accumulator pressure is high, indicating a heavy load on the drivetrain, apply oil takes much longer to bottom the piston. As a result, most of the pressure in the apply circuit is immediately available to apply the band much more firmly. In both cases, once the accumulator piston is bottomed, full mainline pressure is present in the circuit to hold the band securely applied.

2-3 shift valve assembly
The 2-3 shift valve controls upshifts from second to third, and downshifts from third to second. The assembly includes the 2-3 shift valve, the 2-3 throttle valve, and a spring, figure 20-29. The 2-3 shift valve is controlled by opposition between governor pressure on one end, and combined spring force and modulated throttle upshift pressure on the other end.

Mainline pressure (RND4-D3 oil) is routed from the manual valve to the 2-3 shift valve where it is blocked by the downshifted valve. When vehicle speed increases enough for governor pressure to overcome spring force and modulated throttle upshift pressure acting on the 2-3 throttle valve, the 2-3 shift valve upshifts. RND4-D3 oil then flows through the upshifted valve to the high-overdrive clutch, the release side of the intermediate-overdrive servo, and 3-2 control valve.

Third gear circuit pressure to the high clutch applies the clutch to place the transmission in third gear. The pressure routed to the release side of the intermediate-overdrive servo combines with spring force to release the servo and intermediate-overdrive band. The stroking action of the servo piston acts as an accumulator

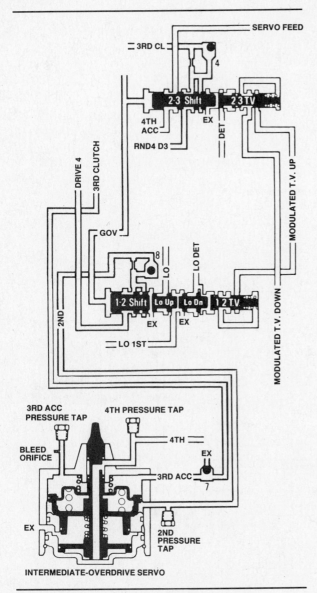

Figure 20-29. The THM 700-R4 2-3 shift valve assembly in its third gear position.

for high clutch application. The time it takes third gear oil to release the intermediate servo provides a delay in high clutch application. Pressure in the high clutch circuit closes the accumulator check valve to seal the exhaust port in the passage at this time.

To prevent hunting between gears after the 2-3 upshift, modulated throttle upshift pressure is blocked from entering the 2-3 shift valve assembly. A closed-throttle 3-2 downshift occurs when governor pressure drops (road speed decreases) to the point where spring force acts on the 2-3 throttle valve to move the 2-3 shift valve against governor pressure.

A part-throttle 3-2 modulated downshift occurs when modulated throttle downshift pressure combines with spring force acting on the 2-3 throttle valve to move the 2-3 shift valve against governor pressure. A full-throttle 3-2 detent downshift occurs at speeds below approximately 60 mph (97 kph) when the accelerator is fully depressed and detent pressure from the throttle valve flows to the 2-3 throttle valve. The combination of detent pressure, modulated throttle downshift pressure, and spring force moves the 2-3 shift valve against governor pressure and causes a 3-2 downshift. If the vehicle speed is below approximately 30 mph (48 kph), detent pressure will also downshift the 1-2 shift valve, resulting in a 3-1 downshift.

3-2 control valve

The 3-2 control valve, figure 20-30, regulates the rate of high-overdrive clutch release and intermediate-overdrive band application during modulated and full-throttle (detent) 3-2 downshifts. The valve is controlled by opposition between spring force and governor pressure. When the vehicle is stopped, spring force holds the 3-2 control valve wide open. As speed rises, governor pressure moves the valve against spring force to restrict the valve opening. Above approximately 50 mph (80 kph), governor pressure closes the valve entirely.

During part-throttle 3-2 downshifts below approximately 50 mph (80 kph), the 3-2 control valve opening regulates the downshift in relation to vehicle speed. Oil exhausting from the high-overdrive clutch flows through a passage to the 2-3 shift valve where it seats a check ball (4). The oil then passes through a restricting orifice and into the shift valve where it exhausts. Oil exhausting from the release side of the intermediate-overdrive servo seats two additional check balls (7 and 2) and flows to the 3-2 control valve. The oil then passes through a restricting orifice and the 3-2 control valve. The size of the opening in the 3-2 control valve allows the high-overdrive clutch to release, and the intermediate-overdrive band to apply, at a rate that provides the best shift feel for a given vehicle speed. From the 3-2 control valve, accumulator release oil flows to the 2-3 shift valve where it exhausts.

During full-throttle 3-2 detent downshifts above approximately 50 mph (80 kph), the 3-2 control valve is closed. As a result, all of the exhausting accumulator oil from the intermediate servo must pass through the restricting orifice before joining the oil exhausting from the high clutch. This causes band application

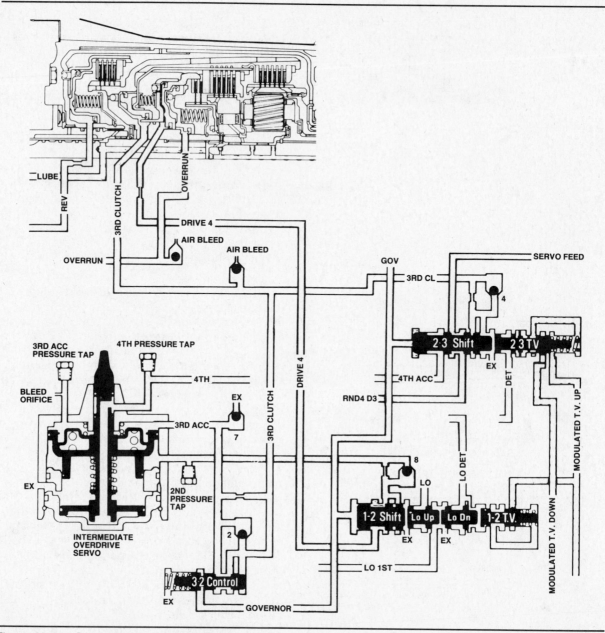

Figure 20-30. Operation of the THM 700-R4 3-2 control valve during a 3-2 downshift.

to be delayed even more, allowing engine speed to increase and better match the higher gear ratio of second gear.

3-4 shift valve assembly

The 3-4 shift valve controls upshifts from third to fourth, and downshifts from fourth to third. The assembly consists of the 3-4 shift valve, 3-4 throttle valve, and a spring, figure 20-31. The 3-4 shift valve is controlled by opposition between governor pressure on one end, and combined spring force and modulated throttle upshift pressure on the other end.

In Overdrive, mainline pressure (DR4 oil) is routed from the manual valve to the 3-4 shift valve where it is blocked by the downshifted valve. When vehicle speed increases enough for governor pressure to overcome spring force and modulated throttle upshift pressure acting on the 3-4 throttle valve, the 3-4 shift valve upshifts. DR4 oil then flows through the upshifted valve to the 3-4 relay valve, making an upshift possible. Exactly how this happens is explained in the next section.

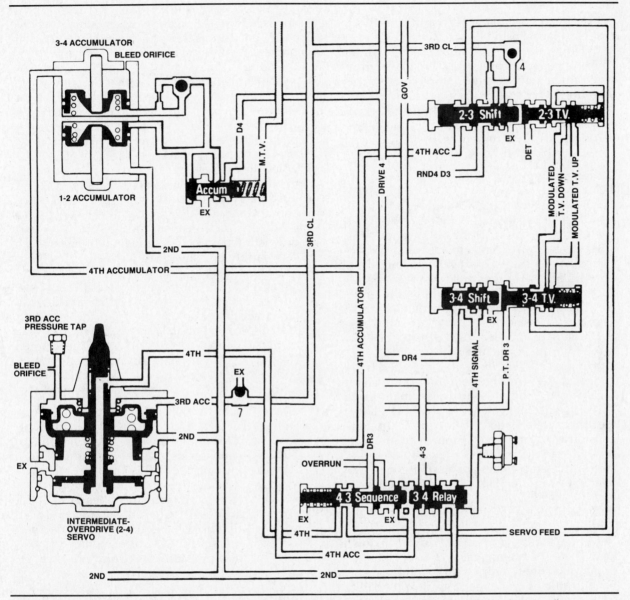

Figure 20-31. The THM 700-R4 3-4 shift valve assembly, 3-4 relay valve, 4-3 sequence valve, intermediate-overdrive servo, and 3-4 accumulator in their fourth gear positions.

To prevent hunting between gears after the 3-4 shift valve has upshifted, modulated throttle upshift pressure is blocked from entering the 3-4 shift valve assembly. A closed-throttle 4-3 downshift occurs when governor pressure drops (road speed decreases) to the point where spring force acts on the 3-4 throttle valve to move the 3-4 shift valve against governor pressure.

A part-throttle 4-3 modulated downshift can be caused by modulated throttle downshift pressure acting on the 2-3 throttle valve. As explained earlier, when throttle pressure rises above 40 psi (276 kPa), the modulated throttle downshift valve opens and sends modulated

throttle downshift pressure to the 3-4 throttle valve. If the combination of spring force and modulated throttle downshift pressure acting on the 3-4 throttle valve becomes great enough, it will move the 3-4 shift valve against governor pressure to cause a 4-3 modulated downshift.

A part-throttle 4-3 downshift is also possible when the accelerator linkage moves the throttle plunger far enough to uncover the part-throttle port in the throttle valve. This routes part-throttle pressure through the part-throttle/DR3 passage to the 3-4 throttle valve. The combination of part-throttle pressure and spring force acting on the 3-4 throttle valve then moves the

3-4 shift valve against governor pressure to cause a 4-3 downshift.

When the gear selector is placed in manual third (3), mainline pressure (DR3 oil) from the manual valve is routed through the part-throttle/D3 passage to the 4-3 throttle valve. The mainline pressure downshifts the 3-4 shift valve against governor pressure, and keeps the valve downshifted regardless of vehicle speed or throttle position.

3-4 relay valve
4-3 sequence valve
The 3-4 relay valve and 4-3 sequence valve, figure 20-31, are located in the same bore in the valve body. A spring acting on the end of the 4-3 sequence valve tends to keep both valves closed. When the 3-4 shift valve upshifts, mainline (4th SIGNAL) pressure from the 3-4 shift valve opens both valves. This allows intermediate-overdrive band (2nd) apply oil to enter the servo feed passage.

Pressure in the servo feed passage passes through the 4-3 sequence valve and exits as fourth gear oil. Fourth gear oil is routed to the intermediate-overdrive servo where it combines with pressure in the second gear circuit to apply the intermediate-overdrive band. Servo feed pressure also flows through the upshifted 2-3 shift valve and exits as fourth accumulator pressure. This pressure is routed to the 3-4 accumulator which helps cushion intermediate-overdrive band application as described earlier. Fourth accumulator pressure is also routed to the area between the 3-4 relay and 4-3 sequence valves.

To regulate the shift feel during a 4-3 downshift, the 4-3 sequence valve controls the rate of intermediate-overdrive servo release. When part-throttle or modulated throttle downshift oil closes the 3-4 shift valve, fourth accumulator pressure closes the 3-4 relay valve, but holds the 4-3 sequence valve open. This opens the 4-3 passage which contains a restricting orifice and leads to the throttle limit valve and an exhaust port.

As a result, exhausting intermediate-overdrive servo apply oil flows back through the 4-3 sequence valve and the upshifted 2-3 shift valve where it is joined by exhausting 3-4 accumulator oil. The combined oil then flows through the area between the 3-4 relay valve and 4-3 sequence valve, and into the 4-3 passage where it exhausts at the throttle limit valve. The restricting orifice in the 4-3 passage slows release of the intermediate-overdrive band for proper shift feel. When fourth accumulator pressure drops low enough, spring force closes the 4-3 sequence valve, opening an exhaust port at the

spring end of the valve. Any remaining servo apply pressure is then quickly exhausted to prevent the intermediate-overdrive band from dragging.

A 4-3 downshift requires that the intermediate-overdrive band be released. A 3-2 downshift requires that the high clutch be released and the intermediate-overdrive band be applied. Thus, a 4-2 downshift requires that the intermediate-overdrive band remain applied while the high clutch is released. Remember that during a 4-3 downshift exhausting intermediate-overdrive servo apply pressure must pass through the upshifted 2-3 shift valve. In a 4-2 downshift, however, the 2-3 shift valve is downshifted and blocks the flow of oil so that servo apply pressure cannot exhaust. At the same time, 3-4 accumulator pressure exhausting at the throttle limit valve holds the 4-3 sequence valve open so that fourth gear intermediate-overdrive servo apply oil cannot exhaust at that location either. Meanwhile, second gear oil fills the apply side of the servo. Then, when 3-4 accumulator pressure drops low enough, spring force closes the 4-3 sequence valve and fourth gear servo apply pressure exhausts.

Converter clutch shift valve assembly
The THM 700-R4 torque converter clutch cannot be applied until the converter clutch shift valve assembly sends a signal pressure to the converter clutch apply valve. The assembly consists of the converter clutch shift valve spool, the converter clutch throttle valve, and a spring figure 20-32. The shift valve is controlled by opposition between governor pressure on one end, and spring force on the other. Modulated throttle upshift pressure also acts on a land at the center of the valve to oppose governor pressure.

When the 1-2 shift valve upshifts, mainline pressure (second gear oil) routed to the converter clutch shift valve provides a source of signal pressure. At lower vehicle speeds and larger throttle openings, the shift valve is closed by spring force and modulated throttle upshift pressure, figure 20-32. This blocks the flow of signal pressure to the apply valve. Once the vehicle speed exceeds a minimum value and the throttle is closed beyond a certain point, governor pressure opens the shift valve against spring force and modulated throttle upshift pressure; this routes signal pressure to the apply valve.

During full-throttle downshifts, detent pressure is routed to the spring end of the converter clutch throttle valve to close the shift valve and exhaust converter clutch signal pressure. This unlocks the clutch so that torque multiplication can take place in the converter.

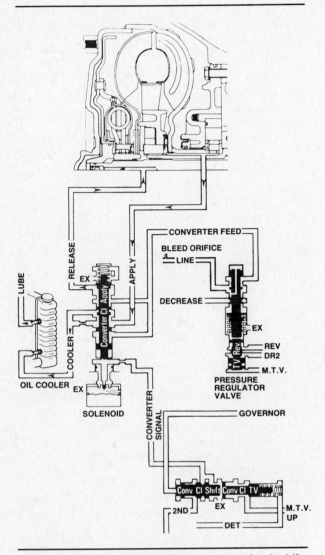

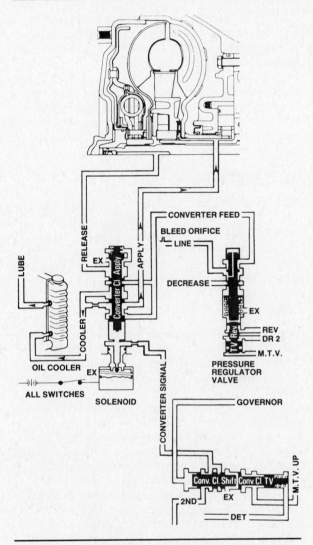

Figure 20-32. The THM 700-R4 converter clutch shift valve assembly, TCC solenoid, and converter clutch apply valve in the release position.

Figure 20-33. The THM 700-R4 converter clutch shift valve assembly, TCC solenoid, and converter clutch apply valve in the apply position.

Converter clutch apply valve

The converter clutch apply valve receives converter feed oil from the pressure regulator valve. It then directs this pressure to the torque converter and transmission cooler through various passages that determine whether the converter clutch is released or applied. Converter clutch apply valve position is controlled by opposition between spring force at one end, and converter clutch signal pressure and the TCC solenoid at the other end.

The electrical and electronic controls used to control the TCC solenoid are described in Chapter 8. The paragraphs below describe the hydraulic operation of the converter clutch system.

When the TCC solenoid is de-energized, figure 20-32, its exhaust port is open and signal oil is exhausted. Thus, spring force holds the converter clutch apply valve in the release position. In this position, converter feed oil is routed to the torque converter through the clutch release passage. Oil enters the converter between the clutch pressure plate and the converter cover; this prevents the clutch from applying. Oil leaving the converter returns to the converter clutch apply valve through the clutch apply passage, and is then routed to the transmission cooler and lubrication circuits.

When the TCC solenoid is energized, figure 20-33, its exhaust port is closed and signal pressure moves the converter clutch apply valve to the apply position against spring force. In this position, converter feed pressure is routed to the torque converter through the clutch apply passage. Oil enters the converter between the clutch pressure plate and the turbine; this forces the pressure plate into contact with the converter cover, locking up the converter. A portion of the converter feed oil is routed through the apply valve and a restricting orifice to the transmission cooler and lubrication circuits.

MULTIPLE DISC CLUTCHES AND SERVOS

The preceding paragraphs examined the individual valves in the hydraulic system. Before we trace the complete hydraulic system operation in each driving range, we will take a quick look at the multiple-disc clutches and servos.

Clutches

The THM 700-R4 transmission has five multiple-disc clutches. Each clutch is applied hydraulically, and released by several small coil springs when oil is exhausted from the clutch piston cavity.

The forward clutch, together with the forward 1-way clutch, is applied in all forward gear ranges to drive the front sun gear. The high-overdrive clutch is applied in third and fourth gears to drive the front ring gear/rear carrier. The low-reverse clutch is applied in manual low and reverse gears to lock the rear carrier to the transmission case. The drive (overrun) clutch is applied in manual third, manual second, and manual low gears to lock the front sun gear to the input shaft. The reverse (input) clutch is applied in reverse gear to lock the rear sun gear to the input shaft.

The low-reverse clutch is designed so that the piston has two surfaces on which hydraulic pressure can act. To apply the low-reverse clutch in manual low, pressure is routed only to the smaller, inner apply area of the piston. When the clutch is applied in Reverse, pressure is routed to the inner apply area of the piston, and also to the larger, outer area. This allows the clutch to develop greater holding force in Reverse, when it is subject to higher torque loads.

Servo

The THM 700-R4 has a single servo, figure 20-5, that is used to apply the intermediate-overdrive band. The servo bore is in the transmission case. Band adjustment is controlled by the length of the servo rod, or apply pin, which is

selected at the time of transmission assembly, using a special tool.

HYDRAULIC SYSTEM SUMMARY

We have examined the individual valves, clutches, and servos in the hydraulic system and discussed their functions. The following paragraphs summarize their combined operations in the various gear ranges. Refer to figure 20-20 to trace the hydraulic system operation.

Park and Neutral

In Park and Neutral, oil flows from the pump to the pressure regulator valve, which regulates the pump pressure to create mainline pressure. The pressure regulator valve also creates converter feed pressure which is routed to the converter clutch apply valve. The converter clutch apply valve routes the feed pressure to the torque converter through the release passage. Return oil from the converter is routed back through the apply valve, and then to the transmission cooler and lubrication circuit.

Pressure in the mainline circuit is routed to the pressure relief valve, manual valve, and throttle limit valve. The pressure relief valve limits mainline pressure to a maximum of 320 to 360 psi (2206 to 2482 kPa). In park, mainline pressure to the manual valve is blocked. In Neutral, the manual valve routes mainline pressure (RND4-D3 oil) to the 2-3 shift valve, and from there into the servo feed circuit where it travels to the 3-4 relay valve and 4-3 sequence valve.

The throttle limit valve limits mainline pressure to a maximum of 90 psi (621 kPa) before sending it to the throttle valve as throttle feed pressure. The throttle valve then regulates throttle feed pressure to create a throttle pressure that is proportional to throttle opening. Throttle pressure is routed to the throttle plunger, modulated throttle upshift and downshift valves, and line bias valve.

Throttle pressure to the throttle plunger acts to increase the amount of spring force on the throttle valve and reduce the amount of force needed to hold the plunger in position. Throttle pressure to the modulated throttle upshift and downshift valves opposes spring force. Above certain pressures, the modulated throttle upshift and downshift valves regulate throttle pressure to created modulated throttle upshift and downshift pressures. These pressures are then applied to the spring ends of the 1-2, 2-3, 3-4, and converter clutch throttle valves where they act to help delay upshifts and force downshifts.

The line bias valve regulates throttle pressure to create a modulated throttle pressure that is routed to the throttle boost valve in the pressure regulator valve assembly to increase mainline pressure. Modulated throttle pressure is also routed to the spring end of the accumulator valve where it combines with spring force to help bottom the valve. All of the above throttle pressure applications are made in all gear ranges.

In Park and Neutral, all clutches and the band are released. Except for those valves just discussed, all valve positions are determined by spring force alone. In Park, a manual linkage engages the parking pawl with the parking gear on the outside of the rear ring gear. This locks the output shaft to the case and prevents the vehicle from moving.

Overdrive Low

When the gear selector is moved to Overdrive, the manual valve routes mainline pressure (DR4 oil) to the apply area of the forward clutch, 1-2 and 3-4 shift valves, accumulator valve, and governor valve. DR4 oil applies the forward clutch to place the transmission in first gear. DR4 oil at the downshifted 1-2 and 3-4 shift valves is blocked. The accumulator valve regulates DR4 oil in relation to throttle opening to create an accumulator pressure that is applied to the spring sides of the 1-2 and 3-4 accumulator pistons.

The governor valve regulates DR4 oil in relation to vehicle speed to create a governor pressure that is applied to the ends of the 1-2, 2-3, 3-4, and converter clutch shift valves, and also to the 3-2 control valve. Governor pressure acting on the shift valves acts to force upshifts. Governor pressure on the 3-2 control valve opposes spring force. These governor pressure applications are made in all forward gear ranges.

In Overdrive low, the forward clutch applied. The forward and low 1-way clutches hold on acceleration, but overrun on deceleration.

Overdrive Second

As vehicle speed and governor pressure increase in Overdrive, governor pressure overcomes spring force and modulated throttle upshift pressure and upshifts the 1-2 shift valve. This allows mainline pressure (DR4 oil) to enter the second gear circuit. Second gear oil then flows through a restricting orifice to the intermediate-overdrive servo, 1-2 accumulator, 3-4 relay valve, and converter clutch shift valve.

Second gear oil applies the intermediate-overdrive servo and band to place the transmission in second gear. Second gear oil to the 1-2 accumulator strokes the accumulator piston to cushion the upshift in relation to throttle opening. Second gear oil at the 3-4 relay valve is blocked.

If the converter clutch shift valve is closed, second gear oil to the valve is blocked. At higher speeds and smaller throttle openings, governor pressure overcomes spring force and modulated throttle upshift pressure to open the shift valve. Second gear oil then flows through the valve and becomes converter signal pressure, which is routed to the converter clutch apply valve. If the TCC solenoid is de-energized, signal pressure is exhausted. When the TCC solenoid is energized, signal pressure moves the converter clutch apply valve to the apply position against spring force. The apply valve then routes converter feed oil to the torque converter through the apply passage to apply the clutch. Some of the converter feed oil is routed through an orifice to the transmission cooler and lubrication circuits. These torque converter clutch pressure applications may be made in second, third, or fourth gear.

In Overdrive second, the forward clutch and the intermediate-overdrive band are applied. The forward 1-way clutch holds on acceleration, but overruns on deceleration.

Overdrive Third

As vehicle speed and governor pressure continue to increase in Overdrive, governor pressure overcomes spring force and modulated throttle upshift pressure and upshifts the 2-3 shift valve. The mainline pressure (RNDR4-D3 oil) that was previously routed into the servo feed circuit is now routed into the third gear circuit, and residual pressure in the servo feed circuit exhausts through the fourth accumulator passage.

Third gear oil flows through a restricting orifice to the apply area of the high-overdrive clutch, intermediate-overdrive servo, and 3-2 control valve. Third gear oil applies the high-overdrive clutch and releases the intermediate-overdrive servo and band to place the transmission in third gear. Third gear oil at the 3-2 control valve has no effect at this time.

In Overdrive third, the forward and high-overdrive clutches are applied. The forward 1-way clutch holds on acceleration, but overruns on deceleration.

Overdrive Fourth

As vehicle speed and governor pressure continue to increase in Overdrive, governor pressure overcomes spring force and modulated throttle upshift pressure and upshifts the 3-4 shift valve. This allows mainline pressure (DR4 oil) to flow through the shift valve where it becomes fourth signal pressure. This pressure is routed to the end of the 3-4 relay valve where it opposes spring force to move the 3-4 relay and 4-3 sequence valves to their open positions.

The open 3-4 relay valve directs second gear oil into the servo feed oil where it travels to the 4-3 sequence valve and the 2-3 shift valve. Servo feed oil at the 4-3 sequence valve passes through the valve and becomes fourth gear oil. This pressure is routed to the servo to apply the intermediate-overdrive band and place the transmission in fourth gear. Servo feed oil to the 2-3 shift valve passes through the valve and becomes fourth accumulator oil. This pressure is routed to the 3-4 accumulator where it strokes the accumulator piston through its range of travel to cushion the band application.

In Overdrive fourth, high-overdrive clutch and intermediate-overdrive band are applied. The forward clutch is also applied, but it is ineffective because the forward 1-way clutch overruns.

Overdrive Range Forced Downshifts

A part-throttle 4-3 downshift occurs when the accelerator linkage moves the throttle plunger far enough to uncover the part-throttle port in the throttle valve. This routes part-throttle pressure to a land on the 3-4 throttle valve. The combination of part-throttle pressure and spring force then moves the 3-4 shift valve against governor pressure to cause a 4-3 downshift. A part-throttle (modulated) 4-3 downshift may also occur at lower speeds when modulated throttle downshift pressure combines with spring force acting on the 3-4 throttle valve to move the 3-4 shift valve against governor pressure.

A full-throttle 3-2 detent downshift occurs at speeds below approximately 60 mph (97 kph) when the accelerator linkage moves the throttle plunger far enough to uncover the detent port in the throttle valve. This routes detent pressure to a land on the 2-3 throttle valve. The combination of detent pressure and spring force then moves the 2-3 shift valve against governor pressure and causes a 3-2 downshift. If the vehicle speed is below approximately 30

mph (48 kph) detent pressure will also downshift the 1-2 shift valve, resulting in a 3-1 downshift. A part-throttle (modulated) 3-2 downshift may also occur at lower speeds when modulated throttle downshift pressure combines with spring force acting on the 3-2 throttle valve to move the 3-4 shift valve against governor pressure.

A full-throttle 2-1 detent downshift occurs at speeds below approximately 30 mph (48 kph) when detent pressure from the throttle valve flows to the 1-2 low downshift or low range valve. The combination of detent pressure and spring force moves the 1-2 shift valve against governor pressure to cause a 2-1 downshift.

Manual Third

When the vehicle is traveling in Overdrive fourth, a forced 4-3 downshift can be made by moving the gear selector to the manual third position (D or 3). The manual valve then routes mainline pressure (DR3 oil) to the 4-3 sequence valve where it closes both the 4-3 sequence valve and the 3-4 relay valve. Servo apply pressure and fourth accumulator pressure then exhaust to release the intermediate-overdrive band.

DR3 oil also flows from the 4-3 sequence valve to apply the drive (overrun) clutch to provide engine compression braking. As long as the gear selector is in the manual third position, the drive clutch remains applied.

In manual third, the forward, high-overdrive, and drive clutches are applied. The forward 1-way clutch holds on acceleration; the drive clutch prevents it from overrunning on deceleration.

Manual Second

A forced 4-2 or 3-2 downshift can be made by moving the gear selector to the manual second position (2). The manual valve then exhausts RND4-D3 oil, which releases the high-overdrive clutch and allows second gear oil to apply the intermediate-overdrive band. The 3-2 control valve acts to regulate the rate of high-overdrive clutch release and intermediate-overdrive band application in relation to vehicle speed. The manual valve also routes DR2 oil to the reverse boost valve to increase mainline pressure. This pressure application is made in manual low as well.

In manual second, the forward and drive clutches, along with the intermediate-overdrive band, are applied. The forward 1-way clutch holds on acceleration; the drive clutch prevents it from overrunning on deceleration.

Manual Low

In manual low, mainline pressure (LOW oil) from the manual valve is routed to the low upshift and downshift valves, or to two lands on the low range valve. This pressure combines with spring force acting on the 1-2 throttle valve to downshift the 1-2 shift valve and keep it downshifted regardless of vehicle speed or throttle position. LOW oil is also routed from the 1-2 shift valve assembly to apply the low-reverse clutch to provide engine compression braking.

If the vehicle is shifted to manual low at a speed above approximately 30 mph (48 kph), governor pressure will be high enough to prevent a 2-1 downshift. Once vehicle speed drops below this level, the transmission will downshift to low and remain there until the gear selector is moved to another position.

In manual low, the forward and low-reverse clutches are applied. The forward and low 1-way clutches hold on acceleration; the drive and low-reverse clutches prevent them from overrunning on deceleration.

Reverse

In Reverse, mainline pressure (REVERSE oil) from the manual valve is routed to the inner and outer apply areas of the low-reverse clutch, the reverse clutch, and the reverse boost valve in the pressure regulator valve assembly. The reverse and low-reverse clutches apply to place the transmission in reverse gear. The reverse boost valve increases mainline pressure, and modulated throttle pressure flows to the throttle boost valve to further boost mainline pressure.

Review Questions

Select the single most correct answer
Compare your answers to the correct answers on page 603

1. Mechanic A says the THM 700-R4 transmission was introduced in the 1982 model year.
   Mechanic B says the THM 700-R4 transmission is used in high-performance applications.
   Who is correct?
   a. Mechanic A
   b. Mechanic B
   c. Both mechanic A and B
   d. Neither mechanic A nor B

2. Mechanic A says all THM 700-R4 transmissions have a torque converter clutch.
   Mechanic B says some shop manuals refer to the THM 700-R4 transmission as the ME-9 gearbox.
   Who is correct?
   a. Mechanic A
   b. Mechanic B
   c. Both mechanic A and B
   d. Neither mechanic A nor B

3. Mechanic A says the THM 700-R4 transmission has four multiple-disc clutches.
   Mechanic B says the THM 700-R4 transmission has two 1-way (sprag) clutches.
   Who is correct?
   a. Mechanic A
   b. Mechanic B
   c. Both mechanic A and B
   d. Neither mechanic A nor B

4. Mechanic A says the THM 700-R4 identification numbers are stamped at the rear of the right-side pan rail.
   Mechanic B says the THM 700-R4 uses a Ravigneaux compound planetary gearset.
   Who is correct?
   a. Mechanic A
   b. Mechanic B
   c. Both mechanic A and B
   d. Neither mechanic A nor B

5. Mechanic A says the THM 700-R4 transmission makes a 4-1 coasting downshift when decelerating to a stop in Overdrive.
   Mechanic B says the THM 700-R4 transmission makes a 3-2 and a 2-1 downshift when decelerating to a stop in manual third.
   Who is correct?
   a. Mechanic A
   b. Mechanic B
   c. Both mechanic A and B
   d. Neither mechanic A nor B

6. Mechanic A says "line oil" is the GM name for one form of THM 700-R4 mainline pressure.
   Mechanic B says "RND4-D3 oil" is the GM name for one form of THM 700-R4 mainline pressure.
   Who is correct?
   a. Mechanic A
   b. Mechanic B
   c. Both mechanic A and B
   d. Neither mechanic A nor B

7. Mechanic A says the THM 700-R4 transmission has a gear-type variable-displacement oil pump.
   Mechanic B says the THM 700-R4 provides engine compression braking in low gear when the gear selector is in the manual third position.
   Who is correct?
   a. Mechanic A
   b. Mechanic B
   c. Both mechanic A and B
   d. Neither mechanic A nor B

8. Mechanic A says "reaction sun gear" is the GM term for the THM 700-R4 rear sun gear.
   Mechanic B says the THM 700-R4 reverse clutch holds the rear ring gear stationary.
   Who is correct?
   a. Mechanic A
   b. Mechanic B
   c. Both mechanic A and B
   d. Neither mechanic A nor B

9. Mechanic A says the THM 700-R4 intermediate-overdrive band is applied in second and fourth gears.
   Mechanic B says the THM 700-R4 intermediate-overdrive band holds the rear sun gear stationary.
   Who is correct?
   a. Mechanic A
   b. Mechanic B
   c. Both mechanic A and B
   d. Neither mechanic A nor B

10. Mechanic A says the THM 700-R4 forward clutch is applied in all gear ranges.
Mechanic B says the THM 700-R4 drive (overrun) clutch is applied in manual low, manual second, and manual third gears.
Who is correct?
a. Mechanic A
b. Mechanic B
c. Both mechanic A and B
d. Neither mechanic A nor B

11. Mechanic A says the THM 700-R4 high-overdrive clutch is applied in third and fourth gears.
Mechanic B says the THM 700-R4 high-overdrive clutch drives the rear ring gear.
Who is correct?
a. Mechanic A
b. Mechanic B
c. Both mechanic A and B
d. Neither mechanic A nor B

12. Mechanic A says the THM 700-R4 rear sun gear input shell is wrapped by the intermediate-overdrive band.
Mechanic B says the THM 700-R4 intermediate-overdrive band is adjusted by the length of the servo apply rod.
Who is correct?
a. Mechanic A
b. Mechanic B
c. Both mechanic A and B
d. Neither mechanic A nor B

13. Mechanic A says the THM 700-R4 forward clutch is applied in fourth gear.
Mechanic B says the THM 700-R4 forward 1-way clutch overruns in fourth gear.
Who is correct?
a. Mechanic A
b. Mechanic B
c. Both mechanic A and B
d. Neither mechanic A nor B

14. Mechanic A says the THM 700-R4 reverse and low-reverse clutches are applied in Reverse.
Mechanic B says the THM 700-R4 low 1-way clutch overruns in Reverse.
Who is correct?
a. Mechanic A
b. Mechanic B
c. Both mechanic A and B
d. Neither mechanic A nor B

15. Mechanic A says the THM 700-R4 has two multiple-disc clutches applied in Overdrive second.
Mechanic B says the THM 700-R4 has two 1-way clutches overrunning during deceleration in Overdrive low.
Who is correct?
a. Mechanic A
b. Mechanic B
c. Both mechanic A and B
d. Neither mechanic A nor B

16. Mechanic A says the THM 700-R4 converter clutch is controlled by the TCC solenoid and the converter clutch apply valve.
Mechanic B says the THM 700-R4 converter clutch is controlled by the converter clutch shift valve.
Who is correct?
a. Mechanic A
b. Mechanic B
c. Both mechanic A and B
d. Neither mechanic A nor B

17. Mechanic A says the THM 700-R4 governor uses check balls to regulate governor pressure.
Mechanic B says the THM 700-R4 governor is located at the left rear of the case and driven by the output shaft.
Who is correct?
a. Mechanic A
b. Mechanic B
c. Both mechanic A and B
d. Neither mechanic A nor B

18. Mechanic A says the 1-2 and 3-4 accumulators are located in the same bore in the THM 700-R4 case.
Mechanic B says the 2-4 servo in the THM 700-R4 transmission also acts on the third gear accumulator.
Who is correct?
a. Mechanic A
b. Mechanic B
c. Both mechanic A and B
d. Neither mechanic A nor B

19. Mechanic A says the THM 700-R4 uses a vacuum modulator to regulate throttle pressure.
Mechanic B says the THM 700-R4 uses a throttle limit valve controlled by a cable to regulate throttle pressure.
Who is correct?
a. Mechanic A
b. Mechanic B
c. Both mechanic A and B
d. Neither mechanic A nor B

20. Mechanic A says the THM 700-R4 3-2 control valve regulates the rate of high-overdrive clutch release during a part-throttle 3-2 modulated downshift.
Mechanic B says the THM 700-R4 3-2 control valve regulates the rate of intermediate-overdrive band application during a full-throttle 3-2 detent downshift.
Who is correct?
a. Mechanic A
b. Mechanic B
c. Both mechanic A and B
d. Neither mechanic A nor B

NIASE Mechanic Certification Sample Test

This sample test is similar in format to the series of eight tests given by the National Institute for Automotive Service Excellence (ASE). Each of these exams covers a specific area of automotive repair and service. The tests are given every fall and spring throughout the United States.

For a technician to earn certification in a particular field, he or she must successfully complete one of these tests, and have at least two years of "hands on" experience (or a combination of work experience and formal automobile technician training). Successfully finishing all eight tests earns the person certification as a Master Automobile Technician.

In the following sample test, the questions follow the form of the ASE exams. Learning to take this kind of test will help you if you plan to apply for certification later in your career. You can find the answers to the questions in this sample exam on page 603.

For test registration forms or additional information on the automobile technician certification program, write to:

National Institute for
AUTOMOTIVE SERVICE EXCELLENCE
1920 Association Drive, Suite 400
Reston VA 22019-1502

1. The sun gear of a single planetary gearset is held while the carrier is driven.

   Mechanic A says the result is that the ring gear is driven in overdrive.

   Mechanic B says the pinion gears and ring gear turn in the same direction.

   Who is right?
   a. Mechanic A
   b. Mechanic B
   c. Both mechanic A and B
   d. Neither mechanic A nor B

2. Mechanic A says that in order to produce gear reduction with a single planetary gearset, the carrier is always the output member.

   Mechanic B says that in order to produce reverse gear with a single planetary gear set, the carrier is always the input member.

   Who is right?
   a. Mechanic A
   b. Mechanic B
   c. Both mechanic A and B
   d. Neither mechanic A nor B

3. A hydraulic pressure of 100 psi is applied to a 10-square-inch servo piston.

   Mechanic A says the result is a band application force of 100 pounds.

   Mechanic B says you should divide the hydraulic pressure by the piston surface area to determine the application force.

   Who is right?
   a. Mechanic A
   b. Mechanic B
   c. Both mechanic A and B
   d. Neither mechanic A nor B

4. A shift valve is a type of switching valve used to change gears in an automatic transmission.

   Mechanic A says shift valves are switched by regulated pressures created by other valves in the hydraulic system.

   Mechanic B says all of the pressures in the hydraulic system are derived from governor pressure.

   Who is right?
   a. Mechanic A
   b. Mechanic B
   c. Both mechanic A and B
   d. Neither mechanic A nor B

5. Automatic transmissions use either gear-type or vane-type oil pumps. Which of the following is not true of *both* pump designs?

   a. They may be driven by the torque converter hub.
   b. They may be driven by a pump driveshaft.
   c. They may provide a variable level of output volume.
   d. They may mount at the front of the transmission case.

6. Mechanic A says mainline pressure is developed and controlled by the pressure regulator valve.

   Mechanic B says higher mainline pressure is necessary during reverse gear operation.

   Who is right?
   a. Mechanic A
   b. Mechanic B
   c. Both mechanic A and B
   d. Neither mechanic A nor B

7. Mechanic A says governor pressure remains constant as vehicle speed increases.

   Mechanic B says throttle pressure decreases as engine load and torque increase.

   Who is right?

   a. Mechanic A
   b. Mechanic B
   c. Both mechanic A and B
   d. Neither mechanic A nor B

8. Mechanic A says throttle pressure controls upshifts and downshifts.

   Mechanic B says that throttle pressure never exceeds governor pressure.

   Who is right?

   a. Mechanic A
   b. Mechanic B
   c. Both mechanic A and B
   d. Neither mechanic A nor B

9. Multiple-disc clutches, 1-way clutches, and bands are the most common types of apply devices. Which of the following statements about these devices is false?

   a. Bands are used as holding devices only.
   b. Multiple disc clutches may be used as either driving or holding devices.
   c. One-way clutches may be used as either driving or holding devices.
   d. All of the above apply devices require hydraulic pressure to operate.

10. Mechanic A says an accumulator in a transmission hydraulic system may be either a spool valve in the valve body or a piston in a servo bore.

    Mechanic B says when an accumulator reaches the end of its stroke, the pressure on the apply device increases to its full value.

    Who is right?

    a. Mechanic A
    b. Mechanic B
    c. Both mechanic A and B
    d. Neither mechanic A nor B

11. Mechanic A says all modern automatic transmission fluids are basically the same, so any type can be used in any transmission.

    Mechanic B says transmission fluid temperature is the

most important factor in determining when to change the fluid.

   Who is right?

   a. Mechanic A
   b. Mechanic B
   c. Both mechanic A and B
   d. Neither mechanic A nor B

12. Mechanic A says a lip-type piston seal is always installed with the lip facing the fluid to be contained.

    Mechanic B says metal seal rings and Teflon seal rings are basically the same except for the materials from which they are made.

    Who is right?

    a. Mechanic A
    b. Mechanic B
    c. Both mechanic A and B
    d. Neither mechanic A nor B

13. Mechanic A says endplay is the back and forth movement of a part along the axis of a shaft.

    Mechanic B says bushings often serve as a type of internal transmission seal.

    Who is right?

    a. Mechanic A
    b. Mechanic B
    c. Both mechanic A and B
    d. Neither mechanic A nor B

14. Mechanic A says the impeller of a torque converter acts as a stator during deceleration to provide engine braking.

    Mechanic B says minimum torque multiplication occurs at the converter stall speed.

    Who is right?

    a. Mechanic A
    b. Mechanic B
    c. Both mechanic A and B
    d. Neither mechanic A nor B

15. Mechanic A says the turbine is locked to the converter cover when a torque converter clutch is applied.

    Mechanic B says a centrifugally locked torque converter clutch slips to provide torque multiplication during hard acceleration at higher speeds.

    Who is right?

    a. Mechanic A
    b. Mechanic B
    c. Both mechanic A and B
    d. Neither mechanic A nor B

16. A car with a properly tuned engine is very sluggish when accelerating from a stop.

    Mechanic A says this could be caused by a bad governor valve.

    Mechanic B says this could be caused by a bad torque converter.

    Who is right?

    a. Mechanic A
    b. Mechanic B
    c. Both mechanic A and B
    d. Neither mechanic A nor B

17. A car with an automatic transmission does not downshift from high gear when the throttle is floored at low speeds.

    Mechanic A says the most likely cause of the problem is a misadjusted linkage.

    Mechanic B says the most likely cause of the problem is a sticking shift valve.

    Who is right?

    a. Mechanic A
    b. Mechanic B
    c. Both mechanic A and B
    d. Neither mechanic A nor B

18. With the transmission pan removed, the best way to pinpoint an oil pressure leak is to:

    a. Check the filter pickup tube seal
    b. Check the mainline pressure
    c. Remove and inspect the valve body
    d. Perform an air pressure test

19. Mechanic A says a stall test can be used to diagnose planetary gear noise.

    Mechanic B says a stall test can be used to check the torque converter stator 1-way clutch.

    Who is right?

    a. Mechanic A
    b. Mechanic B
    c. Both mechanic A and B
    d. Neither mechanic A nor B

20. All of the following can cause an automatic transmission to slip *except:*

    a. Hardened servo seals
    b. Plugged filter
    c. Worn planetary gears
    d. Faulty 1-way clutch

21. Mechanic A says that in a transmission with a vacuum modulator, high intake manifold vacuum causes high throttle pressure.

    Mechanic B says that in a transmission with a vacuum modulator, low intake manifold vacuum causes upshifts to occur at higher vehicle speeds.

    Who is right?
    a. Mechanic A
    b. Mechanic B
    c. Both mechanic A and B
    d. Neither mechanic A nor B

22. Mechanic A says torque multiplication in a torque converter occurs because the stator unlocks and redirects fluid returning to the impeller.

    Mechanic B says torque multiplication in a torque converter occurs because the impeller drives the turbine faster than engine speed.

    Who is right?
    a. Mechanic A
    b. Mechanic B
    c. Both mechanic A and B
    d. Neither mechanic A nor B

23. Mechanic A says throttle pressure controls upshifts and downshifts.

    Mechanic B says General Motors often calls throttle pressure modulator pressure.

    Who is right?
    a. Mechanic A
    b. Mechanic B
    c. Both mechanic A and B
    d. Neither mechanic A nor B

24. Torque converter stall speed is the:
    a. Maximum engine speed with a locked stator
    b. Minimum engine speed with a locked stator
    c. Maximum engine speed with a locked turbine
    d. Maximum engine speed with a locked impeller

25. At a certain point during normal torque converter operation, the stator unlocks and torque multiplication ends.

    Mechanic A says the converter enters the coupling phase at this time.

    Mechanic B says the engine speed at this time indicates the torque converter capacity.

    Who is right?
    a. Mechanic A
    b. Mechanic B
    c. Both mechanic A and B
    d. Neither mechanic A nor B

26. When the torque converter speed ratio is low, torque multiplication is:
    a. Low
    b. Variable in cycles
    c. High
    d. Zero

27. A car with an automatic transmission creeps at idle.

    Mechanic A says the governor may be faulty.

    Mechanic B says the idle speed may be too high.

    Who is right?
    a. Mechanic A
    b. Mechanic B
    c. Both mechanic A and B
    d. Neither mechanic A nor B

28. The fluid on an automatic transmission dipstick is milky pink.

    Mechanic A says the fluid has been contaminated with engine coolant.

    Mechanic B says the fluid has been contaminated by engine oil.

    Who is right?
    a. Mechanic A
    b. Mechanic B
    c. Both mechanic A and B
    d. Neither mechanic A nor B

29. A road test reveals that an automatic transmission is upshifting late.

    Mechanic A says the first step in diagnosis is to adjust the transmission bands.

    Mechanic B says the first step is to perform a hydraulic pressure test on the transmission.

    Who is right?
    a. Mechanic A
    b. Mechanic B
    c. Both mechanic A and B
    d. Neither mechanic A nor B

30. Mechanic A says you can install a transmission clutch piston by working the seal into the bore with a feeler gauge.

    Mechanic B says all transmission seals should be lubricated with fresh ATF prior to installation.

    Who is right?
    a. Mechanic A
    b. Mechanic B
    c. Both mechanic A and B
    d. Neither mechanic A nor B

31. An automatic transmission needs to be removed from the vehicle for repairs.

    Mechanic A says you should mark the flexplate and torque converter for proper reassembly before the transmission is removed.

    Mechanic B says the torque converter may be left attached to the engine when the transmission is removed.

    Who is right?
    a. Mechanic A
    b. Mechanic B
    c. Both mechanic A and B
    d. Neither mechanic A nor B

32. Mechanic A says low hydraulic pressure may be caused by fluid aeration resulting from an overfilled transmission.

    Mechanic B says the indicated fluid level in a THM 125/125C transaxle drops as the transaxle warms to operating temperature.

    Who is right?
    a. Mechanic A
    b. Mechanic B
    c. Both mechanic A and B
    d. Neither mechanic A nor B

33. During an automatic transmission overhaul, extensive bushing and friction material particles are discovered in the gearbox.

    Mechanic A says torque converter must be flushed before the rebuilt transmission is installed.

    Mechanic B says the transmission cooler does not need to be flushed because the transmission filter will trap any debris remaining in the lines.

    Who is right?
    a. Mechanic A
    b. Mechanic B
    c. Both mechanic A and B
    d. Neither mechanic A nor B

34. Mechanic A says an automatic transmission fluid leak resulting from case porosity will require replacement of the case casting.

    Mechanic B says a transmission case porosity problem can be repaired by welding the affected area.

    Who is right?
    a. Mechanic A
    b. Mechanic B
    c. Both mechanic A and B
    d. Neither mechanic A nor B

35. The starter on a vehicle with an automatic transmission will not operate when the gear selector is in Park, but it will crank when the selector is in Reverse.

    Mechanic A says the most likely cause of the problem is a defective neutral start switch.

    Mechanic B says the most likely cause of the problem is a misadjusted gear selector linkage.

    Who is right?
    a. Mechanic A
    b. Mechanic B
    c. Both mechanic A and B
    d. Neither mechanic A nor B

36. Mechanic A says the intermediate (kickdown or front) band requires adjustment on all Torqueflite transmissions and transaxles.

Mechanic B says the low-reverse (rear) band also requires adjustment on all Torqueflite transmissions and transaxles.

Who is right?
a. Mechanic A
b. Mechanic B
c. Both mechanic A and B
d. Neither mechanic A nor B

37. An automatic transmission with a vacuum modulator has abrupt and harsh upshifts. A hand vacuum pump connected to the modulator is unable to maintain a constant vacuum level.

    Mechanic A says the modulator is operating properly and the problem is in the vacuum line from the intake manifold.

    Mechanic B says the abrupt and harsh upshifts are the result of high throttle pressure caused by a defective modulator.

    Who is right?
    a. Mechanic A
    b. Mechanic B
    c. Both mechanic A and B
    d. Neither mechanic A nor B

38. Mechanic A says the bands in a THM 350/350C transmission may be adjusted by turning adjusting screws on the outside of the case.

    Mechanic B says you must remove the oil pan to reach the band adjusting screws for these transmissions.

Who is right?
a. Mechanic A
b. Mechanic B
c. Both mechanic A and B
d. Neither mechanic A nor B

39. A transmission with a lockup torque converter is being rebuilt.

    Mechanic A says the torque converter can be reused provided there were no friction material particles in the oil pan.

    Mechanic B says the torque converter can be reused provided that the transmission was not overheated.

    Who is right?
    a. Mechanic A
    b. Mechanic B
    c. Both mechanic A and B
    d. Neither mechanic A nor B

40. An electronically controlled torque converter clutch is failing to lockup.

    Mechanic A says the problem may be a faulty sensor in the electronic engine control system.

    Mechanic B says the problem may be a faulty torque converter clutch (TCC) solenoid.

    Who is right?
    a. Mechanic A
    b. Mechanic B
    c. Both mechanic A and B
    d. Neither mechanic A nor B

Glossary of Technical Terms

Accumulator: A device that absorbs the shock of sudden pressure surges within a hydraulic system. Accumulators are used in transmission hydraulic systems to control shift quality.

Annulus Gear: Another name for the internal ring gear in a planetary gearset.

Apply Device: Hydraulically operated bands and multi-disc clutches, and mechanically operated 1-way clutches, that drive or hold the members of a planetary gearset.

Axial Motion: Movement along, or parallel to, the centerline (axis) of a shaft. A dynamic seal is required to contain fluids where axial motion is present.

Axial Play: Movement along, or parallel to, the centerline (axis) of a shaft. Also called end thrust or endplay.

Axis: The centerline around which a gear, wheel, or shaft rotates.

Balance Valve: A spool valve and spring combination used to regulate hydraulic system pressure. Some balance valves also use auxiliary lever force or fluid force in doing their jobs.

Belleville Spring: A diaphragm-type spring sometimes used to help apply and release a multiple-disc clutch.

Bevel Gear: A gear with teeth cut at an angle on its outside surface. Bevel gears are commonly used to transmit power between two shafts at an angle to one another.

Booster Valve: A valve that increases mainline pressure when driveline loads are high to prevent apply devices from slipping.

Cam-Cut Drum: A 1-way roller clutch drum whose inner surface is machined with a series of angled grooves into which rollers are wedged.

Cantilever: A lever that is anchored and supported at one end by its fulcrum, and provides an opposing force at its opposite end.

Centrifugal Force: The natural tendency of objects, when rotated, to move away from the center of rotation.

Clutch Pack: The assembly of clutch plates and pressure plates that provides the friction surfaces in a multiple-disc clutch.

Clutch Plates: A generic term for the friction discs and steel discs used in a multiple-disc clutch.

Coefficient of Friction: A numerical value expressing the amount of friction between two surfaces. The coefficient of friction is obtained by dividing the force required to slide the surfaces across one another by the pressure holding the surfaces together.

Compressibility: The ability of a gasket to conform to surface irregularities. "Soft" gaskets are more compressible than "hard" gaskets.

Control Valving: Any devices that restrict, direct, or regulate the flow of fluid in a hydraulic system.

Copolymer: A substance made of giant molecules formed from the smaller molecules of two or more unlike substances.

Coupling Phase: The period of torque converter operation when there is no torque multiplication and rotary flow causes the stator to unlock and rotate with the impeller and turbine at approximately the same speed.

Depth Filter: A filter that traps contaminants within the matrix of the filter material. A felt filter is a type of depth filter.

Differential: The assembly of a carrier and spider gears that allows the drive axles to rotate at different rates of speed as the car turns a corner.

Direct Drive: A 1:1 gear ratio in which the engine, transmission, and driveshaft (on RWD vehicles) all turn at the same speed.

Downshift Valve: An auxiliary shift valve that increases throttle pressure to force a downshift under high driveline loads. Also called a kickdown valve or detent valve.

Dynamic Friction: The coefficient of friction between two surfaces that have relative motion between them. Also called kinetic friction.

Dynamic Seal: A seal that prevents fluid passage between two parts that are in motion relative to one another.

Early Fuel Evaporation (EFE): A system that preheats the incoming air-fuel mixture to improve driveability when the engine is cold.

Elastomer: A synthetic polymer or copolymer that has elastic properties similar to those of rubber.

Endplay: The total amount of axial play in an automatic transmission. Endplay is typically measured at the input shaft.

Exhaust Gas Recirculation (EGR): A pollution control system that injects a small quantity of exhaust gasses into the air-fuel mixture to lower combustion temperatures and limit the formation of oxides of nitrogen (NO_x).

Final Drive: The last set of reduction gears the powerflow passes through on its way to the drive axles.

Flexplate: The thin metal plate, used in place of a flywheel, that joins the engine crankshaft to the fluid coupling or torque converter.

Fluid Shear: The internal friction that occurs within a fluid when its rate or direction of flow is suddenly changed. Fluid shear in the torque converter is the major source of transmission fluid heat buildup.

Force: A push or pull acting on an object. Force is usually measured in pounds or Newtons. Force = Pressure × Area.

Friction Modifiers: Additives that enable lubricants to maintain their viscosity over a wide range of temperatures.

Fulcrum: The support and pivot point of a lever.

Gear Pump: A positive-displacement pump that uses an inner drive gear and an outer driven gear, separated on one side by a crescent, to produce oil flow. Gear pumps may use either helical or spur gears, and are sometimes called gear-and-crescent pumps.

Gear Ratio: The number of turns, or revolutions, made by a drive gear compared to the number of turns made by a driven gear. For example, if the drive gear turns three times for one revolution of the driven gear, the gear ratio is three, 3 to 1, or 3:1. Also, the ratio between the number of teeth on two gears.

Gear Reduction: A condition in which the drive gear rotates faster than the driven gear. Output speed of the driven gear is reduced, while output torque is increased. A gear ratio of 3:1 is a gear reduction ratio.

Geartrain: A series of two or more gears meshed together so that power is transmitted between them.

Governor Pressure: The transmission hydraulic pressure that is directly related to vehicle speed. Governor pressure increases with vehicle speed and is one of the principle pressures used to control shift points.

Governor Valve: The valve that regulates governor pressure in relation to vehicle road speed. Commonly called a governor.

Helical Gear: A gear on which the teeth are at an angle to the gear's axis of rotation.

Hydraulic Circuit: An arrangement of an input source, fluid passages, control valves, and an output device that is used to transmit motion and force to do work.

Hydraulics: The study, or science, of liquids or fluids, and their use to transmit force and motion.

Hydraulically Operated Switching Valve: A spool valve whose position is determined by hydraulic pressures acting on its faces. The shift valves that control most gear changes in an automatic transmission are hydraulically operated switching valves.

Hydrodynamics: The study, or science, of the mechanical movement and action of liquids or fluids in motion.

Hypoid Gearset: A combination of a ring gear and pinion gear in which the pinion meshes with the ring gear below the centerline of the ring gear. Commonly used in automotive final drives.

Hysteresis Bands: The area between two curves on a graph that indicate when a torque converter clutch locks and when it unlocks. Hysteresis bands also exist between graph curves that indicate when a transmission upshifts and when it downshifts between two gears.

Idler Gears: Gears that transmit movement between the drive and driven gears, but do not affect the speed relationship of those gears.

Input Source: The piston or pump that supplies the input force in a hydraulic system.

Input Member: The drive member of a planetary gearset.

Intake Manifold Vacuum: The low pressure area (vacuum) located below the throttle butterfly valve in the engine intake manifold.

Internal Ring Gear: A gear with a hole through its center and teeth cut on the inner circumference. Also called an "annulus gear".

Land: The large, outer circumference of a valve spool that slides against the valve bore. Most spool valves have several lands that are used to block fluid passages.

Mainline Pressure: The pressure developed from the fluid output of the pump and controlled by the pressure regulator valve. Mainline pressure operates the apply devices in the transmission and is the source of all other pressures in the hydraulic system.

Manually Operated Switching Valve: A spool valve whose position is controlled by a mechanical lever. Automatic transmission gear selector levers usually connect to a manually operated switching valve.

Manual Valve: The valve that is moved manually, through the shift linkage, to select the transmission drive range. The manual valve directs and blocks fluid flow to various hydraulic circuits.

Miscibility: The property that allows one fluid to blend with other fluids.

Multiple-Disc Clutch: A clutch that consists of alternating friction discs and steel discs that are forced together hydraulically to lock one transmission part to another.

One-Way Check Valve: A type of switching valve that allows fluid to pass in one direction only, and then only when the pressure is sufficient to unseat the valve.

One-Way Clutch: A mechanical holding device that prevents rotation in one direction, but overruns to allow it in the other. One-way clutches are either roller clutches or sprag clutches.

Orifice: A small opening or restriction in a line or passage that is used to regulate pressure and flow.

Output Device: A piston or motor that transmits output force created by the pressure in a hydraulic system.

Output Member: The driven member of a planetary gearset.

Overall Top Gear Ratio: The ratio obtained when the gear ratio of the transmission's highest forward speed is multiplied times the final drive gear ratio.

Overdrive: A condition in which the drive gear rotates slower than the driven gear. Output speed of the driven gear is increased, while output torque is reduced. A gear ratio of .70:1 is an overdrive gear ratio.

Oxidation: The process in which one element of a compound is combined with oxygen in a chemical process that produces another compound. Oxidation generates heat as a byproduct.

Pinion Gear: A smaller gear that meshes with a larger gearwheel or toothed rack.

Planetary Gearset: A system of gears consisting of a sun gear, an internal ring gear, and a planet carrier with planet pinion gears.

Planet Carrier Assembly: One of the members of a planetary gearset. The carrier, or bracket, on which the planet pinions are mounted.

Planet Pinions: The pinion gears mounted on the planet carrier assembly in a planetary gearset. The planet pinions rotate on the carrier and revolve around the sun gear.

Polymer: A substance made of giant molecules formed from smaller molecules of the same substance.

Ported Vacuum: The low pressure area (vacuum) located just above the throttle butterfly valve in a carburetor or fuel injection throttle body.

Positive-Displacement Pump: A pump that delivers the same volume of fluid with each revolution. The gear and rotor pumps used in automatic transmissions are positive-displacement pumps.

Power: The rate at which work is done or force is applied. In mechanics, power is measured as torque times speed and expressed in units such as horsepower or kilowatts.

Pressure: The force exerted on a given unit of area. Pressure is usually measured in pounds per square inch or kilopascals. Pressure = Force ÷ Area.

Pressure Regulator Valve: The valve that regulates mainline pressure by creating a variable restriction.

Pressure-Relief Valve: A valve that limits the maximum pressure in a hydraulic system or circuit. Also called a pressure-limiting valve.

Radial Play: Movement at a right angle to the axis of rotation of a shaft, or along the radius of a circle or shaft. Also called side thrust or sideplay.

Ravigneaux Gearset: A compound planetary gear system consisting of two sun gears and two sets of planet pinions that share a common ring gear.

Reaction Member: The member of a planetary gearset that is held in order to produce an output motion. Other members react against the stationary, held member.

Reference Voltage: The basic operating voltage in an electronic engine control system. The reference voltage is often modified by sensors to provide the computer with information on engine and vehicle operation.

Reservoir: A tank, pan, or other container that stores fluid for use in a hydraulic system.

Resultant Force: The combined force and oil flow direction produced by rotary flow and vortex flow in a fluid coupling or torque converter.

Rotary Flow: The oil flow path, in a fluid coupling or torque converter, that is in the same circular direction as the rotation of the impeller.

Rotational Motion: Movement that occurs when a shaft turns (rotates) on its axis. A dynamic seal is required to contain fluids where rotational motion is present.

Rotor Pump: A positive-displacement pump that uses an inner drive rotor and an outer driven rotor to produce oil flow. Lobes on the rotors create fluid chambers of varying volumes, and eliminate the need for a crescent as used in a gear pump.

Servo: A hydraulic piston and cylinder assembly that controls the application and release of a transmission band.

Shift Program: The set of instructions that the computer in an electronic shift control system uses to decide when to upshift and downshift the transmission. Many systems offer a choice of two or more shift programs that provide different performance characteristics.

Shift Valve: A spool valve acted on by throttle pressure and governor pressure to time transmission shifts. Also called a "snap" valve or a timing valve.

Simpson Gearset: A compound planetary gear system consisting of two ring gears and two planet carrier assemblies that share a common sun gear.

Sintered: Metal particles fused together under high heat and pressure. Sintering is used to make the porous metal discs in some vacuum delay valves.

Solenoid Ball Valve: An assembly consisting of a ball and seat check valve controlled by an electrical solenoid. When the solenoid is not energized, fluid flows through the check valve. When the solenoid is energized, fluid flow through the check valve is blocked.

Speed Differential: The difference in speed between two objects.

Speed Ratio: The number of revolutions made by the turbine for each revolution of the impeller. Turbine (output) speed is divided by impeller (input) speed and expressed as a percentage.

Spool Valve: A type of sliding hydraulic control valve, made up of lands and valleys, that resembles a sewing thread spool.

Sprag: A figure-eight-shaped locking element of a 1-way sprag clutch.

Spur Gear: A gear on which the teeth are parallel with the gear's axis of rotation.

Stall Speed: The maximum possible engine and torque converter impeller speed, measured in rpm, with the turbine held stationary and the engine throttle wide open.

Static Friction: The coefficient of friction between two surfaces that are in fixed, or nearly fixed, contact with one another.

Static Seal: A seal that prevents fluid passage between two parts that are in fixed positions relative to one another.

Sump: The oil pan, or reservoir, that contains a supply of fluid for the transmission hydraulic system.

Sun Gear: The central gear of a planetary gearset around which the other gears rotate.

Surface Filter: A filter that traps contaminants on its surface. Screen and paper filters are types of surface filters.

Temperature Differential: The difference in temperature between two objects or areas. The greater the temperature differential, the greater the rate at which heat will be transferred between the objects or areas.

Thermodynamics: The study, or science, of the relationship between mechanical and heat energy, and the laws governing the conversion of energy.

Throttle Pressure: The transmission hydraulic pressure that is directly related to engine load. Throttle pressure increases with throttle opening and engine torque output, and is one of the principle pressures used to control shift points.

Throttle Valve: The valve that regulates throttle pressure based on throttle butterfly opening or intake manifold vacuum.

Torque: Twisting or turning force measured in terms of the distance times the amount of force applied. Commonly expressed in foot-pounds, inch-pounds, or Newton-meters.

Torque Converter: A type of fluid coupling used to connect the engine crankshaft to the automatic transmission input shaft. Torque converters multiply the available engine torque under certain operating conditions.

Torque Converter Capacity: The ability of a torque converter to absorb and transmit engine torque in relation to the amount of slippage in the converter.

Torque Curve: A graphic depiction of the amount of torque available at different engine speeds.

Torque Multiplication Phase: The period of torque converter operation when vortex oil flow is redirected through the stator to accelerate impeller flow to the turbine and increase engine torque.

Transmission Band: A flexible steel band lined with friction material that is clamped around a circular drum to hold it from turning.

Two-Way Check Valve: A type of switching valve that controls fluid flow in two separate hydraulic circuits, and allows them to share a single fluid passage.

Vacuum Modulator: A metal housing divided into two chambers by a flexible rubber diaphragm. One chamber is open to atmospheric pressure, the other is connected to intake manifold vacuum and contains a spring. Changes in manifold vacuum cause movement of the diaphragm against spring tension. In automatic transmissions, this movement is often used to control a throttle valve.

Vacuum Pump: A mechanically or electrically driven pump that provides a source of vacuum. Vacuum pumps are commonly used with diesel engines which have little or no intake manifold vacuum.

Valley: The depressions formed between the lands of a spool valve. Valleys allow the passage of fluid into and/or through the valve depending on the position of the spool.

Valve Body: The casting that contains most of the valves in a transmission hydraulic system. The valve body also has passages for the flow of hydraulic fluid.

Vane Pump: A pump that uses a slotted rotor and sliding vanes to produce oil flow. The vane pumps used in automatic transmissions are variable-displacement pumps.

Variable-Displacement Pump: A pump whose output volume per revolution can be varied to increase or decrease fluid delivery. The vane pumps used in automatic transmissions are variable-displacement pumps.

Viscosity: The tendency of a liquid, such as an oil or hydraulic fluid, to resist flowing.

Vortex Flow: The oil flow path, in a fluid coupling or torque converter, that is at a right angle to the rotation of the impeller and to rotary flow.

Work: The transfer of energy from one system to another, particularly through the application of force.

Worm Gear: A shaft with gear teeth cut as a continuous spiral around its outer surface. A worm gear meshes with an external gear and is used to change the axis of rotation.

Index

Answers to Review and NIASE Questions

Chapter 1: Gears
1-D 2-B 3-C 4-D 5-C 6-C 7-D
8-B 9-C 10-A 11-C 12-C 13-D
14-A 15-B 16-C 17-C

Chapter 2: Hydraulic Fundamentals
1-C 2-A 3-B 4-B 5-C 6-A 7-B
8-D 9-D 10-B 11-A 12-B

Chapter 3: Transmission Hydraulic Systems
1-D 2-C 3-C 4-C 5-D 6-B 7-D
8-C 9-A 10-D 11-B 12-D 13-C
14-C 15-D 16-A 17-B 18-C

Chapter 4: Apply Devices
1-D 2-B 3-C 4-C 5-B 6-D 7-D
8-C 9-A 10-D 11-B 12-A 13-B
14-D

Chapter 5: Transmission Fluids, Filters, and Coolers
1-C 2-C 3-A 4-D 5-B 6-A 7-B
8-A 9-A 10-D 11-D 12-C 13-D
14-C

Chapter 6: Gaskets, Seals, Bushings, Bearings, Thrust Washers, and Snaprings
1-B 2-A 3-C 4-D 5-B 6-C 7-A
8-C 9-C 10-D 11-B 12-D 13-C
14-D 15-C 16-B 17-B

Chapter 7: Fluid Couplings and Torque Converters
1-C 2-D 3-B 4-D 5-A 6-A 7-B
8-A 9-B 10-D 11-A 12-C 13-A
14-A 15-D 16-B 17-B 18-A 19-C
20-C 21-D 22-C 23-A

Chapter 8: Electrical and Electronic Transmission Controls
1-A 2-D 3-C 4-C 5-A 6-B 7-C
8-C 9-D 10-B 11-C 12-B 13-C
14-A 15-A 16-A 17-D 18-B 19-C
20-D 21-C 22-D 23-A 24-A 25-B

Chapter 9: Torqueflite Transmissions and Transaxles
1-D 2-B 3-A 4-C 5-D 6-C 7-B
8-C 9-A 10-B 11-A 12-B 13-C
14-B 15-D 16-A 17-C 18-D 19-B
20-D 21-C 22-D 23-B 24-A 25-A

Chapter 10: Ford C3, C4, C5, C6, A4LD, and Jatco Transmissions
1-D 2-C 3-B 4-A 5-C 6-D 7-C
8-A 9-A 10-C 11-C 12-D 13-D
14-C 15-D 16-B 17-C 18-C 19-B
20-B

Chapter 11: The Ford ATX Transaxle
1-A 2-B 3-C 4-D 5-B 6-A 7-A
8-C 9-B 10-A 11-D 12-C 13-B
14-C 15-C

Chapter 12: The Ford AOD Transmission
1-C 2-B 3-A 4-A 5-C 6-D 7-B
8-B 9-A 10-B 11-C 12-A 13-B
14-D 15-C 16-B 17-C 18-A 19-A
20-C

Chapter 13: The Ford AXOD Transaxle
1-B 2-C 3-A 4-B 5-D 6-D 7-C
8-A 9-B 10-C 11-A 12-D 13-C
14-B 15-C 16-A 17-C 18-A 19-D
20-C

Chapter 14: Turbo Hydra-matic 125 and 125C Transaxles
1-C 2-B 3-C 4-C 5-A 6-C 7-D
8-A 9-B 10-C 11-A 12-A 13-C
14-D 15-B 16-A 17-D 18-C 19-C
20-A

Chapter 15: Turbo Hydra-matic 180 and 180C Transmissions
1-B 2-A 3-B 4-A 5-D 6-B 7-A
8-D 9-C 10-A 11-C 12-B 13-C
14-B 15-A 16-D 17-A 18-A 19-C
20-B

Chapter 16: Turbo Hydra-matic 200, 200C, 200-4R, 325, and 325-4L Transmissions
1-B 2-D 3-C 4-D 5-A 6-A 7-C
8-D 9-A 10-A 11-C 12-B 13-C
14-D 15-D 16-A 17-D 18-B 19-C
20-D

Chapter 17: Turbo Hydra-matic 250, 250C, 350, 350C, and 375B Transmissions
1-A 2-C 3-D 4-D 5-A 6-B 7-B
8-A 9-B 10-A 11-B 12-C 13-C
14-D 15-B 16-C 17-D 18-C 19-A
20-B

Chapter 18: Turbo Hydra-matic 375, 400, 425, and 475 Transmissions
1-B 2-A 3-C 4-C 5-D 6-C 7-B
8-C 9-C 10-A 11-B 12-A 13-C
14-B 15-B 16-A 17-B 18-A 19-C
20-A

Chapter 19: The Turbo Hydra-matic 440-T4 Transaxle
1-C 2-B 3-D 4-D 5-B 6-A 7-A
8-B 9-A 10-C 11-B 12-C 13-C
14-D 15-B 16-C 17-B 18-A 19-C
20-B

Chapter 20: The Turbo Hydra-matic 700-R4 Transmission
1-C 2-A 3-D 4-A 5-B 6-C 7-D
8-A 9-C 10-B 11-A 12-B 13-C
14-A 15-B 16-C 17-B 18-C 19-D
20-C

NIASE Mechanic Certification Sample Test
1-C 2-A 3-D 4-A 5-C 6-C 7-D
8-A 9-D 10-C 11-B 12-C 13-C
14-D 15-C 16-B 17-A 18-D 19-B
20-C 21-B 22-D 23-C 24-C 25-A
26-C 27-B 28-A 29-B 30-C 31-A
32-C 33-A 34-D 35-B 36-A 37-D
38-D 39-C 40-C